Climate Change and Critical Agrarian Studies

Climate change is perhaps the greatest threat to humanity today and plays out as a cruel engine of myriad forms of injustice, violence and destruction. The effects of climate change from human-made emissions of greenhouse gases are devastating and accelerating, yet are uncertain and uneven both in terms of geography and socio-economic impacts. Emerging from the dynamics of capitalism since the industrial revolution—as well as industrialization under state-led socialism—the consequences of climate change are especially profound for the countryside and its inhabitants.

This book interrogates the narratives and strategies that frame climate change and examines the institutionalised responses in agrarian settings, highlighting what exclusions and inclusions result. It explores how different people—in relation to class and other co-constituted axes of social difference such as gender, race, ethnicity, age and occupation—are affected by climate change, as well as the climate adaptation and mitigation responses being implemented in rural areas. This book in turn explores how climate change—and the responses to it—affects processes of social differentiation, trajectories of accumulation and in turn agrarian politics. Finally, this book examines what strategies are required to confront climate change and the underlying political-economic dynamics that cause it, reflecting on what this means for agrarian struggles across the world.

The 26 chapters in this volume explore how the relationship between capitalism and climate change plays out in the rural world and, in particular, the way agrarian struggles connect with the huge challenge of climate change. Through a huge variety of case studies alongside more conceptual chapters, this book makes the often-missing connection between climate change and critical agrarian studies. This book argues that making the connection between climate and agrarian justice is crucial.

The chapters in this book were originally published in *The Journal of Peasant Studies.*

Ian Scoones is Professor at the Institute of Development Studies at the University of Sussex, Brighton, UK. He is an agricultural ecologist by original training but today works on questions of policy around land, agriculture and agrarian change, mostly in Africa. He is the principal investigator of the ERC-funded PASTRES programme (http: //pastres.org).

Saturnino M. Borras Jr. is Professor of Agrarian Studies at the International Institute of Social Studies (ISS) of Erasmus University Rotterdam, the Netherlands; Distinguished Professor at China Agricultural University, Beijing, China; and Associate of the Amsterdam-based Transnational Institute.

Amita Baviskar is Dean, Faculty and Professor of Environmental Studies and Sociology and Anthropology at Ashoka University, Sonipat, India. Her research and teaching address the cultural politics of environment and development in rural and urban India. She focuses on the role of social inequalities and identities in natural resource conflicts.

Marc Edelman is Professor of Anthropology at Hunter College and the City University of New York Graduate Center, USA. His latest book is *Peasant Politics of the Twenty-first Century: Transnational Social Movements and Agrarian Change* (2024).

Nancy Lee Peluso is Henry J. Vaux Distinguished Professor of Forest Policy in the Department of Environmental Science, Policy and Management and Chair of the Center for Southeast Asia Studies at the University of California, Berkeley, USA.

Wendy Wolford is Robert A. and Ruth E. Polson Professor of Global Development in the Department of Global Development and Vice Provost for International Affairs at Cornell University, Ithaca, USA. Her research includes work on international development, land use and distribution, social mobilization, agrarian societies and critical ethnography.

Critical Agrarian Studies

Critical Agrarian Studies is the accompanying book series to *The Journal of Peasant Studies*. It publishes selected special issues of the journal and, occasionally, books that offer major contributions in the field of critical agrarian studies. The book series builds on the long and rich history of the journal and its former accompanying book series, the *Library of Peasant Studies* (1973–2008), which published several important monographs and special issues as books.

Series Editor: *Saturnino M. Borras Jr.*

Gender and Generation in Southeast Asian Agrarian Transformations
Edited by Clara Mi Young Park and Ben White

De-centring Land Grabbing
Edited by Peter Vandergeest and Laura Schoenberger

Soy, Globalization, and Environmental Politics in South America
Edited by Gustavo de L. T. Oliveira and Susanna B. Hecht

An Endogenous Theory of Property Rights
Edited by Peter Ho

Global Land Grabbing and Political Reactions 'from Below'
Edited by Marc Edelman, Ruth Hall, by Saturnino M. Borras Jr, Ian Scoones and Wendy Wolford

Critical Perspectives in Rural Development Studies
Edited by Saturnino M. Borras Jr.

The Politics of Biofuels, Land and Agrarian Change
Edited by Saturnino M. Borras Jr., Philip McMichael and Ian Scoones

New Frontiers of Land Control
Edited by Nancy Lee Peluso and Christian Lund

Outcomes of Post-2000 Fast Track Land Reform in Zimbabwe
Edited by Lionel Cliffe, Jocelyn Alexander, Ben Cousins and Rudo Gaidzanwa

Green Grabbing: A New Appropriation of Nature
Edited by James Fairhead, Melissa Leach and Ian Scoones

The New Enclosures: Critical Perspectives on Corporate Land Deals
Edited by Ben White, Saturnino M. Borras Jnr., Ruth Hall, Ian Scoones and Wendy Wolford

For more information about this series, please visit:
https://www.routledge.com/Critical-Agrarian-Studies/book-series/CAG.

Climate Change and Critical Agrarian Studies

Edited by
Ian Scoones, Saturnino M. Borras Jr., Amita Baviskar, Marc Edelman, Nancy Lee Peluso and Wendy Wolford

Routledge
Taylor & Francis Group
LONDON AND NEW YORK

Cover image: "Kaibigang puno", watercolour on paper (2017)
© Image by Boy Dominguez

First published 2024
by Routledge
4 Park Square, Milton Park, Abingdon, Oxon, OX14 4RN

and by Routledge
605 Third Avenue, New York, NY 10158

Routledge is an imprint of the Taylor & Francis Group, an informa business

Funded by the European Union

British Library Cataloguing-in-Publication Data
A catalogue record for this book is available from the British Library

ISBN13: 978-1-032-74165-9 (hbk)
ISBN13: 978-1-032-74167-3 (pbk)
ISBN13: 978-1-003-46796-0 (ebk)

DOI: 10.4324/9781003467960

Typeset in Myriad Pro
by codeMantra

Contents

Citation Information

The following chapters were originally published in various volumes and issues of *The Journal of Peasant Studies*. When citing this material, please use the original page numbering for each article, as follows:

Chapter 1

Climate change and agrarian struggles an invitation to contribute to a JPS Forum
Saturnino M. Borras Jr., Ian Scoones, Amita Baviskar, Marc Edelman, Nancy Lee Peluso and Wendy Wolford
The Journal of Peasant Studies, volume 49, issue 1 (2022) pp. 1–28

Chapter 2

The environmentalization of the agrarian question and the agrarianization of the climate justice movement
Zehra Taşdemir Yaşın
The Journal of Peasant Studies, volume 49, issue 7 (2022) pp. 1355–1386

Chapter 3

Violent silence: framing out social causes of climate-related crises
Jesse Ribot
The Journal of Peasant Studies, volume 49, issue 4 (2022) pp. 683–712

Chapter 4

Climate change and class conflict in the Anthropocene: sink or swim together?
Murat Arsel
The Journal of Peasant Studies, volume 50, issue 1 (2023) pp. 67–95

Chapter 5

The political life of mitigation: from carbon accounting to agrarian counter-accounts
Shaila Seshia Galvin and Diego Silva Garzón
The Journal of Peasant Studies, volume 50, issue 6 (2023) pp. 2259–2282

Chapter 6

Imagined transitions: agrarian capitalism and climate change adaptation in Colombia
Alejandro Camargo
The Journal of Peasant Studies, volume 49, issue 4 (2022) pp. 713–733

Chapter 7

Beyond bad weather: climates of uncertainty in rural India
Tanya Matthan
The Journal of Peasant Studies, volume 50, issue 1 (2023) pp. 114–135

Chapter 8

Climate rentierism after coal: forests, carbon offsets and post-coal politics in the Appalachian coalfields
Gabe Schwartzman
The Journal of Peasant Studies, volume 49, issue 5 (2022) pp. 924–944

Chapter 9

Up in the air: the challenge of conceptualizing and crafting a post-carbon planetary politics to confront climate change
Alistair Fraser
The Journal of Peasant Studies, volume 50, issue 1 (2023) pp. 96–113

Chapter 10

Power for the Plantationocene: solar parks as the colonial form of an energy plantation
Ryan Stock
The Journal of Peasant Studies, volume 50, issue 1 (2023) pp. 162–184

Chapter 11

Oro blanco*: assembling extractivism in the lithium triangle*
Daniela Soto Hernandez and Peter Newell
The Journal of Peasant Studies, volume 49, issue 5 (2022) pp. 945–968

Chapter 12

Adapting to climate change among transitioning Maasai pastoralists in southern Kenya: an intersectional analysis of differentiated abilities to benefit from diversification processes
Edwige Marty, Renee Bullock, Matthew Cashmore, Todd Crane and Siri Eriksen
The Journal of Peasant Studies, volume 50, issue 1 (2023) pp. 136–161

Chapter 13

Advocating afforestation, betting on BECCS: land-based negative emissions technologies (NETs) and agrarian livelihoods in the global South
Pamela McElwee
The Journal of Peasant Studies, volume 50, issue 1 (2023) pp. 185–214

Chapter 14

Food, famine and the free trade fallacy: the dangers of market fundamentalism in an era of climate emergency
Matias E. Margulis, Kristen Hopewell and Edi Qereshniku
The Journal of Peasant Studies, volume 50, issue 1 (2023) pp. 215–239

Chapter 15

Uneven resilience and everyday adaptation: making Rwanda's green revolution 'climate smart'
Nathan Clay
The Journal of Peasant Studies, volume 50, issue 1 (2023) pp. 240–261

Chapter 16

Rethinking 'just transitions' from coal: the dynamics of land and labour in anti-coal struggles
Amod Shah
The Journal of Peasant Studies, volume 50, issue 6 (2023) pp. 2145–2164

Chapter 17

Rescaling the land rush? Global political ecologies of land use and cover change in key scenario archetypes for achieving the 1.5°C Paris agreement target
Jevgeniy Bluwstein and Connor Cavanagh
The Journal of Peasant Studies, volume 50, issue 1 (2023) pp. 262–294

Chapter 18

Producing nature-based solutions: infrastructural nature and agrarian change in San Martín, Peru
Will Lock
The Journal of Peasant Studies, volume 50, issue 6 (2023) pp. 2361–2380

Chapter 19

Climate refugees or labour migrants? Climate reductive translations of women's migration from coastal Bangladesh
Camelia Dewan
The Journal of Peasant Studies, volume 50, issue 6 (2023) pp. 2339–2360

Chapter 20

Certificated exclusion: forest carbon sequestration project in Southwest China
Jun He and Jiping Wang
The Journal of Peasant Studies, volume 50, issue 6 (2023) pp. 2165–2186

Chapter 21

Resilience and conflict: rethinking climate resilience through Indigenous territorial struggles
Noémi Gonda, Selmira Flores, Jennifer J. Casolo and Andrea J. Nightingale
The Journal of Peasant Studies, volume 50, issue 6 (2023) pp. 2312–2338

Chapter 22

Resisting, leveraging, and reworking climate change adaptation projects from below: placing adaptation in Ecuador's agrarian struggle
Megan Mills-Novoa, Rutgerd Boelens, Jaime Hoogesteger and Jeroen Vos
The Journal of Peasant Studies, volume 50, issue 6 (2023) pp. 2283–2311

Chapter 23

Linking climate-smart agriculture to farming as a service: mapping an emergent paradigm of datafied dispossession in India
S. Ali Malik
The Journal of Peasant Studies, volume 50, issue 6 (2023) pp. 2187–2209

Chapter 24

Prefiguring buen sobrevivir*: Lenca women's (e)utopianism amid climate change*
Benjamin C. Fash, Betty del Carmen Vásquez Rivera and María Sojob
The Journal of Peasant Studies, volume 50, issue 6 (2023) pp. 2232–2258

Chapter 25

Forest as 'nature' or forest as territory? Knowledge, power and climate change conservation in the Peruvian Amazon
Maritza Paredes and Anke Kaulard
The Journal of Peasant Studies, volume 50, issue 6 (2023) pp. 2210–2231

Chapter 26

Whose security? Politics, risks and alternatives for climate security practices in agrarian-environmental perspectives
Corinne Lamain
The Journal of Peasant Studies (2023) DOI: 10.1080/03066150.2023.2246386

Notes on Contributors

Murat Arsel, International Institute of Social Studies, Erasmus University Rotterdam, The Hague, Netherlands.

Amita Baviskar, Environmental Studies and Sociology-Anthropology, Ashoka University, Sonipat, India.

Jevgeniy Bluwstein, University of Fribourg, Department of Geosciences, Switzerland.

Rutgerd Boelens, Department of Environmental Sciences, Wageningen University, Netherlands; Centre for Latin American Research and Documentation (CEDLA), University of Amsterdam, Netherlands; Faculty of Agronomy, Universidad Central de Ecuador, Quito, Ecuador.

Saturnino M. Borras Jr., International Institute of Social Studies, Erasmus University Rotterdam, The Hague, Netherlands.

Renee Bullock, International Livestock Research Institute, Nairobi, Kenya.

Alejandro Camargo, Department of History and Social Sciences, Universidad del Norte, Barranquilla, Colombia.

Matthew Cashmore, Norwegian University of Life Sciences, Norway.

Jennifer J. Casolo, Research and Development Institute Nitlapan, Central American University, Managua, Nicaragua; Institute of Development Policy, University of Antwerp, Belgium; Institute of Ch'orti' Maya Science and Technology Upejkna'r e Ja', Camotán, Guatemala.

Connor Cavanagh, University of Bergen, Department of Geography, University of Bergen, Bergen, Norway.

Nathan Clay, Stockholm Resilience Center, Stockholm University, Stockholm, Sweden

Todd Crane, International Livestock Research Institute, Nairobi, Kenya.

Camelia Dewan, Department of Social Anthropology, University of Oslo, Norway.

Marc Edelman, Anthropology Department, Hunter College-CUNY, USA.

Mercedes Ejarque, Collective of Agrarian Scholar-Activists from the South.

Siri Eriksen, Norwegian University of Life Sciences, Norway.

Benjamin C. Fash, Graduate School of Geography, Clark University, Worcester, USA; Cine Bolomchon, Mexico.

Selmira Flores, Research and Development Institute Nitlapan, Central American University, Managua, Nicaragua.

Jennifer Franco, Transnational Institute, Amsterdam, Netherlands.

Alistair Fraser, Department of Geography, Maynooth University, Ireland.

Noémi Gonda, Department of Urban and Rural Development, Swedish University of Agricultural Sciences, Uppsala, Sweden.

Jacobo Grajales, *The Journal of Peasant Studies*; University of Lille, France.

Ruth Hall, *The Journal of Peasant Studies*; PLAAS, University of the Western Cape, South Africa.

Jun He, National Centre for Borderland Ethnic Studies in Southwest China, Yunnan University, Kunming, People's Republic of China; School of Ethnology and Sociology, Yunnan University, Kunming, People's Republic of China.

Jaime Hoogesteger, Department of Environmental Sciences, Wageningen University, Netherlands; Department of Geomatic and Hydraulic Engineering, Universidad de Guanajuato, Mexico; Instituto de Investigaciones en Ecosistemas y Sustentabilidad (IIES), Universidad de Autónoma de México, Mexico.

Kristen Hopewell, Faculty of Land and Food Systems and School of Public Policy and Global Affairs, University of British Columbia, Vancouver, Canada.

Ricardo Jacobs, *The Journal of Peasant Studies*; University of California, Santa Barbara, USA.

Anke Kaulard, Pontificia Universidad Católica del Perú, Social Sciences (Sociology), Lima, Peru.

Sinem Kavak, Collective of Agrarian Scholar-Activists from the South.

Corinne Lamaine, International Institute for Social Studies, Erasmus University Rotterdam, The Hague, Netherlands.

Will Lock, School of Global Studies, University of Sussex, Brighton, UK.

S. Ali Malik, Law, Politics and Society, Drake University, Des Moines, USA.

Matias E. Margulis, Faculty of Land and Food Systems and School of Public Policy and Global Affairs, University of British Columbia, Vancouver, Canada.

Edwige Marty, International Livestock Research Institute, Nairobi, Kenya; Norwegian University of Life Sciences, Norway.

Tanya Matthan, Department of Geography, University of California, Berkeley, USA.

Pamela McElwee, Department of Human Ecology, Rutgers, The State University of New Jersey, New Brunswick, USA.

Megan Mills-Novoa, Department of Environmental Science, Policy, and Management/Energy & Resources Group, University of California, Berkeley, USA.

Peter Newell, Department of International Relations, University of Sussex, Brighton, UK.

Andrea J. Nightingale, Department of Urban and Rural Development, Swedish University of Agricultural Sciences, Uppsala, Sweden; Department of Sociology and Human Geography, University of Oslo, Norway.

Maritza Paredes, Pontificia Universidad Católica del Perú, Social Sciences (Sociology), Lima, Peru.

Nancy Lee Peluso, Department of Environmental Science, Policy and Management, University of California, Berkeley, USA.

Edi Qereshniku, Faculty of Land and Food Systems and School of Public Policy and Global Affairs, University of British Columbia, Vancouver, Canada.

Jesse Ribot, School of International Service, American University, Washington, DC, USA.

Katie Sandwell, Transnational Institute, Amsterdam, Netherlands.

Sergio Sauer, *The Journal of Peasant Studies;* University of Brasilia, Brazil.

Gabe Schwartzman, Geography, Environment and Society, University of Minnesota Twin Cities, Minneapolis, USA.

Ian Scoones, Institute of Development Studies, University of Sussex, Brighton, UK.

Shaila Seshia Galvin, *The Journal of Peasant Studies*; Department of Anthropology and Sociology, Geneva Graduate Institute, Switzerland.

Amod Shah, International Institute of Social Studies, Erasmus University Rotterdam, The Hague, Netherlands.

Annie Shattuck, *The Journal of Peasant Studies*; University of Indiana, USA.

Diego Silva Garzón, Centre for International Environmental Studies, Geneva Graduate Institute, Switzerland.

María Sojob, Cine Bolomchon, Mexico.

Daniela Soto Hernandez, Department of Anthropology, University of Sussex, Brighton, UK.

Ryan Stock, Department of Earth, Environmental and Geographical Sciences, Northern Michigan University, Marquette, USA.

Zehra Taşdemir Yaşın, Department of Sociology, Social Sciences University of Ankara, Turkey.

Betty del Carmen Vásquez Rivera, Movimiento Ambientalista SantaBarbarense, Santa Bárbara, Honduras.

Jeroen Vos, Department of Environmental Sciences, Wageningen University, Netherlands.

Jiping Wang, National Centre for Borderland Ethnic Studies in Southwest China, Yunnan University, Kunming, People's Republic of China; School of International Chinese Language Education, Yunnan University, Kunming, People's Republic of China.

Wendy Wolford, Department of Global Development, Cornell University, Ithaca, USA.

Acknowledgements

The open access publication of this book has been made possible through the PASTRES (Pastoralism, Uncertainty, Resilience: Global Lessons from the Margins) project, www.pastres.org, which has received Advanced Grant funding from the European Research Council (ERC) under the European Union's Horizon 2020 research and innovation programme (Grant agreement No. 740342).

Preface: Climate Change and Critical Agrarian Studies

In rural worlds, the violence of climate change is both fast and slow. Inflicted by the sudden devastation of floods, storms or fire; drawn out over months or years-long drought; cruelly felt by individuals, within families and wider communities as abrupt displacement or, more gradually, as deepening, protracted crises of social reproduction; experienced bodily as acute exhaustion from unrelenting extreme heat, as illness due to longer-term shifts in disease ecologies or through chronic mental and emotional distress.

As global average air and ocean temperatures rise to new levels, accompanied by ever more frequent and dire warnings from international agencies, climate emergency and climate crisis have become established bywords of these times. Too often, though, terms like emergency and crisis gloss over the deeper historical and political drivers of climate change. In doing so, they also isolate climate change from a much wider set of challenges experienced by people for whom security and survival, across the many senses of these terms, are already persistent or recurrent struggles.

As scholars and activists of agrarian questions have noted, very often climate change is not a word on the lips of those who experience its harshest effects, nor is it necessarily at the forefront of their concerns. In everyday life and organised political action in rural worlds, the problems of accelerating climate change are folded into enduring issues related to land, labour, livelihoods and lifeways. As such, they are inseparable from gender and social reproduction; from inequalities and forms of oppression sedimented through race, class, religion and caste among other differences; from persisting forms of coloniality and imperialism and from capitalism as well as state socialist industrialism.

Despite global narratives and agendas of justice, equality and rights, interventions by the state and international organizations have not consistently contributed to mitigating the effects of such intersecting oppressions and at times have instead exacerbated them. We see this, for example, in the grim persistence of world hunger, which affects 2.9 billion people or nearly one-third of the world's population—a figure of staggering magnitude that demands an equally forceful reckoning with the ways that this and related vulnerabilities are differentiated socially, geographically and historically. Sustained attention to the deeply political and world-historical underpinnings of climate change, and their intersection with older, more enduring crises, is necessary for just climate action and for connecting the problems of climate change to the ongoing movements and struggles of those who continue to experience injustice and oppression.

The long tradition of agrarian studies tells us that the notion and the experience of crisis can never be taken for granted—as crises reveal tensions and intensifying contradictions, they demand analytical attention and inspire political thought and action. In many ways, this volume, *Climate Change and Critical Agrarian Studies*, maintains this tradition, spurred by the vital need to grapple better with the implications of climate change, as well as approaches to addressing it, for people and politics across diverse rural worlds.

Beginning with a 2021 essay by then members of *The Journal of Peasant Studies* Editorial Collective that invited contributions to a *JPS* Forum, a momentum gathered. A large number of submissions were received and an international online conference was convened to deepen the discussions among scholars and with social movements and activists. The conference was jointly organised in September 2022 by the Collective of Agrarian Scholar-Activists from the South, *The Journal of Peasant Studies* and the Transnational Institute. Inviting exchanges across the breadth of different scholar-activist backgrounds and experiences, the conference comprised four plenary conversations and 63 presentations and attracted around 2,000 attendees from 105 countries over four days of intensive plenary and panel discussions.

The conference highlighted the intersections between the climate crisis and energy, food and other forms of crisis in rural worlds; the multiplier effect of social differentiation generated by climate change; the urgency of climate justice and the risks of its co-optation and the dangers of green extractivism and false technocratic or political solutions. Powerful messages emerging from the conference emphasized that discussions of climate change cannot be separated from critical attention to capitalism, patriarchy and political authoritarianism, while scholar activism can contribute to powering, incubating, co-creating and disseminating emancipatory alternatives. Across what can be fragmented climate, environmental and agrarian justice movements, the conference foregrounded a vital need to strengthen grounded alliances of people whose lives and work are marked by enduring, and often multiple and converging, forms of insecurity and precarity.

The chapters in this volume go a considerable way to speak to these imperatives. In them, we see the persistence of class, race, indigeneity and gender as central keys to understanding the causes and effects of climate change within a wider crucible of crises experienced by working people; new approaches to, or new reads of, older theories anchored in environmental and agrarian justice; radical perspectives on just transitions and the articulation of food sovereignty and climate action; critical interrogations of dominant framings of climate action that centre labour, including social and reproductive labour and its gendered and racial dimensions; critical studies of climate adaptation projects and of resilience discourse; close studies of the dynamic interaction of climate change and capitalism, especially with respect to indigenous territorial struggles, capitalist agro-industries and finance, and the financialization of the climate change response itself; sharp analyses of the agrarian implications of the latest IPCC reports (AR6) and climate-neutral and net zero pledges made in connection with COP or SDG goals; the political economy of mining and extractive industry expansion associated with the International Energy Agency's prescription for a transition away from fossil fuels and political responses to decarbonization in rural regions of fossil fuel extraction, the rise of right-wing populism and 'fossil fascism'. Through critical analysis of current mainstream modes of environmental action, these articles collectively shine a light on the urgent need to continue developing, refining, expanding and advancing genuinely transformative visions including those being elaborated by agrarian and environmental justice movements.

As formal, institutional processes ostensibly aimed at addressing the climate crisis stagnate or deliver solutions geared towards capital accumulation through the creation of new (predominantly rural) sacrifice zones, the need for global movements and robust scholarship to drive transformative political-ecological change is increasingly urgent.

Climate, agrarian and environmental justice struggles are exploring and building alliances in response to this need, as signified by initiatives like the process towards a new Nyéléni global gathering and the formation of other transnational alliances. Activist-scholars can play a critical role in accompanying movements as they build new alliances and syntheses necessary to genuinely address the ongoing violence of climate change.

These new convergences offer a fertile ground for critical agrarian studies scholarship and scholar activism to develop in generative directions. Critical agrarian studies can make vital contributions to deepen the understanding of the situation and struggles of working people positioned within and outside climate, agrarian and environmental justice movements. Engaging with these movements, while also recognizing and grappling with their own internal complexities, can further enable the interrogation of false solutions that extend processes of extractivism and accumulation. These efforts can also support the revitalization of rural struggles to sustain peasant and indigenous livelihoods and lifeways and resist new burdens compounded by climate change.

Although these movements may follow distinctive trajectories, the inter-relatedness of climate, environmental, energy, labour, food and agrarian problems remains an inescapable reality, one that is also rooted in what feminist movements and scholars in the global south have identified as a wider crisis of care and the ability to establish or reproduce conditions for a secure and dignified life. In this regard, critical agrarian studies have an important role to play in discussing, examining and amplifying possibilities and sites for transformation and emancipatory alternatives. It is precisely through such reckonings that the trenchant analyses, provocations and close studies undertaken in these pages make meaningful contributions to the ongoing work for agrarian and rural worlds beyond crisis.

Shaila Seshia Galvin*, Mercedes Ejarque**, Jennifer Franco***, Jacobo Grajales*, Ruth Hall*, Ricardo Jacobs*, Sinem Kavak**, Katie Sandwell***, Sergio Sauer* and Annie Shattuck*

**Journal of Peasant Studies, **Collective of Agrarian Scholar-Activists from the South, ***Transnational Institute*

OPEN ACCESS

Climate change and agrarian struggles

Saturnino M. Borras Jr., Ian Scoones, Amita Baviskar, Marc Edelman, Nancy Lee Peluso and Wendy Wolford

ABSTRACT

This essay introduces and invites contributions to a new *Journal of Peasant Studies* Forum on 'climate change and critical agrarian studies'. Climate change is inextricably entwined with contemporary capitalism, but how the relationship between capitalism and climate change plays out in the rural world requires deeper analysis. In particular, the way agrarian struggles connect with the huge challenge of climate change is a vital focus for both thinking and action. In this essay, we make the connections between climate change and critical agrarian studies and identify competing, although overlapping, narratives. These narratives frame climate change debates and the way that the dynamics of climate change shape and are shaped by the rural world, whether through state policies, international governance, corporate influence, or agrarian struggles. We use a simple framework to examine different logics and strategies for anti-capitalist struggles that might connect climate change and agrarian mobilisations. We conclude with some overall reflections and suggestions for broad, guiding questions for future inquiry as part of the *JPS* Forum.

Introduction

Climate change is an existential threat to humanity and the planet and a cruel engine of myriad forms of injustice, disruption and destruction. The effects of climate change from human-made emissions of greenhouse gases are devastating and accelerating, yet are uncertain and uneven, both in terms of geography and socio-economic impacts. Emerging from the dynamics of capitalism since the industrial revolution — as well as industrialisation through state-led socialism — the consequences of climate change are especially profound for the countryside and its inhabitants. In this essay, we ask: what are the implications of climate change (and climate change mitigation efforts) for rural areas around the world, and how have rural people responded politically to these challenges?

Climate change is deeply entangled with the functioning of contemporary capitalism, as well as industrialism associated with state socialism.[1] Discussions range from Karl Marx's identification of the 'metabolic rift'[2], to wider debates about the 'Anthropocene' (Reisman and Fairbairn 2021), to how capitalism generates 'climate apartheid' and the contradictions that lead to climate chaos (Mwenda and Bond 2020; Rice, Long, and Levenda 2021). But how the relationship between capitalism and climate change plays out on the ground in the rural world has received less attention. In particular, the way agrarian struggles — led by peasants, pastoralists, fisherfolk, rural workers and others — connect with the challenge of climate change, linking to and going beyond the already widespread challenges to expropriation and extraction in rural areas, is a vital focus for both thinking and action.

This essay presents a set of notes and ideas from the *Journal of Peasant Studies* editorial collective and introduces a new *JPS* Forum on 'climate change and critical agrarian studies'. The essay is an invitation to contribute to the Forum, responding to the questions posed, or coming up with new ones. Contributions combining wider theoretical reflections with empirical analyses are welcomed.

The essay proceeds as follows. In the next section, we make the connections between climate change and critical agrarian studies through three themes. After presenting this background we identify four competing and overlapping narratives that frame climate change debates and influence how they play out in the rural world. These narratives in turn shape the nature of climate politics and the formation of agrarian struggles. We then use a framework to explore different logics and strategies for anti-capitalist and anti-state struggles that might connect climate change and agrarian mobilisations. We conclude with some overall reflections and suggestions of broad, guiding questions for future inquiry as part of the *JPS* Forum.

Climate change and critical agrarian studies: making the connections

In this first section, we explore the connections between climate change and critical agrarian studies, moving from debates about capitalism and 'nature', to situating climate change issues in rural contexts to discussions of climate change and agrarian politics.

Capitalism and 'nature'

Over the past few decades, the fields of ecological Marxism and political ecology have intensively explored relationships between capitalism and 'nature'. These concerns

[1]Many authors explore the connections between climate change and capitalism from different angles (e.g., O'Connor 1998; Newell and Paterson 2010; Moore 2015, 2017; Klein 2015; Malm 2016, 2018; Millar and Mitchell 2017; Wainwright and Mann 2018; Gonzalez 2020; Newell 2021), but to assert such connections does not mean minimising the historical responsibility of what used to be called 'actually existing socialism' and of societies, notably China, that evolved in the direction of state capitalism. Both used (and China and Russia still use) carbon-intensive practices with major environmental consequences (Rogers 2015; Smith 2020). In recent years, China has been the number one national emitter of greenhouse gases and Russia number seven; the substantial emissions of the Soviet Union (until its dissolution in 1991) are evident in the data analysed by Griffin (2017, 8).

[2]In the third volume of *Capital*, Marx argues that 'capitalist property relations provoke an irreparable rift in the interdependent process of social metabolism, a metabolism prescribed by the natural laws of life itself" (Marx 1992, 949).

overlap with some of those of critical agrarian studies (Edelman and Wolford 2017), but we argue that we must go beyond past emphases.

The environmental dimensions of the interrelations between production, circulation, exchange, consumption and waste have been a key theme in ecological Marxism, specifically its focus on the 'metabolic rift' (Foster 1999; Foster, Clark, and York 2011). As highlighted in a significant literature, the metabolic rift occurred at a specific historical moment in the development of capitalism, and it continues wherever these conditions pertain: as capitalism penetrates a previously agrarian society, more commodities are produced that are circulated and consumed in distant places, the natural cycle of localised nutrient use and re-use is broken and the divide between agriculture and industry, as well as between rural and urban zones, widens (Schneider and McMichael 2010).

Jason Moore (2017) takes issue with the 'metabolic rift' argument, pointing out that humans and nature are not separated, but humans act in nature, part of a systemic 'world ecology'. Meanwhile, identifying multi-species realities, where nature–society separations are dissolved into hybrid assemblages, can offer deeper insights into the realities of the Anthropocene (Latour 2004; Haraway 2015; Haraway et al. 2016; Galvin 2018; Latour et al. 2018; Tsing, Mathews, and Bubandt 2019). However, Andreas Malm (2016) objects, and advocates a dialectical approach between humans and the natural world, in which interactions, relations and contradictions are central (cf. Soper 1995). This is a stance that echoes Raymond Williams (1980) and other cultural Marxists on the material and ideological entanglement of labour and nature. Radical, dialectical polarisations, including elaborations of the 'metabolic rift', can help expose the contours of politics around which mobilisation can occur, 'recovering a theoretical basis for ecological militancy' (Malm 2016, 156).

Emphasising the relations between capitalism and nature, James O'Connor (1998) identifies the 'second contradiction of capitalism' as the tendency for capitalism to utilise the natural resources upon which it depends at an unsustainable rate. The logic of capital in search of endless profit requires a continuous supply of cheap or free inputs (nature, labour, energy, food and so on), particularly on capitalism's frontiers, where inputs are mobilised, often violently and with little compensation (Peluso and Lund 2011; Patel and Moore 2017). This generates uneven development of capitalism across geographic spaces and societies over time (Harvey 2003; Smith 2008), providing the basis for colonial and imperial relationships.

In a similar vein, Nancy Fraser (2021, 120) argues that 'capitalism harbours a deep-seated ecological contradiction that inclines it *non-accidentally* to environmental crisis; [...] those dynamics are inextricably entwined with other, "non-environmental" crisis tendencies and cannot be resolved in isolation from them'. She argues that: 'The political implications are conceptually simple if practically challenging: an eco-politics capable of saving the planet must be *anti-capitalist* and *trans-environmental*' (ibid.; original emphasis). By 'trans-environmental' she means going beyond merely an environmental focus, as climate change is deeply entwined with the systemic crises generated by contemporary capitalism. She concludes that, 'Anti-capitalism is the piece that gives political direction and critical force to trans-environmentalism. If the latter opens eco-politics to the larger world, the former trains its focus on the main enemy' (Fraser 2021, 126).

Many of the arguments in ecological Marxism are foundational to the field of political ecology, although political ecology was in part a reaction to the ahistorical, functionalist frameworks of the sub-discipline of cultural ecology that grew out of 1960s anthropology and that emphasised how cultural practices were 'adaptive' and reconciled imbalances between humans and their environment.[3] In the 1970s and 1980s, the presence of militant peasant uprisings and radical groups, from Colombia to the Philippines and Vietnam, pushed political ecologists to argue that all life, human and non-human, produces value through labour. Society is shaped by the uneven ways in which labour was expended, extracted and distributed. Political ecologists have therefore focused on the material, relational and symbolic manifestations of power in agrarian settings, alongside a non-hierarchical, dialectical focus on scale and interactions between local and global dynamics (Rangan and Kull 2009; Sayre 2015). Rejecting the Malthusian implications of some interpretations of the 'Anthropocene' (Yusoff 2021; Malm and Hornborg 2014), political ecologists have instead increasingly engaged with the 'Capitalocene' (Moore 2017), the 'racial Capitalocene' (Vergès 2019) and the 'Plantationocene' (Haraway 2015; Tsing, Mathews, and Bubandt 2019; Wolford 2020; Carney 2021), all of which centre capitalism or the world economy in understandings of environmental change.[4] Political ecology, like critical agrarian studies, recognises that capitalism, as with climate change, is not a global process that happens *to* local communities; rather, capitalism and climate change are social and ecological processes that are both produced and experienced at multiple sites and scales.[5]

These now-large bodies of work highlight how capitalism and climate change mutually constitute each other, and how this can lead to catastrophic consequences. Responses range from facilitating local action for climate justice (Temper et al. 2018), focusing on what Martinez-Alier (2002) calls 'the environmentalism of the poor', to revolutionary action against climate polluters. How then should climate change be addressed specifically in agrarian settings? Can the long tradition of critical agrarian studies (Edelman and Wolford 2017) draw inspiration from ecological Marxism, political ecology and other fields, and shed new light on this urgent problem?

In an important contribution, Henry Bernstein (2010, 300) highlights an 'inherited weakness' in many materialist conceptions of the 'development of the productive forces in capitalist agriculture' that embrace such development as 'forever progressive', ignoring the ecological havoc it wreaks. He suggests, as we do here, that many critical scholars critique the market while implicitly believing in its ability to manage climate change through adaptation and technological innovation. It is however necessary to grapple with the very real limits imposed by climate change and industrial production systems, and so envisage major transformations. There is a scalar logic at work here too: many scholars of political ecology and agrarian studies have analysed localised relations with the understanding that these take shape in the context of broader, global capitalist relations. With climate change, though, we are forced to examine such relations in the context of broader *environmental* relations as well. Just as capitalism is everywhere, so too is climate change.

[3]For example, Peluso (1992); McCarthy (2002); Davis (2002); Zimmerer and Bassett (2003); Hecht and Cockburn (2011); Robbins (2011); Watts (2013 [orig. 1983]); Barnes et al. (2013); Perreault, Bridge, and McCarthy (2015) and Peluso and Vandergeest (2020), among many other important contributions.

[4]For a recent discussion of the 'Plantationocene' concept by a range of scholars and activists, including many in political ecology and agrarian studiės, see the recordings from a recent conversation, https://einaudi.cornell.edu/research/global-research-priorities/conversation-plantationocene.

[5]Thanks to Kasia Paprocki for this point.

We argue that climate change needs to be seen in its wider, historical context, and not just as a technical phenomenon emerging from anthropogenic greenhouse gas emissions. One challenge of climate change emerges from long-running patterns of 'uneven ecological exchange' and consequent 'ecological debt', which result in part from histories of colonial and imperial relations between the core and periphery (Roberts and Parks 2009; Foster and Holleman 2014; Ajl 2021). Contemporary economic and political relations that result in forms of enclosure and extraction — particularly in the Global South — are often justified in the name of meeting net-zero commitments, offsetting targets or providing technologies for low-carbon transitions, and are the direct consequence of such historical processes and patterns of uneven exchange. This applies equally to the depredations of capitalist firms as well as to state-backed industrialisation led by state-owned enterprises and others. Any reflections to address climate change, therefore, must employ a more expansive, historically informed analysis that situates 'climate' within a wider set of environmental struggles in agrarian settings.[6]

We argue that new work in critical agrarian studies needs to retain the focus on local material histories and power relations, while embedding both in long-term analysis of global environmental change and understandings of the way in which this new historical moment and the phenomenon of climate change are shaped by both material limits and the legacies of colonial and imperial inequality. This requires thorough conceptualisations of the relations between capitalism (in its many forms, whether transnational, state-led or local) and diverse forms of 'natures' and 'socio-natures', including the climate, for agrarian settings.

Climate change and the rural world

How then do climate change and the rural world intersect? The rural world is the site where forests are both protected and exploited; where huge mining interests compete with small-scale miners and where rural industrial activity pollutes the air and destroys the land (Peluso 1992, 2017; Peluso and Watts 2001; Hecht and Cockburn 2011). Yet industrial capitalism has often treated nature as inexhaustible and rural inhabitants as disposable through the extraction of vast quantities of resources, from crops to oil, natural gas, minerals and more (Alonso-Fradejas 2021). Neoliberal, state socialist and populist governments alike have thus predicated national development on cheap natural resources and this tendency is likely to be aggravated in the context of climate change.

Climate change exacerbates the uncertainty and amplifies the risks attached to capitalist agriculture, thereby increasing the vulnerability of rural populations. Responses to climate change range from migration to locally based practices that respond to increased variability of rainfall (Mehta et al. 2019) to more institutionalised 'adaptation' and 'mitigation' schemes, often the centrepiece of rural development projects today. There is a plethora of labels, including 'climate-smart', 'nature-positive' or 'resilient' development, but how such concepts and programmes are constructed, through what forms of knowledge and practice, requires further scrutiny.[7]

---

[6]We are grateful to Max Ajl for this important point.

[7]For example, see: Agrawal (2008), Pelling (2010), Ribot (2014), Nightingale (2017), Nightingale et al. (2020), Eriksen et al. (2021), Mehta et al. (2021), Paprocki (2021).

Too often, technocratic approaches frame interventions, obscuring more cosmopolitan 'civic epistemologies' and the intersecting uncertainties and diverse local understandings of climate change in particular settings (Szerszynski and Urry 2010; Wynne 2010; Jasanoff 2011). Such interventions in turn generate a new technocratically-driven politics of climate change, particularly in marginalised areas of the world where climate threats are deemed most pressing (Paprocki 2019). Further, this can replicate the colonial relations that overshadow or subordinate local perspectives, knowledges and 'technologies of humility' (Jasanoff 2007). Central to questions of climate and rural development are relations of power, with external interventions based on forms of accredited science often reinforcing the dominant visions of the powerful (Forsyth 2012). They thus may act to protect the inequitable *status quo* from the threat of climate change, shoring up exclusionary, sovereign notions of place, state rule and citizenship (Potter 2013).

Policy responses to climate change have an impact on social, economic and political relationships in the rural world. These include climate-financing, carbon-offsetting and sequestration schemes that transform rural landscapes through various forms of enclosure.[8] Climate responses centred on shifts to low-carbon alternatives also involve the extraction of resources from rural areas to produce renewable energy and infrastructure, whether biofuels, hydropower or solar and wind farms (Franco et al. 2010; Borras Jr et al. 2016; Barnes 2017; Dunlap 2018; Stock and Birkenholtz 2021; Torres Contreras 2021). Prompted by the climate challenge, and backed by donor and private finance, national plans the world over are full of investments in biofuel, hydropower, REDD+ carbon forestry and Bio-energy with Carbon Capture and Storage (BECCS) projects (Corbera 2012; Leach and Scoones 2015; Turnhout et al. 2017).

Such interventions require restructuring access to and use of resources, and the reframing of what are represented as climate-acceptable practices. Technocrats and investors continue to disparage and scapegoat agrarian production systems, such as mobile pastoralism, swidden agriculture and artisanal fishing, as destructive, wasteful and polluting, without deeper knowledge of the actual impacts of such practices on the environment and climate (Dressler et al. 2017; Franco and Borras Jr 2019; Houzer and Scoones 2021).[9] Meanwhile, dominant actors tolerate extractive and 'productive' corporations and state enterprises without question.

Climate change and agrarian politics

The impacts of climate change and responses to them are highly differentiated. Many suffer, while others prosper and accumulate (Watts 2013 [orig. 1983]). Very often the consequences of climate change are not spectacular, as presented in the form of a major drought, famine, hurricane or other disaster, but emerge as patterns of 'silent' or 'slow' violence (Watts 2013 [orig. 1983]; Peluso and Watts 2001; Nixon 2011; see also Benjaminsen et al. 2012), occurring incrementally, over time and hidden from view. These

[8]Even though such policies have not been implemented in many parts of the world on a significant scale, and state and market agents continue with business as usual, they are significant indicators of the direction of future interventions.

[9]See also the many contributions on 'green grabbing' and related processes, including, Brockington and Duffy (2011), Arsel and Büscher (2012), White et al. (2012), Fairhead, Leach, and Scoones (2012), Mehta, Veldwisch, and Franco (2012), Rulli, Saviori, and D'Odorico (2013), Hunsberger et al. (2017), Dell'Angelo et al. (2017), Borras Jr, Franco, and Nam (2020) and Liao et al. (2021).

differentiated consequences of both climate change itself and institutionalised responses to it have given rise to a new climate-related politics in rural areas.

Contemporary agrarian politics in many parts of the world have roots in the upheavals of the last century, including communist-inspired revolutions and anti-colonial struggles. While peasants did not lead these wars, they played an important role, often providing the mass base for the insurgent people's armies of the 'peasant wars' of the twentieth century (Wolf 1969). These struggles in turn shaped patterns of national development and underdevelopment and generated tensions between the imperial core and the colonised periphery (Rodney 1972; Amin 1974; Cooper and Stoler 1997; Carney 2011; Moyo, Jha, and Yeros 2013), resulting in historically embedded patterns of unequal ecological exchange that frame the current climate crisis (Ajl 2021).

Rural peoples have historically focused on four overlapping arenas of political contestation. These include: (1) changing social relations around property (especially access to the means of production, including access to land, forests, grazing or water); (2) labour regimes and relations; (3) income (profit or wages) and (4) consumption and social reproduction. The peasantry, alongside the rural labour force, differentiated by social class — along with co-constitutive social relations of race, ethnicity, caste, gender, religion and generation, among other elements — shape agrarian politics. From classic studies of agrarian societies, provocative questions arise, like those in Marx's *Eighteenth Brumaire*, about the conservative politics of some smallholding peasants (Marx 1982 [orig. 1852]). How do rural people become revolutionary, form alliances and create conditions for transformation? (Wolf 1969; Huizer 1972; Paige 1975)? What kinds of class alliances and agrarian transformations lead to which kinds of state and modes of political rule and institutions (Moore Jr. 1967)? Why and when do peasants – and other rural peoples - revolt (Scott 1977; but see Popkin 1979)?

While these classic concerns remain relevant, the context for agrarian struggles has shifted over the past century (Bernstein 2006; Akram-Lodhi and Kay 2009; Levien, Watts, and Yan 2018). Contemporary struggles remain firmly linked to global capitalism, but since the early 1980s autonomous agrarian social movements have often supplanted communist or socialist parties as the main protagonists. Many of these are not national in scope, but sectoral, subnational or transnational, single-issue campaigns, mobilising alongside stand-alone localised initiatives (Fox 1992; Edelman 1999; Moyo and Yeros 2005; Wolford 2010; Hall et al. 2015; Edelman and Borras 2016). Some combine class politics with identity politics around race, ethnicity, gender, religion or advocacy such as rural villagers dispossessed by large hydropower projects (Baviskar 1995). Others are rural environmental justice struggles against mining, pollution, 'fortress conservation', energy investments, concentrated animal production operations and industrial monoculture plantations.[10]

In addition to these localised conflicts, transnational agrarian movements have been prominent in struggles against neoliberal globalisation and against the World Trade

[10]'Fortress conservation' is the practice of barring people from forest and other environments that they traditionally used to conserve biodiversity and habitats. Most 'fortress' policies rest on a flawed notion of pristineness that sees 'natural' environments as apart from and uninfluenced by low-impact human activities. See the many contributions on resource extractivism and enclosure from different settings (e.g., Peluso and Watts 2001; Bebbington et al. 2008; Hecht and Cockburn 2011; Weis 2013; Martinez-Alier et al. 2016; Arsel, Hogenboom, and Pellegrini 2016; Adaman, Arsel, and Akbulut 2019; Scheidel et al. 2020; Shah et al. 2021; Kröger 2021).

Organisation (WTO) (Desmarais 2007; Edelman and Borras 2016). Asserting that 'another world is possible', they have advocated food sovereignty and agroecology as an alternative to the corporate, carbon-intensive industrial agri-food system.[11] More broadly, food sovereignty and agroecology have become central dimensions of social justice movements' critique of neoliberalism and a political compass for the construction of positive, alternative futures (Gibson-Graham 1997, 2008).

Nancy Fraser argues that environmental justice movements are too narrowly focused and 'fail to pay sufficient heed to the underlying structural dynamics of a social system that produces not only disparities in outcomes but a general crisis that threatens the well-being of all, not to mention the planet' (Fraser 2021, 125). This a useful and pointed critique, but how does it connect with the agrarian contexts of concern here? If capitalism and climate change are linked, then class and co-constitutive social relations of race, ethnicity, caste, gender and generation must be put front and centre of any analysis of the causes and conditions of climate change, as well as of climate actions.

One consequence of the penetration of neoliberal capitalism into rural areas and the violence of states in processes of enclosure, extraction and exclusion is the appeal of nationalist, authoritarian and right-wing movements, offering populist solutions to protect 'the people' from both the state and the market. Authoritarian populist appeals have tapped into the disenfranchisement and long-term neglect of rural populations and sometimes articulate concerns around environmental protection and climate change, arguing for populist, non-interventionist, local responses.[12] In the same way, movements that are partly religion-based, such as Zero Budget Natural Farming in India, may project anti-science and exclusionary narratives, emphasising a mythical, golden pre-colonial past (Khadse et al. 2018).

An urgent question is whether contemporary agrarian movements have internalised climate change politics as a key context for and object of political struggle, and if so, how and to what extent? Conversely, we must ask whether environmental and climate justice movements take agrarian justice seriously and if so, how and to what extent?[13] And further we must ask, how has the worldwide rise of various combinations of authoritarianism and populism, in which the rural world plays a significant role, influenced such processes? These questions suggest important issues for empirical research, exploring the connections between agrarian and environmental/climate struggles in different political contexts across the world.

Framing the climate challenge: contrasting narratives

Climate change and climate action have assumed growing urgency in recent years, whether in UN deliberations around the 2015 Paris Agreement through the Conference of the Parties (COP) process, scientific analyses of the International Panel on Climate Change (IPCC), or commitments of governments, corporations and financiers to 'low-

---

[11]See for example important contributions from, Patel (2009), Weis (2010), Perfecto and Vandermeer (2010), Wittman, Desmarais, and Wiebe (2010), Rosset and Altieri (2017), Bezner Kerr et al. (2018), Anderson et al. (2019), Holt-Giménez, Shattuck, and Van Lammeren (2021) and Akram-Lodhi (2021), among others.

[12]See, Scoones et al. (2018), McCarthy (2019), Neimark et al. (2019), McKay, Oliveira, and Liu (2020), Mamonova and Franquesa (2020), and Roman-Alcalá, Graddy-Lovelace, and Edelman (2021).

[13]See related discussion on the notion of 'agrarian climate justice' by Borras Jr and Franco (2018).

carbon' or 'net-zero' futures. Agrarian movements have increasingly engaged with these political spaces and international platforms (Chatterton, Featherstone, and Routledge 2013; McKeon 2015; Tramel 2016; Claeys and Delgado Pugley 2017; Routledge, Cumbers, and Derickson 2018), opening up debates to rural issues.

Inevitably, there are competing explanations for the causes and consequences of the climate crisis and for the course of appropriate climate action. As Mike Hulme (2009, 251) points out, 'one of the reasons we disagree about climate change is because we understand development differently' (see also Gupta 2010). Given the hegemony of capitalism in contemporary development, this means that when we disagree about climate change, it is often because we disagree about capitalism — and thus also about the appropriate role for agriculture and land use under capitalism, including wider patterns of ownership and control.

At the risk of reifying 'ideal types', we suggest that there are four main competing narratives about climate change and agrarian struggle. They all overlap, there are multiple strands within each, and they are often combined. But in thinking about how they emerge and are responded to in diverse agrarian settings, it is important to explore how different actors and their wider movements frame the climate challenge and the role of capitalism in particular, as this informs how actions are conceived and struggles are defined.

Corporate-driven, technological narratives

Corporate-driven, technological narratives frequently link business and philanthropic interests, connected through think-tanks, NGOs and elite organisations like the World Economic Forum. The basic assumption, sometimes implicit, is that there is nothing inherent in corporate capitalism that has led to the climate crisis. Proponents of this premise see capitalism as a self-correcting system that can simultaneously generate unbridled expanded reproduction of capital while pursuing effective climate change mitigation and adaptation. While corporate capitalism might have helped cause climate change in the past, they view this as largely accidental and suggest it can be reversed through an open market-place with the right commitments and incentives.

At the core of this narrative is the idea that the crisis is exogenous to the system of production: the market did not create the problem, therefore corporations and the technologies they can deliver can solve the climate crisis. This presumes new systems of incentives/disincentives, involving the 'marketising' of nature. In turn, a 'circular economy' is envisaged that allows for continuous capital accumulation, even in a low-carbon economy, where profits can be made from everything, from cultured meat to cool roofs to renewable energy installations. The primary goal is a win-win situation where corporations continue as profit-making enterprises under an emissions-reduction regime. The vision of the 'great transition', whereby capitalism is saved, patterns of accumulation are redefined and the worst of climate change is averted is, of course, a class project of Global North elites, in which offsets and fantasies of 'net-zero' are central, even though critics increasingly lambast these as 'dangerous traps' (Dyke, Watson, and Knorr 2021).

The technological fixes central to these approaches are fundamentally about achieving efficiency in production, circulation, exchange and consumption of commodities globally, or 'geoengineering' the planet to slow warming (Surprise 2018; Pearce 2019). This implies technical competence and distributive and allocative *efficiency* (rather than justice), all with unbridled expanded reproduction of capital and economic growth without limits, celebrating a peculiarly Western vision of modernity and progress (Isenhour 2016). The controversial 'ecomodernist manifesto' published by the Breakthrough Institute (Asafu-Adjaye et al. 2015) encapsulates this view, with an argument for technology-led decoupling of economy and environment that has provoked many critiques (Caradonna et al. 2015; Hickel 2020; Albert 2020).

Promises of technological fixes generate market and policy expectations and in turn investment, powerfully framing and influencing the climate discussion (McLaren and Markusson 2020). Financial actors are always looking for new opportunities (Clapp and Isakson 2018). Techno-fix advocates point to intensification of production and circulation as 'land-sparing' alternatives that release land for protective conservation uses and 'half-earth' solutions (Wilson 2016). Some suggest conserving 30 percent of the planet's surface by 2030 (Dinerstein et al. 2020; Waldron et al. 2020). These and similar 'solutions' have been targets of both scientific and ethical scepticism (Büscher et al. 2017).[14]

Pro-corporate approaches frequently incorporate a 'techno-spatial fix' (Harvey 2003), with offsetting schemes that sell far-off carbon sinks, typically forest areas or monocrop tree plantations (Bumpus and Liverman 2008; Lovell, Bulkeley, and Liverman 2009; Huff 2021). Why, for instance, constrain the aviation industry from maintaining its operations when some communities of poor rural villagers in the Global South can instead benefit from a carbon sequestration scheme to create a 'net-zero' balance? Resources captured in this way also need to be protected from those assumed to be ecologically destructive users, such as poor villagers living in and around these areas, mostly in rural parts of the Global South. This may be the ultimate metabolic rift.

The logic of repair and restoration through market-led and technological interventions extends to rural settings in the form of 'climate-smart agriculture'. Taking many forms, the broad approach of 'climate-smart agriculture' (as well as 'sustainable intensification' or 'digital agriculture') aims to increase efficiency and productivity in agriculture and to reduce emissions. Effected through market-driven systems of incentives and disincentives, and via a plethora of projects, 'climate-smart' approaches are now widespread (FAO 2013; World Bank 2016). In this vision, capitalist farms, including large agribusinesses, using precision technologies, labour-displacing artificial intelligence and automation, and genetic engineering, constitute climate-smart farming systems, while traditional swidden agriculturalists or mobile pastoralists receive the blame for ecologically destructive farming and livestock-keeping practices (Taylor 2014; Clapp, Newell, and Brent 2018; Newell and Taylor 2018).

The corporate-led narrative champions market mechanisms, voluntary guidelines, corporate social responsibility, codes of conduct and business-led sustainability initiatives, but state enforcement and reliance on subsidies and regulatory frameworks are still very much part of the picture. To 'adapt' to climate change, public-sector investments protect capitalist interests and maintain business as usual. Fashionable proposals include huge publicly funded efforts around everything from geoengineering in space to building sea walls. Private philanthropy is also, ironically, called upon to restore resources destroyed by prior wealth accumulation (Morrison 2019; Pearce 2019; Ribeiro

[14]https://openlettertowaldronetal.wordpress.com/

2019). And in the cases of enclosure for climate-led initiatives, proponents of this narrative are quick to call on the military, police, paramilitaries and courts to enforce various kinds of 'fortress conservation' (Brockington 2002; Dunlap 2018; Verweijen and Marijnen 2018). Technological approaches to mitigating the climate crisis are no doubt essential (Hawken 2017), but everything depends on who controls the technologies and how they are inserted in a social and economic system and regime of accumulation.

Corporate-driven, technological narratives have significant promoters among commercial producers, such as the World Farmers' Organisation (WFO). For example, during the 2019 Madrid COP25, Theo de Jager, former president of Agri South Africa (AgriSA), and current president of the World Farmers' Organisation, declared that 'smallholder farmers need to be exposed to climate smart agriculture for sustainability' (Spore 2019). This echoes the commitments of large philanthropic organisations and foundations, such as the Bill and Melinda Gates Foundation, as well as the corporate business community, most prominently through the World Economic Forum (Schurman and Munro 2013; Wise 2019).

Climate emergency narratives

There are two main types of climate emergency narrative. On the one hand, scientists and even some national security experts point out that the widely accepted assessments, projections, targets and claims of the IPCC are negotiated findings subject to political influence that likely understate the severity and urgency of the crisis (Spratt and Dunlop 2019). More radical critics add that the IPCC is largely silent on capitalism's role in the crisis. On the other hand, other climate emergency narratives argue that disaster is imminent and that this justifies unusual, aggressive and sometimes undemocratic measures, including 'states of exception' where 'sovereign power' would sideline citizens' rights, agency and knowledge (Agamben 2005; Anderson 2017; Gills and Morgan 2020; Paprocki 2021). The two types sometimes overlap and the boundaries between them shift. While we applaud those who point to the gravity of the crisis and stress the imperative of radical action, we have deep reservations about the anti-democratic and authoritarian premises of the 'state of exception' advocates.

Proponents of authoritarian 'emergency' interventions typically stress projected increases in global temperature and identify thresholds and deadlines. These justify urgent measures, even though some critics argue that such 'deadline'-focused action may be dangerous (Asayama et al. 2019; Hulme 2019). Climate emergency narratives feature apocalyptic futures based on current trends and projections: melting glaciers, thawing permafrost, rising sea levels, droughts and heat waves, more severe El Niño and La Niña phenomena, irreparably damaged nature, rampant pollution and eventual civilisational collapse (Skrimshire 2010). They focus on the need for concerted, urgent, rapid action, no matter what, and adamantly reject any gradualist reformism. Meanwhile, grassroots, localised efforts at mitigation or adaptation are frequently rejected as piecemeal, too slow or insufficient.

There are increasing calls to declare 'climate emergencies', from global to local levels (Ruiz-Campillo, Castán Broto, and Westman 2021). Many cities, for example, have made such declarations, calling for changes in individual choices in food consumption and for a rethinking of transport, housing infrastructure and planning. The idea of 'degrowth', involving a shift in patterns of economic production and consumption, has become a popular rallying point (Demaria and Latouche 2019; Gerber 2020; Hickel 2020; Kallis et al. 2020; Jackson 2021), although this too has attracted critique (Milanovic 2017).

A diverse political coalition links to these narratives, with ambivalent, sometimes contradictory ideas about the role of the state and wider democratic processes. In some cases, there remains a contradiction between calls for urgent, emergency action and suspension of democratic accountabilities, and calls for wider citizen involvement through 'citizens' assemblies' or other deliberative fora as a way forward. Others advocate something like radical 'war communism' (Malm 2020) through a revolutionary mobilisation of forces in the face of impending catastrophe, drawing inspiration from Vladimir Lenin and Rosa Luxemburg, amongst others (Bensaïd 2002a).

Climate justice narratives

There are multiple, sometimes competing, climate justice narratives, with contrasting politics around the relationship between climate change and capitalism in agrarian settings. Climate justice narratives start from the observation that inequalities and injustice are at the root of the causes and impacts of climate change (Adger et al. 2006; Marino and Ribot 2012; Swilling and Annecke 2012; Agostino 2015; Klinsky et al. 2017; Boyce 2018). They call for a 'just transition' (Swilling 2019) or the creation of just 'pathways to sustainability' (Leach, Stirling, and Scoones 2010). These approaches identify a range of injustices related to knowledge (whose knowledge counts?), procedure (who is involved in deciding?), distribution (who gets which benefits and who suffers what costs/risks?), and correctives (how are past wrongs addressed?). Proponents argue that a focus on different dimensions of justice is key to addressing climate change (Gardiner 2011). Climate change goes beyond the biophysical and technical and must be seen through the lenses of inequality and injustice (Newell and Mulvaney 2013; Lynch et al. 2019; Tuana 2019; Newell et al. 2020; Sultana 2021), extending beyond the rights of people to those of the living world, with a multi-species conception of environmental and climate justice (Tsing et al. 2020; Tschakert et al. 2021).

Although calls for 'climate justice' or 'just transitions' have become commonplace, the political implications are sometimes unclear (Schlosberg 2009). Some embrace a liberal rights-oriented notion of justice, underscoring allocation and compensation issues. Many of the corporate-led solutions through market and technological fixes have highly variable outcomes (Eriksen et al. 2021); in the now-pervasive discussion of 'planetary boundaries' (Rockström et al. 2009), ideas of justice are combined with perspectives on 'safe spaces' within boundaries. For example, Carl Folke and colleagues (2021, 834) indicate that 'the Anthropocene reality of rising system-wide turbulence calls for transformative change based on emerging technologies, social innovations, shifts in cultural repertoires, and a diverse portfolio of active stewardship of human actions in support of a resilient biosphere'. This, of course, raises big questions about the meaning and politics of transformative change, planetary stewardship and what technologies and social innovations are desired by whom.

Those who take a more radical approach to justice emphasise the rights of those already structurally marginalised, including the poor and future generations (Tschakert and Machado 2012), highlighting transformations within capitalism as the major challenge (Malm 2016; Wainwright and Mann 2018). Still others focus on the rights of non-humans, invoking Andean ideas of '*buen vivir*', '*Pachamama*' and the 'rights of nature', celebrating indigenous and non-Western constructions of inseparable human–nature relations, as highlighted by the Cochabamba declaration of 2010.[15] Such approaches challenge conventional forms of knowledge production and underscore basic issues of epistemic justice (Temper and Del Bene 2016; Whyte 2018). Competing notions of fairness, justice, restitution, reparations and ethics, including non-Western ones, can provide a useful compass in contemporary climate politics (Baer 2011; Gardiner 2011; Schlosberg and Collins 2014; Harris 2016). More radical climate justice narratives highlight the historical injustices of unequal exchange and ecological debt, whereby climate challenges in the periphery are the direct consequences of long histories of exploitation and unequal relations of global power.[16]

These different framings of climate justice are manifested in contrasting approaches to capitalism. In some, if the disadvantaged are protected and have rights, then capitalist relations can be part of the solution. Like the other narratives, the climate justice narrative is also significantly differentiated, ranging from positions that are liberal and social democratic in orientation to more transformative approaches that seek to control capitalism and rein in its worst depredations and effects. Climate justice advocates include reformist positions around 'green new deals' and those associated with mainstream green parties (Ajl 2021; Newell 2019), while others take more radical positions on rights and justice and consider the critique of capitalism more fundamental. Large international coalitions and movements – such as La Via Campesina – often find themselves and their members in the various currents within these climate justice narratives, navigating between more reformist and radical positions.

Structural transformation narratives

For structural transformation narratives, the fundamental problem is that growth in contemporary economies is dependent on fossil capital (Mitchell 2011) and plantation production (Wolford 2020), producing wealth that is maldistributed across classes, regions and economies (Hickel and Kallis 2020). The solution is not to tinker at the margins, temporarily easing the crises of capitalism through technological, market or state welfare fixes, but to transform the relations of production that generate climate change in the first place, through reparation, redistribution and decolonisation (Watts 2004; Cadieux et al. 2019; Ajl 2020). This is a more radical vision of a 'green new deal' that restructures economies in favour of a low-carbon future under people's control (Ajl 2021; Mastini, Kallis, and Hickel 2021; Selwyn 2021).

From an agrarian perspective, structural transformation would include a radical shift from capitalist, industrial agriculture to a different model. This would transform the four fundamental dimensions of agrarian political economy: property, labour, income and consumption/reproduction. It would entail a three-pronged approach to food system transformation: dismantling the global food system controlled by large

---

[15]https://www.therightsofnature.org/universal-declaration/

[16]See, https://www.academia.edu/9167899/Calculating_Climate_Debt_A_Proposal and as highlighted in the People's Agreement of Cochabamba in 2010, https://pwccc.wordpress.com/2010/04/24/peoples-agreement/

corporations; taking over the state and developing new systems of deliberative governance and building something new, drawing from food sovereignty and agroecological ideas and practices, ushering in a new food regime (McMichael 2009; Weis 2010). All of these require an accompanying restructuring of access to and control over natural resources, including through redistributive agrarian reforms. Proponents of structural transformation approaches are not always or necessarily socialist in long-term perspective, although many draw on 'eco-socialism' (Löwy 2005; Le Quang and Vercoutére 2013; Fraser 2021) and feminist ideals of care (Agostino 2015; Fraser 2016; Klein 2020; Mehta and Harcourt 2021).

* * *

Many actors' and movements' positions of course span these narratives, shaping agrarian struggles around climate in different ways. Broad international coalitions – for example, La Via Campesina and its key members – can be seen as both a 'single actor' and as an 'arena of contestation' (Edelman and Borras 2016). Within the movement, narratives of climate emergency, climate justice and structural transformation may combine. This contrasts with positions of rival movements, such as the defunct International Federation and Agricultural Producers (IFAP) and its informal successor, the World Farmers' Organisation (WFO), which emphasise a corporate-driven technological narrative centred on the promotion of biofuels and climate smart agriculture. How the climate challenge is framed and what narratives, with varied inflections and combinations, are pushed then has important implications for how we understand climate politics, and in turn agrarian struggles.

A politics of agrarian struggle for the climate change era?

These positions on climate change, capitalism and agrarian change are not forged solely by climate politics but have been moulded within wider political relations (Desmarais 2007). Not all responses by agrarian movements are explicitly anti-capitalist, as there are many tactical and strategic positions that evolve as alliances are forged. Anti-capitalist struggles also take various forms, from very localised 'do-it-yourself' transformations to movement actions engaged in wider contentious politics at the national and international levels.

In understanding agrarian struggles to confront climate change, oppositional choices between, for example, a village level seed saving-sharing campaign against a corporate seed business versus a national militant agrarian movement that demands for land reform and an end to industrial monoculture plantations, are misplaced. The challenge instead is to examine whether different groups straddle various narratives, and if so how and why; and whether groups coalesce, and if so, how and with what impact?

In discussing the politics of climate change and linking this to agrarian struggles, we find Wright's (2019) typology of 'strategic logics' of anti-capitalist struggles useful. These may be against corporate capitalism as well as wider forms of industrialism, whether notionally socialist or capitalist in character, and may involve alliances with foci of contention that go beyond climate change and may be in collaboration with other players who may not have an explicitly anti-capitalist orientation. Our focus is to enquire further into the diversity, scale and form of agrarian struggles that are addressing climate change or confronting the negative consequences of climate change mitigation interventions in rural settings. Necessarily schematic, the typology identifies five broad 'strategic logics' that historically animated anti-capitalist struggles: 'smashing capitalism', 'dismantling capitalism', 'taming capitalism', 'resisting capitalism' and 'escaping capitalism' (Wright 2019, 38–64).[17]

'Smashing capitalism' is the classic logic of revolutionaries, following Marx and Lenin. To destroy in order to build requires seizing state power, as in the classic peasant revolutions of the past. For Wright, the twentieth-century experience demonstrated that destruction of the old system through revolution did not necessarily result in a truly emancipatory new system, and he questions the 'the plausibility of a strategy that attempts to destroy in a ruptural manner the dominance of capitalism' (ibid., 42), particularly as transitions to state socialism associated with different forms of modernising industrialism have not generated the basis for confronting climate change either.

'Dismantling capitalism' shares the fundamental goals of revolutionaries but accepts the scepticism about the ruptural overthrow of capitalism, while being firmly committed to democratic socialism. Instead, the idea is to have a 'gradual dismantling of capitalism and the building up of the alternative through the sustained action of the state' (ibid., 43). Many of the rural social movements that formed in Latin America in the 1980s and 1990s moved into this space as their initial gains included a deepening of the democratic state. Movements like the MST in Brazil sought to work within and beyond state institutions, building alliances in a Gramscian-style 'war of position' (Wolford 2010). Seizing state power occurs through 'a broad, mass-based socialist party capable of winning elections and staying in power for a sufficiently long time' (Wright 2019, 43). 'Smashing capitalism' and 'dismantling capitalism' both aspire to the 'ultimate possibility of replacing capitalism with a fundamentally different kind of structure, socialism' (ibid., 44).

By contrast, Wright argues that logics focused on 'taming capitalism' see capitalism as causing fundamental harm in society. Reformist 'social democracy' is emblematic of this approach informed by liberal, justice narratives. Capitalism, it is argued, can be 'tamed by well-crafted state policies', including through regulation and redistribution. Wright argues that 'to accomplish this requires popular mobilisation and political will; one can never rely on the enlightened benevolence of elites' (ibid., 45). Through such processes, more fundamental structural transformations may emerge, driven by 'mission-led' initiatives and 'entrepreneurial' state policies (Mazzucato 2021), but at the same time pushed by citizen action and mobilisation (cf. Scoones, Newell, and Leach 2015; Scoones et al. 2020). This might result, for example, in land redistribution and agrarian reform, alongside structural shifts in food systems and land use supported by state regulation and incentives to enhance climate mitigation.

'Resisting capitalism', for Wright, refers to 'struggles that oppose capitalism from outside of the state but do not themselves attempt to gain state power' (ibid., 49). Direct-to-consumer farm movements and other solidarity campaigns fit this description, as does the promotion of local economies involving a prefiguring of food sovereignty

[17]As with any typology, there are nuances and complexities that are overlooked, but as a provocation for analysis Wright's framework helps in thinking through the diversity of 'anti-capitalist' approaches, with potentially important insights into the diversity of agrarian struggles. There are of course other diagnostic typologies exploring the politics of climate change, such as that produced by Wainwright and Mann in *Climate Leviathan* (2018). There is no sense that any of these should necessarily be a starting point for analysis in this *JPS* Forum.

alternatives to capitalism. Resistance may be galvanised through urgent, 'emergency' narratives, as a route to people's mobilisation. This logic overlaps with 'escaping capitalism'. If capitalism is too powerful to fight and defeat, the best hope is to insulate from its damaging effects. Within this logic, according to Wright, 'the "lifestyle" of voluntary simplicity can contribute to broader rejection of the consumerism and preoccupation with economic growth in capitalism' (ibid.). Here alternative movements focused on regenerative agriculture, agroecology and food sovereignty, supporting local economies and food systems, are examples of where agrarian struggles are located.

Wright argues that a strategic combination of four of the five logics — dismantling, escaping, taming and resisting — offers the most effective route forward towards an anti-capitalist transformation that can confront climate and environmental change. It is not a matter of choosing which one logic is somehow 'correct' or devising a singular strategy he argues; rather, combining multiple logics and linked to different narratives, each centred on challenging capitalism and – we would add – other forms of climate-damaging industrialism. He calls this 'eroding capitalism', representing the case for progressive social change both 'from above' through state-oriented actions and through mobilisations 'from below' that create new, emancipatory social relations.

Who, though, are the potential social forces behind such political struggles? In classic agrarian studies, landless rural labourers and poor and middle peasants are the class fractions that are the most reliable forces for transformative change (Wolf 1969; Paige 1975). The processes of enclosure and extraction that neoliberalism accelerated (McCarthy and Prudham 2004), however, have changed agrarian class dynamics, as the recent wave of global land and green grabbing indicates. Today, there is a staggering rise in the number of people who originated from rural areas but are now partly or fully separated from their means of production and social reproduction. This includes those who remain in rural settings but are unable to construct a livelihood sufficient for their survival. These are the rural 'surplus populations', 'working people', 'precariat', 'footloose labour', 'semi-proletariat' and 'fractured classes of labour' (respectively, Li 2010; Shivji 2017; Standing 2014; Breman 1996; Moyo 2005; Bernstein 2006).

As rural class formations fracture under late capitalism, it is rare to find agrarian struggles of the classic types that are consistently anti-capitalist and categorically class-oriented. A combined force that, following Fraser (2021), is 'sufficiently' anti-capitalist, trans-environmental and agrarian and that acts to 'erode capitalism' (Wright 2019) may not be widespread at present. But social forces, political movements and struggles are built over time. As Mike Davis (2020, xviii) argues, class capacities emerge conjuncturally, in the confluence of struggles and within class antagonism, and this is where the most radically transformative organising occurs. Generating an anti-capitalist politics to confront climate change in the context of diverse, sometimes competing, non-class forms of identity is a major challenge, reflecting a 'non-linear' version of Marx for our times (Bensaïd 2002b).

This conjuncture offers a politically aspirational opportunity for bringing rural movements, interests and identities together. Such a focus is nonetheless fraught with contradictions: some actions may satisfy the demands of environmental justice campaigners, but undermine the immediate interest of agrarian justice movements. Much will depend on context, but empirical investigation of diverse experiences may shed light on how a coalition of cross-class forces can form and under what terms.

In classical studies of agrarian politics, traditional allies of agrarian movements ranged from the ranks of the 'enlightened' educated middle class in small towns and their institutions (teachers/students and schools, religious practitioners and organisations) to political parties, usually communist and socialist parties, some of which had armed components (Huizer 1972; Wolford 2003). Alliances in this context often involved class-conscious politics, with an organising focus on landless labourers and poor and middle peasants, distanced from rich farmers and the agrarian bourgeoisie. Such alliance formation is less evident today. For better or for worse, a bewildering array of NGOs has supplanted such traditional allies for agrarian struggles (Edelman and Borras 2016).

Changing rural class formations under neoliberalism alter how we ask questions about agrarian struggles today. Does the erosion of classic 'peasants' amid the rise of rural 'working people' lead to a decrease in the potential for agrarian struggles? If so, what does this imply for broader anti-capitalist and climate/environmental struggles? Does the upsurge in rural–urban and international migration undermine place-based struggles? And how does the rise of populist and often authoritarian politics rooted in rural settings change the character of struggles that connect agrarian and environmental issues? These are, of course, all empirical questions that need to be investigated in different settings.

Towards a research and action agenda linking climate change and agrarian struggles

How does all of this translate into an agenda for thinking and action that analyses and connects climate change and agrarian struggles? In laying out an agenda for future work linking climate change to critical agrarian studies, we draw on the frameworks introduced in this essay and identify three overlapping clusters of questions. These might apply to multiple settings, as individual cases, or as part of global, regional or local analyses:

- How and in what specific, local and global ways, does climate change differ from past environmental exclusions or threats? What combinations of narratives and strategies frame climate change and the institutionalised responses to it in agrarian settings? What exclusions and inclusions result from this?
- How are different people — in relation to class and other co-constituted axes of social difference such as gender, race, ethnicity, age, occupation — affected by climate change and the institutionalised responses to it in agrarian settings? How does this affect processes of social differentiation, trajectories of accumulation and in turn agrarian politics?
- What political logics and strategies can together act to 'erode capitalism' and so the causes of climate change? How can these be central to agrarian struggles now and in the future? How might these operate in contexts of 'authoritarian populism' and what progressive, emancipatory coalitions and alliances can be forged?

In sum we ask: can we envision a sufficiently anti-capitalist, trans-environmental and agrarian approach to confront climate change in rural settings, and what would this look like in practice? This essay has offered a few pointers for grappling with this core question, with some heuristic frameworks drawn from diverse sources that pose questions and help structure thinking and potential action. These are not intended to be a definitive or prescriptive guide to future work, but merely a provocation to encourage it.

In concluding, we encourage contributions to this *JPS* Forum that speak to the competing perspectives mapped out in this essay, as well as new questions and themes that emerge from the essay and indeed challenge it. Contributions will involve new empirical material, with different conceptual starting points and diverse methodologies focusing on agrarian and rural settings anywhere in the world, as well as more global, international reflections. Connecting concerns around climate change and critical agrarian studies, and so deepening debates around agrarian struggles, is long overdue, and this essay is an invitation to others to contribute to the debate.

Acknowledgements

We would like to thank the three external reviewers for their constructive and helpful comments, as well as Paula Bownas for her excellent copyediting assistance.

Disclosure statement

No potential conflict of interest was reported by the author(s).

Funding

Borras' and Scoones' work on this paper has received support from the European Research Council through Advanced Grants, No. 834006 and No. 740342.

ORCID

Ian Scoones http://orcid.org/0000-0001-8547-4464
Marc Edelman http://orcid.org/0000-0001-7359-1858

References

Adaman, F., M. Arsel, and B. Akbulut. 2019. "Neoliberal Developmentalism, Authoritarian Populism, and Extractivism in the Countryside: The Soma Mining Disaster in Turkey." *Journal of Peasant Studies* 46 (3): 514–536.

Adger, W. N., J. Paavola, S. Huq, and M. J. Mace. 2006. *Fairness in Adaptation to Climate Change*. Cambridge: MIT press.

Agamben, G. 2005. "State of Exception." *Nova srpska politička misao* 12 (01 + 04): 135–145.

Agostino, A. 2015. "Climate Justice and Women's Agency: Voicing Other Ways of Doing Things." In *The Oxford Handbook of Transnational Feminist Movements*, edited by R. Baksh, and W. Harcourt, 815–837. Oxford: Oxford University Press.

Agrawal, A. 2008. *The Role of Local Institutions in Adaptation to Climate Change*. Washington DC: World Bank. https://openknowledge.worldbank.org/handle/10986/28274.

Ajl, M. 2020. "The Hidden Legacy of Samir Amin: Delinking's Ecological Foundation." *Review of African Political Economy* 48 (167): 82–101.

Ajl, M. 2021. *A People's Green new Deal*. New York: Pluto Press.

Akram-Lodhi, H. 2021. "The Ties That Bind? Agroecology and the Agrarian Question in the Twenty-First Century." *Journal of Peasant Studies* 48 (4): 687–714.

Akram-Lodhi, A. H., and C. Kay. eds. 2009. *Peasants and Globalization: Political Economy, Agrarian Transformation and Development*. London: Routledge.
Albert, M. J. 2020. "The Dangers of Decoupling: Earth System Crisis and the 'Fourth Industrial Revolution'." *Global Policy* 11 (2): 245–254.
Alonso-Fradejas, A. 2021. "'Leaving no one Unscathed'in Sustainability Transitions: The Life Purging Agro-Extractivism of Corporate Renewables." *Journal of Rural Studies* 81: 127–138.
Amin, S. 1974. *Accumulation on a World Scale: A Critique of the Theory of Underdevelopment*. New York: Monthly Review Press.
Anderson, B. 2017. "Emergency Futures: Exception, Urgency, Interval, Hope." *The Sociological Review* 65 (3): 463–477.
Anderson, C. R., J. Bruil, M. J. Chappell, C. Kiss, and M. P. Pimbert. 2019. "From Transition to Domains of Transformation: Getting to Sustainable and Just Food Systems Through Agroecology." *Sustainability* 11 (19): 2–28.
Arsel, M., and B. Büscher. 2012. "Nature™ Inc.: Changes and Continuities in Neoliberal Conservation and Market-Based Environmental Policy." *Development and Change* 43 (1): 53–78.
Arsel, M., B. Hogenboom, and L. Pellegrini. 2016. "The Extractive Imperative and the Boom in Environmental Conflicts at the end of the Progressive Cycle in Latin America." *The Extractive Industries and Society* 3 (4): 877–879.
Asafu-Adjaye, J., L. Blomquist, S. Brand, B. W. Brook, R. DeFries, E. Ellis, C. Foreman, D. Keith, M. Lewis, and M. Lynas. 2015. *An Ecomodernist Manifesto*. Oakland, CA: Breakthrough Institute.
Asayama, S., R. Bellamy, O. Geden, W. Pearce, and M. Hulme. 2019. "Why Setting a Climate Deadline is Dangerous." *Nature Climate Change* 9 (8): 570–572.
Baer, P. 2011. "International Justice." In *The Oxford Handbook of Climate Change and Society*, edited by J. S. Dryzek, R. B. Norgaard, and D. Schlosberg, 323–337. Oxford: Oxford University Press.
Barnes, J. 2017. "The Future of the Nile: Climate Change, Land use, Infrastructure Management, and Treaty Negotiations in a Transboundary River Basin." *Wiley Interdisciplinary Reviews: Climate Change* 8 (2): 1–18. doi:10.1002/wcc.449.
Barnes, J., M. Dove, M. Lahsen, A. Mathews, P. McElwee, R. McIntosh, et al. 2013. "Contribution of Anthropology to the Study of Climate Change." *Nature Climate Change* 3 (6): 541–544.
Baviskar, A. 1995. *In the Belly of the River: Tribal Conflicts Over Development in the Narmada Valley*. New Delhi: Oxford University Press.
Bebbington, A., D. H. Bebbington, J. Bury, J. Lingan, J. P. Muñoz, and M. Scurrah. 2008. "Mining and Social Movements: Struggles Over Livelihood and Rural Territorial Development in the Andes." *World Development* 36 (12): 2888–2905.
Benjaminsen, T. A., K. Alinon, H. Buhaug, and J. T. Buseth. 2012. "Does Climate Change Drive Land-use Conflicts in the Sahel?" *Journal of Peace Research* 49 (1): 97–111.
Bensaïd, D. 2002a. "'Leaps! Leaps! Leaps!': Lenin and Politics." *International Socialism* 2: 95. https://www.marxists.org/archive/bensaid/2002/07/leaps.htm.
Bensaïd, D. 2002b. *Marx for our Times: Adventures and Misadventures of a Critique*. New York: Verso Books.
Bernstein, H. 2006. "Is There an Agrarian Question in the Twenty-First Century?" *Canadian Journal of Development Studies/Revue Canadienne D'études du Développement* 27 (4): 449–460.
Bernstein, H. 2010. "Introduction: Some Questions Concerning the Productive Forces." *Journal of Agrarian Change* 10 (3): 300–314.
Bezner Kerr, R., H. Nyantakyi-Frimpong, L. Dakishoni, E. Lupafya, L. Shumba, I. Luginaah, and S. S. Snapp. 2018. "Knowledge Politics in Participatory Climate Change Adaptation Research on Agroecology in Malawi." *Renewable Agriculture and Food Systems* 33 (3): 238–251.
Borras Jr, S. M., and J. C. Franco. 2018. "The Challenge of Locating Land-Based Climate Change Mitigation and Adaptation Politics Within a Social Justice Perspective: Towards an Idea of Agrarian Climate Justice." *Third World Quarterly* 39 (7): 1308–1325.
Borras Jr, S. M., J. C. Franco, S. R. Isakson, L. Levidow, and P. Vervest. 2016. "The Rise of Flex Crops and Commodities: Implications for Research." *Journal of Peasant Studies* 43 (1): 93–115.
Borras Jr, S. M., J. C. Franco, and Z. Nam. 2020. "Climate Change and Land: Insights from Myanmar." *World Development* 129: 3–36.

Boyce, J. K. 2018. "Carbon Pricing: Effectiveness and Equity." *Ecological Economics* 150: 52–61.

Breman, J. 1996. *Footloose Labour: Working in India's Informal Economy*. Cambridge: Cambridge University Press.

Brockington, D. 2002. *Fortress Conservation: The Preservation of the Mkomazi Game Reserve, Tanzania*. Bloomington: Indiana University Press.

Brockington, D., and R. Duffy. Eds. 2011. *Capitalism and Conservation*. Hoboken, NJ: John Wiley & Sons.

Bumpus, A. G., and D. M. Liverman. 2008. "Accumulation by Decarbonization and the Governance of Carbon Offsets." *Economic Geography* 84 (2): 127–155.

Büscher, B., R. Fletcher, D. Brockington, C. Sandbrook, W. M. Adams, L. Campbell, C. Corson, W. Dressler, R. Duffy, and N. Gray. 2017. "Half-earth or Whole Earth? Radical Ideas for Conservation, and Their Implications." *Oryx* 51 (3): 407–410.

Cadieux, K. V., S. Carpenter, A. Liebman, R. Blumberg, and B. Upadhyay. 2019. "Reparation Ecologies: Regimes of Repair in Populist Agroecology." *Annals of the American Association of Geographers* 109 (2): 644–660.

Caradonna, J., I. Borowy, T. Green, P. A. Victor, M. Cohen, A. Gow, and R. Heinberg. 2015. "A Degrowth Response to an Ecomodernist Manifesto." *Resilience. Org*. https://www.resilience.org/stories/2015-05-06/a-degrowth-response-to-an-ecomodernist-manifesto/; downloaded 16 July 2021

Carney, J. 2011. *In the Shadow of Slavery*. Berkeley: University of California Press.

Carney, J. A. 2021. "Subsistence in the Plantationocene: Dooryard Gardens, Agrobiodiversity, and the Subaltern Economies of Slavery." *Journal of Peasant Studies* 48 (5): 1075–1099. doi:10.1080/03066150.2020.1725488.

Chatterton, P., D. Featherstone, and P. Routledge. 2013. "Articulating Climate Justice in Copenhagen: Antagonism, the Commons, and Solidarity." *Antipode* 45 (3): 602–620.

Claeys, P., and D. Delgado Pugley. 2017. "Peasant and Indigenous Transnational Social Movements Engaging with Climate Justice." *Canadian Journal of Development Studies/Revue Canadienne D'études du Développement* 38 (3): 325–340.

Clapp, J., and S. R. Isakson. 2018. *Speculative Harvests: Financialization, Food, and Agriculture*. Rugby, UK: Practical Action Publishing.

Clapp, J., P. Newell, and Z. W. Brent. 2018. "The Global Political Economy of Climate Change, Agriculture and Food Systems." *Journal of Peasant Studies* 45 (1): 80–88.

Cooper, F., and A. L. Stoler1997. *Tensions of Empire: Colonial Cultures in a Bourgeois World*. Berkeley: University of California Press.

Corbera, E. 2012. "Problematizing REDD+ as an Experiment in Payments for Ecosystem Services." *Current Opinion in Environmental Sustainability* 4 (6): 612–619.

Davis, M. 2002. *Late Victorian Holocausts: El Niño Famines and the Making of the Third World*. New York: Verso Books.

Davis, M. 2020. *Old Gods, new Enigmas: Marx's Lost Theory*. New York: Verso Books.

Dell'Angelo, J., P. D'Odorico, M. C. Rulli, and P. Marchand. 2017. "The Tragedy of the Grabbed Commons: Coercion and Dispossession in the Global Land Rush." *World Development* 92: 1–12.

Demaria, F., S. Latouche, et al. 2019. "Degrowth." In *Pluriverse: A Post-Development Dictionary*, edited by A. Kothari, 148–151. New Delhi: Tulika Books and Authorsupfront.

Desmarais, A. 2007. *La Via Campesina*. Halifax, NS: Fernwood.

Dinerstein, E., A. R. Joshi, C. Vynne, A. T. L. Lee, F. Pharand-Deschênes, M. França, S. Fernando, T. Birch, K. Burkart, and G. P. Asner. 2020. "A 'Global Safety Net' to Reverse Biodiversity Loss and Stabilize Earth's Climate." *Science Advances* 6 (36): 1–13.

Dressler, W. H., D. Wilson, J. Clendenning, R. Cramb, R. Keenan, S. Mahanty, & R. D. Lasco. 2017. "The Impact of Swidden Decline on Livelihoods and Ecosystem Services in Southeast Asia: A Review of the Evidence from 1990 to 2015." *Ambio* 46 (3): 291–310.

Dunlap, A. 2018. "Counterinsurgency for Wind Energy: The Bíi Hioxo Wind Park in Juchitán, Mexico." *Journal of Peasant Studies* 45 (3): 630–652.

Dyke, J., R. Watson, and W. Knorr. 2021. Climate scientists: concept of net zero is a dangerous trap. 22 April. https://theconversation.com/climate-scientists-concept-of-net-zero-is-a-dangerous-trap-157368.

Edelman, M. 1999. *Peasants Against Globalization: Rural Social Movements in Costa Rica*. Redwood, CA: Stanford University Press.

Edelman, M., and S. M. Borras. 2016. *Political Dynamics of Transnational Agrarian Social Movements*. Fernwood: Practical Action Publishing.

Edelman, M., and W. Wolford. 2017. "Introduction: Critical Agrarian Studies in Theory and Practice." *Antipode* 49 (4): 959–976.

Eriksen, S., E. L. F. Schipper, M. Scoville-Simonds, K. Vincent, H. N. Adam, N. Brooks, B. Harding, L. Lenaerts, D. Liverman, and M. Mills-Novoa. 2021. "Adaptation Interventions and Their Effect on Vulnerability in Developing Countries: Help, Hindrance or Irrelevance?" *World Development* 141: 1–16.

Fairhead, J., M. Leach, and I. Scoones. 2012. "Green Grabbing: A new Appropriation of Nature?" *Journal of Peasant Studies* 39 (2): 237–261.

FAO. 2013. *Climate-smart Agriculture Sourcebook*. Rome: FAO.

Folke, C., S. Polasky, J. Rockström, et al. 2021. "Our Future in the Anthropocene Biosphere." *Ambio* 50: 834–869.

Forsyth, T. 2012. "Politicizing Environmental Science Does Not Mean Denying Climate Science Nor Endorsing It Without Question." *Global Environmental Politics* 12 (2): 18–23.

Foster, J. B. 1999. "Marx's Theory of Metabolic Rift: Classical Foundations for Environmental Sociology." *American Journal of Sociology* 105 (2): 366–405.

Foster, J. B., B. Clark, and R. York. 2011. *The Ecological Rift: Capitalism's war on the Earth*. New York: NYU Press.

Foster, J. B., and H. Holleman. 2014. "The Theory of Unequal Ecological Exchange: a Marx-Odum Dialectic." *Journal of Peasant Studies* 41 (2): 199–233.

Fox, J. 1992. "Democratic Rural Development: Leadership Accountability in Regional Peasant Organizations." *Development and Change* 23 (2): 1–36.

Franco, J. C., and S. M. Borras Jr. 2019. "Grey Areas in Green Grabbing: Subtle and Indirect Interconnections Between Climate Change Politics and Land Grabs and Their Implications for Research." *Land Use Policy* 84: 192–199.

Franco, J., L. Levidow, D. Fig, L. Goldfarb, M. Hönicke, and M. L. Mendonça. 2010. "Assumptions in the European Union Biofuels Policy: Frictions with Experiences in Germany, Brazil and Mozambique." *Journal of Peasant Studies* 37 (4): 661–698.

Fraser, N. 2016. "Contradictions of Capital and Care." *New Left Review* 100: 99–117.

Fraser, N. 2021. "Climates of Capital: For a Trans-Environmental eco-Socialism." *New Left Review* 127: 94–127.

Galvin, S. S. 2018. "Interspecies Relations and Agrarian Worlds." *Annual Review of Anthropology* 47 (1): 233–249. doi:10.1146/annurev-anthro-102317-050232.

Gardiner, S. 2011. "Climate Justice." In *The Oxford Handbook of Climate Change and Society*, edited by J. S. Dryzek, R. B. Norgaard, and D. Schlosberg, 309–322. Oxford: Oxford University Press.

Gerber, J. F. 2020. "Degrowth and Critical Agrarian Studies." *Journal of Peasant Studies* 47 (2): 235–264.

Gibson-Graham, J. K. 1997. "The end of Capitalism (as we Knew it): A Feminist Critique of Political Economy." *Capital & Class* 21 (2): 186–188.

Gibson-Graham, J. K. 2008. "Diverse Economies: Performative Practices for Other Worlds." *Progress in Human Geography* 32 (5): 613–632.

Gills, B., and J. Morgan. 2020. "Global Climate Emergency: After COP24, Climate Science, Urgency, and the Threat to Humanity." *Globalizations* 17 (6): 885–902.

Gonzalez, C. G. 2020. "Racial Capitalism and the Anthropocene." In *The Cambridge Handbook of Environmental Justice and Sustainable Development*, edited by S. A. Atapattu, C. G. Gonzales, and S. L. Seck, 72–85. Cambridge: Cambridge University Press.

Griffin, P. 2017. The carbon majors database: CDP Carbon Majors Report 2017. Climate Accountability Institute. https://6fefcbb86e61af1b2fc4-c70d8ead6ced550b4d987d7c03fcdd1d.ssl.cf3.rackcdn.com/cms/reports/documents/000/002/327/original/Carbon-Majors-Report-2017.pdf?1501833772.

Gupta, J. 2010. "A History of International Climate Change Policy." *Wiley Interdisciplinary Reviews: Climate Change* 1 (5): 636–653.

Hall, R., M. Edelman, S. M. Borras, I. Scoones, B. White, and W. Wolford. 2015. "Resistance, Acquiescence or Incorporation? An Introduction to Land Grabbing and Political Reactions 'from Below'." *Journal of Peasant Studies* 42 (3–4): 467–488.

Haraway, D. 2015. "Anthropocene, Capitalocene, Plantationocene, Chthulucene: Making kin." *Environmental Humanities* 6 (1): 159–165.

Haraway, D., N. Ishikawa, S. F. Gilbert, K. Olwig, A. L. Tsing, and N. Bubandt. 2016. "Anthropologists are Talking – About the Anthropocene." *Ethnos* 81 (3): 535–564.

Harris, P. G. 2016. *Global Ethics and Climate Change*. Edinburgh: Edinburgh University Press.

Harvey, D. 2003. *The new Imperialism*. Oxford: Oxford University Press.

Hawken, P.2017. *Drawdown: The Most Comprehensive Plan Ever Proposed to Reverse Global Warming*. London: Penguin Books.

Hecht, S. B., and A. Cockburn. 2011. *The Fate of the Forest*. Chicago: University of Chicago Press.

Hickel, J. 2020. *Less is More: How Degrowth Will Save the World*. New York: Random House.

Hickel, J., and G. Kallis. 2020. "Is Green Growth Possible?" *New Political Economy* 25 (4): 469–486.

Holt-Giménez, E., A. Shattuck, and I. Van Lammeren. 2021. "Thresholds of Resistance: Agroecology, Resilience and the Agrarian Question." *Journal of Peasant Studies* 48 (4): 715–733.

Houzer, E., and I. Scoones. 2021. *forthcoming. Are Livestock Always bad for the Planet? Rethinking the Protein Transition and Climate Change Debate*. Brighton: PASTRES Programme, Institute of Development Studies.

Huff, A. 2021. "Frictitious Commodities: Virtuality, Virtue and Value in the Carbon Economy of Repair." *Environment and Planning E: Nature and Space* 25148486211015056: 1–26.

Huizer, G. 1972. *The Revolutionary Potential of Peasants in Latin America*. Lanham, MD: Lexington Books.

Hulme, M. 2009. *Why we Disagree About Climate Change: Understanding Controversy, Inaction and Opportunity*. Cambridge: Cambridge University Press.

Hulme, M. 2019. "Climate Emergency Politics is Dangerous." *Issues in Science and Technology* 36 (1): 23–25.

Hunsberger, C., E. Corbera, S. M. Borras Jr, J. C. Franco, K. Woods, C. Work, & C. Vaddhanaphuti. 2017. "Climate Change Mitigation, Land Grabbing and Conflict: Towards a Landscape-Based and Collaborative Action Research Agenda." *Canadian Journal of Development Studies/Revue Canadienne D'études du Développement* 38 (3): 305–324.

Isenhour, C. 2016. "Unearthing Human Progress? Ecomodernism and Contrasting Definitions of Technological Progress in the Anthropocene." *Economic Anthropology* 3 (2): 315–328.

Jackson, T. 2021. *Post Growth: Life After Capitalism*. Hoboken, NJ: John Wiley & Sons.

Jasanoff, S. 2007. "Technologies of Humility." *Nature* 450 (7166): 33–33.

Jasanoff, S. 2011. "Cosmopolitan Knowledge: Climate Science and Global Civic Epistemology." In *The Oxford Handbook of Climate Change and Society*, edited by J. S. Dryzek, R. B. Norgaard, and D. Schlosberg, 129–143. Oxford: Oxford University Press.

Kallis, G., S. Paulson, G. D'Alisa, and F. Demaria. 2020. *The Case for Degrowth*. Oxford: John Wiley & Sons.

Khadse, A., P. M. Rosset, H. Morales, and B. G. Ferguson. 2018. "Taking Agroecology to Scale: The Zero Budget Natural Farming Peasant Movement in Karnataka, India." *Journal of Peasant Studies* 45 (1): 192–219.

Klein, N. 2015. *This Changes Everything: Capitalism vs. the Climate*. New York: Simon and Schuster.

Klein, N. 2020. "Care and Repair: Left Politics in the age of Climate Change." *Dissent* 67 (1): 97–108.

Klinsky, S., T. Roberts, S. Huq, C. Okereke, P. Newell, P. Dauvergne, K. O'Brien, H. Schroeder, P. Tschakert, and J. Clapp. 2017. "Why Equity is Fundamental in Climate Change Policy Research." *Global Environmental Change* 44: 170–173.

Kröger, M. 2021. *Iron Will: Global Extractivism and Mining Resistance in Brazil and India*. Michigan: University of Michigan Press.

Latour, B. 2004. *Politics of Nature*. Cambridge, MA: Harvard University Press.

Latour, B., I. Stengers, A. Tsing, and N. Bubandt. 2018. "Anthropologists are Talking – About Capitalism, Ecology, and Apocalypse." *Ethnos* 83 (3): 587–606.
Le Quang, M., and T. Vercoutére. 2013. *Ecosocialismo y Buen Vivir: Diálogo Entre dos Alternativas al Capitalismo*. Quito: Instituto de Altos Estudios Nacionales, Quito.
Leach, M., and I. Scoones. eds. 2015. *Carbon Conflicts and Forest Landscapes in Africa*. London: Routledge.
Leach, M., A. C. Stirling, and I. Scoones. 2010. *Dynamic Sustainabilities: Technology, Environment, Social Justice*. London: Routledge.
Levien, M., M. Watts, and H. Yan. 2018. "Agrarian Marxism." *Journal of Peasant Studies* 45 (5–6): 853–883.
Li, T. M. 2010. "To Make Live or let die? Rural Dispossession and the Protection of Surplus Populations." *Antipode* 41: 66–93.
Liao, C., K. Nolte, J. A. Sullivan, D. G. Brown, J. Lay, C. Althoff, and A. Agrawal. 2021. "Carbon Emissions from the Global Land Rush and Potential Mitigation." *Nature Food* 2 (1): 15–18.
Lovell, H., H. Bulkeley, and D. Liverman. 2009. "Carbon Offsetting: Sustaining Consumption?" *Environment and Planning A* 41 (10): 2357–2379.
Löwy, M. 2005. "What is Ecosocialism?" *Capitalism Nature Socialism* 16 (2): 15–24.
Lynch, M. J., M. A. Long, P. B. Stretesky, and K. L. Barrett. 2019. "Measuring the Ecological Impact of the Wealthy: Excessive Consumption, Ecological Disorganization, Green Crime, and Justice." *Social Currents* 6 (4): 377–395. doi:10.1177/2329496519847491.
Malm, A. 2016. *Fossil Capital: The Rise of Steam Power and the Roots of Global Warming*. New York: Verso Books.
Malm, A. 2018. *The Progress of This Storm: Nature and Society in a Warming World*. New York: Verso Books.
Malm, A. 2020. *Corona, Climate, Chronic Emergency: War Communism in the Twenty-First Century*. New York: Verso Books.
Malm, A., and A. Hornborg. 2014. "The Geology of Mankind? A Critique of the Anthropocene Narrative." *The Anthropocene Review* 1 (1): 62–69. doi:10.1177/2053019613516291.
Mamonova, N., and J. Franquesa. 2020. "Populism, Neoliberalism and Agrarian Movements in Europe. Understanding Rural Support for Right-Wing Politics and Looking for Progressive Solutions." *Sociologia Ruralis* 60 (4): 710–731.
Marino, E., and J. Ribot. 2012. "Adding Insult to Injury: Climate Change, Social Stratification, and the Inequities of Intervention." *Global Environmental Change* 22 (2): 323–398.
Martinez-Alier, J. 2002. *The Environmentalism of the Poor: A Study of Ecological Conflicts and Valuation*. Northampton, MA: Edward Elgar Publishing.
Martinez-Alier, J., L. Temper, D. Del Bene, and A. Scheidel. 2016. "Is There a Global Environmental Justice Movement?" *Journal of Peasant Studies* 43 (3): 731–755.
Marx, K. 1982 [orig. 1852]. "*The Eighteenth Brumaire of Louis Bonaparte*." In *Marx/Engel Selected Works in one Volume*, 96–179. Moscow: Progress Publishers.
Marx, K. 1992. *Capital: Volume III*. London: Penguin UK.
Mastini, R., G. Kallis, and J. Hickel. 2021. "A Green new Deal Without Growth?" *Ecological Economics* 179: 1–9.
Mazzucato, M. 2021. *Mission Economy: A Moonshot Guide to Changing Capitalism*. Westminster, UK: Allen Lane.
McCarthy, J. 2002. "First World Political Ecology: Lessons from the Wise Use Movement." *Environment and Planning A* 34 (7): 1281–1302.
McCarthy, J. 2019. "Authoritarianism, Populism, and the Environment: Comparative Experiences, Insights, and Perspectives." *Annals of the American Association of Geographers* 109 (2): 301–313.
McCarthy, J., and S. Prudham. 2004. "Neoliberal Nature and the Nature of Neoliberalism." *Geoforum; Journal of Physical, Human, and Regional Geosciences* 35 (3): 275–283.
McKay, B. M., G. D. L. Oliveira, and J. Liu. 2020. "Authoritarianism, Populism, Nationalism and Resistance in the Agrarian South." *Canadian Journal of Development Studies/Revue Canadienne D'études du Développement* 41 (3): 347–362.

McKeon, N. 2015. "La Via Campesina: The 'Peasants' Way' to Changing the System, not the Climate." *Journal of World-Systems Research* 21 (2): 241–249.
McLaren, D., and N. Markusson. 2020. "The co-Evolution of Technological Promises, Modelling, Policies and Climate Change Targets." *Nature Climate Change* 10 (5): 392–397.
McMichael, P. 2009. "A Food Regime Genealogy." *Journal of Peasant Studies* 36 (1): 139–169.
Mehta, L., and W. Harcourt. 2021. "Beyond Limits and Scarcity: Feminist and Decolonial Contributions to Degrowth." *Political Geography*, 1–3. doi:10.1016/j.polgeo.2021.102411.
Mehta, L., S. Srivastava, H. N. Adam, S. Bose, U. Ghosh, and V. V. Kumar. 2019. "Climate Change and Uncertainty from 'Above' and 'Below': Perspectives from India." *Regional Environmental Change* 19 (6): 1533–1547.
Mehta, L., S. Srivastava, S. Movik, H. N. Adam, R. D'Souza, D. Parthasarathy, L. O. Naess, and N. Ohte. 2021. "Transformation as Praxis: Responding to Climate Change Uncertainties in Marginal Environments in South Asia." *Current Opinion in Environmental Sustainability* 49: 110–117.
Mehta, L., G. J. Veldwisch, and J. Franco. 2012. "Introduction to the Special Issue: Water Grabbing? Focus on the (re) Appropriation of Finite Water Resources." *Water Alternatives* 5 (2): 193–207.
Milanovic, B. 2017. The illusion of 'degrowth' in a poor and unequal world. GlobalInequality 18 November 2017. http://glineq.blogspot.be/2017/11/the-illusion-of-degrowth-in-poor-and.html.
Millar, S. W., and D. Mitchell. 2017. "The Tight Dialectic: The Anthropocene and the Capitalist Production of Nature." *Antipode* 49: 75–93.
Mitchell, T. 2011. *Carbon Democracy: Political Power in the age of oil*. London: Verso.
Moore, J. W. 2015. *Capitalism in the Web of Life*. New York: Verso Books.
Moore, J. W. 2017. "The Capitalocene, Part I: On the Nature and Origins of our Ecological Crisis." *Journal of Peasant Studies* 44 (3): 594–630.
Moore Jr., B. 1967. *Social Origins of Dictatorship and Democracy: Lord and Peasant in the Making of the Modern World*. Boston: Beacon Press.
Morrison, J. 2019. Who will Pay for the Huge costs of holding back Rising Seas?. Yale Environment 360. 5 August 2019. https://e360.yale.edu/features/who-will-pay-for-the-huge-costs-of-holding-back-rising-seas.
Moyo, S. 2005. "The Land Question and the Peasantry in Southern Africa." In *Politics and Social Movements in an Hegemonic World: Lessons for Africa, Asia and Latin America*, edited by A. Boron, and G. Lechini, 147–163. Buenos Aires: CLACSO Books.
Moyo, S., P. Jha, and P. Yeros. 2013. "The Classical Agrarian Question: Myth, Reality and Relevance Today." *Agrarian South: Journal of Political Economy* 2 (1): 93–119.
Moyo, S., and P. Yeros. eds. 2005. *Reclaiming the Land: The Resurgence of Rural Movements in Africa, Asia and Latin America*. London: Zed Books.
Mwenda, M., and P. Bond. 2020. "African Climate Justice: Articulations and Activism." In *Climate Justice and Community Renewal*, edited by B. Tokar, and T. Gilbertson, 108–127. London: Routledge.
Neimark, B., J. Childs, A. J. Nightingale, C. J. Cavanagh, S. Sullivan, T. A. Benjaminsen, S. Batterbury, S. Koot, and W. Harcourt. 2019. "Speaking Power to 'Post-Truth': Critical Political Ecology and the new Authoritarianism." *Annals of the American Association of Geographers* 109 (2): 613–623.
Newell, P. 2019. *Global Green Politics*. Cambridge: Cambridge University Press.
Newell, P. 2021. *Power Shift: The Global Political Economy of Energy Transitions*. Cambridge: Cambridge University Press.
Newell, P., and D. Mulvaney. 2013. "The Political Economy of the 'Just Transition'." *The Geographical Journal* 179 (2): 132–140.
Newell, P., and M. Paterson. 2010. *Climate Capitalism: Global Warming and the Transformation of the Global Economy*. Cambridge: Cambridge University Press.
Newell, P., R. Srivastava, L. O. Naess, G. A. Torres Contreras, and R. Price. 2020. *Towards Transformative Climate Justice: Key Challenges and Future Directions for Research*. Brighton: Institute of Development Studies.
Newell, P., and O. Taylor. 2018. "Contested Landscapes: The Global Political Economy of Climate-Smart Agriculture." *Journal of Peasant Studies* 45 (1): 108–129.

Nightingale, A. J. 2017. "Power and Politics in Climate Change Adaptation Efforts: Struggles Over Authority and Recognition in the Context of Political Instability." *Geoforum; Journal of Physical, Human, and Regional Geosciences* 84: 11–20.
Nightingale, A. J., S. Eriksen, M. Taylor, T. Forsyth, M. Pelling, et al. 2020. "Beyond Technical Fixes: Climate Solutions and the Great Derangement." *Climate and Development* 12 (4): 343–352.
Nixon, R. 2011. *Slow Violence and the Environmentalism of the Poor*. Cambridge, MA: Harvard University Press.
O'Connor, J. 1998. *Natural Causes: Essays in Ecological Marxism*. New York: Guilford Press.
Paige, J. M. 1975. *Agrarian Revolution*. New York: Simon and Schuster.
Paprocki, K. 2019. "All That is Solid Melts Into the bay: Anticipatory Ruination and Climate Change Adaptation." *Antipode* 51 (1): 295–315.
Paprocki, K. 2021. *Threatening Dystopias: The Global Politics of Climate Change Adaptation in Bangladesh*. Ithaca, NY: Cornell University Press.
Patel, R. 2009. "Food Sovereignty." *Journal of Peasant Studies* 36 (3): 663–706.
Patel, R., and J. W. Moore. 2017. *A History of the World in Seven Cheap Things: A Guide to Capitalism, Nature, and the Future of the Planet*. Berkeley: University of California Press.
Pearce, F. 2019. Geoengineer the planet? More scientists now say it must be an option. Yale Environment 360. 29 May 2019. https://e360.yale.edu/features/geoengineer-the-planet-more-scientists-now-say-it-must-be-an-option.
Pelling, M. 2010. *Adaptation to Climate Change: From Resilience to Transformation*. London: Routledge.
Peluso, N. L. 1992. *Rich Forests, Poor People: Resource Control and Resistance in Java*. Berkeley: University of California Press.
Peluso, N. L. 2017. "Plantations and Mines: Resource Frontiers and the Politics of the Smallholder Slot." *Journal of Peasant Studies* 44 (4): 834–869.
Peluso, N. L., and C. Lund. 2011. "New Frontiers of Land Control: Introduction." *Journal of Peasant Studies* 38 (4): 667–681.
Peluso, N. L., and P. Vandergeest. 2020. "Writing Political Forests." *Antipode* 52 (4): 1083–1103.
Peluso, N. L., and M. Watts. eds. 2001. *Violent Environments*. Ithaca, NY: Cornell University Press.
Perfecto, I., and J. Vandermeer. 2010. "The Agroecological Matrix as Alternative to the Land-Sparing/Agriculture Intensification Model." *Proceedings of the National Academy of Sciences* 107 (13): 5786–5791.
Perreault, T., G. Bridge, and J. McCarthy. eds. 2015. *The Routledge Handbook of Political Ecology*. London: Routledge.
Popkin, S. L. 1979. *The Rational Peasant*. Berkeley: University of California Press.
Potter, E. 2013. "Climate Change and non-Indigenous Belonging in Postcolonial Australia." *Continuum* 27 (1): 30–40.
Rangan, H., and C. A. Kull. 2009. "What Makes Ecology 'Political'?: Rethinking 'Scale' in Political Ecology." *Progress in Human Geography* 33 (1): 28–45.
Reisman, E., and M. Fairbairn. 2021. "Agri-Food Systems and the Anthropocene." *Annals of the American Association of Geographers* 111 (3): 687–697.
Ribeiro, S., 2019. "Geo-engineering." In *Pluriverse: A Post-Development Dictionary*, edited by A. Kothari et al., 53–56. New Delhi: Tulika Books and Authorsupfront.
Ribot, J. 2014. "Cause and Response: Vulnerability and Climate in the Anthropocene." *Journal of Peasant Studies* 41 (5): 667–705.
Rice, J., J. Long, and A. Levenda. 2021. "Against Climate Apartheid: Confronting the Persistent Legacies of Expendability for Climate Justice." *Environment and Planning E: Nature and Space* 2514848621999286: 1–21.
Robbins, P. 2011. *Political Ecology: A Critical Introduction*. Hoboken, NJ: John Wiley & Sons.
Roberts, J. T., and B. C. Parks. 2009. "Ecologically Unequal Exchange, Ecological Debt, and Climate Justice: The History and Implications of Three Related Ideas for a new Social Movement." *International Journal of Comparative Sociology* 50 (3-4): 385–409.
Rockström, J., W. Steffen, K. Noone, A. Persson, F. S. Chapin III, E. Lambin, T. M. Lenton, M. Scheffer, C. Folke, and H. J. Schellnhuber. 2009. "Planetary Boundaries: Exploring the Safe Operating Space for Humanity." *Ecology and Society* 14 (2): 1–33.

Rodney, W. 1972. *How Europe Underdeveloped Africa*. New York: Verso Trade.
Rogers, D. 2015. *The Depths of Russia: Oil, Power, and Culture After Socialism*. Ithaca, NY: Cornell University Press.
Roman-Alcalá, A., G. Graddy-Lovelace, and M. Edelman. 2021. "Authoritarian Populism and Emancipatory Politics in the Rural United States." *Journal of Rural Studies* 82: 500–504.
Rosset, P. M., and M. A. Altieri. 2017. *Agroecology: Science and Politics*. Rugby, UK: Practical ActionPublishing.
Routledge, P., A. Cumbers, and K. D. Derickson. 2018. "States of Just Transition: Realising Climate Justice Through and Against the State." *Geoforum; Journal of Physical, Human, and Regional Geosciences* 88: 78–86.
Ruiz-Campillo, X., V. Castán Broto, and L. Westman. 2021. "Motivations and Intended Outcomes in Local Governments' Declarations of Climate Emergency." *Politics and Governance* 9 (2): 17–28.
Rulli, M. C., A. Saviori, and P. D'Odorico. 2013. "Global Land and Water Grabbing." *Proceedings of the National Academy of Sciences* 110 (3): 892–897.
Sayre, N. F. 2015. "Scales and Polities." In *The Routledge Handbook of Political Ecology*, edited by T. Perreault, G. Bridge, and J. McCarthy, 504–515. Oxon: Routledge.
Scheidel, A., D. Del Bene, J. Liu, G. Navas, S. Mingorría, F. Demaria, & J. Martínez-Alier. 2020. "Environmental Conflicts and Defenders: A Global Overview." *Global Environmental Change* 63: 1–12.
Schlosberg, D. 2009. *Defining Environmental Justice: Theories, Movements, and Nature*. Oxford: Oxford University Press.
Schlosberg, D., and L. B. Collins. 2014. "From Environmental to Climate Justice: Climate Change and the Discourse of Environmental Justice." *Wiley Interdisciplinary Reviews: Climate Change* 5 (3): 359–374.
Schneider, M., and P. McMichael. 2010. "Deepening, and Repairing, the Metabolic Rift." *Journal of Peasant Studies* 37 (3): 461–484.
Schurman, R., and W. A. Munro. 2013. *Fighting for the Future of Food: Activists Versus Agribusiness in the Struggle Over Biotechnology*. Minneapolis: University of Minnesota Press.
Scoones, I., M. Edelman, S. M. Borras Jr, R. Hall, W. Wolford, and B. White. 2018. "Emancipatory Rural Politics: Confronting Authoritarian Populism." *Journal of Peasant Studies* 45 (1): 1–20.
Scoones, I., P. Newell, and M. Leach. 2015. *The Politics of Green Transformations*. London: Routledge.
Scoones, I., A. Stirling, D. Abrol, J. Atela, L. Charli-Joseph, H. Eakin, A. Ely, P. Olsson, L. Pereira, and R. Priya. 2020. "Transformations to Sustainability: Combining Structural, Systemic and Enabling Approaches." *Current Opinion in Environmental Sustainability* 42: 65–75.
Scott, J. C. 1977. *The Moral Economy of the Peasant: Rebellion and Subsistence in Southeast Asia*. New Haven, CT: Yale University Press.
Selwyn, B. 2021. "A Green new Deal for Agriculture: For, Within, or Against Capitalism?" *Journal of Peasant Studies* 48 (4): 778–806.
Shah, E., J. Vos, G. J. Veldwisch, R. Boelens, and B. Duarte-Abadía. 2021. "Environmental Justice Movements in Globalising Networks: A Critical Discussion on Social Resistance Against Large Dams." *Journal of Peasant Studies* 48 (5): 1008–1032. doi:10.1080/03066150.2019.1669566.
Shivji, I. G. 2017. "The Concept of 'Working People'." *Agrarian South: Journal of Political Economy* 6 (1): 1–13.
Skrimshire, S. 2010. *Future Ethics: Climate Change and Apocalyptic Imagination*. New York: Bloomsbury Publishing.
Smith, N. 2008. *Uneven Development: Nature, Capital, and the Production of Space*. Atlanta, GA: University of Georgia Press.
Smith, R. 2020. *China's Engine of Environmental Collapse*. London: Pluto Press.
Soper, K. 1995. *What is Nature?* Hoboken, NJ: Blackwell.
Spore. 2019. "World Farmers Organization pushes climate-smart agriculture for Africa." 6 December. https://spore.cta.int/en/press-reviews/article/world-farmers-organization-pushes-climate-smart-agriculture-for-africa-sid06db03be7-b3e4-43b7-8416-a73035b38144.
Spratt, D., and I. Dunlop. 2019. Existential climate-related security risk: A scenario approach. BT Policy Paper. Melbourne, Australia: Breakthrough - National Centre for Climate Restoration. https://docs.wixstatic.com/ugd/148cb0_b2c0c79dc4344b279bcf2365336ff23b.pdf.

Standing, G. 2014. "The Precariat." *Contexts* 13 (4): 10–12.
Stock, R., and T. Birkenholtz. 2021. "The sun and the Scythe: Energy Dispossessions and the Question of Labor in Solar Parks." *Journal of Peasant Studies* 48 (5): 984–1007. doi:10.1080/03066150.2019.1683002.
Sultana, F. 2021. "Climate Change, COVID-19, and the co-Production of Injustices: A Feminist Reading of Overlapping Crises." *Social & Cultural Geography* 22 (4): 447–460.
Surprise, K. 2018. "Preempting the second contradiction: Solar geoengineering as spatiotemporal fix." *Annals of the American Association of Geographers*, 108 (5): 1228–1244. doi:10.1080/24694452.2018.1426435.
Swilling, M. 2019. *The Age of Sustainability: Just Transitions in a Complex World*. London: Routledge.
Swilling, M., and E. Annecke. 2012. *Just Transitions: Explorations of Sustainability in an Unfair World*. Cape Town: Juta and Company (Pty) Ltd.
Szerszynski, B., and J. Urry. 2010. *Changing Climates: Introduction*. Newbury Park, CA: SAGE Publications.
Taylor, M. 2014. *The Political Ecology of Climate Change Adaptation: Livelihoods, Agrarian Change and the Conflicts of Development*. London: Routledge.
Temper, L., and D. Del Bene. 2016. "Transforming Knowledge Creation for Environmental and Epistemic Justice." *Current Opinion in Environmental Sustainability* 20: 41–49.
Temper, L., F. Demaria, A. Scheidel, D. Del Bene, and J. Martinez-Alier. 2018. "The Global Environmental Justice Atlas (EJAtlas): Ecological Distribution Conflicts as Forces for Sustainability." *Sustainability Science* 13 (3): 573–584.
Torres Contreras, G. A. 2021. "Twenty-five Years Under the Wind Turbines in La Venta, Mexico: Social Difference, Land Control and Agrarian Change." . *Journal of Peasant Studies*, 1–19. doi:10.1080/03066150.2021.1873293.
Tramel, S. 2016. "The Road Through Paris: Climate Change, Carbon, and the Political Dynamics of Convergence." *Globalizations* 13 (6): 960–969.
Tschakert, P., and M. Machado. 2012. "Gender Justice and Rights in Climate Change Adaptation: Opportunities and Pitfalls." *Ethics and Social Welfare* 6 (3): 275–289.
Tschakert, P., D. Schlosberg, D. Celermajer, L. Rickards, C. Winter, M. Thaler, M. Stewart-Harawira, and B. Verlie. 2021. "Multispecies Justice: Climate-Just Futures with, for and Beyond Humans." *Wiley Interdisciplinary Reviews: Climate Change* 12 (2): 1–10.
Tsing, A. L., J. Deger, A. Keleman Saxena, and F. Zhou. 2020. *Feral Atlas: The More-Than-Human Anthropocene*. Redwood, CA: Stanford University Press.
Tsing, A. L., A. S. Mathews, and N. Bubandt. 2019. "Patchy Anthropocene: Landscape Structure, Multispecies History, and the Retooling of Anthropology: An Introduction to Supplement 20." *Current Anthropology* 60 (S20): S186–S197.
Tuana, N. 2019. "Climate Apartheid: The Forgetting of Race in the Anthropocene." *Critical Philosophy of Race* 7 (1): 1–31.
Turnhout, E., A. Gupta, J. Weatherley-Singh, M. J. Vijge, J. De Koning, I. J. Visseren-Hamakers, M. Herold, and M. Lederer. 2017. "Envisioning REDD+ in a Post-Paris era: Between Evolving Expectations and Current Practice." *Wiley Interdisciplinary Reviews: Climate Change* 8 (1): 1–13.
Vergès, F. 2019. "Capitalocene, Waste, Race, and Gender." *e-flux Journal* 100: 1–13.
Verweijen, J., and E. Marijnen. 2018. "The Counterinsurgency/Conservation Nexus: Guerrilla Livelihoods and the Dynamics of Conflict and Violence in the Virunga National Park, Democratic Republic of the Congo." *Journal of Peasant Studies* 45 (2): 300–320.
Wainwright, J., and G. Mann. 2018. *Climate Leviathan: A Political Theory of our Planetary Future*. New York: Verso Books.
Waldron, A., V. Adams, J. Allan, A. Arnell, G. Asner, S. Atkinson, A. Baccini, J. Baillie, A. Balmford, and J. Austin Beau. 2020. Protecting 30% of the planet for nature: costs, benefits and economic implications.
Watts, M. 2004. "Resource Curse? Governmentality, oil and Power in the Niger Delta, Nigeria." *Geopolitics* 9 (1): 50–80.
Watts, M. J. 2013 [orig. 1983]. *Silent Violence: Food, Famine, and Peasantry in Northern Nigeria*. Atlanta, GA: University of Georgia Press.

Weis, T. 2010. "The Accelerating Biophysical Contradictions of Industrial Capitalist Agriculture." *Journal of Agrarian Change* 10 (3): 315–341.
Weis, T. 2013. *The Ecological Hoofprint: The Global Burden of Industrial Livestock*. London: Zed Books.
White, B., S. M. Borras Jr, R. Hall, I. Scoones, and W. Wolford. 2012. "The new Enclosures: Critical Perspectives on Corporate Land Deals." *Journal of Peasant Studies* 39 (3–4): 619–647.
Whyte, K. P. 2018. "Indigenous Science (Fiction) for the Anthropocene: Ancestral Dystopias and Fantasies of Climate Change Crises." *Environment and Planning E: Nature and Space* 1 (1–2): 224–242.
Williams, R. 1980. "Ideas of Nature." In *Problems of Materialism and Culture: Selected Essays*, edited by R. Williams, 67–85. New York: Verso Books.
Wilson, E. O. 2016. *Half-earth: Our Planet's Fight for Life*. New York: W.W. Norton & Company.
Wise, T. A. 2019. *Eating Tomorrow: Agribusiness, Family Farmers, and the Battle for the Future of Food*. New York: The New Press.
Wittman, H., A. Desmarais, and N. Wiebe. 2010. *Food Sovereignty*. Halifax, NS: Fernwood.
Wolf, E. R. 1969. *Peasant Wars of the Twentieth Century*. Oklahoma: University of Oklahoma Press.
Wolford, W. 2003. "Producing Community: The MST and Land Reform Settlements in Brazil." *Journal of Agrarian Change* 3 (4): 500–520.
Wolford, W. 2010. *This Land is Ours now: Social Mobilization and the Meanings of Land in Brazil*. Durham, NC: Duke University Press.
Wolford, W. 2020. "The Plantationocene: A Lusotropical Contribution to the Theory." *Annals of the American Association of Geographers* 1–18. doi:10.1080/24694452.2020.1850231.
World Bank. 2016. World Bank Group climate change action plan. 7 April 2016. World Bank.
Wright, E. O. 2019. *How to be an Anti-Capitalist in the Twenty-First Century*. New York: Verso Books.
Wynne, B. 2010. "Strange Weather, Again." *Theory, Culture & Society* 27 (2–3): 289–305.
Yusoff, K. 2021. "The Inhumanities." *Annals of the American Association of Geographers* 111 (3): 663–676. doi:10.1080/24694452.2020.1814688.
Zimmerer, K. S., and T. J. Bassett. eds. 2003. *Political Ecology: An Integrative Approach to Geography and Environment-Development Studies*. Guilford Press.

The environmentalization of the agrarian question and the agrarianization of the climate justice movement

Zehra Taşdemir Yaşın

ABSTRACT
This paper makes two central arguments: i) The world-historical generalization of the capital-nature relation in the context of the global extractivist turn has reconfigured the twenty-first century agrarian question as *the agrarian question of nature* as a critical component of the broader socio-ecological question. ii) The historical context of the agrarian question of nature, in turn, gave birth to not only environmental-agrarian movements, but also *agrarianization of the politics and movements of environmental/ climate justice*. The agroecology movement signifies the emergence of a contemporary form of political agrarianism emerged from within the context of the environmentalization of the agrarian question.

Introduction

This paper addresses a pivotal issue raised by Borras et al. (2021) in the invitation article to this forum: situating climate change in its broader historical context of agrarian-environmental transformation and struggles. They posit a central question: whether and how we can 'envision a sufficiently anti-capitalist, trans-environmental and agrarian approach to confront climate change in rural settings' (17). In this paper, I will respond to this question from the theoretical lens of 'the value theory of nature' (Taşdemir Yaşın 2017). The value theory of nature recognizes the social relation of capital and nature as the increasingly predominating relation through which the contradiction of capitalist value production is reconstituted (conceived as 'the second contradiction' by James O'Connor [1988]) and deepened in historically and geographically complex and differentiated interaction with capital-labor relation.

Eco-Marxism has located the socio-ecological outcomes of capitalist expansion in what Marx conceptualized as *the metabolic rift* (Schneider and McMichael 2010; Taşdemir Yaşın 2017). The metabolic rift is about the unsustainable relation of capital to the soil-based in the *systematic disruption of the nutrient cycle* through the separation between food production and food consumption, i.e. *the rural-urban divide*. Although the metabolic rift might have predated capitalism (Duncan 1996; Moore 2008), the historical context or temporality of its world-historical generalization and intensification is the formation of the capital-labor relation, i.e. proletarianization. While the metabolic rift underpins critically the soil-based ecological contradictions of capitalist development, as Phillip McMichael notes, 'it does not address the biospheric rift' in broader context (2012, 145). With the realized potential to outstrip all other past and present environmental problems, 'climate change has introduced a certain temporality into the environment' concentrating the very long-term cyclical trend of the biospheric change into the short-term secular trend of the twenty-first century (McMichael 2012, 138, 142–144). This has manifested as *the biospheric rift* that concerns the *systematic disruption of the carbon cycle*, i.e. the rupture of the long-term balance between carbon (energy) storage and carbon emission capacity of the earth that sustains the balance of gases in the atmosphere and the earth temperature (Clark and York 2005). Clark and York (2005, 395) locate the biospheric rift in *the human-nature divide* created by the global expansion and deepening of capital accumulation and its exploitation of nature. They argue that 'mining the earth to remove stored energy to fuel' the technology-based capitalist industrial expansion gave rise to a new form of human-nature interchange based on a constant increase of the material and energy throughput and the override of the self-life-support systems of nature (404–406). Capitalist industrialization based on the development of supposedly time and space-saving technologies has displaced environmental burdens to the extractive sectors in the context of a global unequal exchange, or unequal time–space appropriation (Hornborg 2007) leading to the carbon rift. I argue that *the historical context or temporality of the deepening biospheric rift is the formation and generalization of the capital-nature relation in commodity production*. This second relation specifies the commodification and an 'abstract form of social domination' (Postone 1993, 3) of nature and its reconstitution as *abstract nature* or value producing nature (Taşdemir Yaşın 2017). The capital-labor relation appropriates in the process of production labor power and time, which can be conceived as a product of consumption of nutrients/energy stored in space, i.e. life-giving (organic) space. The capital-nature relation appropriates in the process of extraction directly the organic space/matter, which can be conceived as a product of accumulation of energy and nutrients through storage of plants and animals in the long climatic time, i.e. the transformation of an energy-storing *longue durée* (McMichael 2012) into a life/energy-giving space.

The value theory of nature employs *abstract nature* as a historically specific *abstract conceptual category* following Marx's historically specific abstraction of *abstract labor* as *the social source of exchange value* (Sayer 1987, 129). In Marx's method of logical exposition of 'the value theory of labor' (Elson 1979), the commodity form is central and, hence, constitutes 'the necessary logical starting point' of his method of presentation (Tomich 2004, 21–31). Through the commodity form as the unity of use-value and exchange-value, Marx discloses how capital becomes a form of social domination exercised through commodity fetishism by which social labor is determined as abstract labor and the abstract aspect of labor dominates its concrete aspect (Elson 1979, 159, 165– 66). Marx conceptualizes capitalist surplus value production as the product of 'the totality of relations comprising the division of labor and world market', i.e. *as a world-historical relation* (Tomich 2015, 361–362). In the contemporary concrete totality of the global division of labor and nature, the world market and the commoditization of everything (Watts, Robbins, and Peet 2010), i.e. the whole lifeworld, it is more apparent that the commodity form is 'simultaneously the historical condition and manifestation of value' structuring not only (social) labor and the socio-historical domination of concrete human labor, as rendered by the value theory of labor; but also nature (or 'socio-nature' [Peluso 2012]) and the socio-historical domination of concrete nature and human-nature interaction, as the value theory of nature aims to reveal (Taşdemir Yaşın 2017, 21–22). 'The simultaneity of substantial and abstract dimensions' (Postone 1993, 269) of not labor but nature creates the dual character of the commodity form and commodity fetishism in socio-natural commodities produced through extractive processes.

From a bioeconomic perspective depicted in Georgescu-Roegen's theory of entropy (Martinez-Alier 2011; Leff 2014), we can situate both the metabolic rift and the biospheric rift as differentiated manifestations of a broader process of entropic degradation of nature or the planet through a *socio-ecological rift* that underpins both the conventional uncritical idea of 'Anthropocene' and its socio-historically based critique through 'Capitalocene' (Moore 2017). *The socio-ecological rift* has not only entailed *the separation of humans* (peasants) *from nature* (their means of agricultural production) in the formation of wage labor as value producing labor and urbanization, but also *the separation of nature from humans* (or culture) in the formation of abstract nature as value producing nature. The latter is about the transformation of matter and energy stored in nature into an unrecyclable form in the commodity form generating carbon emissions, irreversible heat and/or waste. The deepening of the socio-ecological rift through the world-historical expansion of capital-nature relation has rendered the emergence of *a socio-ecological question* with increasing historical visibility and urgency articulated in differentiated yet interdependent forms: 'the agro-environmental crisis' (Van der Ploeg 2008), 'the food question' (McMichael 2008, 2009), 'the ecological question' (McMichael 2013) or 'the environmental question' (Leff 2014, 2021), the question of 'capitalist world-ecology' (Moore 2015, 2017) and the climate question (or the climate change). This paper focuses particularly on the relationship between the agro-environmental question and the climate question by locating them in the wider context of the socio-ecological question from the conceptual lens of the value theory of nature.

Van der Ploeg (2008, 11) characterizes the 'agro-environmental crisis' as a product of organization of agriculture based on a systematic destruction and contamination of the wider ecosystems. The contribution of agro-industrial food production to climate change and global environmental crisis has been vastly demonstrated: The global food system is responsible for the sixth mass extinction of wildlife, a third of global greenhouse gas emissions, the uses of 70% of global fresh water and 40% of global land, the transformation of 33% of global croplands into feed crop frontiers and 26% of the world's ice-free land into livestock grazing frontiers (Weis 2007, 2010; Baird and Barney 2017; Mahowald et al. 2017; Crippa et al. 2021; Selwyn 2021). These sources of global socio-ecological instability materialize in the forms of biodiversity loss, soil degradation, overexploitation of water, food waste, high price volatility of food, malnourishment, obesity, rural displacement, labor casualization, migration and overurbanization (Weis 2010; Araghi 2009b; McMichael 2016; Altieri and Nicholls 2020). 'Green' solutions to the agro-environmental crisis and the introduction of 'climate-smart' technologies, as constituents of neoliberal climate politics, have further exacerbated the override of the biophysical foundations of both agriculture and climate and the growth of the global population of environmental refugees (Weis 2010; Borras and Franco 2018; Akram-Lodhi 2021, 696).

In this paper, I would like to broaden this agro-environmental perspective and specify both the agro-environmental crisis and climate change as globally manifesting historical constituents of the broader socio-ecological question. The urgency of the socio-ecological question increases also the urgency of Enrique Leff's (2021, 136–138, 148–149, 150–152) call for 'a new theory' by taking into account the contribution of nature to (surplus) value formation: a critical qualitative 'theory of production grounded on the potentials and conditions of Nature' rendering possible an alternative 'productive rationality that accounts for the incorporation of natural processes in the general conditions of production and the construction of an environmental rationality' based upon ecological resiliency, territorial rights and environmental justice (167–168). In this paper, I draw on the conceptual perspective of *the value theory of nature* to grasp the historically distinctive socio-ecological relationship between the agro-environmental crisis and the climate change/justice, grounded in an interdisciplinary engagement of critical agrarian studies and political ecology. From this perspective, the following account explores *the agro-environmental question* or what I call *the environmentalization of the agrarian question* by specifying it as *an agrarian question of nature* with deepened and expanded consequences for the carbon/energy cycle and climate change. In turn, it also explores how this relationship has reconfigured the climate change politics leading to *the agrarianization of the climate justice movement* as a part of a wider anti-systemic socio-ecological movement from the lens of a new possible understanding of class enabled by the value theory of nature.

The environmentalization of the agrarian question

Colin Duncan's (1996) historical account of 'the centrality of agriculture' in understanding both English modernization in the eighteenth and nineteenth century and its socio-ecological destabilization from the late nineteenth century onwards disturbs the presumption of the 'inevitable fate of proletarianization' of the peasants underlining many classical accounts of the original agrarian question. His analysis of the historical shift in England from 'high farming' that characterizes a 'long regime of preindustrial agrarian modernity' to antiecological methods 'more akin to mining than to proper farming' through the opening of the English countryside to the world market provides an important lens to recognize the missing socio-ecological dimensions of the agrarian question in the classical accounts (1996, 63–102). It unsettles the prevailing understanding of peasant farming as anti-modern in the classical narratives that were centered on the problem of the formation and reproduction of capital/labor relation within the geographical context of Europe. Thereby, it reveals their ecological blind spots as outcomes of their preoccupation with the subordination of landed property as opposed to the subordination of land by the soil-exhaustive capitalist agriculture leading to the ecological crisis which Marx had already underpinned as the metabolic rift (McMichael 2013).

This English instance as 'the classical cradle of capitalist development' displays the possible contradictions between historical constructions of agrarian transformation from the distinct epistemological lenses based on economic rationality (underlying the value theory of labor) and environmental rationality (Schneider and McMichael 2010; McMichael 2013; Leff 2014, 2016). I propose, in this section, the *environmentalization of the agrarian question* in the contemporary historical context. This argument has three sub-components that will be presented in an integrated way in the analysis: The first one is a historical/ontological argument proposing that the agrarian question has become a central component of the socio-ecological question as an agro-environmental question. Put differently, the agrarian question has transformed into a question of socio-ecological sustainability in relation to labor, land/nature and food. The second one is a theoretical/epistemological argument, with political implications, proposing that without the theoretical lens of value theory of nature we cannot completely comprehend the historical nature of the agro-environmental question, which I formulate as *the agrarian question of nature*. Finally, a reconsideration of the contemporary debates of what has been expressed as 'the twenty-first century agrarian question' (Bernstein 2006; McMichael 2007; Petras and Veltmeyer 2014; Akram-Lodhi 2021) with a socio-ecological lens, particularly in relation to the value theory of nature, demonstrates the environmentalization of the conceptual analysis in the contemporary formulations of agrarian question as well.

Akram-Lodhi and Kay (2010, 264–267) undertook a comprehensive and systematic survey of seven distinct conceptualizations of the contemporary agrarian question in relation to rural production, accumulation and politics. Among these accounts, 'the class forces agrarian question' as mainly developed by the works of Terence Byres and 'the path-dependent agrarian question', as developed by the work of Bill Warren, can be located as two differentiated extensions of the classical accounts of the agrarian question (with a focus on agrarian transition, agrarian capital and rural capitalism) in the spatio-temporal context of the twentieth century rural geographies of the Global South. In the context of the twenty-first century, Henry Bernstein's 'agrarian question of labor', which I will re-denote as CAQ1 (i.e. contemporary agrarian question), moves beyond these accounts. He argues that the classical Marxist accounts of the agrarian question of capital or capitalist transition in agrarian societies subsumed the question of labor within the context of the politics of socialist transition. Yet, the socio-historical elimination of the landed property and the complete internalization in the countrysides of capitalist social relations of production and reproduction, if not the collectivized social relations under socialist/communist transitions, by the end of 1970s marked both the end of the agrarian question of capital and the peasant question on a world scale, sustaining only diversified agrarian classes of capital and labor (2006, 449–450). With the crisis of socialism, the collapse of state-led developmentalism and the global expansion of neoliberalism, the working poor in the Global South confronted a dual crisis of reproduction and fragmentation moving across rural and urban spaces between diverse low-wage, precarious or informal forms of work (Levien, Watts, and Hairong 2018, 867). The agrarian question of labor concerns, thereby, how the capital-labor relation in the neoliberal era reorganizes, differentiates, respatializes and makes vulnerable the rural populations and livelihoods as components of heterogeneous classes of labor. Hence, rural politics becomes a necessary component of a world-historical class struggle that has become a livelihood struggle for access to a living wage.

The conception of agrarian question from the value theory of labor presumes land as a relation of rent, i.e. as a 'factor' of commodity production. Bernstein's (2006, 452) thesis of the historical disappearance of the peasant question through the vanishing of the *'predatory landed property as a significant economic and political force by the end of the 1970s'* thanks to the post-war peasant rebellions, the subsequent land reforms and capitalist restructuring reproduces this presumption. It conceives contemporary farming as already only capitalist production based on a capital-labor relationship, leaving no room for a complex 'simultaneity of different temporalities' cohabiting in geographical space, as expressed by Milton Santos (cited in Leff 2016, 261). This approach situates the category of 'the peasant' and its relation to land in a 'pre-capitalist' social context and delinks at the conceptual level the rural livelihoods of diverse agrarian classes of labor from land. It is not land, but wages/profits in capitalist agriculture which become a political terrain of (semi-)rural struggles.

From a socio-ecological perspective based in 'environmental rationality' (Leff 2014), this approach underestimates how land and nature, instead of wages, has become a central political terrain of rural struggle in the context of the world-historical expansion of extractivist agrarian accumulation. As Leff has argued, the value theory (of labor) has unvalued the productive yet entropic or destructive consumption of nature and 'the ecological conditions for the expanded reproduction of capital' in time–space. It, thereby, could not locate the economic poverty and social differentiation of the peasantry and indigenous peoples inhabiting rural areas in the 'historical process of entropic degradation of their environment and their livelihoods'. Nor could it locate their persistence in their resistance to dispossession, their cultural resilience rooted in nature, and their adoption of a socio-ecological ontology that has challenged the capitalist value relations degrading their nature (Leff 2016, 248–249, 256–259; 2021, 145–147, 155). From the theoretical lens of the value theory of nature, the agrarian question of labor conceived in the form of the fragmentation and increased precarity of the social conditions and livelihood sources of the world of 'unskilled' labor is still relevant. Yet, it becomes an outcome of the dispossessory processes of extractivist accumulation, i.e. the world-historical capital-nature relation.

In this respect, the environmentalization of the twenty-first century agrarian question concerns how the world-historical temporality of capitalist agrarian transformation through the intensification of value production based in extractivist processes configures the ecologically embedded local temporalities of peasants, pastoral and indigenous people living based on Earth. In specifying the environmentalization of the contemporary agrarian question, I will engage in the following conceptual formulations of agrarian question: (i) Farshad Araghi's *agrarian question of dispossession* (or *agrarian question of depeasantization*) (CAQ2); (ii) Philip McMichael's *agrarian question of food* (CAQ3); (iii) McMichael's (2016, 649) reformulation of '*agrarian question as a general socio-ecological question*' (CAQ4) as developed by him and Tony Weis (2007, 2010); (iv) Akram-Lodhi's (2021) recent formulation of *ecological agrarian question* (CAQ5); (v) Finally, what I propose in concurrence with much of these reformulations, but situate as a central component of the broader socio-ecological question with the conceptual and methodological lens of the value theory of nature as an *agrarian question of nature* (CAQ6). Table 1 maps out broadly each of these approaches. As distinct from CAQ1 centered on the capital-labor relationality, the environmentalization of the contemporary agrarian question through the respective accounts of CAQ2, CAQ3, CAQ4, CAQ5 and CAQ6 demonstrates: Not only rural socio-ecological transformation and the agro-environmental crisis, but also urban-based problems of social reproduction and livelihood struggles as well as increasing rural and urban vulnerability to climate change can be specified in relation to the twenty-first century agrarian question configured by a global extractivist accumulation based on land and nature.

The agrarian question of dispossession or *depeasantization* (CAQ2) is centered on an understanding of how the global conquest for increasing relative surplus value in the

Table 1. Environmentalization of the Contemporary (the twenty-first Century) Agrarian Question.

Conception of the contemporary agrarian question	Underlying social/ socio-ecological processes	Socio-ecological outcomes	Epistemological perspective	Relevance from the value theory of nature	Anti-systemic political potentials
CAQ1: The agrarian question of labor	Fragmentation of rural classes of labor in relation to global capital No underlying socio-ecological process	Environmental costs of the high productivity levels of modern capitalist farming	Capital/labor relation Class analytic approach	Still relevant but through the operation of extractivist accumulation	Alliance of fragmented classes of labor to struggle for improved social conditions and wages
CAQ2: The agrarian question of dispossession/ depeasantization	The enclosure food regime Overconsumption and under reproduction of nature Dispossession by displacement of the world peasantries	The interconnected food, environmental and malnutrition crises	'Labor in nature' perspective or value-producing nature via labor	Links depeasantization to the capital-nature perspective Recognizes extractivist accumulation based on 'underreproduction of nature'	Alliance of the world of depeasantized and deproleterianized people
CAQ3: The agrarian question of food	Commoditization of food and agriculture under the global corporate food regime Dispossession of the peasant-farmers Agro-industrial destruction of the land/nature and socio-ecological rights	Food and social reproduction crisis Ecological crisis	Critique of class analytic approach and value episteme based on capital/labor relation Revaluing land, ecology and the peasant	Implies the capital/nature theoretical lens	Peasants as world-historical subjects: the food sovereignty movement
CAQ4: The socio-ecological agrarian question	The extractive food/fuel/biomass regime Land/resource/green grabbing	Food crisis Climate crisis Socio-ecological crisis	Critique of capital/labor analytic approach A socio-ecological perspective based on the human-nature unity	Implies and leads to the capital/ nature theoretical lens	The emergence of a new peasantry reclaiming food and land sovereignty to reconstitute a new social-ecological order
CAQ5: The ecological agrarian question	The increasing throughput of matter and energy through extractive agriculture under the corporate food regime	Exhaustion of resources and entropic degradation of the earth	Capital/labor analytic lens in the context of accumulation by appropriation and unpaid nature approach	Explicates the material processes of entropic degradation through extractive accumulation	Agro-ecological transition through the agro-ecology movement
CAQ6: The agrarian question of nature	The world-historical expansion of capital-nature relation through agro-extractivism, land/resource/ green grabbing and urbanization of nature	The socio-ecological question as a unity of multiple interlinked questions/crises	Value theory of nature Capital/nature analytic lens and a new conception of class based in this lens	Proposes the value theory of nature lens to understand the socio-ecologically distinctive historical character of the agrarian question	Broadened perspective of anti-systemic socio-ecological justice movements

neoliberal context has dispossessed the world peasantries or the rural populations especially in the Global South (Araghi 2000, 2009a, 2009b). Araghi (2000) specifies the historical nature of the contemporary process of depeasantization as a process of *dispossession through displacement* based on the 'mobilization of agricultural land', in Polanyi's words, rather than *dispossession through differentiation*, which underlies the agrarian questions of capital and labor. In contradistinction to CAQ1, the notion of dispossession through displacement/enclosures and the formation of a global reserve army of labor in CAQ2 centers the land and the relationship between agrarian production, commoditization and the world market. Dispossession by displacement links the agrarian question via land to the socio-ecological question by specifying 'the accumulation of "surplus nature"' (2009b, 114).

What he calls 'the standpoint of "labor in nature"' discerns the reproductive fertility of nature as value producing via labor 'in that it determines the amount of necessary labor time and thus directly affects the availability of surplus labor time' (2009b, 121). This understanding of nature as a 'life giving, labor producing, and productive relation of social life' distinguishes the socio-ecological dimensions of CAQ2 on especially three axes of 'accumulation by displacement': (i) in terms of labor, simultaneous depeasantization and deproleterianization and production of surplus value through forced underreproduction of either (underconsuming) urban surplus labor or migrant/informal agrarian labor; (ii) in terms of ecology, forced overconsumption, appropriation and contamination of deruralized environments that are transformed into global surplus nature (ecological enclosures) of the agro-industrial capital as the basis for the global food regime; (iii) the enclosure food regime characterized by forced underconsumption for the surplus populations living in the world's hyperurbanized cities through an overdependence on the market mechanism for access to commoditized food alongside a subsidized consumption and overconsumption among an urban minority (2009a, 113, 118–119, 2009b, 124–127, 134–136; Patel 2007). The neoliberal global value regime, therefore, operates through a continuous and destructive process of redistribution of value from global surplus labor and surplus nature to the global capital. This process entails underreproduction of labor power, 'under-reproduction of nature' (Moore 2011, 28) and the enclosure food regime leading to its multiple systemic crises: an environmental crisis in the form of a deepening 'metabolic rift', a food crisis especially with the end of a food regime in 2006 and a widespread crisis of malnutrition and hunger (Araghi 2009b). CAQ2, thus, temporarily and spatially links the rural and urban manifestations of the socio-ecological crisis by locating the world-historical relationship between labor, nature and food.

While the 'labor in nature' standpoint recognizes 'nature as directly value producing' (2009b, 121), in my view, nature as the 'past and future reproductive context' of labor becomes still indirectly value producing via surplus labor time, or the capital-labor relation. McMichael argues that the 'labor in nature' perspective 'collapses the society/nature binary, positing a unity in the exploitation of human labor and its natural component'. He argues further that the concept of value allows this perspective 'to demystify price (and payment for ecosystem services) as a fetishized representation of the social and ecological relations inherent in commodity production' (2013, 135). The recognition of the unity of labor and nature at the world-historical level as value producing is invaluable for understanding the historical nature of capitalist value relations, but not sufficient. It is also necessary to recognize the historical nature of this unity as well as their historically specific separation through the global value relations. From the conceptual lens of the value theory of nature, the value producing capital-nature relation in the socio-ecologies of the displaced peasantries (ecological enclosures) operates as another and historically predominating basis for the underreproduced nature/labor and the enclosure food regime. To put in Araghi's terms, the 'labor in nature' standpoint has its equivalence in the 'nature in labor' standpoint. The former recognizes *surplus nature* as an empirical category in the concrete form of 'the spaces of the existence of the world peasantries'. The latter, following Marx's method in his value theory of labor (Marx 1976; Elson 1979; Sayer 1987; Tomich 2004), specifies the historically distinctive objectification of (concrete/surplus) nature as *abstract nature* as the condition of the equivalence of ('socio-natural' [Peluso 2012]) commodities in the extractivist capital accumulation processes. While *concrete nature* is the direct biophysical source (in the form of energy/carbon or nutrient/organic matter/space) of the use-value, abstract nature is the historical source of exchange value and surplus value (Taşdemir Yaşın 2017). The extractivist industrial agriculture is an ecologically specific historical moment of these processes. From this inverted standpoint, not only the reproduction of labor depends on the reproduction of nature, but also the reproduction of nature depends on the socio-ecological relation between nature and labor. I will elaborate this point further in the context of the subsequent CAQs as well as of the rise of the agroecology movement in the second section. Yet, it is necessary to note that Araghi's CAQ2 renders the possibilities for linking the processes of depeasantization to the global value relations based on the capital-nature relation.

The agrarian question of food (CAQ3) specifies the CAQ through the lens of the corporate food regime (McMichael 2007, 2008, 2009, 2013). The food regime lens enables resituating the CAQ in relation to political ecology and rural politics of food, land and nature, rendering how 'the classical, capital-centric approach discounts landed ecology, and discounts farmers/peasants as historical subjects' (2013, 62). Accordingly, the corporate food regime has catalyzed 'a new agrarian question' (2007) that re-placed the peasants at the center of the contemporary agrarian politics, in contradistinction to the thesis of CAQ1. The epistemic challenge created by the peasant/agrarian resistance against capitalist value relations by reclaiming the centrality of the peasant relation with the land 'constitutes the twenty-first-century variant of the agrarian question' in the form of CAQ3 (2007, 61). As opposed to the modernist categories perceiving the peasantry as anachronic or residual, the contemporary peasant resistance demonstrates the possibilities for alternative epistemologies based on modern yet anti-capitalist visions of agrarianism. Thereby, it unsettles the hegemonic value episteme of capitalist development as well as its Marxist critiques based on the value theory of labor (2013, 65–83; 2007).

As a conceptual, epistemological and political reflection of the ecologically based critique and struggle of the food sovereignty movement, CAQ3 shifts the substantive focus of the CAQ from *rural production* to sustainable *social reproduction*. Accordingly, the classical accounts of agrarian question enclosed the meaning of 'social reproduction' to 'the reproduction of labor power through the wage relation' as they focused on the expanded reproduction of industrial capital and, thereby, affirmed the 'disappearance of the traditional peasant'. In the context of the CAQ, the food sovereignty movement problematizes and opposes the 'food security' politics of the corporate food regime that encloses the meaning of 'social reproduction' with a rhetoric of 'feeding the world'. This rhetoric conceals the reconfiguration of the food production/circulation through its separation from the peasants/small farmers at the world-scale and (re)integration in the neoliberal world market under the domination of the agro-industrial capital. The food sovereignty movement unravels how the commoditization and 'the accelerated movement of food' relates to 'the accelerated dispossession of the peasantry', cheapening the cost of both production of food and reproduction of wage labor worldwide. It demonstrates the reconfiguration of the rural socio-ecologies of social reproduction as the agro-industrial (commodity) frontiers of 'agriculture without farmers' who, in turn, transform into environmental refugees in the instable and transitory urban social-ecologies of the slums. In this context, it transforms the CAQ into a question of 'the reappearance of "a new peasantry"' reclaiming their rights to connect with land and produce their food without dependence on the world market (McMichael 2008, 45–48; 2009, 298, 307–308; 2013, 81).

The rise of the claims for food sovereignty challenges the capitalist domination of land and agriculture from an *ecological standpoint* unsettling the domination of economic values over ecological values. Likewise, 'its advocacy of revaluing small farming' poses agriculture not only as a material form of food production but also as a distinctive multifunctional form of socio-ecological relationship that is central and necessary for restabilizing human-nature relation. In this respect, CAQ3 poses the agrarian question of food from an understanding of food as a product of distinctive social, cultural and ecological values and relations that are central to sustainable forms of social reproduction, addressing not only the food crisis of neoliberal capitalism, but also its socio-ecological crisis. It conceives the food sovereignty movement as a precursor of a new agrarian ontology as expressed by Hannah Wittman's notion of 'agrarian citizenship' and Duncan's proposition for 'the centrality of agriculture', emphasizing the revitalization of a socio-ecologically sustainable agrarian culture and biophysically diversified local food ecologies through the stewardship of land (2008, 46, 49–50; 2009, 300–308).

In my view, CAQ3 manifests the socio-ecologically based conceptual perspective of the value theory of nature and its later reformulation (2013, 2016) as *a broader socio-ecological question* as a due outcome. McMichael argues that the CAQ3 posed by the food sovereignty movement is conceptually 'unthinkable' from the capital/labor analytic lens that ends up with a characterization of 'the dispossessed (peasant) as unemployed labor' (2008, 49, parenthesis added). This lens does not enable a critique of commoditization of food and fetishism of capitalist food production, which is, by contrast, restored by the food sovereignty movement (2009, 298). I would like to further qualify this argument by arguing that the food sovereignty movement's epistemological challenge and ontological vision restores the critique of fetishism through a new conceptual/epistemological perspective that is specified by the value theory of nature, i.e. the capital/nature analytic lens. 'Abstraction of food production' (2009, 304) is increasingly realized as an extractive-industrial process, i.e. based on *the social abstraction of land/nature,* as the predominating source of food as a 'socio-natural commodity' (Peluso 2012), enabled by dispossession by displacement. In this respect, the historical connection of the crisis of social reproduction to the ecological crisis underlined by CAQ3 opens the substantive space for exploring the world-historical relationality and unity between the capital-labor relation and the capital-nature relation.

McMichael (2013, 2014, 2016) broadens the substantive and political focus of CAQ3 by prioritizing its socio-ecological character in relation to the life-threatening climate crisis (2012) reformulating it 'as a general socio-ecological question' (2016, 649) with its political implications for 'the question of socio-ecological survival in a post-industrial-agricultural era' (2013, 18). I will express this later reformulation of CAQ together with the contributions of Tony Weis (2007, 2010) as *the socio-ecological agrarian question* (CAQ4), although it continues to comprise CAQ3 as well. McMichael writes following Weis: 'The industrial agriculture underwritten by energy and agribusiness subsidies is reproduced by methods of "biophysical override" (Weis 2007), which substitute short-term financial gains (value-override) for long-term ecological sustainability. *Value override by the food regime has critical ecological consequences requiring a reformulation of the agrarian question*' (McMichael 2013, 63 *emphasis added*). Accordingly, the growing global agrarian crisis of industrial agriculture entailing the twin crises of climate and food/malnourishment recenters the agrarian politics on the peasant question (2013, 80–81). CAQ4 transforms the food question into a 'civilizational question' in the context of the conjunction of food, energy, climate and financial crises that can be resolved by a post-industrial socio-ecological farming practice based on land stewardship and renewing ecological cycles (2013, 65, 82–83; 2016, 651–652). Thus, CAQ4 shifts the conceptual and political focus from 'labor in nature' standpoint to what I call 'nature in labor' standpoint.

At the same time, it centralizes in the analysis how industrial agriculture 'as a relatively safe investment haven for the relatively long-term' has become a crucial moment and medium of increasing dependence of commodity production on 'capitalization of non-human nature' (Moore 2010), 'triggering the "global land grab"' (McMichael 2013, 117). A striking moment of this is the agrofuel project developed as a 'green' solution to the 'absolute exhaustion' of fossil fuels. The invention of interchangeable 'flex crops' (food, feed and fuels) such as agrofuels (Borras et al. 2012) manifests how agriculture has become a flexible environmental frontier of commodity production 'converting all crops to exchange value par excellence'. Agrofuels production commences a crucial moment of 'relative exhaustion' or underreproduction of nature, recycling the problem of greenhouse gas emissions as a 'green' solution (2013, 114–117; 2016, 660). Therefore, the essence of CAQ4 is not only about the tension between global abstraction of food production ('food from nowhere') based on commodified inputs and generalized commodification of food, on one hand, and the peasant countermovement against it reclaiming socio-ecologically based local food production ('food from somewhere'), on the other. But also, it is about 'the displacement of agriculture' as a form of production based on land/nature 'from social provisioning and (socio-ecological) multi-functionality' and its remodeling as a site of multiple/flexible commodity production in the form of 'expanding bioeconomic, transgenic and meatification complexes' at the expense of the land use for socio-ecological reproduction (2016, 649, 660, parenthesis added; Weis 2007).

CAQ4 situates the development of bioeconomy or biocapitalism as the highest stage of commodification based on crop substitutability and the global land/resource/green grabbing driving 'a spatio-sectoral shift in capital accumulation toward *a new extractive food/fuel/biomass regime*' through financial speculation. As neoliberal climate change politics presents bioeconomy as a sustainable techno-science intensive model, monopoly use of smart agriculture technologies by agro-TNCs in enabling 'sustainable intensification' of monocrop production 'technologizes' the extractive regime of food/fuel/biomass (McMichael 2013, 119–125, 132, *emphasis added*). CAQ4 articulates, thereby, a substantive and political manifestation of a shift of emphasis from the corporate food regime to the extractive food/fuel/biomass regime. This shift corresponds to the historical expansion of (capitalist) frontier of agriculture or farming beyond the spatio-temporality defined by the neoliberal/corporate 'food security' politics to include the spatio-temporality defined by the 'climate security' politics through a combination of 'bioeconomy' and 'green growth' models based on bio-technological methods.

Akram-Lodhi's (2021) recent formulation of CAQ as *ecological agrarian question* (CAQ5) can be conceived as a conceptually differentiated substantive extension of CAQ4 with more focus on the extractivist character of fossil-fuel driven, large-scale and capital-intensive capitalist agriculture within the context of the corporate food regime. The substantive focus of CAQ5 is the extractive nature of the 'industrial grain-oilseed-livestock' agro-food complex (Weis 2013) understood through the material/energy flows. It mainly argues that extractivist capitalist agriculture is less productive than small-scale agriculture or does not develop productive forces in terms of energy returns relative to energy investment, i.e. 'the EROI of agriculture' (Martinez-Alier 2011). Yet, epistemologically, CAQ5 is governed by the capital-labor theoretical lens and how this relation 'shapes and is shaped by the prevailing ecology' (2021, 701). From this lens, extraction is understood based on Moore's (2015) notion of 'extractive appropriation' of 'Cheap Natures' that is translated in agriculture as appropriation 'by capital of stocks of energy in natural resources such as hydrocarbons, soils and water, which capital in turn transforms into flows of energy-carrying materials, in the form of material farm inputs' (Akram-Lodhi 2021, 699). As such, the subordination of small-scale farming under the corporate food regime to extractivist capitalist agriculture increases the throughput of matter and energy, exhaustion of resources and entropy.

The cheap or unpaid nature approach to the process of extraction or extractive agriculture has one consequence: as it recognizes the exhaustion of nature as a material process or flow of energy, it reproduces the economic rationality underlying the classical accounts of agrarian question from the lens of value theory of labor. It conceives of the increase of material throughput in relation to the labor productivity and nature's depletion in the production of surplus value. The thesis that low EROI of industrial agriculture increases the entropic demise of the ecology constitutes a crucial material aspect of the CAQ. Yet, as underlined by CAQ4 and the environmental rationality perspective of Leff (2014, 2016, 2021), this is a one-sided concrete moment of a broader process of capitalist domination and transformation of socio-ecological relations defining peasant farming (with cultural and territorial consequences). This broader socio-ecological lens has critical political implications as well for the second thesis of CAQ5, which I will further discuss in the subsequent section on the *agrarianization of climate justice*: Small-scale agroecological farming promises to increase the EROI of agriculture and the resolution of the CAQ can be possible by an 'agroecological agrarian transition' as it can 'develop the productive forces necessary to lay the foundation of a post-capitalist future' (Akram-Lodhi 2021, 688).

From the value theory of nature lens, the global expansion of extractive industrial agro-food/agroforestry frontiers (of flex crops) articulates a distinctive socio-ecological relationship between capital and nature as a value relation and a distinctive process of 'commodity frontier' formation (Moore 2000) based on extractivism at the expense of locally specific socio-ecological relations and valuations of nature (Martinez-Alier 2002). This relationship constitutes the socio-historical condition/context of the entropic material flows. This lens shifts the centrality of the CAQ from the commodification and incorporation of farming into the capitalist world market to the commodification and incorporation of nature into the world-historical relation of capital. The key to the unpaid nature approach is Moore's (2015) notion of 'abstract social nature' in the context of his account of 'capitalism's Cheap Nature strategy', as distinct from the *abstract (surplus) nature* as the source of value in the value theory of nature. According to his formulation, 'abstract social nature' is a product of colonialism, enclosures and accumulation by dispossession in their enhancement of labor productivity 'without the costs and risks associated with M–C-M' (capitalization)' (2015, 98). He defines it as 'a systemic family of processes aimed on simplifying, standardizing, and otherwise mapping the world in service to the quantitative expansion of abstract labor'. Therefore, as I pointed out in Araghi's use of 'surplus nature', 'abstract social nature' is a *concrete category* signifying 'those spatio-temporal practices that identify and facilitate the appropriation of unpaid work' or Cheap Nature (2015, 126–127). These concrete processes of mapping, identifying, quantifying, measuring, and coding (2015, 122) partake in the social domination of nature in the extractive frontiers as constituents of the historical process of the formation of abstract nature, i.e. the capital-nature relation as a world-historical value relation, contributing mainly to the 'productivity' of nature.

The example of bioeconomic models of extractive agriculture based on technologically controlled environments is demonstrative of how capital can produce value through extractive accumulation based in *abstract surplus nature* rendering labor redundant. Historically, value production based on the socio-material unity of abstract surplus labor and abstract surplus nature in industrial agriculture became a vital source of the metabolic rift in the form of soil exhaustion. Nonetheless, as Weis (2010) shows through the instance of factory-farmed meat, global value production through abstract surplus nature has become an unprecedented source of the biospheric rift (Clark and York 2005; McMichael 2012). In this respect, the CAQ concerns not only the rural-urban divide underlying the metabolic rift and its reversal. It increasingly concerns a systemic world-historical *socio-ecological rift* expanding and deepening through the contemporary extractivist moment of commodity fetishism based in the 'social domination' (Postone 1993) of land/nature through processes of global land/resource/green grabbing, extractivist shift in industrial agriculture and monopolization of high-technology vis-à-vis the small producers. The development of what the World Economic Forum propagates as 'the fourth industrial revolution' based on digital/automation technologies 'fusing the physical, digital and biological worlds' (Schwab 2017) reconfigures capitalist frontiers of agriculture increasingly based on a global socio-ecological rift. This indicates a highest stage of what Marx states as the 'urbanization of the countryside' that can be re-framed as *urbanization of nature* corresponding to the *environmentalization of agriculture* (for food, feed or fuel) through extractivist industrial shift and (bio)technological domination of nature, i.e. 'technologization of nature' (McMichael 2013, 121).

In this context, from the prism of the value theory of nature, I propose reformulating the CAQ as *an agrarian question of nature* (CAQ6). This formulation concurs with and builds upon the substantive constituents of the formulations of CAQ2,

CAQ3, CAQ4 and CAQ5 and methodological-epistemological critique of the formulations of CAQ3 and CAQ4. Nonetheless, it situates the CAQ in relation to a broader process of the world-historical expansion of the capital-nature relation and, thereby, conceives it as a central component or constituent of the socio-ecological question. The CAQ6 presents a critical demystification of commodity fetishism underlying the CAQ through objectification of nature as well as simplification and elimination of concrete agro-ecological and socio-ecological diversity and complexity. It links the four distinct socio-ecological formulations of the CAQ to the capital-nature analytical lens and broadens their substantive analysis in the face of the world-historical expansion of the extractive environmental frontier of capital accumulation. In this respect, Petras and Veltmeyer's (2014) conception of the twenty-first century agrarian question in the form of 'agro-extractivism' constitutes not only but a central historically specific climate and planet threatening socio-ecological dynamic of the agrarian question of nature.

In situating the historical specificity of extractive industrial agriculture or agro-extractivism, its historical differentiation from petro-chemicals based industrial agriculture is fruitful. Agriculture dependent on the "mining' of soils elsewhere on the planet' (Duncan 1996, 97) has featured as an aspect of the world-economic integration of agrarian production with the first (settler) food regime since the late nineteenth century. This regime underlined wheat imports from the colonial frontiers with the long-term consequences of ecological collapse of family farms and the soil exhaustion in the colonies (Friedmann and McMichael 1989; McMichael 2013, 72, 102). Soil mining intensified in the post-war era in the context of US-centered second food regime with agro-chemical industrialization grounded on the 'cheap' oil regime in relation to the post-colonial expansion of the nation-state system (Araghi 2009b, 126; Friedmann and McMichael 1989). This was based on the 'the substitution of (artificial) fertilizers for labour in soil maintenance' and, thereby, on the decay of ecologically sensitive soil management. This labor-saving shift can be viewed as a *relative increase of abstract surplus nature* vis-à-vis abstract surplus labor in surplus-value production. This increased both the efficiency of soil mining and the productivity of labor 'antiecologically' by erasing the agronomic necessity to adapt to local ecological peculiarities and by eroding the soil, its organic fertility and water-storage capacity (Duncan 1996, 38–39, 70, 113, 119, 122). Therefore, we can argue that the socio-ecological shift to petro-chemicals based industrial agriculture already commenced the historical formation of *an agrarian question of soil* (and subsoil water) in the post-war era yet concealed it with overproduction of food (leading to what McMichael identifies as 'the food-aid regime'). It, thereby, set the historical ground for the agrarian questions of depeasantization, food, energy and nature in the neoliberal era that can be characterized by the intensification of global capital accumulation based on extractivism, a process distinguished by the notion of 'new extractivism' (Bebbington 2009).

Agro-extractivism or 'agrarian extractivism' (Petras and Veltmeyer 2014; McKay 2017, 2020) can be specified by an *absolute increase of abstract surplus nature* at the expense of the abstract surplus labor in commodity/value production rendering labor increasingly redundant, thanks to the fusion of petro-chemicals based agro-industrial accumulation with biotechnology, genetics and digital (smart) technologies. The historical shift in industrial production towards agro-extractivism corresponds to the renovation of the corporate food regime in the form of an extractive food/fuel/biomass regime (McMichael 2013). To state in concrete terms, it is a shift from the intensive chemical–mechanical practice of agricultural/food production as a biophysical process accelerated by industrial fertilizers, drugs and machines (Duncan 1996, 131) to a biologically-genetically modified and technologically controlled practice of multi-functional commodity production (precision agriculture). In the latter, the biophysical override/underreproduction of not only soil but the wider ecosystem (forests, animals, water, biosphere) is maximized through the elimination of any socio-ecologically-assisted organic nutrient/waste recycling and ecosystem (self-)repair/regeneration services/practices.

A predatory and political process of incorporation of nature, local agrarian/industrial food production complexes and markets (McMichael 2014) into the monopoly of agro-input and/or agro-food TNCs (Weis 2010) forms or expands the agro-extractive plantation frontiers, i.e. the capitalized and commoditized nature (Leff 2016, 259–260), through large-scale land acquisitions, transformation of local agrarian practices and the acquisitions of local agents of food production/circulation chains. The expansion of monoculture biofuel and flex-crop (food, feed, fuel) production in Latin America (Petras and Veltmeyer 2014) exemplifies a geographically shaped moment of agro-extractivist frontier expansion in distinctive processes: institutional changes enabling corporate acquisitions of local industries leading to economic concentrations (monopolies/oligopolies); the domination of the TNCs over the key stages of the agro-food chain including the provision of seeds and inputs; large-scale land grabbing and land ownership; industrial processing of the produce and global marketing; the generalization of the transgenic seeds in the market with increased local dependency on TNCs; massive ecological and social degradation including increased rural unemployment. In the particular instance of Bolivia, McKay (2020) demonstrates a similar process of what he specifies through triple moments of *land control*, *state control* and *value-chain control*.

As such, a neoliberal and neo-extractive reconfiguration of plantations is being concretely realized as pointed out by the concept of *Plantationocene* (Wolford 2021). This concept indicates a specific biophysically and technologically manipulated-controlled socio-ecological space or modality of extractive production/accumulation. Unlike colonial plantations, which were dependent on social domination of labor (the formation of abstract labor) through enslavement or enforcement (with biophysical override as 'ecological externality'), modern 'Plantationocene' is dependent on social domination of land/nature (the formation of abstract nature) through encroachment and penetration based on displacement-ridden processes (with social override as 'social externality'). These processes consist mainly of land/nature grabbing (Borras and Franco 2012, 2018); the modification, surveillance and control of land/nature (Peluso 2017) through the use of 'genetically resilient', 'labor-saving' and 'climate smart' technologies; and financialization/financial speculation in different phases of the agri-food chain (Clapp and Isakson 2018). They eventually enforce a 'planetary urbanization' (Brenner and Schmid 2012) as a distinguishing aspect of the emerging global socio-ecological rift. As a contradiction at first sight, the expansion of monocrop plantations takes place within a context of hegemonic climate change politics and market-based climate change mitigation and adaptation initiatives that it reinforces. As the studies on the historical instances of Myanmar and Cambodia (Borras and Franco 2018, 1318–1319; Borras, Franco, and Nam 2020) show, these 'sensitized' neoliberal initiatives through climate smart agriculture legitimize and fortify the global land rush by recasting and reconfiguring this agro-extractivist capitalist penetration and control of the countryside.

Land/nature control in the context of agro-industrial/agro-extractivist shift should be viewed not only as a spatio-temporal expansion of the capitalist frontier, but also as a historically specific process of socio-ecological deepening of capitalist accumulation or value relations, which is not distinguished by Harvey's concept of 'accumulation by dispossession'. The land acquisitions do not target only 'idle' or 'remote' lands, but mostly diverse complex socio-natures based on the peculiarities of land cover, soil structure and population density (Messerli et al. 2014). The enclosures and control of local territories entail domination, forced destruction and transformation of the existing place-based socio-ecological relationships and ecologically and culturally based value forms and reproduction patterns in the formation/ expansion of value forms and relationships based on capital-nature relation. Their conversion into the new commodity frontiers with emerging social-ecologies of the respective commodities is an active and contentious process in which local people and hegemonic actors struggle over the material, spatial, legal and institutional reconfiguration of these territories based on the biological and physical 'materiality' of the commodity, of its extraction/production and of the ecology (Peluso and Lund 2011; Tomich 2004; Sneddon 2007).

The instance of Senegal's agro-industrial transformation (Benegiamo 2020) shows how the grabbing and control of the last open-access pastoral lands in a protected reserve in the context of an agro-industrial investment in the production of agrofuel and sunflower seeds for the biomass power plants eroded and marginalized in the longer term local (semi-)nomadic pastoral land rights, local knowledges and livelihoods producing diversified forms of agro-pastoral resistance. Benegiamo (2020, 538) portrays the historical distinctiveness of the forced enclosures and conversion of this agro-pastoral reserve into an agro-extractive farmland and the 'key elements of rupture' from the colonial period in this instance as follows: a distinct moment of reorganization and reordering of the landscape, the water resources and 'the biological formation of the territory' through 'the saturation of the marginal areas' and 'the establishment of a parallel agro-industry poorly linked to peasant farming', thus 'creating new practices among local populations'. In the context of *the agrarian question of nature*, I have reconceived what she does not explicitly state as the historical distinctiveness of the contemporary moment of agro-extractive expansion: the ruling out of locally specific diverse socio-ecological relationships and practices of material, social and cultural reproduction and forms of valuation by *the world-historical generalization of the capital-nature relation*. Agro-extractivism, in turn, has been a central constituent of the global (neo)extractivist value accumulation and the formation of a global nature frontier together with 'natural resource' extractivism, which even in specific contexts turn into competing forms of frontiers (Peluso 2017), with an outcome of the socio-ecological rift.

The agrarianization of the climate justice movement

The increasing urgency of the socio-ecological question in the twenty-first century carried not only the agro-environmental question but also the question of climate change/justice to the center-stage, increasing the emphasis on 'the climate-changing environment' (Schlosberg and Collins 2014). The climate justice movement emerged during the early 2000s mainly from the urban-based environmental justice organizations (EJOs) and NGOs and politicized in late 2000s through the advocacy movements (*Climate Justice Now!* in 2007 and *Reclaim Power* in 2009 in Copenhagen) contesting the mainstream climate politics. Yet, the consumption-centric carbon focus of global warming and climate change politics centered the core concerns of the climate justice movement around reducing carbon emissions through renewable energy and reformed systems of agriculture, public transport, waste disposal and production (Bond 2013; Martinez-Alier et al. 2016). This focus undermined the broader socio-ecological origins of the unbalanced carbon cycle or the biospheric rift rooted in global extractivist expansion and the consequent socio-ecological rift.

In an engagement to situate climate justice in relation to the world-historical context of the agrarian question of nature, I have two main arguments in this section. First, I argue that *an anti-systemic climate justice politics* capable to locate climate change/justice in a world-historical perspective has been developed through *the agrarianization of the climate justice (movement)* in relation to *the environmentalization of the agrarian question and agrarian movements*. Second, I argue that the agrarianization of the climate justice movement reveals *a reformulation of 'class' (or class relations) within the capitalist world system*, representing the agrarian question of nature. The reformulation indicates and specifies the significant ontological shift at work in the climate justice movement. From the perspective of a new understanding of class based on the value theory of nature, the rise of agrarian-based or agrarianized climate justice movements strengthens the already-proposed thesis of the re-emergence of agrarian people/movements as one of 'the most potent' world-historical subjects or anti-systemic social forces (McMichael 2008, 57; Weis 2010; Borras et al. 2018, 1233–1234). At the same time, it enables to transcend the existing 'limits of theory' (Tomich 2015) that bound the understanding of the concept of class to the capital/wage labor relation.

In developing these arguments, I make an analytical distinction between the *environmentalization of agrarian movements*, which has been already noted in political ecology and critical agrarian studies (Guha and Martinez-Alier 1997; Martinez-Alier 2002, 2011; McMichael 2008, 2009, 2013, 2014, 2016), and the *agrarianization of the climate/environmental justice movements*, which has been indicated by Borras and Franco's (2018) notion of 'agrarian climate justice'. The ontological distinction between environmental agrarian movements and agrarian climate/environmental justice movements is more blurred in both rural and transnational contexts, as their development is situated in the world-historical context of the agrarian question of nature and the expansion of extractive frontiers. In this paper, I specify them as internally linked constituents of what we can call the emergence of a broader *anti-systemic socio-ecological justice movement*. The integration of the claims for food sovereignty/food justice with land sovereignty and agroecology and the development of agroecology both as field and as a movement (in both rural, semi-rural and urban contexts) in the last decade manifests the agrarian basis of anti-systemic solutions to the socio-ecological contradictions of the capitalist world economy.

The concept of 'environmental/ecological agrarianism' (or 'ecological neo-Narodnism') was developed from 1985 onwards in the work of Ramachandra Guha and Joan Martinez-Alier (1997) to signify the emergence and development of the peasant resistance in India, such as the Chipko movement, as an environmental struggle against the historical shift towards industrial agriculture and 'scientific' forestry. It portrayed the consequent transformation of local peasants and indigenous people, who were becoming displaced agrarian and 'ecological refugees', into environmental defenders merging agrarianism and environmentalism. Environmental agrarianism fused the politics of land with the politics of resources (e.g. forest, water, mine), as their relation to social reproduction was central. Environmental agrarianism also expanded, as in the instance of 1980s Indonesia, through the engagement between agrarian and grassroots environmental struggles politicizing and challenging the expansion of the nature frontiers of agrarian and mining capital in the 'not-yet-capitalized' local socio-ecologies of the Global South (Peluso, Afiff, and Rachman 2008). In such instances, the shifting alliances between 'new agrarian' and environmental movements depended on 'the temporal and political economic origins and histories of the respective movements, on the types of land contested, on the politics of access to those lands and on the emergence of "discourse coalitions", that dominate discursive spaces' (2008, 209). The lines between agrarian and environmental justice movements were 'often blurred by their specific and common histories in opposing state expropriations', although their rhetorical strategies obscured the common articulations and 'convergence' between them (2008, 232). The changing historical context of political economy and political field (shaped by alliances with diverse/wider movements and by changing priorities/strategies) and their increasing interdependence in the context of land control and land management forced them over time to find a common ground. The formation and expansion of place-specific environmental-agrarian struggles constituted local moments of the globally spreading environmental justice movements centered on environmental distribution and valuation (Guha and Martinez-Alier 1997; Martinez-Alier 2002; Martinez-Alier et al. 2016).

Yet, each of these categories, i.e. agrarianism and environmentalism, were mostly studied, documented and contextualized distinctively especially until 2010s by 'critical agrarian studies' (Edelman and Wolford 2017) and political ecology, respectively. This analytical distinction and separation produced the issue of alliance building and 'convergence' between agrarian and environmental movements as a 'fertile subject for debate' at the transnational level, between transnational agrarian movements (TAMs) and transnational environmental movements (Borras, Edelman, and Kay 2008, 31). There are both ontological-historical and conceptual sources of this distinction and its waning through the idea of 'convergence' between the two sets of movements from 2010s onwards. We can argue that the manifestation of the food crisis and the rise of food sovereignty politics, the intensification of land/resource/green grabbing and the rise of land sovereignty politics and, finally, the manifestation of climate crisis and the rise of climate justice politics have constituted the most influential moments of 'contact', 'common grounds' and/or 'confluence' at the global level between 'new'/transnational agrarian/peasant movements and 'the global environmental justice movement' and transnational EJOs (Martinez-Alier 2002; Martinez-Alier et al. 2016).

In relation to *the environmentalization of agrarian movements*, McMichael (2008, 42) demonstrated this common ground in the context the agrarian question of food by arguing that the food sovereignty movement (such as IPC for Food Sovereignty and Via Campesina) transformed 'the terms of the question' from agrarian transition to the *agrarian revaluation as central to socio-ecological sustainability*. This corresponds to a critical shift from the politicization of the property relations for land, distinguishing the classical agrarian movements to the politicization of 'the socio-ecology of the property relations' (2008, 46). By replacing the 'class-analytic approach' (conceived through capital-labor relationality) with a 'food regime analytic', McMichael situates the 'cross-class political alliances' between diverse peasant movements, from MST in Brazil to KRRS in India, in the emergence of 'peasantry' as a new world-historical political force in the twenty-first century agrarian question (2008, 57; 2016, 651–654). He notes that 'the food sovereignty movement is not simply about peasants, or food; rather … it is about reorganizing international political economy, modeling social struggle around democratic principles, gender equity, producer rights, ecological practices and rebalancing the urban/rural divide' (2016, 649). It builds a modern yet post-capitalist agrarian identity and politics of 'agrarian citizenship' based on alternative value relations and valuations of agriculture as well as alternative socio-ecological subjectivities/practices such as socio-ecological stewardship and 'revitalization of local food ecologies' (2008, 46–48; 2013, 138; 2014). Thereby, it internalizes or reclaims environmentalism for an alternative socio-ecological ontology in which peasants and agriculture have centrality in a 'multifunctional' form based on ecological restoration, regeneration and biodiversity in addition to self-provisioning (2013, 138–140).

Borras and Franco's (2012) contribution of the political frame of 'land sovereignty' as a broader and more embracive alternative to the limited (in the present context) political concepts of 'land reform' and 'land (tenure) security' grasps the historical essence of contemporary struggles for land, rendering another critical aspect of *'environmentalizing' agrarian movements*. They demarcate it as '*the right of working peoples to have effective access to, use of, and control over land and the benefits of its use and occupation, where land is understood as resource, territory, and landscape*'. Leff (2014, 2016, 2021) has shown such 'land sovereignty' perspective as an already organic constituent of contemporary environmental struggles of peasants and indigenous people. 'Environmental struggles do not only claim a piece of land, *but a territory*', he notes, 'as the habitat where they can experience their habitus, where they can deploy their imaginaries and practices in order to preserve livelihoods, *envision their sustainable life-worlds*', and 'where *the new political actors and environmental movements are emerging to reappropriate their bio-cultural heritage and reconstruct their life-territories*' (2021, 159–160 emphasis added). As a countermovement against the commodification of their life-worlds in the historical context of the agrarian question of nature, therefore, Leff (2021, 14–15) suggests that 'the peoples of Earth' '*are territorializing sustainability*' as opposed to global '*geopolitics of sustainable development*'. Land sovereignty in the form 'territorial sovereignty', thus, implies a new form of revaluation and reclamation of land entailing the restoration of its biology, of the rights to land and food and territorial/cultural identity based on a new conception of agrarian citizenship (McMichael 2013, 149). We can argue that food sovereignty and land sovereignty together build both an anti-capitalist politics and post-capitalist ontology of *socio-ecological sovereignty* or *autonomy* that cannot be captured with the conventional theoretical explanation of the relationship between capital and land. To put in theoretical terms, the former agrarian/land reform movements exposed and resisted the historically specific relationship between capital and land via rent, as conceptualized by the value theory of labor. The contemporary agrarian-environmental food and land sovereignty movements expose and resist the historically specific socio-ecological relationship between capital and nature, as exposed by the value theory of nature. As MST leader João Pedro Stédile puts it:

> From the time of Zapata in Mexico, or of Julião in Brazil, the inspiration for agrarian reform was the idea that the land belonged to those who worked it. Today we need to go beyond this. It's not enough to argue that if you work the land, you have proprietary rights over it ... We want *an agrarian practice that transforms farmers into guardians of the land, and a different way of farming, that ensures an ecological equilibrium and also guarantees that land is not seen as private property*. (cited in McMichael 2009, 297–298, emphasis added)

Thereby, contemporary environmental agrarianism both exposes and builds 'a new peasantry of the twenty-first century' from 'the ecological ground' or 'environmental rationality' constituting the central force for the future persistence of peasantry and agrarianism (Van der Ploeg 2008, 2010; McMichael 2013; Leff 2014, 2016, 2021). These movements do not view land only as a physical 'factor' of agricultural production, but as a 'territory' that signifies a broader complex comprising socio-ecological relations of reproduction as well as biological and cultural heritage/diversity of ecology. Capitalization and commodification of their life territories collapse 'the complex connection between locality, place and identity' (Martinez-Alier 2014) and, therefore, their resistance calls for recognition, restitution and regeneration (Borras and Franco 2018).

The rise of climate justice politics has realized a more apparent and comprehensive moment of environmentalization of TAMs or 'confluence' between agrarian and environmental justice movements as well as between critical agrarian studies and political ecology. The food sovereignty movement's involvement in the climate justice debate (especially with their active presence in the Copenhagen Climate Change Conference in 2009) with the thesis of 'sustainable peasant agriculture is cooling down the earth' was noted by Martinez-Alier (2011), who already conceived by then Via Campesina as 'the most important transnational *socio-environmental* movement in the world' (151, *emphasis added*). He situates its deeper involvement in climate justice in the spread and politicization of the perception of agriculture as system of energy transformation, especially in the wake of the global agro-extractivist shift, with the argument that industrial agriculture is 'a major contributor to global warming and climate change by ... *transforming agriculture from an energy producer into an energy consumer*' (146). This argument enabled Via Campesina to situate itself not only as a movement centered on food sovereignty, but also as a socio-environmental defender of small-scale peasant agriculture on the basis of its greater energy efficiency, fewer emission of greenhouse gases, higher capacity to capture carbon and greater ability for conservation of seeds and biodiversity. Martinez-Alier characterizes this broader repositioning as a contemporary example of 'ecological neo-Narodnism' with its real solutions to climate change based in 'a genuine agrarian reform to strengthen peasant agriculture' (2011, 149–157). In a more recent account, Tramel (2018, 1302) argues that resource grabbing and its fusion with climate change mitigation in the production of a global carbon complex have been influential processes in the inclusion of environmental and climate concerns into the repertoires of contention of agrarian movements. Through the instance of West Africa, she confirms the historical shift in the last decade in the common ground for confluence between transnational agrarian and environmental justice movements, such as La Vía Campesina and Friends of the Earth International, from the struggle for socio-ecologically sustainable food production and food sovereignty to the struggle over climate-related resource capture (such as for biofuel production) and climate-smart agriculture.

We can recapitulate the environmentalization of the agrarian movements in conceptual terms in relation to the CAQ in two historical axes: (i) By centralizing the food sovereignty movement and politics, the agrarian question of food (CAQ3), first, unbound the understanding of the new peasantry from the conceptual limits of the capital-labor/wage-land/rent trio that marginalizes the peasant and foresees its (semi-)proletarianization and, second, defetishized the world-scale commodification of food through destabilization of local social-ecological relations (i.e. the politicization of *'the socio-ecology of property relations'*); (ii) The rise of climate crisis and climate justice politics/movement re-articulated this historical-ontological ground in the context of the environmentalization of the agrarian question (CAQ4-5-6) leading to the environmentalization of the transnational agrarian movements as well. The latter has indicated the repositioning of the new peasantry as an anti-systemic force politicizing also *the socio-ecology of the production relations* defetishized by the conceptual lens of capital-nature relation.

At the epistemic level, the environmentalization of the (transnational) agrarian movements denotes a perspective of comprehending agrarianism with a political ecological lens or, to put differently, integrating political ecology in critical agrarian studies. Twisting this interdisciplinary engagement and comprehending the climate/environmental justice movements from the lens of critical agrarian studies, i.e. integrating critical agrarian perspective in political ecology, I argue, will reveal the *agrarianization of the climate/environmental justice movement* as the flipside of the process of environmentalization of agrarian movements in the historical context of the agrarian question of nature. In other words, simultaneously, grassroot rural/indigenous and transnational environmental movements have transformed into the defenders of agrarian-based solutions to climate change. Martinez-Alier (2011) already noted this, as he argued that not only Via Campesina but also transnational climate/environmental justice movements such as the World Rainforest Movement adopted the claims for strengthening peasant agriculture and food sovereignty as necessary constituents of their climate justice-centered political agenda. This has been expressed as a 'convergence of issues and problems' between agrarian and climate justice movements (Scoones et al. 2018, 10). Comparing transnational peasant (La Via Campesina, GRAIN) and indigenous peoples movements (Indigenous Peoples (IP) Forum on Climate Change-IIPFCC), Claeys and Pugley argued that despite their distinctive political positions, strategies and framings of the climate issue, both kinds of movements integrated climate justice within their political discourse and struggles as a common ground especially 'in reaction to the market-based and state-led strategies advanced by the international community to mitigate' climate change (2016, 2). Accordingly, they repositioned themselves in relation to climate change from 'major drivers of climate change' to 'modern actors' proposing alternative post-capitalist peasant and indigenous community-based models/solutions. TAMs prioritized food and land sovereignty and agroecology as a simultaneous solution to 'feeding the world' and 'cooling the planet', while IP proposed territorial autonomy, biodiversity and local forest management to keep forests alive (Claeys and Pugley 2016, 6–8).

A present updated look to IIPFCC's political agenda reveals a visible shift towards 'agrarianization' of their agenda. As they made it clear in the 2021 UN Climate Change Conference (COP26) in Glasgow, IP prioritizes food sovereignty as a constituent of their climate solution model through agroforestry and sustainable agriculture (http://www.iipfcc.org/). Friends of the Earth International integrates 'ecological peasant farming' in their political vision proposing that it 'can preserve biodiversity and local cultures, cool the planet, provide healthy food and livelihoods for all' (https://www.foei.org/). Climate Justice Alliance argues that the industrial model of agribusiness is major contributor to the climate crisis and proposes local food systems working with nature (https://climatejusticealliance.org/). Thus, concurring with Tramel (2018, 1292) who situates agroecology as 'a strategic frame bridging food and climate struggles', we can argue that food sovereignty and agroecology have become a central constituent of the political agendas of transnational climate/environmental/justice movements.

Borras and Franco (2018) suggest the notion of 'agrarian climate justice' as a normative framework to specify the increasing historical convergence of issues between agrarian justice and climate justice and to create a broader political field that can unify diverse progressive struggles. They point out the historically necessary interdependence between agrarian justice and climate justice as follows:

> The kinds of social movement that are required in the current context are those that are capable of building on and going beyond conventional stand-alone peasant struggles or peasant struggles that subsume other identities or strategies, e.g. ethnic nationalities, indigenous peoples, internally displaced peoples, people displaced by mines and dams. There are two sides of this argument. First, climate change advocacy work by state and social forces ... will be stronger only *if deeply sensitized to agrarian issues and movements* and intimately rooted in and engaged with questions of redistribution, recognition and restitution of rights and claims by poor villagers over their land, water, forest and territory. Second, the only way the urgency and relevance of land policies around redistribution, recognition, restitution and regeneration in the era of global land rush can (re)gain traction in today's world *if it is sensitised to and embedded within the broader climate justice struggles.* (Borras and Franco 2018, 1320, *emphasis added*)

If our point of departure becomes the issue of 'convergence' (Tramel 2018; Scoones et al. 2018; Claeys and Pugley 2016; Borras and Franco 2018), we can argue, following Martinez-Alier (2011), that climate justice has been the most influential point of confluence between the agrarian and environmental/climate justice movements. If our departure point becomes the possibility of creating a broader anti-systemic political field as indicated by Borras and Franco (2018), then, we can situate both the *environmentalization* of the agrarian/food sovereignty movements/politics and the *agrarianization* of the transnational climate justice movements /politics as a relational process of recognition and adaptation of a broader socio-ecological perception of justice, sovereignty, diversity and rights in the face of the (agro-)extractivist turn of the global value relations prone to multiple interlinked soci-ecological crises. Food justice and climate justice are dependent on each other, because as GRAIN states 'the climate crisis and the food crisis are intimately linked' through the industrial food system. As FoodFirst states, agriculture has become the 'battleground for climate justice', as 'how we produce and consume food

Table 2. Social composition of world-wide rural-based environmental conflicts (number of selected cases = 2167, *n* = 3687)

Mobilized groups	Number of cases	Frequency (in all cases, %)
Farmers	1181	32,03
Fisher people	475	12,88
Landless peasants	315	8,54
Pastoralists	184	4,99
Artisanal miners	103	2,79

Source: EJATLAS, https://ejatlas.org/, as of June 2022.

is contributing to climate change'. This link is both an outcome and a manifestation of the agrarian question of nature. If we approach from a broader lens of the socio-ecological question, both climate justice politics and food sovereignty politics become already differentiated expressions of a unified anti-systemic politics situated in the socio-ecological question.

If we move from the level of transnational organizations to the local grassroot conflicts and movements, we can see more concretely that the lines between place-based agrarian and climate/environmental justice movements are blurred, as the earlier historical studies argued (Peluso, Afiff, and Rachman 2008). Constructed based on Environmental Justice Atlas, Table 2 shows that a considerable majority of place-specific environmental conflicts is composed of rural struggles of which more than half composed of struggles of farmers and landless peasants, confirming either environmental agrarianism or agrarian environmentalism depending on the nature and the source of the conflict and the claims of the mobilizing groups. Although the sources of these conflicts are diverse, entailing such as biodiversity conservation, biomass and land conflicts, water management, waste management, mining or fossil fuel extraction projects aimed at value accumulation based on socio-natural commodities, the affected communities are rural-agrarian based communities with shared claims to protect the agro-diversity and biodiversity of their territories, their access/relation to nature (land, forests, water) and agrarian forms of socio-ecological reproduction. As Table 3 shows, more than half of environmental conflicts in the category of 'fossil fuels and climate justice' already emerge in rural-agrarian contexts (399 cases) against projects involving oil and gas exploration and extraction (174), coal extraction and processing (73), oil and gas refining (42), large scale wind energy plants (35), REDD/CDM (12) and large scale solar plants (11). In an important number of these

Table 3. Social composition of reported cases of worldwide environmental conflicts in the category of 'fossil fuels and climate justice' (number of selected cases = 668, Number of selected cases with rural population type: 399, *n* = 3687).

Mobilized groups	Number of cases	Frequency (in all selected cases, %)
Farmers	307	45,96
Fisher people	196	29,34
Landless peasants	82	12,28
Pastoralists	40	5,99
Artisanal miners	7	1,05

Source: EJATLAS, https://ejatlas.org/, as of June 2022.

Table 4. Major types of conflicts in the category of 'the biomass and land conflicts' among rural populations (number of selected cases = 426, $n = 3687$).

rural populations (Number of selected cases= 426, n=3687)

The type of the conflict	Number of cases	Frequency (in all selected cases, %)
Land acquisition	260	61,03
Deforestation	178	41,78
Plantations	169	39,67
Intensive monoculture and livestock	131	30,75
Logging and non-timber extraction	76	17,84
Agro-fuels and biomass energy plants	46	10,80
Water access rights and entitlements	33	7,75
REDD/CDM	18	4,23
Dams and water distribution	16	3,76
Mineral ore exploration	14	3,29

Source: EJATLAS, https://ejatlas.org/, as of June 2022.

conflicts, land acquisition (98), water access rights and entitlements (28) and deforestation (22) figure as the underlying motive. These figures substantiate globally the idea of agrarian environmentalism/climate justice as well as the arguments of Tramel (2018) and Borras and Franco (2018) on the triggering impact of climate change mitigation policies/practices on the spread of agrarian-based climate justice movements. Finally, if we look, from the opposite pole, at the biomass and land conflicts among rural populations, we can see that land acquisition (131) and deforestation (69) based on flex crops (specified with palm oil, corn, maize, sugarcane, soybeans), agro-fuels and biomass energy plants (46), REDD/CDM (18), dams and water distribution (16) projects figure as the underlying context in the specified cases (see Table 4 for significant types of conflicts in this category). This, in turn, substantiates the argument of Borras, Franco, and Nam (2020) on the entanglement between the climate change politics and the rural-agrarian conflicts over access to nature (land, forest, water) operating through land grabbing, green grabbing and resource grabbing. In all these conflicts, the peasant/pastoral/indigenous identities of local-rural people get politicized and gain a more complex political-ecological character merging agrarianism and environmentalism (Taşdemir Yaşın 2019).

Both the environmentalization of agrarian movements, which accelerated at the transnational level in the twenty-first century, and the agrarianization of climate/ environmental justice movements, which became visible in the twenty-first century, in complex/dynamic historical relation to one another *uncover and imply the agrarian question of nature (or the environmentalization of the agrarian question) as a major historical source of the socio-ecological question* (entailing the food/ land/ displacement/ climate crises). They also *manifest the anti-systemic solutions to the climate change and the wider socio-ecological question* in agrarian-based practices, which has become more evident with the agrarianization of climate justice. In this respect, the emergence of 'agrarian-climate justice' signifies an anti-systemic field of politicizing the social-ecology of relations of production, in stark contrast to diverse urban-based/carbon-centered climate justice politics/movements politicizing the socio-ecology of relations of consumption. The rise of the agroecology movement (with diverse articulations), the integration of the claims for food sovereignty and climate justice with the propositions of agroecology, and the spread of agroecological forms of food production (in complex and interlinked rural, semi-rural and urban spatial contexts) represent the solutions within this anti-systemic political field to the socio-ecological contradictions of the capitalist world economy. These solutions are developed in the form an 'anti-systemic praxis', i.e. a union of practices as 'catalysts for cultural transformation' in creating a 'circulatory system of society' (Duncan 1996, 177). To historically specify Holt-Giménez, Shattuck, and Lammeren's (2021) argument, 'agroecology engages the agrarian question' but particularly the agrarian question of nature.

In this context, Akram-Lodhi's second thesis of CAQ5 is critical: 'an agroecological agrarian transition' requires 'transcending those (capitalist) social conditions and relations and developing a post-capitalist agrarian – and non-agrarian – alternative' in order to resolve 'the ecological agrarian question' (2021, 688, 710). Akram-Lohdi locates agroecology in contradistinction to capitalist property relations within the context of the value theory of labor and the conceptual perspective of capital-labor/wage-land/rent relationality. With the value theory of nature lens, I suggest a broader historically-situated theoretical perspective to the understanding of a post-capitalist agrarian socio-ecological alternative. The most critical aspect of this conceptual shift is that it settles the deep-seated impasse to position the food sovereignty movement and the agrarian-climate justice movement within a class-analytic perspective of Marxist agrarian studies framed within the limits of the value theory of labor (McMichael 2008, 2009, 2013, 2016; Tramel 2018; Borras 2020; Akram-Lodhi 2021). This can be viewed as a tension between agrarian political economy and agrarian political ecology. As Borras (2020, 13–14) argues, unifying in a singular way diverse contemporary agrarian movements under the category of 'agrarian (neo)populism' 'undermine[s] the potency of the Marxist framework in advancing urgent and necessary critiques of contemporary agrarian movements' and the anti-capitalist political possibilities developed by them.

However, a class-analytic approach is still possible from the perspective of a value theory of nature that opens a new realm of anti-systemic politics confronting the ecological and socio-ecological contradictions inherent in the capital-nature relation, whose most disruptive long-term and large-scale manifestation has become the planetary climate crisis (McMichael 2012). Rethinking the social category of class in the form of socio-ecological differentiations produced by the world-historical capital-nature relation repositions the environmental agrarian and agrarian climate justice movements as the historical manifestations of the reconfiguration of class formations within the capitalist world system. Thereby, they form already an anti-systemic class movement aiming to disclose, transform and transcend commodity fetishism based in extractive value accumulation. Increasing accumulation of dispossessed but not proletarianized/or 'semi-proletarianized' rural-based redundant populations in the urban planet of slums at the present moment indicates this second class dynamic (evoking 'the second contradiction of capitalism'). The vocabularies of protest used by the transnational coalitions (such as 'industrial agriculture heats up the planet, small-scale agriculture cools down the earth' or 'small-scale farmers feed the world') represent this new, socio-ecologically based political field of class struggle.

The reformulation of the agro-environmental (justice) movements with a new understanding of class re-centers the agrarian/rural people as an *anti-systemic* world-historical social force, proposing sustainable and locally embedded agro-ecological practices as *anti-capitalist praxis* of an emerging *post-capitalist temporality* (McMichael 2013, 135–137; Akram-Lodhi 2021; Leff 2021). It conceptually enables 'the new peasantry', representing a progressive, emancipatory and post-capitalist agrarianism (Van der Ploeg 2008, 2010) and building 'the social and material bases for an equitable and sustainable world order, grounded on the forces and the conditions of life' (Leff 2021, 205). In this respect, *socio-ecological justice/ autonomy/ sovereignty* can be viewed as a unifying anti-systemic theme of various anti-capitalist 'sectors' emphasized by Borras et al. (2018).

From this perspective, the rise of agroecology both as a globally spreading socio-ecological movement/praxis and as a field of study that can unite agronomy, political ecology and critical agrarian perspectives can be situated as a critical component of a broader anti-systemic resolution to the agrarian question of nature and to the broader socio-ecological question. Highly relevant for this perspective is van der Ploeg's argument that 'agroecology is changing the social relations of production in agriculture' (2021, 275), as 'a permanent, material and highly visible critique of the logic of capital ... , a comprehensive and convincing critique that speaks through successfully applying alternative practices and obtaining results that show that agroecology performs better' (2021, 292). It embodies, reaffirms and revitalizes a wide set of socio-ecological mechanisms such as self-provisioning, co-production, the autonomy of nature, agricultural multifunctionality, integration of different agricultural (crops and animals) and non-agricultural production, and the creation of new marketplaces and networks as well as social relations of mutual help, cooperation and solidarity (2021, 279). It, thereby, 'extends the agrarian frontier far beyond the self-imposed limits of entrepreneurial and capitalist farming', promising simultaneously to 'feed the world' (Van der Ploeg 2010, 24), to increase the EROI of agriculture (Martinez-Alier 2011) and to build 'a sustainable mode of production based in the negentropic potentials of life' (Leff 2021, 158).

This potential of agroecology has already been shown in several works as an effective model to increase the future food production and to create equitable food systems as well as sources of skillful employment (Selwyn 2021, 797; Altieri and Nicholls 2020, 893; Weis 2010). According to Altieri and Nicholls (2020, 893–894), 'agroecology has grown into a global movement' of rural producers (peasants and farmers) and activists 'seeking to insure food sovereignty, agrarian reform, the establishment of cooperative models, and the protection of biodiversity' envisioning 'a shift from a market economy to a solidarity economy'. It is, thereby, a *post-capitalist practice* of 'both ecologically sustainable and socially-just farming' (Rosset and Altieri 2017, 9) with a 'very different conception of productivity' based in a labor-intensive, biodiverse, decentralized, ecological stewardship-oriented model of production (Weis 2010, 334), with the potential to build different cultural forms embedded in 'environmental rationality' (Leff 2021, 17) and to return agriculture 'to its rightful, central place in culture, but on both a new ecological basis and a new socioeconomic basis' (Duncan 1996, 181–182).

Conclusion

This paper is an engagement of 'going *against the grain of Marx's classical theoretical presentation'* of the value theory of labor (Tomich 2004, 28) and *reclaiming Marx's method of theory* (Elson 1979; Sayer 1987; Tomich 2004, 2015) *in understanding the concrete socio-ecological complexity of the twenty-first century* through the value theory of nature. The value theory of nature enables grasping the socio-ecologically distinctive nature of the contemporary agrarian question as *the agrarian question of nature* and situates it as a critical component of the broader socio-ecological question. It also enables a reformulation of class and recognition of new class configurations based on the relation of capital to nature. The reformulation locates the rural/agrarian people and environmental-agrarian movements at the center of the socio-ecologically based anti-systemic political field, catalyzing the *agrarianization* of the politics and movements of environmental and climate justice. Particularly, the agroecology movement signifies the emergence of *a contemporary form of political agrarianism* developed within the context of the agrarian question of nature.

In the last three years, I have been involved in various rural and (semi)urban-based local-regional initiatives, organizations and networks of permaculture, a specific practice of agroecology, in Ankara, Istanbul and Bursa, Turkey. These networks are examples of a 'growing global movement' of agroecology (Altieri and Nicholls 2020), practicing redesigning not only food production but a lifeworld. They indicate the socio-material reemergence of a twenty-first century agrarianism as a new catalyzer/progressive social force for a post-capitalist socio-ecological transformation and for re-grounding agrarian production in nature as opposed to the world-historical temporality of capital that transforms living nature into abstract nature. In Harriet Friedmann's (2016, 681–683) visionary expressions, the growth of the agroecology movement indicates and realizes the 'emerging possibilities for a different complexity' entailing new ways of agro-ecologically based living: 'post-industrial (and possibly post-capitalist) … knowledge-intensive farming' as a 'part of a wider (re)invention of meaningful, solidary, unalienated work', 'informed by earth sciences and information technologies, and pioneered by a multiplicity of farmers in different parts of the world', pointing 'not simply to a 'localized' past, but also a to (possible) future that is cosmopolitan, translocally networked and land-centered … emerging in the interstices of decaying capitalist societies'. The perspective presented in this paper specifies the agroecology movement as a socio-historical outcome of the agrarian question of nature and as a constituent of a broader anti-systemic strategy to resolve the socio-ecological question. As such, its relationship with other solutions (such as *the degrowth movement* as a prominent example) situated in the historical context of the socio-ecological question and calling for a post-capitalist transition emerge as a critical research theme in building a broader unified future agenda of an anti-systemic politics and praxis and in further integrating critical agrarian studies and political ecology.

Disclosure statement

No potential conflict of interest was reported by the author(s).

ORCID

Zehra Taşdemir Yaşın http://orcid.org/0000-0002-5823-7145

Acknowledgements

This paper draws on long-term work in building the value theory of nature perspective grounded in historical-empirical research on the geographies of extraction in the Middle East and Turkey. I am deeply grateful to Dale Tomich and Philip McMichael for their generous and inspiring intellectual and companionate guidance in this long journey and for their reading of this paper with very encouraging and helpful comments towards its revision. I am also thankful to Joan Martinez-Alier for sincerely expressing his supportive comments for the proposed theoretical perspective in another context, which became stimulating for this paper as well. I thank the JPS editors for creating with this special issue a critical interdisciplinary scholarly platform that enabled me to compose this paper in this particular way and for giving me the opportunity to be its part. Finally, I would like to express my gratitude to the three anonymous reviewers for their reassuring comments and critical suggestions in refining and making the paper more accessible to a wider audience.

References

Akram-Lodhi, A. H. 2021. "The Ties That Bind? Agroecology and the Agrarian Question in the Twenty-First Century." *Journal of Peasant Studies* 48 (4): 687–714. doi:10.1080/03066150.2021.1923010.

Akram-Lodhi, A. H., and C. Kay. 2010. "Surveying the Agrarian Question (Part 2): Current Debates and Beyond." *Journal of Peasant Studies* 37 (2): 255–284. doi:10.1080/03066151003594906.

Altieri, M. A., and C. I. Nicholls. 2020. "Agroecology and the Reconstruction of a Post-COVID-19 Agriculture." *Journal of Peasant Studies* 47 (5): 881–898. doi:10.1080/03066150.2020.1782891.

Araghi, F. 2000. "The Great Global Enclosure of Our Times: Peasants and the Agrarian Question at the End of the Twentieth Century." In *Hungry for Profit: The Agribusiness Threat to Farmers, Food and the Environment*, edited by F. Magdoff, F. H. Buttel, and J. B. Foster, 145–160. New York: Monthly Review Press.

Araghi, F. 2009a. "The Invisible Hand and the Visible Foot Peasants, Dispossession and Globalization." In *Peasants and Globalization: Political Economy, Rural Transformation and the Agrarian Question*, edited by A. H. Akram-Lodhi and Cristóbal Kay, 111–147. New York: Routledge.

Araghi, F. 2009b. "Accumulation by Displacement: Global Enclosures, Food Crisis, and the Ecological Contradictions of Capitalism." *Review (Fernand Braudel Center)* 32 (1): 113–146.

Baird, I., and K. Barney. 2017. "The Political Ecology of Cross-Sectoral Cumulative Impacts: Modern Landscapes, Large Hydropower Dams and Industrial Tree Plantations in Laos and Cambodia." *Journal of Peasant Studies* 44 (4): 769–795. doi:10.1080/03066150.2017.1289921.

Bebbington, A. 2009. "The New Extraction: Rewriting the Political Ecology of the Andes?" *NACLA Report on the Americas* 42 (5): 12–20. doi:10.1080/10714839.2009.11722221.

Benegiamo, M. 2020. "Extractivism, Exclusion and Conflicts in Senegal's Agro-Industrial Transformation." *Review of African Political Economy* 47 (166): 522–544. doi:10.1080/03056244.2020.1794661.

Bernstein, H. 2006. "Is There an Agrarian Question in the 21st Century?" *Canadian Journal of Development Studies* 27 (4): 449–460. doi:10.1080/02255189.2006.9669166.

Bond, P. 2013. *"Justice" in Critical Environmental Politics, Edited by C. Death, 133–145*. London: Routledge.

Borras, S. M., Jr. 2020. "Agrarian Social Movements: The Absurdly Difficult but not Impossible Agenda of Defeating Right-Wing Populism and Exploring a Socialist Future." *Journal of Agrarian Change* 20 (1): 3–36. doi:10.1111/joac.12311.

Borras, S. M., Jr., M. Edelman, and C. Kay. 2008. "Transnational Agrarian Movements: Origins and Politics, Campaigns and Impact." In *Transnational Agrarian Movements Confronting Globalization*, edited by S. M. Borras, M. Edelman, and C. Kay, 1–36. Malden, MA: Wiley-Blackwell.

Borras, S. M., Jr., and J. C. Franco. 2012. "A 'Land Sovereignty' Alternative? Towards a People's Counter-Enclosure." *TNI Agrarian Justice Programme Discussion Paper*, July.

Borras, S. M., Jr., and J. C. Franco. 2018. "The Challenge of Locating Land-Based Climate Change Mitigation and Adaptation Politics Within a Social Justice Perspective: Towards an Idea of Agrarian Climate Justice." *Third World Quarterly* 39 (7): 1308–1325. doi:10.1080/01436597.2018.1460592.

Borras, S. M., Jr., J. C. Franco, S. Gómez, C. Kay, and M. Spoor. 2012. "Land Grabbing in Latin America and the Caribbean." *Journal of Peasant Studies* 39 (3–4): 845–872. doi:10.1080/03066150.2012.679931.

Borras, S. M., Jr., C. Franco, and Z. Nam. 2020. "Climate Change and Land: Insights from Myanmar." *World Development* 129. doi:10.1016/j.worlddev.2019.104864.

Borras, S. M., Jr., T. Moreda, A. Alonso-Fradejas, and Z. W. Brent. 2018. "Converging Social Justice Issues and Movements: Implications for Political Actions and Research." *Third World Quarterly* 39 (7): 1227–1246. doi:10.1080/01436597.2018.1491301.

Borras, S. M., Jr., I. Scoones, A. Baviskar, M. Edelman, N. L. Peluso, and W. Wolford. 2021. "Climate Change and Agrarian Struggles: An Invitation to Contribute to a JPS Forum." *Journal of Peasant Studies*. doi:10.1080/03066150.2021.1956473.

Brenner, N., and C. Schmid. 2012. *"Planetary Urbanization." in Urban Constellations, Edited by M. Gandy, 10-13*. Berlin: Jovis.

Claeys, P., and D. D. Pugley. 2016. "Peasant and Indigenous Transnational Social Movements Engaging with Climate Justice." *Canadian Journal of Development Studies* 38 (3): 325–340. doi:10.1080/02255189.2016.1235018.

Clapp, J., and S. R. Isakson. 2018. *Speculative Harvests: Financialization, Food, and Agriculture*. Rugby, UK: Practical Action Publications.

Clark, B., and R. York. 2005. "Carbon Metabolism: Global Capitalism, Climate Change, and the Biospheric Rift." *Theory and Society* 34: 391–428. doi:10.1007/s11186-005-1993-4.

Crippa, M., E. Solazzo, D. Guizzardi, F. Monforti-Ferrario, F. N. Tubiello, and A. Leip. 2021. "Food Systems are Responsible for a Third of Global Anthropogenic GHG Emissions." *Nat Food* 2: 198–209. doi:10.1038/s43016-021-00225-9.

Duncan, C. A. M. 1996. *The Centrality of Agriculture: Between Humankind and the Rest of Nature*. Montreal: McGill-Queen's University Press.

Edelman, M., and W. Wolford. 2017. "Introduction: Critical Agrarian Studies in Theory and Practice." *Antipode* 49: 4. doi:10.1111/anti.12326.

Elson, D. 1979. *Value: The Representation of Labour in Capitalism*. London: CSE Books.

Friedmann, H. 2016. "Commentary: Food Regime Analysis and Agrarian Questions: Widening the Conversation." *Journal of Peasant Studies* 43 (3): 671–692. doi:10.1080/03066150.2016.1146254.

Friedmann, H., and P. McMichael. 1989. "Agriculture and the State System: The Rise and Decline of National Agricultures, 1870 to the Present." *Sociologia Ruralis* 29 (2): 93–117.

Guha, R., and J. Martinez-Alier. 1997. *Varieties of Environmentalism: Essays North and South*. London: Earthscan Publications.

Holt-Giménez, E., A. Shattuck, and I. van Lammeren. 2021. "Thresholds of Resistance: Agroecology, Resilience and the Agrarian Question." *Journal of Peasant Studies* 48 (4): 715–733. doi:10.1080/03066150.2020.1847090.

Hornborg, A. 2007. "Footprints in the Cotton Fields: The Industrial Revolution as Time-Space Appropriation and Environmental Load Displacement." In *Rethinking Environmental History, World-System History and Global Environmental Change*, edited by A. Hornborg, J. R. McNeill, and J. Martinez-Alier, 259–272. Lanham, MD: Atamira Press.

Leff, E. 2014. "Environmental Rationality: The Social Re-Appropriation of Nature." *Alternautas* 1 (1): 88–99.

Leff, E. 2016. "From the Persistence of the Peasantry in Capitalism to the Environmentalism of Indigenous Peoples and the Sustainability of Life." In *Peasant Poverty and Persistence in the Twenty-First Century: Theories, Debates, Realities and Policies*, edited by J. Boltvinik and S. Mann, 247–268. London: Zed Books.

Leff, E. 2021. *Political Ecology: Deconstructing Capital and Territorializing Life*. Cham, Switzerland: Palgrave MacMillan.

Levien, M., M. Watts, and Y. Hairong. 2018. "Agrarian Marxism." *Journal of Peasant Studies* 45 (5-6): 853–883. doi:10.1080/03066150.2018.1534101.

Mahowald, N. M., D. S. Ward, S. C. Doney, P. G. Hess, and J. T. Randerson. 2017. "Are the Impacts of Land use on Warming Underestimated in Climate Policy?" *Environmental Research Letters* 12: 9. doi:10.1088/1748-9326/aa836d.

Martinez-Alier, J. 2002. *The Environmentalism of the Poor: A Study in Ecological Conflicts and Valuation*. Cheltenham, UK: Edward Elgar Publishing.

Martinez-Alier, J. 2011. "The EROI of Agriculture and Its Use by the Via Campesina." *Journal of Peasant Studies* 38 (1): 145–160. doi:10.1080/03066150.2010.538582.

Martinez-Alier, J. 2014. "The Environmentalism of the Poor." *Geoforum; Journal of Physical, Human, and Regional Geosciences* 45: 239–241.

Martinez-Alier, J., L. Temper, D. Del Bene, and A. Scheidel. 2016. "Is There a Global Environmental Justice Movement?" *Journal of Peasant Studies* 43 (3): 731–755. doi:10.1080/03066150.2016.1141198.

Marx, K. 1976. *Capital I: A Critique of Political Economy*. London: Penguin Books.

McKay, B. M. 2017. "Agrarian Extractivism in Bolivia." *World Development* 97: 199–211. doi:10.1016/j.worlddev.2017.04.007.

McKay, Ben M. 2020. *The Political Economy of Agrarian Extractivism: Lessons from Bolivia*. Rugby, UK: Practical Action Publishing.

McMichael, P. 2007. "Reframing Development: Global Peasant Movements and the New Agrarian Question." *Canadian Journal of Development Studies* 10 (4): 57–71. doi:10.1080/02255189.2006.9669169.

McMichael, P. 2008. "Peasants Make Their Own History, But Not Just as They Please ... " In *Transnational Agrarian Movements Confronting Globalization*, edited by S. M. Borras, M. Edelman, and C. Kay, 37–60. Malden, MA: Wiley-Blackwell.

McMichael, P. 2009. "Food Sovereignty, Social Reproduction and the Agrarian Question." In *Peasants and Globalization: Political Economy, Rural Transformation and the Agrarian Question*, edited by A. H. Akram-Lodhi, and C. Kay, 288–312. New York: Routledge.

McMichael, P. 2012. "In the Short Run Are We All Dead? A Political Ecology of the Development Climate." In *The Longue Durée and World-Systems Analysis*, edited by R. E. Lee, 137–160. Albany: SUNY Press.

McMichael, P. 2013. *Food Regimes and Agrarian Questions*. 2021 ed. Rugby, UK: Practical Action Publishing.

McMichael, P. 2014. "Historicizing Food Sovereignty." *Journal of Peasant Studies* 41 (6): 933–957. doi:10.1080/03066150.2013.876999.

McMichael, P. 2016. "Commentary: Food Regime for Thought." *Journal of Peasant Studies* 43 (3): 648–670. doi:10.1080/03066150.2016.1143816.

Messerli, P., M. Giger, M. B. Dwyer, T. Breu, and S. Eckert. 2014. "The Geography of Large-Scale Land Acquisitions: Analysing Socio-Ecological Patterns of Target Contexts in the Global South." *Applied Geography* 53: 449–459. doi:10.1016/j.apgeog.2014.07.005.

Moore, J. W. 2000. "Sugar and the Expansion of the Early Modern World-Economy: Commodity Frontiers, Ecological Transformation, and Industrialization." *Review* 23 (3): 409–433.

Moore, J. W. 2008. "Ecological Crises and the Agrarian Question in World-Historical Perspective." *Monthly Review* 60 (6): 54–63.

Moore, J. W. 2010. "The End of the Road? Agricultural Revolutions in the Capitalist World-Ecology, 1450– 2010." *Journal of Agrarian Change* 10 (3): 389–413. doi:10.1111/j.1471-0366.2010.00276.x.

Moore, J. W. 2011. "Transcending the Metabolic Rift: A Theory of Crises in the Capitalist World-Ecology." *Journal of Peasant Studies* 38 (1): 1–46. doi:10.1080/03066150.2010.538579.

Moore, J. W. 2015. *Capitalism in the Web of Life: Ecology and the Accumulation of Capital*. London, New York: Verso.

Moore, J. W. 2017. "The Capitalocene, Part I: On the Nature and Origins of Our Ecological Crisis." *Journal of Peasant Studies* 44 (3): 594–630. doi:10.1080/03066150.2016.1235036.

O'Connor, J. 1988. "Capitalism, Nature, Socialism: A Theoretical Introduction." *Capitalism, Nature, Socialism* 1 (1): 11–38.

Patel, R. 2007. *Stuffed and Starved: The Hidden Battle for the World Food System*. Brooklyn, NY: Melville House Publishers.

Peluso, N. L. 2012. "What's Nature Got to Do With it? A Situated Historical Perspective on Socio-Natural Commodities." *Development and Change* 43 (1): 79–104. doi:10.1111/j.1467-7660.2012.01755.x.
Peluso, N. L. 2017. "Plantations and Mines: Resource Frontiers and the Politics of the Smallholder Slot." *Journal of Peasant Studies* 44 (4): 834–869. doi:10.1080/03066150.2017.1339692.
Peluso, N. L., S. Afiff, and N. F. Rachman. 2008. "Claiming the Grounds for Reform: Agrarian and Environmental Movements in Indonesia." In *Transnational Agrarian Movements Confronting Globalization*, edited by S. M. Borras, M. Edelman, and C. Kay, 209–238. Malden, MA: Wiley-Blackwell.
Peluso, N. L., and C. Lund. 2011. "New Frontiers of Land Control: Introduction." *Journal of Peasant Studies* 38 (4): 667–681. doi:10.1080/03066150.2011.607692.
Petras, J., and H. Veltmeyer. 2014. *Extractive Imperialism in the Americas: Capitalism's New Frontier.* Leiden: Brill.
Postone, M. 1993. *Time, Labor, and Social Domination: A Reinterpretation of Marx's Critical Theory.* New York: Cambridge University Press.
Rosset, P., and M. Altieri. 2017. *Politics and Science of Agroecology*. London: Practical Action Books.
Sayer, D. 1987. *The Violence of Abstraction*. Oxford: Basil Blackwell.
Schlosberg, D., and L. B. Collins. 2014. "From Environmental to Climate Justice: Climate Change and the Discourse of Environmental Justice." *WIRES Climate Change*. doi:10.1002/wcc.275.
Schneider, M., and P. McMichael. 2010. "Deepening, and Repairing, the Metabolic Rift." *Journal of Peasant Studies* 37 (3): 461–484. doi:10.1080/03066150.2010.494371.
Schwab, K. 2017. *The Fourth Industrial Revolution*. London: Portfolio Penguin.
Scoones, I., M. Edelman, S. M. Borras Jr, Ruth Hall, W. Wolford, and B. White. 2018. "Emancipatory Rural Politics: Confronting Authoritarian Populism." *The Journal of Peasant Studies* 45 (1): 1–20. doi:10.1080/03066150.2017.1339693.
Selwyn, B. 2021. "A Green New Deal for Agriculture: For, Within, or Against Capitalism?" *Journal of Peasant Studies* 48 (4): 778–806. doi:10.1080/03066150.2020.1854740.
Sneddon, C. 2007. "Nature's Materiality and the Circuitous Paths of Accumulation: Dispossession of Freshwater Fisheries in Cambodia." *Antipode* 39 (1): 167–193. doi:10.1111/j.1467-8330.2007.00511.x.
Taşdemir Yaşın, Z. 2017. "The Adventure of Capital with Nature: From the Metabolic Rift to the Value Theory of Nature." *Journal of Peasant Studies* 44 (2): 377–401. doi:10.1080/03066150.2016.1246435.
Taşdemir Yaşın, Z. 2019. "Contextualizing the Rise of Environmental Movements in Turkey: Two Instances of Anti-Gold Mining Resistance." In *Transforming Socio-Natures in Turkey: Landscapes, State and Environmental Movements*, edited by B. Inal and E. Turhan, 137–165. London: Routledge.
Tomich, D. 2004. *Through the Prism of Slavery: Labor, Capital, and World Economy*. Lanham: Rowman and Littlefield Publishers.
Tomich, D. 2015. "The Limits of Theory: Capital, Temporality, and History." *Review (Fernand Braudel Center)* 38 (4): 329–368.
Tramel, S. 2018. "Convergence as Political Strategy: Social Justice Movements, Natural Resources and Climate Change." *Third World Quarterly* 39 (7): 1290–1307. doi:10.1080/01436597.2018.1460196.
Van der Ploeg, J. D. 2008. *The New Peasantries: Struggles for Autonomy and Sustainability in an Era of Empire and Globalization*. London: Earthscan.
Van der Ploeg, J. D. 2010. "The Peasantries of the Twenty-First Century: The Commoditisation Debate Revisited." *Journal of Peasant Studies* 37 (1): 1–30. doi:10.1080/03066150903498721.
Van der Ploeg, J. D. 2021. "The Political Economy of Agroecology." *Journal of Peasant Studies* 48 (2): 274–297. doi:10.1080/03066150.2020.1725489.
Watts, M., P. Robbins, and R. Peet, eds. 2010. *Global Political Ecology*. London: Routledge.
Weis, T. 2007. *The Global Food Economy: The Battle for the Future of Farming*. London: Zed Press.
Weis, T. 2010. "The Accelerating Biophysical Contradictions of Industrial Capitalist Agriculture." *Journal of Agrarian Change* 10 (3): 315–341. doi:10.1111/j.1471-0366.2010.00273.x.
Weis, T. 2013. *The Ecological Hoofprint: The Global Burden of Industrial Livestock*. London: Zed Books.
Wolford, W. 2021. "The Plantationocene: A Lusotropical Contribution to the Theory." *Annals of the American Association of Geographers* 111 (6): 1622–1639. doi:10.1080/24694452.2020.1850231.

Violent silence: framing out social causes of climate-related crises

Jesse Ribot

ABSTRACT

Climate change is a problem of unimaginable scope and magnitude – in cause, implication and responsibility. Predominant and ostensibly scientific frames for evaluating climate-related loss and damage focus on the climate events as the primary cause. This approach clouds out and silences the many non-climatic, social and political-economic, causes of crises. Framing the social back in highlights a fuller range of causes and potential solutions. It is also contentious as it locates cause in decisions, policies and institutions – indicating responsibility and blame. Choosing a social and political-economic analytic has implications for action and ethics as it broadens response abilities and responsibility.

No one would say that a lack of money in the world is the reason there are poor people; yet, many blithely suggest that a lack of food is the reason a billion go hungry. (Beyond the scarcity scare, Lappé 2013, 227)

Whither causality?

It may seem that starvation is caused by a lack of food, yet Sen (1980) showed that modern-era food crises occur where there is more than enough food for everyone.[1]

[1]It is important to note that Sen's analysis is based on a neoclassical notion that takes initial conditions (i.e. assets and endowments) as given – without tracing histories of where they came from (Ribot 1995, 2014). The other analysts I cite go further to locate the causes of crisis in the very causes of the distribution of those assets. As Fine (1997, 630) states, 'the micro-foundations of the entitlement approach are to be rejected because of their inability to address satisfactorily

Sen (1980, 620) expressed '... disquiet about this focus on food, and the importance attached to it to the exclusion of other variables,' opening our eyes to a broader array of proximate and structural causes of hunger. Today, there are reasons for disquiet about the focus on climate change as the cause of crises, and the importance it is now afforded to the exclusion of other variables (Hulme 2011; Lahsen and Ribot 2021; Cottier et al. 2022). Sen, among many others in famine and agrarian studies (de Castro 1952; Watts 1983; Blaikie 1985; Fine 1997, 633; de Waal 1997; Devereux 2000, 27; Wisner, Blaikie, and Cannon 2004), understood how simplistic explanations of famine misguided and hindered famine prevention and response. Simplistic assumptions that crises following climate events are caused by those events have the same effect. They cloud out the many causes of, and, thus, potential protections against, crises. How do we conduct a full analysis of the causes of climate-related disasters? We do so by understanding the fragilities that crises reveal and the causes of these fragilities.

The crises that follow extreme climate events expose pre-existing vulnerabilities of the agrarian world and farm households. The precarious are plunged into hunger, famine, economic loss or dangerous dislocation while the secure are barely affected (Watts and Bohle 1993; Blaikie et al. 1994; Ribot 1995). Indeed, Marshall Sahlins (1972, 111, 114–130) labels disasters 'revelatory crises,'[2] as they make underlying vulnerabilities visible – yanking the vail off of present but hidden structures of unequal power, wealth, protection, exposure. Soloway (1994) applied this notion to drought, and Mostafanezhad (2020) and Achiume, Gammeltoft-Hansen, and Spijkerboer (2020) to Covid19. Further back, even our term 'apocalypse' comes from the Greek word for an uncovering, disclosure, or revelation.[3] In cases of climate change, Covid19, or any other hazard, damages are socially stratified by pre-existing vulnerabilities (Turner 2016). The distribution of damages has social and political-economic causes (Sen 1981; Watts 1983; Nixon 2011). In short, there are social and political actions and structures responsible for the differentiated vulnerabilities that turn hazard into crisis for some (Blaikie et al. 1994; Wisner, Blaikie, and Cannon 2004). This essay explores the veiling and unveiling of the causes of vulnerability and crisis while attending to the attribution of responsibility. It also includes responsibility and its absence in the causal analytic of vulnerability.

While *triggered*[4] by climate stress, the evidence is overwhelming that 'climate-related' damages are stratified by historical, social and political-economic arrangements

---

the social relations and structures through which famines are fundamentally caused.' Limits to entitlements approaches are not taken up in this article.

[2]Sahlins (1972, 114) frames crises as revelatory, citing Firth (1959) who wrote of what a famine can 'reveal.'

[3]https://www.merriam-webster.com/dictionary/apocalypse#learn-more.

[4]Viewing hazards as 'triggers' rather than cause enables analysts to include the causes of the fragilities that enable hazards to result in damages. Watts (1991, 15) states, 'Climate, "over-population" and war, while potentially significant as proximate or trigger factors, have been substantially discredited as primary factors.' Fine (1997, 637) writes of 'proximate events that trigger and distribute its [entitlement failure's] incidence.' Pelling (2003, 47) explains, 'The challenge today is to integrate agency and structure in examinations of the production of vulnerability, in specific places, whilst also acknowledging the importance of physical systems in generating hazard that can trigger disaster.' Wisner, Blaikie, and Cannon (2004, 61) point to how '... most government agencies charged with such responsibilities as "environment," "health and welfare" and "public safety" generally still deal with disasters as though they are *equivalent to* the hazards that trigger them.' In an 'aetiology of hunger,' Nally (2011, 4) considers 'droughts, floods, and crop failures are "trigger factors," though not necessarily an "underlying cause," of famine' and quotes Arnold (1988) who says famines are 'a symptom rather than a cause of social weakness.' Temudo and Cabral (2021) also describe climate change as a trigger of rural conflict in Guinea Bissau. While all crisis can be explained by the vulnerabilities, a definitional matter, the stratification of damages is certainly a function of stratified vulnerabilities.

that render some people secure and others vulnerable. The social basis of crises, the vulnerabilities, are often evident. In Bangladesh, cyclone fatalities declined by 150-fold (from over 500,000 to 3400 deaths) between the cyclones Bohla and Sidr that hit the same coastline with similar intensities and surges in 1970 and 2007. The decline was due to planning and reforms on the ground (Bern 1993; CEDMHA 2007; Batha 2008; Government of Bangladesh 2008; MFDMB 2008). In 2020, Cyclone Amphan, the strongest cyclone on record in the Bay of Bengal, registered fewer than 30 deaths – due to effective government action (Kelman and Ahmed 2020). The ability to respond was identified and acted upon. Vulnerability and damage were reduced.

When weather – ordinary or extreme – meets precarity, damages can follow (Blaikie et al. 1994; Wisner, Blaikie, and Cannon 2004) – precarity is social and political-economic. The 1943 West Bengal famine, blamed on crop disease, was caused by well-functioning markets that allocated food away from the hungry; there was enough to feed all and no absolute shortage (Sen 1981).[5] The 1959–1960 famine in China was produced by a depravedly extractive administration, not drought (Jisheng 2012). The 2011 Somali famine was a product of 'interplay of livelihoods, clan and politics,' not drought (Majid and McDowell 2012, 37). In 2005, 'the disaster in New Orleans after Katrina was unnatural and man-made,' (Bullard and Wright 2009, 2); the 1300, disproportionately African-American, fatalities resulted from a long history of government negligence, not the hurricane (White House 2006; Hayes 2009; also see Somers 2008). Damages were a function of conditions on the ground – conditions that don't fall from the sky.

Yet, weather and associated damage often seem to drop randomly from nowhere – as we always face uncertainties. Indeed, as some argue, uncertainty 'defines our times' (Scoones 2019, 5; also see Taddei 2008; Beck 1992). Uncertainty, implies unpredictability such that knowledge, planning and control over the future are indefinite. But this indefinite future does not absolve society and its planners of responsibility, it also does not imply that the past did not have a definite causal logic. Future uncertainty does not mean we cannot trace back the origins of already manifest outcomes. Uncertainty is forward looking. It is about futures. Yet, the past is a certainty, as we can observe what has happened and we can often uncover why. The past can be known as it is completed, inscribed upon the present, and often legible. Vulnerabilities, as they are about a predisposition in the present, contribute to the sense of uncertainty about futures – even when those vulnerabilities and their causes are well understood. The vulnerable are prone in known ways.

Critical realist Bob Jessop (1982, 2014) has characterized this relation between future uncertainty and clear readings of past cause as 'contingent necessity.' Uncertainties make planning and control over the future difficult. They do not, however, override or erase the causes of what has come to pass. We may not have been able to predict the convergence of historical trends, but their convergence, which could have happened or not, explains the outcomes we observe. The implication for responsibility and blame derives from the fact that it is often possible, even in an uncertain world, to identify who or what caused a given outcome – even when that outcome was caused by known or ignored uncertainties. Indeed, we can identify who ignored or poopooed

[5]As Fine (1997, 635) observed '... there is many a slip 'twixt cup and lip when it comes to the access to, as opposed to the supply of, food.... '

conventions like the precautionary principle. We can understand causality by the ignoring, invention or misreadings of (un)certainties and of wisdom.[6]

In an essay about ignoring precaution, Russell (1924) argued that the hubris of science causes damage – as Daedalus learned when he gave Icarus wings. Icarus – burnt by the sun – provides a good precautionary tale for climate change. Indeed, given uncertainties, 'greater humility and vigilance are required' (Scoones 2019, 5); and, when they are not applied, their absence is cause, and, there is fault. This article, however, is not concerned with causes of the future. It focuses on past causes. I explore the causes of crises – which can include, concretely, how uncertainty is misread, generated and abused, and what was known and unknown. The study of causality is not about planning – despite having planning implications. It is about explanation of that which is immutably inscribed that can then, in a forward-looking moment, inform action or planning. The past is a complex multifaceted composite, and at times cause cannot be discerned, but even when unknown it is not uncertain (Nally 2011, viii–ix). It is, simply, what happened – no matter how much we quibble over it.

Since de Castro's (1952) exploration of the political basis of hunger in Brazil, and Sen's (1981) India famine studies, we have had more and more nuanced approaches to vulnerability studies. Why, then, are these vulnerability approaches, that point to social and political-economic causes of crisis, not more widely used? Why is there continuous slippage back into hazard-oriented explanations of damage? 'Why,' as Oliver-Smith (2013, 1) asks, 'are the problems I saw 40 years ago after the Peruvian earthquake of 1970 still with us when we have learned so much about their drivers?' In this article I suggest there are many reasons – not the least of which can be found in the threatening nature of social causality. Cause is social and thus political. The ability to trace a line from damage back to individuals, institutions, or their actions, inactions, ideologies or beliefs, indicating possible responsibility, makes it contentious.

Uncertainty too can evoke relations of blame. Giddens (1999) argues that the idea of 'risk' is '… bound up with the aspiration to control and particularly with the idea of controlling the future.' He poses that hazard and danger were traditionally taken as givens – acts of nature and God – but risk is new. It is a preoccupation with the future. Here nature and God are external to society. Giddens (1999, 8) calls human-induced uncertainties 'manufactured risk' – responded to by insurance or the welfare state and argues that 'the transition from external to manufactured risk is bringing about a crisis of responsibility… .' It shifts the cause of risk itself to human action – and so does the ability to control, or even insure against, external uncertainties and hazard. The ability to act aligns cause with action and inaction, bringing about responsibility.

In all societies, people seek cause for pain and suffering, and in the analysis of cause lies responsibility (Calebresi 1975; Douglas and Wildavsky 1982; Wilkinson 2010). Any analytic frame that shows cause in human action can be used to evoke blame and liability, and is, therefore, avoided by some and sought by others. The choice or avoidance of a frame depends on purpose, position, alliances. Causality is contentious, about interest, so choosing frames that place it in the external and blameless is the easy way out – and helps maintain the appearance of neutrality in 'scientistic' endeavors – like modeling the relation between a storm and a subsequent disaster. Because there is purpose in

[6]Of course, there is even causality in lies and misinformation (Badrinathan 2021).

analyzing cause, and in how we place responsibility, there is also normative choice in and of the frames we bring to this analytic endeavor (see Calebresi 1975; Forsyth 2008, 756 on Blaikie 1985; Cronon 1992; Lakoff 2010). Sticking to a biophysical hazard-driven causal model is very different from choosing a social-causal model to explain disaster (Cottier et al. 2022).

In this article, I explore why causality is such a contentious and morally complex arena, how it embeds social norms and values, and the role of social values in the choice of causal frames. I evoke human exceptionalism and the nature of human nature as the root of moral values (*a la* Arendt), arguing that it is with values – moral or other – that analysts choose analytic frames to interrogate causality. This choice of frames, and the causes they steer us to, establishes or occludes potential responses and responsibilities – with responsibility then recursively turning back to shape the choice of frame. I ask what causes (and responsibilities) *hazard* and *vulnerability* analytic frames for the analysis of crises can reveal and what they might hide – bringing us to the violent silences (see Kashwan and Ribot 2021) that are generated by predominant frames of hazard-based 'climate-impact' analytics. These frames tend to concentrate on the hazard, thus silencing the agrarian histories that push so many farmers and pastoralists to the edge of subsistence; histories that produce the precarities that make it easy for a mere storm or drought to push them over the edge. I end with a call for a 'sociodicy' of climate change – a fuller social analysis of the causes of the vulnerabilities that enable hazards to trigger damage.

Contentions of causality: bringing 'should' back in

> In risk society there is a new moral climate of politics, one marked by a push-and-pull between accusations of scaremongering on the one hand and of cover-ups on the other. (Giddens 1999, 5)

> Responsibility is a contested site, with partisans of particular normative outlooks arguing for attributions of responsibility, while their opponents deny or reassign the attributions. (Jamieson 2015, 36)

Cause, especially of crisis, points to responsibility.[7] Thus, its attribution is constantly disputed. The analysis of cause always has a purpose – a human objective – or we would not seek to know it. As legal scholar Guido Calebresi (1975, 106) explains, '… in law the term "cause" is used in different guises but always to identify those pressure points that are most amenable to the social goals we wish to accomplish…. Where goals differ, so does the practical definition of causation.'[8] Using what is now a timely example, Calebresi (1975, 105) explained that '… so far as legal language is concerned, the "cause" of a disease would depend on how, at any given time, it could be most easily controlled.' Thus, he explains, the cause of tuberculosis in the nineteenth century would include inadequate exposure to sunlight and poor living conditions. Today the causes would include the failure to be inoculated. Or, it is due to inequalities – 'a

[7]Of course, responsibility is contentious too – as it is the contentious content of causality. 'Responsibility is a contested site, with partisans of particular normative outlooks arguing for attributions of responsibility, while their opponents deny or reassign the attributions' (Jamieson 2015, 36). 'In the accountability field, our choice of words also informs broader narratives about the reasons for accountability failures. These "causal stories" are relevant for guiding action because they point the finger at who is responsible for specific problems' (Fox 2022, 9 citing Stone 1989).

[8]The very notion of 'problem' is inherently an ethical issue (Gardiner 2011, 20–21).

health-care system that millions can't access, a public-health system that's been rotting for decades, and extreme inequities that leave large swaths of society susceptible to a new virus' (Young 2021). Causality cannot be separated from goals – in this case, improved health and wellbeing. It also cannot be separated from possibilities – the ability of people to manage the disease. What could have been or can be done becomes causes when neglected.

The word 'responsibility' is key – it links our response abilities (what we can do) to moral purpose – and shapes what we see as cause. Death, when protections are possible, is no longer due to a pathogen that we can protect against; it is now squarely due to negligence (failure to get vaccinated) or deprivation (inability to isolate or quarantine).[9] The causes of vulnerability are social and the analysis of causality can identify potential solutions by identifying treatable causes and by indicating responsible parties or structures. The potential solutions or preventative actions that *could* be, or *could* have been, create responsibility. The *ability* to protect ourselves and others (i.e. to *respond*) carries obligation where protection is a moral goal. This ability, combined with contract obligation (*contract* being a root of the word responsibility[10]), indicates blame and liability when obligation is unmet.

Could, in a moral world, is a pre-condition for *should*. It is only when one could get vaccinated that failure to vaccinate became a cause of disease (Calebresi 1975, 105). *Should*, a social or moral judgement that, when agreed upon or viewed as legitimate[11] (in law or less formally via custom or convention), generates a legal or social contract; it establishes some of the obligations we call responsibility. Moral principles of action – the 'shoulds' of our contracts – have many roots (from platinum or golden rules to shared vulnerability or shared humanity – a la Arendt 1963; Butler 2009; Hobbes in Ferrarese 2016, 5; Mill's harm principle as noted by Jamieson 2015, 26; Nyerere as represented in Shivji 2020; to Samafal, *a la* Carruth 2021). Along with such moral tenets, the analysis of the causes of vulnerability, and of the damages that vulnerabilities enable, is always a first step in establishing responsibility – in both senses: identifying the ability to respond, the 'could,' and the contract of 'should.'

Causality linked to human suffering or wellbeing is never normatively or morally neutral. First, suffering and wellbeing are *judged* to be relevant – this is a normative act. Second, how we frame the cause of a benefit or a damage also has implicit moral judgment, as different analytic frames include different causal variables. If we start with an outcome, a damage, and ask what caused it, a natural science frame will identify biophysical elements that impinge on that outcome – while normative in motivation, the analysis itself provides no indication of what could or should have been. Yet a social science approach, whose subject is 'the social' and cannot be separated from norms, will include a broader set of causes – beyond the biophysical. Causality involving social

[9]True for earthquakes too: ' … the case of people dying from earthquakes today would not warrant an analysis in terms of violence, but the day after tomorrow, when earthquakes may become avoidable, such deaths may be seen as the result of violence' (Galtung 1969, 168–169).

[10]Rooted in latin for a contract, pledge,or vow. https://etymologeek.com/eng/responsible. In this sense it is a fundamental element of social relations.

[11]Weber (1968) viewed 'legitimate' (although he used multiple definitions) as that which is not resisted. So, one may not agree with a law, but one submits to it – often for reasons that have to do with subordination and an inability to resist a monopoly on violence – as that held by the state.

actions is always contingent on normative *latent* elements of human interaction unaccounted for in, and unaccountable by, the natural sciences.

The latent is invisible and cannot be measured with calipers. Normative content, for instance, can be clearly discerned through events that did not occur. It is the moral context of expected action, i.e. the role of individual, social or political *expectations*, that make a non-action into a cause. Moral expectation, 'should,' distinguishes social from natural science causality – as a moral[12] judgment, the content of 'should,' can make the expectation that renders a non-event causal. What is not done can take on the label 'negligence' or 'turpitude' and can even be viewed as an act of 'malice' – due to moral expectation. For example, peasants judge failed reciprocities – that lead to hunger in times of drought – to be unacceptable, indeed, to be sufficient cause for rebellion (Scott 1976). Failure of the Army Corps of Engineers to maintain the levies of New Orleans' Lower 9th Ward can be judged as the cause of Katrina-related deaths (Hayes 2009) – behind which one might even place the malice of racism (Harden, Walker, and Akuno 2007; Bullard and Wright 2009, 38; Thomas and Haynes 2020). Thus, both what is and is not done can shape cause in the social world – yet, only what does happen is made visible by the calipers of natural sciences. In non-reductionist social analytics, these non-events can be causal – this is never the case in a strictly biophysical causal frame (Ennis 2012). In this sense, the choice of analytic frame applied to the causes of crisis is itself a moral judgment – as it will reveal different causes. For more on the choice of frames see Ribot (1995, 2014; Cottier et al. 2022).

The social sciences shine light on elements of cause in other ways that differ from the natural sciences. Lund (2014, 225) points out,

> The social sciences are the empirical science of historical reality. A discipline is essentially historical when its statements cannot be completely severed from the context from which data were drawn. It is the distinguishing feature of social sciences, which contrary to the natural science, cannot 'control for context.'

Social sciences are not inferior – as context, including histories, norms and values, is an empirical fact.[13] 'But,' as Lund (2014, 225) aptly states, social science 'is a *different* science' (italics in original). In short, social sciences make sense of causal chains within social context – which always includes the norms and values that establish expectations. It is worth recalling, that values are 'real' even if they are not measurable in biophysical terms (no calipers). Their reality is malleable and changes with time and context, but that does not make them un-real. These are the social and moral contracts – sometimes encrusted in law – that constitute society. These are the basis of the *shoulds* that link cause to responsibility.[14] These are also the basis of actions with real biophysical and social consequence.

---

[12] I include the 'legal' within moral judgment – as the legal is a codified form of mores. Actions are guided by law, custom and convention, or the 'contracts' of expectation established by one or more people – all of these create expectations.

[13] Indeed, 'There is no risk that can be described without reference to value' (Giddens 1999, 5). Further, 'Risk only exists when there are decisions to be taken … The idea of responsibility also presumes decisions. … someone takes a decision having discernable consequences' (Giddens 1999, 8). The normative is always implicit.

[14] I agree broadly with Jamieson (2015, 37) that '…moral responsibility encompasses "contributing to an outcome" or being "complicit in sustaining a state of affairs," even if these are not causal notions.' But, would still, on the grounds of social contract, consider these notions 'causal.'

Norms and values are causal social facts

> The contention that famine results from a kind of natural law has no basis in scientific knowledge. (de Castro 1952, 12)

> Science ... cannot create ends and, even less, instill them in human beings; science, at most, can supply the means by which to attain certain ends. (Einstein 1949)

> ... you cannot be rational without emotions. Without emotion, you would not know what to want, since like and not-like would be meaningless to you. When there is neither like or not-like, nor any judgment of the emotional reactions of others, you cannot make rational decisions. ("Why It Matters How We Frame the Environment," Lakoff 2010, 72)

As disasters, such as famines, can be avoided (or generated) by social and political action, they are social and political events – they are allowed to happen; while some are, indeed, made to happen (see Devereux 2000, 27 in Edkins 2002, 15). Alex de Waal views famines that are allowed to occur as 'famine crimes' – where, if, for example, there is a social contract between rulers and the people against allowing famine to happen, they become political scandals (de Waal 1997 in Edkins 2002, 15; also see Scott 1976). It is because of responsibility and related liabilities (individual, social or political) that vulnerability – and how we frame analysis of its causes – is contentious, and, itself, political. Vulnerability in the social sciences is defined as a predisposition to damage (Blaikie et al. 1994; IPCC 2019). As society has agency and history, attempting to identify vulnerability's causes always points to the social arrangements that make for this predisposition. In tort law, cause, traced to intention or negligence, must be established to demonstrate responsibility, liability or blame (Calebresi 1975; Hart and Honoré 1959).[15] Causality, thus, can imply guilt, whether in law, the humanities (as in history) or in science – as the 'could' that is demonstrated will be read socially and can thus locate obligation and negligence – or merely identify a social/political decision on the protections to establish or forego. In analyzing climate-related disasters, responsibility – which combines the ability to respond and the moral or legal obligation to do so – follows from an understanding of the causes of vulnerability.

It may seem logical and self-evident to natural scientists that a storm or drought ('natural' or anthropogenic) causes damages that follow, but it is equally obvious in the social sciences that the vulnerabilities, without which damage would not have happened, are the causes, just as well. Causal models differ – and so do the responsibilities they indicate. Causal models, and the choice of causal frames, related to damages are never neutral – as those choosing a causal frame have some interest – 'a goal' in Calebresi's (1975) words – in the kinds of outcomes their analysis will indicate. Clearly, the natural sciences' tendency to cordon off 'context' is shaped by such purposes and constitutes a manifest exclusion of relevant empirical (as norms and values are empirical social *facts*; Durkheim 1982 [1895]; Douglas 1992; Lund 2014) driving forces of climate-related risks. These facts cannot simply be bracketed out without distorting the picture. The choice to do so is normative, and thus political, rather than 'scientific.' All scientists – natural and social – live in a

[15]Believing that cause is rooted in instinct, rather than will, Nietzsche (2003, 64) would disagree with this thesis, stating:

> Everywhere accountability is sought, it is usually the instinct of punishing and judging which seeks it. One has deprived becoming of its innocence if being in this or that state is traced back to will, to intentions, to accountable acts: the doctrine of will has been invented essentially for the purpose of punishment, that is of finding guilty.

For Nietzsche (2003), 'free will' is the invention of theologians as a means attributing discipline and punishment. Cause of events, in turn, is sought to comfort people by reducing the uncertainties of the unknown.

social world, and their choices of research problem, framing, methods, and data all carry (and influence) normative content.

The weight of choice: polluter or exploiter pays?

> ... international environmental law has developed distinctive approaches to standard-setting, institutions, and compliance.... It promotes compliance through transparency and forward-looking, non-adversarial procedures, aimed at improving effectiveness, rather than through traditional international dispute settlement, which takes a backward-looking approach, focusing on the issue of state responsibility. (Bodansky 2020, 3)

> The framings of risk lead scientists and decision makers to ask different questions, which inevitably lead to the implementation of different solutions. And those solutions have significant material impacts on people's lives. (Colette 2016, 44)

> I am talking of millions of men who have been skillfully injected with fear, inferiority complexes, trepidation, servility, despair, abasement. (Aimé Césaire, *Discours sur le Colonialisme* in Fanon 1986, 9)

If the view from 'environmental law' cannot attend to causes, then it is inadequate to the crises at hand – a crisis of responsibility (Giddens 1999, 8). Of course, there is some looking back when attributing the cause of climate change and the ensuing hazards. Clearly, a choice to look back, or not, has multiple implications for policy and practice. So does the choice of frames for attributing causality if and when one does look back. A specific idea of responsibility for damages in climate change, for example, is built into the United Nations Framework Convention on Climate Change (UNFCCC) procedures via the *Polluter Pays Principle* (PPP). UNFCCC uses an analytic frame that places cause in the hazard itself, and, thus, traces responsibility for damages to the parties whose effluents intensified its force. In their model, damages are 'impacts' of climate events – and their reports repeat the words 'climate impact' and 'climate-change impact' over and over (IPCC 2014). Increased damages are attributed to, seen as 'impacts' of, the increase in the intensity of climate events. We call this a hazards model (or *Environmental-Drivers* model, see Cottier et al. 2022) in which the hazard is the cause of the damages that follow, perhaps mediated by some static social elements.

The very different frame used in vulnerability analysis (also characterized as a *social-causal* model, Cottier et al. 2022), however, identifies the sources of fragility in social and human-created arrangements that the hazard finds in place. The damages, and the way they scale with the force or duration of climate events, are due to social vulnerabilities in place. Hazard analysis, implicit in UNFCCC's frame, places causes of damage within the hazard (Bassett and Fogelman 2013).[16] Vulnerability analysis places causes of damage and responsibility in society – so, for example, history may show that patterns of labor exploitation, price fixing, or lack of representation have produced the fragilities in current infrastructures and social arrangements (Ribot 2014; Ribot, Faye, and Turner 2020). The vulnerability model examines the multiple causes (among which the hazard

[16]Bassett and Fogelman (2013) analyzed four IPCC reports and adaptation-focused articles in the scholarly journals *Global Environmental Change, Climatic Change, Climate and Development*, and *Mitigation and Adaptation Strategies for Global Change*. They state,

> Our content analysis shows the dominance (70%) of "adjustment adaptation" approaches, which view climate impacts as the main source of vulnerability. A much smaller percentage (3%) of articles focus on the social roots of vulnerability and the necessity for political-economic change to achieve 'transformative adaptation.' A larger share (27%) locates risk in both society and the biophysical hazard.

may be one) of a single outcome (e.g. economic loss or dislocation) – rather than the multiple outcomes of a single presumed cause (the hazard).

Tracing the origins of vulnerabilities, thus, might lead to a different principle – such as the *Exploiter Pays Principle* (EPP – see Kashwan and Ribot 2021). PPP places cause and responsibility in the hazard and its distant cause in the generation of greenhouse gases, EPP places cause in the fragilities in place – and their historic social and political-economic antecedents. PPP poses a paradox. It attributes the increase in damages that follow a normal versus an intensified hazard (storm or drought) to the climate change – to the climate-change effects of the 'pollution.' Then, following UNFCCC's PPP, the damages are viewed as being commutatively 'caused' by that increment of hazard and are thus the responsibility of those who caused climate change (see Lahsen and Ribot 2021) – the industrial nations (of course, identification of who the polluters are is also very dicey business, see Agarwal and Narain 1991). The PPP idea is to pay reparations, or create protections, to bring those affected back to the state they would have been in had there been no climate change. This return to pre-climate change levels of security (or insecurity), however, is the wrong target. This is equivalent to saying, 'let's provide protections that maintain affected people in their original pre-climate-change state of vulnerability – that is, restore them to an already unacceptable state of precarity.

Not surprising, to many social scientists and historians, the precarious state of the adaptation-fund-eligible populations in 'developing countries' is also largely caused by the same metropoles that generate the climate change. Indeed, the very ability to industrialize, hog and burn carbon fuels was a product of colonial domination and exploitation that made these populations vulnerable (see Fanon 1986 [1952]; Wolf 1981; Mamdani 1996, 2020; Guldi and Armitage 2014; Patel and Moore 2017; Davis et al. 2019). Yet, this precarity is hidden by the hazards focus (which frames the hazard as cause). So, calling the climate increment the cause of observed suffering obscures the broader responsibility to insure people's overall wellbeing – and not just to restore the state of misery they were enjoying in the absence of climate change. PPP is clouding out EPP. People who are vulnerable in the face of ordinary climate events, and vulnerable along many other lines, should not merely be protected from the increased intensity of storms. They need to be compensated for a history of colonization, exploitation, extraction and marginalization. Climate proofing is simply not enough (Morrissey 2014; Brottem and Brooks 2018; Smucker et al. 2015, 40).

Of course, PPP and EPP are integrally interlinked. In law PPP and EPP can easily be combined to establish responsibility – in cases where the action ('pollution' or 'exploitation') or force (a climate event) is traceable to human agency or intentionality. In law, actions without which the observed damage would not have occurred are called 'but-for causes' (essential or *sine qua non* conditions). The link from damage to necessary conditions is considered legally adequate to establish responsibility. With multiple but-for causes, the law does not have a calculus to parse and apportion damages among intervening variables – but it does so through normative judgment; as there are no measurable or quantifiable proportionalities. Each (and every) but-for cause, without which there would be no damage, can, thus, be one-hundred percent causal.

Indeed, while hazard and vulnerability work together, a root-cause analysis cannot attribute a specific portion of damage to a hazard (Cottier et al. 2022). Such quantifiable attribution is often made by analysts, but only by controlling for context (varying the hazard and calculating different damages for a fixed set of conditions in place). This attribution is specious, as context is a cause and has causes. So, this form of analysis is tantamount to controlling for (or taking as fixed) the very causes we are aiming to identify. Causality lies in the conditions that enable a hazard to trigger damages – and these must be accounted for. The relative weight of the causal contribution of each remains indeterminate – even though it is easy to see that hazards have causes and vulnerabilities have causes. Causes on each side of the hazard-vulnerability relation can be accounted for but not quantified for relative contribution.

Violent silences in the choice of frame

> The risks that we choose to see or not see for ourselves or for others are deeply embedded in the way our societies are organized and in the beliefs that sustain and perpetuate that organization. (Oliver-Smith 2013, 1)

> The scientists kept a pointed silence about the living conditions of the world's hungry masses; consciously or unconsciously, they became accomplices in the conspiracy. The social reality of hunger stayed outside their laboratory walls. (de Castro 1952, 8)

> The contention that famine results from a kind of natural law has no basis in scientific knowledge. (de Castro 1952, 12)

> News that is going to be accepted as true information has to wear a badge of loyalty to the particular political regime which the person supports; the rest is suspect, deliberately censored or unconsciously ignored. From this standpoint, the proper way to organize a programme of studying risk is to start with studying institutional design. (Douglas 1992, 19)

> Shifting frames and changing practice are not easy, and are very challenging to incumbent institutions, disciplines and professional practices. (Scoones 2019, 28)

> In understanding contemporary environmental crises, it is crucial to remain attuned to the ways in which 'destruction in the colonial era becomes visible in the postcolonial era'. (Davis et al. 2019, 3 citing Vergès 2017)

> We … need to examine … the erasure of historical memory and other forms of desocialization as enabling conditions of structures that are both 'sinful' and ostensibly 'nobody's fault.' (Farmer 2004, 307)

> Erasing history is perhaps the most common explanatory sleight-of-hand relied upon by the architects of structural violence. Erasure or distortion of history is part of the process of desocialization necessary for the emergence of hegemonic accounts of what happened and why. (Farmer 2004, 308)

A whole set of institutions support analytic approaches that locate the cause of crises in climate hazards. This is not surprising as it is good and necessary to convince the world that anthropogenic climate change is real and dangerous. So, because scientists and the media are under pressure to demonstrate the dangers of climate change, they seek to show how every environmental crisis has a climate change signature (Lahsen and Ribot 2021; Lahsen, de Azevedo Couto, and Lorenzoni 2019). Hulme (2011, 247) calls this attribution of crises to climate alone 'climate reductionism' – a reduction to a set of variables recognized by natural sciences. He explains that, 'In seeking to predict a climate-shaped future, proponents of this logic reduce the complexity of interactions between climate, environments and societies, and a new variant of climate determinism emerges.' He continues, 'Once isolated, climate is then elevated to the role of dominant predictor variable.' Yet, many other variables make climate-related damage possible.

When attributing causality of damages, however, whether to trace cause through the hazard or through histories of place is a *choice* (certainly embedded in institutional and social pressures) – it is the choice of what to hold constant; a common and highly problematic act in natural sciences.[17] Which variables to privilege and which to hold constant is, moreover, a moral choice, as it has implications for potential response and responsibility. Climate modelers resolve this by separating biophysical phenomena from the social conditions – they choose to 'control for context' – and then calculate changes in damages from a controlled social starting point (Cottier et al. 2022). Only then could modelers state that five percent of migrants are leaving Latin America due to climate change-induced drought (Lustgarten 2020). But this number is specious, as that five-percent increase in dislocation is caused by the many local conditions and histories (security or vulnerability) that converge with the drought. It is only caused by the hazard if one occludes the history and causes of the vulnerability that enabled calculating that five-percent effect. Clearly, even a quantitative climate model produces a normatively laden theoretically derived number. So, while an amount of damage would be followed by a specific increase in hazard in a given place and time, the increase is still due to the pre-existing vulnerabilities – and their causes – not to the hazard alone. It is due to histories behind the conditions that were taken as fixed.

Choice of theories and frames of analysis matter – as frames of analysis, like moral tenets, carry human values, as well as implications for action, and are thus always, at least implicitly, normative. Hazards models, attribute blame and responsibility to that event or through that event to the causes of climate change – as in PPP. In the event of blaming ordinary, or 'natural,' weather, the frame blames nobody. It is akin to calling a crisis an 'act of God' or an 'act of nature.' In the event of blaming anthropogenic climate change, the frame traces responsibility to the scads of scattered sources for carbon emissions or to the social and political-economic causes of those many sources. The chain of causality is from climate to carbon emissions back to social acts and structures. Yet, such blame, in the event of climate change, is relatively diffuse, even if social – rather than blaming someone, or specific economic structures, it often blames everyone (e.g. 'anthro-,' the human species) or whole countries or overly broad economic activities (Agarwal and Narain 1991; Castree 2014; Rudiak-Gould 2015; Schwartz 2019). Here, even if vague, the social component of blame articulates through the sky via the human agency transported by the transformed climate event. The focus on cause in and of the hazard, however, entirely misses, indeed occludes, the deep causes in place – the social and historical causes of vulnerability. Reflecting Watts's *Silent Violence* (1983), Kashwan and Ribot

[17]Others writing on the origins of crisis have argued a more general set of Western occlusion via the reductionist sciences. de Castro (1952, 12) argued that

> [t]he tremendous impact of scientific progress produced a fragmentation of culture and pulverized it into little grains of learning. Each scientific specialist seized his granule and turned it over and over beneath the powerful lens of his microscope, striving to penetrate its microcosm, with marvelous indifference to and towering ignorance of everything around him.

In Europe and the US, he continues, universities have created a 'specialists' civilization,' 'directed by men whose scientific outlook is rigorous but who suffer from deplorable cultural and political myopia.' These are, following Ortega y Gasset (1940), quoted by de Castro, the 'new barbarians – men ever more and more learned, and less and less cultured.' Similarly, Margaret Somers (2008, 9), drawing on Foucault, speaks of social histories and causes of crisis as '... subjugated knowledges,' meaning 'those ways of seeing and understanding the world which have been disqualified for their supposed lack of rigor or "scientificity," those knowledges that have been present but which are often made invisible.'

(2021) call this silencing of history a *violent silence* – as it does harm by hiding the factors that cause damage and thus hiding potential responses and responsibility.[18]

The nature of human-nature

> To say that humanity's physical and mental life is linked to nature simply means that nature is linked to itself, for humans are part of nature. (Marx 1894)

> ... laissez-faire was not only 'planned,' as Karl Polanyi famously insisted; its imposition required an increase in repressive measures, as labourers, peasants, and smallholders were forced to bear the cost of *market regulation*. The assumption that markets are 'natural systems' operating outside of power and politics is itself an invention of the nineteenth century. (*Human Encumbrances*, Nally 2011, 8)

> There is an ecology of bad ideas, just as there is an ecology of weeds, and its characteristic of the system that basic error propagates itself. (*Pathologies of Epistemology*, Bateson 1971, 489)

> It is remarkable ... how the (not so) dreaded comparison between human and animal slavery is brandished about in the field of animal studies and how black liberation struggles serve as both the positive and negative foil for making a case for the sentience and therefore emancipation of nonhuman beings. (*Habeas Viscus: Racializing Assemblages, Biopolitics, and Black Feminist Theories of the Human*, Weheliye 2014, 10)

> ... the physical mixing of nature and society does not warrant the abandonment of their *analytical* distinction. (Malm and Hornborg 2014, 64)

It is not a new idea that frames matter for the study of climate-related crises. Almost half a century ago, O'Keefe, Westgate, and Wisner (1976) wrote 'Taking the Naturalness out of Natural Disasters.' Their impulse was practical and moral – directing analysts to locate cause, and thus responsibility and response, in society. Their argument does not imply that nature cannot wreak havoc. Rather, they argue that the social stage must be set – exposure and vulnerability established – for nature to take a toll; they assert that the causes of disaster *can*, and *should*, be framed as social. The frame of causal analysis we bring to disaster is a moral *choice* that shapes how and where we locate responsibility and identify possible response. In some frames – which I would not choose – nature is cause; in others it is not (de Castro 1952; Ribot 1995; Cottier et al. 2022). For any given event and damage, analysts can, if conscious, *choose* among different analytic frames, each of which can rigorously locate causality in totally different arenas. Taking naturalness out of disasters is a *decision* to place cause, and thus responsibility, squarely within society – enabling us to leverage social response. In this sense, the choice of analytic frame is a moral choice – as it shapes, based on the analyst's position or sensibilities, where we locate, via cause, responsibility. This choice of causal analytic shapes responsibility by identifying the culpable and by showing what is possible.

The late Smith (2006) wrongly asserted, however, that 'there's no such thing as a natural disaster.' His argument was based on the idea that the disaster was rooted in social precarities on the ground – he was not concerned with how natural or anthropogenic the hazard was. Yet, for those who take a hazards approach, there can be such

[18]Guldi and Armatage (2014, 83) point to many factors that militated a shallowing of history from the 1970s to 2000s. These factors pushed historians and other social scientists to stop writing for the policy makers as economists replaced them; science envy also steered social scientists toward modelling and '... a focus on game theory and rational actors – ... a retreat to individual and the abstract, not the collective and the concrete.'

thing (assuming that the hazard modeler believes the storm is, at least partly, 'natural'). But, the claim that nature, or even the anthropogenic storm, is the cause of a disaster is, even if true to the assumptions made, a dubious moral stance as it is a choice that can obscure social causes of the vulnerabilities in place. So, it is more accurate to state (for those who believe the storm named Katrina[19] was a natural event) that 'we *should* not attribute disaster to "natural" events.' We *should* choose social-causal frames. We *should* choose responsibility. Any crisis can be shown to have cause in social agency – even if the triggering hazard emerges from something we call nature, or even a seemingly random event – human-induced or not.

Understanding the nature of 'nature' is important for avoiding confusion in causal analysis. Everything, even our tools, machines and artefacts, derives from nature – as do we. Calling disaster an act of God or of nature has long been a way for governments and insurance companies to avoid responsibility (Wilkinson 2010) – as people accept crises as fate and God's will (Scoones 2019, 23). But, as soon as influenced by human action, nature is cultured,[20] it is transformed into something that carries human agency, human influence, or 'dead labor' (*a la* Marx 1894). Climate change cultures nature. And would seemingly place responsibility back with society – those specific actors and structures that generated it. This is not 'species thinking' implicit in terms like 'anthropocene,' which blames the whole human species, absolving the responsible specific social arrangements and sub-groups (Malm and Hornborg 2014; Guldi and Armitage 2014).

Climate change, however, brings in a second and different crisis of responsibility, as when in 2008 government agents or journalists in the State of Santa Catarina, Brazil avoid local responsibility by claiming that disasters following flooding are due to climate change. Climate scientists and other journalists, however, pointed out that the 110 deaths, 78,000 displaced people, 6000 destroyed properties, and 2.5 billion in damages were due to government policies and practices in the region. They cite deforestation of reserves and lack of respect for the Forest Code concerning hilltops, hillsides and riparian forests, poor land use and water management, that enabled erosion and silting, poorly enforced laws that allowed settlement in reserve areas, and policies that left poor people living in precarious areas prone to landslides and floods. One journal article suggested the need was to unclog drains and relocate those living on dangerous slopes and close to the waterline. In this case, however, politicians chose to attribute the disaster to climate change – despite that local people and even climate scientists knew this crisis was socially generated (Lahsen, de Azevedo Couto, and Lorenzoni 2019, 5–6). In this case, there is a disingenuous attempt to place blame somewhere up in the sky – in what may have been anthropogenic rains, but which fell on socially produced vulnerabilities.

While some want to blame pure nature (or God), it is questionable whether there is any pristine nature – environment that has not been shaped by human action – whether in a productive or destructive manner (Ribot 2014).[21] Levis et al. (2018) view even the Amazon

[19]Note this anthropomorphizing with human names like 'Katrina.' Perhaps this is part of directing blame to them – even before climate change – as if they were agential forces.

[20]From Latin *colere* 'tend, cultivate, inhabit' (Oxford Languages online dictionary).

[21]Giddens (1999, 3) places 'the end of nature' at the point when '… we stopped worrying so much about what nature could do to us, and we started worrying more about what we have done to nature.' This is part of our entry into what Ulrich Beck called 'risk society.'

as a domesticated space – with a mix of species shaped by human action (also see Hecht 1990). All nature, since the advent of humans – as we act in the world and even our breathing transforms the atmosphere – has been influenced by humans. This is why Bauer and Ellis (2018) have argued that it is impossible to find a precise starting point for the 'anthropocene' (also see Malm and Hornborg 2014; Davis et al. 2019). Indeed, human action has always shaped the world. As poet Francis Thompson (1913) wrote, ' … thou canst not stir a flower without troubling of a star.' Clearly influenced by his contemporary Isaac Newton, Thompson saw that all things in the world – indeed, all things in the universe – are inter-dependent. Human action has, thus, shaped everything, intentionally or not. So, finding pure nature is not possible. And, it is also not necessary or relevant.

The link that matters is between human action and what humans judge to be good or bad; gains or losses, benefits or damages. These human actions are relevant to attaining human aims and desires. And, it is possible to trace a causal link from a given human-valued outcome, a gain or loss, benefit or disaster, to intentionality or negligence (*a la* Calabresi 1975; Hart and Honoré 1959). We are not concerned, like Thompson, with the effects of all human action, just the agency that causes benefits or damages and that we thus have reason to attended to, care about, encourage or prevent – and this includes the agency in reshaping the biophysical world. So, while human influence may be everywhere, agency,[22] which derives from consciousness and the ability to think and to judge our actions (*a la* Arendt 1963), is the element from which we derive morality and responsibility.

We humans are not separate from nature. We are nature. But, as Arendt (1961, 170–171) pointed out, the miracles of evolution are authored by probability, whereas we know the author of the even more frequent miracle of political change through women and men '… who because they have received the twofold gift of freedom and action can establish a reality of their own.' She recognizes our exceptional position in nature – and some argue it is language and thought that are unique elements of our – human – nature. These give us the freedoms and possibility for action that do not appear to be the nature of other creatures – these are unique to us (see Arendt 1963; Chomsky 2018). We are able to freely think, frame, plan and manage, while other entities and 'creatures' cannot. And, even if they are sentient or communicate or even have 'agency' in *some*, rather absurd, definitions (as Latour 2004 [1999]; Bennett 2010; Simard 2021 assert), the elephants, dolphins, whales, octopodes, scallops, trees, mushrooms, mosquitos, viruses, rivers or rocks are not going to intentionally manage human wellbeing or save us from ourselves. They will not steer us toward a just and sustainable future. As Malm and Hornborg (2014, 68) say, ' … they lack the capacity to scrutinize and stand up to human actions.' That said, if we consider that thought (along with abilities that free us to translate it into action) is our nature, then our actions on the world are also nature and so we can invert this whole argument by saying that everything is natural – even disasters. But, we can say it in a way that does not absolve us of responsibility.

The key responsibility issue, then, is not what is or is not 'natural.'[23] The key is why we, humans, care and are in a unique (exceptional) position to do something about our effect

[22]Latour's (2004) and Bennett's (2010) 'destributive agency' – a claim that all things and all acts are results of agency and that agency is thus everywhere – is a serious diversion from the quest for responsibility.

[23]In some sense, the search for pure nature is like the search for a Garden of Eden – that time before knowledge (and attendant responsibility) that the Judeo-Christian tradition viewed as pure (Ribot 2018). This kind of search for purity

on the each other and the world – for human purposes such as security, sustainability or wellbeing. Thought is our nature, and because of it we are responsible for what we do (*a la* Arendt 2003; Nietzsche 1882); with thought, and being capable, or able to do, we are unique. We have agency *and* can use it for what we value – what we judge to be good or bad. This responsibility is the essence of human exceptionalism. This makes us quite unique potential stewards of the earth and all nature – even our own. This *humanism* does not have to be patronizing. This can be practical and just. We must exercise our very-human agency – as diverse and politically complex as it is. If not us, then who (or what)? Of course, we do so with humility (see Russell 1924). Thought allows us to reflect on whether we can live with our own actions (Arendt 2003) – requiring of us a deep understanding of difference and a value of justice (*a la* Ghandi, in Kashwan et al. 2020).

Agency is not in every act or influence – human or otherwise (despite Latour's 2004 [1999]; Bennett's 2010 flat 'distributive' assertions). As storms or droughts may no longer be 'natural,' as they are profoundly influenced by humans, an element of what they transmit can be attributed to human agency – the element of human influence that can be traced to our ability to think and thus to our ability to be moral beings and to act accordingly. Attribution sciences (Trenberth, Fasullo, and Shepherd 2015), the science of identifying the element of climate events due to anthropogenic climate change, are estimating the climate increment or change in a given event that is due to human agency – the agency made explicit by our ability to know and reflect on our contribution of greenhouse gasses to changing climate. Through this attribution they are establishing particular lines of responsibility. They are establishing human action as the cause of this biophysical change. This change may also be stopped or reversed, via human intentionality. If that is not done, it is our own fault – belonging to no other being. Responsibility belongs to those with agency. Agency belongs to a moral thinking being with the ability to act and intervene. Agency is natural – a unique element of human nature.

Scoones (2019, 17) explains, stepping into slippery actor-network territory, that

> ... interactions between human and non-human natures, bound up in extended actor networks, must necessarily become central to our understandings of how uncertainties are lived with (Law 1999; Latour 2005). In this view, multiple agencies and diverse practices link human–nature networks in ways that both generate and confront uncertainties as horizons of possibilities.

Perhaps this applies to uncertainties ahead, 'horizons,' but not to the past. On the contrary, if we intend to trace responsibility to human intention and agency, the 'actor-network' approach falls short; as it denies antecedents, and the distinctness of human agency (Latour 2004 [1999], 2005; Bennett 2010), it diffuses any notion of causality and human responsibility (*a la* Bryant 2011; *cf.* Hornborg 2017).[24] Rather, in a grounded approach, Borras et al. (2022, 5) observe,

too is an origin myth that has many dangerous (Nazi, Proud Boys ...) implications we won't delve into here (see Huq and Mochida 2018; Kashwan 2020). Nobody tainted us with knowledge – no God or serpent anointed us with insight.

[24]Marx did not blur the distinction between humans and nature, but rather acknowledged interdependencies and embeddedness. These relations are material, as well as ideological, dialectics. They are not vague 'assemblages' or 'actor-networks,' but, rather causal recursive dialectical relations that we can observe and unpack. So, I do not understand how 'identifying multi-species realities, where nature-society separations are dissolved into hybrid assemblages, can offer deeper insights into the realities of the Anthropocene' (Borras et al. 2022, 3).

> ... work in critical agrarian studies needs to retain the focus on local material histories and power relations, while embedding both in long-term analysis of global environmental change and understandings of the way in which this new historical moment and the phenomenon of climate change are shaped by both material limits and the legacies of colonial and imperial inequality.

We must trace the present to human actions of the past.

Disjuncture – the commutative gap

> Sometimes problems are not solved but dissolved in favour of another way of posing the problem. (Jessop 2014)

We have established that not all human influence is tied to agency or responsibility. We have not established whether the human agency that is responsible for changing the climate is also responsible for the subsequent damages that unfold on the ground. Paradoxically, establishing a human influence, an anthropogenic signature, on a climate event does not establish whether or how that anthropogenic change is related to any damages that follow (Lahsen and Ribot 2021). Responsibility for a change in climate does not directly translate into change in subsequent damages or responsibility for those damages.[25] These changes in damage are still caused by other factors – the social and political-economic dynamics and histories that shape precarity.

In the era when de Castro, Sen and Watts were first studying hunger and famine, climate events were viewed as natural backdrop.[26] They were not seen as anthropogenic (although they often were without us knowing – see Sahel discussion below). Here it was easy to state that the causes of crisis were in the vulnerabilities in place – as storms were seen as acts of nature and all possible social response was in action in the locale. People were protected or they were not. Responsibility was squarely on the ground within society (also see Douglas and Wildavsky 1982; Douglas 1992). There was no 'anthropogenic' climate-change increment that could be assessed and to which increased damages could be attributed. With the advent of this increment came calculations of 'additionality'[27] – anthropogenically augmented intensity seemed to imply a corresponding increase in damage.

As we have seen, however, the relation between climate increment and damages is not commutative (see Ribot 1995; Hulme 2011; Lahsen and Ribot 2021; Cottier et al. 2022). As weather events now have human agency (in addition to many other incidental human influences) built into their intensity and frequency, the increment in that intensity and frequency can arc to responsibility and can be a meaningful social cause of that change. So,

---

[25]Except in a 'but-for' sense.

[26]Giddens (1999) points out that

> ... there doesn't seem in fact to be a notion of risk in any traditional culture. The reason for this is that dangers are experienced as given. Either they come from God, or they come simply from a world which one takes for granted.

He goes on, advent of the idea of 'risk' is '... bound up with the aspiration to control and particularly with the idea of controlling the future.' Here hazard and danger are givens, risk is new – it is a preoccupation with the future. I would argue for a moral economy predicated on the realization that we generate our own dangers – and thus we are responsible for them; explaining the rise of insurance and the welfare state (Giddens 1999, 4). Giddens (1999, 8) calls this 'manufactured risk,' and argues 'The transition from external to manufactured risk is bringing about a crisis of responsibility ... .'

[27]See Lemos and Boyd (2010) for an excellent discussion of 'additionality' politics.

responsibility for the cause of the changed climate event can be identified, yet this responsibility for climate change cannot be extended to responsibility for a precise damage increment – as the cause of damage remains, at least in part, in the vulnerability that the climate event finds in place. Nonetheless, climate events now link to human value as they carry human agency.[28] Made by human labors, anthropogenic climate, like objects of use or tools and machines, has embedded agency – the work that keeps on working. So, today, in what is being called the Anthropocene,[29] there is human agency in the sky, which can indicate responsibility. But responsibility for what? The climate event; not necessarily the damages that follow.

Full causal analysis

> The framing of climate events as sudden natural shocks can redirect the attentions away from long-term capacity building and social transformation towards short-term emergency planning .... (Huq and Mochida 2018, 36)

> The challenge today is to integrate agency and structure in examinations of the production of vulnerability, in specific places, whilst also acknowledging the importance of physical systems in generating hazard that can trigger disaster. (Pelling 2003, 47)

> Even brand-new forms of menace, without having been anticipated, can be labeled and slotted into existing categories of responsibility. (Douglas 1992, 54)

> ... the new commonsense must avoid reductive 'ecologism.' Far from treating global warming as a trump card that overrides everything else, it must trace that threat to underlying societal dynamics that also drive other strands of the present crisis. (Fraser 2021, 96)

Sen (1981) revolutionized thinking on famine by debunking the simple food availability decline (FAD) theory of famines and provided a basis for a causal analysis of food deprivation rooted in proximate and some distal political factors (Sen 1981; Drèze and Sen 1989) – including capabilities and democracy. Fine (1997, 645), challenged this critique, bringing FAD back in as one causal element in need of explanation (in addition to entitlement failures), arguing that 'it is essential both that socioeconomic factors be analysed as causally prior and that famine be specifically tied to food rather than to a general theory of access to life's capabilities.' For Fine (1997), FAD cannot be dismissed in analyzing the causal structure of famines. Food is itself something that can fall short – supply can be disrupted or diminished. This itself can be a weapon or factor used by or generated within a larger political economy. So, the ability to shape the political economy that shapes food availability (Fine 1997, 645), in addition to the ability to shape the political economy that shapes entitlements (Watts 1991), is part of the causal structure of famines.

Without resolving the dilemma of how to weigh food decline versus entitlement failure, hazard versus vulnerability, a full social causal analysis of damage can be accomplished by separately analyzing both human causes of and responsibilities for climate change (*a la* Fraser 2021, 100; Borras et al. 2022, 8) and the human causes and responsibilities for vulnerabilities in place. For a given instance of crisis, such as the Sahelian

[28]I.e. Marx's (1894) dead labor at work in machines and infrastructures, or John Locke's (1960 [1689]), transformative labors by which we 'own' the sky – and owning it we are linked to its effects in the world. It is now ours. Being ours, we are also responsible for it, and all the externalities that are coming back to bite everyone.

[29]There are many reasons to avoid a term like 'Anthropocene' (see Bauer and Ellis 2018; Castree 2014; Malm and Hornborg 2014; Davis et al. 2019). Nonetheless, I will use it at times to evoke the era in which we are living under a changing climate.

droughts of the 1970s and 80s, this allows us to attribute responsibility for the hazard (the drought driven by aerosol forcing in Europe and the United States; see Biasutti and Alessandra 2006; Marvel, Biasutti, and Bonfils 2020) with the systematic colonial production of vulnerability (Franke and Chasin 1980; Watts 1983). Here, interestingly enough, we find that many causes of the climate increment and of vulnerability trace to the same Northern origins. They give us, in essence, a PPP-EPP analytic. Both polluter and exploiter pay – and in the case of 1970s and 80s droughts in West Africa, they indicate in aggregate (although not in the specifics of scale and timing) the same responsible parties. Thus, both the hazard and precarity had cause in Northern colonial and industrial activities. Effectively, we have two 'but-for' causes that point to the same responsible parties.

Of course, there are also many situated chains of causality that will show proximate and distal cause of place-based vulnerabilities – that indicate roles from the structure of gender inequalities in the household to local identity politics to structures of rights and representation shaped by local rules and regulations, infrastructures, the structure of markets, and to national and international politics and policies (Sen 1980; Watts 1983; Bassett and Fogelman 2013; Ribot 2014).[30]

Nonetheless, despite there being local, national and international causes of vulnerability, as well as a mix of causes contributing to climate change, focusing on vulnerability by holding hazard as the disaster trigger (rather than as cause), does not take polluters off the hook. It identifies the arena of intervention where, as in the Sahel case, the industrial world and colonial metropole are largely responsible for damages and, thus, reparations – the common cause in both hazard and vulnerability make responsibility unambiguous. This indicates a moral obligation to address the vulnerabilities and not merely those related to a climate increment – as the UNFCCC would have it (see Ribot 2014; Khan and Roberts 2013; Kashwan and Ribot 2021). The overall analysis of causes of vulnerability and of damage will always require a nuanced situated historical causal analysis – and can, likely often, show common root causes.

However, this dual analysis of causality – tracing out of the causes of changed frequency and intensity of climate events and the causes of vulnerabilities – cannot provide a quantitative weighting of the relative roles in damage production of the hazard versus the vulnerability. Neither element can cause damage by itself (Blaikie et al. 1994). Further, in a given place and time, for a different level of hazard there is a different level of damage, as for a different level of vulnerability damage will vary. So, were a climate event to strike a secure village, there may be no damage. Were the intensity of that event to be doubled and to strike this secure village, there still may be no damage. But were the village highly vulnerable, the damage might be quadrupled by a doubling of intensity.

The vulnerability transforms mere events, at least when on the scale of human experience, into hazards – without vulnerability they remain mere events (in this sense, vulnerability defines hazard).[31] While for a specified (controlled-for) vulnerability we can attribute an increment of damage to an increment of hazard, even a fixed vulnerability

[30]As de Sherbinin (2020) notes, 'Social vulnerability is a function of the population's sociodemographic characteristics such as age, sex, ethnicity, race, education, and major livelihoods, as well as its access to financial and other capitals and adaptive capacity.' And, each of these indicators and proximate causes, has a chain of causality behind it – what might be called the structural and political-economic root causes.

[31]When that hazard is on a human scale. But, even with extreme events, such as gigantic meteorites, one can simply say we are vulnerable in the face of meteorites as we do not have the protections.

has causes too, thus the causes of the damage cannot be located in the fixed conditions – they are located in the causes of these conditions. This follows from the observation by Lund (2014, 225), made above, that the social sciences do not control for or set context aside – as context always plays a causal role. In this case, context, both social and biophysical, shapes the entire possibility of crisis. There will never be a fixed or neutral quantified apportionment of cause or responsibility between the hazard and the vulnerability.[32] The question at hand becomes a moral one. Which causes lead us to places where we are *able* to *respond* – in a timely, long- and short-term, and meaningful manner – to reduce pain and suffering associated with climate events? This also evokes the question of who is able, or has the means, to respond – and thus who 'should' respond.

Toward a sociodicy of crisis

> To remember history is not to lament it. Rather it is to purposefully take stock of the magnitude of damages wrought by various forms of discrimination and to devise interventions that redress the sources of that discrimination. … such an approach would push environmental scholarship to internalize 'the deep spatial historical logics' – the plantation, the colony, the reservation – that mark contemporary racialized environments. (Ranganathan 2017, 5)

> To become counter-hegemonic … a new commonsense must transcend the 'merely environmental.' Addressing the full extent of our general crisis, it must connect its ecological diagnosis to other vital concerns – including livelihood insecurity and denial of labour rights; public disinvestment from social reproduction and chronic undervaluation of carework; ethno-racial imperial oppression and gender and sex domination; dispossession, expulsion and exclusion of migrants; militarization, political authoritarianism and police brutality. (Fraser 2021, 96)

> More radical climate justice narratives highlight the historical injustices of unequal exchange and ecological debt, whereby climate challenges in the periphery are the direct consequences of long histories of exploitation and unequal relations of global power. (Borras et al. 2022, 13)

> As climate scientists have investigated future climate scenarios – and potential social responses to environmental changes – they have become, *ipso facto*, social scientists. (Wainwright 2010, 983)

> There are clearly alternative worlds of political critique, parallel to the dominant one, that need to be taken seriously if we are to restore the contingency of history and see the past anew as a site of possibility. (Nally 2011, xiii)

> If we continue to frame adaptation in apolitical terms, our efforts may well end up being futile. Worse than that, in cases where new resources are made available to existing elites they may end up buttressing exclusion and therefore entrenching vulnerability. (Morrissey 2014)

> … social processes and the risks they represent are all outcomes of human decision-making about how resources (including places) are used and by whom they are used. (Oliver-Smith 2013, 1)

Climate is changing. There are stronger storms, more droughts and rising seas. These forces have to be contended with as they *trigger* all kinds of damages. There are

[32]Of course, a total destruction of earth by a meteorite could make preparation irrelevant. But, the effects will still be stratified. Presidents of powerful nations will probably blast off into space. Yes, it will be a bummer to die alone in space and watch the earth vanish, but the vulnerability will still be stratified – by have rocket ships and have not rocketships. In short, any human-scale crisis is stratified by vulnerabilities.

dangerous biophysical forces out there. We are generating them and could stop them. This 'could' is part of the social force we are not using. A different set of conditions, vulnerabilities, *enable* those new forces to trigger damages. What would allow us to withstand and protect against them? This is also a set of social forces that we must work with. The first set of social forces is called mitigation. The second is spoken of as 'adaptation' – although I call them 'vulnerability reduction.'[33] Despite being interlinked they remain distinct. Both condition and are conditioned by the systems we create and live in.

Indeed, we live in a system with built-in contradictions, predicated on damaging growth. Fraser (2021, 96) makes it quite clear that our economic system[34] '… represents the socio-historical driver of climate change, and the core institutionalized dynamic that must be dismantled in order to stop it.' Because of this system, Bourdieu (1997) has argued 'today precarity is everywhere.' The production of hazard and of vulnerabilities, while they could be treated through transition to another kind of economy, when and if that is possible, they are also in need of much less-sweeping social action, resistance and reform. There is a need to fight the production of greenhouse gasses on many fronts – to mitigate. We must also be fighting exploitation and enabling people to keep more of the wealth they generate – to 'adapt' or to reduce their vulnerabilities. It means not allowing people, no matter how wealthy, to live in oversized and over-cooled/heated houses. It means treating social causes, proximate and structural, violences direct and indirect.

Both mitigation and 'adaptation' are social acts – as are the conditions that cause climate change and vulnerability. In 2013, I was happy to see Connelly (2013) accuse social scientists of practicing 'sociodicy' – tracing all cause to the social. Connelly was comparing social science practice to Leibnitz's (1710) *Theodicy*, the justification of the goodness of God in the face of pain and suffering (see Wilkinson 2010; cf. Voltaire's 1759 [1931] lucid critique of theodicy). But Connelly's attempt to insult the social sciences is quite useful, as it accurately labels what we *should* do. For me it is a moral imperative to locate causality where there is a proactive or retroactive response ability, the ability to act – by individuals and society or the institutions that we have built. Thus, we need a sociodicy for the current era of climate change (Ribot 2014, 2019). One can make other analytic choices – to ignore or avert moral responsibility – but, I'd rather not.

God and nature, non-social causes, have always been convenient foils for responsibility. But, having killed God, Nietzsche (1882) indicates we must shoulder responsibility all by our little selves. Perhaps we, people, have killed nature – by transforming it, mixing our labor with it, culturing it, via various social acts (from cultivation to climate change). This view might seem to place us outside of nature – it is implicitly a stance of human exceptionalism. Were it not, then all our actions and the changes or things we make in the world (including machines) are also mere nature – elements of being. We humans are part of that whole – and perhaps we have put our smudge on all 'nature.' In this sense, while nature too is often a foil for responsibility, we have also foiled this foil – by humanizing it, thus pointing back to ourselves. We stand responsible again.

We are *also*, however, not part of that whole – we are, indeed, exceptional. Consciousness and thought mean that we know that we depend on nature (cultured or not), other

[33]See Ribot (2011) for a critique of 'adaptation.'

[34]In their fabulous book, Patel and Moore (2017, 3) wrote that 'taking capitalism seriously [means] understanding it not just as an economic system but as a way of organizing the relation between humans and the rest of nature.'

than ourselves, which we transform for our valued beings and doings. The separation of us from other nature is our own creation via consciousness. But, our own security and survival, of which we are conscious, depends on recognition of this separation – as we are uniquely able to manage that nature so as to sustain or destroy ourselves. Because of consciousness and value, we have intention. Because of ability to realize that intention, and to reflect upon and judge our potential acts, we have responsibility (Arendt 2003). If 'the greatness of this deed' is 'too great for us' we must indeed 'become gods simply to appear worthy of it' (Nietzsche 1882). We have a choice. We can manage or not. I choose to manage. I choose life. To manage is our exceptional role – call it patronizing or matronizing; it does not have to be. It is humanism and should be done inclusively with and for care (regardless of race, ethnicity, gender, caste, class, age, place of origin, orientation, etc.!). Not to manage is a shirking.

A sociodicy – a social-causal approach – is a conscious means of identifying the ability to respond and responsibility. It does not mean ignoring the biophysical causes of force and change or the human effects on the biophysical world. The social depends on the biophysical as *should* depends on *could*. We operate within the possible – a world where the resources at hand and the principles that shape what we can do with them and what they can, in turn, do for or to us are physical entities with their own limits and possibilities. $F = ma$; $E = mc^2$; $PV = nRT$. What matters is not whether the biophysical exists or has effects – it certainly does. What matters is what we do with and make of these realities.[35] These formulas and the biophysical world establish a realm of the biophysical possible. Society operates within and on that possible. The social still shapes their effects on us and our effects on the physical shape conditions that we respond to as well.[36] We read and act in the naked world with social categories and objectives.

The way forward is to practice a conscious sociodicy of climate-related pain and suffering – a problem-oriented approach to crisis that asks which causes of the beings and doings we care about are socially and political-economically generated and which are amenable to social and political-economic solutions. Vulnerability analysis is the entry point into the most immediate solutions to those elements of loss and damage that are with us today. We have methods of analysis of climate-related vulnerabilities (de Castro 1952; Sen 1981; Watts 1983; Blaikie et al. 1994; Wisner, Blaikie, and Cannon 2004; Nally 2011; Ribot 1995, 2014). All start with the problem – identifying something that affects people's ability to do and be (to function) with health and wellbeing. Then they trace chains of causality outward from that problem – from a moment of loss or damage. For such social-causal analyses related to diseases, such as Covid-19, see Farmer (2004), Dzingirai et al. (2017) and

[35]As Oliver-Smith (2013, 2) reminds us:

> There is no cultural equivalent of the law of gravity. There is no social physics. Even our most basic biological realities are culturally framed and structured. Thus, market logics and structural constraints are ultimately cultural products, the outcome of decisions and choices made by people.

[36]Another important nuance in the relation between hazard and vulnerability (the biophysical and social) is that crises triggered by hazards depend on vulnerabilities, but they also change vulnerabilities. Hazard, by triggering crisis, can change existing infrastructures and the availability of resources. Thus crises, still made possible by vulnerabilities, feed back into vulnerabilities in the face of subsequent events. See Turner et al. (2003a, 2003b), Turner 2010; Swift (1989) has called this a ratchetting down. Also see Horton et al. (2021, 1280) on how climate change can change the resource base. They aptly label these 'habitability changes.' In this sense, vulnerability can include the vulnerability to become more vulnerable – a kind of vulnerability trap. Yet the vulnerability to be made more vulnerable also has social causes. Also see Beymer-Farris, Bassett, and Bryceson 2012 on the recursive relation between social and biophysical factors.

Auerbach and Thachil (2021). On tracing causes of xenophobic violence see Achiume (2014). The choices of frame – one that focuses on social and political-economic causes – is made explicitly to identify what society can do in response, society's responsibility.

Li (2007, 267) brilliantly dissects the tensions between solution and problem orientations, stating 'Community forest management ... begins ... from a proposed solution rather than a unified specification of a problem.' By focusing on solutions, one asks 'how' to change and not 'why' the contentious or challenging situation arises. Interests align around goals, not around problems. This solution orientation is common in climate change circles – 'adaptation' is entirely a 'how' kind of beast (Ribot 2011, 2014). Bodansky (2020, 3) shows how '... international environmental law ... promotes compliance through transparency and forward-looking, non-adversarial procedures, aimed at improving effectiveness, rather than through traditional international dispute settlement, which takes a backward-looking approach, focusing on the issue of state responsibility.' This is precisely the 'non-confrontational' strategy that occludes causes (the 'why') and, thus, misses many potential solutions.

The solutions that causal analyses uncover may indicate blame and liability as well as reparations – something some parties will want to avoid. Nonetheless, they provide a full range of possible interventions as well. In some senses, by occluding some causes, then, the forward-oriented view is part of the causes of risk in the first place. By not attempting to reveal or address root causes it allows them to continue or deepen. It also side steps the real problem, which for forest users, in Li's (2007, 266) case, may be self-determination and not getting these people to 'do as they ought' – by governing their conduct of forest management. The 'will to improve,' is likely the will of one party – the foresters, those who govern – rather than the communities living in these forests (for an adaptation case illustration see Beymer-Farris, Bassett, and Bryceson (2012). A problem-oriented approach starts with the problem and traces back – it then asks what can be done to treat the full range of causes. It is a choice to see, and to address, the non-contentious and contentious social causes.

The framings we choose and caveats we insert shape our world. As Lund (2014, 226) observed, 'By not questioning the concepts and categories with which we read the "naked facts," it is easy to make a set of facts look speciously unequivocal and pervasive.' We also must state the caveats so that those observing our work know what we assume. So, there is nothing wrong with stating that a given increase in climate intensity in a given place and time may trigger specific additional damages, depending on the vulnerabilities in place. Still, a climate event, a more-intense storm, or its incremental intensity or frequency, triggers damages as a function of the vulnerabilities that it finds on the ground. So, we can show there is a role of the biophysical force – the climate event. Yet, coping with and failing to cope (via mitigation or so-called adaptation) remains a social act with social responsibility. We may fail to cope, but the whole situation – vulnerability with (and defining of) hazard (Blaikie et al. 1994; Wisner, Blaikie, and Cannon 2004) – is social.

In a responsible frame, the cause of the damages and of the intensification of those damages with a more-intense event remains a function of the vulnerabilities that enable the event to trigger damage. A responsible frame identifies where we (individuals, groups and societies) are able to respond. As it is also possible for the social world to now shape this trigger, the hazard must also be part of the risk equation and enters into response ability. In short, we are weighing, or rather we should pragmatically weigh, the importance of cause against our ability to respond (see Jamieson 2015).[37] Accounting for all (biophysical and social) causes (at least all we have time and resources to account for) helps us to identify those that most matter – those we can plausibly reshape in order to reduce damages and increase wellbeing. In the world of rural political economy and agrarian studies, this approach to crisis is not entirely new (Polanyi 1944; de Castro 1952; Scott 1976; O'Keef et al. 1976; Wolf 1981; Sen 1981; Watts 1983; Blaikie 1985; Blaikie et al. 1994; Fine 1997; Davis 2001; Nally 2011; Turner 2016). Histories matter.

Acknowledgements

Many thanks to Tom Bassett, Daniela P. G. Calmon, Timothy Forsyth, Alessandra Giannini, Ali Kaba, Prakash Kashwan, Christian Lund, Barbara Nakangu Bugembe, Joanne Pascale, Yukari Sekine and Matthew Turner, and two external reviewers, for their insightful constructive comments on this text.

Disclosure statement

No potential conflict of interest was reported by the author(s).

ORCID

Jesse Ribot http://orcid.org/0000-0002-5164-9315

References

Achiume, E. Tendaye. 2014. "Beyond Prejudice: Structural Xenophobic Discrimination Against Refugees." *Georgetown Journal of International Law* 45 (3): 323–381.

Achiume, E. Tendaye, Thomas Gammeltoft-Hansen, and Thomas Spijkerboer. 2020. "Introduction to the Symposium on COVID-19, Global Mobility and International Law." *AJIL Unbound* 114: 312–316.

Agarwal, Anil, and Sunita Narain. 1991. *Global Warming in an Unequal World: A Case of Environmental Colonialism*. New Delhi: Centre for Science and Environment.

Arendt, Hannah. 1961. *Between Past and Future: Six Exercises in Political Thought*. London: Faber & Faber.

Arendt, Hannah. 1963. *Eichmann in Jerusalem: A Report on the Banality of Evil*. New York: Viking Press.

Arendt, Hannah. 2003. *Responsibility and Judgment*. New York: Schocken.

Arnold, David. 1988. *Famine: Social Crisis and Historical Change*. Oxford: Basil Blackwell.

Auerbach, Adam M., and Tariq Thachil. 2021. "How Does Covid-19 Affect Urban Slums? Evidence from Settlement Leaders in India." *World Development* 140: 1–11.

Badrinathan, Sumitra. 2021. "Educative Interventions to Combat Misinformation: Evidence From A Field Experiment in India." *American Political Science Review* 115 (4): 1325–1341.

Bassett, Thomas, and Charles Fogelman. 2013. "Déjà vu or Something New? The Adaptation Concept in the Climate Change Literature." *Geoforum* 48: 42–53.

Batha, E. 2008. "Cyclone Sidr Would Have Killed 100,000 Not Long Ago." *AlertNet*, November 16, 2007. Accessed 19 May 2022. http://alertnet.org/db/blogs/19216/2007/10/16-165438-1.htm.

Bateson, Gregory. 1971. *Steps to an Ecology of Mind*. Northvale, NJ: Jason Aronson. Inc.

[37]As Jamieson (2015, 23) pragmatically suggests,

> we should be pluralists about responsibility and shape whatever conceptions can help to explain, guide, and motivate our responses to climate change. His conceptions is forward looking and powerful. The sociodicy frame looks backward to understand causes so as to be able to identify them in the present and future. Histories matter and give us insight into the structures of vulnerability and the crises they enable.

Bauer, Andrew M, and Erle C. Ellis. 2018. "The Anthropocene Divide: Obscuring Understanding of Social-Environmental Change." *Current Anthropology* 59: 209–227.
Beck, Ulrich. 1992. *Risk Society: Toward a New Modernity*. London: Sage.
Bennett, Jane. 2010. *Vibrant Matter: A Political Ecology of Things*. Duke University Press. doi:10.2307/j.ctv111jh6w
Bern, C. 1993. "Risk Factors for Mortality in the Bangladesh Cyclone of 1991." *Bulletin of World Health Organization* 71 (1): 73–78.
Beymer-Farris, Betsey A., Thomas J. Bassett, and Ian Bryceson. 2012. "Promises and Pitfalls of Adaptive Management in Resilience Thinking: The Lens of Political Ecology." In *Resilience and the Cultural Landscape*, edited by T. Plieninger and C. Bieling, 283–299. Cambridge: Cambridge University Press.
Biasutti, Michela, and Giannini Alessandra. 2006. "Robust Sahel Drying in Response to Late 20th Century Forcings." *Geophysical Research Letters* 33: L11706. doi:10.1029/2006GL026067.
Blaikie, Piers. 1985. *The Political Economy of Soil Erosion in Developing Countries*. London: Longman Press.
Blaikie, Piers, Terry Cannon, Ian Davis, and Ben Wisner. 1994. *At Risk: Natural Hazards, People's Vulnerability, and Disasters*. London: Routledge.
Bodansky, Daniel. 2020. "Top 10 Developments in International Environmental Law: 1990–2020." In *Yearbook of International Environmental Law*. Oxford: Oxford University Press. doi:10.2139/ssrn.3644557.
Borras, Saturnino M., Ian Scoones, Amita Baviskar, Marc Edelman, Nancy Lee Peluso, and Wendy Wolford. 2022. "Climate Change and Agrarian Struggles: An Invitation to Contribute to a JPS Forum." *Journal of Peasant Studies* 49(1): 1–28.
Bourdieu, Pierre. 1997. La précarité est aujourd'hui partout: Intervention lors des Rencontres européennes contre la précarité. Grenoble. 12–13 décembre 1997. Contre-feux, Ed. Liber Raisons d'agir, Grenoble.
Brottem, Leif, and Bonnie Brooks. 2018. "Crops and Livestock Under the Sun: Obstacles to Rural Livelihood Adaptations to Hotter Twenty-First-Century Temperatures in Eastern Senegal." *Land Degradation & Development* 29 (1): 118–126.
Bryant, Levi R. 2011. *The Democracy of Objects*. London: Open Humanities Press.
Bullard, Robert D., and Beverly Wright, eds. 2009. *Race, Place, and Environmental Justice After Hurricane Katrina: Struggles to Reclaim, Rebuild, and Revitalize New Orleans and the Gulf Coast*. Boulder: Westview Press.
Butler, Judith. 2009. *Frames of War: When Is Life Grievable?* New York: Verso.
Calebresi, Guido. 1975. "Concerning Cause and the Law of Torts: An Essay for Harry Kalven, Jr." *The University of Chicago Law Review* 43: 69–108.
Carruth, Lauren. 2021. *Love and Liberation: Humanitarian Work in Ethiopia's Somali Region*. Ithaca: Cornell University Press.
Castree, Noel. 2014. "The Anthropocene and Geography I: The Back Story." *Geography Compass* 8 (7): 436–449.
CEDMHA (Center for Excellence in Disaster Management and Humanitarian Assistance). 2007. "Cyclone Sidr Update." November 15. http://www.coedmha.org/Bangladesh/Sidr11152007.htm.
Chomsky, Noam. 2018. *What Kind of Creatures Are We*. New York: Columbia University Press.
Colette, April L. 2016. "The Politics of Framing Risk: Minding the Vulnerability Gap in Climate Change Research." *World Development Perspecitves* 1: 43–48.
Connelly, William. 2013. *The Fragility of Things*. Durham: Duke University Press.
Cottier, Fabien, Marie-Laurence Flahaux, Jesse Ribot, Richard Seager, and Godfreyb Ssekajja. 2022. "Cause and Effect in Climate-related Migration," *World Development*. Article Number 106016. Accepted for publication 22 June 2022.
Cronon, William. 1992. "A Place for Stories: Nature, History, and Narrative." *The Journal of American History* 78 (4): 1347–1376.
Davis, Mike. 2001. *Late Victorian Holoausts: El Niño Famines and the Making of the Third World*. London: Verso.
Davis, Janae, Alex A. Moulton, Levi Van Sant, and Brian Williams. 2019. "Anthropocene, Capitalocene, Plantationocene?: A Manifesto for Ecological Justice in an Age of Global Crises." *Geography Compass* 13: e12438.

de Castro, Josué. 1952. *The Geography of Hunger*. Boston: Little, Brown and Company.

de Sherbinin, Alex. 2020. "Impacts of Climate Change as Drivers of Migration" *Migration Information Source* October 23 Issue.

de Waal, Alex. 1997. *Famine Crimes: Politics and the Disaster Relief Industry in Africa*. Oxford: African Rights and International African Institute/James Currey.

Devereux, Stephen. 2000. *Famine in the Twentieth Century*. IDS Working Paper 105. Brighton: Institute for Development Studies.

Douglas, Mary. 1992. *Risk and Blame: Essays in Cultural Theory*. London: Routledge.

Douglas, Mary and Aaron Wildavsky. 1982. *Risk and Culture: An Essay on the Selection of Technological and Enviornmental Dangers*. Berkeley: University of California Press.

Drèze, Jean and Amartya Sen. 1989. *Hunger and Public Action*. Oxford: Clarendon Press.

Durkheim, Émile. 1982. [1st pub. 1895]. Lukes, Steven (ed.). *The Rules of Sociological Method and Selected Texts on Sociology and its Method*. W. D. Halls (trans.). New York: Free Press.

Dzingirai, Vupenyu, Lindiwe Mangwanya, Melissa Leach, Annie Wilkinson, and Sally Bukachi. 2017. "Structural Drivers of Vulnerability to Zoonotic Diseases in Africa." *Philosophical Transactions B* 372: 20160–20169.

Edkins, J. 2002. "Mass Starvations and the Limitations of Famine Theorising." *IDS Bulletin* 33 (4): 12–18.

Einstein, Albert. 1949. "Why Socialism?" *Monthly Review*. May. Viewed 5 Jan 2022 at https://www.update.uu.se/~fbendz/library/why_soc.htm.

Ennis, Robert H. 2012. "Analysis and Defense of Sole Singular Causal Claims." Unabridged Presentation Version for the PSA, University of Illinois, UC 13 November.

Fanon, Frantz. 1986 [1952]. *Black Skin, White Masks*. London: Pluto Press.

Farmer, Paul. 2004. "An Anthropology of Structural Violence." *Current Anthropology* 45 (3): 305–325.

Ferrarese, Estelle. 2016. *Vulnerability and Critical Theory*. Leiden: Brill.

Fine, Ben. 1997. "Entitlement Failure." *Development and Change* 28: 617–647.

Firth, Raymond, ed. 1959. *Man and Culture: An Evaluation of the Work of Brownislaw Malinowski*. London: Routledge and Kegan Paul.

Forsyth, Timothy. 2008. "Political Ecology and the Epistemology of Social Justice." *Geoforum* 39: 756–764.

Fox, Jonathan. 2022. *Accountability Keywords. Accountability Working Paper 11*. Washington, DC: Accountability Research Center.

Franke, Richard W., and Barbara H. Chasin. 1980. *Seeds of Famine: Ecological Destruction and the Development Dilemma in the West African Sahel*. Montclair, NJ: Allanheld, Osmun.

Fraser, Nancy. 2021. "Climates of Capital: For a Trans-Environmental Eco-Socialism." *New Left Review* 127: 94–127.

Galtung, Johan. 1969. "Violence, Peace, and Peace Research." *Journal of Peace Research* 6: 167–191.

Gardiner, Stephen M. 2011. *A Perfect Moral Storm: The Ethical Tragedy of Climate Change*. Oxford: Oxford University Press.

Giddens, Anthony. 1999. "Risk and Responsibility." *The Modern Law Review* 62 (1): 1–10.

Government of Bangladesh. 2008 April. "Cyclone Sidr in Bangladesh: Damage, Loss, and Needs Assessment for Disaster Recovery and Reconstruction." *Prepared by the Government of Bangladesh Assisted by the International Development Community with Financial Support from the European Commission*.

Guldi, Jo, and David Armitage. 2014. *The History Manifesto*. Cambridge: Cambridge University Press.

Harden, M., N. Walker, and K. Akuno. 2007. *Hurricane Katrina: Racial Discrimination and Ethnic Cleansing in the United States in the Aftermath of Hurricane Katrina*. A Response to the 2007 Periodic Report of the United States of America.

Hart, H. L. A., and A. M. Honoré. 1959. *Causation in the Law*. Oxford: Oxford University Press.

Hayes, Ashley. 2009. Army Corps of Engineers Liable for Katrina Flooding. CNN. http://www.cnn.com/2009/US/11/18/louisiana.katrina.lawsuit/index.html.

Hecht, Susanna. 1990. *The Fate of the Forest Developers, Destroyers, and Defenders of the Amazon*. Chicago: University of Chicago Press.

Hornborg, Alf. 2017. "Dithering While the Planet Burns: Anthropologists' Approach to the Anthropocene." *Reviews in Anthropology* 46 (2-3): 61–77. doi:10.1080/00938157.2017.1343023.

Horton, Radley M., Alex de Sherbinin, David Wrathall, and Michael Oppenheimer. 2021. "Assessing Human Habitability and Migration: Integrate Global Top-Down and Local Bottom-Up Analyses." *Science* 372 (6548): 1279–1283.

Hulme, Michael. 2011. "Reducing the Future to Climate: A Story of Climate Determinism and Reductionism." *Osiris* 26: 245–266.

Huq, Efadul, and Henry Mochida. 2018. "The Rise of Environmental Fascism and the Securitization of Climate Change." *Projections* 30 March.

IPCC. 2014. *Climate Change 2014: Impacts, Adaptation, and Vulnerability. Part A: Global and Sectoral Aspects.* Contribution of Working Group II to the Fifth Assessment Report of the Intergovernmental Panel on Climate Change [Field, C. B., V. R. Barros, D. J. Dokken, K. J. Mach, M. D. Mastrandrea, T. E. Bilir, M. Chatterjee, K. L. Ebi, Y. O. Estrada, R. C. Genova, B. Girma, E. S. Kissel, A. N. Levy, S. MacCracken, P. R. Mastrandrea, and L. L.White (eds.)]. Cambridge University Press, Cambridge, United Kingdom and New York, NY, USA.

IPCC (Inter-governmental Panel on Climate Change). 2019. *Climate Change and Land: an IPCC Special Report on Climate Change, Desertification, Land Degradation, Sustainable Land Management, Food Security, and Greenhouse Gas Fluxes in Terrestrial Ecosystems* [P. R. Shukla, J. Skea, E. Calvo Buendia, V. Masson-Delmotte, H.-O. Pörtner, D. C. Roberts, P. Zhai, R. Slade, S. Connors, R. van Diemen, M. Ferrat, E. Haughey, S. Luz, S. Neogi, M. Pathak, J. Petzold, J. Portugal Pereira, P. Vyas, E. Huntley, K. Kissick, M. Belkacemi, J. Malley, (eds.)]. In press.

Jamieson, Dale. 2015. "Responsibility and Climate Change." *Global Justice: Theory Practice Rhetoric* 8/2: 23–42.

Jessop, Bob. 1982. *The Capitalist State: Marxist Theories and Methods.* By Bob Jessop. New York: New York University Press.

Jessop, Bob. 2014. "The Strategic-Relational Approach: An Interview with Bob Jessop." December 2. WordPress Interview. Accessed 15 December 2021. https://bobjessop.wordpress.com/2014/12/02/the-strategic-relational-approach-an-interview-with-bob-jessop/.

Jisheng, Yang. 2012. *Tombstone: The Great Chinese Famine 1959–1962.* New York: Farrar, Straus and Giroux.

Kashwan, Prakash. 2020. "American Environmentalism's Racist Roots have Shaped Global Thinking About Conservation." *The Conversation*, September 2. https://theconversation.com/american-environmentalisms-racist-roots-have-shaped-global-thinking-about-conservation-143783.

Kashwan, Prakash, Frank Biermann, Aarti Gupta, and Chukwumerije Okereke. 2020. "Planetary Justice: Prioritizing the Poor in Earth System Governance." *Earth System Governance* 6: 1–5.

Kashwan, Prakash, and Jesse Ribot. 2021. "Violent Silence: Erasing History and Justice in Global Climate Negotiations & Action." *Current History* 120 (829): 326–331.

Kelman, Illan, and Bayer Ahmed. 2020. "Bangladesh has Saved Thousands of Lives from a Devastating Cyclone – Here's How." *The Conversation* June issue. https://theconversation.com/bangladesh-has-saved-thousands-of-lives-from-a-devastating-cyclone-heres-how-139903.

Khan, M. R., and J. T. Roberts. 2013. "Adaptation and International Climate Policy." *WIRES Climate Change* 4 (3): 171–189.

Lahsen, Myanna, G. de Azevedo Couto, and I. Lorenzoni. 2019. "When Climate Change is Not Blamed: The Politics of Disaster Attribution in International Perspective." *Climatic Change* 158: 213–233.

Lahsen, Myanna, and Jesse Ribot. 2021. "Politics of Attributing Extreme Events and Disasters to Climate Change." *WIRES Climate Change* 13: 1–12.

Lakoff, George. 2010. "Why It Matters How We Frame the Environment." *Environmental Communication* 4 (1): 70–81.

Lappé, Francis Moore. 2013. "Beyond the Scarcity Scare: Reframing the Discourse of Hunger with an Eco-Mind." *Journal of Peasant Studies* 40 (1): 219–238. doi:10.1080/03066150.2012.708859.

Latour, Bruno. 2004 [1999]. *Politics of Nature. How to Bring the Sciences Into Democracy, Catherine Porter (Trans.).* Cambridge, MA: Harvard University Press.

Latour, Bruno. 2005. *Reassembling the Social: An Introduction to Actor-Network-Theory.* Oxford: Oxford University Press.

Law, John. 1999. "After ANT: Complexity, Naming and Topology." *The Sociological Review* 47: 1–14.

Leibniz, Gottfried Wilhelm. 1710. *Théodicée [Theodicy]: Essays on the Goodness of God, the Freedom of Man and the Origin of Evil.*
Lemos, Maria Carmen, and Emily Boyd. 2010. "The Politics of Additionality Across Scales: The Implications of Additionality to Policy Choice and Development." In *The Politics of Climate Change: A Survey*, edited by Maxwell Boykoff, 96–110. London: Rutledge.
Levis, Carolina, Bernardo M. Flores, Priscila A. Moreira, Bruno G. Luize, Rubana P. Alves, Juliano Franco-Moralse, Juliana Lins, et al. 2018. "How People Domesticated Amazonian Forests." *Frontiers in Ecology and Evolution: Agroecology* 5 (171): 1–21.
Li, Tania Murray. 2007. "Practices of Assemblage and Community Forest Management." *Economy and Society* 36 (2): 263–293.
Locke, John. 1960 [1689]. *Second Treatise of Government*. Cambridge: Cambridge University Press.
Lund, Christian. 2014. "Of What is This a Case? Analytical Movements in Qualitative Social Science Research." *Human Organization* 73 (3): 224–234.
Lustgarten, Abraham. 2020. "The Great Climate Migration." Accessed 30 November 2020. https://www.nytimes.com/interactive/2020/07/23/magazine/climate-migration.html.
Majid, Nisar, and Stephen McDowell. 2012. "Hidden Dimensions of the Somalia Famine." *Global Food Security* 1 (1): 36–42. doi:10.1016/j.gfs.2012.07.003.
Malm, Andreas, and Alf Hornborg. 2014. "The Geology of Mankind? A Critique of the Anthropocene Narrative." *The Anthropocene Review* 1 (1): 62–69.
Mamdani, Mahmood. 1996. *Citizen and Subject: Contemporary Africa and the Legacy of Late Colonialism*. London: Routledge.
Mamdani, Mahmood. 2020. *Neither Settler nor Native: The Making and Unmaking of Permanent Minorities*. Cambridge: Harvard University Press.
Marvel, Kate, Michela Biasutti, and Céline Bonfils. 2020. "Fingerprints of External Forcings on Sahel Rainfall: Aerosols, Greenhouse Gases, and Model Observation Discrepancies." *Environmental Research Letters* 15: 084023.
Marx, Karl. 1894. *Capital Volume III: The Process of Capitalist Production as a Whole*. New York: International Publishers.
MFDMB (Ministry of Food and Disaster Management of Bangladesh). 2008. *Super Cyclone Sidr 2007: Impacts and Strategies for Interventions*. Dhaka: Bangladesh Secretariat. http://www.ecologyandsociety.org/vol15/iss3/art11/.
Morrissey, James. 2014. "Addressing the Political Deficit in Climate Adaptation." *Aljazeera*, 19 May. http://www.aljazeera.com/indepth/opinion/2014/05/climate-adaptation-africa-2014519103948475231.html.
Mostafanezhad, Mary. 2020. "Covid-19 is an Unnatural Disaster: Hope in Revelatory Moments of Crisis." *Tourism Geographies* 22 (3): 639–645.
Nally, David P. 2011. *Human Encumbrances: Political Violence and the Great Irish Famine*. Notre Dame: University of Notre Dame Press.
Nietzsche, Friedrich. 1882. *The Gay Science*, edited by Walter Kaufmann, 181–182. New York: Vintage.
Nietzsche, Friedrich. 2003. *Twilight of the Idols and The Anti-Christ*. New York: Penguin Books.
Nixon, Rob. 2011. *Slow Violence and the Environmentalism of the Poor*. Cambridge: Harvard University Press.
O'Keefe, Phil, Ken Westgate, and Ben Wisner. 1976. "Taking the Naturalness Out of Natural Disasters." *Nature* 260: 566–567.
Oliver-Smith, Anthony. 2013. "A Matter of Choice (Editorial)." *International Journal of Disaster Risk Reduction* 3: 1–3.
Ortega y Gasset, José. 1940. *El Libro de las Misiones*. Buenos Aires.
Patel, Raj, and Jason W. Moore. 2017. *A History of the World in Seven Cheap Things: A Guide to Capitalism, Nature and the Future of the Planet*. Berkeley: University of California Press.
Pelling, Mark. 2003. *The Vulnerability of Cities: Natural Disasters and Social Resilience*. London: Earthscan.
Polanyi, Karl. 1944. *The Great Transformation*. New York: Rinehart.
Pope Francis. 2015. *Encyclical Letter Laudato Si' of the Holy Father Francis on Care for our Common Home*. Rome: Vatican Press.

Ranganathan, Malini. 2017. *The Environment as Freedom: A Decolonial Reimagining*. Social Science Research Council. https://items.ssrc.org/just-environments/the-environment-as-freedom-a-decolonial-reimagining/.

Ribot, Jesse. 1995. "The Causal Structure of Vulnerability: Its Application to Climate Impact Analysis." *GeoJournal* 35 (2): 119–122.

Ribot, Jesse. 2011. "Vulnerability Before Adaptation: Toward Transformative Climate Action' From Affirmative to Transformative Climate Action." *Global Environmental Change* 21: 4.

Ribot, Jesse. 2014. "Cause and Response: Climate Vulnerability in the Anthropocene." *Journal of Peasant Studies* 4 (5): 667–705.

Ribot, Jesse. 2018. "Ontologies of Occlusion in the Anthropocene." *Current Anthropology* 59: 2. Comment on Andrew M. Bauer and Erle C. Ellis "The Anthropocene Divide: Obscuring Understanding of Social-Environmental Change."

Ribot, Jesse. 2019. "Social Causality of our Common Climate Crisis: Toward a Sociodicy for the Anthropocene." In *Commons in a Glocal World: Global Connections and Local Responses*, edited by Tobias Haller, Thomas Breu, Tine De Moor, Christian Rohr, and Heinzpeter Znoj, 34–53. London: Routledge.

Ribot, Jesse, Papa Faye, and Matthew Turner. 2020. "Climate of Anxiety in the Sahel: Emigration in Xenophobic Times." *Public Culture* 32 (1): 45–75.

Rudiak-Gould, Peter. 2015. "The Social Life of Blame in the Anthropocene." *Environment and Society* 6: 48–65.

Russell, Bertrand. 1924. *ICARUS or The Future of Science*. New York: E. P. Dutton & Company.

Sahlins, Marshall. 1972. *Stone Age Economics*. London: Routledge.

Schwartz, John. 2019. What Makes a Catastrophic Flood? And Is Climate Change Causing More of Them? *The New York Times*. https://www.nytimes.com/2019/03/22/climate/what-causes-floods.html.

Scoones, Ian. 2019. *What is Uncertainty and Why Does it Matter?* STEPS Center Working Paper 105, IDS Sussex.

Scott, James C. 1976. *The Moral Economy of the Peasant: Rebellion and Subsistence in Southeast Asia*. New Haven: Yale University Press.

Sen, Amartya K. 1980. "Famines." *World Development* 8 (9): 613–621.

Sen, Amartya K. 1981. *Poverty and Famines: An Essay on Entitlement and Deprivation*. Oxford: Oxford University Press.

Shivji, Issa G. 2020. "Trying Times." *Naked Punch*. Accessed 17 October 2021. http://nakedpunch.com/articles/338.

Simard, Suzanne. 2021. *Finding the Mother Tree: Uncovering the Wisdom and Intelligence of the Forest*. New York: Alfred Knopf.

Smith, Neil. 2006. "There is No Such Thing as a Natural Disaster." SSRC. http://understandingkatrina.ssrc.org/Smith/.

Smucker, Thomas A., Ben Wisner, Adolfo Mascarenhas, Pantaleo Munishi, Elizabeth Wangui, Sinha Guarav, Daniel Weiner, Charles Bwenge, and Eric Lovell. 2015. "Differentiated Livelihoods, Local Institutions, and the Adaptation Imperative: Assessing Climate Change Adaptation Policy in Tanzania." *Geoforum* 59: 39–50.

Soloway, Jacqueline S. 1994. "Drought as a 'Revelatory Crisis': An Exploration of Shifting Entitlements and Hierarchies in the Kalahari, Botswana." *Development and Change* 25: 471–495.

Somers, Margaret R. 2008. *Genealogies of Citizenship, Markets, Statelessness, and the Right to Have Rights*. Cambridge: Cambridge University Press.

Swift, Jeremy. 1989. "Why are Rural People Vulnerable to Famine?" *IDS Bulletin* 20 (2): 8–15.

Taddei, Renzo. 2008. "Blame: The Hidden (and Difficult) Side of the Climate Change Debate." *Anthropology News*, pp. 45–6.

Temudo, Marina Padrão, and Ana I. R. Cabral. 2021. "Climate Change as the Last Trigger in a Long-Lasting Conflict: The Production of Vulnerability in Northern Guinea-Bissau, West Africa." *Journal of Peasant Studies*. https://www.tandfonline.com/eprint/3ZVQ24AZRMFPSSER3IK4/full?target=10.1080/03066150.2021.1996355

Thomas, Adelle, and Rueanna Haynes. 2020. "Black Lives Matter: The Link Between Climate Change and Racial Justice." *The Climate Analytics Blog*. 22 June. https://climateanalytics.org/blog/2020/black-lives-matter-the-link-between-climate-change-and-racial-justice/.

Thompson, Francis. 1913. *The Mistress of Vision*. Accessed 16 October 2021. https://quotepark.com/quotes/1894388-francis-thompson-thou-canst-not-stir-a-flower-without-troubling-o/.

Trenberth, Kevin E., J. T. Fasullo, and T. G. Shepherd. 2015. "Attribution of Climate Extreme Events." *Nature Climate Change* 5: 725–730.

Turner, B. L. 2010. "Vulnerability and Resilience: Coalescing or Paralleling Approaches for Sustainability Science?" *Global Environmental Change* 20 (4): 570–576.

Turner, B. L., P. A. Matson, J. J. McCarthy, R. W. Corell, L. Christensen, N. Eckley, G. K. Hovelsrud-Broda, et al. 2003a. "A Framework for Vulnerability Analysis in Sustainability Science." *Proceedings of the National Academy of Sciences US* 100: 8074–8079.

Turner, B. L., P. A. Matson, J. J. McCarthy, R. W. Corell, L. Christensen, N. Eckley, G. K. Hovelsrud-Broda, et al. 2003b. "Illustrating the Coupled Human-Environment System for Vulnerability Analysis: Three Case Studies." *Proceedings of the National Academy of Sciences US* 100: 8080–8085.

Turner, Matthew D. 2016. "Climate Vulnerability as a Relational Concept." *Geoforum* 68: 29–38.

Voltaire. 1931. *[1759]. Candide: Ou, L'optimisme*. Paris: André Morize.

Wainwright, Joel. 2010. "Climate Change, Capitalism, and the Challenge of Transdisciplinarity." *Annals of the Association of American Geographers* 100 (4): 983–991.

Watts, Michael J. 1983. *Silent Voice: Food Famine and Peasantry in Northern Nigeria*. Berkeley: University of California Press.

Watts, Michael. 1991. "Entitlements or Empowerment? Famine and Starvation in Africa." *Review of African Political Economy* 51: 9–26.

Watts, Michael J., and Hans Bohle. 1993. "The Space of Vulnerability: The Causal Structure of Hunger and Famine." *Progress in Human Geography* 17 (1): 43–68.

Weber, Max. 1968. *Economy and Society: An Outline of Interpretive Sociology*. New York: Bedminster Press.

Weheliye, Alexander G. 2014. *Habeas Viscus: Racializing Assemblages, Biopolitics, and Black Feminist Theories of the Human*. Durham: Duke University Press.

White House. 2006. *The Federal Response to Hurricane Katrina*. Accessed on 5 December 2008. http://www.whitehouse.gov/reports/katrina-lessons-learned.pdf.

Wilkinson, Iain. 2010. *Vulnerability in Everyday Life*. London: Routledge.

Wisner, Ben, Piers M. Blaikie, and Terry Cannon. 2004. *At Risk: Natural Hazards, People's Vulnerability and Disasters*. London: Routledge.

Wolf, Eric. 1981. *Europe and the People Without History*. Los Angeles: University of California Press.

Young, Ed. 2021. "We're Already Barreling Toward the Next Pandemic." *The Atlantic* September Issue. Accessed 17 October 2021. https://www.theatlantic.com/health/archive/2021/09/america-prepared-next-pandemic/620238/.

OPEN ACCESS

Climate change and class conflict in the Anthropocene: sink or swim together?

Murat Arsel

ABSTRACT

Class is key to understanding the genesis and impacts of climate change. Nevertheless, it is commonly argued that 'we are all in the same ship', suggesting that emerging climate politics will not be conflictual along class lines. This paper demonstrates that (agrarian) political economy and political ecology scholars have not adequately scrutinized the relevance of class to contemporary environmental politics to counteract such claims. It also briefly considers two questions – can there be progress without conflict? and can there be conflict without an enemy? – before calling for the development of a Marxist theory of environmental conflicts in the Anthropocene.

I. The same ship

While recent criticisms have shown that the Anthropocene is not merely a geological concept and that it needs to be apprehended within a historical political economy framework (Bonneuil and Fressoz 2016), contemporary discussions around how best to transition to sustainability continue to appeal for united action from humanity as a whole. This is especially the case within the context of climate change. Since dramatic sea level rises are one of its most spectacular impacts, it is not surprising that the metaphor of humanity traveling 'in the same ship' has emerged as a common image. This trope is often deployed by prominent international actors such as Kofi Annan to argue that 'we are all at risk' and that we will 'sink or swim together' (Sherriff 2015).

The 'same ship' metaphor has been challenged forcefully for two related reasons which show that climate change cannot be understood without explicit reference to socio-economic inequalities (Schmitz and Scoones 2015). On the one hand, the passengers have differing responsibilities in terms of their roles in creating the problem. This has both a historical dimension (much of the existing accumulation is due to the greenhouse gas emissions of already industrialized countries) and one of wealth (a billionaire from Beijing is responsible for much more emissions than a homeless person from London). On the other, adaptation to climate change is 'intrinsically spatial' (Shi et al. 2016, 132).

A poor Dutch woman might be better protected than a rich Bangladeshi one given the unevenness in the capabilities of their respective countries. It is of course not only one's national location that matters – gender, race and ethnicity contribute to the inequalities in the vulnerability to the effects of climate change (Adger 2006), as demonstrated in the experience of poor African Americans in New Orleans during Hurricane Katrina (Daniels et al. 2006). As recent work in agrarian political economy and political ecology has documented extensively (e.g. Andreucci and Zografos 2022; Bratman, Auch, and Stinchfield 2022; Camargo 2022; Newell 2022), the ship might be one but it is structured along class lines. Poor and marginalised groups who have done little to create the problem and stand to suffer most significantly from the effects are in lower decks. Affluent and powerful groups, who have not only played the leading role in bringing about climate change but can also adapt more freely and dictate climate policies to their benefit, can be found in first class cabins.

If the genesis of climate change and the distribution of its impacts need to be understood by reference to class positions, what about societal responses? More precisely, is it possible to read social conflicts over the environment in general and climate change in particular from a similar Marxist perspective? This is far from clear because there is a fundamental difference between 'traditional' and 'ecological' distribution conflicts, which emanates from the materiality of climate change in the Anthropocene. Whereas there is no material interest for the members of the bourgeoisie to fight against most of the myriad ills of the capitalist mode of production, they do have a vested interest in overcoming its environmental blind spots because, without genuine and dramatic changes, the negative impacts of climate change are likely to be catastrophic at a planetary scale (Beck 2015). Similarly, contemporary anti-capitalist movements do not have the luxury of 'cheering on a superstorm' the way they might a strike or a sit-in (Malm 2018, 207). To take the trope of the common ship one final step, is it possible to expect that the passengers could act (or are already acting) beyond class lines since, in the words of the autonomist Marxist Amadeo Bordiga, 'if the third class and the crew are not safe, the superior class, that paid stupendous passage fares, is not safe either' (Bordiga 1956; also see Swyngedouw 2013)?

For Swyngedouw, thinking beyond class within the context of the environment is unrealistic as starkly illustrated by the example of the Titanic, where a 'large number of the first-class passengers found a lifeship; the others were trapped in the belly of the beast' (2013, 17). In other words, the ship might be one but the politics and possibilities of the passengers will differ along existing lines of socio-economic inequality, obviating the possibility of cross-class alliances. Chakrabarty, however, has suggested that such a reading is limited by its over-reliance on outdated concepts and behavioural patterns drawn from the study of societal conflicts relating to capitalism. In the Anthropocene, cognizance of the gravity of climate change could lead humanity to see past the cabin structure of the ship, to 'think of humans on different scales and in different contexts' (2017, 25) and possibly transcend the lines of socio-economic inequality to tackle the problem collectively. Thus, Chakrabarty asserts that 'there are no lifeships here for the rich and the privileged' (Chakrabarty 2009, 221) and that the 'politics of climate change is more than the politics of capitalism' (Chakrabarty 2017, 25). The tension between these two positions points toward an important yet surprisingly undertheorised question: what is the relationship between climate change politics and class?

This question in many ways is a precursor to the one asked in the framing paper of this special issue concerning the five types of strategic logics of anti-capitalist struggle codified by Wright (2019, 38–64) and the concrete shape they will take in practice: 'Who ... are the potential social forces behind such political struggles?' (Borras et al. 2022, 16). The primary ambition of this paper is not to answer these questions as this would require a much larger and empirically grounded analysis. Rather, by critically rereading (agrarian) political economy and political ecology literatures as well as a more disparate collection of environmental studies research, it analyses to what extent and how the class nature of environmental conflicts in general and political responses to climate change in particular has been investigated and theorised.

It argues that neither stream of literature has pursued this line of inquiry with sufficient rigour and that we lack a Marxist theory of conflict that can explain environmental conflicts in general and emerging climate conflicts in particular. Agrarian political economy literature, which has long been the *locus classicus* of class-centred analyses in the social sciences, has not fully grappled with the implications of the Anthropocene in general and the political implications of catastrophic climate change. Political ecology, another scholarly tradition that is particularly attune to inequality and marginalisation, has not advanced our understanding of class politics in a systematic way either. This is partly because the primary focus of first-generation political ecology has been to demonstrate that the experience of (rather than responses to) environmental change is class-based and the central thrust of second generation has been to show that not just class but also other dimensions of socio-political and cultural differentiation also matter in understanding how the impacts are manifested. In both cases, the relationship between class and political mobilization in response to climate change (including but potentially going beyond) conflict has been left undertheorised.

A third body of work, which draws primarily from sociology and political science, has in fact engaged far more directly with the class nature of environmental politics (including climate change) but very little of this is from a Marxian understanding class. Taken together, the paper argues that there remains work to be done in understanding if and how class dynamics shape political responses to climate change. While this is a fascinating intellectual puzzle – arguably the most important one facing Marxist and Marxian scholarship in the era of the Anthropocene – it is also massively important for political action. As Bernstein put it cogently, 'understanding class dynamics should always be a point of departure and a central element of' (2010, 123) transformative activism.

In the next section, the paper situates this challenge within a broader intellectual context by discussing the re-emergence of natural limits as a significant conceptual and empirical factor and the overall decline of class analysis in social sciences and environmental studies' relationship with this decline. It also provides a very broad framework for a Marxist approach to class, which forms the basis of the subsequent section. The focus there is on three related streams of scholarship: ecological Marxism; political economy of nature; and environmental politics. Building on a critical re-reading of this vast terrain of scholarship, the paper articulates two questions that can guide further thinking on class and climate change: Can there be progress without social conflict? And can there be conflict without an enemy? The brief final section relates these questions back to the theoretical puzzle set out in the introduction, concluding on a question of urgent theoretical and practical significance.

II. Capitalism, limits and conflict

The theoretical tension regarding if and how class dynamics shape climate change politics is best apprehended within the context of the Anthropocene, or more specifically, in the context of the growing awareness that human activities on the planet can now be apprehended in geological terms. It is clear that capitalism has been defining humanity's relationship with nature for at least a few centuries. The most important implication of the growing awareness of the Anthropocene is arguably the re-emergence of ecological limits as a key political signifier. Shorn off its 1970s neo-Malthusian associations, the concept has now regained its significance in various guises. Martinez-Alier's work on the intensification of the social metabolism (2009), the concept of 'planetary boundaries' (Rockström et al. 2009) and Kallis' (2019) recent and more direct attempt to reclaim limits for progressive purposes all point towards the acceptance that humanity's relationship with nature is (and has been for some time) in a crisis circumscribed by what is physically available for human use.

This crisis signals the bankruptcy of capitalist modernity's cornucopian promise of material prosperity that can be shared with an ever-increasing share of humanity. Looking past the smokescreen of the idea of reforming capitalism – whether it is couched in the language of ecological modernization (Jänicke 2008), circular economy (Korhonen, Honkasalo, and Seppälä 2018), or the Singularity (Morris 2010) – would show that distributional concerns can only be displaced but not fundamentally eliminated. In other words, while the global expansion of capitalism was made possible through various political, economic, and cultural maneuvers – colonialism, class-compromise of social democracy; spatio-temporal fixes of footloose capital; glorification of consumerism; etc. – the challenge posed by the concrete reality of natural limits in the Anthropocene resignifies the centrality of distributional conflicts. Consequently, the value of the environment in its various guises – as resources, as services, as focus of spirituality – has come to be refracted through its finiteness in the Anthropocene and struggles over it can either replace or supplement 'traditional' conflicts, those that are over the distribution of surplus value created by labour. To this end, Martinez Alier and O'Connor's designation of contemporary environmental politics as ecological distribution conflicts is as accurate as it is prescient (Martinez Alier and O'Connor 1995).

But are these ecological distribution conflicts analytically similar enough to traditional conflicts so that they too can be subjected to the same type of class analysis? While there are some obvious parallels between the uneven and unjust appropriation of value produced by human labour and the value produced by the labour of nature (Parenti 2015; also see Battistoni 2017), this is primarily true in the sense of 'natural resources' such as timber, oil, and extractive goods in general. However, the picture becomes far more complicated if we also consider ecological 'bads' (ecological services are yet another complication). Some of the 'bads', for example, the health impacts of an oil spill from a broken pipeline, do accrue to a specific group such as an indigenous community in the Amazon that could perhaps be collectively conceptualized in class-specific terms (e.g. Arsel, Pellegrini, and Mena 2019). However, others might defy spatial and temporal boundaries to such an unprecedent extent, as in the case of nuclear power or biotechnology, that their impact can potentially also evade boundaries of class in terms of who experiences them (Goldblatt 1996).

However, the real difference between traditional distributional conflicts and ecological distribution conflicts in the Anthropocene might emerge when looking at political responses to potentially existential threats. Chakrabarty's argument resonates here since even according to the dictates of neoclassical economics, rational individuals – including members of the bourgeoisie – would be expected to prioritize the survival of essential planetary ecosystems (if not human species as a whole) over the survival of capitalism so as to ensure their own personal safety. Thus, the joint involvement of distinct classes – including those who are organically implicated in the creation of the very environmental crisis in question – or the potential obsolescence of a Marxist notion of class in such instances cannot by definition be explained away either as green washing, bourgeois false consciousness, altruism or as a form of environmentalism that is a vehicle to achieve various other political goals (Arsel, Akbulut, and Adaman 2015). The analytical utility of class in explaining political struggles is, therefore, very much in question when it comes to environmental problems such as climate change that are, or can be construed as being, linked to planetary survival.

It is important to note that environmental politics is not the only domain in which the currency of class as an analytical concept has been questioned. There is an ongoing if slow-burning discussion regarding the diminished prevalence of class analysis in the social sciences as a whole that aims to articulate how the concept can be revitalized (e.g. Wright 1996; Grusky and Sorensen 1998; Crompton and Scott 1999; Davis 2013). Responses to the decline of class analysis, on which there seems to be little doubt, register several commonalities. For instance, there is incredulity that at this stage of late capitalism, where patterns of inequality especially within nations but across the globe as a whole have deepened and seem to be on the verge of ossifying, the analytical tool specifically and normatively built around the elimination of inequality has waned in influence. While debates regarding socioeconomic inequality have reached the mainstream (Deaton 2013; Saez and Zucman 2016; Piketty 2017), attempts at recognizing it structurally as part of the logic of capitalism seem to be lagging far behind.

Just as perplexing to the participants of this debate is the type of analytical approaches that have come to dominate instead of Marxist class analysis. On the one hand, the hegemony of neoclassical thought has resulted in the imperialism of economics. No longer content to be practicing one of the social sciences, mainstream economists aim to subsume all social scientific inquiry under their methodological and ideological dictates. This unfortunately effective strategy has been possible by perpetrating the dominance of the fictitious and consciously apolitical social actor called the rational individual (Akbulut, Adaman, and Madra 2015). On the other, post-modern and post-structural approaches, despite paying lip service to broader and historical inequalities unleashed by the rise of capitalism, are unable to replace the task carried out by Marxist class analysis. This is not to argue that race, gender, and ethnicity are unimportant but that their fundamental malleability makes them unsuitable for grounded political economy analysis on their own (Chibber 2006). This is especially so since the persistent focus of post-modern and post-structural theory on the limitations of structural explanations unwittingly contributes to accounts that privilege if not the individual then the potentially just as vacuous concept of community. Put differently, the type of approach that has come to fill the void left by class analysis is ultimately apolitical regardless of its provenance since it fails to apprehend social change through the transformation of underlying economic structures.

While decline of class analysis overall and its limited deployment in environmental conflicts, in particular, is notable, the irony of the fact that the acceptance of the environmental problematique as a central concern for political economy was only possible because of the challenges to the validity of orthodox class analysis is inescapable. Concern for the environment, as with concern for gender, race and ethnicity, was often written off from 1960s onwards if not as false consciousness then as being a second-order concern that could be dealt with once the struggle between the proletariat and the bourgeoisie was transcended. To the extent that environmental studies in general and political ecology, in particular, were able to rise to scholarly significance, this was necessarily achieved at the expense of orthodox Marxism that left little room for anything but 'traditional' class conflict. It is not surprising, therefore, that 'first generation' political ecology (Biersack and Greenberg 2006), which crystallized in the first edition of Peet and Watts' 'Liberation Ecologies' (Peet and Watts 1996), while aiming to put capitalism squarely within its analytical viewfinder, emerged from a broader attempt at reconciling radical political economy with post-modernist and post-structuralist approaches. Having established the political nature of environmental change by illuminating the ways in which capitalism's relationship with nature feeds off and reproduces socio-economic inequalities, political ecology has since moved toward asking more pointed questions about pathways to constructing an alternative to capitalism. This 'second generation' political ecology, developing within a broader context in which decoloniality has come to be seen as the vanguard of critical inquiry, has moved further away from the concerns of orthodox Marxism. This is mainly because orthodox Marxism and its conceptual toolkit have been charged as being a constituent part of the logic of Eurocentric dominance – economically, politically as well as racially – that need to be systematically dismantled. As such, for much of contemporary political ecology literature, anti-capitalist struggle is no longer expected to be waged through class struggle and analytical attention is consequently focused on socio-cultural markers of difference rather than a materially grounded conceptualisation of class.

As the preceding discussion shows, the decline of the primacy of class analysis itself is overdetermined and was important for the growth of environmental inquiry. Therefore, the critique that follows is not intended as a call for purity of dogma. To the extent that it suggests the absence of rigorous class analysis in the study of environmental change, this is not argued from the vantage point of a strict definition. In fact, given the breadth and depth of debates on the meaning of class within traditional, analytical and post-Marxists (as well those that don't easily fit into these categories), articulating a single definition and treating it 'as if it were the only or the ultimate or the universally agreed kind of class analysis' (Resnick and Wolf 2003, 8) would be a Procrustean exercise.

Nevertheless, it would be useful to sketch out several building blocks of Marxist class analysis that can be used, broadly, to develop the proceeding critique of (agrarian) political economy and political ecology (the discussion of the third stream of literature is centred around other, competing approaches to class, including Weberian and Bourdieusian ones). First, political consciousness (class-in-itself) and action (class-for-itself) emerge from material foundations, though this cannot be reduced simply to the economic and can be conceptualised to include larger societal power relations that can contain social and cultural bases of difference as well. Second, inequality between groups in the distribution of these assets has structural implications to political processes in general and reproduction of the prevailing mode of production in particular. Third, further emphasizing the centrality of inequality, the concept of class is fundamentally relational and experiential. Fourth, capitalism as a mode of production structures all contemporary political relations. Finally, the relationship between different classes is overall and in the long durée, though not necessarily at every given moment, characterised by antagonism.

These contours are offered not as an hortatory effort for definitional purity and more in the spirit of delineating a broadly Marxist reading. As such, they are largely in line with a type of conceptualisation that can also be found in the Communist Manifesto though with certain modifications that draw from more recent interventions, drawing particular inspiration from E.P. Thompson (1971; Sewell 1990) and, to a lesser extent, the contributions of E.O. Wright (1997, 1999). While they are presented in an abstract form, they are intended to reflect the challenges posed by environmental degradation in general and climate change in particular within the context of contemporary global capitalism.

III. Theorizing class and environment

It is within these intellectual and ecological contexts that the relevance of class to societal responses to climate change in particular and environmental degradation in general need to be scrutinized. This is undoubtedly a massive and rapidly expanding body of work. What follows engages primarily with contributions that explicitly focus on the relationship between capitalism (and, in certain cases, capitalism together with modernity) and nature with a view to critically reading their (non-)utilization of class to understand societal responses. For the sake of analytical clarity, the critique of this literature is presented under three headings. While they are handled separately, these streams do of course cross and build on each other.

The first focuses primarily on the logic of capitalism as a mode of production and how its relationship with nature should be conceptualized, especially in relation to the growing awareness that nature can no longer be treated as infinitely abundant (e.g. Burkett 1999). This approach can be described as the political economy of nature. The second is more directly concerned with the impact of capitalism on nature and society and can be subsumed under the joint rubric of agrarian political economy and political ecology (e.g. Pye 2021; Svarstad and Benjaminsen 2020). It focuses especially on the differential impacts of capitalism on specific spaces, natures and communities as well as how these communities respond or fight back (actor network theory influenced variants of this work also focuses on the agency of non-human entities, e.g. Bennett 2009). While this is a somewhat arbitrary distinction, the first can be seen as a study of capitalism's internal workings vis-à-vis nature and the second of the distribution of its impacts. In different ways, both literatures are concerned with understanding how nature (and society, or for some, socio-natures, e.g. Swyngedouw 1999) is made more amenable for the continuation and deepening of processes of accumulation and how peasant and marginalized communities resist and articulate alternatives. Both of these two streams are built around a commitment to Marxian political economy. The third stream explicitly focuses on societal reactions to environmental change, including but going beyond, conflicts. It is in this stream where class is explicitly and consistently discussed and, perhaps surprisingly, a substantial portion of these contributions come from environmental sociologists that do not write within a Marxist framework.

A. Political economy of nature

This first stream of explicitly Marxist literature on environmental studies can also be described as 'ecological Marxism' (Kovel 1995; Benton 1996; Burkett 1999). The main thrust of this literature was to discover the 'green Marx' by re-reading his writings with the fresh set of eyes provided by the post-World War II boom in environmental consciousness and to rework Marxist concepts to suit explicitly environmentalist analytical ends. James O'Connor's 'second contradiction of capitalism' is emblematic in this regard (1988). The first contradiction is essentially one of overproduction of goods, creating an ever-widening gulf between labour's shrinking ability to consume and capitalism's ever-expanding ability to produce. As this gulf widens, it is expected that the tension would lead to revolutionary change that will result in labour assuming full control of the means of production. O'Connor's second contradiction is one of underproduction, namely the ability of capital to reproduce the natural conditions (which includes ecological as well as human foundations of capitalism) that it needs to thrive. As capitalism continues to demand more and more resources both for the production of goods and absorption of 'bads', it undermines ecosystems' ability to reproduce themselves (which is a crisis that is also intimately linked to the crisis of social reproduction, as richly argued by socialist eco-feminism, e.g. Salleh 1995). O'Connor argues, therefore, that 'there may exist a contradiction of capitalism which leads to an 'ecological' theory of crisis and social transformation' (O'Connor 1988, 14).

O'Connor's contribution was more in terms of the development of a theoretical postulate, whose precise machinations were left for other scholars to describe. How the crisis would come about was best described by the work of John Bellamy Foster, who revived the concept of 'ecological rift', which was coined by Marx (Foster 1999). What Foster and his collaborators have effectively done is to materially illustrate O'Connor's somewhat nebulous theoretical formulation, giving it analytical purchase by linking actual ecological concepts with economic dynamics. To the extent that Foster's conceptual-methodological breakthrough is an essential component in the ecological Marxist toolbox, it concerns the emergence of the material conditions upon which the contradiction would emerge rather than how it would be resolved. To wit, in the eponymous book of 544 pages there are only a handful of direct references to class, none of which actually deal with the social transformation question (Foster, Clark, and York 2010). In short, Foster's approach illustrates well the analytical ambitions of the political economy of nature literature, which does not theorize class-based action in relation to environmental degradation. Nevertheless, O'Connor himself is clearly attuned to the significance of how the second contradiction of capitalism might unleash revolutionary forces. After laying out the analytical foundations of the second contradiction, he makes this bold but problematic assessment:

> The combination of crisis-stricken capitalism externalising more costs, the reckless use of technology and nature for value realisation in the sphere of circulation, and the like, must sooner or later lead to a 'rebellion of nature', that is, to powerful social movements demanding an end to ecological exploitation. (1988, 32)

Putting aside the fundamental anthropocentricism of the argument that 'rebellion of nature' is in fact human social movements speaking on behalf of nature [a critique that the concept of the Anthropocene, where the earth itself emerges as an actant, seems to anticipate Davies (2016)], O'Connor does not at all specify who would comprise these 'powerful social movements'. Demonstrating perhaps that this puzzle is particularly thorny, his concluding thoughts bypass the question as to who comprises these 'powerful social movements' to focus instead on who should not be defining the character of the environmental backlash against capitalism. Responding to the seeming contradiction that the emerging political response to capitalism's crisis has taken the shape of the worryingly 'post-class' new social movements, O'Connor is interested mainly in critiquing the post-Marxist thought of Laclau and Mouffe (1985) as well as Offe (1985). Summarily dismissing the significance of new social movements (as discussed below) by likening them to 'other fringe movements', he predicts that they are 'bound to self-destruct' (1988, 32). What remains from his analysis is the implicit assumption that the 'powerful social movements' brought up by the second contradiction would conform to the class-based features anticipated by the first contradiction. However, as Martinez Alier has argued 'environmental conflicts … do not always correspond to fights between workers and capitalists. Sometimes they do, like pollution in a factory. But quite often the actors are different … who are the protagonists of what O'Connor called the second contradiction of capitalism?' (Pellegrini 2012, 349).

Another problematic aspect of O'Connor's second contradiction thesis is its failure to anticipate that capital (with the assistance of the state) could convert its own crisis into a new accumulation strategy and in so doing mollify emerging societal responses. This is mainly a failure to anticipate that global ecological crisis would become an undeniable fact (the US experience in climate denialism notwithstanding), one that is more convenient to respond to than to ignore. Capital's embrace of the environmental problematic is itself a function partly of the availability of scientific and technological alternatives [e.g. the successful global response to the Ozone layer owes much to the fact that DuPont already had a compound lined up to replace the CFCs, (Maxwell and Briscoe 1997)] and partly of the regulatory possibilities made possible by the neoliberal turn, whose dominance starting from the late 1970s dovetails the mainstreaming of environmentalism.

The literature on neoliberal conservation (e.g. Büscher et al. 2012), responds precisely to the innovative environmental mechanisms that began to emerge in the neoliberal era as nature – *pace* Polanyi (1957) – is transformed into a highly valuable commodity that goes beyond the traditionally traded goods such as oil and timber. While its debt to O'Connor is rarely expressed, neoliberal conservation can be read as an attempt to update his work, primarily by showing how the second contradiction fails to materialize (Arsel and Büscher 2012). The update is needed precisely because social forces had begun – as discussed in the third part of this section – to rise up to demand meaningful action against an increasingly undeniable global ecological crisis. To a lesser but certainly not an insignificant extent, the mounting of environmental problems also came to be a barrier to further accumulation in certain sectors, demonstrating that just as social actors, capitalists too can display a degree of internal heterogeneity in relation to environmental politics (see, for e.g. Paterson 2001 and also the conclusion of this article). Thus, given these economic and political imperatives to respond to the environmental crisis, capital (and the state) moved from deregulation (of the economy so that environmental impacts could be externalized) to reregulation (of nature so that environmental impacts could be internalized as profit opportunities), a transition explicitly and forcefully endorsed by the Stern report (2007) within the context of climate change. This transition made it possible to convert the ability of the atmosphere into a tradable commodity by creating (with the help of the state) tradable emissions markets but also to achieve the commodification of environmental services. In all such instances, the creation of new structures of ownership were decisive in the realization of such 'innovations', which, in a climate of neoliberal multiculturalism fed off existing demands for new forms of territorial sovereignty by marginalized communities such as peasants and indigenous groups. The sum of these transformations made environmentalism a source of profit, creating a new breed of entrepreneurs singing the gospel of win-win-win solutions (Arsel and Büscher 2012). Where these solutions failed to achieve results (which was most often) or failed to convince that the solutions were meaningful, the promise of future scientific breakthroughs continue to be dangled as talismans as exemplified by the massive investment in geoengineering technologies (Buck 2019).

The labour of neoliberal conservation scholars has focused primarily on how these mechanisms were articulated, justified and implemented. Societal responses to them, however, have not been studied systematically. This is not to suggest that conflicts have not been tackled. However, they are treated as case studies demonstrating the unevenness once again of capitalist market mechanisms in creating winners and losers. To the extent that class enters these discussions, it is to suggest (accurately) that neoliberalism is a class project to create renewed domination of labour, peasantry, indigenous people, and other marginalized actors. However, the protagonists are not theorized beyond recording their (usually) negative experiences and describing the shape of their fight to stop or slow down the commodification of their life spaces. More significantly, they are not studied as movements whose class positions and strategies are unpacked but as examples of the fundamental destructiveness of neoliberal conservation. In other words, they assume that the resistance of a given marginalised community is a class struggle since it challenges the expansion and deepening of capitalist dynamics without concretely theorising what the concept of class might signify within the context of, for instance, an indigenous community in the Ecuadorian Amazon or poor farmers in the Turkish countryside. In the absence of careful Marxist engagement with social movements, mainstream political science approaches have come to dominate (Silver and Karatasli 2015). Even a cursory glance at this literature shows that not only class but the concept of capitalism itself is largely absent.

Whereas the neoliberal conservation literature aims to unearth how capitalism's relationship with nature has been evolving to create new and more intensive ways in which the environment could be integrated into the class-project of neoliberal capitalism, Jason Moore's *suis generis* Marxism has opened up an entirely new way to conceptualize the capitalism-nature link. His approach harks back to the earlier generation of ecological Marxism in the sense that Moore is fundamentally concerned with understanding how ecology forms a barrier to capital's future reproduction (Moore 2015). In other words, Moore, contrary to the neoliberal conservation literature whose post-structural undercurrents prevent it from acknowledging the material limits of accumulation, is squarely concerned with them and how these are continuously challenged by transformative social processes. However, departing from first-generation ecological Marxists who saw nature as an external constraint on capital, his innovative move is to reject what he sees as a Cartesian dualist separation between ecology and economy. Rather than capital vs. nature, his construction is capital-in-nature, which is itself a class-relationship since labour (and humans as well as other living and non-living components of nature) are implicated on both sides of this unity.

Moore clearly sees a role for class struggle to overthrow the unity of capital-in-nature. This seems to be transmitted via a Polanyian reclamation of the autonomy of society to give nature as well as society's relationship with it meanings that defy the capitalist logic of value, which sees 'all elements of human and extra-human nature [as] effectively interchangeable' (Moore 2018). This is as clear a description of contemporary struggles – from food sovereignty to land grabbing to commoning – as one could make. Yet, Moore's formulation still does not go far enough in terms of placing class conflict centrally into attempts to transcend capitalism. Moore sees all these conflicts through capitalism's dependence on 'cheap nature' [a theme he develops more extensively with Raj Patel (Patel and Moore 2017)]. Here his acknowledgement of ecological limits comes back to hobble his argument because what he ultimately anticipates the fall of capitalism to be is the ecological limits themselves. To the extent that class conflict figures prominently in his non-dualist vision, he seems to be arguing that class conflict will not only be functional to limits being confronted but, ontologically, animated by the limits themselves. While this technically shows a certain loyalty to an orthodox approach to materialism, it neglects the fact that many of the contemporary environmental struggles are not simply about the material availability of resources but the specific constellation of meanings they are imbued with. Put differently, food sovereignty, for example, is not strictly about the availability of sufficient amount of calories but the specific shape those calories take and authority over how they are created, transported and consumed.

This leads to two related problems for Moore's formulation. His vision of a 'rebellion' against the 'value/monoculture nexus of modernity' is too rigid to include symbolic struggles over nature. More significantly, his over-reliance on limits to be the driving force not only smacks of unwarranted optimism that capital is unable to find ever new ways to displace its impacts and/or co-opt sufficiently powerful communities to secure enough legitimacy to extend the exploitation of nature in novel ways. It also is a type of reverse neo-Smithianism in the sense that as dependency scholars falsely saw the rise of capitalism as a function of intensified trade (Brenner 1977), Moore expects its fall to be an outcome of the inability of the system to keep trading more and more cheap goods.

The contemporary terrain of ecological distribution conflicts, however, are far more complicated than Moore's formulation can capture. Not only the struggles are not necessarily at the edges of systemic limits as such but they also do not necessarily seem to take a class-specific shape. Moore implicitly acknowledges this when he theorizes these struggles as 'the struggle over the relation between humans and the rest of nature is necessarily a class struggle. (But not *just* a class struggle)'(2015, 150; original italics). The thought in brackets works less as a clarification and more as an attempt to postpone serious reckoning with the question as to who exactly will lead the rebellion against capitalism's colonization of nature.

B. Agrarian political economy and political ecology of capitalism

Arguably, the stream of social scientific research that has shown most sustained and serious investment into class analysis in recent decades is agrarian political economy.

This intellectual position is to a large extent a function of this group of scholars' commitment to understanding (and defending) the peasantry, both in terms of its mode of production but also as a socio-political force (Borras 2009; Bernstein and Byers 2001; Bernstein 2010; Friedmann 2019). Given the centrality of the concept of class, there has been much work providing both conceptual nuance to how to differentiate and identify (peasant) classes as well as those enriching these discussions in terms of specific geographic contexts (Cousins 2011; Harriss-White 2018; Lerche and Shah 2018; Schneider 2015).

Space limitations and the *a priori* commitment to the survival of the peasantry [except, in dissenting traditions, such as the Warrenites (Kiely 2009)] are not the only reasons why an extensive discussion of these mostly intra-community debates cannot be fully synthesized here. It is also possible to argue that the ecology question has received scant attention in this body of work (beyond those concentrating around the question of metabolic rift as discussed above). This is not to suggest that environmental change, land degradation and other cognate problems have not been tackled. But these have either focused on the environment primarily as a productive resource (as land, water, etc.) or as another dimension along which marginalisation takes places or is exacerbated rather than class-based *environmentalist* action. This also broadly applies to the agrarian political economy of climate change, which foregrounds the impact of climate change and adaptation processes aimed at mitigating it (e.g. rise of biofuels) rather than its mobilising potential [Raju Das's nearly 700-page entreaty on class analysis, for instance, only mentions climate change once in a foonote (Das 2017)!].

There are of course exceptions to this overall subsumption of ecological concerns under historic lines of defence by the peasantry. The most significant one is centred around the 'ecological agrarian question', the seventh and final one of the contemporary agrarian questions proposed by Akram-Lodhi and Kay. The concept 'requires paying close attention to the ways in which the capital-labour relation shapes and is shaped by the prevailing ecology and as such is about how rural labour processes and ecological processes are intertwined' (Akram-Lodhi 2021, 701). While this certainly gestures in the direction of centralising the link between class and environmentalism, the focus, even when it is explicitly on how environmental problems create 'the preconditions for a set of social and political struggles that promote farming systems' (701) remains inwardly focused and does not question if and how the threat of climate change in the Anthropocene affects the political consciousness of peasants and agriculture producers. Consequently, Akram-Lodhi points towards agro-ecology as a potential solution, which, notwithstanding the strengths of agro-ecology, demonstrates the closed-loop nature of theorising the environmental challenge in the agrarian political economy literature. A similar posture is also evident in Saguin (2016), who is primarily interested in 'articulating the place of nature in agrarian change', Neimark and Healy (2018), who are concerned with how ecological transition dynamics (in this case, the European Union's bioeconomy) created gendered outcomes, and Latorre (2021), who focuses on a broader sets of processes of marginalization. In effect, this more ecological turn in agrarian political economy does not so much theorise the link between class and nature as bringing the literature closer in synch with the political ecology literature (discussed below).

Beyond the ecological agrarian question, agrarian political economy research is gradually taking up the question of class positions and political struggles. Perhaps surprisingly, much of this literature is pointing towards not necessarily the diminishing of the centrality of class positions nor the uniqueness of the peasant class as a bulwark against capitalism's excesses. Rather there is considerable soul searching regarding the need to form cross-class alliances as a precondition for achieving sustainable development. For example, Scoones' expansive review article on the politics of sustainability and development underscores the salience of alliance building (2016). More recently, writing about the protests of Dutch farmers against climate change mitigation measures, van der Ploeg (2020) points in a similar direction by highlighting the multi-class nature of the movement but also warning that such a force could fall prey to the appeals of regressive populism and ultimately fail to serve the needs and interests of peasant producers.

Ultimately, there are signs in the agrarian political economy literature that class and climate change require some reconsideration of certain long-cherished intellectual dogmas (though, of course, the valuable instinct to resist the watering down of key concepts is also alive and well as demonstrated by Pye's (2021) strong defence of the proletariat as the revolutionary class). In their special issue, Levien, Watts, and Yan (2018) highlight the very question at the heart of this paper as one of the generative questions for future work in agrarian political economy, drawing attention to the Marxist connections between climate change and the Anthropocene. Echoing Bernstein's warning that the 'social locations and identities the working poor inhabit, combine and move between make for ever more fluid boundaries and defy inherited assumptions of fixed and uniform notions of 'worker', 'farmer', 'petty trader', 'urban', 'rural', 'employed', and 'self-employed'' (2010, 111), Scoones argues that any future transformation would have to 'draw on an unruly politics, involving diverse knowledges, and multiple actors'(2016, 293). The paper framing this special issue also points in the same direction when it suggests that 'the way agrarian struggles – led by peasants, pastoralists, fisherfolk, rural workers and others – connect with the challenge of climate change, linking to and going beyond the already widespread challenges to expropriation and extraction in rural areas, is a vital focus for both thinking and action' (Borras et al. 2022, 2). In short, recent agrarian political economy has opened the door to rethinking (and reconstructing) the link between class, climate change and agrarian political economy but this work is in its infancy.

Arguably, class is a more explicit component of the now vast literature on political ecology of climate change, though this applies primarily to the distribution of environmental impacts and does not yield a class-centred study of resulting conflicts. Its proponents have built on the ground prepared by early ecological Marxists as well as scholars from other cognate (sub)fields, be it cultural geography, social anthropology, or environmental history. Explicitly targeting power relations, political ecology concerns itself with more than the 'environmental' in the sense that the literature captures the complex interrelations between ecological change and the political economic dynamics surrounding them. The goal of this approach is ultimately to demonstrate how the creation and maintenance of environmental inequalities within capitalism are fundamentally political and interrelated with myriad other spheres such as health, gender, indigeneity, and race.

Increased attention to these attributes within the context of a Marxist framework has certainly done much to correct orthodox Marxism's blind spots. Perhaps in part because of the need to empirically demonstrate how factors beyond class do matter in apprehending the unevenness of capitalist economic dynamics and that they deserve analytical and political support in terms of the conflicts they generate, political ecology has excelled in delivering fine-grained documentation of various conflagrations at the local level. This variegated understanding of capitalism's impact on the ground did also attend to class in addition to race, gender, and ethnicity. However, rather than seeing class as a meta-structure interpenetrating all these attributes, political ecology literature has succumbed to a facile intersectional approach that treats class as simply another marker of difference with no analytical priority.

This reluctance to theorize capitalism's link with nature through class terms also continued in terms of understanding movements of resistance. Here the impact of political ecology's entanglement with both Foucauldian post-structuralism and, more problematically, with post-development thinking has come at the expense of not only willingness to theorize broadly but also ability to recognize structural conditions that animate movements. Instead, political ecology literature has taken community-level acts of resistance as ontologically coherent entities rather than as part of a greater whole (though there are some notable exceptions, e.g. Martinez Alier et al. 2016). To this end, it is possible to see the common refrain of 'fine-grained' inquiry not simply as a methodological choice but an ideological posture that celebrates each movement as unique and context-specific. As such, the analytical terrain for understanding counter-movements in the ecological sphere were ceded to the new social movements and resource mobilization theories as already mentioned above.

This admittedly broad-brush picture of political ecology's relationship to class needs to be qualified in two ways. One concerns Martinez-Alier and Guha's 'environmentalism of the poor' (1997), which is discussed in the next section. The other is the transposition of the concept of class from relationships of production to international distribution of wealth and power. Specifically, in the absence of class as a driving leitmotif in terms of struggles against capital in ecological distribution conflicts, political ecology has transposed it to the global arena. By so doing, rather than apprehending class positions of individuals, political ecology has recognized a class relationship between rich and poor countries in a manner consonant with dependency theory's spatial ordering of the world along centre and periphery. While this view does have much merit in terms of flows of financial resources, corporate ownership structures, and 'othering' of communities and spaces, it cannot fulfil the function of the class analysis that takes social classes as its starting point.

Furthermore, while a robust defence of the concept of periphery is very much possible (Fischer 2015), the division is nevertheless coming under increasing critical scrutiny [as with most divisions that are built around the notion of 'developed' and 'developing' (Horner and Hulme 2017)]. For instance, the literature on environmental justice does show that race as well as class in developed countries correlate with certain exposure to environmental harms in much the same way as it does in the developing world (Nixon 2011). Similarly, the arrival on the development scene of countries such as China whose position in this global class relationship are much harder to capture within existing political ecology conceptualizations (Henderson, Appelbaum, and Ho 2013). Ultimately, it is still possible to assert that while political ecology has demonstrated the unevenness of capitalism's environmental impacts, leading to the possibility of empirically seeing how class is a central concern, it has not done sufficient analytical labour to show how inequality and class in terms of environmental struggles connect together.

This might not simply a problem of misaligned intellectual posture. The environmental conflicts that have been emerging from around the world themselves have been shy of using a language of class themselves. This could be due to various reasons. In certain contexts, the language of environmentalism was deployed consciously and deliberately instead of left-wing repertoires of dissent. For instance, some of the most prominent environmental movements in Turkey were built by former leftist activists, who had suffered heavily under military repression during the 1970s and 1980s (Arsel 2003; Arsel, Akbulut, and Adaman 2015). In others, instead of domestic political considerations, environmentalism was the chosen vehicle of activists because of the expected gains that can be made in the international sphere from relatively mainstream campaigners and non-governmental organizations. The Ogoni struggle in Nigeria remains the epitome of such strategic positioning (Agbonifo 2018). The overall intellectual climate too can shape the way environmental conflicts can downplay their class-specific demands. As mentioned already, within the contemporary political landscape, neoliberalism (and neoliberal multiculturalism, see Hale 2005) from the right and decoloniality from the left create the conditions for the ascension of identity-based demands to public prominence. It is not, surprising, therefore that many Latin American ecological distribution conflicts congregate around the language of indigenous rights (leading to problematic cases where certain groups suddenly and without much cultural or historical foregrounding discover their own indigeneity; see for example, Fernandez-Salvador 2018; Fontana 2014). Faced with such struggles, political ecologists have chosen to replicate the overall political climate by privileging indigenous and other non-class based identities in their conceptualization of ecological distribution conflicts even in cases where class-specific demands do enter the picture (Pellegrini and Arsel 2018).

C. Class dimensions of ecological distribution conflicts

The failure of ecological Marxism as well as agrarian political economy and political ecology to tackle class directly is likely to due to the fact that it is 'much easier to celebrate class struggle than to analyze it' (Moore 2015, 38). In the absence of a theoretically consistent treatment of how class struggle fits into environmental movements confronting capitalism, there has nevertheless emerged a vibrant discussion regarding class and environmentalism in disparate parts of environmental politics and (environmental) sociology. While some of these have developed directly in response to one another (e.g. Martinez Alier and Guha's 'environmentalism of the poor' as a critique of Inglehart's post-materialism), others have emerged as part of sociological debates that arrived at environmental politics by way of a broader questioning of modernity and capitalism (e.g. Beck's reflexive materialism). As such, rather than forming distinct intellectual traditions (such as ecological Marxism or political ecology), they represent disparate strands of inquiry that can be compartmentalized between two opposing camps: those who argue class is not important or at least is marginal to contemporary environmental politics, and those who argue that it is, in different ways. A related third stream of literature that builds on these traditions recognizes the class composition of contemporary struggles but does not necessarily conceive of this dimension in a classically Marxist lens that is inherently conflictual.

i. Post-class environmentalism

The argument that class is not a salient factor in understanding contemporary politics of societal transformation in relation to environmental problems itself can be read in terms of two separate literatures. While both of these focus on the changes wrought by what can be termed 'high modernity' and the consequent rise of a 'postmodern world' (Pichardo 1997), one focuses on changes in social structures ranging from individual self-perception to family ties to state-society dynamics (e.g. Offe 1985) and the other is primarily interested in how both the materiality of nature and its societal perceptions have been transformed (Beck 1992). Both literatures point to the rise of 'new social movements' as a decisive moment for political subjectivities under capitalism (Melucci 1988). In so doing, they bundle ecological politics with other critical political processes that similarly challenge the authoritative scripts of Western modernity on various grounds of identity, be it race, ethnicity or gender. Even if it's not explicitly acknowledged, these analyses suffer from the linearity of much of Western social theory (resulting in such distinctions as 'first modernity' and 'second modernity'). They apply primarily to the industrialized countries in the West, though the popularity of 'new social movements' and the 'reflexive modernisation' literatures mentioned above have travelled beyond their original geographical settings in Western Europe and the United States (e.g. Veltmeyer 1997; Dwivedi 1999).

The first stream of criticism is part of an intellectual tradition borne out of the intensifying signs that state-centred management of society via bureaucratic and technological interventions had entered a regressive phase, with the idea of progress losing its cachet and coming to stand for 'an awful desolation, insecurity and simple nullity' (Latouche 1993, 13). Given the disillusionment with the overall project of modernity, its key institutions, not only the state but labour unions and other established channels for political action, had become decentred from political life. Part of this fall from grace of course relates to the inability of the political institutions of modernity to take seriously concerns such as gender equality, persistent racial discrimination, growing ecological degradation and the obstinately centripetal forces of identity formation and fragmentation. Failure to respond to these challenges was seen as systemic – rather than episodic failures of the state or the market – and thus all major political concepts undergirding this sphere came to be discredited. A corollary of this view was a rejection of the idea that 'a single political economic transformation would solve the whole range of social ills' (Calhoun 1993). The diminishing of class and the elevation of various types of identity-based formations can be understood within this context. To reiterate, within this reading, the rise of the environmental challenge to modernity is only one aspect of a process of disenchantment with modernity. The rise of these critiques not only discredited class as a central concept but also opened up new political avenues for political action, which came to be known as 'new social movements' that organized themselves differently both in terms of composition and action, perceived their relationship to modern institutions reflexively, and made demands that were not defined by material redistribution.

The novelty of new social movements, therefore, emerges not simply from the newness of their demands – e.g. an end to environmental degradation – but also the manner in which they are conceptualized as part of the Western political sphere. As such, their main forms of solidarity are expected to go beyond class both because the onset of a postindustrial economic landscape rendered some of the class-based concerns reduntant but also because the subjectivity of the political actors affected by these issues went across class lines, even if they were to be accepted to persist to a certain extent. These two arguments form a unity when applied to the context of environmental change. The first is tantamount to asserting that to the extent that inequalities in material attainment remain in the post-industrial landscape, these are either not so grave to be a central organizing principle for social actors or that their consequences in terms of the achievement of life satisfaction are not especially salient. The second suggests that the environmental changes experienced in these contexts cut across class lines and manifest their impacts in a class-less fashion. Putting these two together would yield an argument that the experience of and, therefore, political responses to environmental degradation is, fundamentally, a post-class dynamic. Beck's famously misguided aphorism that 'poverty is hierarchic, smog is democratic' (1992, 36) summarizes the putative irrelevance of class to environmental political action and, therefore, to social theory.

As the discussion on political ecology has demonstrated, smog and most other manifestations of environmental degradation are not at all democratic and do fall along lines of class as well as race and gender (Curran 2018a). However, Beck's contribution to this debate goes beyond this unfortunate assertion. Focusing also on the manner in which environmental questions arise rather than on simply how they affect the world, Beck has argued that there is a fundamental qualitative difference between environmental problems in the contemporary era, which he designates as 'reflexive modernity' or 'second modernity'. Many of the environmental problems characteristic of this era defy the geographic, temporal and, indeed, class barriers to which environmental problems in first modernity adhered. As such 'this means that the divisions and tensions that are inherent in the risk society cannot be analyzed properly using traditional analytical prisms of class, gender, and age' (Mythen 2021, 537). For instance, in Beck's smog example, air pollution from a coal power plant settle on a reasonably small and clearly delineable area, affect mostly communities in the current or a few future generations, and impact on those who do not have the means to relocate to a healthier location (e.g. labour working in the plant itself). Radioactivity from a nuclear power plant, however, affects a vastly greater geographic area, lasts for countless generations and makes it much harder (or at least much more costly) for the affluent to escape from its (long-term) path. That said, it is important to recognize that the differences between these two types of risks – emerging from different type of technologies characterizing different phases of modernity – are more useful as stylized facts rather than ecological truths (after all, impact of smog can in fact stay in the ground for several centuries).

Beck's argument is essentially a commentary on the institutions underpinning the creation and societal rollout of advanced scientific and technological innovations. Beck argues that given their massive complexity, they defy the institutions that were created to regulate capitalist modernity. Technologies such as nuclear radioactivity and genetic modification operate at such a rarified scientific sphere that existing bureaucratic and political mechanisms to assess their viability, desirability and perimeters of operation simply cease to function in a meaningful manner. For instance, elected members of a national parliament are unlikely to be equipped with the necessary scientific training to be able to design effective and realistic regulation to deal with the potential impacts of biotechnology. The burden of knowledge to assess these technologies would be so high as to render meaningful communication from scientific expert bodies to 'laypersons' extremely unfeasible (Wynne 1994). Taking the argument further, Beck suggests that given the vast time horizons in question and the total and complete annihilation of humanity and all life on earth emerging as a distinct potentiality for the first time in human history, society might not be equipped with the institutional infrastructure and the necessary moral horizon required with dealing with the problems created by 'reflexive modernity' itself.

In practice, therefore, regulation of science and technology under reflexive modernity is primarily a theatre of regulation rather than actual regulation, a situation described by Beck as 'organized irresponsibility'. Within this climate, with science and technology acting as a runaway train, noone is deemed to be safe as the magnitude of risks are far too great to respect class lines. In effect, while Beck's aphorism does not work with smog, his argument holds more appeal if phrased as 'radioactivity is democratic' since it and other similar risks (e.g. risks from 'runaway biotechnology' and extreme manifestations of climate change) are expected to defeat the potential material and spatial barriers the affluent classes can erect to protect themselves. Nevertheless, Beck's seemingly total rejection of class (along with ethnicity, nuclear families, etc.) as a 'zombie' category is, as Curran aptly describes it, bombastic (2018b, 30).

Rather than seeing him as an essentially post-class social theorist (as he himself had adamantly claimed), Curran has attempted to reconcile Beck's otherwise significant highlighting of risk (and risk exposure) with Marxist and, to a lesser extent, with Bourdieusian class analysis. This has taken two different analytical tracks. First, Curran has argued that there exists remarkable analytical symmetry between Marx's and Beck's problematization of the modern world (Curran 2016). This symmetry can be found not only in the way they both focus on the implications the development of productive forces have on nature-society relationships but also in their theoretical focus on the necessity of bringing key aspects of modern society under democratic control. Their resulting analysis is certainly symmetrical in the sense that in 'Beck's theory of risk society, risk occupies the same *structural* position that class occupies in Marx's historical materialism' (281). The fundamental difference between the experience of risk and exploitation, however, limits the effectiveness of this line of defence of Beck's work, mainly serving to highlight that he – as with most Western European social theorists – owes an immense debt to Marx's analytical method.

Curran's second approach focuses on the elaboration of the concept of risk-class (see, among others, 2013, 2016, 2018a, 2018b). He effectively argues – with more patience and attention to detail that can be found in Beck's own writing – that increased exposure to risk under second modernity does not materialise in a political economic context devoid of class relationships but in one where class carves the pathways that structure the intensity of individuals' exposure to environmental risks' and their capability to afford 'escape routes' from them (2017). Ultimately, Curran's develops 'risk-class' with a view to providing a 'potential framework to begin to move beyond goods-centric political economy to place the production and distribution of risks alongside goods on an equal footing' (304). While he is right to correct Beck's approach which claimed that the rise of risk meant the irrelevance of class, recognising the analytical – and political – significance of risk in and of itself does not necessarily mean the two are of equal significance. More importantly, Curran's reconciliation of risk and class ultimately replicates the shortcoming of political ecology literature in the sense that class-structured exposure to risks or experience of environmental degradation does not necessarily mean that conflicts arising in response to them will be similarly class centred. In other words, whereas Marx's theory of class is not only – or, in fact, primarily – intended to account for the creation of inequalities but to account for the mechanism through which they would be transcended (through class conflict), Beck's theory of risk, even with Curran's substantial corrections and elaboration, does not come with a similar theory of conflict.

ii. Environmentalism as class politics

On the other side of the equation it is once again possible to identify two streams arguing this time that class is an important factor in environmental politics. The first and arguably the most influential one is also the one that utilizes a non-Marxist conceptualization of class, the post-materialism thesis of Inglehart (1981; Inglehart and Flanagan 1987). The post-materialism literature, which makes extensive use of quantitative data sets such as the World Values Survey and the International Social Survey Program. Given its roots in mainstream political science literature (of the United States), inequalities are not addressed through class but categories such as 'affluence'. Initially formulated by Inglehart (1981), the central premise of this literature is that concern for the environment only begins to be expressed once more fundamental, 'material' needs have been fulfilled. It is asserted that an empirical relationship has been uncovered which shows that '... beginning in the 1960s there has been increasing evidence of a shift in the basic value systems of citizens of advanced industrialized nations. Traditional materialist values have been gradually replaced by higher order, non-economic concerns. These post-materialist values involve appreciation for social equality, participation in decision-making, freedom of expression, and the improvement of the quality of life in general' (Goksen, Adaman, and Zenginobuz 2002). The clear implication of this position is of course that it is necessary to be rich to be green.

A spate of empirical studies seems to support this thesis, arguing that 'environmental concern is closely correlated with the wealth of the nations' (Franzen and Vogl 2013, 1001) and that '[p]rogress in economic development results in declining perceived risk and increasing propensity to act' (Lo 2016, 874). Despite such assertions, the post-materialism thesis has come under rigorous critique, which can be summarized under three headings. First, the empirical studies purporting to show this relationship do not always measure the same effect: 'environmental concern' and 'propensity to act' are substantially different concepts. Second, the nature of the data – often national or international datasets – obscures the implications of inequalities at the local level on environmental politics as well as the specificities of various environmental issues. Third, and most importantly, however, regardless of the robustness of the empirical tests utilized, these studies fail to account for the countless instances of environmental action by poor communities, not only in the developing but also developed world.

It is this last critique that the 'varieties of environmentalism' literature developed by Joan Martinez Alier and his colleagues tackles directly (Martinez Alier and Guha 1997; Anguelovski and Martinez Alier 2014; Martinez Alier 1995). They do not challenge Inglehart's conceptualization on Marxist grounds *per se*. Rather they seek to make his concept of environmental politics more complicated. They argue that what Inglehart describes as environmentalism is only one type, which obscures another, arguably more progressive movement. Repositioning Inglehart's definition of environmentalism as the 'environmentalism of the rich', they coin another type, which is the 'environmentalism of the poor'. The basic argument is that the defence of productive resources mounted by mostly though not exclusively rural communities, peasants as well as indigenous peoples, is also a form of environmentalism. Their argument can be considered to be the most promising theoretical advance to unite class and environmentalism since O'Connor's articulation of the second contradiction.

In this formulation, the peasantry and indigenous people engage in fights to control their forests, lands or other natural resources can be said to be fighting for the control of the means of production. However, unlike traditional class conflicts where labour struggle to control the means of production (e.g. factory), here the peasantry and indigenous communities are too fighting to control their means of production which happen to be bound up with and comprised of nature itself (an argument that anticipates Moore's critique of dualism). It is necessary to note that the environmentalism of the poor, with its green on the outside but red inside formulation, is very much a constitutive process of the class that wages it. Put differently, the class position of these actors begins to emerge only when confronted by capital (or its agents in the state) with the threat of transforming existing (possibly non-capitalist) relationship with nature (as it gets transposed into 'natural resources' or 'ecosystem services'). Until this encounter, many such actors exist on the peripheries of capitalism or lead a dual life, where intra-communal relationships are governed by one logic (non-capitalist) and external with another (capitalist). Thus, resistance against the intrusion of capital is not only a resistance against the destruction of nature but also against the logic of capitalism itself.

iii. Bourgeois environmentalism and the 'new class'

Just as Inglehart is primarily interested in the political consequences of increased affluence, 'environmentalism of the poor' is similarly narrowly focused on political responses from a specific segment of society. Despite their theoretical significance, both approaches have been used primarily to document the empirical validity of their perspectives (it is necessary to be rich to be green vs. the poor are also environmentalists) without going further to theorize class broadly. For the post-materialism thesis, this would require also thinking about class relations rather than class positions and for environmentalism of the poor there is a need to go beyond conflictual outcomes. Outside of these, there is a third approach that explicitly engages with the idea of class, though more flexibly. This approach can be seen in at least three different types of arguments that can loosely be seen as economic, cultural and political constructions of class.

The first is demonstrated by the concept of 'bourgeois environmentalism', which builds on the Martinez Alier framework. This line of argument has been pursued most effectively by Amita Baviskar (2003; also see Mawdsley, Mehra, and Beazley 2009), who has argued that urban environmental politics in Delhi have come to be dominated by the interests of the 'bourgeois' or the 'upper class'. While the struggle for green spaces might be dominated by bourgeois activists, Baviskar nevertheless maintains that progress of a green agenda is not necessarily 'antagonistic to working-class interests' (2003, 95). In other words, bourgeois environmentalists might be seen to be either as playing a temporary leadership role or, alternatively, harking back to the post-class arguments outlined above, certain issues can be expected to result in meaningful alliances between classes even within the vastly unequal contexts as in Delhi's urban politics.

The second interpretation centres on the political responses that the 'new social movements' literature questions (Eckersley 1989). While maintaining the concept of class but looking not at material economic indicators but the cultural forces unleashed by certain constellations of productive relations, it is argued that there is a direct relationship between the rise of environmentally significant behaviours, especially in Western Europe, and the emergence of a 'new class' comprised of 'educated individuals employed in teaching, creative, or caring occupations' (Giugni and Grasso 2015). To reiterate, class is certainly seen as an important variable but this 'new class' occupies an awkward position, reminiscent of Wright's (1996) understanding of the shifting political positionalities of the middle classes, where they are neither directly in charge of the means of production nor are they (cap)able to move in step with traditional labouring classes.

The third and final formulation, the 'environmentalism of the malcontent', aims to make sense of the key qualities of both 'bourgeois environmentalists' and the members of the 'new class'. For Arsel, Akbulut, and Adaman (2015), it is possible to identify a group of activist who arrive at transformative environmental politics not because of their economic class positions but because of their political subjectivities. More specifically, they find that certain activists approach environmental struggles as arenas in which the alliance between state and capital can most effectively targeted. More importantly, however, while their ability to engage in environmentalist struggles is made possible by their class positions (as 'bourgeois' or 'new class' actors), their willingness emerges from their previous experience in progressive politics for traditional Marxian causes.

While the environmentalism of the malcontent explicitly requires crass-class collaborations (for the 'malcontent' do not have sufficient political legitimacy from a strictly ecological stand point), the other two formulations too leave an analytical open door to the possibility of multi-class struggles either because they assume that multiple, otherwise antagonistic classes can be desirous of certain environmental goals or because they do not perceive class antagonisms to play out in their traditional script. Regardless of their formulation, these three approaches are noteworthy because they take the concept of class seriously without an *a priori* conclusion that transformative environmental struggles need to be zero sum conflicts between competing class interests.

IV. Conflict, progress and the enemy

What, then, is the relationship between climate change conflicts and class in the Anthropocene? More specifically, how do we make analytical sense of the protagonists of emerging and almost certain to intensify conflicts over how best to deal with climate change? The preceding re-reading of broad swaths of literatures demonstrates the analytical difficulty in identifying the protagonists of environmental conflicts in a way that can have purchase across the various spatial, geographical, political and ecological contexts in which conflicts continue to conflagrate. This difficulty in systematizing contemporary (and emerging) environmental conflicts is especially challenging in terms of a Marxist understanding of class positions. Early examples of ecological Marxism, for example the work of O'Connor, have essentially evaded this question. More recently, Moore's interventions on nature and capitalism has pointed towards the unfeasibility of sustaining the cheap provision of certain ingredients that power global capitalism as the source of an eventual systemic transition. However, since protagonists still need to be social actors (rather than 'actants', including nature itself), it is not possible to see natural limits themselves as the drivers of socioeconomic change. Whereas political ecology, especially in its second generation, has focused on actors in a way to prioritise their diversity, agrarian political economy has steadfastly focused on the primacy of the peasantry as the vanguard of resistance against capitalism's environmental excesses. These are analytically and politically valuable contributions on their own. However, in terms of constructing a broadly applicable theory of environmental conflicts, their sum is less than the parts. The third stream of literature has taken class seriously but without developing a theory of conflict.

In other words, what has remained missing is the linkage between these specific struggles and a broader transformative movement. Absence of a rigorous theory of environmental conflict makes it difficult to conceptualise the pathways away from the environmental crisis of capitalism. In their absence, 'in the same ship' narratives come to dominate the intellectual and political landscape and their main utility for their proponents is to advance a win-win ideology that short-circuits the type of drastic changes required to address climate change. The 'in the same ship' is negatory in two very different ways, both of which, independently as well as together, stand in stark contrast to a Marxist understanding of social progress whose engine of change is class conflict.

One way of reading the exhortation 'we are in the same ship' is to suggest that progress in climate change mitigation and adaptation can be made without conflict. However, even the most cursory reading of history would suggest that the notion of progress has been intrinsically linked to conflict. To take examples from modern capitalist history, the process of decolonisation, fight for racial equality, emancipation of women, and struggle for workers rights have all emerged from conflictual dynamics (that are ongoing). Could it be that climate politics defies this seemingly iron-clad law of history? For the proponents of the 'in the same ship' approach, this would either require a *deus ex machina* technological breakthrough – e.g. CO2 capture mechanisms – that can give a new lease of life for fossil-fuel dependent capitalism. Or the exceptional nature of climate change in the Anthropocene would compel potentially conflictive social groups to put aside their differences to focus single-mindedly on climate change.

The problem with the technocentric solution is not only that investing so much hope into a 'moonshot' technological breakthrough is risky in and of itself lest it fails to materialise. It is also that partial advancement of technological breakthroughs do little in terms of dealing with the overall problem and they substantially expand and deepen the capitalism's environmental limitations since they are guided by the logic of accumulation not sustainability. It is also important to add that taking the 'moonshot' analogy literally would also reveal that not only the Soviet but also the US space programme was built by state entities. Incipient attempts at making a Green New Deal in the US or European Green Deal show no such commitment to massive infusion of public funds into responding to climate change, except in terms of defensive gestures aimed primarily to keep anticipated flow of climate refugees out of the European Union and the United States.

The latter – the case in which potentially conflictive social groups would make an (temporary) alliance to battle a shared existential risk – is similarly problematic because those with the grievances are exactly the ones housed in the 'third class' of the putative ship. In other words, the 'same ship' metaphor contains within itself the implicit expectation that the poor and marginalized communities who have done the least in generating the problem of climate change and who experience some of its worse impacts to suspend their otherwise justified demands for socio-economic justice in order to stave off planetary catastrophe. It is precisely because this is an unlikely scenario, which makes the coupling of 'same ship' ideology with climate alarmism dangerous. Accepting the existential threat of climate change without genuine transformative political economic changes can lead to calls for putatively benign authoritarianism as illustrated clearly by no less than the intellectual father of the modern Gaia thesis James Lovelock who argued that since climate change can be seen as analogous to a war, democratic decision making processes would not deliver us the necessary results because another 'IPCC report won't be enough. We would argue over it like now' (Hickman 2010). Instead, he argues, we need 'a more authoritative world. We've become a sort of cheeky, egalitarian world where everyone can have their say. It's all very well, but there are certain circumstances – a war is a typical example – where you can't do that. You've got to have a few people with authority who you trust who are running it' (Hickman 2010). In short, in the absence of a major technological breakthrough to diffuse the risks of climate change, the poor and marginalised could end up facing a Hobson's choice between suspending their claims for socio-economic justice or having their rights usurped by a war-like approach to climate change.

The second negation does not presuppose the elimination of conflict but instead blurs or completely eliminates the notion of an 'enemy'. In other words, the recognition of the supposed fact of humanity being on the 'same ship' could be expected to unleash a transformative struggle to respond to climate change by undertaking the types of political, economic, social and cultural changes required. If humanity is seen as a unity when it comes to its experience of and respond to climate change, it could perhaps struggle with itself. Such an argument is in fact consonant with Beck's 'reflexive modernisation' and, if taken to this logical conclusion, could call for the articulation of novel conceptual scaffolding as suggested by Chakrabarty. However, a call for universal reflexivity would need to go dramatically beyond Beck's notion of a 'second modernity' and instead result in a 'collective reimagining of our world, or our worlds' (Wainwright 2020, 211) with a view to achieving global climate justice by building non-capitalist political economies (Mann and Wainwright 2019, 57). The possibility of undertaking such a humanity-wide transformative self-struggle could in a timely and effective manner is supported neither by the trajectory of climate politics so far or by a Marxist reading of history.

When thinking through these two questions – can there be struggle with conflict and can there be conflict without an enemy – within the context of a transition away from climate change in the Anthropocene, it would be useful to note two caveats. The first concerns the ontological unity of the 'enemy', especially within an orthodox Marxist view of conflict. While the argument in this paper has been that (agrarian) political economists and political ecologists have not sufficiently theorised the identify of the chief protagonists to reckon with the possibility of their comprising a class unity, there has been even less effort by these scholars on the identity of the putative class enemy. The analytical purchase of an orthodox two-class model could be limited not only because the 'proletariat' is not sufficiently homogenous. Even if they were, it could be possible that their enemy, the 'bourgeoisie', is too fragmented within the context of climate change in the Anthropocene. Recognizing that the sole objective of capital is to reproduce itself, it would be possible to expect fundamental conflict amongst different economic sectors and capitalists operating within them.

This leads to the second caveat that to the extent that questioning the class-nature of climate conflicts within the Anthropocene makes it possible to highlight some key analytical dimensions, a full reckoning with the question needs to be situated within a much broader empirical terrain. In other words, reducing the Anthropocene to its arguably most spectacular instance can blur away the broader tensions inherent in the relationship between nature, society and capital. To this end, it is also important to recognize that climate change might not be an ontological unity and instead the summation of the changes of the myriad interconnected ecological dynamics with extremely difficult to predict feedback loops. In inquiring into the class nature of climate conflicts, therefore, it would be important to look beyond isolated instances in the short term and develop generative analytical approaches that can focus on the global and the long-duree.

V. Who will blow up the pipeline?

In conclusion, although (agrarian) political economy and political ecology have done much to advance our understanding of environmental conflicts including those concerned with climate change, the two provocations by O'Connor and Chakrabarty remain unanswered. On the one hand, fidelity to Marxist class analysis – broadly construed but anchored in concrete materialism – would help retheorize some of the existing experiences of resistance especially in the developing world, not just by rethinking how to apprehend the identity of the protagonists but also by reconsidering the nature of their demands by recognizing that struggle against capitalist development and its environmental excesses are not necessarily tantamount to the abandonment of emancipatory material development nor a retreat into a type of eco-medievalism. Such a task would therefore necessitate a broader rethinking of the analytical function of class also in cognate fields, especially development studies (Arsel and Dasgupta 2015). On the other hand, it bears repeating that fidelity to Marxist class analysis does not require a rigid two-class framework, the erasure of genuine cross-class formations, or the possibility that additional conceptual tools might need to be developed to respond to the exigencies of climate change in the Anthropocene.

The increasing recognition that climate change is no longer a distant and potential outcome but very much concrete reality in our time coupled with mounting political angst concerning the resilience of business as usual approaches to national and global governance has recently, and perhaps unsurprisingly, engendered discussions around the necessity of more direct and potentially violent actions to bring about the needed socio-economic transformations (for e.g. Malm 2021; Wainwright 2020). In a recent commentary Battistoni (2022) has argued that Malm's 'How to blow up a pipeline?' is not so much a discussion of 'how' but 'why'. It could be added that 'who will blow up the pipeline?' is also an important question awaiting an answer. Doing so – in both practical as well as analytical terms – requires the urgent development of a Marxist theory of environmental conflict in the Anthropocene.

Acknowledgements

I am grateful for the productive feedback of the anonymous referees and JPS editorial team. I would also like to thank colleagues who have read previous drafts or engaged with the core arguments of the paper, including Fikret Adaman, Bill Adams, Anirban Dasgupta, Wendy Harcourt, Jessica Hope and Lorenzo Pellegrini.

Disclosure statement

No potential conflict of interest was reported by the author(s).

References

Adger, W. N. 2006. "Vulnerability." *Global Environmental Change* 16 (3): 268–281.

Agbonifo, J. 2018. *Environment and Conflict: The Place and Logic of Collective Action in the Niger Delta*. London: Routledge.

Akbulut, B., F. Adaman, and Y. M. Madra. 2015. "The Decimation and Displacement of Development Economics." *Development and Change* 46 (4): 733–761.

Akram-Lodhi, A. H. 2021. "The Ties That Bind? Agroecology and the Agrarian Question in the Twenty-First Century." *Journal of Peasant Studies* 48 (4): 687–714.

Andreucci, D., & Zografos, C. 2022. Between Improvement and sacrifice: Othering and the (bio) Political Ecology of Climate Change. Political Geography, *92*, Article 102512.

Anguelovski, I., and J. Martinez-Alier. 2014. "The 'Environmentalism of the Poor' Revisited: Territory and Place in Disconnected Glocal Struggles." *Ecological Economics* 102: 167–176.

Arsel, M. 2003. Risk Society at Europe's Periphery? The Case of the Bergama Resistance inTurkey1. Integrating and articulating environments: A challenge for northern and southern Europe, 3, 29.

Arsel, M., B. Akbulut, and F. Adaman. 2015. "Environmentalism of the Malcontent: Anatomy of an Anti-Coal Power Plant Struggle in Turkey." *Journal of Peasant Studies* 42 (2): 371–395.

Arsel, M., and B. Büscher. 2012. "Nature™ Inc.: Changes and Continuities in Neoliberal Conservation and Market-Based Environmental Policy." *Development and Change* 43 (1): 53–78.

Arsel, M., and A. Dasgupta. 2015. "Critique, Rediscovery and Revival in Development Studies." *Development and Change* 46 (4): 644–665.

Arsel, M., L. Pellegrini, and C. Mena. 2019. "Maria's Paradox and the Misery of Living Without Development Alternatives in the Ecuadorian Amazon." In *Immiserizing Growth: When Growth Fails the Poor*, edited by R. Kanbur, R. Sandbrook, and P. Shafer, 203–225. Oxford: Oxford University Press.

Battistoni, A. 2017. "Bringing in the Work of Nature: From Natural Capital to Hybrid Labor." *Political Theory* 45 (1): 5–31.

Battistonni, A. 2022. "Is Sabotage a Pipe Dream?" In *Property Will Cost Us the Earth Direct Action and the Future of the Global Climate Movement*, edited by J. Kindig, 85. Verso Books.

Baviskar, A. 2003. "Between Violence and Desire: Space, Power, and Identity in the Making of Metropolitan Delhi." *International Social Science Journal* 55 (175): 89–98.

Beck, U. 1992. *Risk Society: Towards a New Modernity*. London: Sage.

Beck, U. 2015. "Emancipatory Catastrophism: What Does it Mean to Climate Change and Risk Society?" *Current Sociology* 63 (1): 75–88.

Bennett, J. 2009. *Vibrant Matter: A Political Ecology of Things*. Durham: Duke University Press.

Benton, T. 1996. *The Greening of Marxism*. New York: The Guilford Press.

Bernstein, H. 2010. *Class Dynamics of Agrarian Change*. Halifax: Fernwood Press.

Bernstein, H., and T. J. Byres. 2001. "From Peasant Studies to Agrarian Change." *Journal of Agrarian Change* 11): 5–56.

Biersack, A., and J. B. Greenberg. 2006. *Reimagining Political Ecology*. Durham: Duke University Press.

Bonneuil, C., and J. B. Fressoz. 2016. *The Shock of the Anthropocene: The Earth, History and Us*. London: Verso.

Bordiga, A. 1956. Drammi gialli e sinistri della moderna decadenza sociale. http://www.quinterna.org/archivio/1952_1970/drammi_gialli.htm

Borras Jr, S. M. 2009. "Agrarian Change and Peasant Studies: Changes, Continuities and Challenges–an Introduction." *Journal of Peasant Studies* 36 (1): 5–31.

Borras Jr, S. M., I. Scoones, A. Baviskar, M. Edelman, N. L. Peluso, and W. Wolford. 2022. "Climate Change and Agrarian Struggles: An Invitation to Contribute to a JPS Forum." *Journal of Peasant Studies* 49 (1): 1–28.

Bratman, E., T. Auch, and B. Stinchfield. 2022. "The Fracking Frontier in the United States: A Case Study of Foreign Investment, Civil Liberties and Land Ethics in the Shale Industry." *Development and Change* 53 (3): 469–494.
Brenner, R. 1977. "The Origins of Capitalist Development: A Critique of neo-Smithian Marxism." *New Left Review* 104: 25–92.
Buck, H. J. 2019. *After Geoengineering: Climate Tragedy, Repair, and Restoration*. London: Verso.
Burkett, P. 1999. "Natural Causes: Essays in Ecological Marxism." *Monthly Review* 50 (9): 47–48.
Büscher, B., S. Sullivan, K. Neves, J. Igoe, and D. Brockington. 2012. "Towards a Synthesized Critique of Neoliberal Biodiversity Conservation." *Capitalism Nature Socialism* 23 (2): 4–30.
Calhoun, C. 1993. ""New Social Movements" of the Early Nineteenth Century." *Social Science History* 17 (3): 385–427.
Camargo, A. 2022. "Imagined Transitions: Agrarian Capitalism and Climate Change Adaptation in Colombia." *Journal of Peasant Studies* 29 (4): 1–21.
Chakrabarty, D. 2009. "The Climate of History: Four Theses." *Critical Inquiry* 35: 197–222.
Chakrabarty, D. 2017. "The Politics of Climate Change is More Than the Politics of Capitalism." *Theory, Culture & Society* 34 (2-3): 25–37.
Chibber, V. 2006. "On the Decline of Class Analysis in South Asian Studies." *Critical Asian Studies* 38 (4): 357–387.
Cousins, B. 2011. "What is a 'Smallholder'? Class-Analytic Perspectives on Small-Scale Farming and Agrarian Reform in South Africa." In *Land and Resource Reform in South Africa: Impacts on Livelihoods*, edited by P. Hebinck and C. Shackleton, 86–111. Abingdon: Routledge.
Crompton, R., and J. Scott. 1999. "Introduction: The State of Class Analysis." *The Sociological Review* 47 (S2): 1–15.
Curran, D. 2013. "Risk Society and the Distribution of Bads: Theorizing Class in the Risk Society." *The British Journal of Sociology* 64 (1): 44–62.
Curran, D. 2016. "Risk Society and Marxism: Beyond Simple Antagonism." *Journal of Classical Sociology* 16 (3): 280–296.
Curran, D. 2018a. "Beck's Creative Challenge to Class Analysis: From the Rejection of Class to the Discovery of Risk-Class." *Journal of Risk Research* 21 (1): 29–40.
Curran, D. 2018b. "Environmental Justice Meets Risk-Class: The Relational Distribution of Environmental Bads." *Antipode* 50 (2): 298–318.
Daniels, R. J., D. F. Kettle, and H. Kunreuther. 2006. *On risk and disaster: Lessons from Hurricane Katrina*. Philadelphia: University of Pennsylvania Press.
Das, R. J. 2017. *Marxist Class Theory for a Skeptical World*. Leiden: Brill.
Davies, J. 2016. *The Birth of the Anthropocene*. Oakland, CA: University of California Press.
Davis, M. 2013. "Rethinking Class: The Lineage of the Socialist Register." *Socialist Register* 50: 287–304.
Deaton, A. 2013. *The Great Escape: Health, Wealth, and the Origins of Inequality*. Princeton, NJ: Princeton University Press.
Dwivedi, R. 1999. "Displacement, Risks and Resistance: Local Perceptions and Actions in the Sardar Sarovar." *Development and Change* 30 (1): 43–78.
Eckersley, R. 1989. "Green Politics and the new Class: Selfishness or Virtue?" *Political Studies* 372: 205–223.
Fernandez-Salvador, C. 2018. 'The Shuar and Large-Scale Mining in Zamora-Chinchipe, Ecuador' PhD Dissertation. International Institute of Social Studies, The Hague.
Fischer, A. M. 2015. "The end of Peripheries? On the Enduring Relevance of Structuralism for Understanding Contemporary Global Development." *Development and Change* 46 (4): 700–732.
Fontana, L. B. 2014. "Indigenous Peoples vs Peasant Unions: Land Conflicts and Rural Movements in Plurinational Bolivia." *Journal of Peasant Studies* 41 (3): 297–319.
Foster, J. B. 1999. "Marx's Theory of Metabolic Rift: Classical Foundations for Environmental Sociology." *American Journal of Sociology* 105 (2): 366–405.
Foster, J. B., B. Clark, and R. York. 2010. *The Ecological Rift*. New York: Monthly Review Press.
Franzen, A., and D. Vogl. 2013. "Two Decades of Measuring Environmental Attitudes: A Comparative Analysis of 33 Countries." *Global Environmental Change* 23 (5): 1001–1008.
Friedmann, H. 2019. "The Awkward Class: A Foundation for Peasant Studies." *Journal of Peasant Studies* 46 (5): 1096–1105.

Giugni, M., and M. T. Grasso. 2015. "Environmental Movements in Advanced Industrial Democracies: Heterogeneity, Transformation, and Institutionalization." *Annual Review of Environment and Resources* 40: 337–361.

Goksen, F., F. Adaman, and EÜ Zenginobuz. 2002. "On Environmental Concern, Willingness to pay, and Postmaterialist Values: Evidence from Istanbul." *Environment and Behavior* 34 (5): 616–633.

Goldblatt, D. 1996. *Social Theory and the Environment*. Cambridge: Polity Press.

Grusky, D. B., and J. B. Sørensen. 1998. "Can Class Analysis be Salvaged?" *American Journal of Sociology* 103 (5): 1187–1234.

Hale, C. R. 2005. "Neoliberal Multiculturalism." *PoLAR: Political and Legal Anthropology Review* 28 (1): 10–19.

Harriss-White, B. 2018. "Awkward Classes and India's Development." *Review of Political Economy* 30 (3): 355–376.

Henderson, J., R. P. Appelbaum, and S. Y. Ho. 2013. "Globalization with Chinese Characteristics: Externalization, Dynamics and Transformations." *Development and Change* 44 (6): 1221–1253.

Hickman, L. 2010. James Lovelock: Humans are too stupid to prevent climate change. 29 March. https://www.theguardian.com/science/2010/mar/29/james-lovelock-climate-change

Horner, R., and D. Hulme. 2017. "From International to Global Development: New Geographies of 21st Century Development." *Development and Change* 50 (2): 495–510.

Inglehart, R. 1981. "Post-materialism in an Environment of Insecurity." *American Political Science Review* 75 (4): 880–900.

Inglehart, R., and S. C. Flanagan. 1987. "Value Change in Industrial Societies." *American Political Science Review* 81 (4): 1289–1319.

Jänicke, M. 2008. "Ecological Modernisation: New Perspectives." *Journal of Cleaner Production* 16 (5): 557–565.

Kallis, G. 2019. *Limits: Why Malthus was Wrong and why Environmentalists Should Care*. Stanford: Stanford University Press.

Kiely, R. 2009. "The Globalization of Manufacturing Production: Warrenite Fantasies and Uneven and Unequal Realities." In *Peasants and Globalization: Political Economy, Rural Transformation and the Agrarian Question*, edited by A. H. Akram-Lodhi and C. Kay, 169–189. London: Routledge.

Korhonen, J., A. Honkasalo, and J. Seppälä. 2018. "Circular Economy: The Concept and its Limitations." *Ecological Economics* 143: 37–46.

Kovel, J. 1995. "Ecological Marxism and Dialectic." *Capitalism Nature Socialism* 6 (4): 31–50.

Laclau, E., and C. Mouffe. 1985. *Hegemony and Socialist Strategy: Towards a Radical Democratic Politics*. London: Verso.

Latorre, S.. 2021. "The Role of Ecuadorian Working-Class Environmentalism in Promoting Environmental Justice: An Overview of the Hydrocarbon and Agricultural Sectors." In *The Palgrave Handbook of Environmental Labour Studies*, edited by N. Räthzel, D. Stevis, and D. Uzzell, 271–294. Cham: Palgrave Macmillan.

Latouche, S. 1993. *In the Wake of the Affluent Society: An Exploration of Post-Development*. London: Zed Books.

Lerche, J., and A. Shah. 2018. "Conjugated Oppression Within Contemporary Capitalism: Class, Caste, Tribe and Agrarian Change in India." *Journal of Peasant Studies* 45 (5-6): 927–949.

Levien, M., M. Watts, and H. Yan. 2018. "Agrarian Marxism." *Journal of Peasant Studies* 45 (5-6): 853–883.

Lo, A. Y. 2016. "National Income and Environmental Concern: Observations from 35 Countries." *Public Understanding of Science* 25 (7): 873–890.

Malm, A. 2018. *The Progress of This Storm: Nature and Society in a Warming World*. London: Verso Books.

Malm, A. 2021. *How to Blow up a Pipeline*. London: Verso Books.

Mann, G., and J. Wainwright. 2019. "Political Scenarios for Climate Disaster." *Dissent* 66 (3): 65–72.

Martinez-Alier, J. 1995. "The Environment as a Luxury Good or "Too Poor to be Green"?" *Ecological Economics* 13 (1): 1–10.

Martinez-Alier, J. 2009. "Social Metabolism, Ecological Distribution Conflicts, and Languages of Valuation." *Capitalism Nature Socialism* 20 (1): 58–87.

Martinez-Alier, J., and R. Guha. 1997. *Varieties of Environmentalism: Essays North and South*. London: Earthscan.

Martinez-Alier, J., L. Temper, D. Del Bene, and A. Scheidel. 2016. "Is There a Global Environmental Justice Movement?" *Journal of Peasant Studies* 43 (3): 731–755.
Martínez-Alier, J., and M. O'Connor. 1995. Ecological and economic distribution conflicts (No. 321.95). Unitat de Fonaments de l'Anàlisi Econòmica (UAB) and Institut d'Anàlisi Econòmica (CSIC).
Mawdsley, E., D. Mehra, and K. Beazley. 2009. "Nature Lovers, Picnickers and Bourgeois Environmentalism." *Economic and Political Weekly* XLIV (11): 49–59.
Maxwell, J., and F. Briscoe. 1997. "There's Money in the air: The CFC ban and DuPont's Regulatory Strategy." *Business Strategy and the Environment* 6 (5): 276–286.
Melucci, A. 1988. Social movements and the democratization of everyday life.
Moore, J. W. 2015. *Capitalism in the Web of Life: Ecology and the Accumulation of Capital*. London: Verso Books.
Moore, J. W. 2018. "The Capitalocene Part II: Accumulation by Appropriation and the Centrality of Unpaid Work/Energy." *Journal of Peasant Studies* 45 (2): 237–279.
Morris, I. 2010. *Why the West Rules – for now: The Patterns of History and What They Reveal About the Future*. New York: Farrar, Straus, and Giroux.
Mythen, G. 2021. "The Critical Theory of World Risk Society: A Retrospective Analysis." *Risk Analysis* 41 (3): 533–543.
Neimark, B. D., and T. M. Healy. 2018. "Small-Scale Commodity Frontiers: The Bioeconomy Value Chain of Castor oil in Madagascar." *Journal of Agrarian Change* 18 (3): 632–657.
Newell, P. 2022. "Climate Justice." *Journal of Peasant Studies* 49 (5): 1–9.
Nixon, R. 2011. *Slow Violence and the Environmentalism of the Poor*. Cambridge, MA: Harvard University Press.
O'Connor, J. 1988. "Capitalism, Nature, Socialism a Theoretical Introduction." *Capitalism Nature Socialism* 1 (1): 11–38.
Offe, C. 1985. "New Social Movements: Challenging the Boundaries of Institutional Politics." *Social Research* 52 (4): 817–868.
Parenti, C. 2015. "The 2013 ANTIPODE AAG Lecture The Environment Making State: Territory, Nature, and Value." *Antipode* 47 (4): 829–848.
Patel, R., and J. W. Moore. 2017. *A History of the World in Seven Cheap Things: A Guide to Capitalism, Nature, and the Future of the Planet*. Berkeley: University of California Press.
Paterson, M. 2001. "Risky Business: Insurance Companies in Global Warming Politics." *Global Environmental Politics* 1 (4): 18–42.
Peet, R., and M. Watts. 1996. *Liberation Ecologies: Environment, Sustainability, and Environment in an Age of Market Triumphalsism*. London: Routledge.
Pellegrini, L. 2012. "Joan Martinez-Alier." *Development and Change* 43 (1): 341–359.
Pellegrini, L., and M. Arsel. 2018. "Oil and Conflict in the Ecuadorian Amazon." *European Review of Latin American and Caribbean Studies/Revista Europea de Estudios Latinoamericanos y del Caribe* 106: 209–218.
Pichardo, N. A. 1997. "New Social Movements: A Critical Review." *Annual Review of Sociology* 23 (1): 411–430.
Piketty, T. 2017. *Capital in the Twenty-First Century*. Cambridge, MA: Harvard University Press.
Polanyi, K. 1957. *The Great Transformation: (The Political and Economic Origin of our Time)*. Boston, MA: Beacon Press.
Pye, O. 2021. "Agrarian Marxism and the Proletariat: A Palm oil Manifesto." *Journal of Peasant Studies* 48 (4): 807–826.
Resnick, S., and R. Wolff. 2003. "The Diversity of Class Analyses: A Critique of Erik Olin Wright and Beyond." *Critical Sociology* 29 (1): 7–27.
Rockström, J., W. Steffen, K. Noone, Å Persson, F. S. Chapin III, E. Lambin, T. M. Lenton, M. Scheffer, C. Folke, and H. J. Schellnhuber. 2009. "A Safeoperating Space for Humanity. Identifying and Quantifying Planetary Boundaries that must not be Transgressed could Help Prevent Human Activities from Causing Unacceptable Environmental Change." *Nature* 461: 472–475.
Saez, E., and G. Zucman. 2016. "Wealth Inequality in the United States Since 1913: Evidence from Capitalized Income tax Data." *The Quarterly Journal of Economics* 131 (2): 519–578.

Saguin, K. 2016. "Blue Revolution in a Commodity Frontier: Ecologies of Aquaculture and Agrarian Change in Laguna Lake." *Philippines. Journal of Agrarian Change* 16 (4): 571–593.

Salleh, A. 1995. "Nature, Woman, Labor, Capital: Living the Deepest Contradiction." *Capitalism Nature Socialism* 6 (1): 21–39.

Schmitz, H., and I. Scoones. 2015. *Accelerating Sustainability: Why Political Economy Matters*. Report. 152, IDS, University of Sussex, Brighton, UK.

Schneider, M. 2015. "What, Then, is a Chinese Peasant? Nongmin Discourses and Agroindustrialization in Contemporary China." *Agriculture and Human Values* 32 (2): 331–346.

Scoones, I. 2016. "The Politics of Sustainability and Development." *Annual Review of Environment and Resources* 41 (1): 293–319.

Sewell Jr, W. H. 1990. "How Classes are Made: Critical Reflections on EP Thompson's Theory of Working-Class Formation." In *EP Thompson: Critical Perspectives*, edited by H. J. Kaye, and K. McClelland, 50–77. Philadelphia: Temple University Press.

Sherriff, L. 2015. One Young World Summit: 'We Will Sink Or Swim Together', Warns Kofi Annan Over Climate Change. Huffington Post UK. Accessed 11 February 2018. http://www.huffingtonpost.co.uk/2015/11/19/we-will-sink-or-swim-together-warns-kofi-annan-climate-change_n_8597362.html

Shi, L., E. Chu, I. Anguelovski, A. Aylett, J. Debats, K. Goh, T. Schenk, K. C. Seto, D. Dodman, and D. Roberts. 2016. "Roadmap Towards Justice in Urban Climate Adaptation Research." *Nature Climate Change* 6 (2): 131–137.

Silver, B. J., and S. S. Karatasli. 2015. "Historical Dynamics of Capitalism and Labor Movements." In *The Oxford Handbook of Social Movements*, edited by D. Della Porta and M. Diani, 133–145. Oxford: Oxford University Press.

Stern, N. 2007. *The Economics of Climate Change: The Stern Review*. Cambridge: Cambridge University Press.

Svarstad, H., and T. A. Benjaminsen. 2020. "Reading Radical Environmental Justice Through a Political Ecology Lens." *Geoforum; Journal of Physical, Human, and Regional Geosciences* 108: 1–11.

Swyngedouw, E. 1999. "Modernity and Hybridity: Nature, Regeneracionismo, and the Production of the Spanish Waterscape, 1890-1930." *Annals of the Association of American Geographers* 89 (3): 443–465.

Swyngedouw, E. 2013. "Apocalypse now! Fear and Doomsday Pleasures." *Capitalism Nature Socialism* 24 (1): 9–18.

Thompson, E. P. 1971. "The Moral Economy of the English Crowd in the Eighteenth Century." *Past & Present* 50: 76–136.

van der Ploeg, J. D. 2020. "Farmers' Upheaval, Climate Crisis and Populism." *Journal of Peasant Studies* 47 (3): 589–605.

Veltmeyer, H. 1997. "New Social Movements in Latin America: The Dynamics of Class and Identity." *Journal of Peasant Studies* 25 (1): 139–169.

Wainwright, J. 2020. "How Not to Think About Climate Change." *GeoHumanities* 6 (1): 205–214.

Wright, E. O. 1996. "The Continuing Relevance of Class Analysis." *Theory and Society* 25: 693–716.

Wright, E. O. 1997. *Class Counts: Comparative Studies in Class Analysis*. Cambridge: Cambridge University Press.

Wright, E. O. 1999. Foundations of class analysis: A Marxist perspective. annual meeting of the American Sociological Association, Chicago.

Wright, E. O. 2019. *How to be an Anti-Capitalist in the Twenty-First Century*. New York: Verso Books.

Wynne, B. 1994. "May the Sheep Safely Graze? A Reflexive View of the Expert-Lay Knowledge Divide." In *Risk Environment and Modernity: Towards an New Ecology*, edited by S. Lash, S. Bronislaw, and B. Wynne, 44–83. London: Sage.

OPEN ACCESS

The political life of mitigation: from carbon accounting to agrarian counter-accounts

Shaila Seshia Galvin and Diego Silva Garzón

ABSTRACT

This article seeks to stimulate analysis of how accounting knowledge, techniques, and practices work to incorporate agriculture and land use into climate mitigation. Accounting plays a significant role in the ways that capitalism inserts itself into, reworks, or reorganises agrarian webs of life. To study these processes, we train our critical gaze on accounting itself – its epistemic foundations, instruments, and narratives, and their implications for agrarian livelihoods and relations. Through the notion of "agrarian counter-accounts," we conclude by considering the potential of alternative methodologies and understandings of account-giving, taking, and holding in struggles for agrarian climate justice.

In recent years, agricultural land, practices, and livelihoods have increasingly captured the attention of climate science, policy, and finance. On the one hand, agriculture is estimated to account for 22% of global greenhouse gas emissions, with projections rising over the coming decades (IPCC 2022a). On the other, it has been seen to hold the possibility of making a significant contribution to climate mitigation (IPCC 2019). Once on the margins of climate mitigation, agriculture along with other forms of land use are now seen to harbor vast 'untapped potential' in the area of reducing emissions as well as in carbon storage and sequestration, often under the banner of 'nature-based solutions' or 'natural capital' approaches (FAO 2022; Iseman and Miralles-Wilhelm 2021). Such enchantment with agriculture has seen photosynthesis framed as a 'low-cost negative emissions technology,' while companies and countries seeking to meet voluntary or compliance commitments to reduce emissions look to agriculture and land use as areas where emissions reductions may be readily achieved in what is often described as a 'cost-effective' manner.[1] The dual position of agriculture and land use as both a source of, and sink for, greenhouse gas emissions also means that it has been identified as key to achieving climate neutrality and net zero targets, terms themselves that demand critical

[1]See, for example, the Natural Climate Solutions Alliance, convened by the World Business Council for Sustainable Development and World Economic Forum. https://www.wbcsd.org/Programs/Climate-and-Energy/Climate/Natural-Climate-Solutions/The-Natural-Climate-Solutions-Alliance. Accessed 18 November 2022.

interrogation (Stabinsky 2021) as they become a frame and focal point for climate mitigation ambitions for entities that range from the European Union to BP, Shell, and Nestlé.

This article sets out a research agenda to stimulate analysis of the ways that accounting knowledge, techniques and practices play a formative role in incorporating agriculture and land use into climate mitigation narratives, strategies, and interventions.[2] In facilitating such processes of incorporation, accounting helps to constitute more broadly the way that the problem of climate change is apprehended and, as a consequence, the ways that responses to it are envisioned, designed, and enacted. In taking up questions posed by Borras et al. (2022, 17) – 'What combinations of narratives and strategies frame climate change and the institutionalized responses to it in agrarian settings? What exclusions and inclusions result from this?' – we probe one of the crucial yet often overlooked features of many contemporary responses: their reliance on epistemologies, narratives, and practices of accounting.

A key point of departure for us is that grappling with climate change and its mitigation within agrarian settings requires developing critical and analytical approaches for engaging with accounting, whose specific power and impact on agriculture and agrarian modes of production and ways of life remains under-examined. With Borras et al. (2022, 5), we concur that 'climate change needs to be seen in its wider, historical context, and not just as a technical phenomenon emerging from anthropogenic greenhouse gas emissions.' Building on this, we argue that not only climate change, but ways of perceiving, understanding, and responding to it must also be historicized. The forms of knowledge and practice that constitute dominant approaches to climate mitigation cannot be separated from the framing and understanding of the problem of climate change itself.

In this article we develop a framework for engaging critically with the way that accounting comes to shape and condition climate mitigation in agrarian settings. Our work draws on on-going research conducted since 2020, using methods that include document analysis, participation in online training courses on environmental and carbon accounting, and ethnographic fieldwork in Argentina, Colombia, and western Canada. In this article and our larger work we approach accounting as a world- or environment-making phenomenon (Tsing 2015; Moore 2015). By this, we mean that in seeking to make visible particular social and ecological phenomena, accounting invariably transforms those environmental and agrarian relations that are the objects of its knowledge and interventions. Accounting is not, therefore, purely a means of measurement and calculation; it is also a novel way of reconfiguring these relations by disembedding greenhouse gas emissions from the socio-environmental conditions in which they are produced, and rendering them quantifiable units of carbon dioxide equivalent (CO_2e). In this way, it works to draw human and nonhuman life and relations, as they are constituted within agrarian worlds and practices, into emergent capitalist networks and technologies of climate mitigation. This reconfiguration, moreover, follows a specific path, as biological and ecological processes in agricultural settings are harnessed in new ways to achieve climate mitigation without halting abiding historical processes of accumulation. Accounting knowledge and practice contributes crucially to the work of opening up new spaces for accumulation in agrarian settings through the production

[2]Shaila Seshia Galvin and Diego Silva Garzón are the joint co-authors of this article.

of data, and the regulation and management of land, human labor, and nonhuman life. This, we show, has profound implications for the scales and modes of agricultural production that are likely to be supported by mitigation interventions, as well as for farmers' productive autonomy. As others map some of the manifestations of these new accumulation pathways (Matthan 2022; Schwartzman 2022; Stock 2022), we train our critical gaze on accounting itself – its epistemic foundations, instruments, and narratives, and their implications for agrarian livelihoods and relations.

Climate mitigation, accounting, and critical agrarian studies

The prominence of agriculture within global climate negotiations grew substantially after 2015, with the adoption of the Paris Agreement. As the first legally binding global agreement on climate change, it committed its signatories to implement Nationally Determined Contributions in an effort to limit global temperature rise to less than a two degrees Celsius increase (and ideally 1.5 degrees) from pre-industrial levels (UNFCCC 2015). With the agreement, agriculture came to occupy a more central place in climate discourse and strategy, both nationally and internationally. Notably, 148 countries included agriculture in their mitigation targets and/or actions submitted with their NDCs (Strohmaier et al. 2016). The subsequent year, at COP22 in Marrakech where signatories committed to the implementation of their NDCs, agriculture emerged as a 'critical site for the adaptation to and mitigation of climate change' (Newell and Taylor 2018). More recently, since the publication of the IPCC special reports on 1.5 degrees and on land, focus on agriculture has intensified with reductions in agricultural emissions deemed necessary to meet targets and deadlines (Leahy, Clark, and Reisinger 2020; IPCC 2018; IPCC 2019). Here, agriculture is framed as a 'sector' or 'set of activities,' necessarily reflecting the way that climate change more broadly is understood as a 'biophysical problem predominantly driven by carbon emissions' (Nightingale et al. 2020, 346). As agriculture draws more attention and investment, it is accounting that offers a discursive and technical framework that casts the agrarian as a biophysical domain within the contemporary science and policy of mitigation, bracketing agricultural emissions off from their social, historical, political, and economic moorings.[3]

Such a feat is not easily achieved. The emission of greenhouse gases, notably methane and nitrous oxide as well as carbon dioxide, accompany agricultural production through processes such as enteric fermentation in ruminant animals, the deposition of synthetic fertilizers or livestock manures on fields and pastures, or the microbial activity associated with soil respiration as well as flooded rice cultivation (IPCC 2006). Agrarian settings have been particularly resistant to climate mitigation initiatives and their associated accounting frameworks, at least in part because of the difficulty of measuring or estimating emissions that are multiple, diffuse, and variable. And because agricultural greenhouse gases (GHGs) emerge in different sites and through biological processes across vast areas above and below ground, the production of accurate, region-specific emissions data is more complex than in many other spheres of activity.

[3]A good example of this is a now widely reproduced image of agricultural sources and sinks for greenhouse gas emissions, which first appeared as 'Figure1.1 The main greenhouse gas emission sources/removals and processes in managed ecosystems' in Volume 4 on Agriculture, Forestry, and Other Land Use as part of the IPCC's 2006 Guidelines for National Greenhouse Gas Inventories. See: (IPCC 2006).

Rethinking how greenhouse gases are understood and analyzed in relation to agriculture is important for developing an agrarian political economy attuned to the ways that pathways of accumulation are enabled by climate mitigation and to the role that accounting frameworks play in establishing these pathways. Unraveling these connections often necessitates treating carbon dioxide and other greenhouse gases as more than chemical compounds, instead probing them as both 'substance and relation' (Dalsgaard 2014). As a relation, a number of scholars have observed how almost every form of human activity along with the foundations of capitalism as well as social and economic life more generally are implicated in the generation of greenhouse gases. In the twenty first century, carbon has become a 'common denominator for thinking about social life in relation to the environment' (Bridge 2011, 821), though this common denominator is also uncommoned by uneven histories of oppression and dispossession that are linked to imperialism and colonialization, ongoing processes of settler colonialism, race, and gender inequality (Yusoff 2021; Davis and Todd 2017).

Pushing back against a narrow conceptualization of agriculture as a sector, source, and sink in relation to the generation of greenhouse gases, Reisman and Fairbairn (2021, 688) observe that agriculture is 'not a set of impacts to be avoided but rather a site of political and economic processes to be accounted for and reimagined.' By enabling analytical purchase on political economic processes underpinning climate change and responses to it, critical agrarian studies has, to date, engaged with climate mitigation from a range of perspectives, including green grabbing (Fairhead, Leach, and Scoones 2012), Reducing Emissions from Deforestation and Forest Degradation or REDD+ (McElwee 2016), climate smart agriculture (Newell and Taylor 2018; Clapp, Newell, and Brent 2018) and carbon offsetting (Cavanagh and Benjaminsen 2014). These studies pose a direct challenge to the way that climate science and policy decouple the problem of climate change from the political, economic, and bio-historical circumstances in which it emerged (see Chakrabarty 2009; Mitchell 2011). In particular, they highlight the tendency of such responses to 'render technical' (Li 2007) the problem of climate change as well as the means of tackling it. Thus, for example, framing climate change in a narrowly scientific sense as simply a 'problem of emissions' has enabled the proliferation of techno-fixes, such as carbon capture and storage which are increasingly recognized as necessary to achieve reductions needed to keep warming within 1.5 or 2 degrees. This elides the ways that climate change is fundamentally embedded within economic systems, activities, and practices, while intensifying asymmetries of power and inequality, opening new avenues of dispossession as well as capitalist accumulation (Clapp, Newell, and Brent 2018; Nightingale et al. 2020; Stock and Birkenholtz 2021). Climate change increasingly represents an opportunity rather than a crisis for capitalism, as climate mitigation becomes an arena of investment and industry through which capitalism finds and establishes new sites and modes of accumulation in rural worlds.

Accounting knowledge and practices, we contend, play a critical role in enabling such novel forms of accumulation because of the ways that they figure in capitalist processes, inserting themselves into, reworking, and reorganizing what Jason Moore (2015, 2017) has described as 'webs of life.' By making visible, quantifiable, and manageable the work of trees, plants and soils in storing or sequestering carbon emissions, accounting helps to incorporate agrarian activities into mitigation efforts. It also opens new avenues for transforming the work/energy of the biosphere into capital by internalizing previously untapped forms of more-than-human work in the service of creating or extending networks of capital accumulation. While this is also achieved by mitigation initiatives in other sectors, such as REDD+, in agrarian settings modes of accounting, and notably carbon accounting, help to embed mitigation within productive activities, responding to the historical need of capitalism to mitigate GHG emissions without halting economic growth.

Yet despite its importance, accounting has received little attention in critical agrarian studies, as attention has frequently been drawn to particular sites (on land, in forests) and the structural processes with which they are associated. Such relatively scant attention to accounting is at least partially attributable to the difficulties posed by the complexity and dynamism of socionatural agrarian environments, where the elements that make up greenhouse gases assume different forms and undergo multiple transformations prior to their production as GHG emissions. But to overlook carbon accounting in agriculture, we suggest, is to neglect an important mechanism through which novel processes of accumulation are beginning to take shape as climate mitigation is enacted in agrarian settings. In what follows, we attune ourselves to accounting as an epistemic and ontological force that animates approaches to mitigation in agriculture.

Climate responses and accounting epistemologies

Contemporary methods of accounting for agriculture's emissions, or its capacities for storage and sequestration, extend a long history of calculative practices such as double-entry book-keeping (Mills 1994) and techniques of legibility, that have been critical in creating the conditions of possibility for capital accumulation and state power (Scott 1998). Historians of science and economics have documented how accounting practices have become authoritative forms of knowledge, melding the form that management and government assume across various domains of life (Porter 1994). Accounting is therefore 'intrinsic to and constitutive of social relations, rather than secondary and derivative' (Miller 2001, 392). More than a form of knowledge, it is replete with tools and instruments that enable concrete interventions in the world (Loft 2021; Fleischman, Funnell, and Walker 2013) and that intertwine modes of measurement with practices of government and accumulation. Processes of appropriation and accumulation, Jason Moore has pointed out, do not only take place through primitive accumulation, expanded reproduction, or accumulation by dispossession, but are importantly enabled in spaces 'in between,' particularly through 'those practices committed to locating, quantifying and rationalizing human and extra-human natures.' (Moore 2018, 249) Accounting, in this respect, enables the government and management of particular ecologies, notably in the service of shaping socio-environmental relations into productive forces that work in the service of capitalism within agriculture.

Accounting practices are deeply embedded in both climate science and policy (Bebbington, Larrinaga, and Thomson 2021; Charnock, Brander, and Thomas 2021; Newell, Boykoff, and Boyd 2012), and they lie at the core of many responses to climate change across a wide range of contexts and scales that extend well beyond agriculture – from the national inventories of greenhouse gas emissions and removals binding signatories to the UNFCCC to personal carbon footprint calculators (Dalsgaard 2014). The ubiquity of accounting epistemologies in mitigation approaches and practices builds on the emergence of much broader subfields of social and environmental accounting over the course of the twentieth century. Though critical scholarship now links accounting with mainstream, neoliberal responses to climate change, of note is that in their early days both social and environmental accounting were subfields which claimed a certain counter-hegemonic position vis-à-vis the wider field in which they were located. While social accounting challenged mainstream financial accounting's claims of being objective and value-free, environmental accounting arose out of the recognition that as capital advances, it destroys its biophysical conditions of reproduction (Gray 1990).

Over time, however, environmental accounting has grown into a field of accounting practice that extends, rather than challenges, traditional accounting principles and frameworks by incorporating elements of nature and the natural environment that have long been considered external to society (Lohmann 2009; Hopwood 2009; Moore 2017). Managerial aspects of traditional accounting are evident in the way that environmental accounting has been applied to the internal management of corporate environmental behavior (Kumarasiri and Jubb 2016; Olsthoorn et al. 2001; Tyteca 1996) or external environmental reporting (Pattberg 2017; Hahn, Reimsbach, and Schiemann 2015; Figge et al. 2002) and, with regard to climate change, internal carbon management accounting and external carbon disclosure at the level of organizations, products and supply chains (Stechemesser and Guenther 2012). As environmental considerations became more mainstream in the corporate world under the banner of sustainability and, more recently, Environmental Social and Governance (ESG) criteria, environmental accounting has lost its once counter-hegemonic status as it becomes incorporated into management, communications, and marketing (Spence, Husillos, and Correa-Ruiz 2010).

While climate change helped to spur the development of carbon accounting as an offshoot of environmental accounting, international discussions were fostering debates on a related global concern – conservation through the valuation of biodiversity. From 2001 to 2005, the work of more than 1360 scientists resulted in the Millennium Ecosystem Assessment (MEA), a scientific appraisal of the 'conditions and trends in the world's ecosystems and the services they provide' (Millenium Ecosystem Assessment 2022). In addition, a ministerial gathering of environment ministers from the G8+6 countries in 2008, resulted in an initiative called The Economics of Ecoystems and Biodiversity (TEEB), which sought to redress the 'invisibility of nature' by recognizing and capturing the value of biodiversity and ecosystem services for business and society (TEEB 2022). Both MEA and TEEB renewed the significance of natural capital approaches – a concept first deployed in the late 1970s to highlight the role of natural resources in the production of economic value (Schumacher 1979) and later mobilized to bridge the world of ecology and accounting (Gray 1990; Rubenstein 1992) – to inform understandings of sustainability among policy makers and the business community (Groot and Braat 2015, 233). By 2013, natural capital approaches had become so popular that a Guardian article stated 'If you are looking for the next big thing in sustainability, you needn't look much further than natural capital accounting' (McGill 2013).

Carbon and natural capital accounting have facilitated the commodification of ecosystem processes, promoting a vision of 'nature' as something that can be valued in monetary terms and substituted for other forms of capital (Gómez-Baggethun and Martin Lopez 2015). The expansion of accounting principles and frameworks into environmental domains is thus a critical force underlying the economization of the environment (Sullivan 2013). The influence of accounting practices is evident, for instance, in the rise of payments for ecosystem services, and in emissions trading systems, which are mediated by international and industry specific accounting frameworks and methodologies such as the IPCC Guidelines, the UN's System of Environmental Economic Accounting (SEEA), the Greenhouse Gas Protocol administered by the World Resources Institute and World Business Council for Sustainable Development, and the Verified Carbon Standard among many others. These accounting frameworks and methodologies have a bearing on agrarian worlds, where carbon accounting and natural capital accounting are applied. For example, based on life cycle assessment methodologies, the Greenhouse Gas Protocol, often referred to as the gold-standard for calculating and reporting GHG emissions (Ratnatunga, Jones, and Balachandran 2011), includes specific guidelines for the agricultural sector (WRI and WBCSD 2014). These guidelines are primarily intended for large producers and companies, such as fruit and crop growers, ranchers and biofuel producers, who want to develop inventories about their greenhouse emissions (WRI and WBCSD 2014, 9), but they can also be used by governmental and non-governmental initiatives.

While environmental accounting has aimed at internalizing the environmental costs of capitalist production, it has also helped to deploy rationalizing techniques to render the environment amenable to ongoing processes of capital accumulation. Although some have argued that climate change presents a crucial challenge for capitalism, it is increasingly the case that the development and deployment of environmental, carbon, and natural capital accounting in ever-widening spheres of society, economy, and daily life helps to create new domains of capital accumulation by finding novel ways of making socionatural and ecological systems work for capital. Crucially, accounting enables greenhouse gases associated with agricultural practices and activities to be disembedded from the socioecological, historical, and political economic conditions in which they are generated. This disembedding renders them objects of measurement, estimation, and calculation that can be re-embedded in commodity and market relations amenable to appropriation and exploitation. Accounting epistemologies, therefore, provide an important foundation and condition from which new accumulation pathways can develop. But, first, the ontological and epistemological positioning of climate change as a technical and economic problem must be discursively amplified.

Amplifying accounting, dis-/re-embedding carbon: instruments and narratives

Accounting is a field of technical and professional knowledge, but it is also one where particular vocabularies, concepts, and metaphors work as epistemic objects to coproduce 'climate change as an issue of political and scientific concern' (Jasanoff 2010; Knorr-Cetina 1999; Lahn 2021, 4). The linguistic field of climate science and policy yields plentiful possibilities in this regard, and it is instructive to ask how accounting informs climate change vernaculars and with what political effects. Here, we focus on three concepts – budgets, balance sheets, and debt – which have acquired particular contemporary salience for the way in which they have amplified accounting as a means of apprehending and responding to climate change.[4] In different ways, the language of budgets, balance

[4]Other accounting terms, such as 'stock' and 'flow,' as well as interest and dividend, also warrant attention but are beyond what we can consider here.

sheets, and debt, and their associated practices, have worked to make the immense complexities of climate change graspable and manageable within existing architectures of management and governance where accounting logics are deeply entrenched. In some instances, they have also generated opportunities for counter-mobilizations and political resistance in struggles for climate justice through the 'lenses of inequality and injustice' (Borras Jr. et al. 2022, 12; Sultana 2022; Tuana 2019; Borras and Franco 2018).

Together with calculative devices that seek to render greenhouse gas emissions quantifiable, and capable of being inventoried and commensurated, this trinity of concepts are among those that participate in the process of disembedding carbon from situated social and ecological agrarian settings and establishing conditions for them to be re-embedded in novel financial and market relations in ways that can facilitate accumulation. One of the starkest ways in which this occurs is through the enunciation of calculative devices that have been developed to make different greenhouse gases commensurable, or equivalent to one another. Commensuration, or 'the comparison of different entities according to a common metric' (Espeland and Stevens 1998, 313), has been identified as a key premise for the commodification of carbon and the constitution of carbon markets. Within scholarship on carbon accounting, several studies explore how different greenhouse gases are 'made the same,' that is, made commensurable through metrics such as carbon dioxide equivalent (CO_2e) and greenhouse warming potential (GWP). A prime example of this is 'Global Warming Potential,' an index which renders different greenhouse gases comparable to (and exchangeable with) carbon dioxide and is used in the calculation of CO_2 equivalents (MacKenzie 2009; Whitington 2016). MacKenzie (2009, 447) goes so far as to liken metrics such as Global Warming Potential and CO_2e to 'exchange rates' which enable fungibility and liquidity in carbon markets by allowing units of CO_2e to be isolated and exchanged across jurisdictions. Instruments such as CO_2e and Global Warming Potential amplify accounting, allowing greenhouse gases to be disembedded from the immediate conditions in which they are produced and repurposed to enable novel forms of accumulation within market relations.

But the mathematical relation brought forth through techno-scientific devices that render greenhouse gases commensurable or fungible neglects the multiple other relational dimensions of these gases. Illustrating this are the comments of the Director of the Colombian branch of the Dutch NGO Tropenbos, dedicated to the conservation of rainforests, who qualified the ways in which equivalence has had the effect of 'making things the same.' (MacKenzie 2009) To the contrary, this director told us, 'not all carbon emissions are made equal.'[5] He went on to explain that a mitigation project promoting the conservation of a biodiverse rainforest is very different from a mitigation project based on reforestation through commercial plantations. While both projects might contribute to removing atmospheric CO_2, the role of a biodiverse rainforest in regulating the climate cannot be reduce to units of CO_2 emitted in a particular period of time.

Devices such as Global Warming Potential and CO_2 equivalent are powerful tools that disembed greenhouse gases from the agrarian relations through which they are produced, but they need to be accompanied by narratives that scale up and extend the

[5]Despite the ways in which they are rendered commensurable, greenhouse gases also harbor different legal, political, and chemical meanings a point that Gökçe Günel (2016) has usefully developed. Such multiplicity is just as important as commensuration for processes of commodification (Günel 2016) and, we would argue more broadly, for accumulation.

reach of accounting logics in mitigation action and finance. It is here that vernacular notions of budgets, balance sheets, and debt have enabled carbon dioxide and its equivalents to be conceptualized as a resource amenable to management and regulation, with the potential to be harnessed in accumulation processes. Such discursive transformations warrant scrutiny, while also calling for attention to be paid to the ways that agrarian relations come to be reconfigured around carbon.

One of the most fundamental ways in which narratives work is to condition the way that the problem of climate change is itself understood. In this, one of the most pervasive accounting metaphors to be taken up in climate science and policy is that of the carbon budget. The notion of the carbon budget, which posits carbon dioxide as a planetary resource to be managed, is a relatively recent one appearing with the publication of the IPCC's Fifth Assessment Report in 2013 (Lahn 2021, 5). This assessment report marked a shift from a prior emphasis on the stabilization of greenhouse gas emissions and atmospheric concentrations of CO_2 to one that established a global carbon budget, fixing cumulative carbon dioxide emissions to a level that was capped by the target of limiting warming to 2 degrees Celsius, and later 1.5 degrees Celsius, above pre-industrial levels (Lahn 2021). By connecting the goal of reducing GHG emissions to the figurative notion of a financial budget, the notion of a global carbon budget has captured the attention of a range of actors and catalyzed responses within policy, finance, corporate, and public sectors. It has also succeeded in mobilizing a broad public imagination, as in the Guardian's Carbon Countdown Clock which counts the time left before the world surpasses the IPCC's 2 degree Celsius target at current emissions levels (Evershed 2017).

To some extent, the notion of the carbon budget, more than that of stabilizing emissions, has unleashed more radical, transformative politics, with climate strikes, direct action, and campaigns to leave fossil fuels in the ground sometimes also appropriating the notion of a budget (see Lahn 2021, 4–5). In this vein, the notion of a global carbon budget can create political and discursive space for distributional questions of how and among whom this budget is to be allocated (Neumayer 2000). Notions of carbon deficits and debt similarly raise questions about how these are to be serviced, although they have also sparked concern because of the discursive openings they create for leveraging concrete interventions including negative emissions technologies such as carbon capture and storage, solar radiation management, and other technologies to not only reduce deficits or debts but even to underwrite overshoots or expansions of the global budget (Asayama, Hulme, and Markusson 2021, 1; Mazzai 2021).

Though agriculture's principal emissions – methane and nitrous oxide – are not captured in strict scientific formulations of carbon budgets which are limited to carbon dioxide, in the wake of the IPCC's Fifth Assessment Reports (AR5), as well as the Special Reports on Global Warming of 1.5°C and Climate Change and Land, published between the fifth and sixth assessment cycles, agriculture has become more central in the way that budgets are reckoned with (IPCC 2019, 2014, 2018). Within scientific communities, it has been noted that remaining within a carbon budget of 580 Gigatonnes demands a 'substantial concurrent reduction of agricultural emissions' and that achieving this will not be possible without such reductions (Leahy, Clark, and Reisinger 2020, 2). Others, however, have sounded a note of caution at the aptness of the carbon budget, particularly in relation to agricultural greenhouse gas emissions, noting that cumulative budgets cannot account for the complexity and dynamism of agricultural emissions (Lahn 2021).

Mitigation action, nonetheless, continues to be configured using accounting terms such as budgets and balance sheets. These vernaculars have, in turn, spawned others – notably they have created the conditions of possibility for notions such as Net Zero and climate neutrality. Promoted by advocacy programs such as the UNFCCC's (2021) 'Race to Zero' campaign, the notion of net-zero and climate neutrality have become a central focus in recent scientific and political discussions (Asayama, Hulme, and Markusson 2021, 2; Tanaka and O'Neill 2018; Rogelj et al. 2021). These ambitions are increasingly translated into programs and interventions at varying scales, gathering force among a wide swath of actors in the scientific, government and policy, as well as private and financial sectors. In a recent presentation, Miguel Taboada from the Argentinean Institute of Agricultural Research (INTA) remarked that carbon neutrality works like an accounting balance sheet 'we have to make a balance between the level of GHG emissions that we are producing ... and the possibility to compensate those emissions' (Taboanada 2021, min 10). Here, we see how in an era of global climate change, the classic accountant's balance sheet, which normally displays net-revenues, is reworked to instead display net carbon emissions. Principles of double-entry book-keeping, part of what was so crucial for the development of early capitalism in northern Italy (Mills 1994), are now applied in the management of carbon accounts and budgets.

Yet, the global framing of the carbon budget and its balance sheet obscures the unequal responsibility that different countries have had in the creation of a 'carbon deficit' (Neumayer 2000; Callahan and Mankin 2022). Environmental activists, as well as heads of state and coalitions of countries of the Global South, argue that industrialized countries are responsible for the largest amount of CO_2 emissions that have been historically produced and that are currently accumulated in the atmosphere. At the same time, the effects of climate change are disproportionately felt in poorer countries that have less financial and technological capacities to respond (Newell 2022; Arsel 2022; Borras Jr. et al. 2022). In struggles for climate justice, the accounting logics implied in the balance sheet have facilitated the emergence of debates about how climate debt is construed, as well as loss and damage and climate reparations. Industrialized countries have a 'climate-debt' with countries of the South and should take the lead reducing carbon emissions and funding climate adaptation programs (Pickering and Barry 2012; Khan et al. 2020). As an illustration, Tom Athanasiou, from the EcoEquity nonprofit, suggests that the U.S has over used its share of the global carbon-budget: 'we can't meet our moral and practical burdens simply by reducing our own emissions; we've already put so much carbon into the air (and hence reduced the space that should rightly go to others) that we need to make amends' (McKibben 2020). Such notions of climate debt have led to the (so far unfulfilled) financial pledge from industrialized countries to commit USD 100 billion towards adaptation and resilience in the South from 2020, and they animate long-standing efforts to secure binding international commitments on loss-and-damage, as was evident most recently in COP27.

Framing climate justice in terms of debt is symptomatic of the power of accounting and finance in climate change discussions, but also signals some of the dangers that this entails. The development of climate vernaculars – budgets, balance sheets, and debts – work to mobilize climate action, frame mitigation, and inform policy responses to climate change, and to orient investment. In so doing, GHG emissions become a 'resource' to be managed and governed. In a similar way, carbon emission reductions are turned into a resource that can expand the carbon budget and be traded as carbon offsets. However, this economic strategy defined by carbon budgets, emission rights, and offsets trading, depends on a perfectly calibrated calculative landscape where GHG emissions are accurately measured, emission rights effectively distributed, and emission reductions certified and verified. Despite its questionable efficacy, this strategy is already producing new streams of revenue linked to the production and trade of GHG emission reductions. Such reductions often result from the conjoined mobilization of human labor and more-than-human capacities (often referred to as natural capital, ecosystem services, or nature-based solutions) to capture and store greenhouse gases, such as photosynthesis and soil carbon sequestration. In achieving this, accounting is instrumental in making visible and measurable new forms of more-than-human work/energy as the basis for still more capital accumulation (Moore 2015, 2017, 2018).

Mitigation, accounting, and agrarian relations

When it comes to agrarian relations, carbon accounting must be understood as more than a technical, calculative exercise but a social and political process that may transform local realities. As discussed in the previous section, carbon accounting simplifies the multiple relations that lie behind the production of GHG emissions, by translating them into commensurable substance–units of CO_2 equivalent. It also helps to frame climate mitigation interventions through economic language of balance-sheets, budgets, and debt, and disembeds greenhouse gas emissions from the socio-environmental relations that produce them. Although some degree of disembedding necessarily accompanies efforts to quantify agricultural emissions, it is equally important to attend to what such disembedding enables: that is to say, what historical and socio-environmental relations are obscured through accounting frameworks, and what kinds of relations are enabled by it. From this perspective, it becomes possible to see that in extending the epistemic and discursive dimensions of accounting into agrarian settings, the deployment of quantifying infrastructures that account for GHG emissions on the ground are reconfiguring local agrarian landscapes, re-organizing ecosystems and economic activities, to render them amenable to quantification. In this manner, GHG emissions are re-embedded in agrarian relations within the parameters of carbon accounting and through novel quantification infrastructures.

Carbon accounting helps to render GHG emissions a resource that can be mined and extracted through quantification protocols and methodologies aimed at their mitigation, either in absolute terms or, as is often the case with net zero and carbon neutrality pledges, in terms of emissions intensity. In the process, new infrastructures and networks of agricultural production in which accounting practices are embedded are transforming agrarian worlds. Particularly notable is the way that carbon accounting revolves around the production of data about GHG emissions, the reductions of which can be monetized in carbon and agricultural markets. However, the need to produce data about mitigation efforts influences the shape of mitigation action in agriculture in ways that can further the uniformization of agrarian practices and promote the scaling-up of industrial agricultural production. Carbon accounting in agriculture also expands the reach that some institutional actors have over agrarian settings, strengthening their authority to reshape and centralize production processes and data collection. In so doing, carbon accounting has an impact over farmers' productive autonomy and can influence the role that farmers' play in the future of agriculture.

Consider two corporate mitigation initiatives led by the US company Indigo Ag and the Argentinian company Bioceres Crop solutions. These agricultural inputs providers have attracted increasing attention in the last decade because of their commitment to reduce the global agricultural carbon footprint. In 2019, Indigo Ag was named the 'World's most innovative company' by CNBC disruptors list (CNBC 2019), while some of the technologies of Bioceres have been deemed of national interest by the Argentinian senate (Silva 2020). Some of the star products of these companies include microbial seed treatments that act as plant growth promoters and that can allegedly reduce the need for agrichemical products, contributing to the reduction of GHG emissions (mainly carbon dioxide, nitrous oxide emitted in the production of fertilizers). These companies also promote their clients' adoption of agricultural practices that can increase soil carbon sequestration, such as the use of cover crops and no-till farming.

To promote their microbial products and the adoption of good agricultural practices, Indigo and Bioceres have created mitigation programs that aim at scaling up their mitigation goals with the help of farmers (Indigo Ag 2022; Bioceres Crop Solutions 2022a). Farmers' who want to reduce their GHG emissions can join these programs but they have to commit to follow specific crop management practices, use particular products and provide data about their agricultural practices, inputs and plots (for example, the history of deforestation, rotation crops, agrochemical use, tilling practices). This data is then used to estimate the carbon footprint of particular crops and to support the certification of crops as low carbon or carbon neutral. In return for farmers' compliance, these programs provide monetary compensation. Bioceres pays up to 30 USD per hectare to farmers that provide data allowing the company to keep product traceability of their products (Bioceres Crop Solutions 2022a). Indigo Ag promises to pay farmers 75% of the profits resulting from mitigation efforts (Indigo Ag 2022). These examples show that as corporate coordination of carbon accounting initiatives scale up mitigation efforts, they can also increase corporate control of agricultural value chains. They can motivate centralized networks of agricultural production managed by corporate actors, in which farmers are instructed, and in some cases even legally obliged, to use certain products and adopt particular agricultural practices.

Clearly, the goal of these corporate mitigation programs is not only to promote GHG emission reductions in agriculture but also to create new streams of revenue linked to the commercialization of carbon credits and low carbon products. The capacity of agrifood companies such as Indigo and Bioceres to turn GHG emission reductions into an economic resource depends not only on their ability to enrol farmers and promote recommended agricultural practices, but also on their capacity to calculate and certify GHG emission reductions according to acceptable standards. Thus, mitigation programs follow accepted standards of carbon accounting ensuring that their mitigation achievements can be certified. For example, Indigo's 'Carbon' program helps farmers to certify their GHG emission reductions through the VERRA VM00042 methodology, so that they can be transformed into tradeable carbon credits (Indigo Ag 2022). The productive networks promoted by corporate mitigation programs are thus shaped around highly technical calculating efforts where companies control the accounting knowledge (even designing their own accounting methodologies as Indigo's VM0042), and where farmers become data providers sharing data about their production processes.

The production of GHG emission reductions as a resource is not, however, equally feasible for all agricultural actors. Compliance with carbon accounting standards is technically complex and the certification process is expensive, which prevents, small and less capitalized farmers from measuring, certifying, and monetizing their mitigation achievements on their own. Even large companies such as Bioceres require assistance when trying to produce certifiable GHG emission reductions. While Bioceres is currently not seeking to produce carbon credits as is Indigo, the company aims at certifying agricultural products to be sold in specialty markets at more attractive prices (or at least to ensure that they comply with increasingly stringent environmental regulations).[6] To reach this goal, Bioceres has joined the 'Programa Argentino Carbono Neutro' (PACN), which supports Argentinian agricultural producers in calculating their GHG emissions and communicating their mitigation achievements to agricultural markets through the labeling of certified carbon neutral products (PACN 2020).[7]

While these corporate programs create new streams of revenue that are shared with farmers, the corporate control of these networks means that these benefits come to farmers at the cost of less productive autonomy. Although not all mitigation programs depend on the coordination of corporate actors, for example public and non-governmental actors are also leading mitigation efforts, all of those we have encountered in our research across Argentina, Canada, and Colombia rely on top-down coordination and centralized networks of production. For example, the Colombian public program *Vision Amazonia* brings together indigenous and peasant communities in the Colombian Amazon to work against deforestation and forest degradation, leading to carbon emission reductions that are later exchanged against international cooperation funds (Visión Amazonía 2022). These funds are then distributed to local projects related to forest conservation and agroforestry. The centralized and top-down associative nature of mitigation efforts is related to the data requirements of carbon accounting frameworks. *Vision Amazonia* has a large pedagogical component to train local communities on the environmental and economic benefits of transforming their land-use and agricultural practices, as well as on the need to monitor and keep records of their progress. Therefore, although this type of program is not run by corporate actors, it is still reliant on the governance of local practices by external actors to fulfill the requirement of carbon accounting methodologies.

A new Colombian program called *Biocarbono* aims at producing low-carbon agrarian landscapes (Biocarbono 2022) with a similar result-based methodology to that of *Vision Amazonia*. The program promotes changes in land-use that increase soil carbon

[6]The production of certified low carbon crops follows a similar logic to that of other goods produced for specialty markets (e.g. organic, clean, fair-trade). However, unlike organic and fair trade certifications that produce particular qualities (Galvin 2021), carbon accounting participates in a substantially different type of capitalization where GHG emission reductions become a commodity in themselves through the process of verification and certification.

[7]The PACN offers three main tools for this purpose: a series of manuals of good practices that can help farmers reduce their GHG emissions; a carbon calculator that can be used to estimate the amount of GHG emissions and emission reductions; and product stamps to differentiate the products that are part of the program in agricultural markets. The strictest of the three stamps offered by the PACN certifies that a farmer's production has achieved carbon neutrality. This stamp must be supported by internationally recognized standards so that any mitigation achievements are also recognized in international agricultural markets.

sequestration, such as the replacement of cattle-ranching for cocoa and palm oil agroforestry projects, in regions heavily affected by deforestation or with a high potential for afforestation such as the Colombian Orinoquía. This country-based program is part of the World Bank Biocarbon Fund's Initiative for Sustainable Forest Landscape (ISFL). The ISFL is financed by international cooperation funds and supports agrarian mitigation programs in five different countries of Latin America, Africa and South East Asia. The initiative has a strong focus on carbon accounting (ISFL 2022), seeking to build the technical capacity of participant countries so that they can carry out the measurement, reporting and verification (MRV) of their programs' GHG emission reductions and removals. The data produced by these MRV systems could then be used to justify the disbursement of Biocarbon funds against certified GHG emission results. These disbursements would compensate local actors and communities for their involvement in these programs.

The objectives of the corporate and national mitigation programs discussed so far are not completely the same, but the production of GHG emission reductions as a resource is conditioned by the implementation of carbon accounting frameworks in both types of initiatives. The standards of carbon accounting produced by VERRA are amongst the most widely used for AFOLU (Agriculture, forestry and land use) projects, including more than thirty methodologies. Both Indigo and the Biocarbon Fund have helped to produce VERRA methodologies for the AFOLU sector (VERRA 2011, 2020), while also being positioned to benefit from them, something that we have also observed in the development of carbon offsets for agricultural emissions in the Canadian province of Alberta.[8] Most of these methodologies are aimed at forestry and afforestation projects such as REDD+. However, six of them are directly related to agricultural production (VERRA 2022a), including projects of crop and land management such as the restoration of grazing land through forest plantations, the replacement of chemical fertilizers by organic compost, and the shifting of continuous and unrestricted grazing for planned rotational grazing, amongst others.

Carbon accounting frameworks can influence the types of agriculture that are promoted by mitigation action and the actors that they target. Judging by the VERRA registry (VERRA 2022b), where all the projects using VERRA methodologies are listed, reforestation projects and methodologies predominate over agricultural projects (out of 260 registered AFOLU projects, only six follow methodologies directed at the agricultural sector). The involvement of small farming communities in these projects is prominent in reforestation and agroforestry interventions (such as Vision Amazonia) where GHG emission reductions are achieved through the improvement of forest coverage. The low number of agricultural projects can be indicative of the challenges posed by carbon accounting methodologies for this sector in general, and for small-scale farming in particular. For example, methodologies that involve life-cycle assessments and that require product traceability (such as those implemented in the mitigation programs of Indigo and Bioceres) are more easily implemented in large-scale monocultures than in polyculture farming, where lack of crop uniformity makes product traceability difficult. Moreover, small-scale farming does not offer the same level of scalability for mitigation initiatives, which means that the

[8]Indigo produced the VM00042 methodology for improved agricultural land management (VERRA 2020) and the Biocarbon Fund produced the VM00017 methodology for the adoption of sustainable agricultural land management (VERRA 2011).

economic benefits of implementing carbon accounting methodologies are less attractive. This explains in part why corporate programs usually target industrial farmers of export-oriented monocrops. For example, farmers involved in the Bioceres program cultivate an average of 229 hectares of land (Bioceres Crop Solutions 2022b).[9]

Besides issues of uniformity and scalability, farmers' active participation in mitigation projects so far remains an important condition for the popularization of carbon accounting methodologies in agriculture. Gathering and sharing the data required by carbon accounting methodologies places an additional burden of work on farmers, discouraging them from joining mitigation programs or from strictly following their guidelines of data collection. To face this challenge, agri-food companies deploy field teams to fill data gaps while also coming up with strategies to 'make it easier for farmers to gather the data they need to earn certified carbon credits' (Bomgardner and Erickson 2021 Q16], 14). Companies such as Bayer are testing pilot projects where data gathering and record keeping is automated with the help of precision agricultural machinery, satellite information and digital platforms. These automation strategies may simplify carbon accounting in agriculture but they can also influence the types of agriculture that carbon accounting promotes. Such strategies are often designed and carried out by agri-food corporations that can afford such heavy investment and risk, but they enable industry concentration in large-scale monocrop agriculture. Automation strategies can also disproportionally attract mitigation finance, which is 'inclined towards sectors or emission sources where quantification and monitoring of the mitigation benefits is relatively easy' (FAO 2019, 6).

Finally, automation strategies deployed to simplify the practice of carbon accounting can promote imaginaries of the future that displace farmers from their central role in agricultural production. The CEO of Okaratech, the company that manages Bioceres' digital platform in Argentina, imagines a type of agriculture where satellites, drones, mechanical planters, harvesters and fumigators communicate with each other in real time through the 'internet of things,' coordinating production and gathering data without farmer involvement. In these imaginaries of the future, farmers play a minimal role because they are perceived as an obstacle to data gathering and record keeping. Thus, if these imaginaries of the future of agriculture materialize, they could have serious implications for the role of farmers in agricultural production, and for food sovereignty and productive autonomy among other things.

Even if these imaginaries do not gain traction, carbon accounting is already affecting the type of agricultural production promoted by mitigation initiatives. Mitigation programs are creating new configurations of institutional power and alliances as corporate, state, and non-governmental actors position themselves as expert intermediaries between farmers' practices and mitigation ambitions. This is leading to the emergence of new forms of agricultural governance that are premised on the production, centralization, and control of GHG emission data to be either exchanged for cooperation funds, or certified and monetized. Such moves threaten farmers' productive autonomy, and they may reproduce perceptions of farmers as dependent on assistance and expert guidance

[9]Bioceres program has grown from involving 25 wheat farmers in 2020 to 225 wheat farmers in 2022 and from 15 soy producers in 2020 to 103 farmers in 2022. These farmers produce soy on 23 thousand hectares of land and wheat on 53 thousand hectares of land.

or as obstacles to agricultural modernization, while also generating new demands on labor as farmers become subjected to the guidelines of carbon accounting frameworks.

Conclusions: reimagining accounting, advancing agrarian counter-accounts

Accounting is a field of knowledge and a form of professional practice; it is also a way of viewing the world, and a force that participates in re-making it. In this article, we have examined its particular force along three different dimensions – as epistemology, instrument, and narrative – and explored how and with what effects it is being operationalized through corporate and public climate mitigation interventions in agrarian settings. What emerges as notable from this study is the particular role that accounting tools and methods play in disembedding carbon dioxide and its CO_2e counterparts from the socio-ecological and political-economic relations in which they are produced, while vocabularies, concepts, and narratives do the work of *amplifying* accounting as a means of apprehending and responding to climate change. Throughout, the tight relation between accounting, governance, and management has helped to keep climate change as a primarily 'problem of emissions.'

In this way, accounting must be acknowledged as a condition of possibility for the expansion and deepening of capitalist accumulation in the midst of, and even through, environmental crises. Accounting comes to shape accumulation pathways in situated and particular ways as agrarian environments and relations are brought into the fold of climate mitigation. In the foregoing discussion we have noted that accounting is a force that must be reckoned with in relation to long-standing concerns in agrarian studies with questions of land and labor, as well as more recent ones such as knowledge, data, and nonhuman life and worlds. What forms of accounting and accounts are demanded and produced in the name of climate mitigation? By whom, for whom, and with what effect? As agrarian peoples and places come to be governed through regimes of climate mitigation and low-carbon agriculture, these are some of the questions that we suggest need to be asked and addressed in grappling analytically and politically with both novel and abiding processes of accumulation.

At the same time however, we also look with interest toward the imaginative promise of what the critical accounting scholar Robert Gray (2010) and several of his collaborators have called 'counter-accounts.' Counter-accounts are those that leverage the power and versatility of accounting vocabularies in order to render or take account differently (Gray, Brennan, and Malpas 2014; Gray 2010). Accounting-giving, account-taking, and account-holding are, in this expanded sense, profoundly social practices that can both open, and foreclose, different political potentials. Indeed, while carbon accounting becomes subsumed by the values and representational forms of financial accounting, some critical accountant scholars have reacted by arguing that 'there is no a priori reason for restricting accounts to any particular form of representation, any particular medium of expression' (Gray, Brennan, and Malpas 2014, 262). Instead, we should recognize that accounts can 'cut across pretty much any other category of existence we might choose – be it (say) language, literature, organizations, families or markets' (2014, 262), and that there are multiple ways in which different collectives produce accounts of their existence. Along these lines, such multiple avenues for giving and holding account may open up other ways of figuring climate change in a way that resists the reductionism of a focus on emissions, and that can be harnessed in struggles for what Borras and Franco (2018) have termed agrarian climate justice. Current examples of counter-accounts can be found in mobilizations around climate debt, loss and damage, and unequal ecological exchange.

Emergent forms of what we call agrarian counter-accounts are located at different sites and scales of social, political, and epistemological practice. Some forms of agrarian counter-accounts may not eschew the quantitative logics of accounting entirely, but instead seek to harness them toward different ends. Along these lines, ecological economists have developed modes of account-taking that seek to make legible and quantifiable global environmental injustices, accounting for the unequal flow of virtual soil (Pengue 2017; Zuberman 2019), virtual water (Dalin et al. 2012) and GHG emissions across different geographies of the world (Van Houtan et al. 2021). Whilst still relying on quantitative methods and in some instances also the conceptual parameters of 'natural capital,' the forms of evidence that are generated by these sorts of counter-accounts may be used to support arguments for environmental reparations, compensation, and loss-and-damage, by demonstrating how the cost of pollution and environmental degradation is unequally distributed. They also contribute to wider efforts to center justice in climate change responses, by 'undoing equivalence' (Carton, Lund, and Dooley 2021) and attending to questions of restitution, recognition, and redistribution (Borras and Franco 2018). Community-based approaches, on the other hand, give and take account of local environmental change in ways that extend far beyond methods of measuring and estimating GHG emissions. Some examples of these approaches include social cartographies about the evolution of local ecosystems (Rodríguez 2010), as well as literary accounts of local environmental change (Fujikane 2021). These are but a few examples that point to a range of possible agrarian counter-accounts that may be given, taken, and held, and which exceed the narrow scope of accounting for climate change by accounting for emissions alone. Cumulatively, they show that as much as carbon accounting may be a powerful tool for enabling new forms of accumulation in agrarian settings, agrarian worlds equally offer an important political location from which to develop accounts that mobilize against longstanding practices extraction, exploitation, and unequal exchange.

At the same time, in creating an opening toward the imaginative political and emancipatory potential of counter-accounts, it is worth bearing in mind Audre Lorde's caution 'the master's Tools will never dismantle the master's house' (Lorde 1983). To the extent that accounting has been crucial in the emergence and expansion of capitalism and capitalist accumulation in new domains, it is uncertain whether it can be re-purposed in struggles for climate justice, agrarian and otherwise. So thoroughly have responses to climate change been 'rendered technical' by accounting knowledges and interventions, that the question arises whether accounting in any form can be a means not only for challenging or identifying processes of exclusion and deepening inequality, but for pursuing alternatives that would give substance to robust and meaningful climate justice. What kinds of reimagination of accounting would this entail? What accounts must be taken, given, and to what ends? For now, these remain open questions. But, given the way that accounting has taken firm hold in modes of responding to and engaging with the climate crisis, these may yet be questions worth asking.

Acknowledgements

The authors are grateful to their research interlocutors in Argentina, Colombia, and Canada, and express thanks to John Paulraj, Michelle Niemann, as well as those who participated in our panel at the conference 'Climate Change and Agrarian Justice,' convened by JPS, the Transnational Institute, and the Collective of Agrarian Scholar-Activists from the South. We much appreciate the comments and suggestions from *JPS*'s anonymous reviewers, which helped us to strengthen our manuscript. Errors and emissions remain ours alone.

Disclosure statement

No potential conflict of interest was reported by the author(s).

Funding

Research conducted for this article took place with the support of a grant from the Swiss National Science Foundation for the project 'Accounting for Nature: Agriculture and Mitigation in the Era of Global Climate Change' (grant number 10001H_197546).

ORCID

Shaila Seshia Galvin http://orcid.org/0000-0002-3228-5448
Diego Silva Garzón http://orcid.org/0000-0002-5441-9937

References

Arsel, Murat. 2022. "Climate Change and Class Conflict in the Anthropocene: Sink or Swim Together?" *The Journal of Peasant Studies*, 1–29. doi:10.1080/03066150.2022.2113390.

Asayama, Shinichiro, Mike Hulme, and Nils Markusson. 2021. "Balancing a Budget or Running a Deficit? The Offset Regime of Carbon Removal and Solar Geoengineering under a Carbon Budget." *Climatic Change* 167 (1–2): 25. doi:10.1007/s10584-021-03174-1.

Bebbington, Jan, Carlos Larrinaga, and Ian Thomson. 2021. "Curating Environmental Accounting Knowledge." In *Routledge Handbook of Environmental Accounting*, edited by Jan Bebbington, Carlos Larrinaga, Brendan O'Dwyer, and Ian Thomson, 3–16. Abingdon: Routledge.

Biocarbono. 2022. "Biocarbono Orinoquía." Accessed 2 December 2022. https://biocarbono.org/proyecto/

Bioceres, Crop Solutions. 2022a. "Generaci6n HB4" Accessed 2 December 2022. https://www.generacionhb4.com.ar/#program.

Bioceres Crop Solutions. 2022b. "Financial and Operating Results: Second Quarter Fiscal Year 2022." Rosario Argentina. Accessed 2 December 2022. https://s26.q4cdn.com/783252186/files/doc_financials/2022/q2/BIOX-2Q22-Financial-Results.pdf

Bomgardner , Melody, and Britt Erickson. 2021. "Carbon Farming Gets off the Ground: Replenishing Depleted Soil Can Combat Climate Change If Science, Policy, and Markets Align." *C&EN* 99 (18): 22–28. https://pubs.acs.org/doi/10.1021/cen-09918-cover.

Borras, Saturnino M., and Jennifer C. Franco. 2018. "The Challenge of Locating Land-Based Climate Change Mitigation and Adaptation Politics within a Social Justice Perspective: Towards an Idea of Agrarian Climate Justice." *Third World Quarterly* 39 (7): 1308–1325. doi:10.1080/01436597.2018.1460592.

Borras Jr., Saturnino M., Ian Scoones, Amita Baviskar, Marc Edelman, Nancy Lee Peluso, and Wendy Wolford. 2022. "Climate Change and Agrarian Struggles: An Invitation to Contribute to a *JPS* Forum." *The Journal of Peasant Studies* 49 (1): 1–28. doi:10.1080/03066150.2021.1956473.

Bridge, Gavin. 2011. "Resource Geographies 1: Making Carbon Economies, Old and New." *Progress in Human Geography* 35 (6): 820–834. doi:10.1177/0309132510385524.

Callahan, Christopher W., and Justin S. Mankin. 2022. "National Attribution of Historical Climate Damages." *Climatic Change* 172 (40): 1-19. doi:10.1007/s10584-022-03387-y.

Carton, Wim, Jens Friis Lund, and Kate Dooley. 2021. "Undoing Equivalence: Rethinking Carbon Accounting for Just Carbon Removal." *Frontiers in Climate* 3: 664130. https://www.frontiersin.org/articles/10.3389fclim.2021.664130.

Cavanagh, Connor, and Tor A. Benjaminsen. 2014. "Virtual Nature, Violent Accumulation: The 'Spectacular Failure' of Carbon Offsetting at a Ugandan National Park." *Geoforum* 56: 55–65. doi:10.1016/j.geoforum.2014.06.013.

Chakrabarty, Dipesh. 2009. "The Climate of History: Four Theses." *Critical Inquiry* 35 (2): 197–222. doi:10.1086/596640.

Charnock, Robert, Matthew Brander, and Schneider Thomas. 2021. "Carbon." In *Routledge Handbook of Environmental Accounting*, edited by Jan Bebbington, Carlos Larrinaga, Brendan O'Dwyer, and Ian Thomson, 353–365. Abingdon: Routledge.

Clapp, Jennifer, Peter Newell, and Zoe W. Brent. 2018. "The Global Political Economy of Climate Change, Agriculture and Food Systems." *The Journal of Peasant Studies* 45 (1): 80–88. doi:10.1080/03066150.2017.1381602.

CNBC. 2019. "CNBC Disruptor 50." Accessed 2 December 2022. https://www.cnbc.com/2019/05/14/indigo-ag-2019-disruptor-50.html

Dalin, Carole, Megan Konar, Naota Hanasaki, Andrea Rinaldo, and Ignacio Rodriguez-Iturbe. 2012. "Evolution of the Global Virtual Water Trade Network." *Proceedings of the National Academy of Sciences* 109 (16): 5989–5994. doi:10.1073/pnas.1203176109.

Dalsgaard, Steffen. 2014. "Carbon Value between Equivalence and Differentiation." *Environment and Society* 5 (1): 86–102. doi:10.3167/ares.2014.050106

Davis, Heather, and Zoe Todd. 2017. "On the Importance of a Date, or Decolonizing the Anthropocene." *ACME* 16 (4): 761–780.

Espeland, Wendy Nelson, and Mitchell L. Stevens. 1998. "Commensuration as a Social Process." *Annual Review of Sociology* 24 (1): 313–343. doi:10.1146/annurev.soc.24.1.313.

Evershed, Nick. 2017. "Carbon Countdown Clock: How Much of the World's Carbon Budget Have We Spent?" *The Guardian*, January 19. https://www.theguardian.com/environment/datablog/2017/jan/19/carbon-countdown-clock-how-much-of-the-worlds-carbon-budget-have-we-spent

Fairhead, James, Melissa Leach, and Ian Scoones. 2012. "Green Grabbing: A New Appropriation of Nature?" *Journal of Peasant Studies* 39 (2): 237–261. doi:10.1080/03066150.2012.671770.

FAO. 2022. *The State of the World's Land and Water Resources for Food and Agriculture 2021 – Systems at Breaking Point*. Rome: Food and Agriculture Organization. doi:10.4060/cb9910en.

FAO. 2019. "Reducing Enteric Methane for Improving Food Security and Livelihoods." Rome: Food and Agriculture Organization. https://www.fao.org/3/i5902e/i5902e.pdf

Figge, Frank, Tobias Hahn, Stefan Schaltegger, and Marcus Wagner. 2002. "The Sustainability Balanced Scorecard - Linking Sustainability Management to Business Strategy." *Business Strategy and the Environment* 11 (5): 269–284. doi:10.1002/bse.339.

Fleischman, Richard K., Warwick Funnell, and Stephen P. Walker, eds. 2013. *Critical Histories of Accounting: Sinister Inscriptions in the Modern Era*. Routledge New Works in Accounting History 11. New York: Routledge.

Fujikane, Candace. 2021. *Mapping Abundance for a Planetary Future: Kanaka Maoli and Critical Settler Cartographies in Hawai'i*. Durham: Duke University Press.

Galvin, Shaila Seshia. 2021. *Becoming Organic: Nature and Agriculture in the Indian Himalaya*. London: Yale University Press.

Gómez-Baggethun, Erik, and Berta Martin Lopez. 2015. "Ecological Economics Perspectives on Ecosystem Services Valuation." In *Handbook of Ecological Economics*, edited by Joan Martinez-Alier and Roldan Muradian, 260–282. Cheltenham, UK: Edward Elgar Publishing. doi:10.4337/9781783471416.

Gray, Rob. 1990. *The Greening of Accountancy: The Profession After Pearce*. Research Report 17. London: The Chartered Association of Certified Accountants.

Gray, Rob. 2010. "Is Accounting for Sustainability Actually Accounting for Sustainability … and How Would We Know? An Exploration of Narratives of Organisations and the Planet." *Accounting, Organizations and Society* 35 (1): 47–62. doi:10.1016/j.aos.2009.04.006.

Gray, Rob, Andrew Brennan, and Jeff Malpas. 2014. "New Accounts: Towards a Reframing of Social Accounting." *Accounting Forum* 38 (4): 258–273. doi:10.1016/j.accfor.2013.10.005.

Groot, Rudolf de, and Leon Braat. 2015. "The Contributions of the Ecosystem Services Paradigm to Sustainability Science, Policy and Practice." In *Handbook of Ecological Economics*, edited by Joan Martinez-Alier and Roldan Muradian, 233–259. Cheltenham, UK: Edward Elgar Publishing. doi:10.4337/9781783471416.

Günel, Gökçe. 2016. "What Is Carbon Dioxide? When Is Carbon Dioxide?" *PoLAR: Political and Legal Anthropology Review* 39 (1): 33–45.

Hahn, Rüdiger, Daniel Reimsbach, and Frank Schiemann. 2015. "Organizations, Climate Change, and Transparency: Reviewing the Literature on Carbon Disclosure." *Organization & Environment* 28 (1): 80–102. doi:10.1177/1086026615575542.

Hopwood, Anthony G. 2009. "Accounting and the Environment." *Accounting, Organizations and Society* 34 (3–4): 433–439. doi:10.1016/j.aos.2009.03.002.

Indigo, Ag. 2022. "Carbon by Indigo." Accessed 2 December 2022. https://www.indigoag.com/carbon

IPCC. 2014. *Climate Change 2014: Mitigation of Climate Change: Working Group III Contribution to the Fifth Assessment Report of the Intergovernmental Panel on Climate Change*. Edited by O. Edenhofer, R. Pichs-Madruga, and Y. Sokona. Cambridge, UK: Cambridge University Press.

IPCC. 2022. "Summary for Policymakers." In *Climate Change 2022: Mitigation of Climate Change. Contribution of Working Group III to the Sixth Assessment Report of the Intergovernmental Panel on Climate Change* , edited by P. R. Shukla, J. Skea, R. Slade, A. Al Khourdajie, R. van Diemen, D. McCollum, and M. Pathak, et al. Cambridge, UK: Cambridge University Press.

IPCC. 2006. *2006 IPCC Guidelines for National Greenhouse Gas Inventories*, edited by H. S. Eggleston L. Buendia, K. Miwa T. Ngara and K. Tanabe. Japan: IGES.

IPCC. 2018. *Global Warming of 1.5°C: IPCC Special Report on Impacts of Global Warming of 1.5°C Above Pre-Industrial Levels in Context of Strengthening Response to Climate Change, Sustainable Development, and Efforts to Eradicate Poverty*. Edited by V.Masson-Delmotte, P. Zhai, H.-O. Pörtner, D. Roberts, J. Skea, P. R. Shukla, A. Pirani, et al. 1st ed. Cambridge, UK and New York: Cambridge University Press. https://doi.org/10.1017/9781009157940

IPCC. 2019. "Summary for Policymakers." In *Climate Change and Land: An IPCC Special Report on Climate Change, Desertification, Land Degradation, Sustainable Land Management, Food Security, and Greenhouse Gas Fluxes in Terrestrial Ecosystems*, edited by P. R. Shukla, J. Skea, H.- O. Pörtner, D. C. Roberts, P. Zhai, R. Slade, and S. Connors, et al. https://doi.org/10.1017/9781009157988.001

Iseman, T., and Miralles-Wilhelm, F. 2021. *Nature-Based Solutions in Agriculture: The Case and Pathway for Adoption*. Virginia, US: Food and Agriculture Organization and The Nature Conservancy. doi:10.4060/cb3141en. https://www.fao.org/3/cb3141en/cb3141en.pdf

ISFL. 2022. "Carbon Accounting". Biocarbon Fund. Accessed 2 December 2022. https://www.biocarbonfund-isfl.org/theme?title=Carbon%20Accounting

Jasanoff, Sheila. 2010. "A New Climate for Society." *Theory, Culture & Society* 27 (2–3): 233–253. doi:10.1177/0263276409361497.

Khan, Mizan, Stacy-ann Robinson, Romain Weikmans, David Ciplet, and J. Timmons Roberts. 2020. "Twenty-Five Years of Adaptation Finance through a Climate Justice Lens." *Climatic Change* 161 (2): 251–269. doi:10.1007/s10584-019-02563-x.

Knorr-Cetina, K. 1999. *Epistemic Cultures: How the Sciences Make Knowledge*. Cambridge, MA: Harvard University Press.

Kumarasiri, Jayanthi, and Christine Jubb. 2016. "Carbon Emission Risks and Management Accounting: Australian Evidence." *Accounting Research Journal* 29 (2): 137–153. doi:10.1108/ARJ-03-2015-0040.

Lahn, Bård. 2021. "Changing Climate Change: The Carbon Budget and the Modifying-Work of the IPCC." *Social Studies of Science* 51 (1): 3–27. doi:10.1177/0306312720941933.

Leahy, Sinead, Harry Clark, and Andy Reisinger. 2020. "Challenges and Prospects for Agricultural Greenhouse Gas Mitigation Pathways Consistent With the Paris Agreement." *Frontiers in Sustainable Food Systems* 4: 69. https://www.frontiersin.org/article/10.3389fsufs.2020.00069.

Li, Tania Murray. 2007. *The Will to Improve: Governmentality, Development, and the Practice of Politics*. Durham: Duke University Press

Loft, Anne. 2021. *Understanding Accounting in Its Social and Historical Context: The Case of Cost Accounting in Britain, 1914-1925*. Abingdon: Routledge, Taylor & Francis Group. https://public.ebookcentral.proquest.com/choice/publicfullrecord.aspx?p=6319863.

Lohmann, Larry. 2009. "Toward a Different Debate in Environmental Accounting: The Cases of Carbon and Cost–Benefit." *Accounting, Organizations and Society* 34 (3–4): 499–534. doi:10.1016/j.aos.2008.03.002.

Lorde, Audre. 1983. "The Master's Tools Will Never Dismantle the Mater's House." In *This Bridge Called My Back: Writings by Radical Women of Color*, edited by Cherríe Moraga, and Gloria E. Anzaldúa, 94–101. New York: Kitchen Table: Women of Color Press.

MacKenzie, D. 2009. "Making Things the Same: Gases, Emission Rights and the Politics of Carbon Markets." *Accounting, Organizations and Society* 34 (3–4): 440–455.

Matthan, Tanya. 2022. "Beyond Bad Weather: Climates of Uncertainty in Rural India." *The Journal of Peasant Studies*: 1–22. doi:10.1080/03066150.2022.2116316.

Mazzai, Alessandra. 2021. "Carbon Budget." *Seeds: Words That Feed the Future* (blog). December 10. https://www.climateforesight.eu/seeds/carbon-budget/.

McElwee, Pamela D. 2016. *Forests Are Gold : Trees, People, and Environmental Rule in Vietnam. Culture, Place, and Nature: Studies in Anthropology and Environment*. Seattle: University of Washington Press.

McGill, Alan. 2013 "Natural Capital Accounting: What's All the Fuss About?" *The Guardian*, September 27. Accessed 15 March 2022. https://www.theguardian.com/sustainable-business/natural-capital-accounting-fuss-about?CMP=twt_gu

McKibben, Bill. 2020. "The Climate Debt the U.S. Owes the World." *The New Yorker*, December 2. https://www.newyorker.com/news/annals-of-a-warming-planet/the-climate-debt-the-us-owes-the-world.

Millenium Ecosystem Assessment. 2022. "Overview of the Millennium Ecosystem Assessment." *Millenium Ecosystem Assessment* (blog). 2022. https://www.millenniumassessment.org/en/About.html.

Miller, Peter. 2001. "Governing by Numbers: Why Calculative Practices Matter." *Social Research* 68 (2): 379–396.

Mills, Geofrey. 1994. "Early Accounting in Northern Italy: The Role of Commercial Development and the Printing Press in the Expansion of Double-Entry from Genoa, Florence and Venice." *Accounting Historians Journal* 21 (1). https://egrove.olemiss.edu/aah_journal/vol21/iss1/4.

Mitchell, Timothy. 2011. *Carbon Democracy: Political Power in the Age of Oil*. London: Verso.

Moore, Jason W. 2015. *Capitalism in the Web of Life: Ecology and the Accumulation of Capital*. 1st ed. New York: Verso.

Moore, Jason W. 2017. "The Capitalocene, Part I: On the Nature and Origins of Our Ecological Crisis." *The Journal of Peasant Studies* 44 (3): 594–630. doi:10.1080/03066150.2016.1235036.

Moore, Jason W. 2018. "The Capitalocene Part II: Accumulation by Appropriation and the Centrality of Unpaid Work/Energy." *The Journal of Peasant Studies* 45 (2): 237–279. doi:10.1080/03066150.2016.1272587.

Neumayer, Eric. 2000. "In Defence of Historical Accountability for Greenhouse Gas Emissions." *Ecological Economics* 33 (2): 185–192. doi:10.1016/S0921-8009(00)00135-X.

Newell, Peter. 2022. "Climate Justice." *The Journal of Peasant Studies* 49 (5): 915–923. doi:10.1080/03066150.2022.2080062.

Newell, Peter, Maxwell T. Boykoff, and Emily Boyd. 2012. *The New Carbon Economy: Constitution, Governance and Contestation*. Chichester, West Sussex: Wiley-Blackwell.

Newell, Peter, and Olivia Taylor. 2018. "Contested Landscapes: The Global Political Economy of Climate-Smart Agriculture." *The Journal of Peasant Studies* 45 (1): 108–129. doi:10.1080/03066150.2017.1324426.

Nightingale, Andrea Joslyn, Siri Eriksen, Marcus Taylor, Timothy Forsyth, Mark Pelling, Andrew Newsham, Emily Boyd, et al. 2020. "Beyond Technical Fixes: Climate Solutions and the Great Derangement." *Climate and Development* 12 (4): 343–352. doi:10.1080/17565529.2019.1624495.

Olsthoorn, Xander, Daniel Tyteca, Walter Wehrmeyer, and Marcus Wagner. 2001. "Environmental Indicators for Business: A Review of the Literature and Standardisation Methods." *Journal of Cleaner Production* 9 (5): 453–463. doi:10.1016/S0959-6526(01)00005-1.

PACN. 2020. "Sello Programa Argentino Carbono Neutro." Accessed 2 December 2022. https://carbononeutro.com.ar/sellos/

Pattberg, Philipp. 2017. "The Emergence of Carbon Disclosure: Exploring the Role of Governance Entrepreneurs." *Environment and Planning C: Politics and Space* 35 (8): 1437–1455. doi:10.1177/2399654417723341.

Pengue, Walter. 2017. *El Vaciamiento de Las Pampas: La Exportación de Nutrientes y El Final Del Granero Del Mundo*. Santiago de Chile: Fundación Heinrich Böll Stiftung.

Pickering, Jonathan, and Christian Barry. 2012. "On the Concept of Climate Debt: Its Moral and Political Value." *Critical Review of International Social and Political Philosophy* 15 (5): 667–685. doi:10.1080/13698230.2012.727311.

Porter, Theodore M. 1994. "Making Things Quantitative." *Science in Context* 7 (3): 389–407. doi:10.1017/S0269889700001757.

Ratnatunga, Janek, Stewart Jones, and Kashi R. Balachandran. 2011. "The Valuation and Reporting of Organizational Capability in Carbon Emissions Management." *Accounting Horizons* 25 (1): 127–147. doi:10.2308/acch.2011.25.1.127.

Reisman, Emily, and Madeleine Fairbairn. 2021. "Agri-Food Systems and the Anthropocene." *Annals of the American Association of Geographers* 111 (3): 687–697. doi:10.1080/24694452.2020.1828025.

Rodríguez, Carlos Alberto. 2010. *Cartografía local*. 1st ed. Monitoreos Comunitarios para el manejo de los recursos naturales en la Amazonia Colombiana. 1: 1–51. Bogota: Tropenbos Internacional de Colombia. file:///Users/diegosilva/Downloads/CARTOGRAFIA-LOCAL-web%20(3).pdf.

Rogelj, Joeri, Oliver Geden, Annette Cowie, and Andy Reisinger. 2021. "Net-Zero Emissions Targets Are Vague: Three Ways to Fix." *Nature* 591 (7850): 365–368. doi:10.1038/d41586-021-00662-3.

Rubenstein, Daniel Blake. 1992. "Bridging the Gap between Green Accounting and Black Ink." *Accounting, Organizations and Society* 17 (5): 501–508. doi:10.1016/0361-3682(92)90044-S.

Schumacher, E. F. 1979. *Small Is Beautiful: Economics as If People Mattered*. New York: Harper and Row.

Schwartzman, Gabe. 2022. "Climate Rentierism After Coal: Forests, Carbon Offsets, and Post-Coal Politics in the Appalachian Coalfields." *The Journal of Peasant Studies* 49 (5): 924–944. doi:10.1080/03066150.2022.2078710.

Scott, James C. 1998. *Seeing Like a State: How Certain Schemes to Improve the Human Condition Have Failed*. Yale Agrarian Studies. New Haven: Yale University Press.

Silva, Diego. 2020. "Trigo HB4: realidades, mitos y relatos sobre los cultivos transgénicos." Accessed 2 December 2022. https://tallerecologista.org.ar/trigo-hb4-realidades-mitos-y-relatos-en-torno-de-los-cultivos-transgenicos/

Spence, Crawford, Javier Husillos, and Carmen Correa-Ruiz. 2010. "Cargo Cult Science and the Death of Politics: A Critical Review of Social and Environmental Accounting Research." *Critical Perspectives on Accounting* 21 (1): 76–89. doi:10.1016/j.cpa.2008.09.008.

Stabinsky, Doreen. 2021. "Net Zero and COP26." Friends of the Earth International. https://www.foei.org/wp-content/uploads/2021/11/Net-Zero-COP26.pdf.

Stechemesser, Kristin, and Edeltraud Guenther. 2012. "Carbon Accounting: A Systematic Literature Review." *Journal of Cleaner Production* 36: 17–38. doi:10.1016/j.jclepro.2012.02.021.

Stock, Ryan. 2022. "Power for the Plantationocene: Solar Parks as the Colonial Form of an Energy Plantation." *The Journal of Peasant Studies*, 1–23. doi:10.1080/03066150.2022.2120812.

Stock, Ryan, and Trevor Birkenholtz. 2021. "The Sun and the Scythe: Energy Dispossessions and the Agrarian Question of Labor in Solar Parks." *The Journal of Peasant Studies* 48 (5): 984–1007. doi:10.1080/03066150.2019.1683002.

Strohmaier, Rita, Janie Rioux, Anika Seggel, Alexandre Meybeck, Martial Bernoux, Mirella Salvatore, Javier Miranda, and Astrid Agostini. 2016. The Agriculture Sector in the Intented Nationally Determined Contributions: Analysis. *Environment and Natural Resource Management Working Paper* 62: 1–67. Rome: Food and Agriculture Organization of the United Nations. https://www.fao.org/3/i5687e/i5687e.pdf

Sullivan, Sian. 2013. "Banking Nature? The Spectacular Financialisation of Environmental Conservation." *Antipode* 45 (1): 198–217. doi:10.1111/j.1467-8330.2012.00989.x.

Sultana, Farhana. 2022. "Critical Climate Justice." *The Geographical Journal* 188 (1): 118–124. doi:10.1111/geoj.12417.

Taboanada, Miguel. 2021. "Oportunidades y Desafíos Que Presenta La Medición y Acreditación de La Captura de Carbono." Presented at the Seminario Balance de Carbono, November 9. https://www.youtube.com/watch?v=fpuwJ57BFNE&ab_channel=CanalAgrositio.

Tanaka, Katsumasa, and Brian C. O'Neill. 2018. "The Paris Agreement Zero-Emissions Goal Is Not Always Consistent with the 1.5 °C and 2 °C Temperature Targets." *Nature Climate Change* 8 (4): 319–324. doi:10.1038/s41558-018-0097-x.

TEEB. 2022. "Our Approach." *The Economics of Ecosystems and Biodiversity* (blog). 2022. https://teebweb.org/about/approach/.

UNFCCC. 2021. "The Future of Race to Zero." United Nations Framework Convention on Climate Change. https://unfccc.int/sites/default/files/resource/Public%20Consultation%20on%20the%20Future%20of%20Race%20to%20Zero%20%5Bexternal%5D.pdf

Tsing, Anna Lowenhaupt. 2015. *The Mushroom at the End of the World: On the Possibility of Life in Capitalist Ruins*. Princeton: Princeton University Press.

Tuana, Nancy. 2019. "Climate Apartheid: The Forgetting of Race in the Anthropocene." *Critical Philosophy of Race* 7 (1): 1–31. doi:10.5325/critphilrace.7.1.0001.

Tyteca, Daniel. 1996. "On the Measurement of the Environmental Performance of Firms— A Literature Review and a Productive Efficiency Perspective." *Journal of Environmental Management* 46 (3): 281–308. doi:10.1006/jema.1996.0022.

UNFCCC. 2015. "The Paris Agreement." Paris: United Nations Framework Convention on Climate Change. Accessed 2 December 2022. https://unfccc.int/sites/default/files/english_paris_agreement.pdf

Van Houtan, Kyle S., Kisei R. Tanaka, Tyler O. Gagné, and Sarah L. Becker. 2021. "The Geographic Disparity of Historical Greenhouse Emissions and Projected Climate Change." *Science Advances* 7 (29): eabe4342. doi:10.1126/sciadv.abe4342.

VERRA. 2011. "Approved VCS Methodology VM0017." *Sectorial Scope 14 Agriculture, Forestry, Land Use*. Accessed 2 December 2022. https://verra.org/wp-content/uploads/2018/03/VM0017-SALM-Methodolgy-v1.0.pdf

VERRA. 2020. "VM0042 Methodology for improved agricultural land management." *Sectorial Scope: 14 Agriculture, Forestry, Land Use*. Accessed 2 December 2022. https://verra.org/methodologies/vm0042-methodology-for-improved-agricultural-land-management-v1-0/.

VERRA. 2022a. "VERRA Methodologies for Agriculture." Accessed 2 December 2022. https://verra.org/methodologies/

VERRA. 2022b. "Verra Registry." Accessed 2 December 2022. https://verra.org/registry-system/

Visión Amazonía. 2022. "Visión Amazonía. Que Hacemos y Que No Hacemos?" Accessed 2 December 2022. https://visionamazonia.minambiente.gov.co/que-hacemos-y-que-no-hacemos/

Whitington, J. 2016. "Carbon as a Metric of the Human." *PoLAR: Political and Legal Anthropology Review* 39 (1): 46–63.

WRI and WBCSD. 2014. "GHG Protocol Agricultural Guidance: Interpreting the Corporate Accounting and Reporting Standard for the Agricultural Sector." World Resource Institute and World Business Council for Sustainable Development. Accessed 2 December 2022. https://ghgprotocol.org/sites/default/files/standards/GHG%20Protocol%20Agricultural%20Guidance%20%28April%2026%29_0.pdf

Yusoff, Kathryn. 2021. "The Inhumanities." *Annals of the American Association of Geographers* 111 (3): 663–676. doi:10.1080/24694452.2020.1814688.

Zuberman, Federico. 2019. "Suelo Virtual y Deuda Ecológica. Un Cálculo Para La Expansión de La Soja En Argentina." *SaberEs* 11 (1). doi:10.35305/s.v11i1.194.

OPEN ACCESS

Imagined transitions: agrarian capitalism and climate change adaptation in Colombia

Alejandro Camargo

ABSTRACT
Climate change has significantly affected rural lives around the world. Adaptation, as a political response to this situation, interacts with longer trajectories of agrarian capitalism and peasant's expectations for the future. Through the concept of imagined transitions, this article explores how peasants of northern Colombia manufacture and project their own transition to an agrarian capitalist future in the aftermath of climate-related floods and in the midst of adaptation interventions. Peasants use adaptation to imagine a future in which they are no longer peasants but have instead become rural entrepreneurs who play a proactive role in the development of capitalism.

'Economic adaptation'

In 2010, heavy floods surpassed historical records and ravaged many places around the world (Kundzewicz et al. 2014). Colombia was one of these places. The global intensification of El Niño Southern Oscillation (ENSO) cold phase, known as La Niña, significantly increased rainfall and caused catastrophic flooding throughout the country. Although some cities were affected, the La Niña floods fundamentally constituted an agrarian disaster. Floodwaters destroyed crops, farms, caused massive loss of livestock, and displaced thousands of peasants who became landless for more than a year. The Colombian president publicly deemed this catastrophe a manifestation of global climate change and, consequently, formulated a climate adaptation plan to reconstruct both the economy and infrastructure of the affected areas. This was a pivotal moment, as adaptation to climate change became a national political concern for the first time in Colombia. In fact, this new focus resulted in the creation of the Adaptation Fund, a national institution through which the state has reconstructed devastated areas and developed risk management and adaptation projects. The 2010 floods thus reconfigured agrarian landscapes and state action in significant ways.

In 2012, the Minister of Agriculture of Colombia visited Southern Atlántico, a rural area in the north of the country which was one of the most seriously affected places in 2010. In his speech, the Minister connected the climate crisis and the prospects of the agricultural economic sector in a particular way. In the years immediately following the La Niña disaster, the Colombian government signed new free trade agreements (FTA henceforth) with the United States (2011), Canada (2011), and the European Union (2012). Especially in the case of the trade agreement with the USA, the Colombian government anticipated that the agricultural sector would become a strategic field for job generation, technological innovation, and greater international competitiveness (Bancoldex 2012). The Minister argued that the reconstruction of Southern Atlántico should be an opportunity to transform a ruined rural economy into an 'agricultural export model' (Ministerio de Agricultura 2012).

Shortly thereafter, the Ministry of Agriculture, together with the Adaptation Fund and the local government, formulated a silvopasture and cattle restocking project to recover dairy farming – a very important activity to hundreds of people in Southern Atlántico. The silvopasture scheme was the main component of a broader 'ecological corridor' intended to connect this region with global carbon markets. Oddly, the Adaptation Fund referred to these cattle projects as 'economic adaptation' initiatives whereby the victims of the disaster would be able to 'economically adapt' to climate change. By framing the recovery of this area in these terms, the Minister of Agriculture also created an expectation among the peasants. From this time on, their economic imaginaries were shaped by the desire to connect this devastated landscape with the world via international FTAs and carbon markets. In this way, the ambiguous discourse of climate change adaptation became an opportunity for them to imagine a transition from a catastrophic present to a promising globalized capitalist economy.

I use the concept of imagined transitions to examine how interventions and narratives about adaptation to climate change intersect with the promises and historical trajectories of global agrarian capitalism, as well as the implications of that conflation for the production of economic expectations among those affected by climate disasters. This concept allows me to bring together longstanding concerns about agrarian transitions in peasant studies and contemporary scholarship on adaptation and disastrous climates in political ecology and critical development studies. In addition, the idea of imagination makes it possible to understand agrarian transitions not simply as the result of class struggles or the external forces of the capitalist mode of production, as it is often depicted in the literature. The idea of imagination also allows an understanding of the agencies, desires, and expectations of those who experience climate disasters and adaptation policies in their everyday life. In this sense, it is important to note that imagined transitions in Southern Atlántico were not entirely forward looking. Peasants nurtured their dreams of global agrarian capitalism together with their memories of a rural modernization project supported by the World Bank between the 1960s and 1970s. Although the main economic goals of that project were never achieved (for reasons to be explained later), for the peasants it was a successful experience, and a transition that they wanted to see replicated in the aftermath of the 2010 disaster. Imagined transitions, however, are not simply epistemic elaborations regarding the past, the present, and future. They also shaped everyday practices, rural identities, and were anchored in the materiality of the environment. Peasants created a concrete agenda in order to facilitate the transition from a devastated landscape and a precarious dairy farming economy (which emerged as the only alternative after the collapse of the World Bank project), to a productive agro-export model; and from a peasant collective identity to a new entrepreneurial rural class. Currently, this agenda has not materialized yet for reasons that would require a deeper analysis beyond the objectives of this paper. My interest, however, is not to assess the eventual outcome of that agenda, but to examine the political, historical, and environmental conditions under which it was manufactured and envisioned.

This particular conflation of past experiences, present devastation, and future expectations posits temporality as a critical axis to understand the convergence of climate change adaptation and agrarian capitalism. The conventional idea of adaptation as 'the process of adjustment to actual or expected climate and its effects' (IPCC 2012, 556), is in itself a temporal consideration of a particular future (which may obscure other alternative futures (Paprocki 2019)) in the face of an increasingly disturbed climate. The case of Southern Atlántico complicates this temporality by exposing how expectations of a resilient and prosperous future are shaped by local experiences of the past and the history of agrarian capitalism. Therefore, this paper examines how climate change adaptation becomes a site for the production and recreation of temporalities connecting expectations of prosperity, discourses on socio-economic adjustment, and histories of agrarian transformation. The temporality of climate change adaptation unfolds as global fears of a catastrophic future inform economic and social policies and interventions in vulnerable areas around the world. Yet vulnerable people are not always acquiescent and disempowered. The case of Southern Atlántico shows that expectations to transform a devastated landscape into a resilient and economically prosperous place are not exclusively a consequence of, or an imposition from, the operation of external, powerful, opportunistic, and abstract economic forces – which often thrive at the expense of local people's interest. Rather, economic prosperity and capital accumulation can be embraced, prioritized, sought, and imagined by individuals who even occupy the most marginal places in the capitalist economy and who are highly vulnerable to climate change.

By bringing together climate change narratives and historical processes of environmental change and agrarian capitalism development, this article also contributes to the JPS Forum on 'climate change and critical agrarian studies' as introduced by Borras Jr et al. (2022). These scholars call special attention to the ways in which narratives about climate change, agrarian struggles, and the role of capitalism emerge, and how different actors in rural settings respond to them. The authors identify four competing, and often overlapping, narratives, two of which resonate with the experience of peasants in Southern Atlántico. The first narrative, framed as corporate-driven, maintains that capitalism is a 'self-correcting system' which can operate and, simultaneously, provide solutions to the climate crisis through adaptation and mitigation. The climate crisis, therefore, is seen as external to the logics of capitalism. The second narrative acknowledges the disturbing role of capitalism in the production of the climate crisis and underscores the need of a radical shift from capitalism to a different system of production.

In Southern Atlántico, the global capitalist system of production is seen as a solution to the local crisis and the need of a radical shift is also present. This case, however, pushes us to expand the scope of these narratives as framed in Borras et al.'s piece, and, therefore, to consider other possibilities, imaginaries, and practices of climate and agrarian politics. In this part of the world, the idea of capitalism as a self-correcting system is not a corporate narrative. It is a local expectation elaborated by poor peasants based on their own experience as subjects and beneficiaries of a past international rural modernization project. The relative failure of this project does not mean the failure of capitalism, but a conjunctural flaw than can be fixed in the present with the support of climate adaptation policies to create a better future in the face of a globalizing industrial agriculture. In this context, for the peasants of Southern Atlántico a structural transformation is needed, but not in the form of a shift from capitalism to a different system of production based on, for instance, agroecological ideas and practices – as highlighted by Borras et al. The peasants of Southern Atlántico envision instead a radical shift from the margins of capitalism to its center. It is radical because it necessarily involves a profound transformation of identities, landscapes, forms and scales of production, technologies, and labor relations.

This article draws on twelve months of qualitative research in Southern Atlántico, where I conducted semi-structured interviews, participated in community and government meetings, and observed the everyday life of the victims of the disaster as they rebuilt their agrarian landscapes and domestic spaces. With this methodology, I analyzed the negotiation of post-disaster adaptation projects between peasants and government officials, and I considered their understandings of the local and the global economy. I also reviewed historical documents produced by the World Bank and some Colombian agencies which were involved in the economic and physical transformation of Southern Atlántico in the 1960s and 1970s. The rest of the article is organized into five sections. In the following section, I situate my arguments within broader discussions about transitions in agrarian studies. In the three subsequent sections, I will develop a narrative on (a) how peasants came to see a failed project of rural modernization as a success; (b) how an agrarian economy centered on dairy farming, which was resuscitated in the aftermath of the disaster as an economic adaptation project, became symbolic of poverty and failure among peasants; and (c) the ways in which peasants propelled themselves as rural entrepreneurs onto a prosperous future. In the final section, I will discuss the implications of the concept of imagined transitions for the study of agrarian capitalism in times of climate change.

Imagined transitions

The Southern Atlántico area is a subregion of the Department of Atlántico in the Colombian Caribbean region (see Figure 1). This area is an alluvial plain with an undulating surface that, as Figure 1 shows, is surrounded by three important bodies of water: the Magdalena River, the Dique Canal, and the Guájaro Reservoir. At the time of the catastrophe, Southern Atlántico was home to nearly 90,000 people, of which around 50% lived with unmet basic needs (UNINORTE 2011, 3). These individuals, seen as the poorest of the Department of Atlántico, are distributed in the towns of Manatí, Santa Lucía, Repelón, Suán, Candelaria, and Campo de La Cruz. Unlike other rural areas in Colombia, people in Southern Atlántico do not identify land concentration, land inequality, or land conflicts as actual problems. In fact, there is a general agreement that the agrarian reform program in the 1960s and 1970s was successful precisely because it guaranteed farmers access to land. Current data on rural land tenure in the Department of Atlántico is inaccurate because almost 61% of the rural property is informally owned. However, the Gini coefficient on land concentration shows that the towns of Southern Atlántico have a more equitable distribution of land (between 0.540 and 0.638, 0 meaning that all agricultural holdings hold the same amount of land) as compared to the Northern part of the Department in the Barranquilla metropolitan area (where it is between 0.773 and 0.876, being 1 the case in which a single agricultural holding holds

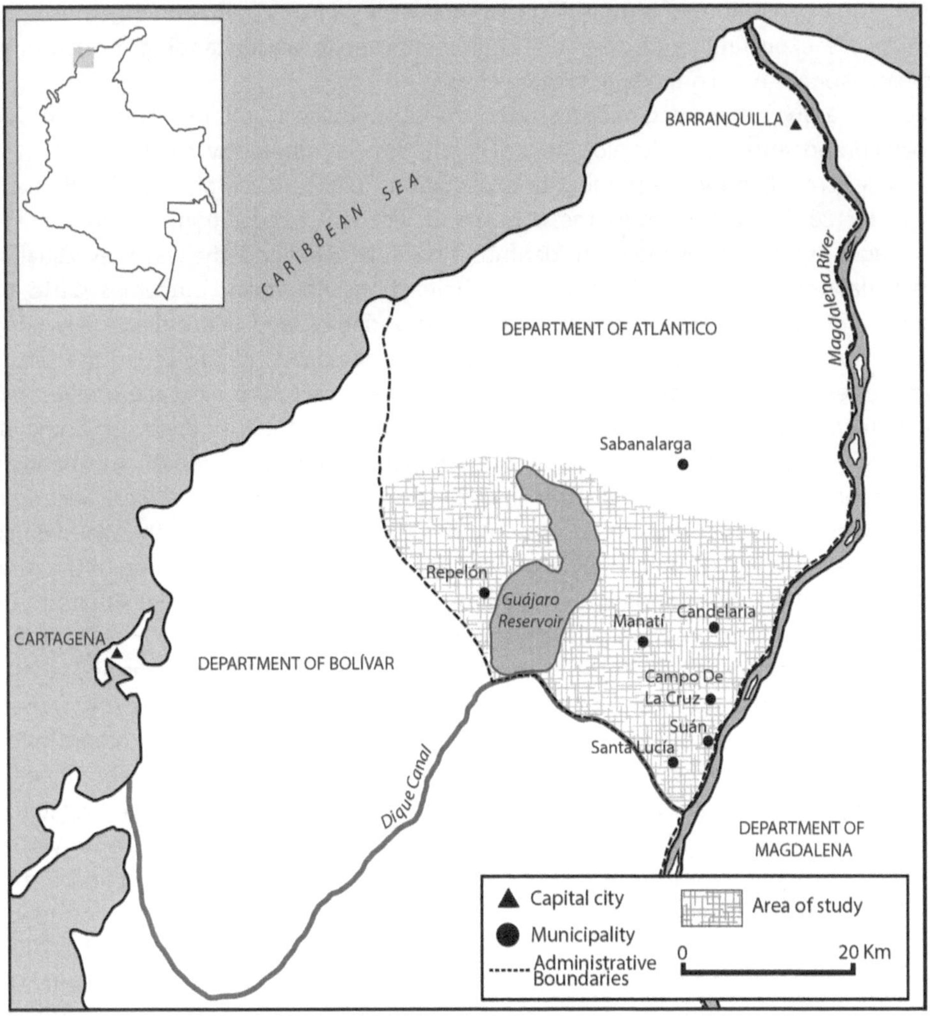

Figure 1. Southern Atlántico in the Colombian Caribbean region.

all the land) (FAO-Minagricultura 2019, 38). By the time of the disaster, small-scale dairy farming was the main economic activity in Southern Atlántico. Small-scale cultivation of corn, cassava, plantains, and fruits played a secondary role. A report on soils and agricultural management (G2C Ingénierie 2011) described dairy farming as highly unproductive and marginal, which barely produced enough for farmer's subsistence and otherwise contributed to soil erosion and degradation. Crop production, for its part, is depicted as lacking any form of technical development or industrial vocation.

This pessimistic view of the agrarian economy of Southern Atlántico was not merely an external observation. During my fieldwork, peasants repeatedly referred to dairy farming and subsistence agriculture as unproductive, valueless, and undesirable practices. These attitudes gained more currency after the disaster of 2010, when the destruction of crops, farms, the massive loss of livestock, as well as the forced displacement of thousands of farmers configured a landscape of devastation. It was upon this material and ideological terrain that expectations of a transition to a prosperous agro-export economy took hold. But why are expectations of desirable futures relevant to understanding the logics of agrarian capitalism in contexts of climate crisis?

Beckert argues that imaginaries of how the economy will unfold in the future are pivotal components of the development of capitalism. Put differently, capitalist dynamics 'are vitally propelled by the shaping of expectations' (2016, 6; see also Phelps, Bunnell, and Miller 2011). Beckert coins the concept of 'fictional expectations' to understand 'the images actors form as they consider future states of the world, the way they visualize causal relations, and the ways they perceive their actions influencing outcomes' (2016, 9). Expectations ascribe value to specific objects and variables, and orient decision-making despite uncertainty and the incalculability of future outcomes. By projecting the future onto the temporal order of capitalism, Beckert observes, actors project themselves and their perceptions of the world. Although Beckert focuses on the future, he also acknowledges that fictional expectations are strongly shaped by the past. In Southern Atlántico, the prospects fueling the project of a high-value crop production scheme were anchored in memories of a bygone prosperous agricultural economy sponsored by the World Bank, the emergence of a precarious dairy economy, and the historical degradation of the rural landscape. In this way, economic expectations in Southern Atlántico were situated in a wider fluctuating history of capitalist prosperity and crisis.

Agrarian transitions, as capitalist phenomena, entail different imaginaries and expectations of the future. The concept of agrarian transition usually refers to the process whereby capitalism becomes the dominant mode of production in the countryside (Byres 1996, 3). This process involves, among other changes, a shift from a predominantly peasant agriculture to an agricultural system characterized by a large rural proletariat (Goodman and Redclift 1982, viii). The idea of agrarian transition has been controversial because in its early formulations it was predicated on a teleological conception of the development of capitalism. According to these early views (mostly from orthodox Marxism), the commodification process is unilinear and irreversible (O'Laughlin 2009, 204), and therefore it follows a pre-determined path (Byres 1996, 3). As capitalism unfolds in the countryside, non-capitalist or non-commodified forms would inevitably disappear (Araghi 2009, 114). Peasants, therefore, were seen as technologically backward and 'doomed by the forces of modernization and industrialization' (Bryceson 2000, 6).

The literature has shown that peasants are far from absent in capitalism (McMichael 2008; Akram-Lodhi and Kay 2010) and that the development of this mode of production does not occur in an unilinear way. Rather, agrarian transitions evolve under specific conditions and in particular times (Goodman and Redclift 1982, 66). In the case of Latin America, land reform and its concomitant agricultural modernization projects have been crucial for the expansion of capitalist relations in the rural world. Within the broader context of development, mainly from the 1950s through the 1970s, these 'state-induced transitions' sought to guarantee land ownership and access to credits and technological inputs in order to transform traditional forms of production into 'modern capitalist enterprise' (Goodman and Redclift 1982, 112–119; see also Bernstein 2009, 245). These expectations of development, as Hegel would say, can be situated in the human capacity to desire and imagine a future significantly different from the present (Cowen and Shenton 1996, xi). Yet, for Escobar (1995), development became a dream that eventually turned into a nightmare. Critical development studies have extensively demonstrated that attempts to create conditions for local accumulation, economic growth, and welfare through development have repeatedly failed (Wainwright 2008, 4; see also Escobar 1995 and Ferguson 1994). In the case of land reform in many Latin American countries, peasants were not necessarily the main beneficiaries despite their constant struggle for land. In fact, their living standards never improved as expected (Teubal 2009; Mondragón 2006). Consequently, Latin America currently has the greatest inequality in the distribution of land. In this context, Colombia is at the top of the list as the most unequal country (Oxfam 2016). Transitions to a modernizing agrarian capitalism through land reform and development occurred in conflicting ways.

Although the idea of agrarian transition has been conceptually and empirically problematic, it continues to be relevant to an understanding of the processes of class formation and capital accumulation in our current neoliberal world (Beban and Gorman 2017). In this era of global climate change, however, the analysis of agrarian transitions demands a reconsideration. In the late nineteenth century, Kautsky observed:

> The peasant was now dependent on the market, which proved to be even more moody and unpredictable than the weather. At least the weather's perfidiousness could be prepared for. Ditches could mitigate the effects of a particularly wet summer, and irrigation could counter the consequences of drought; smoking fires could protect vines from spring frost and so on. But the peasant had no means of arresting a collapse in prices, or of selling unsellable corn. (Kautsky [1899]1988, 16)

Today, we need to recalibrate Kautsky's articulation of weather and the market in the face of global climate disasters and adaptation governance. These global phenomena have made increasingly evident that peasants are hardly prepared for 'weather's perfidiousness,' and that ditches and irrigation are insufficient to withhold the devastating effects of extreme weather events. For the peasants of the twenty-first century, weather is as 'moody and unpredictable' as the market. On the one hand, abrupt climate disruptions (such as floods and droughts) inevitably bring about profound devaluations in the built environment and the provision of services (Johnson 2015). Agricultural investments are particularly prone to devaluation because of their sensitivity to shifting climatic conditions which increase the likelihood of crisis (Sayre 2010). On the other hand, climate disasters and crises create the conditions for 'profitable spatial fixes' (Johnson 2015). In fact, addressing the effects of climate change on the built environment necessarily requires a 'capital switch' (Castree and Christophers 2015, 380) to redesign and rebuild infrastructures. Climate change adaptation, as a regime to govern people and landscapes (Paprocki 2018), precisely provides the bureaucratic, epistemic, institutional, and financial apparatus to envisage a transition to a future in which capital and climate crisis come together (Taylor 2014; Klepp and Chavez-Rodriguez 2018).

Imagined transitions, as they play out in Southern Atlántico, invite a further elaboration of some of what has been discussed above. First, imagined transitions are teleological, not in an evolutionary and unilinear sense, but as open-ended orientations of everyday life to plan, hope for, imagine, or give up possible futures (Bryant and Knight 2019). Expectations, as a sense of how things ought to be (Bryant and Knight 2019, 58), are key orientations in which the 'future is made present' (Koselleck 2004, 259). Secondly, although the thesis of the disappearance of the peasantry has been highly criticized, in Southern Atlántico people envisioned a future in which they are no longer peasants, but rural entrepreneurs instead. Peasants are not merely the vulnerable victims of a disturbing conflation of climate and capitalism. They make vulnerability into agency in order to designate how capitalism and climate adaptation should work in their favor. Thirdly, according to the World Bank, the agricultural modernization project in Southern Atlántico was a failure, which mirrors the academic criticism of development projects at large. But for the peasants, it was a success. Adaptation, from their perspective, would be considered as success as long as it resuscitates that prosperous past.

Prosperous times

Government officials and peasants alike imagined commercial crop production as the path towards economic prosperity in the aftermath of the disaster, but they did so from different temporal perspectives. The former spoke exclusively in the context of the recently signed FTAs, but the latter also built their expectations on the grounds of their memories of a prior transition from a mainly subsistence economy to a large-scale agricultural project. During the late 1960s and 1970s, the Colombian government created in Southern Atlántico one of the most ambitious programs of agrarian reform and agricultural modernization in the country with the financial and technical support of the World Bank. The goal of this program, known as the Atlántico No 3 Project, was to overcome poverty and increase productivity and wealth via development. To this end, the 'introduction of modern agriculture' (World Bank 1967, ii) was conceived of as the engine to foster progress in what was formerly a floodplain where peasants combined seasonal, small-scale agriculture and fishing.

This project occurred at a very critical moment. Since the 1950s, a relatively modern capitalist agriculture began to emerge in Colombia mainly as a consequence of subsidized credits and massive imports of fertilizers, pesticides, and agricultural machinery (Palacios and Abel 1991, 630). One of the goals of this modernization project was to diversify national production and incentivize exports. Between 1951 and 1960, coffee crop represented 79.2 percent of the total export trade in Colombia, which made the national economy vulnerable to an eventual crisis of that particular crop (Kofas 1986, 17). In this context, products such as cotton, rice, bananas, and sorghum broadened and diversified the spectrum of national crop production. For instance, cotton exports (58,222 tons) surpassed national consumption (54,751tons) in 1968, and rice yields increased from 1.94 to 4.26 tons per hectare between 1966 and 1975 (Kalmanovitz and Enrique 2006, 274). These changes occurred at the same time as the central government enacted an Agrarian Reform Law in 1961. The complex history behind this law is beyond the scope of this paper. But it is sufficient to say that the impetus for agricultural modernization, as well as the mobilization of peasants against land concentration and rural inequality, had an important influence on this legal transformation.

Southern Atlántico presented an attractive area for the expansion of agricultural modernization projects and, therefore, for the development of capitalist agriculture. To that end, the Colombian government requested a loan of 1966 from the International Bank for Reconstruction and Development (World Bank hereafter). This loan was intended to finance agricultural development, irrigation, and flood control works in 35,500 ha of Southern Atlántico (World Bank 1967, 1). By the time this loan was negotiated, some initial agricultural projects were already underway in the region. In the mid-1950s, the

Colombian government built a 70 km levee surrounding the triangle of Southern Atlántico to protect this area against flooding. This levee interrupted the various connections between marshes, the Dique Canal, and the Magdalena river. As a consequence, the marshes progressively disappeared and more land was available for agriculture (INCORA 1970, 13). Mixed cultivations of corn, plantains, vegetables, and yucca grew in an area of nearly 1500 ha with the help of fertilizers and pesticides (World Bank 1967, 4). Cotton was already taking hold in response to the growing demand from the emerging textile industry in Barranquilla. In 1957, for instance, Southern Atlántico produced 638,00 kg of cotton, which amounted to 83% of the total production in the Department of Atlántico (Arteta 2012, 157). Yet in the eyes of the World Bank, this was just an incipient modernization project, as agriculture was still considered 'backward and underdeveloped' (1967, 4). Furthermore, the disappearance of marshes fostered an economic crisis among fishers. In a survey carried out in 1962 in areas that were partially desiccated, government officials found that the number of fishers had decreased from 1130 to 18 (Mejía Gutierrez 1966, 5). A World Bank report observed that because of the fishing crisis 'the area was considered one of the most economically depressed in the country' (World Bank 1982, 2). From its outset, agricultural modernization involved serious challenges.

The World Bank project intended to fight poverty and 'underdevelopment' in Southern Atlántico through modern agriculture on 35,500 ha, which would directly benefit 1800 families and employ 650 heads of households as farm laborers (World Bank 1972, 10). By this time, the central government had already begun an agrarian reform program which secured access to land to around 1252 families (World Bank 1983, 26). In addition to the encouraging prospect of national agricultural modernization, the project envisioned new markets in Europe and the US for the crops produced under the new scheme. The local market also appeared promising, as the Barranquilla-based processing company Conservas California S. A. (a subsidiary of Grace & Co. Ltd from the US) had expressed its willingness to purchase the tomato produced in Southern Atlántico for its canning operations (World Bank 1967, 10). With new irrigation and drainage infrastructures, agricultural production started in the early 1970s with tomato, cotton, sorghum, soybeans, and peanuts crops as the main commercial products (World Bank 1972).

What seemed to be a promising beginning in the march towards an agricultural emporium, however, soon followed an unfortunate path. First, the project never adequately considered the high salinity and alkalinity of soils and groundwater. This factor rendered large extensions of land unsuitable for crop cultivation and, consequently, had a considerable effect on the project's main goals:

> because of the unsuitable soils, the plan to produce high value export crops had to be abandoned and was replaced by a traditional annual crops program; consequently, there was no need to construct facilities for the storage, grading, packing and processing of export crop products. (World Bank 1982, 4)

Secondly, flooding was never successfully controlled. After catastrophic floods in 1975 and 1978, the 'violent waters' of the Magdalena River breached a levee in 1984 and flooded around 10,000 ha of arable land, 'leaving behind a wake of desolation and misery' (Diario del Caribe 1984, 8). A local newspaper succinctly observed that: 'irrigation districts, worth more than a billion pesos and built several years ago [...] are completely destroyed, making it almost impossible to recover what has been called the Southern triangle, or the breadbasket of this part of the country' (Diario del Caribe 1984, 8). The drainage system within the project area did not receive proper maintenance and many drainage canals silted up (World Bank 1982, 8).

And finally, agricultural production through cooperatives failed to generate substantial profit because of insufficient harvesting equipment, flood-related losses, and problems with the accounting system (World Bank 1982, 10). Many peasants fell into debt and decided to quit agriculture altogether because it was no longer profitable. Subsequently, this crisis accelerated a wave of immigration to Venezuela and the concomitant erosion of the land tenure system (World Bank 1982, 11). By 1983, about 250 families who had been beneficiaries of the project had left their lands abandoned (World Bank 1983, iii), others lent them to someone else, and still others sold them by way of informal transactions according to local interlocutors. One of the most dramatic indicators of this crisis was the rapid fall in the area devoted to agriculture. In 1970, the cash crop area was 7000 ha, while in 1980 it dropped to 1000 ha (World Bank 1982, 12). In an evaluation report, World Bank officials concluded: 'With the benefit of hindsight it can be said that the Government's decision to transform the economy of the depressed Atlantico area through agrarian reform and agricultural development was correct.' Yet, it continued, the project was 'overly optimistic and complex,' and therefore from an economic perspective it 'must be termed a failure' (World Bank 1982, 20).

At first glance, the story of the World Bank development project in Southern Atlántico can be seen as demonstrative of 'the inevitable gap between what is attempted and what is accomplished' (Li 2007, 1). But this is not the full picture. Paradoxically, for the farmers who participated in this project, this is a story of success and prosperity. In the nine meetings I attended in 2013 where government officials met the communities of Southern Atlántico, there was always at least one person who publicly expressed the importance of replicating the agricultural project of the 1960s and 1970s in the present. Martín,[1] a peasant from Manatí who participated in the World Bank project, explained to me the reasons why they see this project as a success:

> The best thing the state has ever done for this region was the Atlántico No. 3 project. It was a very important project widely known both nationally and internationally. Almost everybody found a job in agriculture. Just ask people here in Manatí and you will see that those who own a big house and who could afford to send their kids to college were beneficiaries of that project [...] not to mention that the land we have today was acquired in that time. (Martín, Manatí, Agosto 15, 2013)

According to Martín Olivero, whose voice is representative of a generalized agreement among farmers, the only way to overcome the post disaster crisis is by 'resuscitating' commercial crop production. Yet achieving this, according to him, would involve the demise of dairy farming as the region's main economic activity. As crop production faded away, the state had sponsored dairy farming as an alternative to the economic crisis. Cattle units in the project area grew from 3558 in 1972 to 17,883 in 1983, while pasture area increased from 11,150 ha in 1972 to 18,567 ha in 1981 (World Bank 1983).[2] Nevertheless, it is very common to hear people like Martín Olivero refer to dairy farming as a form of 'slavery.'

[1] I use pseudonyms throughout the article.

[2] The Atlántico No. 3 Project had envisioned that because of the hypothetical success of commercial agriculture, pasture area would be around 2600 ha by 1986 (World Bank 1983, 27).

Peasants constantly argued, in meetings and during interviews, that dairy farming is a precarious activity that has forced them to work for no profit and provides no way out. Dairy farming is, for them, the main cause of poverty and therefore it is an obstacle to economic prosperity. Paradoxically, in the aftermath of the 2010 floods, initial efforts to provide a quick response to the post-disaster economic crisis involved the restoration of dairy farming. The Adaptation Fund, the Government of Atlántico, and the Ministry of Agriculture attempted to do this through the allocation of five thousand cows to those who had lost their cattle as part of the 'economic adaptation' project.

Five thousand cows

As floodwaters advanced throughout Southern Atlántico, hordes of peasants, with their families and livestock, started a forced and uncertain transhumance to the highlands. But the loss of hundreds of cows was inevitable. A month after the collapse of the Dique Canal, 69,414 head of cattle had been lost (Sánchez 2011, 15). Those who managed to save their cattle, however, faced a scarcity of water, pasture, and money that would have been necessary to keep their animals alive. For these peasants, the only way out was to sell their cattle in an improvised, unequal, an opportunistic market that formed amid the disaster. Word of desperate peasants trying to sell their cattle spread rapidly throughout the region. This situation attracted buyers from surrounding areas who exploited this opportunity to purchase cattle at a very cheap price. Cows that were regularly worth COP$2,000,000, for instance, were sold for COP$200,000, according to an interview with a member of one of the two dairy farming cooperatives of Manatí. Peasants did not have the opportunity to renegotiate this price or to refuse an offer because their only alternative was to let the cattle die. Furthermore, the scant money peasants made in these unequal transactions was desperately needed to cover quotidian needs, and not to invest in other permanent assets. To some extent, the convergence of this opportunistic market, and the difficult environmental conditions to sustain the cattle, constituted a 'disaster after the disaster' (Schuller 2008, 18). Yet those who capitalized on this disaster were not external capitalist agents, as in the widely known 'disaster capitalism' (Klein 2007) narrative, but instead the very neighbors of the flood victims.

Although the Minister of Agriculture promoted large-scale commercial crop production as a strategic sector to connect Southern Atlántico with the global food economy, the recovery of dairy farming became a priority because of this massive loss of cattle. The project allocated five thousand cows in the most affected municipalities to reactivate milk production. In 2009, a year before the catastrophe, 255,225 ha of land in the Department of Atlántico were devoted to cattle raising for dairy and beef, while only 5638 ha were intended for crop production. Four out of seven municipalities where cattle raising is most concentrated are located in Southern Atlántico. In spite of this, the contribution of Atlántico to the total national milk production was only 1.02% (Lombana et al. 2012). When peasants refer to dairy farming as a form of 'slavery,' they underscore the absurdity of undertaking, against their will, an economic activity which offers nothing beyond what they consider a meager daily income, and only because there are no alternative livelihoods. Even more disturbing for them was the fact that in the aftermath of the disaster, and while they planned and imagined a prosperous agro-export economy, the most immediate need was to demand a solution to the crisis of dairy farming. According to the records of the two dairy cooperatives in Manatí compiled before the catastrophe, their members produced around 20,000 liters of milk a day before the floods. Production stopped in the aftermath of the disaster and resumed one year later. By then, milk production was no more than 500 liters a day.

The Adaptation Fund partially funded the cattle restocking project and conceptualized it as an economic mechanism to adapt to climate change. Its goal was to 'improve the quality of life' for the families of rural producers through the development of dairy farming (Fondo de Adaptación 2013, 5). It was also framed within the 'Productive Alliances' model which has been financially and technically supported by the World Bank in Latin America, Asia, and Africa (Fondo de Adaptación 2013, 3). In addition, the project acknowledged the eventual negative effects of the US FTA on the national dairy sector. Nevertheless, it deemed this threatening conjuncture an opportunity to adjust the sector on the basis of efficiency, higher quality, and innovation (Fondo de Adaptación 2013, 9–10). The project included the implementation of a 3000-hectare silvopastoral system, the reconstruction and maintenance of corrals and wells, technical assistance for cattle raising, and training workshops to infuse peasants with the capitalist principles of economic productivity. Silvopastoral systems are a form of agroforestry intended to provide economic and environmental benefits through the integration of trees, tree products, forage, and livestock. The integration of those components 'simultaneously produces timber commodities, a high-quality forage resource, and efficient livestock production' (USDA 1997, 1). This system was also presented as a potential opportunity to join the world carbon market through the construction of a large ecological corridor. In sum, the silvopastoral system was a means to turn dairy farming into a profitable activity.

The program started with multiple training workshops and meetings in which discourses about economic productivity and the insertion of local production into global markets were used to support the silvopastoral project. This rhetoric, however, also included disdainful narratives about the backwardness of local knowledge and economy. For instance, in a training workshop on soil quality held in Manatí in March 2013, a government official argued that dairy farming was backward and inefficient in Southern Atlántico because peasants lacked a profit-oriented 'consciousness' and the technical knowledge to make soils productive. She emphasized that nothing would change if peasants do not change their 'mentality.' While some of the approximately forty attendees did not respond to those comments, others found them offensive because the official neglected the experience they built during the World Bank project. Even though government officials and peasants equally embraced the idea of a future prosperous agro-economy, they had conflicting notions as to how to achieve that goal.

In November 2013, the Minister of Agriculture visited Southern Atlántico to meet with peasant leaders who expressed their concern regarding the delays in the dairy farming project. Almost three years had passed since the floods and milk production was still inactive. After reinforcing the discourse on economic prosperity that other government officials had regularly disseminated in the area, the Minister promised he would intervene to speed up the allocation of cattle. A few weeks later the first set of cattle arrived in Southern Atlántico. But this happened before the silvopastoral project began, contrary to what the project managers had originally planned. No trees had been planted and many farms had no assistance whatsoever from the technical component of the project. Further, the idea of 'economic adaptation to climate change' was never fully addressed or explained during the workshops. In fact, government officials never clarified why or how allocating cattle in the same area that was flooded would help people adapt to climate change. In an interview with a high-ranking government agent, I insistently asked about the meaning of adaptation in this project, to which he ultimately answered: 'adaptation is also a way to fund projects like this.' Ideas and narratives about economic growth eclipsed other concerns regarding future climate catastrophes and soon the language of climate adaptation disappeared altogether from official discourses on the economic recovery of the area. The government vision of a prosperous future for the people of Southern Atlántico commenced in an unexpected and contradictory way.

Bauman argues that economic progress 'renders once effective modes of making a living unviable and impracticable' (2004, 5). The truth of this statement might be demonstrated in the case of government officials promising prosperity and, at the same time, deeming peasant livelihoods as backward, unprofitable, and therefore unviable. But what is actually unviable and impracticable for peasants is precisely a project to restore the dairy farming economy. Furthermore, what was once an effective mode of making a living for the peasants was not a 'traditional' rural economy, but an internationally supported development project based on large- production. One could even argue that ideas of progress disseminated by government officials were also far from 'effective,' as demonstrated by the way in which the silvopastoral system project began to crystalize. What was initially conceived of as a project connecting ecological corridors, economic productivity, and the green economy of global carbon markets, resulted instead in the mere replacement of the cattle that peasants lost during the floods. The dreams of a transition towards an agrarian capitalist economy in the aftermath of disaster found itself on a very shaky ground where conflicting notions of welfare and productivity collided.

Yet the shortcomings of the initial silvopasture project did not lead to the end of the fictional expectations of a prosperous agrarian economy. Quite the contrary, the flaws of this project reaffirmed what peasants had consistently thought about dairy farming: that this activity is not the right route to progress. In other words, the tensions involved in the silvopastoral system project strengthened the idea that large-scale commercial crop production was the most effective way to overcome the crisis and transform a devastated landscape into a prosperous area. For the peasants, however, this transition requires more than a shift in the agrarian landscape of Southern Atlántico. Embracing such a promising accumulation project also requires a transformation of themselves as economic subjects. Together with the negative aspects of the dairy farming economy, the very condition of being a peasant became synonymous with a historical process of economic decay and impoverishment. The people's prospects of an export-oriented agricultural scenario, therefore, required a transition from peasants to entrepreneurs.

Rural entrepreneurs

For decades, scholars in agrarian studies have been concerned about the fate and future prospects of peasants in global capitalism. In the case of Colombian, for instance, scholars have exposed how capitalism in the form of tourism and conservation (Ojeda 2012), industrial agriculture (Cely-Santos and Hernández-Manrique 2021), and neoliberal policies aimed at making peasants into entrepreneurs (Tobasura Acuña 2011), has adversely affected rural livelihoods. Other scholars beyond Colombia have focused more widely on the development of peasant struggles for a future in which food sovereignty, equal access to land and resources, and environmental sustainability are the ideal conditions for the social and economic reproduction of their livelihoods (McMichael 2006; Schneider and Niederle 2010; Scoones 2008). Although these visions about the future increasingly inform the political action of many peasants across the globe, the people of Southern Atlántico imagined their transition to a better life, as well as their place in a globalizing food economy, in a different way.

Peasants were convinced that the silvopastoral system was not a viable solution to the crisis. Accordingly, in 2013, leaders from five affected municipalities created a committee to actively participate in the 'economic adaptation' project. Specifically, this committee sought to present various economic proposals to state agencies, and to demand a more inclusive process in the decisions regarding the economic future of their region. As a part of this endeavor, the committee wrote a report describing the state of the agricultural sector in the aftermath of the disaster. The report proposed some solutions to overcome the crisis and begin the transition to a prosperous economy:

> The peasantry cannot be seen anymore as a marginal sector providing cheap labor. It is important to envisage investment in rural development as a step towards competitiveness, but in order to do so, peasants need to become rural entrepreneurs and leaders in the development of agro-industry.[3]

Peasant aspirations to become rural entrepreneurs were reinforced by rumors, news, and discourses – such as that of the Minister of Agriculture – on the development of FTAs. Initially, these ambitions were met with concern and uncertainty. As a peasant told me after a meeting with the Minister in July, 2013:

> The world is changing and we need to keep up with it. If we want progress, we need to adapt to FTAs and their requirements. Dairy farming here is very unproductive and if we keep working that way the FTAs will kill us. We do not want to be poor peasants anymore; we want to be rural entrepreneurs.

By the time these opinions about FTAs were taking shape among peasants, the FTA with the US was already having adverse consequences. Exports of cheap dairy products flooded Colombian markets and, consequently, the national dairy sector was severely hit (Contexto Ganadero 2012). For example, in 2012, 2898 tons of powdered milk arrived in Colombia from the US, whereas in 2013, this figure grew to 3872 tons (Contexto Ganadero 2015). The total amount of imported powdered milk (including the US market) in 2014 was 15,048 tons and in 2016 it was 41,040 (Contexto Ganadero 2017). In June 2017, this situation worsened further as dairy farmers in the Santander region of Colombia dumped thousands of liters of milk on the ground to protest the lack of competitiveness and access to markets (Vanguardia 2017). Moreover, a report published in 2015, revealed that after two years of the US-Colombia FTA, the income of small-farmers decreased, and the deficit of the agricultural trade balance increased (Suárez and Barberi 2015). For those

---

[3]Comité central del sur del Departamento del Atlántico, Octubre de 2013, *Diagnóstico económico del sur del Departamento del Atlántico y propuestas para la reactivación del sector agropecuario.* Manatí, Unpublished report. Page 8.

who faced the catastrophe of 2010 and managed to save a few cows, this crisis was rather discouraging.

Although the negative effects of FTAs on dairy farming caused concern among peasants, rumors also spread that agriculture was about to become very profitable because the US would eventually demand more fruits and vegetables from Colombia. Specifically, people heard that the demand for mangoes would growth considerably. Roberto, a dairy farmer from Manatí, for instance, gave up his hopes with milk production and instead decided to plant mango trees to be prepared for the eventual export market. When I visited his farm in May 2013, he explained to me that dairy farming in Southern Atlántico would eventually disappear because of the FTA situation, the poor competitiveness, and the lack of technological innovation. The mango, according to him, would soon gain a higher value in international markets. These rumors, which depicted agriculture as a key component of FTAs, had a twofold influence on their expected transition from peasants to rural entrepreneurs. First, it reinforced the idea that dairy farming was an ill-suited solution to the crisis, and that being a dairy farmer would constitute a condemnation to economic ruin. Second, fictional expectations of a prosperous future found a possible path to materialization in a concrete agricultural project. This distinction was particularly clear in 2013, when farmers from different regions and with heterogeneous claims blocked roads across the country to protest the adverse consequences of FTAs. Although peasants of Southern Atlántico actively joined this strike, their position was not against the FTAs but instead to demand a more equitable, inclusive, and efficient participation in them. Between uncertainty and optimism, peasants positioned themselves as economic subjects seeking to move from the margins of capitalism towards its center.

Even though peasants projected themselves as rural entrepreneurs in an idealized future, this imagined transition was also anchored in a concomitant transformation of the landscape. Older peasants witnessed and benefited from commercial crop production during the World Bank project, which according to their testimonies, changed their lives in a positive way. That experience endowed them with enough historical authority to maintain that commercial crop production already proved successful, and to show younger generations that this agricultural project should be the main goal of their mobilization in the aftermath of the 2010 floods. But they also knew that resuscitating the bygone 'times of the agrarian reform,' as they call that epoch, would not be an easy endeavor. The history of catastrophic floods has exposed the fragility and flaws of water infrastructure, as well as the vulnerability of agriculture to both floods and droughts. In Manatí, for instance, peasants rely exclusively on rainfall for crop production. But during extended dry periods this technique presents serious challenges. The demise of the World Bank project left behind a ruined agricultural landscape of decaying irrigation and drainage infrastructure. Yet this landscape continues to nurture the hope that something can be done to resuscitate commercial crop production. The peasant committee's report mentions the urgent need to reconstruct and expand abandoned water infrastructures as a mechanism to guarantee year-round agricultural production. Peasants see the problems of the past as something that can be addressed in the present, not as definitive limitations inherent to the failure of past agricultural modernization and capitalist agriculture.

Becoming a rural entrepreneur, therefore, involved profound transformations in the agricultural landscape, labor relations, technological arrangements, rural identities, and in the connections with the market at different levels. Consequently, this imagined transition ultimately would lead to – or require – the disappearance of the peasantry as a cultural and economic category. When people say that they do not want to be 'poor peasants' anymore, but rural entrepreneurs instead, they are actually overthrowing an identity that constrains their economic expectations. Peasants and entrepreneurs are, therefore, mutually exclusive 'agrarian allegories' (Tsing 2003) that shaped images, discourses, and assumptions of prosperous futures.

Malleable adaptation

The analysis of how and under what political, historical, and environmental conditions imagined transitions to global agrarian capitalism are manufactured in moments of climate crisis has different implications for the ways in which we understand rural lives and livelihoods in the present.

First, the conflation of the historical dynamics of agrarian capitalism and climate change adaptation does not necessarily manifest itself as an external force imposed upon the people who were directly affected by these economic and environmental processes. Decisions, aspirations, agencies, and expectations, also play a crucial role in the formation and expansion of capitalist frontiers and the operation of adaptation governance. In other words, as Tania Li (2014) argues, capitalist relations are not always imposed. However, the same can be said about the 'adaptation regime' (Paprocki 2018). The case presented here demonstrates that in the process of adopting, embracing, and actively shaping accumulation and adaptation projects, individuals also reinvent themselves and project a transformation of their identities and practices onto desirable futures. In the origins of political ecology, the dialogue between agrarian and disaster studies exposed the multiple ways in which capitalist forces may worsen the impacts of climatic hazards and disasters (i.e. Watts 1983). More recently, the narrative of disaster capitalism has helped us understand how disasters may become opportunities for the expansion of neoliberalism (Gunewardena and Schuller 2008). Reactions to the disaster capitalism narrative have also provided important insights into the production of capitalist subjects (i.e. Faas 2018). Yet a better understanding of these subjective and quotidian processes will also help us to grasp the role of capitalism in the production and aftermath of climate disasters beyond conventional dualisms such as domination versus resistance, and external capitalist forces against disempowered individuals.

The story presented here is not simply a story about marginalized individuals demanding interventions. Critics of post-development theory have long argued that 'popular struggles in the South are about access to development [...] rather than rejection of it' (Storey 2000, 42). This is a story of how the people of Southern Atlántico manufactured a transition project based on their interpretations of past agrarian transitions, the global capitalist economy, and their own identities as agrarian subjects. This elaboration of the future positioned them beyond the place of disempowered people seeking access to the benefits of adaptation projects, but instead as subjects willing to play an active role in the historical development of capitalism.

Secondly, the study of expectations of economic prosperity in a context of climate crisis has an important analytical value for our broader understanding of how climate governance operates within capitalism. Since the floods of 2010 were explained as a consequence of global climate change, the programs to recover the affected areas were formulated as climate adaptation interventions. Yet the idea of adaptation soon conflated with the idea of economic prosperity. Authors such as Kelman, Gaillard, and Mercer (2015) have argued that development agendas are increasingly being dominated by climate change and disaster risk reduction concerns since disasters can set development back years or decades. In this context, disaster studies have shifted from a predominant focus on vulnerability to an increasing interest in building resilience (Grove 2014a, 614). Resilience, as a form of governance, has at times been highly criticized because it is consonant with a neoliberal rationality that makes individuals responsible for their own adaptation (Grove 2014b, 242; see also Evans and Reid 2014). Moreover, Kelman et al. observe that conventional concepts of resilience envision a return to a pre-disaster state, which is not necessarily a favorable or desirable one (2015, 22). The Silvopastoral system project was precisely an initiative to tackle both the climate and the economic challenges that peasants face in the post-disaster moment. Although government officials envisioned an economically prosperous future through this project, they dissonately returned to a pre-disaster economy based on dairy farming, the same economy that peasants disdained and sought to overcome. Furthermore, once peasants received their cows and some initial technical assistance, they were supposed to start over and be responsible for the successful recovery of their domestic economy. Adaptation was at best a blurry idea that, at least during my fieldwork, never found a place in peasant expectations of transitioning to a capitalist large-scale economy. It was not even clear why allocating cows to people would help them adapt to climate change. Yet it would be misleading to conclude that this idea of adaptation was a deformation of the standard concept that otherwise circulates worldwide. Adaptation is a malleable idea and the experience of Southern Atlántico accounts for the ways in which peasants and government officials adopt, resignify and repurpose this global concept for their own benefit.

Imagining a transition to what peasants saw as a better life brought hope in the midst of a climate crisis. This was the opportunity for peasants to manufacture their own future, to ascribe value to what they considered important, and to imagine themselves as their own economic agents who can be leaders in the evolution of a globalizing agrarian capitalism. In sum, by creating their own expectations of the future, peasants were able to envision, in their own terms, how to turn their tragedy into a market opportunity.

Disclosure statement

No potential conflict of interest was reported by the author(s).

Funding

This work has been supported by a doctoral scholarship from Departamento Administrativo de Ciencia, Teconologia y e Innovacion (COLCIENCIAS), currently known as Ministerio de Ciencia, Tecnologia e innovacion [grant number 79956272].

ORCID

Alejandro Camargo http://orcid.org/0000-0002-5812-8416

References

Akram-Lodhi, A. Haroon, and Cristóbal Kay. 2010. "Surveying the Agrarian Question (Part 2): Current Debates and Beyond." *Journal of Peasant Studies* 37 (2): 255–284.

Araghi, Farshad. 2009. "The Invisible Hand and the Visible Foot: Peasants, Dispossession and Globalization." In *Peasants and Globalization: Political Economy, Rural Transformation and the Agrarian Question*, edited by A. Haroon Akram-Lodhi and Cristóbal Kay, 111–147. New York: Routledge.

Arteta, Cecilia. 2012. "Una experiencia inconclusa: El cultivo de algodón en el departamento del Atlántico, 1935-1965." *Historia Caribe* VII (21): 143–179.

Bancoldex. 2012. *Tratado de libre comercio entre Colombia y los Estados Unidos. Fascículo 3: Agroindustria.* Bogotá: Bancoldex. http://www.bancoldex.com/documentos/3758_TLC_eltiempo_03.pdf.

Bauman, Zygmunt. 2004. *Wasted Lives: Modernity and its Outcasts*. Oxford: Polity.

Beban, Alice, and Timothy Gorman. 2017. "From Land Grab to Agrarian Transition? Hybrid Trajectories of Accumulation and Environmental Change on the Cambodia-Vietnam Border." *Journal of Peasant Studies* 44 (4): 748–768.

Beckert, Jens. 2016. *Imagined Futures: Fictional Expectations and Capitalist Dynamics*. Cambridge, MA: Harvard University Press.

Bernstein, Henry. 2009. "Agrarian Questions from Transition to Globalization." In *Peasants and Globalization: Political Economy, Rural Transformation and the Agrarian Question*, edited by A. Haroon Akram-Lodhi and Cristóbal Kay, 239–261. New York: Routledge.

Borras Jr, Saturnino, Ian Scoones, Amita Baviskar, Marc Edelman, Nancy Peluso, and Wendy Wolford. 2022. "Climate Change and Agrarian Struggles: An Invitation to Contribute to a JPS Forum." *Journal of Peasant Studies* 49 (1): 1–28.

Bryant, Rebecca, and Daniel M. Knight. 2019. *The Anthropology of the Future*. Cambridge: Cambridge University Press.

Bryceson, Deborah. 2000. "Peasant Theories and Smallholder Policies: Past and Present." In *Disappearing Peasantries? Rural Labour in Africa, Asia and Latin America*, edited by Debprah Bryceson, Cristóbal Kay, and Jos Mooij, 1–36. London: Practical Action.

Byres, Terence. 1996. *Capitalism from Above and Capitalism from Below: An Essay in Comparative Political Economy*. New York: St. Martin's Press.

Castree, Noel, and Bett Christophers. 2015. "Banking Spatially on the Future: Capital Switching, Infrastructure, and the Ecological fix." *Annals of the Association of American Geographers* 105 (2): 378–386.

Cely-Santos, Marcela, and Olga Lucía Hernández-Manrique. 2021. "Fighting Change: Interactive Pressures, Gender, and Livelihood Transformations in a Contested Region of the Colombian Caribbean." *Geoforum; Journal of Physical, Human, and Regional Geosciences* 125: 9–24.

Contexto Ganadero. 2012. *Carne y leche colombiana siguen en desventaja con el TLC*, November15. Accessed January 18, 2013. http://www.contextoganadero.com/agricultura/carne-y-leche-colombiana-siguen-en-desventaja-con-el-tlc.

Contexto Ganadero. 2015. TLC con EE.UU.: *Colombia sigue sin exportar carne e inundada de leche por importaciones*. Accessed May 15. http://www.contextoganadero.com/economia/tlc-con-eeuu-colombia-sigue-sin-exportar-carne-e-inundada-de-leche-por-importaciones.

Contexto Ganadero. 2017. *En menos de 40 días de 2017, Colombia está llena de leche importada*, February 8. Accessed July 15, 2017. http://www.contextoganadero.com/economia/en-menos-de-40-dias-de-2017-colombia-esta-llena-de-leche-importada.

Cowen, Michael, and Robert Shenton. 1996. *Doctrines of Development*. London: Routledge.

Diario del Caribe. 1984, December 11. "El río entra en Santa Lucía." *Diario del Caribe*, 8.

Escobar, Arturo. 1995. *Encountering Development: The Making and Unmaking of the Third World*. Princeton, NJ: Princeton University Press.

Evans, Brad, and Julian Reid. 2014. *Resilient Life: The Art of Living Dangerously*. Cambridge and Malden, MA: Polity Press.

Faas, A. J. 2018. "Petit Capitalisms in Disaster, or the Limits of Neoliberal Imagination: Displacement, Recovery, and Opportunism in Highland Ecuador." *Economic Anthropology* 5 (1): 32–44.

FAO and Minagricultura. 2019. *Plan integral de Desarrollo agropecuario y rural con enfoque territorial, Departamento del Atlántico, Tomo II*. Bogotá: FAO-Minagricultura.

Ferguson, James. 1994. *The Anti-Politics Machine: 'Development,' Depoliticization, and Bureaucratic Power in Lesotho*. Minneapolis: University of Minnesota Press.

Fondo de Adaptación. 2013. *Ejercer la interventoría integral del contrato # 121 de 2013, cuyo objeto consiste en ejercer la gerencia del proyecto reactivar: 'reactivación de la cadena lechera con sistemas silvopastoriles en seis municipios del sur del Atlántico para beneficio de 1.000 familias de productores agropecuarios afectados por el fenómeno de La Niña 2010-2011,' como parte del programa de reactivación económica- PRREA REACTIVAR*. Bogotá: Fondo de Adaptación.

G2C Ingénierie. 2011. *Informe sobre manejo agrícola y conservación de suelos*. Venelles: G2C Ingénierie.

Goodman, David, and Michael Redclift. 1982. *From Peasant to Proletarian: Capitalist Development and Agrarian Transitions*. New York: St. Martin's Press.

Grove, Kevin. 2014a. "Adaptation Machines and the Parasitic Politics of Life in Jamaican Disaster Resilience." *Antipode* 46 (3): 611–628.

Grove, Kevin. 2014b. "Agency, Affect, and the Immunological Politics of Disaster Resilience." *Environment and Planning D: Society and Space* 32 (2): 240–256.

Gunewardena, Nandini, and Mark Schuller. 2008. *Capitalizing on Catastrophe: Neoliberal Strategies in Disaster Reconstruction*. Lanham: AltaMira Press.

INCORA. 1970. *Proyecto Atlántico 3*. Bogotá: INCORA.

IPCC. 2012. *Managing the Risks of Extreme Events and Disasters to Advance Climate Change Adaptation: Special Report of the Intergovernmental Panel on Climate Change*. Cambridge: Cambridge University Press.

Johnson, Leigh. 2015. "Catastrophic Fixes: Cyclical Devaluation and Accumulation Through Climate Change Impacts." *Environment and Planning A* 47 (12): 2503–2521.

Kalmanovitz, Salomón, and López Enrique. 2006. *La agricultura colombiana en el siglo XX*. Bogotá: Fondo de Cultura Económica.

Kautsky, Karl. [1899] 1988. *The Agrarian Question, Volumen 1*. London: Zwan Publications.

Kelman, Ilan, J. C. Gaillard, and Jessica Mercer. 2015. "Climate Change's Role in Disaster Risk Reduction's Future: Beyond Vulnerability and Resilience." *International Journal of Disaster Risk Science* 6 (1): 21–27.

Klein, Naomi. 2007. *The Shock Doctrine: The Rise of Disaster Capitalism*. New York: Metropolitan Books/Henry Holt.

Klepp, Siljia and Libertad Chavez-Rodriguez, eds. 2018. *A Critical Approach to Climate Change Adaptation: Discourses, Policies and Practices*. London: Routledge.

Kofas, Jon. 1986. *Dependence and Underdevelopment in Colombia*. Tempe, AZ: Arizona State University, Center for Latin American Studies.

Koselleck, Reinhart. 2004. *Futures Past: On the Semantics of Historical Time*. New York: Columbia University Press.

Kundzewicz, Zbigniew W., Shinjiro Kanae, Sonia I. Seneviratne, John Handmer, Neville Nicholls, Pascal Peduzzi, Reinhard Mechler, et al. 2014. "Flood Risk and Climate Change: Global and Regional Perspectives." *Hydrological Sciences Journal* 59 (1): 1–28.

Li, Tania. 2007. *The Will to Improve: Governmentality, Development, and the Practice of Politics*. Durham: Duke University Press.

Li, Tania. 2014. *Land's end: Capitalist Relations on an Indigenous Frontier*. Durham: Duke University Press.

Lombana, Jair, Dorien Martínez, Miguel Valverde, Jesús Rubio, Jaime Castrillón, and William Marino. 2012. *Caracterización del sector ganadero del Caribe colombiano*. Barranquilla: Universidad del Norte.

McMichael, Philip. 2006. "Peasant Prospects in the Neoliberal age." *New Political Economy* 11 (3): 407–418.

McMichael, Philip. 2008. "Peasants Make Their own History, but not Just as They Please." *Journal of Agrarian Change* 8 (2-3): 205–228.

Mejía Gutierrez, Mario. 1966. *Exploraciones pesqueras: La pesca en el Canal del Dique*. Medellín: Secretaría de Agricultura.

Ministerio de Agricultura. 2012. "Plan de choque para recuperación agrícola del sur del Atlántico anuncia Minagricultura." September 14. Accessed September 15, 2012. http://www.minagricultura.gov.co/inicio/noticias.aspx?idNoticia = 1670.

Mondragón, Héctor. 2006. "Colombia: Agrarian Reform – Fake and Genuine." In *Promised Land: Competing Visions of Agrarian Reform*, edited by Peter Rosset, Raj Patel, and Michael Courville, 165–176. Oakland, CA: Food First Books.

Ojeda, Diana. 2012. "Green Pretexts: Ecotourism, Neoliberal Conservation and Land Grabbing in Tayrona National Natural Park, Colombia." *Journal of Peasant Studies* 39 (2): 357–375.

O'Laughlin, Bridget. 2009. "Gender Justice, Land and the Agrarian Question in Southern Africa." In *Peasants and Globalization: Political Economy, Rural Transformation and the Agrarian Question*, edited by A. Haroon Akram-Lodhi and Cristóbal Kay, 190–213. New York: Routledge.

Oxfam. 2016. *Unearthed: Land, Power and Inequality in Latin America*. https://www.oxfam.org/en/research/unearthed-land-power-and-inequality-latin-america.

Palacios, Marco, and Christopher Abel. 1991. "Colombia." In *The Cambridge History of Latin America*, Vol VIII, edited by Leslie Betthel, 629–686. Cambridge: Cambridge University Press.

Paprocki, Kasia. 2018. "Threatening Dystopias: Development and Adaptation Regimes in Bangladesh." *Annals of the American Association of Geographers* 108 (4): 955–973.

Paprocki, Kasia. 2019. "All That is Solid Melts Into the Bay: Anticipatory Ruination and Climate Change Adaptation." *Antipode* 51 (1): 295–315.

Phelps, Nicholas A., Tim Bunnell, and Michelle Ann Miller. 2011. "Post-disaster Economic Development in Aceh: Neoliberalization and Other Economic-Geographical Imaginaries." *Geoforum; Journal of Physical, Human, and Regional Geosciences* 42 (4): 418–426.

Sayre, Nathan. 2010. "Climate Change, Scale, and Devaluation." *Washington and Lee Journal of Energy, Climate, and the Environment* 1 (1): 93–105.

Sánchez, Andrés. 2011. *Después de la inundación. Documentos de trabajo sobre economía regional, N. 150*. Cartagena: Centro de Estudios Económicos Regionales.

Schneider, Sergio, and Paulo Niederle. 2010. "Resistance Strategies and Diversification of Rural Livelihoods: The Construction of Autonomy among Brazilian Family Farmers." *Journal of Peasant Studies* 37 (2): 379–405.

Schuller, Mark. 2008. "Deconstructing the Disaster After the Disaster: Conceptualizing Disaster Capitalism." In *Capitalizing on Catastrophe: Neoliberal Strategies in Disaster Reconstruction*, edited by Nandini Gunewardena and Mark Schuller, 17–27. London: Altamira Press.

Scoones, Ian. 2008. "Mobilizing Against GM Crops in India, South Africa and Brazil." *Journal of Agrarian Change* 8 (2-3): 315–344.

Storey, Andy. 2000. "Post-development Theory: Romanticism and Pontius Pilate Politics." *Development* 43 (4): 40–46.

Suárez, Aurelio, and Fernando Barberi. 2015. *Efectos del TLC Colombia-EE.UU. sobre el agro: los rostros y las cifras*. Bogotá: Oxfam and Planeta Paz.

Taylor, Marcus. 2014. *The Political Ecology of Climate Change Adaptation: Livelihoods, Agrarian Change and the Conflicts of Development*. London: Routledge.

Teubal, Miguel. 2009. "Peasant Struggles for Land and Agrarian Reform in Latin America." In *Peasants and Globalization: Political Economy, Rural Transformation and the Agrarian Question*, edited by A. Haroon Akram-Lodhi and Cristóbal Kay, 148–166. New York: Routledge.

Tobasura Acuña, Isaías. 2011. "De campesinos a empresarios: la retórica neoliberal de la política agraria en Colombia." *Espacio abierto* 20 (4): 641–658.

Tsing, Anna. 2003. "Agrarian Allegory and Global Futures." In *Nature in the Global South: Environmental Projects in South and Southeast Asia*, edited by Anna Tsing Paul Greenough, 124–169. Durham: Duke University Press.

UNINORTE. 2011. "Salud pública: así estaban los municipios antes del desastre." *Informativo Uninorte* 65: 3.

USDA. 1997. "Silvopasture: An Agroforestry Practice." *Agroforestry Notes*, Note 8, 1. Accessed July 22, 2015. http://nac.unl.edu/documents/agroforestrynotes/an08s01.pdf.

Vanguardia. 2017. *Por crisis, productores de Santander botan la leche*, June 9. Accessed October 12, 2017. http://www.vanguardia.com/economia/local/400236-por-crisis-productores-de-santander-botan-la-leche.

Wainwright, Joel. 2008. *Decolonizing Development: Colonial Power and the Maya*. Malden, MA: Blackwell Pub.

Watts, Michael. 1983. "Hazards and Crises: A Political Economy of Drought and Famine in Northern Nigeria." *Antipode* 15 (1): 24–34.

World Bank. 1967. *Atlantico no. 3 Irrigation Project (Technical Operations Projects Series No. 589)*. Washington, DC: World Bank.

World Bank. 1972. *Report and Recommendation of the President to the Executive Directors on a Proposed Loan to the Republic of Colombia for the Second Atlántico Development Project*. Washington, DC: The World Bank.

World Bank. 1982. *Impact Evaluation Report. Colombia: Atlantico Irrigation Project*. Washington, DC: The World Bank.

World Bank. 1983. *Project Completion Report: Second Atlantico Development Project*. Washington, DC: The World Bank.

Beyond bad weather: climates of uncertainty in rural India

Tanya Matthan

ABSTRACT

Drawing on research in central India, this paper argues that emerging climatic risks are inseparable from the systemic risks of capitalist production, particularly in relation to India's ongoing agrarian crisis. Drawing from classic debates in peasant studies around agrarian risk, subsistence, and moral economy, I argue that the intertwined effects of climatic variability and agrarian capitalism have produced a generalized 'climate of uncertainty' in rural India. The paper explores how encounters with agrarian risk reproduce social inequalities. It concludes that a deeper reckoning with regional histories of agrarian change is critical to forging just and secure rural futures in the face of the global climate crisis.

1. Introduction

The rainy season of 2019 was a tumultuous one across central India. In Malwa, a predominantly agricultural region in the state of Madhya Pradesh, the south-west monsoon – which brings a majority of the region's rainfall – arrived promisingly in late June, perfectly in sync with the sowing period. The steady downpour was a relief to farmers who had been reeling from several years of poor rainfall. Within a few weeks, however, the rain came to a sudden and alarming halt, replaced with scorching heat and clear skies at precisely the period when rain was crucial to the growth of soybean, the primary monsoon crop. When the prolonged dry spell ended, the rain returned with a fury not seen in several years.[1] But farmers' initial elation dissipated when they returned to their waterlogged fields to find wilted stalks and empty pods in the leguminous plant. With few or no other crops to compensate for the devastating season, cultivators could now only hope for a good winter harvest. Once again, it seemed that the weather had not favored their fields and fortunes.

While it is difficult to causally attribute this particular season's weather events to global climate change, meteorologists and climate scientists confirm that climate change is making the South Asian monsoon more erratic, marked by the increased incidence of localized extreme rainfall events, including in central India (Krishnan et al. 2020). Not only will climate change exacerbate this pattern of variable and uneven rainfall, the pattern itself is becoming harder to accurately predict. In rural Malwa, however, the

---

[1]This heavy rainfall was attributed to a positive Indian Ocean Dipole event (also known as the Indian Nino) which, in 2019, was one of the strongest ever on record.

discourse of climate change is largely absent in everyday conversations among rural residents. Rather, most farmers understood the season to be a more extreme and intense form of the weather variability they had experienced for generations – from poor rain, untimely frost, and intense cold spells to droughts and floods. In the current historical moment, however, all discussions of extreme weather inevitably segued into debates around commodity prices, input costs, and debt (see also Eakin 2006). Indeed, for farmers, 'bad weather' is viewed neither as an isolated or novel problem nor as the most crucial issue for cultivators already facing serious agrarian distress.

Ethnographically examining assessments of and engagements with agrarian risk in contemporary India, this paper discusses how the emerging climate crisis intersects with and exacerbates the ongoing agrarian crisis in the Indian countryside. This terminology of 'crisis' is fraught and contentious, and scholars have productively interrogated how and when crisis narratives are deployed, by whom, and with what effects (Roitman 2013). Labels such as 'climate crisis' and 'agrarian crisis' highlight the urgency of the problem at hand but also run the risk of flattening the plurality of farmers' experiences (van de Meerendonk 2020) and legitimizing authoritarian and anti-democratic measures (Borras et al. 2022). Nonetheless, as Marx argued, crises are not aberrations but rather caused by specific historical developments that often become 'normalized'. Crises, then, are revelatory since they expose the internal dynamics and contradictions of social formations (Watts 1983, 33). Analyzing how these crises conjoin and compound each other gets at the core of the current model of fossil-fueled industrial agriculture in India as well as to the structural transformations essential to addressing them.

While the agrarian crisis is sociologically complex and regionally variegated (Lerche 2011; Reddy and Mishra 2009; Shah 2012), in general, many farmers are experiencing the declining profitability of agrarian livelihoods as the viability of Green Revolution technologies – high-yielding varieties of seed, chemical inputs, private tube-wells – decline in tandem with reduced government spending on agricultural subsidies and infrastructures that undergirded increased food production and productivity in the mid-twentieth century.[2] Deepening agricultural risk undergirds this crisis. Farmers now face a range of intersecting economic and ecological risks: rising costs of production financed by debt alongside volatile and often unremunerative prices, degraded lands, and dwindling yields (Vasavi 2012). These risks have tragically culminated in the suicides of over 300,000 farmers over the past three decades. As more farmers are ever more deeply entwined within commercial production and deregulated markets, it is evident that rural India is facing the 'convergence of economic and ecological contradictions in such a way that the material conditions of society as a whole are undermined' (Foster 2013, 1).

As the JPS editorial to this special forum (Borras et al. 2022) highlights, climate change is inextricably tied to capitalist development, but the specific ways in which they intersect in rural worlds needs further examination. As conjoined socio-ecological processes, climate change and capitalism unfold in a similarly uneven and unequal manner in the

[2]Since 1991, the agricultural sector in India has largely stagnated. For instance, its contribution to the GDP was about 15% as of 2011–12 even though nearly 60% of the country's population depended on the sector for their livelihood. Over half of the country's farmers are in debt, and for 85% of farmers, there are no savings in agriculture – that is, they make less than they spend on farming on average. Alongside, landholdings are small and fragmented with over 70 per cent of farmers classified as 'marginal', owning less than one hectare of land.

countryside. How they intertwine to produce a generalized climate of uncertainty is the focus of this article. For instance, while water scarcity and insecurity is likely to worsen with climate change-fuelled droughts, the roots of this crisis in rural India can be traced back to the water-intensive model of industrial agriculture promoted by the state, largely through privatized and mechanized irrigation technologies. Climate change is not the sole driver of environmental exclusion but rather 'articulates with underlying processes of economic and agrarian transformation' (Kennedy 2022, 675).

In studying the micro-dynamics of agrarian change in Malwa, this paper underscores climate change as an often mundane and unspectacular process 'occurring incrementally, over time and hidden from view' (Borras et al. 2022, 6) in ways that make it impossible to neatly separate its patterns from long-standing ecological shifts. As science and technology studies scholars have argued, the scientific fact of 'climate change' as an abstract and planetary phenomenon often sits uneasily with localized experiences and subjective meanings of climate among ordinary people as they interact with the natural world (Jasanoff 2010). Indeed, for the individual farmer and at the level of the village, climate change-induced extreme weather is always perceived and experienced through already transforming socio-natures and political economies, often appearing as declining yields, drying wells, and mounting debt.[3] Moreover, uneven trajectories of development and climate change demand greater attention to 'the vagaries of intra-community inequalities and injustices' (Paprocki and Cons 2014, 1113). The rural world is a 'stratified human landscape in which the risks and opportunities presented by climatic change are unequally distributed' (Taylor 2015, 319), with important implications for social differentiation. Rather than assuming how this plays out, we need to observe how different social groups respond and adapt to these shifts.

Building on these crucial premises, this paper historically locates and ethnographically grounds climate uncertainties squarely within – rather than external to – longer trajectories of political-economic and environmental change. Against these interlocking crises, my focus is on the ways in which climate change intersects with and transforms long-standing agrarian risks as well as emerging uncertainties. In doing so, the paper reflects on the relevance of classic debates in peasant studies for the climate change era. The next section (Section 2) outlines the core argument of the paper in relation to central concepts of risk, subsistence, and moral economy, and situates this study within the region of Malwa in central India. Then, I present the main case study in four parts. Section 3 examines existing assessments of risk in state policies on climate change, showing how these documents reduce climate risk to a problem of 'bad weather', and focus solely on its impacts on yield and productivity to the near complete exclusion of its socio-ecological implications. Section 4 counters the narrative presented by these policies through a historical overview of agrarian transformations in Malwa with the introduction of soybean. It shows how extreme weather intersects with socially and historically

[3]Even if aggregate rainfall remains constant, it is expected that localized extreme rainfall events will increase as will variability in patterns of rainfall dispersal. According to the Government of India's Economic Survey (2017-18), temperatures in South Asia are likely to increase by about 3–4 degrees by the end of the 21st century. As a result, farmer incomes could decline by 15–18 percent, or as high as 20 per cent for unirrigated areas. In central India where Malwa is located, there has already been a three-fold increase in extreme rainfall events in central India since 1950. Here, a 10 per cent decline in average rainfall has been accompanied by a 75 per cent increase in extreme rainfall events (daily rainfall exceeding 150 mm) over the same period. Among the major crops affected by temperature increases, soybean and chickpea rank among the highest, both important crops in Malwa.

produced risk of input-intensive monocropping. The next two sections discuss how modes of adapting to climatic and environmental risks – both formal and informal – have maladaptive outcomes through the acceleration of social differentiation. In Section 5, I analyze a national crop insurance program to demonstrate how this increasingly popular mode of managing climate risk widens social inequalities. Finally, Section 6 focuses on the ubiquitous practice of groundwater extraction, showing how farmers must actively engage groundwater uncertainty in order to continue cultivation, a practice which demonstrates the transformation of agriculture into a precarious gamble. While these four axes – assessing, producing, managing, and engaging – are hardly exhaustive of the range of rural engagements with risk and uncertainty, they offer a glimpse into the intertwined contours of climatic and capitalist transformations in the Indian countryside.

2. Climates of uncertainty

The intertwined dynamics of neoliberal state policy, climatic variability, and capital-intensive production have produced a 'climate of uncertainty' for rural communities. In using the term 'climate', I aim to draw attention to climate (change) as a shifting set of weather patterns (evident in temperature, wind, and precipitation) reshaping rural worlds while also pointing to the broader meaning of climate as the prevailing set of *conditions* characterizing a group, period or place. The climate of uncertainty which pervades agrarian life today is certainly a product of climatic changes but not exclusively so. For most farmers in central India, climate change manifests in subtle shifts which map onto and compound existing threats to their livelihoods – hence, 'climate change' as a specific and identifiable cause of harm simply does not reflect their everyday reality.

Similarly, the analytical lens of risk and uncertainty enables attention to the emerging *and* endemic hazards of agrarian production. 'Risk' itself has been a central and contested concept within critical agrarian studies since at least the 1970s. Scholars have shown how the expansion of colonialism and capitalism greatly amplified the inherent riskiness of peasant livelihoods (Watts 1983). Drawing from E.P. Thompson's (1971) discussion of 'moral economy' in relation to food access during shortages in eighteenth century England, James Scott (1976) argued that pre-capitalist peasant societies in Southeast Asia were organized around a 'subsistence ethic', which firmly prioritized risk-aversion over profit-maximization. Here, subsistence security was protected through norms of reciprocity, redistribution and patronage among peasants and with landlords and states. Subsistence, Scott notes, was not just an economic claim but also a moral one, undergirded by notions of justice and fairness – excessive claims on peasant produce by elites during times of dearth were deemed unjust, and therefore, met with protest and rebellion. The risks of agriculture were, therefore, minimized and managed both through agricultural practices such as intercropping and planting drought-resistant seed varieties and social arrangements of work sharing, gift-giving, and patronage.

Relatedly, in tracing the origins of famine in drought-prone Hausaland in West Africa, Michael Watts (1983) located the region's crisis of food production and distribution squarely within the colonial intensification of commodity production. Cash-cropping coupled with volatile commodity prices and heavy colonial tax burdens destroyed the adaptive capacity and moral economy of peasant communities, thus rendering farming into an increasingly hazardous livelihood. As a result, risks that were agronomically minimized and socially managed were now heightened and individualized – even a small shock such as a poor harvest or fluctuating prices could produce a crisis of subsistence for peasant producers.

By contrast, Samuel Popkin (1980) insisted that the 'moral economists' had an overly romanticized view of pre-capitalist village relations and institutions. Focusing on individual decision-making, Popkin argued that peasants did in fact make risky short- and long-term investments in order to raise their subsistence levels. Moreover, he pointed out that social and moral norms of reciprocity and redistribution are dynamic and shifting in light of power relations and strategic calculations. In his view, the extension of markets did not destroy an existing moral economy but rather offered new opportunities for gain as well as long-term protections for the 'rational' peasant. However, just as Popkin critiqued the moral economy approach for romanticizing village communities, his perspective similarly presented a overly positive view of the role of markets in the lives of peasants.

Given the centrality of these debates around risk, subsistence, and moral economy to the field of peasant studies, it is important to consider their relevance for agrarian politics in the climate change era. In what ways does climate change transform these concepts and the socio-ecological relations to which they refer? At present, agriculturalists confront 'subsistence crises' in the form of droughts, floods, pests and disease at an unprecedented and unpredictable frequency and intensity. But it is not just the climate that has dramatically transformed in the decades since these canonical texts were written. For many farmers across the world, 'the subsistence crisis has become a permanent state' (Edelman 2005, 336). Indeed, farmers face new risks associated with industrialized monocropping (eg: water scarcity, rising input costs) alongside the impacts of economic liberalization (eg: declining subsidies and extension services). Local institutions ensuring subsistence security have been weakened while state relief is limited and directed through financial technologies.[4] At the same time, rural expectations for accumulation and consumption have risen, with concomitant strains on household economies.

At this conjuncture, risk itself has dramatically transformed. The expansion of highly capital-intensive farming has 'injected risk into the operating system of the agricultural order in a newly intensified way' (Sethi 2021, 1405). Rising costs of inputs – from seeds and fertilizers to machinery and wells – combine with unremunerative prices and erratic weather to make it impossible for farmers – even the poorest peasants – to completely minimize risk if they are to continue working the land. Farmers are now compelled not just to manage risks but also to actively make highly risky investments simply in order to survive (cf. Sethi 2018). Contrary to Popkin's argument, however, they do so not because markets offer any security but because it is increasingly difficult to divest from them. These risks – climatic and otherwise – are not (only) externally imposed, but rather differentially and unevenly produced, distributed and engaged within deeply hierarchical agrarian landscapes. If agriculture in India has always been a risky enterprise, it is now a highly precarious gamble on both the monsoon and the market (Aga 2018; Gupta 2017).

Further, with climate change, historically managed agrarian risks have transformed into chronic uncertainty (Gupta 2017). By definition, risk refers to a situation wherein

[4]The large-scale protests (2020-21) by farmers, particularly from the northern Indian states of Punjab and Haryana, against the proposed deregulation of the sale and purchase of agricultural commodities can be read as rebellion against the state's violation of the moral economy of fair prices.

probabilities of various outcomes are known and calculable while uncertainty connotes a situation wherein the probabilities are largely unknown (Knight 1921). While this distinction is rarely clear in real-life situations, it is apparent that climate change is fueling and exacerbating agrarian uncertainty, or the number and kinds of 'unknown unknowns'. Farmers now face novel and unpredictable weather patterns which confound existing knowledge and adaptive mechanisms vis-à-vis extreme events (cf. Srivastava, Mehta, and Adam 2022). While capitalism has firmly implanted risk into the soils and waters of rural Malwa, climate change grafts new uncertainties onto these existing risks. The focus of this paper then is largely but not solely on experiences of uncertainty. I use both the terms 'risk' and 'uncertainty' throughout the paper, but try to clarify the reason for using one over the other in each case.

The climate challenge, therefore, calls for a return to and reconsideration of the core concepts of early political ecology and agrarian studies to show that humans live in environments constituted by natural processes but also 'in one of their own making, constituted by human practice, and subject to ongoing change and historical transformation' (Watts 1983, 25). Specifically, critical agrarian studies offers a framework to analyze changing relations of production as well as diverse trajectories and dynamics of agrarian capital accumulation. Political ecology's attunement to biophysical processes highlights the shifting 'ecological relations of production', or the ways in which capital accumulation reorganizes natural processes and reshapes the land (Beban and Gorman 2017, 751). Bringing these frameworks together, this paper highlights their combined strength in examining the specificities of agrarian change in relation to wider environmental and economic transformations.

These social conditions and transformations fundamentally shape how climate change reconfigures the agrarian landscape of Malwa. Located in the western part of the central Indian state of Madhya Pradesh, Malwa is a fertile area, well-known for its moisture-retentive deep black cotton soil. In agro-climatic terms, it is classified as a semi-arid region marked by undulating and hilly terrain and dry deciduous shrub lands. As a predominantly agricultural region, Malwa has long been entwined in regional and global markets through the cultivation of crops such as opium, cotton, and sugarcane. Since the 1970s, cropping patterns have changed dramatically – soybean now dominates in the *kharif* (monsoon/rainy) season, followed by a second crop of irrigated wheat and gram in the *rabi* (winter) season, made possible by the spread of tube wells and pumps fueled by state-subsidized electricity. In the past two decades, wealthy farmers have taken to the large-scale cultivation of high-value, high-risk horticultural crops such as onions, potatoes, and garlic. The average size of operational landholdings in Madhya Pradesh is 1.4 hectares and is unevenly distributed. While Rajputs (former landlords) are a major landowning caste, it is traditional peasant castes such as Patidars, Khatis, and Jats who constitute the class of agrarian capitalists, engaging in intensive commercial cultivation and investing agricultural surplus in mechanization and land purchases as well as diversifying into agri-commodity trade and input retailing (cf. Aga 2018). Over 75 per cent of farmers are small and marginal landholders with less than 2 hectares of land. Given these small and shrinking landholdings, most rural households currently rely on both petty agri-commodity production and various forms of non-agricultural wage labor and trade (Lerche 2021).

This paper draws on 14 months of ethnographic research conducted in the Malwa region between 2018 and 2019. During this time, I was primarily based in a single village that I call Pipliya, but my research involved travel to and interviews with a range of people in neighboring villages, market towns and government offices across three districts of Ujjain, Dewas and Shajapur. Methods included a household socio-economic survey with 60 rural households in one village, close observation of the agricultural decisions and practices of rural households (stratified by class, caste, and gender) across three agricultural seasons, semi-structured interviews with a range of key actors in the rural economy (farmers, traders, agricultural extension agents, policy-makers, insurers) in Malwa as well as the cities of Indore, Bhopal, Mumbai and New Delhi, and detailed analysis of national and state-level policy documents on agriculture and climate change in central India.

3. Assessment

Assessments of climate risk – what it is, how it manifests, and with what impacts – are central to climate action, shaping decisions about where responsibility lies and what must be done. In this section, I examine the framing of climate risk in key government documents which shape agrarian climate policy in India. Specifically, I look at how risk is understood – its causes, contours, and consequences – in these accounts as well as how this imagination shapes proposed solutions. Across these documents, I show risk is framed as changes to 'temperature' and 'precipitation' with little attention to the intersections between and imbrication of climatic factors and historical political ecologies. The dominant concern of policy-makers are risks posed to 'yield' and 'productivity', rather than the threats posed to rural livelihoods and social equity. Through a reading of the National Action Plan on Climate Change (2008), Indian Economic Survey (2017-18), and the Madhya Pradesh State Action Plan on Climate Change (2014),[5] I argue that this particular constitution and delimitation of risk not only elides a range of existing and intersecting socio-ecological risks, but relatedly, produces adaptation and mitigation 'solutions' that circumvent – and indeed, worsen – the structural roots of risk.

The 2008 National Action Plan on Climate Change outlines eight national missions that act on different aspects of the climate challenge. Among them is the National Mission For Sustainable Agriculture which identifies four priority areas, namely dryland agriculture, access to information, risk management, and use of biotechnology. The focus is on technological solutions such as improved methods of water and soil conservation, development of drought – and pest-resistant crop varieties, and agro-climatic information sharing. For example, the plan calls for increased irrigation for rain-fed areas and innovative agricultural research and technology to increase yields and crop resilience. Insurance is also presented as a key area of development, but, as I show in detail below, it is framed as a way to further enable capitalist agriculture by covering risks incurred to enhance productivity. Across these arenas, there is no discussion about adapting to changing ecological conditions or addressing differential social vulnerabilities to climate impacts. Rather, 'sustainable agriculture' implies sustaining the current input-intensive model in the

[5]These reports are prepared by different state agencies: the National Action Plan by the Prime Minister's Council on Climate Change, Government of India, the Economic Survey by the Ministry of Finance, and the MP State Plan by the regional government's Housing and Environment Department.

face of climatic variability rather than interrogating and transforming agricultural paradigms that produce and exacerbate adverse weather conditions.

The Economic Survey deploys an explicitly Malthusian framework in calling for increased productivity 'against the backdrop of increasing resource constraints' (2018, 82).[6] Indeed, the report frames climate change as the harbinger of a new agrarian transition – in order to increase productivity and ensure prosperity, people must be moved out of agriculture. It insists that farming cannot remain a viable livelihood – the ultimate goal of agrarian policy must therefore be 'to ensure that tomorrow there are fewer farmers and farms but more productive ones' (2018, 83). In this imaginary, there is no future for the rural smallholder in the climate change era. Instead, climate change portends – and indeed, necessitates – urban migration and depeasantization in order to secure the country's food and economic security (see also Paprocki 2020 on similar imaginaries in coastal Bangladesh).

Thus, the primary policy recommendations hinge on new technologies. For instance, it pushes for a renewed focus on irrigation technologies such as drip and sprinkler systems as emphasized by the 'more crop per drop' slogan. While it recognizes the problem of groundwater depletion, it fails to identify the fundamental roots of this water crisis in unsustainable cropping patterns as well as unequal use and distribution of water among and between rural and urban populations. Across the report, the discourse around climate change serves to renew and strengthen calls for a 'New Green Revolution', despite the considerable failures and ongoing legacies of the first one (see Patel 2013).

Although the MP State Action Plan adopts a more ecologically grounded approach, this report too relies heavily on technical and market-oriented modes of adaptation: mechanization, market access, improved seed and cropping techniques, and integrated climate information hubs. Complex principles of agroecology are reduced to organic farming and agro-climatic suitability with little attention to institutional support, resource redistribution, and market access. While the plan recognizes, for example, monocropping and attendant soil damage as an important area of concern, it recommends the 'management of risks for sustainable productivity' through intercropping and use of indigenous seed varieties. While these are crucial restorative measures at the farm level, they do not redress vulnerabilities at 'the scale of the market, of territory, and access to state power' (Holt-Gimenez, Shattuck, and Van Lammeren 2021, 720).

Although the report acknowledges the particular vulnerabilities faced by small and marginal farmers, recognizing inequality is not equivalent to analyzing power (Taylor 2015). It is the difference between recognizing vulnerability as a *state* of being as against '"vulnerabilization" as a relational *process* in which vulnerability is produced and reproduced over time between social groups' (Taylor 2015, 8, emphasis mine). As a result of this framing, the policy suggestions remain limited to short-term and farm-level adaptations rather than structural transformations that address severe and socially produced disparities in access to resources or the decades of state disinvestment from agricultural development.

In distilling the complex phenomenon of climate change into a singular measure of temperature and rainfall, these documents dwell on a 'narrow framework of interactions

[6]While the report discusses a range of issues, I focus here on Chapter 6 on 'Climate, Climate Change and Agriculture' (p. 82-101).

between farmers and climatic stresses', although smallholders themselves recognize that a range of political-economic risks make their livelihoods insecure and unstable (Taylor and Bhasme 2021, 432). As such, they preclude discussion of the connections between climate and the current mode of production. Moreover, the causal attribution of climate change to greenhouse gas emissions from the developed world (while not inaccurate) serves to elide the ways that the agrarian landscapes of India too are unevenly implicated in the climate crisis, as I discuss in the next section.

Given that adverse climate outcomes are exclusively interpreted in terms of effects on 'yield' and 'productivity', it is not surprising that the proposed solutions are located squarely within the 'adaptation-modernisation nexus' (Taylor 2015),geared toward the maintenance of 'mainstream mono-crop cereal-based input intensive production systems' (Raina 2012, 321). Particularly glaring is the lack of concrete discussion of the ongoing agrarian crisis, which forms a key driver of the vulnerabilization of farmers. This crisis is itself transformed into a technical problem of costs and outputs rather than a complex political-economic, social, environmental, and psychic catastrophe. Ignoring these deep roots naturalizes and backgrounds the 'pre-existing misery of precarity' which is a product of specific state policies (Ribot 2014, 672).

At best, the focus is on improving the efficiency of agricultural production through technological innovation and for the protection of farmers through financialized risk-management tools. At worst, these policies advocate for the demise of the 'unproductive' peasantry, citing their incapacity to ensure productivity and food security in a climate-changed agrarian economy. Overall, these solutions broadly align with a 'corporate-driven technological narrative' (Borras et al. 2022, 9) or 'a green economy discourse that is associated with business and market-friendly principles, technological optimism and neglect of issues concerning power, inequity, marginalisation and resource distribution' (Srivastava, Mehta, and Adam 2022, 5). However, as I argue, the specific risks of bad weather are always already shaped by existing environmental relations that are historically and socially produced. It is to this production of risk that I now turn.

4. Production

This paper opened with an especially tumultuous monsoon in Malwa, which caused significant damage to crops across the region. But this was not an entirely unique season. Over the last five years, residents of Pipliya village and the broader region have faced multiple poor soybean harvests. Each year, distinct but intersecting phenomena damaged the crop: in 2017, it was the yellow mosaic virus; in 2019, a long dry spell followed by heavy rain damaged the seed; and in 2020, infestations of white fly and stem fly was followed by excess rainfall during harvest time to ruin the crop. In the latter case, heavy rain and high temperatures not only harmed the crop but also formed the perfect weather for the proliferation of pests. The harvests were so poor that farmers were unable to save even a small amount of good seed for the following year's (2021) sowing, a practice that all farmers follow.[7]

[7]Most farmers were then forced to purchase seeds, often from wealthy farmers who had managed to save and stock good seed. During sowing (late June), seed prices skyrocketed, leading many farmers into debt to buy seed or to reduce their cultivation altogether. As a result, soybean acreage declined by 10% in the 2021 season compared to 2020.

Across Malwa, farmers despaired about the erratic weather. Interestingly, they cursed soybean as a 'weak' crop, one that required steady, if minimal, rainfall to survive and lamented their failure to diversify their fields by growing maize or peanuts. Indeed, as an agricultural extension agent succinctly pointed out, 'It is not the rain that is the problem. The problem is soybean monocropping.'

This section outlines a broader understanding of risk through and beyond climate, specifically outlining how climate change 'exacerbates the uncertainty and amplifies the risks attached to capitalist agriculture, thereby increasing the vulnerability of rural populations' (Borras et al. 2022, 5). As weather patterns transform, farmers face not only heightened risks but also limited (albeit highly differentiated) capacities to manage these risks owing to the intersecting legacies of industrialized monocultures and weakened state supports. That is, farmers' experiences of erratic rainfall, scorching temperatures, and the pests that follow are entangled with and inseparable from existing socio-ecological relations of production which are, in turn, shaped by state policy, global markets, and rural social structure.

Although the farmers of Malwa have long been integrated with global trade networks through crops such as cotton and opium, the past five decades have led to enormous and intense shifts in cropping patterns and production relations. Soyabean lies at the center of these transformations. In the 1960s, agronomists from the University of Illinois visited India as agricultural advisors, but brought with them expertise on the soybean crop which would, it was believed, address India's protein deficiency problem (see Kumar 2016 for a history of soybean in Malwa). Although the Indian palate never quite took to the bean, its cultivation spread in the 1970s and 80s as a crop that was processed into oil and deoiled cake (DOC). Soybean was promoted not as much by the government but by oil mills, processing industries, and commodity trading companies which saw soybean as a lucrative source of foreign exchange. In a few decades, this region came to be popularly known as 'Soya Pradesh' (Soya State).

Soybean quickly became the dominant cash crop, displacing staple monsoon crops such as cotton, maize and sorghum. Its production largely eliminated existing subsistence crops. Indeed, the proportion of cropped area dedicated to food crops declined sharply from 86 per cent in 1960–61 to 72 per cent in 2000-03, largely due to this shift from cereals to oilseeds (Vijay Shankar 2005).[8] This reflects global trends in soybean production. Global soybean production has increased over a 1000 times in the last 60 years, and land under soybean cultivation has quadrupled during this same period. As a flex crop, soybean is valued less for soy itself and far more for soybean meal and oil, which are used as cattle feed, edible oil and a range of industrial products (see de L.T. Oliveira and Schneider 2016).[9] In contributing directly to the agro-industrial livestock sector, the landscapes of Malwa are directly tied to the industry that produces at least 15 percent of greenhouse gas emissions within the agri-food sector.

[8]India currently ranks fifth in global soybean production but its output is far lower than other major producer countries such as Brazil, USA, and Argentina. Much of India's soybean exports are to the US and Southeast Asian countries such as Taiwan and Vietnam, but exports fluctuate depending on global prices, especially since Indian soybean varieties are seen as 'non-competitive' (higher priced) than those produced in places such as Brazil and Argentina.

[9]In 2020-21, soybean production in India was estimated at roughly 11 million metric tons, and about 2 million metric tons was exported as soybean meal, though annual exports vary considerable from year to year.

This has had dramatic social and ecological effects. The introduction of soybean, a short-duration monsoon crop, created a new pattern of double-cropping in Malwa – soon after soybean is harvested, farmers plant wheat, gram and a number of other winter crops. This cropping pattern enabled wealthy, large landowner farmers to further accumulate capital through multiple cropping cycles and reinvest in mechanization technologies, tube-wells, and economic diversification. For small farmers, the coming of soybean enhanced their cash incomes and opened up the possibility of cultivating across seasons which made their small plots of land more viable. But it also undermined their food sovereignty and nutritional security. Soybean cultivation displaced sorghum (the primary food crop) in the monsoon season which was, in turn, replaced by irrigated winter wheat. This meant that smallholders would either have to drill a well to cultivate wheat or leave their land fallow for the winter, hampering their ability to meet basic subsistence needs from their own land (cf. Paprocki and Cons 2014). Marginal farmers who lack access to irrigation are therefore left food insecure, relying wholly on market purchases and (often substandard) grain disbursed through the state's public distribution system.

This 'yellow revolution' (of soybean) has quickly transformed into an ecologically disastrous 'soybean-wheat complex', as farmers are now compelled to drill more and deeper tubewells to cultivate wheat and other winter crops, leading to declining water tables and considerable debt burdens (Kumar 2016). As yields decline in the wake of decades of soybean monocropping, farmers are now confronting falling yields and soil degradation which is met with greater investments in chemical fertilizers and pesticides. Overall, this pattern has 'promoted yield anxiety and further pushed farmers onto a technological treadmill and into a constant state of crisis' (Kumar 2016, 9). In many ways, these socio-ecological transformations resemble earlier shifts toward intensive cash-cropping and commodity production. In the Sahel region of Africa, for instance, Michael Watts (1983) described how the colonial state-driven 'groundnut revolution' decreased the area devoted to food crops and subjected communities to volatile commodity prices, thus transforming seasonal cycles of hunger into full-scale famine.[10]

It is only through this recent history of agrarian change that we can fully grasp the implications of the strange and erratic weather of the past few years. Many of these events – intense rainfall, extreme heat, long dry spells, and frequent pest attacks – are only likely to rise as the climate crisis worsens. But the vulnerability of farmers across these seasons arises at the conjunction of soybean monocropping, chemicalized agriculture, and climate change. As the agricultural extension agent introduced at the beginning of the section explained, the distress faced by farmers cannot be attributed solely to inclement weather. Broader structural forces that have pushed farmers into these destructive monocultures wherein high yields and high prices are the sole objectives of farming regardless of long-term sustainability.

[10]The expansion of shrimp aquaculture in coastal Bangladesh reveals similar dynamics. As Kasia Paprocki (2021) notes, even though development practitioners view shrimp farming as a solution to climate change, it is actually the violent and dispossessive legacies of shrimp and the displacement of rice cultivation that produced new social and environmental threats for peasant cultivators. Commercial shrimp production caused waterlogging and soil salinity, leading to the destruction of trees, grazing lands and gardens - in other words, damaging the means of production and reproduction in the region.

This socio-ecological destruction unfolds at the intersection of climatic and capitalist temporalities. According to agricultural scientists, much of the damage in 2020 could potentially be traced to specific varieties of seed: JS-9560, JS-2029, JS-9305. The JS-9560 variety, for instance, is a locally developed seed variety whose primary quality is its early ripening. With this variety, the crop can be harvested in under 90 days, a huge advantage for farmers who wish to plant potatoes, garlic and onions on the same land over the winter season. However, this variety – despite its temporal benefits – is believed to be less tolerant of pests and extreme weather conditions (Niazi 2020). The push to produce more from the land within shorter periods of time runs up against specific material limits. Moreover, as established weather patterns shift, central India is receiving heavy rainfall even in the months of September and October (when the monsoon typically recedes) which hampers the harvest period for these short-duration varieties. Not only is this quasi-industrial model depleting the soil and water on which it is built but it creates new ecological crises (such as reduced resistance to pests) which then intersect with and are compounded by the climate crisis. Capitalist agriculture – in particular the obsession with yield and productivity that permeates down to the smallholder farm – destabilizes its own conditions of possibility (Fraser 2021).

It is clear that climate change is exacerbating the vulnerabilities of farmers in rural India – but it does not act alone or in isolation. Through an empirical account of soybean, I have demonstrated that climate change is always experienced through specific capitalist socio-natures of crops, soil, water, and seed. In doing so, it offers 'a more expansive, historically informed analysis that situates "climate" within a wider set of environmental struggles in agrarian settings' (Borras et al. 2022). Yet, as the previous section showed, assessments of risk often narrowly decouple climate from these struggles and offer techno-managerial solutions of risk-management, including crop insurance, whose dynamics I now examine.

5. Management

Cultivators have managed and minimized long-standing agrarian risks in multiple ways – from crop diversification to out-migration. A range of coping strategies exist that are both preventive and adaptive, including: state food-for-work programs, public food distribution systems, wage labor, migration, prayer, changing seed and crop varieties, selling off livestock, taking on loans, and so on (see Peterson 2012). These modes of managing and mitigating risk continue to be important in rural contexts. However, as noted above, centuries of colonialism and capitalist development have eroded deep-rooted modes of minimizing risk as well as moral economies of reciprocity and redistribution at the heart of the peasant subsistence economy (Scott 1976; Watts 1983). In Malwa, for instance, this can be seen in the spread of soybean farming which displaced local cropping practices that maintained agro-diversity and agronomic variation on farms to avoid subsistence crises.

In response to emerging climate change, states, corporations, and development institutions across the globe have promoted tools of financialized risk-management. Among these, crop insurance is now a popular mechanism of climate change adaptation and mitigation across agrarian contexts in the global south. In India, risk financing through extension of credit and insurance to enable adaptation to climate risks (and likely crop failure) is posited as a priority area of the National Action Plan on Climate Change and risk management through insurance is highlighted within the National Mission for Sustainable Agriculture, echoing similar emphases in other parts of Asia and Africa (Clapp and Isakson 2018).

Specifically, crop insurance is operationalized through the Pradhan Mantri Fasal Bima Yojana (Prime Minister's Crop Insurance Program, hereafter PMFBY). The program, launched in 2016, follows on the heels of a number of crop insurance programs that have been introduced (and often failed) in various parts of the country since the 1970s.[11] Under this state-funded and privately-operationalized program, insurance premiums, highly subsidized by central and state governments, are collected by insurance companies, who then conduct seasonal crop surveys and make payouts to farmers. In its current form, it marks the massive transfer of public wealth to private companies through premium subsidies. In effect, it facilitates capital accumulation by the insurance industry by enrolling rural smallholders into global circuits of finance in the name of climate mitigation and income security through a range of microfinance products (da Costa 2013; Isakson 2015).

More broadly, the way that insurance is framed as a climate adaptation strategy works to naturalize and de-historicize agrarian risk itself. Overall, the risks against which it protects all appear to be naturally occurring events, rather than historical outcomes of socio-natural processes – from soybean monocropping to unequal land ownership. The program is advertised as protecting farmers from 'contingencies' and 'unforeseen events'. Although this is crucial especially in light of shifting and unpredictable weather patterns, it entirely elides the ways that structural inequities, political-economic transformations, and policy decisions interact with inclement weather to produce adverse outcomes for farmers. Marcus Taylor (2016) notes a similar process in his analysis of a livestock insurance scheme for Mongolian pastoralists. There, the World Bank aided the Mongolian government in setting up a livestock insurance scheme for pastoralists, appearing to be a neutral outsider facilitating adaptation to climate risk. However, Taylor demonstrates that it is the Bank's structural adjustment policies that produced these risks in the first place – through the privatization of herding and the undermining of collective risk-pooling institutions. In response, herders increased their livestock population as a mode of survival, which turned out to be disastrous both ecologically and socially, propelling degradation of pastures and decimation of herds by extreme weather. The crisis which the Bank intended to solve with insurance was one partially of their own creation.

Moreover, insurance necessarily delimits what is classified as 'risk' (and therefore, what losses can be monetarily compensated) while also actively promoting the same risky practices that produce vulnerability to climate hazards. The PMFBY covers risks such as yield losses (caused by adverse weather) and prevented sowing as well as post-harvest losses and localized calamities such as hailstorms, landslides, and inundation. While this scope is fairly broad, it is premised on numerous elisions and erasures, revealing a 'gap between real-world dangers on the one hand and insurable risks on the other' (Aguiton 2019, 285).

[11]While the program is now voluntary, until 2020, it was mandatory for all loanee farmers (farmers availing of crop loans from public and private banks). That is, any farmer who had taken a crop loan was automatically insured - in part as a way for banks to hedge against risk of loan default (Sheth 2017).

For instance, certain crops cannot be insured at all if they are not 'notified' crops, leaving farmers who grow them entirely outside the scheme's purview and certain risks are entirely unrecognized such as damage caused by frost or by stray cows and wild animals. These pre-defined risks do not adequately account for emerging climate uncertainties – for example, the potato crop is only protected against extremely low temperatures (being a winter crop sown in October), but this does not safeguard against untimely rainfall which is increasingly expected to occur well into the sowing period and beyond.

More fundamentally, the goal of these programs – as also outlined in the policy documents reviewed earlier – is to maintain the status quo of capitalist agriculture. The rationale of programs such as these is not to adapt and transform farming to suit an increasingly precarious present and future but rather to locate ways to continue and intensify current models of production while minimizing loss of income. In its own words, the program aims to provide financial support and income stability in the aftermath of natural calamities, to encourage 'innovative and modern' agricultural practices, and to ensure agricultural credit flows. Climate change is deployed as a discursive tool to further bolster an input-intensive and debt-financed model of agrarian development by offering a (minimal) buffer against weather shocks. Indeed, despite the lack of empirical evidence for this, policy-makers and the insurance industry insist that access to insurance will 'embolden farmers to abandon time-tested practices and technologies that generally ensure stable yields in favour of riskier yet potentially more lucrative activities' (Clapp and Isakson 2018, 57). Better risk management is, therefore, associated with growing riskier and more profitable crops – regardless of its social and ecological consequences.

Further, access to insurance has become another mechanism of differentiation among farmers. The program directly excludes the most marginalized of cultivators from its fold – including small and marginal farmers, tenant farmers, sharecroppers, and agricultural laborers. Although these classes of cultivators can in theory take out an insurance policy, field data shows that it is immensely difficult for them to do so. Even subsidized premiums are a considerable sum of money, an impossible investment for smallholders, especially at the beginning of the agricultural season when sowing expenses (seed, fertilizer, tractor costs) are considerable. Often, bank officials are often unwilling to issue policies to smallholders. As one marginal farmer insisted, 'They don't give loans and insurance to *people like us*, those with just one *bigha* [of land].'

Although the program itself is faulty and limited, being excluded from its purview is not entirely inconsequential. Direct relief to farmers in the wake of crop losses is shrinking, andpoor farmers are chided by state representatives for not taking out insurance to protect their crops and livelihoods. Nonetheless, when claims payouts are made, they further enable insured farmers – mostly large- and medium-scale farmers with access to credit and cash reserves – to withstand climatic extremes. Their crops are insured, but they also have the savings to weather crop failures and have diversified into non-farm trades and occupations which afford more secure and steady incomes. Insurance itself becomes another route through which the state subsidizes and supports wealthy landowners while largely abandoning smallholders. If anything, insurance only protects those who already have the means to secure their livelihoods. Meanwhile, in the wake of climate-induced calamites and losses, small farmers are further proletarianized, pushed into exploitative wage labor and forms of 'distress diversification' (Srivastava, Mehta, and Adam 2022). They take on more debt and agricultural labor, often from and on the fields of large landowners, which locks them into cycles of debt, dependency and even dispossession as when they sell parcels of land to pay off loans or withstand repeated crop failures.

At their core, then, adaptive mechanisms such as insurance are primarily geared toward enabling the continuation of technologies, practices, and relations of production that perpetuate the current model of commercial farming, regardless of their sustainability in the face of the climate challenge. In reducing systemic socio-ecological risks to singular occurrences of bad weather, measuring loss solely in terms of yield, and serving as a buffer for the dangers of industrialized farming, it elides the production of risk and, more significantly, amplifies these risks. Similarly, differentiated access means that the poorest and most vulnerable farmers are entirely excluded from even the meager social safety net that insurance offers. Climate change, as well as dominant adaptation mechanisms, thus heightens processes of social differentiation and widens rural inequalities, as I show in more detail below.

6. Engagement

As the JPS editorial highlights, the differentiated impacts and responses to climate change in particular and environmental risk more broadly mean that some suffer while others prosper (Borras et al. 2022, 6). Certain agrarian classes are better able to manage emerging climate uncertainties – to survive and in some cases, thrive despite and through them. This is true of India's agrarian crisis as well. Scholars have shown that experiences of this crisis are socially and spatially uneven (Jodhka 2012; Sinha 2020). For large landowners, new uncertainties can become opportunities for accumulation such as when they sell soybean seed at exorbitant rates in the aftermath of poor harvests or have access to storage infrastructures that enable speculation on harvested crops through spells of heavy rainfall (Matthan 2022). At the same time, nearly all farmers, including smallholders, are compelled to actively *engage* risk, albeit to different extents and with varied consequences. The focus of this section is on how they do so in relation to groundwater, a resource under particular stress in the current moment. Under changing socio-environmental conditions, land-water must be considered as a 'coupled resource' (Chung 2019) whose use, extraction and enclosure cannot be disconnected. Land is now increasingly valued 'as the topographical basis on which to gain access to water' (Taylor and Bhasme 2021, 436). That is, while farmers insist that land can be easily leased or bought, it is becoming far more difficult and lucrative to secure access to water.

Water scarcity is, however, a relatively recent historical and social phenomenon in many parts of India. This is made evident, for example, in historical accounts of droughts and famine in western India which show that shortages of food were common but not shortages of water (Bhatia 1992). The problem of water depletion emerged in the latter part of the twentieth century with changing cropping patterns and expansion of mechanized extraction technologies. Scarcity is not natural in a physical sense, but rather, socially produced through capitalist farming and over-extraction. In Malwa, the soybean-wheat complex is undergirded – quite literally – by groundwater to irrigate water-intensive crops such as wheat and increasingly, capital-intensive vegetables.

Water is now vital to commercial and subsistence cultivation. It is also an enterprise plagued with uncertainty for several reasons: first, with uneven groundwater mapping, the location, depth and quantity of water in subsurface aquifers is itself unclear, causing many wells to fail or quickly run dry; second, drilling tube-wells is an expensive endeavor often financed by debt (Taylor 2015; Vasavi 2020); and third, Malwa's hard rock aquifers have a low replenishment rate, making over-extraction especially dangerous in the long-term (Vijay Shankar 2005). These risks of access are not evenly distributed and experienced. Access to water, as with other environmental resources, is reflective and productive of social inequality (Swyngedouw 2009). The topography of Malwa is such that villages comprise fertile valleys surrounded by rocky hills. In general, while wealthy and upper-caste households own land in the valleys, poor and Dalit ('low-caste') households own land in the rocky and arid hills where groundwater is scant. Given the connected nature of subsurface aquifers, this also means that those who can drill more and deeper wells can deplete water resources to the detriment of smallholders who must also draw water from the same hydrogeological system. As a result, a common property resource is effectively privatized by wealthy farmers through their larger land parcels, deeper wells, stronger pumps, and bigger storage tanks. With limited state investment in public water infrastructures and the absence of a moral economy of 'just' water use, these risks are entirely individualized with serious impacts on the subsistence security of small farmers.

Groundwater depletion and attendant inequities are well-documented, exceeding and preceding the impacts of climate change (see Bhatia 1992; Moench 1992; Shah 2008). Indeed, these inequities of water access are not directly tied to climatic variability but rather to intensive cash-cropping over the past half-century. Nonetheless, scholars point out that climate change will act as a 'force multiplier' to further threaten secure access to water for millions of farmers (Shah 2009). For India, hydro-climatic change has multiple effects: monsoon crops face greater risk of both flooding and drought, while late winter and summer crops will experience higher evapo-transpiration and lower soil moisture due to excess heat, thus requiring more frequent irrigation. Given the highly vulnerable irrigation economy, addressing these hazards and their uneven social impacts will necessitate not just technical interventions such as better storage facilities and efficient irrigation systems but also a reckoning with the current unsustainable and inequitable model of water-intensive farming.

For small farmers, investment in irrigation is already incredibly risky. Yet, these cannot always be avoided and indeed, must be engaged. For instance, Gopal, a Dalit smallholder farmer who owns about half an hectare of unirrigated land, drilled four tube-wells over three years. A good supply of water would enable him to cultivate an additional winter crop on the land. None of the wells yielded water, preventing him from planting wheat or onions, as he had hoped. Following these failed attempts, he leased out this land to a neighbor for the winter season and took up factory work in a nearby town, insisting that farming without access to water was unviable. In this situation, more farmers like Gopal will be pushed to lease or even sell their land, potentially undermining their food security, autonomy, and leading to greater concentration of land and water in the hands of rural elites.

Indeed, farmers who have ready access to capital, credit and social networks reported drilling or deepening wells every year as wells run dry (Dubash 2002). This is not just for their own use but also for sale. In villages in Malwa and elsewhere, it is common for water merchants to sell water through tankers at a high cost during the summer months or enter into share-cropping contracts based on their contribution to irrigation (cf. Bhatia 1992; Taylor 2015, 90). Further, it is these big farmers who can more swiftly adapt to and capitalize on short-term fluctuations in water availability. In 2019, rainfall was plentiful, if devastating to the soybean crop. Streams were full and wells overflowed. While one might have expected most farmers to take advantage of this rare situation, small farmers proceeded to plant water-intensive but low-investment wheat to recover their losses from the heavy rain. Wealthy farmers, however, had the capital to invest in water-intensive but also high-yielding and profitable vegetables, thereby garnering substantial profits despite and through this extreme weather, thus furthering processes of stratified accumulation.

Moreover, in Malwa, capitalist farmers of the dominant Patidar caste are now leasing land in the neighboring Nimar plains (south of Malwa) through which the Narmada river runs. Water, they insist, is more plentiful in this region given its proximity to the river and land is relatively inexpensive. This region comprises both wealthy Patidar farmers and, importantly, a substantial adivasi tribal population. It is the land of these latter communities that farmers from Malwa tend to lease (non-adivasis cannot buy land here due to constitutional laws protecting adivasi land rights).[12] What is unfolding then is a form of 'water grabbing' since land leases are primarily driven by the desire to capture water resources (Mehta, Veldwisch, and Franco 2012). Moving into new lands has become a crucial (mal)adaptation strategy among wealthier households. Although this form of water appropriation is relatively small-scale, it nonetheless reallocates water away from poor, Dalit and adivasi peasants and undermines the quality of their land.

Formal adaptation programs can also become a key mechanism of differentiation. In the water sector, their overall focus is on improving the efficiency of the existing system of production. The solutions offered largely fall into one of two broad categories: techno-managerial and traditionalist. Techno-managerial solutions from drought-resistant seed varieties to drip irrigation are presented as panaceas to climate impacts. Traditionalist solutions call for a return to older modes of conservation and adaptation such as water harvesting techniques and drought-coping mechanisms. While certain technologies may be necessary and welcome, unless they are tied to farmers' needs and address rural hierarchies, they will end up reproducing unequal social and ecological relations of production. For instance, a valorization of traditional water-management systems that does not confront caste-based ritual segregations that exclude Dalits from access to common water resources will only worsen enduring hierarchies.

Watershed development, a cornerstone of climate and development policy in India, is a prime example of this. It generally involves the construction of concrete run-off structures, check dams, and individual farm ponds for the purpose of rainwater harvesting and aquifer recharge. As a study of climate resilience initiatives in a drought-prone region of India (Taylor and Bhasme 2021) demonstrates, while these water infrastructures are publicly funded, the uneven geography of land ownership and water flows as well as the large investment demands of ponds, pumps, wells, and tanks required to utilize harvested water makes it far more difficult for small and marginal farmers to access its

[12]This practice is further justified through dominant discourses that castigate adivasi farmers for their 'unproductive' and 'wasteful' use of the land (and water) which are, in turn, tied to racialized stereotypes of adivasis as 'lazy'.

benefits. In this way, project dynamics are remarkably similar to earlier watershed development programs which concentrate resources in the hands of already privileged individuals in order to ensure the achievement of project 'success', thus reproducing structures of inequality (Baviskar 2007). In Malwa's villages too, it was primarily upper-caste farmers who had the capital required (over and above minimal state subsidies) to invest in water-efficient technologies that ensure their water security and enable agrarian accumulation.

Thus, 'the impacts of climatic change combine with the commodification of water and new forms of market-entrepreneurship' (Taylor 2015, 90) to deepen rural inequality. On the one hand, big farmers respond to water uncertainties by expanding their search for water (both above- and under-ground) and through taking advantage of short-term fluctuations in water availability. On the other hand, Dalit and adivasi smallholders tend to be cornered into two (im)possible situations: they can either continue to make these uncertain investments in water (and other inputs) or abandon cultivation altogether. As grave risks morph into perilous uncertainties, agriculture itself turns into a gamble, but one whose odds are firmly stacked against the small farmer.

7. Conclusion

While the tumult of the 2019 monsoon, with which I began this paper, is not quite a direct outcome of the climate crisis, it portends an emergent present and near future of volatility and uncertainty for farmers. Across South Asia, cultivators are expected to increasingly face extreme and unpredictable weather in the form of intense bouts of heavy rainfall, longer dry spells, and frequent heat waves. While farmers have always confronted erratic weather, they now do so under dramatically changed conditions of production, ones which often preclude risk aversion and demand everyday engagements with economic and environmental uncertainties simply in order to remain cultivators of the land. This 'climate of uncertainty' is produced at the conjuncture of anthropogenic climate change and capitalist modes of cultivation.

In the context of industrialized monocropping, an erratic monsoon is not only felt as more or less rainfall but equally as additional investment in expensive inputs (seeds for re-sowing or chemicals to deal with new pests), delayed harvests, extensions on loan repayments, or the need to drill a new well. Ordinary risks of inclement weather turn into puzzling and treacherous uncertainties, upending established agricultural practices and raising new dilemmas: how to cultivate lucrative short-duration varieties when rainfall now continues into harvest time? Can the standing wheat crop withstand the extreme heat that seems to arrive earlier every year? Will there be enough water to irrigate a field of onions? When the costs of production are so high and returns already minimal, even incremental shifts in weather patterns can engender devastating losses (cf. Gupta 2017; Sethi 2021).

And while these uncertainties structure the everyday decisions and practices of all classes and castes of farmers, it has particularly devastating impacts on small farmers who are unable to consistently engage with these uncertainties or repeatedly withstand losses. For smallholders, farming itself becomes unviable, leading to distress diversification, land dispossession, and worsening poverty. Big landowners are not only better able to withstand and capitalize on these uncertainties owing to their access to capital, credit, and knowledge, but they are also often the primary beneficiaries of state-sponsored climate adaptation programs from crop insurance to watershed development and beyond.

This 'climate of uncertainty' that envelopes rural lives and livelihoods has important theoretical and political implications. In the current moment, the intertwining of capitalist and climate risk does not generate practices of risk minimization. As this essay has shown, most farmers are now compelled to engage in risky practices in order to remain as agriculturalists, achieve basic subsistence standards, and enact upward class mobility. This marks a partial but significant shift in peasant orientation toward agrarian risk. Moreover, previously known risks have mutated into hard-to-predict uncertainties which can disrupt long-standing modes of prediction and adaptation. This calls for a deeper understanding of quotidian experiences of risk and uncertainty among farmers in order to better grasp the desires and compulsions, knowledge and practices, generated at the intersection of climate change and agrarian capitalism. Doing so is crucial to imagining and building more socially and ecologically just modes of engaging with uncertain agrarian futures.

Relatedly, the study of agrarian risk holds value for theorizing and enacting struggles for subsistence security and agrarian justice in the climate change era. Importantly, risk draws our attention to questions of responsibility and redress. Writing half a century prior, scholars working at the intersection of peasant studies and political ecology argued that peasants found some succor in social norms and institutions which minimally ensured the right to subsistence, especially in times of crisis. Today, in the villages of Malwa, there are few village-level safeguards against subsistence crises, and entrenched hierarchies of caste and class often preclude communal sharing and collective action. Without simplistically romanticizing traditional village institutions, global markets, or even state welfare programs, we might ask what moral economies (understood as notions and practices of justice, fairness and security), both extant and emergent, are available and adequate to the current rural predicament.

Crop insurance, discussed above, is one such site that might be retooled for more progressive and transformative ends. As a mechanism of risk-management and social security, insurance programs can be socialized and expanded in ways that are inclusive, accessible and redistributive – as against its current exclusionary, technocratic, and financialized form. The successful protests by north Indian farmers in 2020–21 against the dismantling of an established system of support prices and regulated markets represents another such site of mobilization, even though this system of security is itself tragically built on an unsustainable model of input-intensive farming. Nonetheless, demands for insurance against crop loss and insistence on fixed prices for produce represent dynamic moral-economic struggles that claim protections against agrarian risk and uncertainty from the Indian state. Often, these unfold in expected spaces and have contradictory effects. The question for contemporary agrarian movements then is how to strengthen, extend and reorient these demands in ways that connect issues of fair prices and resource redistribution with broader concerns around climate resilience as well as stable employment, food security, and human dignity.

Forging viable agrarian futures in the face of climate change demands an expansive, relational and historical understanding of risk and uncertainty, one that roots climatic stresses firmly within longer histories of uneven development. Indeed, analyses of the social and historical roots of risk demonstrate how experiences of climate change are always refracted through existing production relations and power structures (Paprocki

2021). As Malwa's farmers well know, the climate of uncertainty that engulfs their livelihoods and life-chances emanates not just from fleeting rain clouds and scorching temperatures, but also from the soybean seed and the tube-well, and the model of industrialized agriculture that brought these technologies to their lands. Placing the seed, the well, and the weather within a unified frame of analysis and action might lead us toward the structural transformation of this model (Borras et al. 2022). Doing so is essential to ensuring that rural producers are able to not simply manage risk or survive uncertainty but also thrive and flourish through a life on the land.

Acknowledgments

I would like to thank four anonymous reviewers for the insightful comments which greatly improved this article. The Wenner-Gren Foundation for Anthropological Research and the UCLA International Institute provided essential funding for this research. The S.V. Ciriacy-Wantrup Post-doctoral Fellowship at the University of California, Berkeley, afforded me the time to write this article. My greatest thanks go to the many people of Malwa with whom I worked. All faults are mine alone.

Disclosure statement

No potential conflict of interest was reported by the author(s).

Funding

This work was supported by Wenner-Gren Foundation: [Grant Number 9661].

References

Aga, A. 2018. "Merchants of Knowledge: Petty Retail and Differentiation Without Consolidation among Farmers in Maharashtra, India." *Journal of Agrarian Change* 18 (3): 658–676.

Aguiton, S. A. 2019. "Fragile Transfers: Index Insurance and the Global Circuits of Climate Risks in Senegal." *Nature and Culture* 14 (3): 282–298.

Baviskar, A. 2007. "The Dream Machine: The Model Development Project and the Remaking of the State." In *Waterscapes: The Cultural Politics of a Natural Resource*, edited by A. Baviskar, 281–313. New Delhi: Permanent Black.

Beban, A., and T. Gorman. 2017. "From Land Grab to Agrarian Transition? Hybrid Trajectories of Accumulation and Environmental Change on the Cambodia–Vietnam Border." *The Journal of Peasant Studies* 44 (4): 748–768.

Bhatia, B. 1992. "Lush Fields and Parched Throats: Political Economy of Groundwater in Gujarat." *Economic and Political Weekly* 27 (51/52): A142–A170.

Borras Jr., S., I. Scoones, A. Baviskar, M. Edelman, N. Peluso and W. Wolford. 2022. "Climate Change and Agrarian Struggles: An Invitation to Contribute to a JPS Forum." *Journal of Peasant Studies*, 49:1, 1-28.

Chung, Y. 2019. "The Grass Beneath: Conservation, Agro-Industrialization, and Land-Water Enclosures in Postcolonial Tanzania." *Annals of the American Association of Geographers* 109 (1): 1–17.

Clapp, J., and S. R. Isakson. 2018. *Speculative Harvests: Financialization, Food and Agriculture*. Rugby, UK: Practical Action Publishing.

da Costa, D. 2013. "The 'Rule of Experts' in Making a Dynamic Micro-Insurance Industry in India." *Journal of Peasant Studies* 40 (5): 845–865.

de L.T. Oliveira, G., and M. Schneider. 2016. "The Politics of Flexing Soybeans: China, Brazil and Global Agroindustrial Restructuring." *The Journal of Peasant Studies* 43 (1): 167–194.

Dubash, N. 2002. *Tubewell Capitalism: Groundwater Development and Agrarian Change in Gujarat*. New Delhi: Oxford University Press.

Eakin, H. 2006. *Weathering Risk in Rural Mexico: Climatic, Institutional and Economic Change*. Tucson: University of Arizona Press.

Edelman, M. 2005. "Bringing the Moral Economy Back in to the Study of 21st-Century Transnational Peasant Movements." *American Anthropologist* 107 (3): 331–345.

Foster, J. B. 2013. "The Epochal Crisis." *Monthly Review* 65 (5): 1.

Fraser, N. 2021. "Climates of Capital: For a Trans-Environmental eco-Socialism." *New Left Review* 127: 94–127.

Gupta, A. 2017. "Farming as a Speculative Activity."." In *The Routledge Companion to Environmental Humanities*, edited by U. Heise, J. Christensen, and M. Niemann, 185–193. New York: Routledge.

Holt-Gimenez, E., A. Shattuck and I. Van Lammeren. 2021. "Thresholds of Resistance: Agroecology, Resilience and the Agrarian Question." *The Journal of Peasant Studies* 48:4, 715-733.

Isakson, S. R. 2015. "Derivatives for Development? Small-Farmer Vulnerability and the Financialization of Climate Risk Management." *Journal of Agrarian Change* 15 (4): 569–580.

Jasanoff, S. 2010. "A New Climate for Society, *Theory, Culture & Society* 27(2-3): 233-253.

Jodhka, S. 2012. "Agrarian Change in the Times of (Neo-Liberal) 'Crises': Revisiting Attached Labour in Haryana." *Economic and Political Weekly* 47 (26–27): 5–13.

Kennedy, S. 2022. "The Power to Stay: Climate, Cocoa, and the Politics of Displacement." *Annals of the American Association of Geographers* 112 (3): 674–683.

Knight, F. H. 1921. *Risk, Uncertainty and Profit*. Houghton: Pantianos Classics.

Krishnan, R., J. Sanjay, C. Gnanaseelan, M. Mujumdar, A. Kulkarni, and S. Chakraborty. 2020. *Assessment of Climate Change Over the Indian Region: A Report of the Ministry of Earth Sciences (MoES), Government of India*. New Delhi: Springer.

Kumar, R. 2016. *Rethinking Revolutions: Soya-Beans, Choupals, and the Changing Countryside in Central India*. New Delhi: Oxford University Press.

Lerche, J. 2011. "Agrarian Crisis and Agrarian Questions in India." *Journal of Agrarian Change* 11 (1): 104–118.

Lerche, J. 2021. "The Farm Laws Struggle 2020–2021: Class-Caste Alliances and Bypassed Agrarian Transition in Neoliberal India." *The Journal of Peasant Studies* 48 (7): 1380–1396.

Matthan, T. 2022. "Speculative Crops: Gambling on the Onion in Rural India." *Geoforum* 130: 115–122.

Mehta, L., G. J. Veldwisch, and J. Franco. 2012. "Introduction to the Special Issue: Water Grabbing? Focus on the (re)Appropriation of Finite Water Resources." *Water Alternatives* 5 (2): 193–207.

Moench, M. 1992. "Chasing the Watertable: Equity and Sustainability in Groundwater Management." *Economic and Political Weekly* 27 (51/52): A171–A177.

Niazi, S. 2020. "India's Soybean Farmers Demand help as Rains and Pests Wipe Out Crops." Thomas Reuters Foundation, 9 October. Available at: https://news.trust.org/item/20201009000315-phbp0/

Paprocki, K. 2020. "The Climate Change of Your Desires: Climate Migration and Imaginaries of Urban and Rural Climate Futures." *Environment and Planning D: Society and Space* 38 (2): 248–266.

Paprocki, K. 2021. *Threatening Dystopias: The Global Politics of Climate Change Adaptation in Bangladesh*. Ithaca: Cornell University Press.

Paprocki, K., and J. Cons. 2014. "Life in a Shrimp Zone: Aqua- and Other Cultures of Bangladesh's Coastal Landscape." *The Journal of Peasant Studies* 41 (6): 1109–1130.
Patel, R. 2013. "The Long Green Revolution." *Journal of Peasant Studies* 40 (1): 1–63.
Peterson, N. 2012. "Developing Climate Adaptation: The Intersection of Climate Research and Development Programmes in Index Insurance." *Development and Change* 43 (2): 557–584.
Popkin, S. 1980. "The Rational Peasant." *Theory and Society* 9 (3): 411–471.
Raina, R. 2012. "Agriculture in the Environment: Are Sustainable Climate Friendly Systems Possible in India?" In *Handbook of Climate Change and India: Development, Politics, and Governance*, edited by Navroz Dubash, 317–327. Oxon: Earthscan.
Reddy, D. N., and S. Mishra, eds. 2009. *Agrarian Crisis in India*. New Delhi: Oxford University Press.
Ribot, J. 2014. "Cause and Response: Vulnerability and Climate in the Anthropocene." *The Journal of Peasant Studies* 41 (5): 667–705.
Roitman, J. 2013. *Anti-crisis*. Durham: Duke University Press.
Scott, J. 1976. *The Moral Economy of the Peasant: Rebellion and Subsistence in Southeast Asia*. New Haven: Yale University Press.
Sethi, A. 2018. The life of debt in rural India. Ph.D. diss., Columbia University.
Sethi, A. 2021. "Terms of Trade and the Cost of Cotton: The Paradox of Commercial Agriculture in India." *The Journal of Peasant Studies* 48 (7): 1397–1408.
Shah, T. 2008. *Taming the Anarchy: Groundwater Governance in South Asia*. London: Routledge.
Shah, T. 2009. "Climate Change and Groundwater: India's Opportunities for Mitigation and Adaptation." *Environmental Research Letters* 4: 035005.
Shah, E. 2012. "'A Life Wasted Making Dust': Affective Histories of Dearth, Death, Debt and Farmers' Suicides in India." *Journal of Peasant Studies* 39 (5): 1159–1179.
Sheth, A. 2017. Cultivating Risk: Weather Insurance, Technology and Financialization in India." PhD diss., Department of Urban Studies and Planning, Massachusetts Institute of Technology.
Sinha, S. 2020. "Betting on Potatoes: Accumulation in Times of Agrarian Crisis in Punjab, India." *Development and Change* 51 (6): 1533–1554.
Srivastava, S., L. Mehta, and H. N. Adam. 2022. "Climate Change and Uncertainty: Politics and Perspectives."." In *The Politics of Climate Change and Uncertainty in India*, edited by L. Mehta, H. N. Adam, and S. Srivastava, 1–26. New York: Routledge.
Swyngedouw, E. 2009. "The Political Economy and Political Ecology of the Hydro-Social Cycle." *Journal of Contemporary Water Research & Education* 142: 56–60.
Taylor, M. 2015. *The Political Ecology of Climate Change Adaptation: Livelihoods, Agrarian Change and the Conflicts of Development*. London: Routledge.
Taylor, M. 2016. "Risky Ventures: Financial Inclusion, Risk Management and the Uncertain Rise of Index-Based Insurance." In *Risking Capitalism*, edited by Susanne Soederberg, 237–266. Emerald Publishing Limited: Bingley.
Taylor, M., and S. Bhasme. 2021. "Between Deficit Rains and Surplus Populations: The Political Ecology of a Climate-Resilient Village in South India." *Geoforum; Journal of Physical, Human, and Regional Geosciences* 126: 431–440.
Thompson, E. P. 1971. "The Moral Economy of the English Crowd in the Eighteenth Century." *Past and Present* 50: 76–136.
van de Meerendonk, T. 2020. "Claiming Crisis: Narratives of Tension and Insurance in Rural India." *Contemporary South Asia* 28 (3): 362–373.
Vasavi, A. R. 2012. *Shadow Space: Suicides and the Predicament of Rural India*. Gurgaon: Three Essays Collective.
Vasavi, A. R. 2020. "The Tiger and the Tube Well: Malevolence in Rural India." *Critical Asian Studies* 52 (3): 429–445.
Vijay Shankar, P. S. 2005. "Four Decades of Agricultural Development in MP: An Agro-Ecological Sub-Region Approach." *Economic and Political Weekly* 40 (48): 5014–5024.
Watts, M. 1983. "Hazards and Crises: A Political Economy of Drought and Famine in Northern Nigeria." *Antipode* 15 (1): 24–34.

Climate rentierism after coal: forests, carbon offsets, and post-coal politics in the Appalachian coalfields

Gabe Schwartzman

ABSTRACT
The central Appalachian coalfields have become a major site of carbon forestry offsets on California's carbon market. I use these coalfields as a vantage point from which to examine the emerging dynamics of climate change and rentier capitalism in the rural Global North. Studying one valley on the Kentucky-Tennessee border where coal mining has largely ended, I document how emergent land uses take the form of rentier capitalism. I conclude that rentier dynamics articulated with deindustrialization have created the conditions for right-wing populism to emerge, in part in response to the experience of becoming 'surplus population,' drawing upon Tania Li's framework.

Introduction

The central Appalachian coalfields have become a major site of carbon offsets on California's carbon market. Spanning from southern West Virginia, eastern Kentucky, southwest Virginia, to northeast Tennessee, the historic coalfields (see Figure 1) are now home to the second-largest concentration of Improved Forestry Management (IFM) offsets outside of California (after Alaska) (California Air Resources Board 2019). Appalachian carbon offset project managers argue that offsets not only provide good return on investment but provide wide social benefits. Researchers have recently documented that many climate investing projects fail in such goals: market-driven climate change mitigation and adaptation schemes remain largely symbolic, failing to create emissions reductions, to generate social benefit, and to turn a profit (Bigger and Carton 2020; Bigger and Millington 2020). Much of this research, however, has originated from an analysis of urban space (green bonds, etc.) (*for exception see* Kay 2018).

Carbon offsets and climate finance in rural Appalachia, on the other hand, are big business. Estimating average prices per-offset from the California market alone,[1] Central Appalachian landowners have netted at least several hundred million dollars from offset sales in the last five years (California Air Resources Board 2019). While the costs associated with certifying the offsets are not insignificant, the offsets can continue paying out over time and the long-term labor costs are very low.

[1]Many landowners sell carbon offsets to voluntary carbon offset markets, e.g., for air travel.

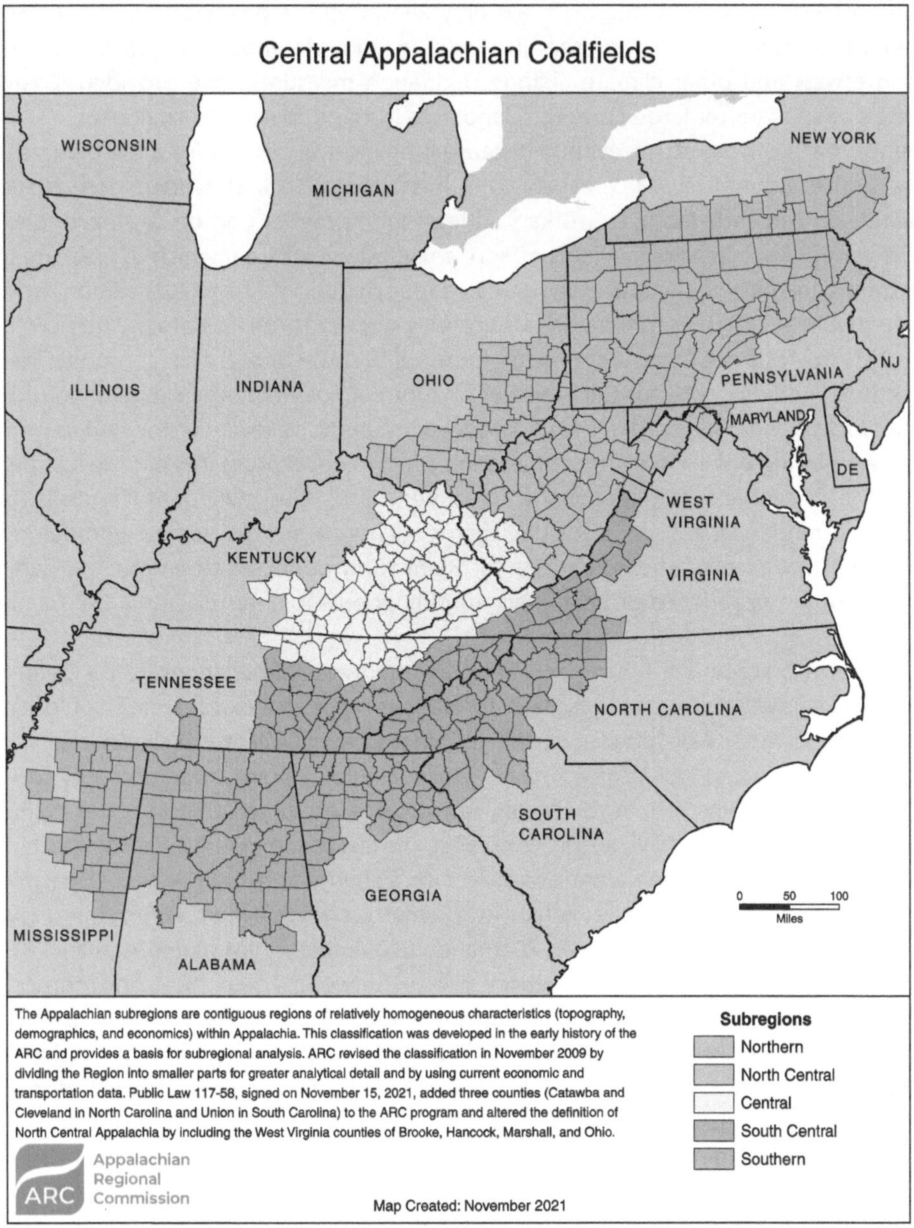

Figure 1. Map of the Central Appalachian region, as defined and mapped by the Appalachian Regional Commission. The coalfields closely map onto the central sub-region.

IFM offsets are a recent addition to the Appalachian landscape. In 2013, the state of California implemented a carbon dioxide emissions 'cap-and-trade' program, a policy that requires large industrial polluters in California (utility companies, oil companies etc.) to reduce emissions by a set amount per year. If a polluter does not meet their emissions targets, they can either purchase credits from another polluter that had reduced emissions below their target, or they can purchase up to eight percent of their target reductions from offsets. Offsets provide emission reductions from actors outside of the carbon market, however, California's law only allows offsets projects to originate from within the United States at present (Hsia-Kiung and Morehouse 2015). Since 2014, carbon offsets and other climate change mitigation measures have provided sizeable new sources of rent for large corporate landlords in the Appalachian coalfields.

Rent is distinct from other revenue because it is, 'income derived from the ownership, possession or control of scarce assets ... ' (Christophers 2019, 6). Understood as such, rents rely on very little labor, or workers, at least in the rentier portion of an economy.

The central Appalachian coalfields offer one vantage point from which to examine the emerging dynamics of climate change and rentier capitalism in the rural Global North. Although climate change mitigation politics may appear to be liberatory, such climate measures in Appalachia are deepening forms of disinvestment and corporate value extraction. I find evidence for this trend in one former coal mining valley that straddles the Kentucky-Tennessee border, the Clearfork Valley. In this valley, two simultaneous processes are playing out. First, deindustrialization leads to novel processes of disinvestment. Second, in this context, climate change investments and other emergent land uses take the form of rentier capitalism. I conclude that rentier dynamics articulated with deindustrialization have created the conditions for Right-wing populism to emerge, in part in response to the experience of becoming 'surplus population,' to use Tania Li's framing (Li 2010).

To make this argument, I trace land use and landownership changes in the Clearfork Valley and I study the effects of these transformations for people in this area. The Clearfork Valley was also the site of John Gaventa's germinal study, *Power and Powerless: Rebellion and Quiescence in an Appalachian Valley* (1982), a helpful point of comparison to the present situation. Dynamics in this valley can provide useful insights into the political economy of the region for two key reasons. First, the last coal mines in this area shuttered in 2020, making the area emblematic of the region's energy transition. Across Appalachia, this transition is well underway. Region-wide coal production and employment have fallen by over forty percent between 2010 and 2020 (Energy Information Agency 2019). Second, in 2019, The Nature Conservancy (TNC), the world's wealthiest environmental organization, purchased nearly all the forested land in this valley. The purchase was part of a half-a-million-acre purchase and market-driven climate change mitigation project that spans the Tennessee, Virginia, and Kentucky coalfields. TNC will continue fossil fuel extraction leases on these lands, including gas leasing in the Clearfork Valley, because they claim they cannot break existing lease agreements with fossil fuel companies. Rather than reducing emissions from fossil fuels, TNC aims to reduce emissions through IFM (improved forestry management): increasing the amount of carbon stored in trees that are destined for the timber industry. However, more than just increasing the amount of carbon stored in trees, TNC promises the land deal will support 'a wide variety of jobs,' and 'increase value to local communities' (Tingley 2021, np). These promises are intimately connected to the decline of the coal industry, as TNC and other investors position their investments as supporting the region's economic transition.

Asking what the effects of emergent climate forest governance regimes are for people living in these forests, I bring a critical agrarian studies approach (Edelman and Wolford 2017) to investigate land use change and the making 'surplus' of much of this population. I draw on fourteen months of ethnographic fieldwork in the Clearfork Valley and over fifty interviews with economic development planners and activists for this analysis.

The research findings contribute to the study of capitalism and climate change in the rural world in two ways. First, I find evidence to support Marxian geographic scholarship that posits that rentierism relies on enclosure to generate value (Felli 2014; Kay 2017; 2018; Karakilic 2022). In this Appalachian valley, climate mitigation rents, or what I term climate rentierism, constitutes a new enclosure of coalfield forests, while other rentier practices, such as leasing the land for recreation, further enclose forests. Yet, these enclosures do not alter the livelihoods of most of the people living in the rentier landscape. Climate change mitigation enclosures build upon ongoing processes of uneven deindustrialization to deepen existing inequalities. New processes of rent seeking, such as climate rentierism, may rely on enclosures to extract value, but that value may not necessarily originate from the people living in the rentier landscape. My second contribution, therefore, is to point to the ways that rentierism renders people living in rentier landscapes 'surplus' and extraneous to processes of value generation. Much of the literature on rentier capitalism has focused on theoretical political economy, with little of it investigating how rentier dynamics affect everyday conditions of human life and the wellbeing of people. I add to this literature with an analysis of the effects of rentier dynamics for people living in the coalfields.

As a contribution to the forum on climate change and agrarian struggles, I respond to provocations in the recent essay in this journal on the subject (Borras et al. 2022). Borras et al. (2022) ask how climate change differs from other environmental exclusions in agrarian environments; how climate change affects trajectories of accumulation in the rural world; and how rural struggles mobilize in the context of climate change and amid authoritarian populist politics. I address these timely questions in three ways: first, I detail how climate change mitigation efforts in the rural Global North mimic and deepen already existing processes of exclusion, reproducing existing trajectories of rentierism, if also providing novel sites of accumulation. Second, as indicated, I suggest that climate rentierism helps produce the conditions for authoritarian populist sentiments to emerge. And third, in this context, I suggest that other politics are possible, presenting possible places to look to consider the future of rural politics in Appalachia and beyond.

In what follows, I first discuss political economic theories of rent, rentierism, and so-called 'surplus population'. Second, I detail how climate investing becomes enrolled in rentier capitalism in the coalfields, a process that has accompanied uneven deindustrialization. Third, I detail how people living in this Appalachian valley experience the effects of emergent rentierism. I then point to the ways that a deepening of inequality stokes 'rural resentment' (Cramer 2016) of perceived unevenness in wealth and prosperity, suggesting how such sentiments play into the politics of right-wing populism in rural America. I conclude with a brief examination of other politics that offer possibilities to reorient political alliances in the current moment, and, perhaps, 'erode capitalism' (Wright 2017) in rural North America.

Climate change and rentier capitalism

Some political economists argue that carbon crediting, offset schemes, and climate-driven conservation easements are novel forms of rent (Felli 2014; Kay 2017). Studies of climate rents are a piece of recent geographic scholarship that has theorized the political economy of rentierism, rentier capitalism, and financialization (A. Gunnoe 2014; Kay 2017; Christophers 2019, 2020; Karakilic 2022).

The effects of rentier capitalism for people living in rentier landscapes, however, has received less explicit attention. Paying attention to the effects of rentier capitalism in Appalachia is important because such scholarship can provide the basis to consider the articulation of political economic transformation and people's political dispositions. Such an analysis helps explain the relationship between rentierism and right-wing populism in Appalachia.

That Appalachia is a rentier landscape has been well documented: in some counties in coal country, land companies own upwards of eighty to ninety percent of total land (Appalachian Land Ownership Task Force 1979). In the 1970s, researchers first documented these dynamics as central to the coal industry's control of the political economy in the region. Dynamics in the Clearfork Valley helped spurr the Land Ownership Task Force to study the dynamics of 'absentee landowners,' an outcome of John Gaventa's study of the valley (1982). In 2013, a group of researchers based out of the West Virginia Center on Budget and Policy conducted a similar study on West Virginia, finding that concentrations of corporate land ownership had stayed relatively constant (Spence et al. 2013). Finer-grained studies of land ownership continue to reveal deepening rentier dynamics, documenting ongoing evictions in southern West Virginia as land companies find tenants to be irrelevant and a nuisance to a rent seeking model (Appalachian Land Study 2022).

Changes in land use and land ownership indicate that the collapse of the coal industry is producing a 'rent gap' (Smith 1979; 1996; Darling 2005). As the price of land and mineral assets fall or slow in growth, investors find new opportunities to purchase land and seek new forms of rent. This new rentierism mirrors processes of financialization of land that geographers have documented across North American resource extraction landscapes and timberlands (A. Gunnoe 2014; Kay 2017; Fairbairn 2020).

'Climate rents' feed into a new rentierism in the coalfields, a rentierism that has largely left the people in the region without work and with diminishing investments in basic infrastructure. Recent debates about rent and rentierism (Felli 2014; Kay 2017; Christophers 2020; Karakilic 2022) highlight how Global North economies increasingly rely on rentierism, as all sectors of the corporate world increasingly monopolize and lease access to scarce assets (e.g. intellectual property, platforms, land etc.) (Christophers 2019; 2020). Marxist geographers have added that not only does rentier capitalism entail monopolization, privatization, and deregulation (products of neoliberal politics), but it also consistently relies upon the enclosure of commons, or commonly accessible resources (A. Gunnoe 2014; Kay 2017; Karakilic 2022).

In North America, these enclosures have accompanied the financialization of land (e.g. Fairbairn 2020). For instance, Kelly Kay's study of Maine timberlands deftly shows how new 'investor-owners' threaten to destroy forests that are vital for the tourist industry as a way to increase financial rents, forcing local communities reliant on the tourist industry to pay for conservation easements (Kay 2017). Kay demonstrates how rent seeking involves obstruction of access to the 'necessary conditions of production,' drawing on the work of Romain Felli (2014, 269) to show that the privatization of nature has relied on obstructing access to aspects of the landscape.

The rentierization of North Atlantic economies, and Appalachia in particular, has accompanied processes of deindustrialization (Winant 2021). Disinvestment has 'hollowed out' rural communities across the United States: industrial employment all-but disappeared, public social infrastructure has been widely abandoned, and people with education, skills, and capital continue to look for better opportunities in cities (Carr and Kefalas 2010; Edelman 2021). In Appalachia, mines mechanized, and then shuttered. The timber industry has taken a similar trajectory, as has manufacturing, which saw growth from the 1970s through the 1990s as plants relocated to southern, less union friendly geographies, only to relocate again offshore in the 2000s (Eller 2008). The growth and decline of data and service oriented companies (e.g. data and call centers) also followed similar trends from the 1990s through the 2000s (Oberhauser 1993; Maggard 1994).

Without industries providing a tax base, funding for schools, roads, water systems, and public services have evaporated. As work becomes scarce and government does not invest in infrastructure (what might otherwise create the conditions for higher quality of life or more diverse economic growth), many people have left deindustrializing communities (Kratzer 2015).

Deindustrialization renders those people that have stayed in Appalachia 'surplus population.' Tania Li defines the term, following Marx's 'relative surplus population,' as people who are not needed for capital accumulation. Li details that there is a growing body of rural poor who have nowhere to go, dispossessed of land or livelihood when new production regimes displace former livelihoods and offer few jobs in return (Li 2014; 2017). The author estimates there are a billion people on the planet whose 'predicament is that their labor is surplus in relation to its utility for capital' (Li 2010, 68). Teleological narratives of development offer, 'the assumption that all these surplus people would find somewhere else to go, and something else to do' (Li 2017, 1250), moving to cities or other jobs. In a world of increasingly deindustrializing cities, however, many people return or stay in the countryside despite dire conditions of poverty. Li draws on the Foucauldian concept of biopolitics to consider state responses to this 'surplus population' of poor and underemployed people in the countryside, biopolitical responses to either 'make live' or 'let die' (Li 2010).

The central Appalachian coalfields exemplify Li's thesis. Many people live without jobs, 'surplus' to the needs of capital, in deeply precarious situations, facing poverty, depression, disease, and despair. On average, less than sixty percent of adults aged eighteen to sixty-five participated in coalfield counties' workforce between 2015–2019, with some counties having lower than forty percent participation rates (Pollard and Jacobsen 2021, 87). The coalfields have some of the lowest and most rapidly falling life expectancies in North America. Compared to the US, 'the region's deficit in life expectancy increased from 0.6 years in 1990–1992 to 2.4 years in 2009–2013' (Singh, Kogan, and Slifkin 2017, 1423). People under the age of sixty-five face disproportionately high rates of premature death from poverty and disease, namely addiction, suicide, and overdose deaths compared to the United States (Rigg, Monnat, and Chavez 2018; Cooper et al. 2020; Meit et al. 2020). An epicenter of the opioid epidemic, many people in the coalfields struggle with opioid addiction and often become entangled in the criminal justice system because of drug criminalization (Ray 2021). This is a decidedly 'let die' biopolitics.

With few public resources available for people to sustain themselves outside of jobs, many have turned to federal disability benefits, Social Security Insurance, as one of the only sources of consistent long-term cash benefits in the United States. Whereas the national average for disability benefits from 2015–2019 was ten percent, coalfield counties in central Appalachia have on average higher than twenty percent of the population age eighteen to sixty-four receiving payments for a disability (Pollard and Jacobsen 2021, 154). Perhaps even more telling about dynamics of 'surplus population,' disproportionate rates of disability benefits extend to children, with seven percent or higher of children in the central Appalachian coalfield counties receiving disability insurance, while the national average was below three percent between 2015 and 2019 (Pollard and Jacobsen 2021, 155). While disability benefits for adults can be partial or due to work related injury, disability in children indicates a life-long inability to work. Clearly, many parents find that, under conditions of poverty and deindustrialization, their children are unlikely to find gainful work for a multitude of reasons. As Alan Maimon points out, disability benefits in coalfield Appalachia are a measure of injury, sickness, and the mental health effects of chronic poverty, *as well as* a measure of people who have given up the hope of ever working given economic and personal circumstances (Maimon 2021).

Rentierism in Appalachia, in the context of deindustrialization, increasingly renders people surplus to capital accumulation. As I conclude, these effects of deindustrialization and deepening rentier relations are intimately tied to emergent 'rural resentment'-driven right-wing populism.

Climate rentierism and enclosures

In the following section, I detail how climate investments offer a business-as-usual approach to forest management, deepening dynamics of rentier capitalism in the coalfields. This deepening of rentier dynamics largely leaves local communities out of the distribution of economic benefit and leaves people increasingly irrelevant to the economic logic.

Corporate and non-profit spokespeople involved with novel climate change mitigation investments in the coalfields, which TNC's land deals typify, are steadfast in their conviction that the marketization of nature can both mitigate climate change and create social goods. TNC wrote in a job advertisement for a staff person for this new project: 'Our vision is based on the conviction that capital markets, businesses and governments must invest in nature as the long-term capital stock of a sustainable, equitable and more efficient economy' (The Nature Conservancy 2021, np). TNC land managers are convinced that these land deals will support community desires for economic development in the collapse of the coal industry. As one staff person explained the goals in starting the project: 'the big overarching goal was how do we manage these properties in a way that support localities, tied to local communities' goals and vision through economic diversification in the wake of the decline in the coal industry' (Interview with non-profit staff. Phone. September 17, 2021). TNC sees their role as maintaining 'working forests' and supporting jobs in a more sustainable timber industry, leasing land for more recreational uses, and protecting lands for conservation.

Interviews with residents near the Cumberland Forest Project, TNC's name for the half-million-acre purchase, however, indicate that these strategies for jobs and community development investment overlook the ways that carbon offset, conservation easement, and recreational leasing programs deepen existing inequalities in the region. What TNC terms community benefits, many in the community understand as business-as-usual.

In fact, TNC's social benefits are so opaque that none of my interviewees in the areas surrounding the Cumberland Forest Project knew that TNC had purchased the property three years prior. That community members are largely unaware of TNC's project indicates that the project has largely unaffected their lives, in part because the project resembles the rest of the landscape. Across multiple interviews, Clearfork Valley residents said that little had changed in terms of economic conditions in recent years. Not only were people unaware that TNC purchased the land bordering their homes; multiple people told me that the land was owned by either Chinese companies or, as one informant stated, 'I heard it was all owned by some Japanese company' (Fieldnotes. January 7th, 2022), neither of which is true for nearby properties. A Chinese firm had, in fact, purchased a local coal company and attendant mineral rights a decade earlier (Boling 2012), however, their last mining operations had shuttered in 2020.

TNC's land deal exemplifies the financialized investments schemes that control much of the coalfield forests. TNC used a corporation they started called NatureVest to acquire the property, raising 'green' venture capital to finance the purchase. NatureVest brings together investors that hope to see large returns on their investments, rather than see these lands in conservation for perpetuity or to predominantly derive rent from timber production. NatureVest intends to sell the property in several years, promising sizeable earnings in the sales price for the investors (Personal correspondence. Email. July 15th, 2020). This format mirrors for-profit financialized timber investment operations that dominate the industry in North America, known at Timber Investment Management Organizations (TIMOs) (Gunnoe and Gellert 2011; Gunnoe, Bailey, and Ameyaw 2018).

As a TIMO-like model, TNC's project joins the region's rentier economy. Gunnoe (2014) illustrates how financialization of timberlands in the United States has ushered in a new kind of rentierism, one where investors aim to extract rents not simply from timber production, but the inflation of the prices of the timberland holdings. Kelly Kay also illustrates this point, noting that the investor-owners of TIMOs are looking for new sources of rent, such as conservation easements, to increase shareholder value outside of timber production (2017).

TNC's project follows this same model. As a TNC staff person familiar with the project explained,

> the limited partnership [NatureVest] is set up much like other timber-oriented investment funds and so it's not meant to be a forever situation … we think plus or minus 10–12 years, we'll need to eventually sell the property. Okay. But before, before we do that, we're going to work to put permanent legal conservation easements in place on as much of the properties as we can. (Interview with the Nature Conservancy staff. Phone. September 17, 2021)

Part of the conservation goals of the project include creating a biodiversity climate migration corridor with conservation easements on these timberlands. The non-profit sees this work as creating the largest species conservation corridor on the east coast of the United States (The Nature Conservancy 2019).

In line with a rentier model for shareholder value, the Cumberland Forest Project is pioneering new sources of rent through both carbon offsets and recreational leasing. What is striking about these new rents is not simply their closure of forest access for nearby communities. Rather, these new rent sources require drastically less local labor or community involvement than resource extraction or other tourist economies. For instance, the carbon offsets on the TNC land derive value from corporations and workers in California, not from the labor or presence of people in Tennessee or Kentucky.

This climate rentierism is not limited to TNC. Many Appalachian land companies and actors the timber industry are getting into offset rents. Business and finance reports indicate that TIMOs have picked up on the trend as a low-cost way to get rent out of forests while still being able to extract timber lease income. Dick Kempka, a spokesperson for Molpus Woodland, a TIMO that owns substantial holdings in the Appalachian coalfields, was quoted in the *Wall Street Journal,* 'We're seeing more and more value from having the trees stay there longer' (Dezember 2020, np). This value comes from carbon offset rents derived in California, as essentially a tax on California polluters, presumably passed on to workers and surrounding communities. To understand the financial scope of these projects, the carbon offset deals on the Cumberland Project lands, although financial details are undisclosed, received at least $80 million dollars from California polluters.[2]

Despite recent research that indicates improved forestry offsets are over-credited for the carbon they mitigate (Badgley et al. 2022), carbon forest offsets are a growing trend across West Virginia, Virginia and Tennessee coalfield counties. IFM offsets comprised 17,880,708 issued offsets in the Central Appalachian coalfields as of 2019 (California Air Resources Board 2019, 18), with the potential for market expansion. As reports on the industry note, land companies in the region are turning to carbon offsets as part of their new revenue strategies at a moment when coal revenue is in free-fall and they have witnessed stagnation in hardwood timber prices during the COVID-19 pandemic (despite refined lumber prices sky-rocketing) (Dezember 2020; Zhang and Stottlemyer 2021).

The map below, Figure 2, shows the extent of coalfield forests already sold on California's offset market as of 2020, data that leaves out any carbon offsets sold on numerous voluntary markets.

Enclosure I: climate rentierism

What are the implications of landownership and land use changes for people living in these forests? Here I argue that climate investing and other rentier activities limit the community's access to forests. I note, however, that most of the effects of climate rentierism largely go unnoticed and unfelt. Carbon offset and conservation economies do not fundamentally alter conditions of employment or livelihood for people in these communities. These new rentier activities do, however, continue to alienate people from local economic activity. Yet, rather than express resentment or anger at rentier corporations, people express more general nostalgia and grievance about current conditions of life in their communities.

Before I consider some of these sentiments, I first examine the ways that climate mitigation investments enclose forest access regimes. Will Bowling, TNC's Central Appalachian Projects Director for Kentucky told TNC's magazine, 'that the plan is not to lock people out of the land, but to prove that conservation can be an economic driver' (Elliston 2019, np). Yet, based on interviews with people living around the forest, TNC's investments deepen on-going trends to make lands accessible to tourists, not residents, and they do not drive increased jobs or business growth in the Clearfork Valley.

[2]Based on the average price of carbon and number of offsets sold on the California market.

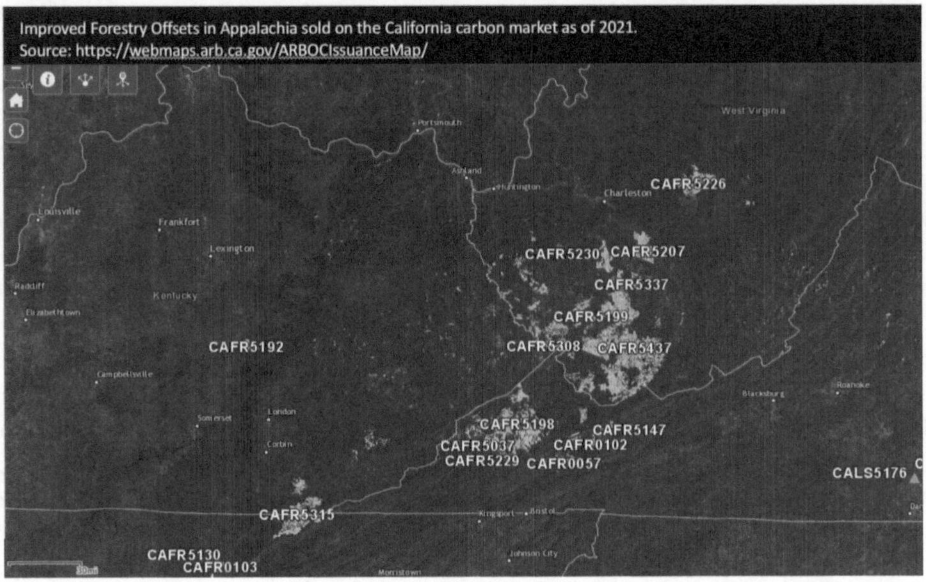

Figure 2. Map of IFM carbon offsets in Central Appalachia that have been sold on the California carbon market as of 2021.

The new carbon offsets join a decades long process of deindustrialization and professionalization of the timber industry in this region. TNC's promise of forestry jobs will likely go to sustainably certified timber management corporations and companies that can conduct carbon accounting on these lands. For small-scale loggers living in the Clearfork Valley and operating with little capital, the new forestry regime means less work. Loggers in the Clearfork Valley said in multiple interviews that many of the small owner-operator mills had closed recently (Fieldnotes. February 7, 2022). Others explained that jobs were declining. In one logger's telling, new restrictions made companies cut less, and he speculated that the local operators were getting fewer contracts too (Interview with coalfield resident. Eagan, TN. February 15, 2022). While the logging industry employment in the coalfields has been marginal in recent decades, employing few people in extraction while most of the wood products manufacturing occurs elsewhere, the move towards sustainable forestry management further limits the amount of work that residents living near these forests will see. These investments may create jobs in the wider region, however, many of them will be higher skilled and higher educated jobs. Timber firms now must hire consultants to count carbon stored in the trees and practice stainable forestry methods, the proof of which often requires certifications and the presences of degree-holding foresters. Such jobs will likely not reach many people in the communities closest to the forests since many of those people have low levels of educational attainment.

Enclosure II: recreational rents

In addition to the new forestry regime, many people in the Clearfork Valley complain about enclosures that occur because landowners, witnessing the decline of coal rents, have leased land to state wildlife and recreation agencies. These public agencies administer the lands as fee-to-access hunting and recreation lands. People living in the forests, again, do not express anger towards the rentier corporations, but instead towards the government bodies that now police their access of forests. While such recreational leasing predates TNC's new land management regime, the Cumberland Forest Project highlights the implication of recreational leasing.

Recreational leasing in the Clearfork Valley began in 2014. The then major landholder, Molpus Woodlands Group, leased their holdings to the Tennessee Wildlife Resources Agency (TWRA) as hunting land, making it a Wildlife Management Area (WMA). While publicly this deal was heralded as making more land publicly accessible, for residents of the valley it was just the opposite.

Across coalfield Appalachia there is a long history of people relying on forest access to collect non-timber forest products for sale (e.g. ginseng); hunting; collecting fuelwood to defray utility bills; and recreation that is a key part of identity (Hufford 1993; 2003). Recreational leasing disrupts these practices, establishing a pay-to-access regime catered to tourists and middle-class users. Discussing how the forest used to be a commons, an interviewee remarked, 'I think TWRA has put an end to all that, because you have to have a permit now, you know' (Fieldnotes. Eagan, TN. January 13th, 2021). Many others expressed similar thoughts.

One ginseng digger in the Clearfork Valley expressed outrage that,

> ... [P]eople that lived here all their life can't even go over on it [TWRA controlled land]' (Interview with ginseng digger. Clairfield, TN. January 25th, 2022). Discussing how ginseng digging practices have changed, the interviewee responded, 'Not as many as there used to [dig it]. Because now there's quite a lot of game wardens. And they've got that land lease, and a lot of people won't pay to travel on it .... (Interview with ginseng digger. Clairfield, TN. January 25th, 2022)

In this area, with additions to make it legal to hunt deer and turkeys, local favorite species, a yearly permit can run in the range of $250 (Interview with ginseng digger. Clairfield, TN. January 25th, 2022).

TWRA's land management has brought very few investments into the valley, and fewer new employment opportunities. Except for a half-dozen campgrounds catering to the all-terrain vehicle (ATV) riders that have flocked to newly established ATV riding trails, the largest of which employees nine year-round employees (Interview with campground manager. Stinking Creek, TN. July 23rd, 2022), there are few indicators of local economic impact from the recreational leasing. Employment figures for the five census tracts in the four counties that comprise the Clearfork Valley have closely tracked job gains and losses in the coal industry over the past decade,[3] indicating little impact from tourism (US Census Bureau 2020).

TWRA's control of this territory has, however, brought a new criminalization of forest access.

> ... [A] lot of people [are] getting in trouble too for it,' said one interviewee. 'If I go 15 feet away from [a permit holder] when I'm on their land, I'll go straight to jail if they pull me over. (Interview with ginseng digger. Clairfield, TN. January 25th, 2022).

[3]The valley lost 1,328 coal jobs from 2010–2020 in a total population of 15,000 in 2020. Numbers of employed persons track this figure.

At the same time, residents report there are more people in the forests than they can remember, with tourists and recreational users accessing the land for hunting and predominantly for recreational all-terrain vehicle (ATV) riding.

The presence of hundreds of ATV riders a weekend is palpable. The hills hum and whine with the echoing sounds of dual-stroke engines as riders enjoy the thrills of off-road trails and rugged abandoned mine features. Of the ATVs, one resident said, 'I hate them' (Fieldnotes. March 10, 2022). Others complained about late night and early morning weekend noise as the riders buzzed past resident's homes.

In 2020, ATV trail riders paid thirty-seven dollars for a permit to ride through the woods, a cost that residents refuse (TN Dept. of Tourism Development 2021). Across Appalachia, ATV riders now pay to ride across thousands of miles of private land company lands (Tennessee Wildlife Resource Agency 2022; Hatfield and McCoy Trails Redevelopment Authority 2011). TWRA and other public agencies collect fees from riders and hunters, sending much of those funds directly to private landowners, such as TNC. Atop the climate rentierism, more traditional forms of rentierism and enclosure, such as recreational leasing, close forests to wider community access.

While enclosure of the forest is felt as a grievance and a wrong, few of my interviewees felt that the closures fundamentally altered their livelihoods or employment. While forest practices are important parts of identity, few people live off of root-digging or hunting in this community, and those that do have found ways to evade TWRA policing. Rather than dispossess people from their livelihoods, these enclosures occur with little economic effect upon local livelihoods.

Nostalgia and grievance in a deindustrialized landscape

Many people in the Clearfork Valley express sentiments of nostalgia for an imagined better past and grievances about economic decline and community dissolution. Despite these grievances, few people talk about leaving the area. The rise of the kind of rentier capitalism that I describe here accompanies processes of uneven deindustrialization. Deindustrialization has left many people across Appalachia without work, in desperate poverty, and with few other places to go. In the Clearfork Valley, many people discuss a sense of being trapped amid attachments to home. Furthermore, rentier capitalism increasingly renders people 'surplus' and relegates their communities to disinvestment and abandonment from capital and the state, dynamics which climate investing only deepens. If this is so, why do people in the Clearfork Valley not go somewhere else?

In previous decades, when people across Appalachia were rendered unnecessary to capital they moved to industrial cities of the North (now the 'Rustbelt') or South (in a period of rapid 'Sunbelt' urban growth) (Obermiller et al. 2009). Many in the coalfields no longer see those as options, and indeed many people that lived in cities have come home to retire or find economic refuge in Appalachian communities. Facing the possibility of dire poverty in deindustrialized cities, many people have come back, and many choose to stay in increasingly deindustrialized and disinvested rural communities. These decisions mirror a trend that Tania Li identified across rural communities around the world (Li 2010; 2014; 2017). Many people in the Appalachia region have nowhere to go.

Understanding these decisions is important to conceptualize how climate politics and capitalist dynamics articulate in Appalachia. The hollowing out of rural communities is intimately tied to distrust, disaffection, and resentment among a rural political base in the US (Carr and Kefalas 2010; Cramer 2016; Ashwood 2018; Edelman 2018; Berlet and Sunshine 2019; Silva 2019; Edelman 2021). As many people in these rural areas have found political affinity with right-wing anti-establishment conservatives, rural voting blocs now offer significant political opposition to legislation and political momentum to regulate fossil fuels and mitigate the effects of a changing climate.

In this section, I explore the question of why people stay in disinvested places, and I consider how they narrate their situations. People voice sentiments of grief, loss, and abandonment, feelings that are intimately connected to deindustrialization, declining rural qualities of life, and a rentierism that operates with nearly no local workers.

Over the past fifty to seventy years, hundreds of thousands of people have left the Appalachian coalfields in search of better opportunities (Obermiller et al. 2009). From 2010 to 2020, most coalfield counties saw more than three percent of their population leave (Pollard and Jacobsen 2021, 13). The figure of three-percent population decline during the recent collapse of the coal industry is perhaps less than some might expect. Coal industry employment declined forty percent region-wide, and up to eighty percent in certain sub-regions (e.g. eastern Kentucky) during that same time period (Fritsch 2019). As one economic development planner in southern West Virginia explained to me, 'the coal industry has collapsed ... [And] you would expect there to be more of an exodus, considering that there are opportunities other places. But so many people who leave come right back' (Interview with economic development professional. Zoom. January 26, 2022).

Despite many people expressing a love for the mountains and the country as why they stay or come back to the coalfields, others are more candid about feeling trapped. In the Clearfork Valley, one informant told me repeatedly, 'If I could leave tomorrow, I would' (Fieldnotes. September 23rd, 2021), discussing his desire to live in northern Michigan if only he did not have to take care of his ailing mother. 'I mean, I love it, but I hate it [the valley],' a young woman in her thirties who had struggled with addiction told me (Fieldnotes. January 7th, 2022).

Many of those that stay in the valley narrate the history of the place through discourses of loss and grief. As one elderly resident told me:

> And that was sad. The day that they pulled all the [coal] trucks out. And they all was in a line and they blowed their horns for the last time. You know, it reminded me a lot of when my brother passed away .... (Interview with coalfield resident. Clairfield, TN. September 19, 2021)

When I asked an elderly couple about how things had changed in their lifetimes, the man said, 'It's gotten pretty bad around here ... Can't find no one to work or do nothing.' He blamed immigrants for taking jobs, parroting conservative talking points. He contradictorily complained about the lack of work ethic in young people and implied that people would have better jobs if immigrants did not take all the existing jobs, a seemingly ludicrous assertation in a place with virtually no jobs or immigrants to speak of. The woman reflected on the 1960s and 1970s, times of poverty, when they did not have running water and people strung up their own electric lines, saying, 'Those were the golden years ... ' (Interview with coalfield residents. Primroy, TN. May 21, 2021), repeating the phrase.

Younger residents discussed wanting to stay in the area because they loved the place but also voiced the knowledge that they lived in a place of poverty and want. High school aged people discussed wanting to leave the Clearfork Valley, most of them hoping to move to small towns nearby in East Tennessee or Eastern Kentucky to get jobs there, such as in nursing. During interviews with working age people, however, when I inquired about whether people thought there was more opportunity elsewhere, most of those over the age of twenty-five did not think so. Most did not feel qualified for jobs that they could find elsewhere, and many expressed a fear of being without resources in place without family to fall back on. Like much of the coalfields, people stay because they love the place, but they also stay because there are few other places that they can easily find a job or are easy to live in for poor and working people: as Tania Li argues, conditions of urban deindustrialization make urban places unaffordable and unlivable.

People in the coalfields are increasingly extraneous for landowners. From losses of coal jobs to timber jobs, to the decline of small manufacturing, employment overall has plummeted in the region. Between 1984 and 2022, the Central Appalachian region has witnessed nearly eighty percent decline in coal employment (Open Source Coal 2022). Appalachia witnessed a twenty one percent decline in manufacturing jobs from 2002 to 2017 (Appalachian Regional Commission 2019, 25), while Central Appalachia lost nearly fifteen percent of its forestry and farming employment in that same fifteen year period (Appalachian Regional Commission 2019, 72). People living in the Clearfork Valley have acutely felt these employment trends. As one interviewee put it, 'it's not a town for a working man. Because if you came in, I don't know where you'd find a job' (Interview with coalfield resident. Clairfield, TN. September 23rd, 2021).

If under the hegemony of coal companies people faced domination, as John Gaventa documented (1982), today they face irrelevance. New rounds of enclosure and processes of rent seeking join decades-long economic trajectories of deindustrialization that have made life in the coalfields difficult, dispossessing people of land, jobs, housing, public infrastructure, and access to the forest. This is the lived condition of Tania Li's 'surplus population;' people are resentful, distrust the government, and feel irrelevant to local economic activity.

Feelings of grievance and nostalgia in rural communities articulate together in what scholars have documented as 'rural resentment' (Cramer 2016; Metzl 2020). Scholars have tied this resentment to Donald Trump's election and the rise of a rural right-wing populism in the United States (Edelman 2021). Across rural America, feelings of abandonment, dispossession, and decline propel some people to give up on political participation, looking to individualized, personal, and moral solutions (Silva 2019). Indeed, Gaventa's recent assessment of the Clearfork Valley indicated very low levels of political participation (Gaventa 2019). Those people that do participate in electoral politics, however, have widely supported far-right anti-establishment candidates (see below), a trend across rural America (Berlet and Sunshine 2019). Operating with possessive investments in whiteness and within the crises of masculinity associated with deindustrialization, the feminization of labor (McDowell 1991; 2014), and downward economic outcomes (Lipsitz 2006; Schwartzman 2013; Moreton-Robinson 2015; Lensmire 2017; White 2017), uneven deindustrialization creates conditions that fuel authoritarian populism in the countryside (Scoones et al. 2018; Gaventa 2019; Edelman 2021).

A new conjuncture

Far-Right politicians now dominate politics in the Clearfork Valley and much of the coalfields, the results of a thirty-year regional political shift towards the Right (Schwartzman 2013; 2015; Young 2022). Among the valley's four counties, Republican presidential candidates garnered between 44 and 56 percent of the vote between 1976 and 2000, collectively electing Democrat Bill Clinton twice. Between 2004 and 2020, however, Republican votes steadily increased from 61 to 82 percent of the total votes cast (Leip 2022). Local county representatives reflect these presidential percentages. One popular commissioner told me, infamous for his failed attempt in 2022 to establish a publicly funded Confederate monuments museum and for saying that people receiving government assistance should not be allowed to vote, discussing post-coal development, 'The people [is] what needs to change, not the government' (Interview with county official. Jacksboro, TN. June 6, 2021). Another county official from the valley opens each commission meeting with a ten-minute invocation to 'Jesus Christ' (Fieldnotes. Tazewell, TN. September 25, 2021). Interviews with valley residents show that people's sentiments of grievance and nostalgia have found a voice in anti-establishment far-Right politics, moralism, and conservatism. Yet, the current politics are not necessarily the politics of the region's future.

The terrain is rapidly shifting in the coalfields. As Mike Davis argues, the most radically transformative organizing happens conjuncturally, only where and when diverse struggles align, link-up, and collide in new ways (Gramsci 2000; Hall 1988; Davis 2020; Hart 2020). Discussing the rise of Margaret Thatcher, Stuart Hall writes, 'When a conjuncture unrolls, there is no 'going back.' History shifts gears. The terrain changes. You are in a new moment' (Hall 1988, 161). The collapse of the coal industry, the crises in neoliberal hegemony (Hall, Massey, and Rustin 2015; Fraser 2019), substantial attacks on neoliberal rule coming from the Right (Slobodian 2021; Cooper 2021), and the emergence of popular discourses about the region's economic transition indicate that the coalfields are in a different conjuncture; a new terrain where struggles emerge.

Perhaps the clearest indication that the coalfields are in this new political and ideological moment is that neoliberal ideology, which has become institutionalized across the coalfields, seems to fail to provide coherent explanation of or substantial answers for the many crises facing people in the coalfields. Across rural America, scholars have noted that neoliberalism faces an ideological crisis (Brown 2019; Fraser 2019; Peck and Theodore 2019; Davies and Gane 2021). Traditional divides between the Left and the Right are eroding in rural places where everyone can agree about their distrust of a state that works for corporate interests (Ashwood 2018). In Appalachia, there is widespread animosity towards mainstream political life that has failed to improve social conditions in the region for fifty years (Martin 2019), neither saving the coal industry nor planning for an economic transition (M. Gunnoe 2019). Political figures that buck a political elite in a Trumpian style are widely popular, such as West Virginia's Democrat-turned-Republican governor, Jim Justice (Young 2022).

Political alliances are reorienting in this moment. For instance, in the Clearfork Valley, a coalition for community-led economic development, the Tennessee Appalachian Community Economies coalition, has garnered sustained community engagement from conservative pastors to anarchist community organizers. There is a growing consensus in the valley and across the region that the current economic development regime will not contend with the crises of disinvestment and deindustrialization. As one non-profit leader assuredly said of the recent wave of federal investments, 'Oh, it'll never make it down to the little guys like us' (Fieldnotes. March 23, 2021), an assessment that has, sadly, so far born true.

No clear ideology or political alliance, however, has usurped neoliberal rule. Instead, reactionary, nostalgic, grievance laden politics that struggle to articulate strong visions of the future (such as those of the county commissioners or of West Virginia's governor) dominate the landscape, indicating that, 'the old is dying and the new cannot be born' (Gramsci 1971, 276). While right-wing anti-elitism animates local politics, such politics fail to offer an imaginable future for the region.

Conclusion: towards a Left rural populism

In this moment, Leftist politics have the potential to galvanize support around brighter imaginable futures for the region. Many organizations and activists in Appalachia are already contesting and, perhaps, 'eroding' (Wright 2017) capitalist rentier relations through efforts to create more livable and democratic communities. These efforts build towards a strategy for rural Left populism.

On a regional scale, several institutions and organizations are working to mobilize grassroots support around economic agendas in Appalachia. These efforts include the Alliance for Appalachia, a coalition of grassroots progressive and environmental justice organizations in the region, advocating, among other things, for federal funding to create jobs remediating mined lands; Appalachian Voices, a regional non-profit advancing various campaigns for economic revitalization and environmental justice; and the ReImagine Appalachia coalition, advocating for a new New Deal and a federal jobs program.

A Left populist strategy, as Chantal Mouffe describes, must aim 'at federating the democratic demands into a collective will to construct a 'we,' a 'people' confronting a common adversary: the oligarchy' (Mouffe 2018, 24). Widespread disaffection with contemporary economic conditions provides an opening for Left populism to craft a new common will, one posed against the political entities of the region's economic status quo.

This article highlights the centrality of land and land ownership in the post-coal political economic trajectory in Appalachia. More research is needed about the potential for struggles over land and land access to become issues that might galvanize new alliances in Appalachia, and to examine what agrarian populism (Borras 2020), however fraught that concept may be (Bernstein 2020), might mean in this context. Further research should consider what a 'make live' (Li 2010) biopolitics would be in Appalachia, and how the post-work politics of welfare, disability, and forest access articulate in emergent political alliances.

In the current moment, many Leftist political programs, such as social democracy, run counter to Right-wing populist rhetoric around individual responsibility and white Christian moralism, the imaginaries articulated to the current hegemonic alliance. Yet, as discussed, there is a palpable crisis of ideology in the region. Political factions that offer renewed political imagination may stand a chance of forging new hegemonic alliances with rural working and non-working people in coalfield Appalachia, alliances that could shift balances of power across the United States.

Political action on climate change requires this type of shift in rural politics in the United States. Rural voters, constituencies, and political representatives currently obstruct any political agendas that might stem fossil fuel use or constitute significant emissions reductions. Reversing this obstruction will require a national political alliance that convincingly promises to improve the quality of life for rural people.

Acknowledgements

The author would like to thank the anonymous reviewers, whose generous and constructive comments have been invaluable in this paper, and to thank the members of the TNACE coalition in the Clearfork Valley, whose work greatly informs this analysis.

Disclosure statement

No potential conflict of interest was reported by the author(s).

Funding

This material is based upon work supported by the National Science Foundation Graduate Research Fellowship [grant number 2019274360].

References

Appalachian Land Ownership Task Force. 1979. *Who Owns Appalachia? Landownership and Its Impact*. Lexington: Univ Pr of Kentucky.

Appalachian Land Study. 2022. "Land Study Mission, Objectives, and Methods." *Appalachian Land Study*. Accessed January 10, 2022. https://www.appalachianlandstudy.org/abou.

Appalachian Regional Commission. 2019. *"Industrial Make-Up of the Appalachian Region: Employment and Earnings 2002–2017."* Division of Research and Evaluation. Washington, DC: Appalachian Regional Commission. https://www.arc.gov/wp-content/uploads/2020/06/IndustrialMakeUpoftheAppalachianRegion2002-2017.pdf.

Ashwood, Loka. 2018. *For-Profit Democracy: Why the Government Is Losing the Trust of Rural America*. New Haven: Yale University Press.

Badgley, Grayson, Jeremy Freeman, Joseph J. Hamman, Barbara Haya, Anna T. Trugman, William R. L. Anderegg, and Danny Cullenward. 2022. "Systematic Over-Crediting in California's Forest Carbon Offsets Program." *Global Change Biology* 28 (4): 1433–1445. doi:10.1111/gcb.15943.

Berlet, Chip, and Spencer Sunshine. 2019. "Rural Rage: The Roots of Right-Wing Populism in the United States." *Journal of Peasant Studies* 46 (3): 480–513. doi:10.1080/03066150.2019.1572603.

Bernstein, Henry. 2020. "Unpacking 'Authoritarian Populism' and Rural Politics: Some Comments on ERPI." *Journal of Peasant Studies* 47 (7): 1526–1542. doi:10.1080/03066150.2020.1786063.

Bigger, Patrick, and Wim Carton. 2020. "Finance and Climate Change." In *The Routledge Handbook of Financial Geography*. New York: Routledge.

Bigger, Patrick, and Nate Millington. 2020. "Getting Soaked? Climate Crisis, Adaptation Finance, and Racialized Austerity." *Environment and Planning E: Nature and Space* 3 (3): 601–623. doi:10.1177/2514848619876539.

Boling, Zhang. 2012. "China Digs into American Coal Mines." *MarketWatch*, May 29, 2012, sec. Industries. https://www.marketwatch.com/story/china-digs-into-american-coal-mines-2012-05-29.

Borras, Saturnino M. 2020. "Agrarian Social Movements: The Absurdly Difficult but Not Impossible Agenda of Defeating Right-Wing Populism and Exploring a Socialist Future." *Journal of Agrarian Change* 20 (1): 3–36. doi:10.1111/joac.12311.

Borras, Saturnino M., Ian Scoones, Amita Baviskar, Marc Edelman, Nancy Lee Peluso, and Wendy Wolford. 2022. "Climate Change and Agrarian Struggles: An Invitation to Contribute to a JPS Forum." *Journal of Peasant Studies* 49 (1): 1–28. doi:10.1080/03066150.2021.1956473.

Brown, Wendy. 2019. *In the Ruins of Neoliberalism: The Rise of Antidemocratic Politics in the West*. New York: Columbia University Press.

California Air Resources Board. 2019. "*U.S. Forest Offset Projects: Overview*." Sacramento: California Air Resources Board.

Carr, Patrick J., and Maria J. Kefalas. 2010. *Hollowing Out the Middle: The Rural Brain Drain and What It Means for America. Illustrated Edition*. Boston: Mass: Beacon Press.

Christophers, Brett. 2019. "The Rentierization of the United Kingdom Economy." *Environment and Planning A: Economy and Space*, September, 0308518X19873007. doi:10.1177/0308518X19873007.

Christophers, Brett. 2020. *Rentier Capitalism: Who Owns the Economy, and Who Pays for It?* London: Verso Books.

Cooper, Melinda. 2021. "The Alt-Right: Neoliberalism, Libertarianism and the Fascist Temptation." *Theory, Culture & Society* 38 (6): 29–50. doi:10.1177/0263276421999446.

Cooper, Hannah LF, David H. Cloud, Patricia R. Freeman, Monica Fadanelli, Travis Green, Connor Van Meter, Stephanie Beane, Umedjon Ibragimov, and April M. Young. 2020. "Buprenorphine Dispensing in an Epicenter of the U.S. Opioid Epidemic: A Case Study of the Rural Risk Environment in Appalachian Kentucky." *International Journal of Drug Policy* 85 (November): 102701. doi:10.1016/j.drugpo.2020.102701.

Cramer, Katherine J. 2016. *The Politics of Resentment: Rural Consciousness in Wisconsin and the Rise of Scott Walker. Illustrated Edition*. Chicago: University of Chicago Press.

Darling, Eliza. 2005. "The City in the Country: Wilderness Gentrification and the Rent Gap." *Environment and Planning A: Economy and Space* 37 (6): 1015–1032. doi:10.1068/a37158.

Davies, William, and Nicholas Gane. 2021. "Post-Neoliberalism? An Introduction." *Theory, Culture & Society* 38 (6): 3–28. doi:10.1177/02632764211036722.

Davis, Mike. 2020. *Old Gods, New Enigmas: Marx's Lost Theory*. London: Verso.

Dezember, Ryan. 2020. "Preserving Trees Becomes Big Business, Driven by Emissions Rules." *Wall Street Journal*, Accessed August 24, 2020, sec. Markets. https://www.wsj.com/articles/preserving-trees-becomes-big-business-driven-by-emissions-rules-11598202541.

Edelman, Marc. 2018. "*Sacrifice zones in rural and non-metro USA: fertile soil for authoritarian populism*." Transnational Institute. Accessed February 21, 2018. https://www.tni.org/my/node/23914.

Edelman, Marc. 2021. "Hollowed out Heartland, USA: How Capital Sacrificed Communities and Paved the Way for Authoritarian Populism." *Journal of Rural Studies* 82 (February): 505–517. doi:10.1016/j.jrurstud.2019.10.045.

Edelman, Marc, and Wendy Wolford. 2017. "Introduction: Critical Agrarian Studies in Theory and Practice." *Antipode* 49 (4): 959–976. doi:10.1111/anti.12326.

Energy Information Agency. 2019. *Annual Energy Outlook 2019 with Projections to 2050*. Washington, DC: Office of Energy Analysis, US Dept. of Energy.

Eller, Ronald D. 2008. *Uneven Ground: Appalachia Since 1945*. 1st ed. Lexington: The University Press of Kentucky.

Elliston, Jon. 2019. "Heart of Appalachia." *The Nature Conservancy Magazine*. Accessed May 19, 2022. https://www.nature.org/en-us/magazine/magazine-articles/cumberland-forest-project/.

Fairbairn, Madeleine. 2020. *Fields of Gold: Financing the Global Land Rush. Cornell Series on Land : Perspectives in Territory, Development, and Environment*. Ithaca, NY: Cornell University Press.

Felli, Romain. 2014. "On Climate Rent." *Historical Materialism* 22 (3–4): 251–280. doi:10.1163/1569206X-12341368.

Fraser, Nancy. 2019. *The Old Is Dying and the New Cannot Be Born: From Progressive Neoliberalism to Trump and Beyond. London* . New York: Verso.

Fritsch, David. 2019. "U.S. *Coal Production Employment Has Fallen 42% since 2011*." Today in Energy. Washington, DC: U.S. Energy Information Administration (EIA). https://www.eia.gov/todayinenergy/detail.php?id = 42275.

Gaventa, John. 1982. *Power and Powerlessness: Quiescence & Rebellion in an Appalachian Valley*. Urbana, IL: University of Illinois Press.

Gaventa, John. 2019. "Power and Powerlessness in an Appalachian Valley – Revisited." *Journal of Peasant Studies* 46 (3): 440–456. doi:10.1080/03066150.2019.1584192.

Gramsci, Antonio. 1971. "*Selections from the Prison Notebooks.*" Edited by Quintin Hoare and Geoffrey Nowell Smith. Reprint, 1989 Edition. New York: International Publishers Co.

Gramsci, Antonio. 2000. *The Antonio Gramsci Reader: Selected Writings 1916–1935*, edited by David Forgacs. New York: New York University Press.

Gunnoe, Andrew. 2014. "The Political Economy of Institutional Landownership: Neorentier Society and the Financialization of Land." *Rural Sociology* 79 (4): 478–504. doi:10.1111/ruso.12045.

Gunnoe, Maria. 2019. "An Open Letter from Fighter against Mountain Top Removal Coal Mining." *People's Tribune (blog)*. Accessed October 11, 2019. http://peoplestribune.org/pt-news/2019/10/letter-from-fighter-against-mountain-top-removal-mining/.

Gunnoe, Andrew, Conner Bailey, and Lord Ameyaw. 2018. "Millions of Acres, Billions of Trees: Socioecological Impacts of Shifting Timberland Ownership." *Rural Sociology* 83 (4): 799–822. doi:10.1111/ruso.12210.

Gunnoe, Andrew, and Paul K. Gellert. 2011. "Financialization, Shareholder Value, and the Transformation of Timberland Ownership in the US." *Critical Sociology* 37 (3): 265–284. doi:10.1177/0896920510378764.

Hall, Stuart. 1988. *The Hard Road to Renewal: Thatcherism And The Crisis Of The Left. London*. New York: Verso Books.

Hall, Stuart, Doreen Massey, and Michael Rustin. 2015. *After Neoliberalism?: The Kilburn Manifesto*. London: Lawrence & Wishart Ltd.

Hart, Gillian. 2020. "Why Did It Take so Long? Trump-Bannonism in a Global Conjunctural Frame." *Geografiska Annaler: Series B, Human Geography* 102 (3): 239–266. doi:10.1080/04353684.2020.1780791.

Hatfield and McCoy Trails Redevelopment Authority. 2011. "*Hatfield and McCoys Trails.*" Charleston, WV: WV Department of Transportation. https://transportation.wv.gov/highways/programplanning/plan_conf/Documents/2011PC/Hatfield_McCoy_Trails.pdf.

Hsia-Kiung, Katherine, and Erica Morehouse. 2015. *Carbon Market California: A COMPREHENSIVE ANALYSIS OF THE GOLDEN STATE'S CAP-AND-TRADE PROGRAM*. Washington, DC: Environmental Defense Fund. https://www.edf.org/sites/default/files/content/carbon-market-california-year_two.pdf.

Hufford, Mary. 1993. "Tending Commons: Ramp Suppers, Biodiversity and the Integrity of 'The Mountains'." *Folklife Center News*. Washington, DC: American Folklife Center, the Libary of Congress.

Hufford, Mary. 2003. "Knowing Ginseng: The Social Life of an Appalachian Root." *Cahiers de littérature orale* 53 (1): 265–291.

Karakilic, Emrah. 2022. "Rentierism and the Commons: A Critical Contribution to Brett Christophers' Rentier Capitalism." *Environment and Planning A: Economy and Space* 54 (2): 422–429. doi:10.1177/0308518X211062233.

Kay, Kelly. 2017. "Rural Rentierism and the Financial Enclosure of Maine's Open Lands Tradition." *Annals of the American Association of Geographers* 107 (6): 1407–1423. doi:10.1080/24694452.2017.1328305.

Kay, Kelly. 2018. "A Hostile Takeover of Nature? Placing Value in Conservation Finance." *Antipode* 50 (1): 164–183. doi:10.1111/anti.12335.

Kratzer, Nate W. 2015. "Coal Mining and Population Loss in Appalachia." *Journal of Appalachian Studies* 21 (2): 173–188. doi:10.5406/jappastud.21.2.0173.

Leip, David. 2022. "*Dave Leip's Atlas of U.S. Presidential Elections.*" 2022. http://uselectionatlas.org/.

Lensmire, Timothy J. 2017. *White Folks: Race and Identity in Rural America. Writing Lives: Ethnographic Narratives*. New York: Routledge, Taylor & Francis Group.

Li, Tania Murray. 2010. "To Make Live or Let Die? Rural Dispossession and the Protection of Surplus Populations." *Antipode* 41 (s1): 66–93. doi:10.1111/j.1467-8330.2009.00717.x.

Li, Tania Murray. 2014. *Land's End: Capitalist Relations on an Indigenous Frontier*. Durham, NC: Duke University Press Books.

Li, Tania Murray. 2017. "After Development: Surplus Population and the Politics of Entitlement." *Development and Change* 48 (6): 1247–1261. doi:10.1111/dech.12344.

Lipsitz, George. 2006. *The Possessive Investment in Whiteness: How White People Profit from Identity Politics, Revised and Expanded Edition*. Philadelphia: Temple University Press.

Maggard, Sally Ward. 1994. "From Farm to Coal Camp to Back Office and McDonald's: Living in the Midst of Appalachia's Latest Transformation." *Journal of the Appalachian Studies Association* 6: 14–38.
Maimon, Alan. 2021. *Twilight in Hazard: An Appalachian Reckoning*. New York: Melville House.
Martin, Lou. 2019. *Appalachia in the Age of Neoliberalism*. Asheville, NC: Marshall University. https://mds.marshall.edu/asa_conference/2019/session6/26.
McDowell, Linda. 1991. "Life Without Father and Ford: The New Gender Order of Post-Fordism." *Transactions of the Institute of British Geographers* 16 (4): 400. doi:10.2307/623027.
McDowell, Linda. 2014. "Gender, Work, Employment and Society: Feminist Reflections on Continuity and Change." *Work, Employment and Society* 28 (5): 825–837. doi:10.1177/0950017014543301.
Meit, Michael, M. P. H. Heffernan, Erin Tanenbaum, Maggie Cherney, and Victoria Hallman. 2020. "*Appalachian Diseases of Despair*." Final Report. Appalachian Diseases of Despair. Chicago: University of Chicago NORC. https://www.arc.gov/report/appalachian-diseases-of-despair/.
Metzl, Jonathan M. 2020. *Dying of Whiteness: How the Politics of Racial Resentment Is Killing America's Heartland. Illustrated Edition*. New York: Basic Books.
Moreton-Robinson, Aileen. 2015. *The White Possessive*. 1 ed. Minneapolis: University of Minnesota Press.
Mouffe, Chantal. 2018. *For a Left Populism. London* . New York: Verso.
The Nature Conservancy. 2019. "Cumberland Forest Project." *The Nature Conservancy*. 2019. https://www.nature.org/en-us/about-us/where-we-work/united-states/kentucky/stories-in-kentucky/cumberland-forest/.
The Nature Conservancy. 2021. "*The Nature Conservancy, Director Cumberland Forest Project*." Accessed September 27, 2021. https://www.impactfinancepro.com/cumberland-forest-the-nature-conservancy/.
Oberhauser, Ann M. 1993. "Industrial Restructuring and Women's Homework in Appalachia: Lessons from West Virginia." *Southeastern Geographer* 33 (1): 23–43. doi:10.1353/sgo.1993.0011.
Obermiller, Phillip, Chad Berry, Roger Guy, J. Trent Alexander, William Philliber, Michael Maloney, and Bruce Tucker. 2009. "Major Turning Points: Rethinking Appalachian Migration." *Appalachian Journal* 36 (3/4): 164–187.
Open Source Coal. 2022. "Coal Production by Mine Type and Employment 1983-2022 in the 'Central Appalachian' Region." *Appalachian Voices*. Accessed May 4, 2022. https://www.opensourcecoal.org.
Peck, Jamie, and Nick Theodore. 2019. "Still Neoliberalism?" *South Atlantic Quarterly* 18 (2): 245–265. doi:10.1215/00382876-7381122.
Pollard, Kevin, and Linda A. Jacobsen. 2021. "The Appalachian Region: A Data Overview from the 2015-2019 American Community Survey. Chartbook." *Appalachian Regional Commission*. https://eric.ed.gov/?id = ED613609.
Ray, Tarence. 2021. "United in Rage." *The Baffler*, July 6, 2021. https://thebaffler.com/salvos/united-in-rage-ray.
Rigg, Khary, Shannon Monnat, and Melody Chavez. 2018. "Opioid-Related Mortality in Rural America: Geographic Heterogeneity and Intervention Strategies." *International Journal of Drug Policy* 57 (July): 119–129. doi:10.1016/j.drugpo.2018.04.011.
Schwartzman, Gabe. 2013. *Where Appalachia Went Right: White Masculinities, Nature, and Pro-Coal Politics in an Era of Climate Change*. Berkeley, CA: Center for Right Wing Studies, University of California. http://escholarship.org/uc/item/4wm02403.
Schwartzman, Gabe. 2015. "How Central Appalachia Went Right." *The Daily Yonder*, Accessed January 13, 2015. http://dailyyonder.com/how-coalfields-went-gop/2015/01/13/.
Scoones, Ian, Marc Edelman, Saturnino M. Borras Jr, Ruth Hall, Wendy Wolford, and Ben White. 2018. "Emancipatory Rural Politics: Confronting Authoritarian Populism." *Journal of Peasant Studies* 45 (1): 1–20. doi:10.1080/03066150.2017.1339693.
Silva, Jennifer M. 2019. *We're Still Here: Pain and Politics in the Heart of America*. New York, NY: Oxford University Press.
Singh, Gopal K., Michael D. Kogan, and Rebecca T. Slifkin. 2017. "Widening Disparities In Infant Mortality And Life Expectancy Between Appalachia And The Rest Of The United States, 1990–2013." *Health Affairs* 36 (8): 1423–1432. doi:10.1377/hlthaff.2016.1571.

Slobodian, Quinn. 2021. "The Backlash Against Neoliberal Globalization from Above: Elite Origins of the Crisis of the New Constitutionalism." *Theory, Culture & Society* 38 (6): 51–69. doi:10.1177/0263276421999440.

Smith, Neil. 1979. "Toward a Theory of Gentrification A Back to the City Movement by Capital, Not People." *Journal of the American Planning Association* 45 (4): 538–548. doi:10.1080/01944367908977002.

Smith, Neil. 1996. *The New Urban Frontier: Gentrification and the Revanchist City. 1st Edition. London* . New York: Routledge.

Spence, Beth, Cathy Kunkel, Elias Schewel, Ted Boettner, and Lou Martin. 2013. *Who Owns West Virginia in the 21st Century*. Charleston, WV: West Virginia Center for Budget and Policy.

Tennessee Wildlife Resource Agency. 2022. "The Nature Conservancy, TWRA Announce Conservation Easement to Benefit North Cumberland WMA's Ed Carter Unit." *Government. Tennesse State Government*. Accessed January 10, 2022. https://www.tn.gov/twra/news/2022/1/10/the-nature-conservancy–twra-announce-conservation-easement-to-benefit-north-cumberland-wma-s-ed-carter-unit.html.

Tingley, Andrew. 2021. "How The Nature Conservancy's Work in the Cumberland Forest Supports Jobs & Climate Mitigation." *Salesforce.Org*. Accessed July 2, 2021. https://www.salesforce.org/blog/the-nature-conservancy-cumberland-forest-project/.

TN Dept. of Tourism Development. 2021. "Tennessee Tourism Outperformed the Nation in 2020 with $16.8 Billion in Visitor Spending Despite COVID-19." *TN Department of Tourism Development*. Accessed August 6, 2021. https://www.tn.gov/tourism/news/2021/8/6/tennessee-tourism-outperformed-the-nation-in-2020-with–16-8-billion-in-visitor-spending-despite-covid-19.html.

US Census Bureau, American Community Survey. 2020. *"American Community Survey 5-Year Estimates, Table S1071."* Generated by Gabe Schwartzman. Accessed May 3, 2022. https://data.census.gov/cedsci/.

White, Dustin. 2017. "The Toxic Masculinity of Fossil Fuels." *Ohio Valley Environmental Coalition (blog)*. Accessed November 11, 2017. https://ohvec.org/toxic-masculinity-fossil-fuels/.

Winant, Gabriel. 2021. *The Next Shift: The Fall of Industry and the Rise of Health Care in Rust Belt America*. Caimbridge, MA: Harvard University Press.

Wright, Erik Olin. 2017. "How to Be an Anti-Capitalist for the 21 St Century." *Theomai* 35: 8–21.

Young, Charles. 2022. "Gov. Jim Justice: 'West Virginia Is in Demand.'" *WV News*. 2022. https://www.wvnews.com/statejournal/news/gov-jim-justice-west-virginia-is-in-demand/article_d4a2c50c-9bd7-11ec-b177-73e9990e2095.html.

Zhang, Xufang, and Anne Stottlemyer. 2021. *Lumber and Timber Price Trends Analysis During the COVID-19 Pandemic*. College Station, TX: Forest Analytics Department.

Up in the air: the challenge of conceptualizing and crafting a post-carbon planetary politics to confront climate change

Alistair Fraser

ABSTRACT
I argue that confronting climate change requires conceptualizing and crafting a post-carbon planetary politics focused on removing carbon from the atmosphere. A focal point for beginning to build this politics should be carbon removal networks. I conceptualize these networks as vehicles that tap diverse knowledge domains (from sciences such as ecology or chemistry to activism and the law) to establish a planetary-wide political alliance which removes carbon while delivering nutrition, shelter, and care to populations in all manner of geographical settings. Such networks would mobilize the cooperative and collaborative power of civil society to establish an extensive, significant, and as yet under-recognized, 'civil society carbon sink.' I then suggest astute political organizing might take advantage of contemporary (and emerging) technologies to build such alliances and amplify the potential impact of organizing to remove carbon from the atmosphere.

Introduction

In response to the question, 'can we envision a sufficiently anti-capitalist, trans-environmental and agrarian approach to confront climate change in rural settings, and what would this look like in practice?' (Borras et al. 2022, 17), the following paper makes two arguments. The first is normative, which is required given a political economic context that tends to produce pessimism and despair. Following Latour (2017, 5), the threat of climate change means human societies must 'start seeking a territory that we and our children can inhabit … ' In turn, I argue the search for an inhabitable territory means that scholars – whether or not their focus is on agrarian change – should understand that confronting climate change will require a long-term *struggle* to create an entirely new planetary-wide geography centred around the act of removing carbon from the atmosphere. An adequate approach to confront climate change will need to pursue far-reaching change, often simultaneously, in numerous settings and with a vast diversity of practices. As Nunes (2021, 289; emphasis in original) argues regarding social change in general, the challenge is to negotiate

> a plurality of timelines and rhythms of change running at variable speeds, an irregular patchwork of continuities and discontinuities that do not miraculously combine to produce

structural transformation but are the object of a *constant, deliberate effort to play them both in support of* (to reinforce) *and against* (to correct the course of) *one another.*

In this paper, I probe what such a 'constant deliberate effort' might entail. Borras et al. (2022) are correct to argue we need a politics that confronts climate change by focusing on transforming society, not least by moving beyond capitalist values. Yet, even if such a transformation can be achieved, human and other forms of life will still face enormous challenges unless new coalitions can be established to find ways of removing (truly, vast quantities of) carbon from the atmosphere. The first part of the following paper expands on this point by shining light on the possibility that the struggle to remove carbon from the atmosphere can and should be led not by start-ups or extant capitalist enterprises, nor even states, but rather by grassroots, community-based organizations and political organizing; in short, by civil society (broadly construed). Climate models ignore the scope for civil society to take the lead in confronting climate change. My focus is on what it might mean if a 'civil society carbon sink' was established and expanded by tapping diverse knowledge domains, from the sciences such as ecology or chemistry to activism and the law. I ask: Using cooperative and collaborative powers latent within civil society, could a planetary-wide political alliance focus on removing carbon?

The paper's second argument builds on the first by engaging with some of the ideas raised in recent scholarship on the impact of digital technologies on agrarian change (e.g. see Bronson and Knezevic 2016; Mooney 2018; Fraser 2019; Prause, Hackfort, and Lindgren 2021; Stock and Gardezi 2021; Visser, Sippel, and Thiemann 2021). I suggest that establishing and expanding the civil society carbon sink might take advantage of technological affordances to amplify the effectiveness of action. Drawing again on Nunes (Nunes 2021, 290), I argue technologies can be employed in political struggle with a view to maximizing 'the structural impact that our limited capacities to act can have … ' Emphatically, my argument is not intended to suggest that technologies are in any way a 'silver bullet.' This is not to fetishize technology, nor is it to claim that the challenge of addressing climate change is just a matter of finding some new technological fixes. There is no 'technical fix' to climate change. Rather, the point is that, especially in this century, agitating, educating, and organizing will be bound up with technology in unpredictable but potentially useful and 'productive' (Ettlinger 2018) ways. With nuanced political debate and bargaining, certain technical barriers in the way of removing carbon could be reduced and/or overcome. I will specifically refer to free/open-source software, open-source intelligence, and blockchain technologies with a view to considering how their affordances might be harnessed by civil society action to remove carbon from the atmosphere.

A civil society carbon sink?

It is well known that the last three hundred years of industrial activity and land-use change, and then the last century of carbon-intensive energy consumption, have created the conditions for rapid climate change. Put differently, humans currently exist in the context of a capitalist mode of production which relies heavily on emitting carbon into the atmosphere (see especially Mitchell 2011). In response, one part of the current challenge is to reduce carbon emissions via 'deep and rapid decarbonization'

(Smith, Davis, and Creutzig 2016, 48) of society, regardless of whether capitalist or other values are hegemonic. Yet, not only must carbon emissions decline rapidly, which seems to be decades away, but according to climate models for this century and beyond (e.g. see Fuss, Canadell, and Peters 2014) there is *also* a need to remove vast amounts of carbon from the atmosphere if we want to create a safer climate in which humans and other species can survive and thrive. In a complete turnaround from the last century of human activity in the wealthiest, most technologically innovative countries and beyond, which to an extraordinary extent relied on extracting and burning fossil fuels, humans must now find ways to bring carbon back from the atmosphere and hold it, somehow, in the soil or elsewhere underground. Removal of carbon from the earth's atmosphere would be the most important material movement of the twenty-first century. At the very least, the carbon removal question is going to matter (on some current debates, see Wallace-Wells 2021).

For those developing climate models, terrestrial carbon dioxide removal (tCDR) tends to be conceptualized as involving processes such as 'bioenergy with carbon capture and storage,' 'biochar' and 'soil carbon sequestration,' 'direct air capture of carbon-dioxide,' 'enhanced weathering,' and 'afforestation and reforestation' to fix carbon in biomass and soils (Rogelj, Popp, and Calvin 2018). All such techniques require land use change or generate impacts on biodiversity and water. They each also yield questions about the *specific* effects: Who will, or won't, be involved; and how should any associated costs be met? Models test the limits of changing land use versus the possible impact on food production systems. One striking feature of these models is the way analysis tends to cohere around the role of two sets of human actors: states and markets. The core assumption underpinning climate models is that entrepreneurs, firms, and perhaps entire industrial sectors will respond to policy, or lead from the front in removing carbon from the atmosphere. The underlying question to be posed is whether climate policy can alter 'economic incentives' (de Coninck, Revi, and Babiker 2018); can technological change alter the 'cost-effectiveness' of one action or another? In essence, the assumption about human life is that climate change in this century will be experienced and negotiated through a state-market nexus, with civil society playing only the role of exerting pressure on governments to act.

I argue that focusing on the climatic effects of moves by states and markets alone is a mistake because it overlooks an alternative and crucial set of agents. Specifically, what is ignored in climate science is the potential for grassroots social movements, community-based organizations, or citizen-led civil society organizations (hereafter, for the sake only of simplicity, 'civil society') to target carbon removal. This is not to suggest the wide range of entities constituting civil society are entirely distinct from states and markets. There are always relations. Neoliberal government policies, for example, operate on the assumption that organizations in civil society, such as charities running food banks, will 'mop up' and soften the blow of decisions that expand inequality. Many firms, meanwhile, will view civil society as a key market for goods and services; and many civil society organizations will have strong relations with local, national, or transnational capital. Nevertheless, in the interpretations and analyses of what the future holds, climate science restricts its conceptual gaze such that only politicians and firms are awarded causal properties.

In opposition to such a view, I argue a more robust interpretation of what faces humanity this century and beyond should recognize that the growth, expansion, and increased capacity of civil society – albeit assisted by carbon-intensive globalization processes equipping them with significant and amplified means of communication and coordination – is one of humanity's greatest (but underrecognized) achievements. It is remarkable that such a wide range and large number of activist organizations, political entities, non-governmental institutions, and even (social) enterprises have been formed by people from diverse backgrounds, sometimes with little to no experience of establishing and operating sustainably, *and often in the context of significant constraints, including violent repression*. Accomplishments vary from achieving incremental changes in public policies regarding human rights; to challenging corporate control over the food system (e.g. see Busscher et al. 2020); and indeed the establishment and expansion of complex entities such as La Via Campesina, a transnational agrarian movement held together via an ongoing struggle to make the case for systemic change in food systems across the world (e.g. see Borras 2008). Given these pointers, which suggest civil society can be a key locus of social change, I want to highlight a new question for our times: What if today's grassroots social movements, community-based organizations, citizen-led civil society organizations, and others not yet established, were to use their cooperative and collaborative powers to build projects and alliances toward a long-term goal of removing carbon from the atmosphere? What if carbon removal in this century and beyond was led by civil society, rather than the state-markets nexus?

Beyond the scope for civil society to push for social change, focusing on this wellspring of change is particularly worthwhile due to what I should like to refer to as the *civil society multiplier*. There are two components here. First, notwithstanding constraints regarding the narrow demographics of volunteers in some civil society organizations or the risk of burnout, civil society action can achieve much more change in a given condition than might be expected because so much of the work is carried out on a voluntary basis, provided for free on a pro bono case, or uses funds from charitable contributions to cover labour inputs. Consider how different this is from the calculation of cost efficiency or from measures of economic incentives used in IPCC reports and supporting literature. When climate modelers imagine how farmers view the land, they assume that actions are measured in relation to cost as defined in neoclassical economics, which underpins economic models of climate change. Planting trees requires labour inputs, for example, which must be paid for. In contrast, although costs certainly matter, the calculations for civil society will be different when labour inputs are provided by volunteers, or if the cost of purchasing labour or materials can be shared across a network (possibly 'crowdsourced'). There is no guarantee that they will, but actors in civil society have scope to consider forms of land use change firms or farmers might not manage to afford. Organizations in civil society can pursue novel, more ambitious pathways toward carbon removal. The multiplier effect here is material: the return, the ongoing achievement, is potentially far greater because grassroots action can operate according to unconventional economic metrics; climate models struggle to capture this reality.

Second, further benefits emerge when openness to difference and diversity (of individuals, opinion, or way of thought) is embraced as a virtue. By no means does this apply necessarily to civil society organizations – there are, of course, many civil society organizations that discriminate or are otherwise exclusionary – but when they *are* 'open' and continuously resolve internal conflicts effectively while navigating the politics of difference, they have a better chance of finding and tapping pertinent sources of knowledge and/or good practice, thereby, increasing their potential to develop astute, effective, and sustainable arrangements or practices (e.g. see Escobar 2004). One core reason why openness should matter here is because the task at hand entails colossal complexity. Astute political organizing to transform space and remove carbon will require tapping expertise on activist practice (for example, to manage political differences while building the necessary alliances and coalitions among diverse populations with competing interests) and the law (for example, to establish legal communal entities such as community land trusts or cooperatives). Moreover, effectively removing carbon from the atmosphere requires tapping into a wide range of knowledge domains in sciences such as ecology or chemistry to calculate best practice regarding matters such as net ecosystem CO_2 exchange. A basic challenge will be defining what carbon removal should mean, in practice. My argument is not that one or another practice will be proven to remove carbon from the atmosphere long enough for it to make a difference; rather, it is that civil society should form networks and alliances capable of beginning to learn what works and then to pursue more of it. Civil society networks to remove carbon can capitalize on openness. Consider now the contrasting likelihood that today's climate models have conceptualized and assessed carbon removal possibilities based only on the results from studies of projects that were overwhelmingly developed, constructed, and rolled out by relatively *closed* scientific teams operating with the limiting parameters set by the parochial imaginations of funding agency steering committees.[1] If, for example, it is clear that creating new, or expanding existing, wetlands can present striking carbon removal opportunities (e.g. see Mitsch, Bernal, and Nahlik 2013), their potential could be amplified by interventions channelled via civil society. Where watercourses might once have followed a path suited to commercial agriculture, perhaps action can identify ways of re-arranging flows, expanding wetlands, and increasing their potential as carbon sinks. Climate models underestimate the potential latent within grassroots organizations and among subaltern populations generally.

To summarize, my normative argument, in response to the framing paper by Borras et al. (2022), which asks how we must confront climate change, is that the confrontation at issue should be about carbon removal. Confronting climate change will require decades of struggle to create an entirely new planetary-wide geography centred around the act of removing carbon from the atmosphere. I have also suggested that such a project can emerge from civil society. Acknowledging the two components of the civil society multiplier suggests grassroots action could make a difference in a desperately difficult situation in which we need to 'change everything' (Felber 2015). The type of political entity needed, therefore, is something we might label 'carbon removal networks.' The point of emphasizing 'networks' here is that humans always occupy positions in overlapping social networks (Ettlinger 2003). The effect is to amplify the opportunities, expand the horizons, and increase the potential of their intended actions because individuals

[1]Laudel (2006), research funding agencies operate on the mistaken assumption their systems support only the highest quality research, whereas evidence from Australia and Germany demonstrate that

> successful acquisition of competitive grants is influenced by a variety of factors such as a country's general investment in research, a scientist's research field, the availability of enabling funds, and the continuity of the research trail. These factors depend either partly or not at all on a scientist's or a proposal's quality. (398)

The parochialism of the 'quality myth' interacts with systemic practices that direct funding toward 'closed clubs' (Enger 2018) of higher education institutions.

within civil society organizations form interlinking webs and chains of social relations with numerous others. The upshot is recognizing that coalitions of carbon removal networks might manage to become a significant and extensive *carbon sink* by mobilizing civil society's cooperative and collaborative power around the objective of increasing the flow of carbon out of the atmosphere. What matters now, then, is probing what might be the transformational prospects of carbon removal networks.

On the transformational prospects of carbon removal networks

What if there was a decades-long project of grassroots political organizing around carbon removal? What transformations might it lead to? Here I want to argue that sustained action of this nature can meet core human needs by delivering nourishment, shelter, and care to populations in all manner of geographical settings at the same time as removing carbon from the atmosphere. I also want to highlight potential tensions emerging from such a vision.

With regards to *nourishment*, carbon removal pursued via agroecological methods could simultaneously aim to improve water management (including reuse to minimize waste) and food supplies. Above ground, carbon removal to increase soil organic carbon can boost crop diversity, which then promotes insect and ruminant biodiversity. Carbon removal can intersect positively with numerous related biodiversity challenges if its ecological practice can work with soil microbes, bacteria, and fungi. The potential for such practices to matter demonstrates that rural spaces are pivotal sites of change. One carbon removal option is likely to involve some form of 'soil carbon sequestration' (SCS), given 'high agreement' and 'robust evidence' that it 'affects nutrients and food security favourably' (de Coninck, Revi, and Babiker 2018, 345). A related practice involves producing and applying biochar to the soil (for a critique, see Leach, Fairhead, and Fraser 2012). Both options raise questions about the permanence of sequestration given varying soil types, although it is necessary to recognize uncertainties and accept that carbon removal networks would need to tap a wide range of knowledge domains to produce and sustain success. Further urgent questions remain about how the rural spaces where SCS could occur *will in fact be used* in this century. For example, although there is evidence to suggest that

> agroecological improvement of the design and management of [traditional farming] systems can make them even more effective carbon sinks, small diversified farms occupy less than 20% of the total arable land, and could not offset the emissions produced by industrial agriculture which occupies 80% of arable land with input intensive monocultures responsible for between 19 and 29% of total greenhouse emissions. (Altieri and Nicholls 2017, 43)

Likewise, insofar as IPCC findings might encourage some optimism that commercial industrial agriculture will develop practices that enable SCS to occur at scale, there are reasons to be alarmed about such a scenario. In relation to approaches such as 'sustainable intensification' or 'climate-smart' agriculture, it is essential to recognize that

> these superficial technical adjustments are ideologically buttressed by intellectual projects to reframe and redefine agroecology by stripping it of its political and social content and promote the wrong notion that agroecological methods can co-exist alongside the aggressive expansion of industrial agriculture, transgenic crops and agrofuels. (Altieri, Nicholls, and Montalba 2017, 3)

Carbon removal grounded in agroecological practice can pursue disruptive innovation that lead to the demise of commercial industrial agriculture. Making the case for carbon removal involves building a policy basis for planetary-wide land reform to ensure growth of the civil society carbon sink. Even without effective land reform policy, however, carbon removal networks could transform the use of rural space and pursue civic-led agrarian reform using agroecological methods that change land and agrarian relations. Literature on transnational agrarian movements demonstrates the extraordinary capacity for organizing and promoting a planetary agenda for agrarian change (Borras 2008; Borras, Edelman, and Kay 2008; Martinez-Torres and Rosset 2010). The challenge now is to ask how the development of activism focused on removing carbon might intersect with existing and emerging tensions and struggles facing food producers around the world. How will arguing for a decades-long struggle to remove carbon from the atmosphere sit with oppressed populations facing problems in the here and now? Is it possible for peasants and their allies to ensure the carbon removal agenda does not become an anti-peasant move (cf. Fairhead, Leach, and Scoones 2012)?

Regarding *shelter*, action to remove carbon could draw on expertise in architecture and design to achieve carbon removal while developing innovative housing designs using renewable materials and integrating living walls and ponds to capture and store carbon. The challenge of climate change intersects with urgent urban questions, especially but not only in global south contexts. Envisioning radical approaches to confronting climate change in rural settings alone will not suffice (Borras et al. 2022). The urban spaces of civil society-led carbon removal might involve re-orienting institutional spaces in cities such as schools, universities, hospitals, or other workplaces to make them carbon sinks. Drawing on pools of voluntary or paid labour funded by the 'gift economy' or new commons-based peer production enterprises (as signposted, for example, by Pazaitis 2020; also see Papadimitropoulos 2020), urban-based networks could create infrastructure to support carbon removal, such as irrigation systems. Collective (and often marginalized) knowledge of urban spaces could identify locations for rolling out appropriate carbon removal infrastructure. Cities can be cooled, food can be produced via *urban* agroecological practice (e.g. Altieri and Nicholls 2018; also Altieri and Toledo 2011, 600–601), materials for creating more adequate forms of shelter can be developed, and carbon can be stored.[2] Rather than an urban scene defined this century by corporate giants and their emerging notions for how a 'smart city' should operate, within grasp is an urbanism defined instead by its potential to be a civil society carbon sink,[3] although it will need to be a type of urbanism that connects fruitfully and equitably with changes in rural settings.

Finally, regarding humanity's need to provide *care* and receive it from others, the therapeutic component of carbon removal deserves attention insofar as the possible regeneration at issue is not simply ecological but rather is also potentially oriented toward

[2]A further consideration is the verticality of urban space: one hectare in a high-rise city could contain many multiples of hectares when roof space and walls are used to grow biomass or food. Capitalists are leading the way in the emergence of urban indoor vertical farms or so-called 'controlled environment agriculture' using LED lights, climate-controls, and carbon dioxide (Benke and Tomkins 2017; van Delden, Sharath Kumar, and Butturini 2021); carbon removal networks might follow suit via experimentation and innovation to improve methodologies, while discovering and sharing new techniques with others.

[3]Such a shift is not quite what has happened with Toronto's move away from Alphabet's smart city vision but there are resemblances (Jacobs 2022).

improving the lives of humans. Mental health can improve with exposure to nature (Beyer, Kaltenbach, and Szabo 2014), for example when planting or harvesting trees, plants, or tending crops. Bonds of solidarity within carbon removal networks could tackle loneliness, conceivably improving the lives of marginalized or aging populations. In addition, if carbon removal becomes a focal point of human life and if such practice can improve nourishment and help to provide shelter, it follows that humans will have more scope to provide care to others. A different planetary-wide life is in view, even if its conceptualization and eventual crafting will be far from easy.

Yet, if the preceding materials suggest we can envision a new planetary-wide geography taking shape when human life is oriented around carbon removal, it is also necessary to consider what the politics of carbon removal networks might involve. To draw parallels with Butler's (2020, 48) notion of 'interdependency among regions and hemispheres,' the required politics should find ways to strategize collectively and then 'think about the threat to the environment, the problem of the global slum, systemic racism, the condition of stateless people whose migration is a common global responsibility, even the more thorough overcoming of colonial modes of power.' Capitalists engage this complex scene for the sake of generating profits; civil society action oriented around carbon removal can engage it for the sake of removing carbon, while addressing human core needs for nourishment, shelter, and care. It is tempting to imagine that such a political move will manage to subordinate the state-market nexus to the task of carbon removal, such that capitalist enterprises will be squeezed out or at least kept on a tight leash. Indeed, to draw inspiration from Wright (2010), maybe political organizing around carbon removal can find ways to create symbiotic, interstitial, and ruptural projects that avoid becoming 'unstable and degenerat[ing] into some form of either a statist or a capitalist economy' (372). A more likely scenario for now is that, for so long as capitalist enterprises exist, the potential for carbon removal networks to achieve success – as measured by lower levels of carbon in the atmosphere or, for example, higher levels of soil organic carbon across vast areas of the planet – will depend to a large extent on whether their actions can contribute to the formation of subaltern *post*-capitalist economies (uneasily) co-existing with the capitalist mainstream (as sensitively explored by Gibson-Graham, Cameron, and Healy 2013; also Gibson-Graham 2006). By thinking laterally about the 'immense and paralyzing question of how to replace capitalism by some other regime' (Latour 2017, 97), a post-capitalist 'community economy' approach to carbon removal might be the most viable form of anti-capitalist action, given that 'what might come after capitalism can only be built from where we stand, using the multiple and messy resources and capacities that present themselves' (Chatterton 2016, 405).

It is appropriate here, therefore, to note that actions *resembling* carbon removal networks already exist. Consider Regeneration International, a non-profit formed in 2017, which aims to 'promote, facilitate and accelerate the global transition to regenerative food, farming and land management for the purpose of restoring climate stability, ending world hunger and rebuilding deteriorated social, ecological and economic systems' (https://regenerationinternational.org/). The International Biochar Initiative (https://biochar-international.org/) and Nature Conservancy (https://www.nature.org/en-us/) have similar targets. A coalition called Grasslands+ (https://grasslandsplus.org.uk/) aims to protect and restore 'grasslands, savannahs, plains, heaths, steppes and meadows' given their capacity to store carbon. Then there is the large international constituency of non-profits and non-governmental organizations within the '4 per 1000' initiative aiming to increase soil carbon storage by 4% per year in the first 30–40 cm of soil (https://www.4p1000.org/). Numerous other entities exist with a national focus, such as the UK's Soil Association (https://www.soilassociation.org/) and Mexico's Vía Orgánica (https://viaorganica.org/). As such, actions that might lay the ground for carbon removal are now taking place, albeit at a pace too slow to make a difference. By no means is it reasonable to suggest that these actions are currently, or will ever become, 'sufficiently anti-capitalist' (Borras et al. 2022, 17), nor even that they fit with respect to the post-capitalist community economy approach (Gibson-Graham, Cameron, and Healy 2013). However, the examples should prompt the question of whether, if carbon removal networks do take shape, they will improve their likelihood of succeeding by explicitly embracing a socially and ecologically constructive mode of production that establishes post-capitalist practices and is organizational and materially robust enough to endure co-existence with capitalist enterprises and values.

Using digital technologies to build carbon removal networks

Recent scholarship of relevance to critical agrarian studies has highlighted the importance of digital technologies (e.g. see Bronson and Knezevic 2016; Mooney 2018; Prause, Hackfort, and Lindgren 2021; Stock and Gardezi 2021; Visser, Sippel, and Thiemann 2021). Capitalists deploy digital technologies to map, survey, record, analyse and strategize about the future of rural space. Practices of agrarian resistance and transformation likewise draw upon and are bound up with digital technologies today (Fraser 2020). More generally, humans in a wide range of contexts exist in sometimes tense relation with digital technologies that they might use to enable communication and coordination while also generating data and potential profits for powerful, distant corporations and their shareholders. Human life in the twenty-first century is (and will become even more so) a digital affair.

Confronting climate change requires the creation of planetary-wide political alliances focused on annually removing enormous quantities of carbon from the atmosphere. I argue here, therefore, that laying the ground for novel yet ambitious and effective political entities such as carbon removal networks objectively will require political organizing that takes advantage of contemporary (and emerging) technologies. There will be no escape from digital life. Start-ups, well-established corporations, and states view carbon removal as an opportunity. They will use all the technical tools at their disposal to pursue their objectives. My argument is that carbon removal from the grassroots, from civil society, should begin to examine how it might *also* use digital technologies. Per my earlier argument regarding the civil society multiplier, there are opportunities for carbon removal networks to exploit technologies in innovative ways, thereby *amplifying* effective action. The challenge, in part, is to move beyond dominant patterns whereby 'misconfigured' innovations, such as those aimed at creating 'smart farming,' are rolled out by capitalist enterprises (Fraser 2021) and instead create new iterative arrangements that effectively remove carbon while delivering nutrition, shelter, and care to populations in all manner of geographical settings. The challenge also is to recognize that the requisite technical skills are unevenly distributed and, crucially for the purposes of this special forum, they tend not to be found among peasants and small-scale farmers searching for an anti-capitalist, trans-environmental and agrarian approach to confront climate change, nor among the most oppressed urban dwellers who might contribute in the struggle to cool the climate. There are clear risks in constructing architectures of resistance that incorporate practices understood only by a technical elite. My argument in response is simply that managing these risks will be germane to the success of carbon removal networks. Effective political organizing will require creating sustainable arrangements or practices that distribute technical understanding far and wide. Failures will occur, for this or for any of many other reasons. The challenge is to debate, to learn, and then to disseminate knowledge about what works. Toward initiating the debate, I now discuss three broad arrays of technology that deserve critical attention, although none of them provide definitive answers about how they should be engaged by carbon removal networks.

Open-source software to discover and coordinate activities

The first is open-source software. Lots of the software embedded in today's digital devices is proprietary, exchanged legally for a fee, and edited or adjusted only by the firm's software developers. A different approach known as open-source software turns everything upside down. The source code is 'open' insofar as anyone (albeit with requisite skills) can read and then proceed to edit it. Communities of volunteers will sometimes work (often on an ongoing basis) to develop such software. Unlike proprietary software, moreover, open-source is typically distributed for free; there is no parent firm claiming intellectual property rights. The relationship between proprietary and open-source software varies across the digital world. Many devices will contain a mix of the two. A universe of opportunity now exists to create devices such as sensors or online environments such as a forum or social networks using open-source software. Communities of practitioners and users focus on eradicating cost-related barriers in the way of even the poorest people accessing software. Social movements or community-based organizations do not need to waste scarce resources on purchasing software products from tech giants such as Microsoft or Apple. Open-source software can underpin, underwrite, and thereby amplify the possibilities of civil society.

The relevance of open-source technologies and practices to scholarship on agrarian studies has been demonstrated by Kloppenburg (2014). More recently, Bronson (2019, 4) has shed light on the development of *FarmOS*, an 'open-source agricultural data platform [in which] all of the code is visible and freely available, anyone can install the platform or host the system, and anyone can contribute by writing code and developing novel features.' Regarding the prospects of crafting carbon removal networks, then, the significance of open-source software affords opportunities to build, expand, and lock in successes. The objective is to learn about and disseminate knowledge regarding practices of carbon removal across a dispersed population of actors connected in a network and thereby taking advantage of digital technologies to coordinate their activities. Like the 'horizontal communication' used in the campesino-to-campesino movement in Cuba (Rosset et al. 2011), carbon removal networks can focus on teaching, and learning from allies, while in this case drawing on and using open-source software. At the same time, for these networks to thrive, practitioners must find ways of developing appropriate forms of *inclusion* which in turn requires effective representation and, it follows, communication across individual and group differences (Young 2000; regarding the centrality of 'mutual social recognition' to innovation in open-source production, see Benkler 2017). Contemporary social life suggests the strong likelihood that digital technologies will play a role. The availability of free and open-source software developed in voluntary communities of dispersed allies should be understood as a potential and likely amplifier that will enable carbon removal networks. Finally, open-source technologies present opportunities for civil society organizations to pursue carbon removal while retaining control over the operation of devices or services and thereby avoid exposure to 'data grabbing' by the largest technology firms.

Open-source intelligence to share and analyse evidence of carbon removal

A second array of technologies concerns open-source intelligence (OSINT). Growth in the availability of digital devices and services in the last decade especially has been a huge contributing factor in the emergence of OSINT communities (e.g. see Gradecki and Curry 2017; Belghith 2021). The objective in such communities is to track, analyse, and report on a wide variety of events and developments, with a major focus on military action in places such as Syria and Ukraine, as well as other humanitarian or environmental issues throughout the world. Participants use digital artefacts such as satellite imagery, cell phone footage uploaded to social media sites such as Twitter and YouTube, or (hacked and open public) database materials to explain actions (such as a drone strike), identify actors (sometimes for the sake of passing information to law enforcement officials), or locate otherwise unidentified places. Sharing their findings with other community members, as well as news media, and sometimes receiving crowdfunded support from otherwise passive allies, OSINT communities construct and conduct wide-ranging investigations. Few other developments in the present era offer quite such striking evidence of the new possibilities and contingencies associated with digital life.

OSINT might prove useful to carbon removal networks. The relevance of OSINT to critical agrarian studies has been emphasized by Fraser (2020). Consider the actions of InfoAmazonia, 'a collaboration between journalists, data scientists, and indigenous communities and land rights activists to respond to displacement and dispossession in the Amazon region' (Fraser 2020, 11; citing Gutierrez and Milan 2017; see also Gutierrez 2019). Referenced in the literature as a form of 'geoactivism' (Gutierrez 2019), the parallels with OSINT are clear. For example, note that:

> communities in the West of Pará, Brazil, linked by 'Rede InfoAmazonia,' share measurements of the quality and quantity of water to be able to make decisions in their daily lives. The initiative connects a network of sensors managed by the communities via mobile phones 'capable of monitoring physical and chemical parameters that help indicate whether the water is contaminated' […] The idea is to produce warnings about the quality of the water, as well as to gather historical data, visualise them on a map and identify trends. (Gutierrez 2019, 105)

With regards to carbon removal networks, OSINT illustrates the possibility that participants and supporters will establish open-source systems to build intelligence about the condition and performance of their projects. Whereas for InfoAmazonia, a collection of sensors is used to monitor water conditions, the focus in carbon removal networks might be on using sensors to measure carbon accumulations or rates of soil respiration.

Thus, if carbon removal networks might look to open-source technology generally to amplify collective action, they might also view open-source intelligence as an additional form of collective practice to improve knowledge about carbon removal, share and analyse evidence of carbon removal, and develop confidence in what works.

Blockchain technologies to support carbon removal networks

Blockchain technologies deserve consideration when constructing carbon removal networks. A blockchain is a 'distributed ledger that records transactions in a transparent, verifiable and permanent manner by storing them into a sequence (or "chain") of blocks' (Poux, De Filippi, and Ramos 2020, 7). Information stored in the blocks – such as a financial transaction – is secured using cryptography; once recorded, permanence is secured by all nodes in the network working to 'guarantee the integrity and authenticity of all data stored into this decentralized database' (8). The decentralized character of blockchain technologies tends to lend them credence to a wide range of non- and indeed anti-state groups and individuals, including many who subscribe to libertarian as well as anarchist and anti-capitalist ideologies. The cryptographic component of blockchain technologies also involves a high degree of technical complexity, which makes them intelligible to people with a background in software or data science but mystifying to many others.

There are numerous debates about blockchain technologies. For some critics, the emergence of blockchain-based cryptocurrencies signposts the demise of fiat currencies, which also means that democratic scrutiny and indeed control over fiscal and financial matters will end (Wolf 2019). Others focus on cryptocurrencies such as Bitcoin and emphasize how they require a significant supply of electrical power to verify transactions, which is especially problematic given the contemporary climate problem (Ross Sorkin 2021). Another focal point is the possible emergence of a technical elite or 'blockocracy' developing and rolling out cryptocurrencies (Kavanagh and Ennis 2020). Then there are critiques focusing on specific blockchain projects, for example where blockchain technology 'is enabling new opportunities for speculative investment [and] the extraction of economic benefits from those suffering the scars of empire in the Global South' (Howson 2020, 5; see also Howson 2019). For supporters, meanwhile, blockchain technologies will yield a progressive world in which financial relations can be established without any input from the state (e.g. see Beller 2017). Other analysts note that cryptocurrencies might eventually require far less electrical power once proof of work protocols that underpin Bitcoin are replaced by proof of stake protocols, which could become the basis for verification in cryptocurrencies such as Ethereum (Howson 2019).

Another pertinent fault line in the blockchain technologies debate is that capitalists in technology, trade, trade finance, and financial sectors more broadly are investing in blockchain technologies. There is a risk, therefore, that blockchain innovations, if their effectiveness can indeed be demonstrated, will eventually be dominated, and conceivably to some extent controlled, by the same financial institutions, firms, and regulatory states that set agendas in today's world. In response, radical blockchain theorists propose creating financial innovations that 'take such tools out of the exclusive hands of banks and states and put them in the hands of migrants, activists, artists, radicals and poets such that we could keep the value that we produce for ourselves ... ' (Beller 2017, 173).

Applauding the 'publicly auditable, anti-state nature of programmable monetary platforms [and their scope to open] the door to transnational currencies and controls that could limit or perhaps eliminate many exploitative relationships,' Beller (2017, 174) has even questioned whether it might be 'possible, through a democratization of financial tools and a decolonization of finance to create new economic spaces and within these spaces utilize subaltern and revolutionary cultural forms and practices to *redefine what is valued*?' (173; my emphasis).

In the context of climate change, a question emerging from Beller's vision is whether it might be possible to use blockchain technologies in defining and supporting the value of carbon removal. There is a chance that leading blockchain capitalists such as ConsenSys, Microsoft, and IBM could move in this direction; such action might even prove successful. Against this backdrop, a challenge is discovering if blockchain technologies might be used by civil society to avoid capitalist control over the carbon removal process. Drawing a parallel between Howson's (2021, 1) argument regarding degrowth activism and my concern with the struggle to establish and expand carbon removal networks, I argue it is conceivable that 'under certain conditions, blockchain might make those struggles more effective.'

One set of clues is provided by SEEDS (https://joinseeds.earth/), which claims to be an emerging 'ecosystem' concerned with 'Sowing Ecological, Equitable and Decentralizing Societies' (SEEDS 2021a). Created by a team of 'entrepreneurs, researchers, technologists, financiers, artists, ecologists and whole-systems designers' (SEEDS 2021b, 3), the purported ambition is to 'build, manage, and evolve voluntary and borderless economic and governance systems, with the transparent and incentivized participation of any person on our planet' (SEEDS 2021c, 2). Blockchain technologies are germane to the project. Using Telos and EOSIO, two blockchain infrastructural platforms, the SEEDS project is organized around a cryptocurrency called 'Seeds.' Participants would join by buying the currency, acquiring voting rights, and even earning an income from the community's 'harvest' (plainly, its surplus) by making pertinent contributions. As such, 'on-chain governance' (De Filippi and McMullen 2018) is intended to be transparent, decentralized, and autonomous. The vision they advance is that the 'people and organizations of SEEDS become investors, owners, users, and partners of their own economic system – thus aligning foundational interests' (SEEDS 2021c, 11).

Nori, which describes itself as a carbon removal organization, offers further clues. Its web-based marketplace matches farmers who agree to remove carbon and lock it in the soil for at least ten years with buyers looking to offset their emissions. To enter the marketplace, farmers in the US, the only country where Nori currently operates, must 'have adopted regenerative agriculture practices in the last ten years' (Nori 2021a) and be willing to satisfy various other conditions, not least that they will provide data on their operations and continue to provide updates in the future. Once enrolled, farmers can receive a Nori Carbon Removal Tonne (NRT) for each tonne of CO2 equivalent they will store, once they have 'entered their farm data into the Nori app, run quantification to determine the incremental soil organic stock (SOC) gained from regenerative practices, verified these by a third party auditor, and signed an NRT agreement' (Nori 2021b). Farmers can then sell NRTs to buyers in the Nori marketplace. Blockchain technologies are part of the picture here by virtue of NORI coin, a cryptocurrency which can be bought and exchanged for one tonne of C02 equivalent 'stored for at least ten years' (Nori 2021c). The vision is that NORI will help create 'a market-driven reference price for carbon removal' (Nori 2021c), with independent verifiers 'carbon quantification tools' sending 'real-world data to Ethereum's blockchain' (Nori 2021c) where data are then stored, securely using blockchain.

Taking stock of these developments regarding blockchain technologies, three points stand out. First, a post-capitalist carbon removal network requires crafting collaborative networks, not direct supplier-buyer market-oriented relationship central to Nori's market-driven practice. Second, blockchain technologies play only a supporting verification role for Nori and SEEDS, whereas I argue carbon removal networks will be likely to need a more direct engagement to achieve verification of carbon removal at scale. A major part of the challenge for carbon removal networks will be creating a democratic, autonomous, and transparent method of verifying that practice and action does in fact deliver carbon removal. Part of the methodological challenge of verification is precisely about the chemistry of carbon removal; but verification also entails humans working with technologies to securely confirm and reach consensus that practices are effective. What blockchain technologies appear to offer is a set of possibilities, rather than a promise or guarantee, that they could become integral when answering the verification question. Nori's approach to verification is indicative of what might take shape, although the notion of 'carbon removal' it advances falls short of what will be needed in this century: removing carbon will require more than token gestures. Finally, only in the case of SEEDS are there opportunities to see how carbon removal networks might be organized precisely *via* blockchain. Thus, what the cases of SEEDS and Nori suggest is that this century's struggles to remove carbon from the atmosphere will lead some practitioners to experiment with blockchain to create new economic spaces that might support carbon removal networks. Beller (2017), it is conceivable that such experiments will manage to 'redefine what is valued,' possibly by advancing the value of carbon removal. The best blockchain technologies remain unwritten; as such, they should not be written off by those who might begin to craft carbon removal networks if they acknowledge the absence of easy solutions and the presence, instead, of a latent civil society carbon sink that might be expanded via astute and critical use of technologies.

Conclusion

To reiterate the overall point of the paper, I have argued that a decades-long struggle to remove carbon from the atmosphere can be led by people and organizations in civil society. Actions therein can create an anti- or at perhaps a *post*-capitalist approach to confront climate change, which would need to be trans-environmental and agrarian, given the intensity of carbon removal's relationship with the land and with soil, and which would need to involve transforming rural *and* urban areas. It is time to turn around the political question about carbon. Carbon removal led by civil society should be at the centre of efforts to create better lives on this planet. There is a strong likelihood that populations will begin to lean toward this perspective when they grasp that states and markets have failed to reduce carbon emissions rapidly enough. As the carbon problem becomes more acute, what other options will we all have? It is, therefore, an apt moment to consider, research, and debate the prospects of creating and expanding the civil society carbon sink.

There are perils to note. Insofar as the need for carbon removal in this century and beyond will intensify, states and corporations might examine what they can do and how they can control what takes place. A civil society carbon sink might value solidarity, cooperation, agroecology, and forms of justice that transform urban life; a corporate-led process, or actions run by an authoritarian state, will operate according to different logics. Likewise, it is important to note that, although civil society can turn to arrays of technologies to amplify the effectiveness of their actions; states, corporations, and start-ups will turn to other emerging technologies, many of them proprietary and building on developments in artificial intelligence, which can already predict the structure of millions of proteins and may begin examining how to edit plant genomes to speed up carbon sequestration. Furthermore, even if a civil society carbon sink can be created and expanded, outcomes will still be uneven, and tensions will emerge over what sort of land-use change under what sort of circumstances will be most successful or tolerable to diverse populations. The upshot: there are going to be proliferating and colliding tensions and struggles when humans confront climate change directly by trying to create a new planetary-wide geography centred around the act of removing carbon from the atmosphere.

I argue finally, then, that an urgent agenda for researchers across the sciences should be to engage, work with, and promote the development of carbon removal networks. Activists working to create (however they define) 'climate justice' must incorporate and then effectively equip these networks. Each year, the atmosphere tells us how much more carbon we need to remove. The response to these signals should be renewed commitments to develop the potential for civil society to cool the planet, while simultaneously improving the quality of human/non-human life. As things stand, civil society's effect on reducing levels of carbon-dioxide in the atmosphere is marginal. But significant potential exists to make a positive difference. Confronting the climate question requires looking beyond states and markets. Grassroots, community-based, citizen-led, civil society organizations *already* conserve, protect, and transform an enormous area and wide variety of forestry, agricultural land, and wetlands. Their potential has not been addressed in carbon removal models; an unfortunate omission. If '[r]egreening the planet through conservation, restoration, and improved land management is a necessary step for our transition to a carbon neutral global economy and a stable climate' (Griscom, Adams, and Ellis 2017), carbon removal networks can play a major role. It is time to build and promote the civil society carbon sink.

Acknowledgements

The author would like to thank Nancy Ettlinger for always reading the numerous versions of this paper with enthusiasm and patience. I also thank Jun Borras and the anonymous reviewers for their insightful and helpful comments on the original and revised version of this manuscript. Any mistakes or errors are mine alone.

Disclosure statement

No potential conflict of interest was reported by the author(s).

ORCID

Alistair Fraser http://orcid.org/0000-0001-8147-7780

References

Altieri, M. A., and C. I. Nicholls. 2017. "The Adaptation and Mitigation Potential of Traditional Agriculture in a Changing Climate." *Climatic Change* 140 (1): 33–45.

Altieri, M. A., and C. I. Nicholls. 2018. "Urban Agroecology: Designing Biodiverse, Productive and Resilient City Farms." *Agro Sur* 46 (2): 49–60.

Altieri, M. A., C. I. Nicholls, and R. Montalba. 2017. "Technological Approaches to Sustainable Agriculture at a Crossroads: An Agroecological Perspective." *Sustainability* 9 (3): 349. doi:10.3390/su9030349.

Altieri, M. A., and V. M. Toledo. 2011. "The Agroecological Revolution in Latin America: Rescuing Nature, Ensuring Food Sovereignty and Empowering Peasants." *Journal of Peasant Studies* 38 (3): 587–612.

Belghith, Y. 2021. "The Social Structures of OSINT: Examining Collaboration and Competition in Open Source Intelligence Investigations." PhD diss., Virginia Tech. https://vtechworks.lib.vt.edu/bitstream/handle/10919/103944/Belghith_Y_T_2021.pdf?sequence=1&isAllowed=y.

Beller, J. 2017. *The Message is Murder: Substrates of Computational Capital*. London: Pluto Press.

Benke, K., and B. Tomkins. 2017. "Future Food-Production Systems: Vertical Farming and Controlled-Environment Agriculture." *Sustainability: Science, Practice and Policy* 13 (1): 13–26.

Benkler, Y. 2017. "Peer Production, the Commons, and the Future of the Firm." *Strategic Organization* 15 (2): 264–274.

Beyer, K., A. Kaltenbach, and A. Szabo. 2014. "Exposure to Neighborhood Green Space and Mental Health: Evidence from the Survey of the Health of Wisconsin." *International Journal of Environmental Research and Public Health* 11 (3): 3453–3472.

Borras, S. M. 2008. "La Via Campesina and its Global Campaign for Agrarian Reform." *Journal of Agrarian Change* 8 (2-3): 258–289.

Borras, S. M., M. Edelman, and C. Kay. 2008. "Transnational Agrarian Movements: Origins and Politics, Campaigns and Impact." *Journal of Agrarian Change* 8 (2-3): 169–204.

Borras, S. M., I. Scoones, A. Baviskar, M. Edelman, N. L. Peluso, and W. Wolford. 2022. "Climate Change and Agrarian Struggles: An Invitation to Contribute to a JPS Forum." *Journal of Peasant Studies* 49 (1): 1–28.

Bronson, K. 2019. "Looking Through a Responsible Innovation Lens at Uneven Engagements with Digital Farming." *NJAS-Wageningen Journal of Life Sciences* 90: 100294.

Bronson, K., and I. Knezevic. 2016. "Big Data in Food and Agriculture." *Big Data & Society* 3 (1): 2053951716648174.

Busscher, N., E. L. Colombo, L. van der Ploeg, J. I. Gabella, and A. Leguizamón. 2020. "Civil Society Challenges the Global Food System: The International Monsanto Tribunal." *Globalizations* 17 (1): 16–30.

Butler, J. 2020. *The Force of Nonviolence: An Ethico-Political Bind*. London: Verso.

Chatterton, P. 2016. "Building Transitions to Post-Capitalist Urban Commons." *Transactions of the Institute of British Geographers* 41 (4): 403–415.

de Coninck, H., A. Revi, and M. Babiker. 2018. "Strengthening and Implementing the Global Response." In: Global Warming of 1.5°C. An IPCC Special Report on the Impacts of Global Warming of 1.5°C Above Pre-Industrial Levels and Related Global Greenhouse Gas Emission Pathways, in the Context of Strengthening the Global Response to the Threat of Climate Change, Sustainable Development, and Efforts to Eradicate Poverty. IPCC-The Intergovernmental Panel on Climate Change.

De Filippi, P., and G. McMullen. 2018. "Governance of Blockchain Systems: Governance of and by Distributed Infrastructure." Doctoral dissertation, Blockchain Research Institute and COALA.

Enger, S. G. 2018. "Closed Clubs: Network Centrality and Participation in Horizon 2020." *Science and Public Policy* 45 (6): 884–896.

Escobar, A. 2004. "Beyond the Third World: Imperial Globality, Global Coloniality and Anti-Globalisation Social Movements." *Third World Quarterly* 25 (125): 207–230.
Ettlinger, N. 2003. "Cultural Economic Geography and a Relational and Microspace Approach to Trusts, Rationalities, Networks, and Change in Collaborative Workplaces." *Journal of Economic Geography* 3 (2): 145–171.
Ettlinger, N. 2018. "Algorithmic Affordances for Productive Resistance." *Big Data & Society* 5 (1): 2053951718771399.
Fairhead, J., M. Leach, and I. Scoones. 2012. "Green Grabbing: A New Appropriation of Nature?" *Journal of Peasant Studies* 39 (2): 237–261.
Felber, C. 2015. *Change Everything: Creating an Economy for the Common Good*. London: Zed Books.
Fraser, A. 2019. "Land Grab/Data Grab: Precision Agriculture and its new Horizons." *Journal of Peasant Studies* 46 (5): 893–912.
Fraser, A. 2020. "The Digital Revolution, Data Curation, and the New Dynamics of Food Sovereignty Construction." *Journal of Peasant Studies* 47 (1): 208–226.
Fraser, A. 2021. "'You Can't Eat Data'? Moving Beyond the Misconfigured Innovations of Smart Farming." *Journal of Rural Studies*, doi:10.1016/j.jrurstud.2021.06.010.
Fuss, S., J. Canadell, and G. Peters. 2014. "Betting on Negative Emissions." *Nature Climate Change* 4: 850–853.
Gibson-Graham, J. K. 2006. *A Postcapitalist Politics*. Minneapolis: University of Minnesota Press.
Gibson-Graham, J. K., J. Cameron, and S. Healy. 2013. *Take Back the Economy: An Ethical Guide for Transforming Our Communities*. Minneapolis: University of Minnesota Press.
Gradecki, J., and D. Curry. 2017. Crowd-Sourced Intelligence Agency: Prototyping Counterveillance." *Big Data and Society* 4 (1): 2053951717693259.
Griscom, B. W., J. Adams, and P. W. Ellis. 2017. "Natural Climate Solutions." *Proceedings of the National Academy of Sciences* 114 (44): 11645–11650.
Gutierrez, M. 2019. "Maputopias: Cartographies of Communication, Coordination and Action – The Cases of Ushahidi and InfoAmazonia." *GeoJournal* 84 (1): 101–120.
Gutierrez, M., and S. Milan. 2017. "Technopolitics in the Age of Big Data." In *Networks, Movements and Technopolitics in Latin America: Critical Analysis and Current Challenges*, edited by F. S. Caballero, and T. Gravante, 95–109. Basingstoke: Palgrave Macmillan. SSRN: https://ssrn.com/abstract=2935141.
Howson, P. 2019. "Tackling Climate Change with Blockchain." *Nature Climate Change* 9 (9): 644–645.
Howson, P. 2020. "Climate Crises and Crypto-Colonialism: Conjuring Value on the Blockchain Frontiers of the Global South." *Frontiers in Blockchain*, doi:10.3389/fbloc.2020.00022.
Howson, P. 2021. "Distributed Degrowth Technology: Challenges for Blockchain Beyond the Green Economy." *Ecological Economics* 184: 107020.
Jacobs, K. 2022. "Toronto Wants to Kill the Smart City Forever." *MIT Technology Review*. https://www.technologyreview.com/2022/06/29/1054005/toronto-kill-the-smart-city/.
Kavanagh, D., and P. J. Ennis. 2020. "Cryptocurrencies and the Emergence of Blockocracy." *The Information Society* 36 (5): 290–300.
Kloppenburg, J. 2014. "Re-Purposing the Master's Tools: The Open Source Seed Initiative and the Struggle for Seed Sovereignty." *Journal of Peasant Studies* 41 (6): 1225–1246.
Latour, B. 2017. *Down to Earth: Politics in the New Climatic Regime*. Cambridge: Polity.
Laudel, G. 2006. "The 'Quality Myth': Promoting and Hindering Conditions for Acquiring Research Funds." *Higher Education* 52: 375–403.
Leach, M., J. Fairhead, and J. Fraser. 2012. "Green Grabs and Biochar: Revaluing African Soils and Farming in the New Carbon Economy." *Journal of Peasant Studies* 39 (2): 285–307.
Martinez-Torres, M. E., and P. M. Rosset. 2010. "La Via Campesina: The Birth and Evolution of a Transnational Social Movement." *Journal of Peasant Studies* 37 (1): 149–175.
Mitchell, T. 2011. *Carbon Democracy. Political Power in the Age of Oil*. London: Verso.
Mitsch, W. J., B. Bernal, and A. M. Nahlik. 2013. "Wetlands, Carbon, and Climate Change." *Landscape Ecology* 28: 583–597.

Mooney, P. 2018. "What's Cooking for Climate Change? Techno-Fixing Dinner for 10 Billion." *Bulletin of the Atomic Scientists* 74 (6): 390–396.

Nori. 2021a. *Growing a Better Future Together*. https://nori.com/for-growers

Nori. 2021b. *Frequently Asked Questions*. https://nori.com/faq

Nori. 2021c. *The NORI Token*. https://nori.com/token

Nunes, R. 2021. *Neither Vertical Nor Horizontal: A Theory of Political Organization*. London: Verso.

Papadimitropoulos, V. 2020. *The Commons: Economic Alternatives in the Digital Age*. University of Westminster Press. https://library.oapen.org/handle/20.500.12657/42987.

Pazaitis, A. 2020. "Breaking the Chains of Open Innovation: Post-Blockchain and the Case of Sensorica." *Information* 11 (2): 104. doi:10.3390/info11020104.

Poux, P., P. De Filippi, and S. Ramos. 2020. "Blockchains for the Governance of Common Goods." In *Proceedings of the 1st International Workshop on Distributed Infrastructure for Common Good*, 7–12.

Prause, L., S. Hackfort, and M. Lindgren. 2021. "Digitalization and the Third Food Regime." *Agriculture and Human Values* 38 (3): 641–655.

Rogelj, J., A. Popp, and K. V. Calvin. 2018. "Scenarios Towards Limiting Global Mean Temperature Increase Below 1.5°C." *Nature Climate Change* 8: 325–332.

Rosset, P. M., B. Machín Sosa, A. M. Roque Jaime, and D. R. Ávila Lozano. 2011. "The Campesino-to-Campesino Agroecology Movement of ANAP in Cuba: Social Process Methodology in the Construction of Sustainable Peasant Agriculture and Food Sovereignty." *Journal of Peasant Studies* 38 (1): 161–191.

Ross Sorkin, A. 2021. "Bitcoin's Climate Problem." *New York Times*, March 9. https://www.nytimes.com/2021/03/09/business/dealbook/bitcoin-climate-change.html.

SEEDS. 2021a. *SEEDS: Constitution*. https://joinseeds.earth/pdf/Official_SEEDS_Constitution_V_1_July_10_2021.pdf.

SEEDS. 2021b. *SEEDS: Conscious Currencies | Regenerative Cultures*. https://joinseeds.earth/pdf/SEEDS-in-a-glance-4-Pager-V6-1.pdf.

SEEDS. 2021c. *SEEDS: An Economic 'Operating System' for Coordinating Equitable & Regenerative Global Societies*. https://joinseeds.earth/pdf/Economic_Renaissance_v5.pdf.

Smith, P., S. Davis, and F. Creutzig. 2016. "Biophysical and Economic Limits to Negative CO2 Emissions." *Nature Climate Change* 6: 42–50.

Stock, R., and M. Gardezi. 2021. "Make Bloom and Let Wither: Biopolitics of Precision Agriculture at the Dawn of Surveillance Capitalism." *Geoforum* 122: 193–203.

van Delden, S. H., M. Sharath Kumar, and M. Butturini. 2021. "Current Status and Future Challenges in Implementing and Upscaling Vertical Farming Systems." *Nature Food* 2 (12): 944–956.

Visser, O., S. R. Sippel, and L. Thiemann. 2021. "Imprecision Farming? Examining the (in) Accuracy and Risks of Digital Agriculture." *Journal of Rural Studies* 86: 623–632.

Wallace-Wells, D. 2021. "Climate Reparations a Trillion Tons of Carbon Hangs in the Air, Put There by the World's Rich, an Existential Threat to its Poor. Can We Remove It?" *New York Magazine*, Nov 1. https://nymag.com/intelligencer/2021/11/climate-change-reparations.html.

Wolf, M. 2019. "The Libertarian Fantasies of Cryptocurrencies." *Financial Times*, February 12. https://www.ft.com/content/eeeacd7c-2e0e-11e9-ba00-0251022932c8.

Wright, E. O. 2010. *Envisioning Real Utopias*. London: Verso.

Young, I. M. 2000. *Inclusion and Democracy*. New York: Oxford University Press.

Power for the Plantationocene: solar parks as the colonial form of an energy plantation

Ryan Stock

ABSTRACT
Solar park development in India constitutes racial regimes of land ownership, as solar-related dispossessions produce a highly racialized (through caste) and gendered surplus population of landless peasants. Conceptualizing the power relations of solar power through the Plantationocene, I argue the highly ordered form of the solar park is a set of neocolonial social relations akin to an energy plantation; an archetype of an imperative, idealized and racialized reordering of nature, economy and society to power a more sustainable world-system. Agrarian climate justice requires intersectional peasant coalitions struggling to transform neocolonial land politics and implementing redistributive and emancipatory solar interventions.

Introduction

> All the required technology is at hand. Adequate land will be available for energy plantations on a large scale without encroachment on land needed for other purposes. Energy plantations offer many potential benefits and no serious drawbacks for our country. (Kemp and Szego 1975)
>
> Plantation regimes create, institutionalize, and manage extreme levels of conflict. (Woods 2017, 40)

Powering all nations through renewable energies will require an estimated land footprint of 119,651,632 km^2 (Jacobson et al. 2017). The imperative to mitigate the climate crisis through renewables has exacerbated a global land rush already under way (Cotula 2012; Dell'Angelo et al. 2017; Franco and Borras 2019; Scheidel and Sorman 2012). The last few decades have seen an increase in foreign land acquisitions for large-scale agricultural investments and the funding of massive infrastructure projects under the auspices of rural development in the periphery by core countries looking to sustain economic hegemony. The rush to enclose vast swathes of rural land has been characterized as *land grabbing* due to the dispossession of arable land and the displacement or partial proletarianization of peasants (Borras and Franco 2012, 2013; Goetz 2021; GRAIN 2008; van der Ploeg, Franco, and Borras 2015).

This land rush is of acute concern in the global South, where *green grabbing* – resource or land dispossession to further environmental or sustainability goals (Fairhead, Leach, and Scoones 2012) – has disrupted agrarian political economy and ruptured peasant moral economies (Franco and Borras 2019; Rocheleau 2015). Specifically, expansive contiguous stretches of agrarian and commons lands have been enclosed or degraded to facilitate the global proliferation of large-scale renewable energy infrastructures (Baird and Barney 2017; Rignall 2016; Siamanta 2019). The enclosure of marginal lands for large-scale renewable energy installations represents historically specific local institutional responses to global socioecological crises. Yet the cordoning off of rural lands to generate energy has exacerbated land and resource conflicts for proximate communities. According to the IPCC (2022, 55),

> Some mitigation options can increase competition for scarce resources including land, water and biomass. Consequently, these can also reduce adaptive capacity, especially if deployed at larger scale and with high expansion rates thus exacerbating existing risks in particular where land and water resources are very limited.

For example, Dunlap and Arce (2021) examine Électricité de France's green grabbing for wind turbines – 'wind factories' – on indigenous held land in Gunaa Sicarú, Oaxaca, Mexico. The firm engaged in various counterinsurgency measures against land defenders whose lives are imperiled, livelihoods are disrupted and lands are dispossessed and degraded by infrastructural colonization (see also Dunlap 2018a). Renewable energy transitions often manifest inequitable land politics, disproportionately impacting marginalized, indigenous and racialized populations.

Utility-scale solar photovoltaic systems are hailed as effective and efficient infrastructures to rapidly mitigate climate change. This is especially true in India, where solar parks are being developed at breakneck pace on marginal public lands to the detriment of smallholding peasants. Altered property relations of new energy regimes disproportionately impact resource-dependent populations and produce new subjectivities. However, it remains unclear how land technologies (global and local) for large-scale renewable energy infrastructures articulate racial subjectivities and the stakes for survival in the Anthropocene.

The solar park form effectively functions as a rationally-ordered, hierarchical, efficient, ideal and imperative social system that is implemented as a solution to the climate crisis. Given these characteristics and an unbridled potential to accumulate private and globalized 'green' capital, I argue that the solar park can be conceived of as an energy plantation. This paper seeks to put plantation studies in conversation with the just transition to conceptualize the power relations of solar power through the *Plantationocene*. Heavily interrogating case studies in India, I demonstrate how solar parks can be situated within the larger genealogy of the plantation, with similar production relations that are repackaged as the charismatic, technologically 'modern,' planetary savior of the Plantationocene epoch. Building from Borras et al.'s (2022) expansive review of climate change and agrarian studies, this paper is a partial response to their provocation to research the narratives and strategies that frame institutionalized responses to climate change and the subsequent production of social differentiation and new exclusions within agrarian milieu. This study emerges from ongoing fieldwork (beginning in 2018) at the Gujarat Solar Park (Gujarat), the Kurnool Solar Park (Andhra Pradesh) and the

Fatehgarh Solar Park (Rajasthan) in India and utilizes data collected using a mixed methods approach that included semi-structured interviews, discourse analysis and participant observation.

This paper contains five additional sections. In the proceeding section, I review literature from critical agrarian studies and political ecology that examines the socioecological impacts of solar parks, specifically focusing on how green grabbing (re)produces social power asymmetries and social differentiation. I then move to interrogate the concept of wastelands, the racialized and contested spaces upon which most solar parks are developed. I assert that dispossessive solar parks built on wastelands constitute racial regimes of ownership, due to their neocolonial production relations and racialization of marginalized peasants. Thereafter, I examine plantation-based economic systems and climate-responsive agrarian transformations in India to make the case that solar parks can be conceived of as energy plantations, given their racial and gendered relations of production. I then situate solar energy plantations in the geological epoch of the Plantationocene, briefly recounting vibrant debates on the Anthropocene concept and its incarnations. Solar parks can be considered postcolonial energy plantations because they are exploitative and racializing social systems that are discursively articulated as imperative and ideal solutions to the climate crisis. I conclude this article by identifying apertures for agrarian climate justice wherein political contestation can manifest alternative land politics and implement equitable alternatives to dispossessive and racializing solar plantations.

Alienation by illumination: land and energy dispossessions for solar power

A solar park is defined as 'a concentrated zone of development of solar power generation projects and provides developers an area that is well characterized, with proper infrastructure and access to amenities and where the risk of the projects can be minimized' (MNRE 2016, 2). Institutionally configured like Special Economic Zones brokered by the state, solar parks are conduits for globalized and financial capital from domestic and multinational firms (see Siamanta 2017). The state operationalizes the investability of land (Fairbairn 2020; Le Billon and Sommerville 2017; Li 2014) – ensuring minimal financial risk for investors (Kennedy and Stock 2021) – in part by blocking and rolling back land redistribution schemes and promoting market-friendly land policies (Franco and Borras 2021). The solar park concept has received widespread acclaim from multilateral and intergovernmental agencies for their ability to generate immense amounts of electricity and capital.

Yet the accumulation of capital via the rapid expansion of utility-scale solar in rural areas of India has upended the agrarian political economy and spawned agrarian transformation at the local-scale, characterized by vast landscapes of rectilinear energy infrastructures upon erstwhile fecund farms (Figure 1). India's many solar parks have an established pattern of negatively impacting the lives of peasants through the enclosure of land and resources, exclusionary planning processes, encroachment into the environment and the entrenchment of social vulnerabilities (Sovacool 2021). The Indian state utilized eminent domain from a colonial-era law to dispossess a marginalized caste group of productive farmlands to build a massive solar park in Gujarat (Yenneti et al. 2016; Yenneti and Day 2016), in addition to numerous procedural injustices such as withholding

Figure 1. Solar park development has dispossessed peasants of land and livelihoods (Photo: Ryan Stock).

information from residents and denying them opportunities to engage in decision-making processes (Yenneti and Day 2015). In Assam, a private firm dispossessed local indigenous peasants of their lands for solar, which also happen to be an ecologically vital elephant corridor (Delhi Solidarity Group 2021). Indigenous groups in Kerala also lost their lands and livelihoods for solar development due to a lack of proper land titles (Bedi 2019). Land dispossessions for solar park development have left local residents without jobs and without any measurable improvement in electrification (Stock and Birkenholtz 2021), while exacerbating water scarcity for dryland farmers (Stock 2021a). The enclosure of marginal public lands for solar parks has also foreclosed poor and lower caste women's access to firewood and grazing lands (Stock and Birkenholtz 2020). To quell women's dissent and discontentment over dispossession and livelihood dislocations, a solar park developer in Gujarat initiated a corporate social responsibility scheme designed for the 'upliftment of Women & Child.' However, the scheme was only offered to women of the dominant caste group and of middle-to-high economic class (relative to the village), thereby reproducing social power asymmetries and marginality for lower caste women (Stock 2021b). Enclosing marginal commons land and green grabbing smallholders' land for solar parks is a modern iteration of rural development practice that reproduces gender and caste disparities in access to land and resources in the postcolonial and post-carbon era.

As the *sine qua non* of techno-managerial solutions to the climate crisis, the solar park concept has been globalized outside of India. Forged in the margins of COP 21 through a partnership between India and France, the International Solar Alliance (ISA) seeks to assist member countries realize their decarbonization commitments through the implementation of solar parks between the Tropic of Cancer and the Tropic of Capricorn.

> Development of large-scale solar generation capacity in various regions and development of bi-lateral, regional and inter-regional transmission inter-connections has the potential to eventually lead to global inter-connection of solar energy resources and solar energy transfer from one part of the world to other. (International Solar Alliance 2022)

The sunrise of India as a geopolitical energy *force majeure* has culminated with the nation leading the ISA, galvanizing global finance from the global North for reproducing dispossessive solar park development throughout the global South. At the recent COP 26 in Glasgow, the ISA forged partnerships with philanthropic organizations to mobilize $1 trillion in solar financing by 2030, the thrust of which will be developing solar parks in the global South (Bloomberg 2021). The global proliferation of solar parks has transformed production relations within agrarian milieu, especially on lands deemed unproductive and under-capitalized in the market economy.

Wastelands and racial regimes of ownership

The institutional juggernaut of the ISA holds immense power in reshaping land regimes in rural areas, targeted by project developers because of vast and contiguous swathes of land and more fungible legal categorizations. For example, marginal public lands – 'wastelands' – are earmarked for solar parks in India. Stipulated in the Ministry of Natural Resources and Environment's *Guidelines for Development of Solar Park:* 'In order to provide for such a large tract of contiguous land with appropriate insolation levels, the State Government may prioritize the use of government waste/non-agricultural land in order to speed up the acquisition process' (MNRE 2016, 5). The term 'wastelands' is an official bureaucratic categorization of marginal lands or degraded lands that are owned by the government or are commons (ICAR 2010), a political construction that obscures the socioeconomic importance and agroecological productivity of these lands for different users (Borras et al. 2010; Jodha 1990). Wastelands function as a politically useful simplification by maintaining intentionally malleable and subjective bureaucratic definitions, often connotating under-capitalized or aesthetically messy lands that lack 'statistical picturing' (Baka 2013; Li 2014; Nalepa and Bauer 2012). In the colonial era, wastelands were lands that did not generate revenue for the British crown (Gidwani 1992; Whitehead 2010). The British Raj seized forest commons from peasants and implemented new property regimes and taxes, exerting a monopoly for export-led commercial timber extraction (Guha 1983). Such dispossessions of forested revenue wastelands were justified through developmentalist discourses as being for the 'common good' (Sivaramakrishnan 2000), despite juridically blocking highly gendered use values. The postcolonial government has maintained continuity in the bureaucratic classification of these spaces (da Corta and Venkateshwarlu 1999). Revenue wastelands gave way to techno-managerial concerns of these spaces after the Green Revolution, focusing on soil degradation and water scarcity (Ariza-Montobbio et al. 2010), and are considered prime locations for globalized investments in new industrial or commercial developments in the neoliberal era (Li 2014).

I assert that differentiating property regimes of energy and agriculture through wasteland technologies for large-scale solar development in India constitute *racial regimes of ownership* (Bhandar 2018), often manifested through caste – a social construct 'inextricably bound with capitalism's dehumanizing impulse' that differentially racializes property and personhood (Cháirez-Garza et al. 2022, 202). Ranganathan's brilliant political-ecological inquiry of environmental unfreedoms in Bengaluru contains this useful definition of caste: 'In a nutshell, modern caste is a form of institutionalized and racialized oppression that inflects hierarchies in labour, land, and capital' (Ranganathan 2022, 6). In the Indian context, '… racialization functions through casteist socio-spatial logics to further racial capitalism' (Cháirez-Garza et al. 2022, 202). The Indian state uses wasteland classifications to maintain neocolonial production relations in agrarian spaces by coercively altering land tenure and differentiating property rights through a gendered racialization of marginalized caste and indigenous groups. Wasteland categorizations represent signifiers for racial politics of property, naturalizing the economy of repair and the improvement of local populations. The British Raj would seize wasteland spaces from peasants, rationalized via the racialization of resource-dependent 'backwards' populations that 'misuse' the land (Gidwani 1992; Baka 2013). The distribution of conditional land titles for wastelands to marginalized groups has a long history in the (post)colonial government, as far back as the colonial-era Madras Presidency giving conditional titles to Dalits (Basu 2008). State-distributed conditional land titles function to subalternize poor citizens (often determined by caste) by differentiating their property rights (Jonnalagadda, Stock, and Misquitta 2021). The Green Revolution drove a redistribution of wastelands in the name of agricultural production, dispossessing commons land fit for foraging and grazing and erasing resource-dependent women's use values (Shiva 2016). The colossal hydroelectric Sardar Sarovar Dam was wrought from the dispossession of 'mismanaged' tribal forests and the internal displacement of the Bhil community (Baviskar 2004), disproportionately affecting women and female-headed households (Mehta 2009).

In the context of solar park development, the conversion of public 'wastelands' – racialized landscapes of often marginal soils tended by marginalized peasants but normatively deemed unproductive for agriculture by elites – has been coupled with the green grabbing of commons (farmed or forested) and private farmlands of smallholders to manifest territorially expansive and extractive sites of 'clean' energy production. The spectacle from above resembles a veritable sea of solar arrays (Figure 2). Large-scale solar parks are charismatic and spectacular examples of global land grabbing of wastelands and smallholder lands for climate interventions (Franco and Borras 2019). There are numerous examples of how states collude with private firms to alter land tenure arrangements, manifesting racial regimes of land ownership in the solar energy transition. The state of Morocco developed a large concentrated solar plant near Ouarzazate utilizing colonial property laws that effectively devalued collectively-owned land of a marginalized ethnic community, allowing the state to acquire the lands despite local resistance (Rignall 2016; Cantoni and Rignall 2019). Wasteland enclosures for solar development in India have dispossessed lower caste women of access to biomass in Gujarat, burdening them to travel further distances to procure firewood (Stock and Birkenholtz 2020). The state of Andhra Pradesh dispossessed Dalits and lower caste peasants of their conditional land titles for a solar park, reconfiguring state-citizen relations (Jonnalagadda, Stock, and Misquitta 2021). The states of Rajasthan and Kerala have also moved to reclassify customarily tenured land as wastelands for the purposes of dispossession for a solar park, disproportionately impacting indigenous groups and marginalized agropastoralists (Chari 2020; Bedi 2019). Conveniently, the central government's Ministry of New and Renewable Energy (MNRE) has bureaucratically erased victims of solar green grabbing. According to clause 1.3 of Annexure-V of

Figure 2. Transformation of wastelands for a solar plantation (Photo: Ryan Stock).

the solar park development guidelines, 'In case waste land is acquired for setting up solar parks, there is no Project Affected Persons (PAPs) or Person Affected Families (PAFs) and hence there is no requirement of Rehabilitation and Resettlement (R&R)' (MNRE 2016, 41). The conversion of 'inefficiently-utilized' and under-capitalized wastelands inhabited by 'backward' rural populations for capital-intensive and efficient 'clean' energy generation is expertly managed by a 'smart' labor force (Stock 2021c), whose infrastructures largely supply energy to urban and industrial centers. Solar parks represent a discursively ideal reconfiguration of socioecological relations within a racial regime of land ownership, an energized iteration of plantation logics in the Anthropocene.

Plantation-based economic systems

Plantation economies are centered upon large-scale export-oriented agriculture dependent upon forced labor, the severing of peasant moral economies and the negation of local use values (Mintz 1985; Stoler 1985; Williams 1944). The colonial form of the plantation remains stubbornly persistent in the neoliberal era (Kenney-Lazar and Ishikawa 2019; Li 2018; Wolford 2021a), embodied in large-scale industrial agriculture and extractive industries that exert racial violence and placelessness for marginalized populations (Carney 2021; McKittrick 2011; Wolford 2021b). Plantation-based economic systems are nature-society relations predicated on the devaluation of labor (through racialization and exploitation) and ecological degradation.

> Plantation systems are a method of colonization that imposes upon social landscapes a distinct regime of political, economic, and ethnic regulation. Central to these regimes is their monopolization ethic: the total elimination, marginalization, or exile of indigenous people and small landowners. (Woods 2017, 41)

I situate my critique of India's solar parks as yet another iteration in the genealogy of dispossessive agrarian transformations, akin to an energy plantation.

To the detriment of smallholding peasants of colonial India growing subsistence crops, plantation[1] agriculture reconfigured agrarian systems for industrial and export production, implemented by the British Raj. As opposed to Atlantic plantation systems that depended on slave labor, Asian plantation systems largely depended on unfree indentured laborers (Behal and Mohapatra 1992). The indentured system of labor was eventually supplanted by a system of free wage labor, though working conditions and production relations remained almost identical (Stoler 1985). During the colonial era, the major export crops grown were indigo, tea, coffee and rubber. Beginning in the 1790s with the East India Company, large plantations cultivating indigo were concentrated in Bihar and deltaic Bengal awash with foreign capital. Indigo plantations in the region were dominated by English planters who had *zamindari* rights with near complete control over the land and exercised coercion to extract labor from peasants on the indigo plantations until they were no longer profitable in the 1920s and ceased to define Bengali commercial agriculture (Gupta 1992). Tea plantations were concentrated in West Bengal and Assam. For the ruling British, tea plantations represented spaces of bucolic nature and industrial production. Plantation labor was heavily exploited and regarded as disposable (Chatterjee 2001). To date, the colonial legacy of tea plantations remains in Darjeeling, where imperial terror is recast as heritage *terroir* and the relations of production devalue Nepali laborers. Plantation capitalists discursively articulate a romanticized and sanitized production process that harkens back to colonial rule, thereby obscuring the persistent violence of plantation production (Besky 2014). Colonial plantations, demonstrated by India's *longue durée* of (post)colonial production relations, necessitate a racialized labor force.

State-financed industrial agriculture was coercively introduced in the global South with the help of philanthropic organizations, discursively articulated as imperative solutions to food crises (Nally and Taylor 2015). The so-called Green Revolution wrought havoc on ecosystems and peasants (Shiva 2016), a biopolitical calculus that further solidified the stratification of agrarian classes and the racialization of farm labor (Patel 2013; Hetherington 2020). Ostensibly for the 'common good,' the Green Revolution was actually a geopolitically imperial project structured through a 'racial hierarchy that equated whiteness with scientific superiority and indigeneity with underdevelopment' (Eddens 2019). Neoliberal food production systems represent a continuity in Green Revolution logics (McMichael 2009; Patel 2013), accelerating and exacerbating the alienation of land and labor for corporate consolidation (McMichael 2012). Articulated through Malthusian discourse (Ojeda et al. 2020), public and private institutions engage in techno-managerial approaches to crises by deploying capital-intensive technologies and infrastructures under the auspice of climate action (Nightingale et al. 2019). Ensuring food security against ecological

[1]Ranajit Gupta (1992, 173) defines the colonial plantation system in India as capitalist enterprises with the following characteristics: 'an agro-industrial enterprise raising one or several crops on a large scale under tropical or semi-tropical climatic conditions; an international- market orientation; the launching and subsequent maintenance of plantations under the ownership and control of foreign capital with the backing of the colonial state; the employment of a large number of producers and labourers (not necessarily wage workers) doing hard manual work under conditions of a primitive labour process; the use of a migrant and/or immigrant labour system; and the mobilisation and control of direct producers through economic and extra-economic coercive methods with the direct and indirect support of the colonial state.'

crises remains at the forefront of climate science and development practice (Mbow et al. 2019). Sustaining the world food order animates recent approaches to improve production yields as the world burns. Criticized as 'old wine, new bottles' (Cavanagh et al. 2017), development agencies promote climate-smart agriculture as the ideal response to food production in the Anthropocene, imperative to ensure food security, climate mitigation and adaptation for peasants of the global South (Gardezi et al. 2022). Yet policy agendas underpinning climate-smart agriculture obscure inequalities and injustices (Shaw and Wilson 2020; Taylor 2018), thereby entrenching social power relations and impeding other pro-poor interventions and modalities of production (Karlsson et al. 2018). Climate-smart agriculture also provides legitimacy to a global land rush and the erasure of smallholding agriculture so long as it mitigates climate change (Borras and Franco 2018; Franco and Borras 2019). Precision agriculture technologies designed to recommend and implement solutions to climate-induced food insecurities and production constraints dispossesses farmers of their data while reconfiguring labor geographies (Fraser 2019; Stock and Gardezi 2021). 'Smart' agricultural solutions to climate change sustain plantation social relations to accumulate globalized and financial capital. Baka (2013, 2017) documents the itinerant wasteland logics that are used for accumulation by dispossession in establishing climate-responsive plantations. The Indian state acquired and dispossessed marginal farmers' land in Tamil Nadu for large-scale Jatropha plantations to produce biofuels by offering them substantially reduced land payments, leveraging bureaucratic processes around land documents to justify expropriation. Land enclosures for Jatropha plantations, along with the attendant obstruction of access to other biomass crops necessary for household reproduction, have reproduced the social vulnerability of smallholding and marginalized farmers in the region (Ariza-Montobbio et al. 2010). These violent, racialized, rationally-ordered plantation logics also animate global renewable energy development, characterized by the proliferation of solar plantations.

Powering the Plantationocene: postcolonial solar energy plantations

Human impacts on planetary systems have been undeniably disruptive. Beyond the climate emergency, human impacts can be measured and material artifacts can be found in geological formations of the Earth. Scientists in the field of stratigraphy are debating as to whether the profoundly transformative human impacts deserve a new geological epoch, separating this moment from the Holocene epoch that began roughly 11,700 years ago in the Neolithic Revolution when humans began domesticating plants and animals. Some have proposed this new geologic epoch to be called the *Anthropocene* (Crutzen and Stoermer 2000; Castree 2014a), and there remains much debate as to when this new geologic epoch began (e.g. Industrial Revolution, Atomic Age) (Zalasiewicz et al. 2021). However, this term has sustained much recent criticism from scholars that rightly assert that human impacts on the planetary system and their consequences for human populations are uneven. We cannot equitably ascribe blame to an undifferentiated *Anthropos*, given the historical and present role of the global North in causing global environmental crises at the expense of the disproportionately vulnerable global South (Castree 2014b). Jason Moore proposes the concept of the *Capitalocene* as an alternative to the Anthropocene to center capitalism's relations of production as responsible for planetary disruptions (Moore 2017). The world-system of markets in the 'age of capital' appropriates the unpaid work/energy of human and non-human labor, and renders nature legible to state-capital-science complexes for management and profit (Moore 2018). And yet extractive and expansive production relations existed before capitalism (Haraway et al. 2016). Centering capital alone obscures the transformative nature-society relations of settler colonialism and chattel slavery, and the racialization of dispossessed, exploited and exterminated populations (Lewis and Maslin 2018; Yusoff 2018).

Accurately accounting for this 'topography of difference' necessitates a critical examination of the plantation form and a centering of racial politics (Haraway et al. 2016; Tsing et al. 2019; Davis et al. 2019; Jegathesan 2021), although vital contributions have been made by scholars racializing the Anthropocene and Capitalocene concepts (Vergès 2017; Whyte 2018; Saldanha 2020; Eichen 2020). Delineating this geological epoch by the plantation form has been called the *Plantationocene*.[2] Reprising Wendy Wolford's (2021b) generative extension of these conversations, 'Plantations are inherently power-laden social structures found in every modern economic system. They embody both racial violence and resistance, straddling or bridging the divide between rural and urban, agriculture and industry, town and country, and local and global' (3). As discussed in the above section with evidence from India, plantation systems are idealized energy and labor-intensive transformations of nature and culture, necropolitically organized for efficient production and governance. Extractive agro-industrial landscapes that characterize the Plantationocene may also involve the alienation of productive land (wastelands or otherwise) and conversion into energy plantations. The plantation form shapes energy regimes (Harrison and Popke 2018), which transforms nature and labor into energy and constructs hierarchical ontologies of 'mattering' for more-than-human populations (Lennon, 2017). Following Wolford's (2021b) characterization of the plantation form as a set of social relations, an imperative and an ideal, I argue that solar parks are energy plantations – harvesting photons to generate photovoltaic electricity – that constitute racial regimes of ownership in the Plantationocene.

Solar parks as a social system

Renewable energy infrastructures are efficient means of generating energy, imperative to mitigating the climate crisis. Solar photovoltaic (PV) infrastructures are being deployed rapidly as part of a worldwide energy transition away from fossil fuels. Large-scale solar PV systems are being rolled out in rural areas, upending agrarian political economies, altering property regimes and reconfiguring labor relations. As discussed previously, peasants who live in villages adjoining wastelands where a solar park is being developed often experience a dispossession of land, energy and water resources. Marginal lands are enclosed for the massive infrastructure, often severing access to biomass and grazing lands for resource-dependent households. A woman from the marginalized Dalit caste of Meghwal had this to say about changes in gendered access to resources at the Fatehgarh Solar Park in Rajasthan:

---

[2]Haraway (2015, 162) defines *Plantationocene* as 'the devastating transformation of diverse kinds of human-tended farms, pastures, and forests into extractive and enclosed plantations, relying on slave labor and other forms of exploited, alienated, and usually spatially transported labor.'

> Women are not allowed to go there. Before the solar park, they were associated with fields, cattle grazing and firewood fetching. Now nothing remains for them except the household. Due to the solar park, deforestation has been done on a larger scale. (F07, 3 January 2022)

As public and private lands become the ground upon which solar arrays are mounted, available productive farmland is reduced which devastates the livelihoods of landholders, laborers and tenant farmers. The change in property relations leads to a proletarianization of much of the peasantry, who have little choice but to seek alternate employment. But due to the highly technical nature of an operating solar park, the solar park does not hire locals who do not have certified skill-sets in electrical engineering. As such, the peasantry suffers a partial proletarianization from solar green grabbing, rendered surplus labor to the solar economy (Li 2011). The solar energy transition highlights the need to critically engage Kautsky's (1988) classic agrarian question, specifically the agrarian question of labor (Bernstein 2004; Akram-Lodhi and Kay 2010). Dispossessed peasants, disproportionately marginalized caste and gender groups, experience a disruption of livelihoods and culture within their communities.

Solar energy plantations establish a new social system predicated on the efficient generation of electricity through large-scale extractive infrastructure within an overdetermined agrarian milieu, exacerbating caste-mediated 'land wars' in the neoliberal era (Nielsen et al. 2020; Levien 2013). Caste structures India's new solar social system, intersecting with other axes of social difference (e.g. gender, religion, class), to differentiate the value of bodies and space for 'green' capitalist accumulation through energy production. State-led solar development attracts domestic and multinational firms to invest in and establish smaller solar arrays within the confines of the larger solar park, enacting a capital switch from the circuit of finance capital to the circuit of fixed capital to resolve crises of overaccumulation of capital and underproduction of nature (Castree and Christophers 2015; Kennedy and Stock 2021). Globalized solar capital circulates through local solar infrastructures, facilitating the generation of electricity that is often routed away from adjacent villages to power regionally proximate urban and/or industrial centers. The new regional social system resembles the solar plantation system itself – a bureaucratically hierarchical regime with a highly racialized labor force and inequitable changes in land tenure arrangements; rationally ordered landscapes that efficiently generate energy resources that supply urban centers with precious electricity at the expense of energy impoverished rural spaces. Although solar electricity and capital flows currently deviate from the traditional agro-export model of plantations insofar as the postcolonial state finances and shepherds private finance for renewable electricity which is consumed domestically, energy and resource access is structured by subnational core-periphery relations.

Solar parks as an imperative

India is striving to achieve 100 gigawatts energy capacity from utility-scale solar plants by the end of 2022 as part of their Nationally Determined Contributions to the Paris Agreement and 500 gigawatts renewable energy capacity by 2030 for commitments made at COP 26 in 2021. India also declared that it will achieve 50% non-fossil-based electricity by 2030, striving for carbon neutrality by 2070. India's ambitious energy transition to meet climate goals is heavily reliant upon the rapid development of large-scale renewable energy infrastructures. So-called 'ultra-mega' solar parks (with ≥500 megawatts capacity) are one of India's primary strategies to attain net-zero. India's solar park concept has gone global, with the International Solar Alliance developing solar parks throughout the global South as an efficient mitigation strategy. The climate crisis has opened new commodity frontiers in agrarian spaces globally. In this sense, the energy itself and that which it is not (i.e. carbon) produce value. Finance capital from state and global firms is invested to convert wastelands into highly efficient energy plantations – useful landscape simplifications for market control and centralized governance within a renewable energy transition. During his launch of India's National Action Plan on Climate Change in 2008, former Indian Prime Minister Manmohan Singh triumphantly declared: 'We will pool our scientific, technical and managerial talents, with sufficient financial resources, to develop solar energy as a source of abundant energy to power our economy and to transform the lives of our people' (MNRE 2016). Climate politics is used to legitimate land grabbing for solar parks (Franco and Borras 2019). The alienation of local land and livelihoods is regarded as acceptable sacrifices to confront global environmental change.

Land enclosures for the utility-scale generation of solar electricity is deemed necessary to mitigate climate change and generate much-needed electricity to combat energy poverty, even by many marginalized and dispossessed peasants. 'The solar park is good for the country but not for us. Our goats cannot go far for grazing. We cannot gather firewood' (FG-F01, 3 January 2022). This dovetails from the (post)colonial state's moralistic and developmentalist push for rural electrification, an imperative for modernization. British and Indian commercial interests, combined with state-led industrialization, drove the colonial state's development of electricity infrastructure. However, India's colonial states approached electrification differently, resulting in uneven development of electricity infrastructure, inequitable access and heterogeneous energy policies. The post-independence era is one of energy federalism (Sareen and Kale 2018), although the central government has sought to standardize electricity policies throughout the nation and has resumed its focus on rural electrification through extensive grid expansions (Kale 2014). The postcolonial state saw rural electrification as a moral project and electricity provisioning was framed as a national good and an entitlement (Chatterjee 2020), yet inequitable access and heterogeneous policies remain (Sareen and Kale 2018). Beginning in 2015, the central government sought to electrify all villages within a span of 1000 days through the Deen Dayal Upadhyaya Gram Jyoti Yojana (DDUGJY) scheme. This goal was ostensibly achieved in 2018, despite 'electrification' being defined as a mere 10% of households having access to electricity. The 2017 Pradhan Mantri Sahaj Bijli Har Ghar Yojana (Saubhagya) scheme was launched in conjunction with DDUGJY to provide free electricity connections to every household in India. Ensuring the energy security for all citizens has become a central development concern for the state. Ironically, rural electrification (a moral and developmental imperative) has been adroitly wielded as a governmental technology of both fossil capitalism and 'green' neoliberalism (Chatterjee 2019). The rapid expansion of solar parks aids in these efforts and are imperative to powering a more sustainable world-system.

Solar parks as an ideal

The solar park form is discursively articulated as the ideal techno-managerial approach to mitigate climate change, with promise to deliver the 'triple bottom line' of ecological protection, economic development and social equity. Even the words we use to characterize large-scale solar infrastructures – solar *parks*, solar *farms*, solar *plants* – serve to naturalize these technologies as ecologically benign, despite the negative impacts they may have on local ecosystems and their inhabitants (Mulvaney 2019; Tanner et al. 2021). Solar energy plantation imaginaries generalize space and place as universally fit for these technological (and profitable) interventions to historically-specific and contingent ecological crises, articulated through postcolonial developmentalist discourses of the global public good to 'save the planet.' Further, they abstract highly complex landscapes and social systems into simplifications that render wastelands and racialized laborers legible to capital. Organizing a post-carbon energy landscape through solar parks in agrarian settings is rationalized through moralistic logics of sustainable ecological modernization and economic development. Smallholding agrarian production systems are framed as ecologically destructive and economically inefficient (Borras and Franco 2018), delegitimizing local climate adaptive and mitigative production systems (Borras et al. 2020). The state and solar developers articulate colonial wasteland discourses to dispossess smallholding and marginalized farmers of their land, under the auspices of modernizing the 'traditional' agrarian economy for energy security and for the remediation of global environmental crises (Cantoni and Rignall 2019; Rignall 2016; Singh 2022; see also Baka 2013, 2017; Franco and Borras 2019). Yet the impetus to technologically repair 'wasted' environments is fraught with racializations that characterize local and marginalized populations as 'backward' people who 'misuse' their lands (Stock 2021d). For instance, the general manager of the project developer for the Gujarat Solar Park confidently asserted: 'It is barren land. That is a backwards area with unusable land. There was no development in Charanka' (GPCL.004, 4 March 2018). When circulated, colonial wasteland discourses function as a politics of knowledge that obscure social power asymmetries and injustices on the ground.

Making value of 'wasted' lands has not brought economic development to solar regions at the local scale. When he inaugurated the Gujarat Solar Park in April 2012, Narendra Modi promised that the project would bring '30,000 jobs including in the manufacturing of solar panels' (Kaur 2022). Modi and project developers also promised other entitlements to peasants, including free electricity, a hospital, ambulance services, drinking water, a high school, paved roads, solar streetlights, a skills development center, a teacher training program and a Sun Temple (Balan 2014; Kaur 2022; MNRE 2014). Project developers of the Kurnool Solar Park also promised thousands of jobs, solar streetlights and functioning drinking water stations (Stock 2022). To date, none of these have been implemented. To the contrary, new solar economies have hollowed out the agrarian political economy and left many peasants landless and jobless, exacerbating their economic precarity (Stock and Birkenholtz 2021). The survival concerns of this Meghwal Dalit man from Fatehgarh demand an urgent response: 'If we lose our livelihoods by the solar park, then what will we do? How will we fill our bellies?' Within plantation logics, solar parks are social systems – constituting racial regimes of land ownership – that are framed as imperative and ideal solutions to economic and ecological crises. Yet the sun

Figure 3. The sun setting on a solar-powered Plantationocene (Photo: Mridul Ganguly).

will set on solar plantations as decolonial and distributed solar interventions rise (Figure 3), illuminating new horizons of agrarian climate justice.

Transcending the energy transition to solar plantations

> For those of us interested in addressing race, space, and premature and preventable death, plantation futures demand decolonial thinking that is predicated on human life. (McKittrick 2013, 3)

Responding to the climate crisis demands swift structural and infrastructural transformations, including a transition away from fossil fuels and towards renewable energy. Large-scale solar infrastructures are proliferating throughout rural landscapes of the global South with assistance from development agencies (e.g. World Bank, Asian Development Bank, International Solar Alliance) to meet this challenge. Yet solar parks developed on rural wastelands often disrupt the agrarian political economy, altering property relations and reconfiguring labor geographies. Worse still, solar parks may reproduce the climate and social vulnerability of smallholding peasants. Alienated from land and livelihoods, solar-related dispossessions have produced a highly racialized and gendered surplus population of landless peasants who are not absorbed into the modern 'smart' economy, a precarious labor class that struggles to withstand another wave of rural development that leaves them behind (Stock 2021c). Adding insult to injury, solar energy generated from the periphery powers the adjacent core urban centers and industry, leaving the dispossessed in the dark.

India's solar parks are an archetype of an idealized and racialized reordering of nature, economy and society to power a more sustainable world-system, a set of colonial social relations akin to an energy plantation that harvests photons to generate photovoltaic electricity. Echoing Clyde Woods, the plantation is not '… a dead, yet still romanticized, aberration killed off by the inevitable march of human progress' (Woods 2017, 4). Plantation relations are resilient, haunting the colonial present as landed interventions to confront the climate crisis. Not only do colonial legacies persist in the era of climate change, they may shape social and climate vulnerabilities as well as structure institutional responses. Solar plantations developed as part of global renewable energy transitions reinstate neocolonial production relations in a climate-altered and carbon-constrained agrarian milieu, which may (re)produce the political economic causes of climate vulnerability and their differentiated and uneven outcomes (Ribot 2014; Sovacool et al. 2019; Sovacool 2021; Thomas et al. 2019). After all, 'vulnerability does not just fall from the sky' (Ribot 2013). Yet solar plantations structure vulnerability from grabbed land, livelihood dislocations and infrastructural violence. Perhaps even more nefarious, massive colonial-capitalist mitigation interventions like solar parks represent a morally-sanctioned continuity of production relations, land politics and social differentiation wrought by petro-racial capitalism as an imperative and ideal solution to the climate crisis. In essence, 'the "solution" is now the "problem"' (Dunlap 2018b).

Such climate coloniality weighs heavy on vulnerable populations suffering in 'sustainable' sacrifice zones (Sultana 2022a). Yet peasants of solar plantations are not passive victims; they demonstrate agency against dispossessive renewable energy infrastructures through various modalities of political resistance from below, ranging from protests and blockades, lawsuits, discursive contestation and embodied and affective resistance (Rignall 2016; Stock and Birkenholtz 2020; Stock, 2022). Political struggle is essential to developing the class consciousness necessary for dispossessed peasants to transform the production relations of solar plantations (Gahman 2020; Pye 2021; Stock 2021c). Solar plantation futures can be dismantled through radical place-making and coalitional land-based politics (McKittrick 2013; Heynen and Ybarra 2021), ushering in decolonized postcarbon futures (Lennon 2020). Scholars and activists alike must foster an intersectional and transnational solidarity praxis of transformative climate justice (Newell et al. 2021) – rooted in 'systemic solutions that account for and financially support care, commoning, repair, and reparation' (Sultana 2022b, 122) – that redresses the gendered and racial injustices of solar plantations (see Lennon 2017). We must confront the stubborn persistence of asymmetrical power relations within market-based techno-managerial fixes to the climate crisis (Nightingale et al. 2019), lest ostensible solutions create a climate apartheid that locks in the suffering of marginalized populations (Rice, Long, and Levenda 2021).

The 'rapidly closing window to secure a liveable future' has been wedged ajar by recent scholarship focused on conceptualizing a truly powerful 'cross-fertilisation' of agrarian justice and climate justice (Newell 2022). Borras and Franco (2018) identify the 5 R's of agrarian climate justice – *redistribution, recognition, restitution, regeneration* and *resistance* – as deep social reforms and responses that are indispensable for equitable changes to land politics in the context of the climate crisis. They further assert that the 5 R's must be sandwiched by a land size ceiling to limit land grabbing and a land size floor to guarantee access for everyone (Borras and Franco 2018). Achieving agrarian climate justice will necessitate a systemic change to land politics in the context of a deeply inequitable renewable energy transition (Franco and Borras 2021). Ergo, just transitions to solar energy must be landed. Alternatives to dispossessive large-scale solar infrastructures –

such as community-based solar cooperatives engaged in prosumption (Shah 2018; Siamanta 2021) – may be equally effective at mitigating the climate crisis while also serving as a form of energy reparations for the violence and deprivations of climate coloniality and racial capitalism (Luke and Heynen 2020). Although socioeconomic and demographic disparities may still exist in distributed generation (Reames 2020; Sovacool et al. 2022), small-scale and medium-scale solar PV initiatives tend to be more equitable in spaces that disproportionately suffer from energy poverty (Nordholm and Sareen 2021; Siamanta 2021). Agroecological practices – with the potential to transform the social relations of food production (van der Ploeg 2021) – can incorporate distributed agrivoltaic regimes on so-called 'wastelands' (Barron-Gafford et al. 2019). If rooted in intersectional and abolitionist praxis (Montenegro de Wit 2021), agroecological approaches to community-based solar agrivoltaics can foster both food and energy sovereignty of peasants dispossessed by solar plantations (Bezner Kerr et al. 2019), with the potential to forge transnational solidarities as social movements that scale (Khadse et al. 2018). Transcending the energy transition to solar plantations requires nothing less than place-based yet transnational, decolonial, intersectional and anti-racist peasant coalitions struggling to transform neocolonial land politics and implementing redistributive and emancipatory solar interventions.

Acknowledgements

Immense gratitude goes to Jun Borras, the JPS Editors and Editorial Collective for their interest in this research and for convening the Special Forum and conference on climate change and critical agrarian studies. I am honored for the opportunity to share this manuscript alongside the works of such brilliant scholars and activists. Thank you to the anonymous reviewers for generative comments and suggested edits. And to Wendy Wolford who inspired these ideas. I am eternally indebted to the intrepid Mridul Ganguly for fieldwork support. Much empowerment and solidarity to the folks in the Illume Lab, especially Isabelle Honkomp, Eli Williams and Joslin Brown. This article is dedicated to Shilpa, Shyan, Lakshay and Raysha.

Disclosure statement

No potential conflict of interest was reported by the authors.

ORCID

Ryan Stock http://orcid.org/0000-0002-6218-3725

References

Akram-Lodhi, A. H., and C. Kay. 2010. "Surveying the Agrarian Question (Part 2): Current Debates and Beyond." *Journal of Peasant Studies* 37 (2): 255–284.

Ariza-Montobbio, P., S. Lele, G. Kallis, and J. Martinez-Alier. 2010. "The Political Ecology of Jatropha Plantations for Biodiesel in Tamil Nadu, India." *The Journal of Peasant Studies* 37 (4): 875–897.

Baird, I. G., and K. Barney. 2017. "The Political Ecology of Cross-Sectoral Cumulative Impacts: Modern Landscapes, Large Hydropower Dams and Industrial Tree Plantations in Laos and Cambodia." *The Journal of Peasant Studies* 44 (4): 769–795.

Baka, J. 2013. "The Political Construction of Wasteland: Governmentality, Land Acquisition and Social Inequality in South India." *Development and Change* 44 (2): 409–428. doi:10.1111/dech.12018.

Baka, J. 2017. "Making Space for Energy: Wasteland Development, Enclosures, and Energy Dispossessions." *Antipode* 49 (4): 977–996.
Balan, P. 2014. "Skill Development Centre to Come Up at Charanka Solar Park." *Business Standard*. https://www.business-standard.com/article/economy-policy/skill-development-centre-to-come-up-at-charanka-solar-park-114101300850_1.html
Barron-Gafford, G. A., M. A. Pavao-Zuckerman, R. L. Minor, L. F. Sutter, I. Barnett-Moreno, D. T. Blackett, and J. E. Macknick. 2019. "Agrivoltaics Provide Mutual Benefits Across the Food–Energy–Water Nexus in Drylands." *Nature Sustainability* 2 (9): 848–855.
Basu, R. S. 2008. "'Rights Over Wastelands' and New Narratives of the Paraiyan Past (1860–1900)." *Studies in History* 24 (2): 265–293.
Baviskar, A. 2004. *In the Belly of the River: Tribal Conflicts Over Development in the Narmada Valley*. Oxford: Oxford University Press.
Bedi, H. P. 2019. "'Lead the District Into the Light': Solar Energy Infrastructure Injustices in Kerala, India." *Global Transitions* 1: 181–189.
Behal, R. P., and P. P. Mohapatra. 1992. "'Tea and Money Versus Human Life': The Rise and Fall of the Indenture System in the Assam Tea Plantations 1840–1908." *The Journal of Peasant Studies* 19 (3-4): 142–172.
Bernstein, H. 2004. "'Changing Before Our Very Eyes': Agrarian Questions and the Politics of Land in Capitalism Today." *Journal of Agrarian Change* 4 (1/2): 190–225.
Besky, S. 2014. "The Labor of Terroir and the Terroir of Labor: Geographical Indication and Darjeeling Tea Plantations." *Agriculture and Human Values* 31 (1): 83–96.
Bezner Kerr, R., C. Hickey, E. Lupafya, and L. Dakishoni. 2019. "Repairing Rifts or Reproducing Inequalities? Agroecology, Food Sovereignty, and Gender Justice in Malawi." *The Journal of Peasant Studies* 46 (7): 1499–1518.
Bhandar, B. 2018. *Colonial Lives of Property: Law, Land and Racial Regimes of Ownership*. Durham: Duke University Press.
Bloomberg. 2021. "Bloomberg Philanthropies Announces New Partnership With International Solar Alliance to Mobilize $1 Trillion in Solar Financing." *Bloomberg Philanthropies*. https://www.bloomberg.org/press/bloomberg-philanthropies-announces-new-partnership-with-international-solar-alliance-to-mobilize-1-trillion-in-solar-financing/
Borras, S. M., and J. C. Franco. 2013. Global Land Grabbing and Political Reactions 'From Below'. *Third World Quarterly*, 34 (9), 1723–1747.
Borras, S. M., and J. C. Franco. 2018. "The Challenge of Locating Land-Based Climate Change Mitigation and Adaptation Politics Within a Social Justice Perspective: Towards an Idea of Agrarian Climate Justice." *Third World Quarterly* 39 (7): 1308–1325.
Borras, S. M., J. C. Franco, and Z. Nam. 2020. "Climate Change and Land: Insights from Myanmar." *World Development* 129: 104864.
Borras, S. M., P. McMichael, and I. Scoones. 2010. "The Politics of Biofuels, Land, and Agrarian Change." *Journal of Peasant Studies* 37: 575–592.
Borras, S. M., I. Scoones, A. Baviskar, M. Edelman, N. L. Peluso, and W. Wolford. 2022. "Climate Change and Agrarian Struggles: An Invitation to Contribute to a JPS Forum." *The Journal of Peasant Studies* 49 (1): 1–28.
Borras., S. M., and J. C. Franco. 2012. "Global Land Grabbing and Trajectories of Agrarian Change: A Preliminary Analysis." *Journal of Agrarian Change* 12 (1): 34–59.
Cantoni, R., and K. Rignall. 2019. "Kingdom of the Sun: A Critical, Multiscalar Analysis of Morocco's Solar Energy Strategy." *Energy Research & Social Science* 51: 20–31.
Carney, J. A. 2021. "Subsistence in the Plantationocene: Dooryard Gardens, Agrobiodiversity, and the Subaltern Economies of Slavery." *The Journal of Peasant Studies* 48 (5): 1075–1099.
Castree, N. 2014a. "The Anthropocene and Geography I: The Back Story." *Geography Compass* 8 (7): 436–449.
Castree, N. 2014b. "Geography and the Anthropocene II: Current Contributions." *Geography Compass* 8 (7): 450–463.
Castree, N., and B. Christophers. 2015. "Banking Spatially on the Future: Capital Switching, Infrastructure, and the Ecological Fix." *Annals of the Association of American Geographers* 105 (2): 378–386. doi:10.1080/00045608.2014.985622.

Cavanagh, C. J., A. K. Chemarum, P. O. Vedeld, and J. G. Petursson. 2017. "Old Wine, New Bottles? Investigating the Differential Adoption of 'Climate-Smart' Agricultural Practices in Western Kenya." *Journal of Rural Studies* 56: 114–123.

Cháirez-Garza, J. F., M. D. Gergan, M. Ranganathan, and P. Vasudevan. 2022. "Introduction to the Special Issue: Rethinking Difference in India Through Racialization." *Ethnic and Racial Studies* 45 (2): 193–215.

Chari, M. 2020. "How Solar Farms Fuel Land Conflicts." *Mint.* https://www.livemint.com/news/india/how-solar-farms-fuel-land-conflicts-11600612526037.html

Chatterjee, P. 2001. *A Time for Tea: Women, Labor, and Post/Colonial Politics on an Indian Plantation.* Durham: Duke University Press.

Chatterjee, E. 2019. "A Climate of Scarcity: Electricity in India, 1899–2016." In *Scarcity in the Modern World: History, Politics, Society and Sustainability,* edited by J. Brewer, N. Fromer, F. A. Jonsson, and F. Trentmann, 1800–2075. Bloomsbury Academic.

Chatterjee, E. 2020. "The Asian Anthropocene: Electricity and Fossil Developmentalism." *The Journal of Asian Studies* 79 (1): 3–24.

Cotula, L. 2012. "The International Political Economy of the Global Land Rush: A Critical Appraisal of Trends, Scale, Geography and Drivers." *Journal of Peasant Studies* 39 (3–4): 649–680.

Crutzen, P. J., and E. F. Stoermer. 2000. "The Anthropocene." *Global Change Newsletter* 41: 17–18.

da Corta, L., and D. Venkateshwarlu. 1999. "Unfree Relations and the Feminisation of Agricultural Labour in Andhra Pradesh, 1970–95." *The Journal of Peasant Studies* 26 (2/3): 71–139.

Davis, J., A. A. Moulton, L. Van Sant, and B. Williams. 2019. "Anthropocene, Capitalocene … Plantationocene? A Manifesto for Ecological Justice in an Age of Global Crises". *Geography Compass* 13 (5): e12438.

Delhi Solidarity Group. 2021. *Anatomy of a Solar Land Grab.* https://delhisolidaritygroup.files.wordpress.com/2021/04/azure-power_mikir-bamuni-grant_ffc_report_compressed.pdf

Dell'Angelo, J., P. D'odorico, M. C. Rulli, and P. Marchand. 2017. "The Tragedy of the Grabbed Commons: Coercion and Dispossession in the Global Land Rush." *World Development* 92: 1–12.

Dunlap, A. 2018a. "Counterinsurgency for Wind Energy: The Bíi Hioxo Wind Park in Juchitán, Mexico." *The Journal of Peasant Studies* 45 (3): 630–652.

Dunlap, A. 2018b. "The 'Solution' is Now the 'Problem:' Wind Energy, Colonisation and the 'Genocide-Ecocide Nexus' in the Isthmus of Tehuantepec, Oaxaca." *The International Journal of Human Rights* 22 (4): 550–573.

Dunlap, A., and M. C. Arce. 2021. "'Murderous Energy' in Oaxaca, Mexico: Wind Factories, Territorial Struggle and Social Warfare." *The Journal of Peasant Studies*: 1–26.

Eddens, A. 2019. "White Science and Indigenous Maize: The Racial Logics of the Green Revolution." *The Journal of Peasant Studies* 46 (3): 653–673.

Eichen, J. R. 2020. "Cheapness and (Labor-) Power: The Role of Early Modern Brazilian Sugar Plantations in the Racializing Capitalocene." *Environment and Planning D: Society and Space* 38 (1): 35–52.

Fairbairn, M. 2020. *Fields of Gold: Financing the Global Land Rush.* New York: Cornell University Press.

Fairhead, J., M. Leach, and I. Scoones. 2012. "Green Grabbing: A New Appropriation of Nature?" *Journal of Peasant Studies* 39 (2): 237–261.

Franco, J. C., and S. M. Borras. 2019. "Grey Areas in Green Grabbing: Subtle and Indirect Interconnections Between Climate Change Politics and Land Grabs and Their Implications for Research." *Land Use Policy* 84: 192–199.

Franco, J. C., and S. M. Borras. 2021. "The Global Climate of Land Politics." *Globalizations* 18 (7): 1277–1297.

Fraser, A. 2019. "Land Grab/Data Grab: Precision Agriculture and its New Horizons." *The Journal of Peasant Studies* 46 (5): 893–912.

Gahman, L. 2020. "Contra Plantation, Prison, and Capitalist Annihilation: Collective Struggle, Social Reproduction, and the Co-Creation of Lifegiving Worlds." *The Journal of Peasant Studies* 47 (3): 503–524.

Gardezi, M., S. Michael, R. Stock, S. Vij, A. Ogunyiola, and I. Ishtiaque. 2022. "Prioritizing Climate-Smart Agriculture: An Organizational and Temporal Review." *WIRES Climate Change.* doi:10.1002/wcc.755.

Gidwani, V. 1992. "'Waste' and the Permanent Settlement in Bengal." *Economic and Political Weekly* 27 (4): PE39–PE46.

Goetz, A. 2021. "Land Grabs." In *Handbook of Critical Agrarian Studies*, edited by A. H. Akram-Lodhi, K. Dietz, B. Engels, and B. M. McKay, 346–356. Edward Elgar.

GRAIN. 2008. "Seized: The 2008 Landgrab for Food and Financial Security." GRAIN. https://grain.org/article/entries/93-seized-the-2008-landgrab-for-food-and-financial-security

Guha, R. 1983. "Forestry in British and Post-British India: A Historical Analysis." *Economic and Political Weekly* 18 (44): 1882–1896.

Gupta, R. D. 1992. "Plantation Labour in Colonial India." *The Journal of Peasant Studies* 19 (3-4): 173–198.

Haraway, D. 2015. "Anthropocene, Capitalocene, Plantationocene, Chthulucene: Making Kin." *Environmental Humanities* 6 (1): 159–165.

Haraway, D., N. Ishikawa, S. F. Gilbert, K. Olwig, A. L. Tsing, and N. Bubandt. 2016. "Anthropologists are Talking–About the Anthropocene." *Ethnos* 81 (3): 535–564.

Harrison, C., and J. Popke. 2018. "Reassembling Caribbean Energy? Petrocaribe, (Post-) Plantation Sovereignty, and Caribbean Energy Futures". *Journal of Latin American Geography* 17 (3): 212–234.

Hetherington, K. 2020. "Agribiopolitics: The Health of Plants and Humans in the Age of Monocrops." *Environment and Planning D: Society and Space* 38 (4): 682–698.

Heynen, N., and M. Ybarra. 2021. "On Abolition Ecologies and Making 'Freedom as a Place'." *Antipode* 53 (1): 21–35.

ICAR. 2010. *Degraded and Wastelands of India: Status and Spatial Distribution*. New Delhi: Indian Council for Agricultural Research.

International Solar Alliance. 2022. *Development of Large-Scale Solar Power Projects Under Solar Park Concept in Cluster/Group of ISA Member Countries*. International Solar Alliance. https://isolaralliance.org/work/solar-park

IPCC. 2022. *Summary for Policymakers. Climate Change 2022: Mitigation of Climate Change*. Intergovernmental Panel on Climate Change. https://report.ipcc.ch/ar6wg3/pdf/IPCC_AR6_WGIII_SummaryForPolicymakers.pdf

Jacobson, M. Z., M. A. Delucchi, Z. A. Bauer, S. C. Goodman, W. E. Chapman, M. A. Cameron, and A. S. Yachanin. 2017. "100% Clean and Renewable Wind, Water, and Sunlight All-Sector Energy Roadmaps for 139 Countries of the World." *Joule* 1 (1): 108–121.

Jegathesan, M. 2021. "Black Feminist Plots Before the Plantationocene and Anthropology's 'Regional Closets'." *Feminist Anthropology* 2 (1): 78–93.

Jodha, N. S. 1990. "Rural Common Property Resources: Contributions and Crisis." *Economic and Political Weekly* 25 (26): A65–A78.

Jonnalagadda, I., R. Stock, and K. Misquitta. 2021. "Titling as a Contested Process: Conditional Land Rights and Subaltern Citizenship in South India." *International Journal of Urban and Regional Research* 45 (3): 458–476. http://dx.doi.org/10.1111/ijur.v45.3.

Kale, S. 2014. "Structures of Power: Electrification in Colonial India. Comparative Studies of South Asia." *Africa and the Middle East* 34 (3): 454–475.

Karlsson, L., L. O. Naess, A. Nightingale, and J. Thompson. 2018. "'Triple Wins' or 'Triple Faults'? Analysing the Equity Implications of Policy Discourses on Climate-Smart Agriculture (CSA)." *Journal of Peasant Studies*. doi:10.1080/03066150.2017.1351433.

Kaur, R. 2022. "A Decade on, India's First Solar Park Has May Promises Left to Fulfil." *Mongabay India*. https://india.mongabay.com/2022/03/a-decade-on-indias-first-solar-park-has-many-promises-left-to-fulfil/

Kautsky, K. 1988. *The Agrarian Question*. London: Zwan Publications.

Kemp, C. C., and G. C. Szego. 1975. "The Energy Plantation." *Energy Sources* 2 (3): 263–274.

Kennedy, S., and R. Stock. 2021. "Alternative Energy Capital of the World? Fix, Risk and Solar Energy in Los Angeles' Urban Periphery." *Environment and Planning E: Nature and Space*, 1–22.

Kenney-Lazar, M., and N. Ishikawa. 2019. "Mega-Plantations in Southeast Asia: Landscapes of Displacement." *Environment and Society* 10 (1): 63–82.

Khadse, A., P. M. Rosset, H. Morales, and B. G. Ferguson. 2018. "Taking Agroecology to Scale: The Zero Budget Natural Farming Peasant Movement in Karnataka, India." *The Journal of Peasant Studies* 45 (1): 192–219.

Le Billon, P., and M. Sommerville. 2017. "Landing Capital and Assembling 'Investable Land' in the Extractive and Agricultural Sectors." *Geoforum; Journal of Physical, Human, and Regional Geosciences* 82: 212–224. doi:10.1016/j.geoforum.2016.08.011.

Lennon, M. 2017. "Decolonizing Energy: Black Lives Matter and Technoscientific Expertise Amid Solar Transitions." *Energy Research & Social Science* 30: 18–27.

Lennon, M. 2020. "Postcarbon Amnesia: Toward a Recognition of Racial Grief in Renewable Energy Futures." *Science, Technology, & Human Values* 45 (5): 934–962.

Levien, M. 2013. "The Politics of Dispossession: Theorizing India's 'Land Wars'." *Politics and Society* 41 (3): 351–394.

Lewis, S., and M. Maslin. 2018. *The Human Planet: How We Created the Anthropocene*. London: Penguin.

Li, T. 2011. "Centering Labor in the Land Grab Debate." *Journal of Peasant Studies* 38 (2): 281–299.

Li, T. M. 2014. "What is Land? Assembling a Resource for Global Investment." *Transactions of the Institute of British Geographers* 39 (4): 589–602.

Li, T. M. 2018. "After the Land Grab: Infrastructural Violence and the 'Mafia System' in Indonesia's Oil Palm Plantation Zones." *Geoforum* 96: 328–337.

Luke, N., and N. Heynen. 2020. "Community Solar as Energy Reparations: Abolishing Petro-Racial Capitalism in New Orleans." *American Quarterly* 72 (3): 603–625.

Mbow, C., C. Rosenzweig, L. G. Barioni, T. G. Benton, M. Herrero, M. Krishnapillai, E. Liwenga, et al. 2019. "Food Security." In *Climate Change and Land: An IPCC Special Report on Climate Change, Desertification, Land Degradation, Sustainable Land Management, Food Security, and Greenhouse gas Fluxes in Terrestrial Ecosystems*, edited by P. R. Shukla, J. Skea, E. Calvo Buendia, V. Masson-Delmotte, H.-O. Pörtner, D. C. Roberts, P. Zhai, R. Slade, S. Connors, R. van Diemen, M. Ferrat, E. Haughey, S. Luz, S. Neogi, M. Pathak, J. Petzold, J. Portugal Pereira, P. Vyas, E. Huntley, K. Kissick, M. Belkacemi, and J. Malley.

McKittrick, K. 2011. "On Plantations, Prisons and a Black Sense of Place." *Social & Cultural Geography* 12 (8): 947–963.

McKittrick, K. 2013. "Plantation Futures." *Small Axe: A Caribbean Journal of Criticism* 17 (42): 1–15.

McMichael, P. 2009. "A Food Regime Genealogy." *The Journal of Peasant Studies* 36 (1): 139–169.

McMichael, P. 2012. "The Land Grab and Corporate Food Regime Restructuring." *The Journal of Peasant Studies* 39 (3–4): 681–701. doi:10.1080/03066150.2012.661369.

Mehta, L. 2009. *Displaced by Development: Confronting Marginalisation and Gender Injustice*. New Delhi: Sage.

Mintz, S. 1985. *Sweetness and Power: The Place of Sugar in Modern History*. New York: Penguin Books.

MNRE. 2014. *Gujarat Solar Park: World's First Multi Developer, Multi Facility, Multi Technology and Multi Beneficiary Solar Park At Charanka Village, Dist Patan*. Ministry of New and Renewable Energy. Government of India.

MNRE. 2016. *Guidelines for Development of Solar Parks*. Ministry of New and Renewable Energy. Government of India.

Montenegro De Wit, M. 2021. "What Grows from a Pandemic? Toward an Abolitionist Agroecology." *The Journal of Peasant Studies* 48 (1): 99–136.

Moore, J. W. 2017. "The Capitalocene, Part I: On the Nature and Origins of Our Ecological Crisis." *The Journal of Peasant Studies* 44 (3): 594–630.

Moore, J. W. 2018. "The Capitalocene Part II: Accumulation by Appropriation and the Centrality of Unpaid Work/Energy." *The Journal of Peasant Studies* 45 (2): 237–279.

Mulvaney, D. 2019. *Solar Power: Innovation, Sustainability, and Environmental Justice*. Oakland, CA: University of California Press.

Nalepa, R. A., and D. M. Bauer. 2012. "Marginal Lands: The Role of Remote Sensing in Constructing Landscapes for Agrofuel Development." *Journal of Peasant Studies* 39 (2): 403–422.

Nally, D., and S. Taylor. 2015. "The Politics of Self-Help: The Rockefeller Foundation, Philanthropy and the 'Long' Green Revolution." *Political Geography* 49: 51–63.

Newell, P. 2022. "Climate Justice." *The Journal of Peasant Studies*, 1–9. doi:10.1080/03066150.2022.2080062.

Newell, P., S. Srivastava, L. O. Naess, G. A. Torres Contreras, and R. Price. 2021. "Toward Transformative Climate Justice: An Emerging Research Agenda." *Wiley Interdisciplinary Reviews: Climate Change* 12 (6): e733.

Nielsen, K. B., S. Sareen, and P. Oskarsson. 2020. "The Politics of Caste in India's New Land Wars." *Journal of Contemporary Asia* 50 (5): 684–695.

Nightingale, A. J., S. Eriksen, M. Taylor, T. Forsyth, M. Pelling, A. Newsham, and R. Bezner Kerr. 2019. "Beyond Technical Fixes: Climate Solutions and the Great Derangement." *Climate and Development*, 1–10.

Nordholm, A., and S. Sareen. 2021. "Scalar Containment of Energy Justice and Its Democratic Discontents: Solar Power and Energy Poverty Alleviation." *Frontiers in Sustainable Cities* 3: 13.

Ojeda, D., J. S. Sasser, and E. Lunstrum. 2020. "Malthus's Specter and the Anthropocene." *Gender, Place & Culture* 27 (3): 316–332.

Patel, R. 2013. "The Long Green Revolution." *The Journal of Peasant Studies* 40 (1): 1–63.

Pye, O. 2021. "Agrarian Marxism and the Proletariat: A Palm Oil Manifesto." *The Journal of Peasant Studies* 48 (4): 807–826.

Ranganathan, M. 2022. "Caste, Racialization, and the Making of Environmental Unfreedoms in Urban India." *Ethnic and Racial Studies* 45 (2): 257–277.

Reames, T. G. 2020. "Distributional Disparities in Residential Rooftop Solar Potential and Penetration in Four Cities in the United States." *Energy Research & Social Science* 69: 101612.

Ribot, J. 2013. "Vulnerability Does Not Just Fall from the Sky: Toward Multi-Scale Pro-Poor Climate Policy." In *Handbook on Climate Change and Human Security*, edited by M. R. Redclift, and M. Grasso, 164–199. Cheltenham: Edward Elgar.

Ribot, J. 2014. "Cause and Response: Vulnerability and Climate in the Anthropocene." *The Journal of Peasant Studies* 41 (5): 667–705.

Rice, J., J. Long, and A. Levenda. 2021. "Against Climate Apartheid: Confronting the Persistent Legacies of Expendability for Climate Justice." *Environment and Planning E: Nature and Space* 5 (2): 625–645.

Rignall, K. E. 2016. "Solar Power. State Power, and the Politics of Energy Transition in pre-Saharan Morocco." *Environment and Planning A: Economy and Space* 48 (3): 540–557.

Rocheleau, D. E. 2015. "Networked, Rooted and Territorial: Green Grabbing and Resistance in Chiapas." *The Journal of Peasant Studies* 42 (3-4): 695–723.

Saldanha, A. 2020. "A Date with Destiny: Racial Capitalism and the Beginnings of the Anthropocene." *Environment and Planning D: Society and Space* 38 (1): 12–34.

Sareen, S., and S. S. Kale. 2018. "Solar 'Power': Socio-Political Dynamics of Infrastructural Development in Two Western Indian States." *Energy Research & Social Science* 41: 270–278.

Scheidel, A., and A. Sorman. 2012. "Energy Transitions and the Global Land Rush: Ultimate Drivers and Persistent Consequences." *Global Environmental Change* 22 (3): 588–595.

Shah, T. 2018. "Kick-Starting the Kisan Urja Suraksha Evam Utthaan Mahabhiyan." *Economic and Political Weekly* 54: 13–16.

Shaw, A., and K. Wilson. 2020. "The Bill and Melinda Gates Foundation and the Necro-Populationism of 'Climate-Smart' Agriculture." *Gender, Place & Culture* 27 (3): 370–393.

Shiva, V. 2016. *The Violence of the Green Revolution: Third World Agriculture, Ecology, and Politics*. Lexington: University Press of Kentucky.

Siamanta, Z. C. 2017. "Building a Green Economy of Low Carbon: The Greek Post-Crisis Experience of Photovoltaics and Financial 'Green Grabbing." *Journal of Political Ecology* 24 (1): 258–276.

Siamanta, Z. C. 2019. "Wind Parks in Post-Crisis Greece: Neoliberalisation Vis-à-Vis Green Grabbing." *Environment and Planning E: Nature and Space* 2 (2): 274–303.

Siamanta, Z. C. 2021. "Conceptualizing Alternatives to Contemporary Renewable Energy Development: Community Renewable Energy Ecologies (CREE)." *Journal of Political Ecology* 28 (1): 258–276.

Singh, D. 2022. "'This is All Waste': Emptying, Cleaning and Clearing Land for Renewable Energy Dispossession in Borderland India." *Contemporary South Asia*, 1–18.

Sivaramakrishnan, K. 2000. "State Sciences and Development Histories: Encoding Local Forestry Knowledge in Bengal." *Development and Change* 31 (1): 61–89. doi:10.1111/1467-7660.00147.

Sovacool, B. K. 2021. "Who are the Victims of Low-Carbon Transitions? Towards a Political Ecology of Climate Change Mitigation. *Energy Research & Social Science* 73: 101916.

Sovacool, B. K., M. L. Barnacle, A. Smith, and M. C. Brisbois. 2022. "Towards Improved Solar Energy Justice: Exploring the Complex Inequities of Household Adoption of Photovoltaic Panels." *Energy Policy* 164: 112868.

Sovacool, B. K., M. Martiskainen, A. Hook, and L. Baker. 2019. "Decarbonization and its Discontents: A Critical Energy Justice Perspective on Four Low-Carbon Transitions." *Climatic Change* 155 (4): 581–619.

Stock, R. 2021a. "Illuminant Intersections: Injustice and Inequality Through Electricity and Water Infrastructures at the Gujarat Solar Park in India." *Energy Research & Social Sciences* 82: 102309. doi:10.1016/j.erss.2021.102309.

Stock, R. 2021b. "Bright as Night: Illuminating the Antinomies of 'Gender Positive' Solar Development." *World Development* 138: 105196. doi:10.1016/j.worlddev.2020.105196.

Stock, R. 2021c. "Praeclariat: Theorising Precarious Labour Geographies of Solar Energy." *Antipode* 53 (3): 928–949. doi:10.1111/ANTI.12698.

Stock, R. 2021d. "*Deus ex Mitigata*: Denaturalizing the Discursive Power of Solar India." *Environment and Planning E: Nature and Space* 4 (2): 354–382. doi:10.1177/2514848620908166.

Stock, R. 2022. "Triggering Resistance: Contesting the Injustices of Solar Park Development in India." *Energy Research & Social Sciences* 86: 102464.

Stock, R., and T. Birkenholtz. 2020. "Photons vs. Firewood: Female (Dis)empowerment by Solar Power in India." *Gender, Place & Culture* 27 (11): 1628–1651. doi:10.1080/0966369X.2020.1811208.

Stock, R., and T. Birkenholtz. 2021. "The Sun and the Scythe: Energy Dispossessions and the Agrarian Question of Labor in Solar Parks." *The Journal of Peasant Studies* 48 (5): 984–1007. doi:10.1080/03066150.2019.1683002.

Stock, R., and M. Gardezi. 2021. "Make Bloom and Let Wither: Biopolitics of Precision Agriculture at the Dawn of Surveillance Capitalism." *Geoforum* 122: 193–203.

Stoler, Ann. 1985. *Capitalism and Confrontation in Sumatra's Plantation Belt, 1870–1979*. New Haven: Yale University Press.

Sultana, F. 2022a. "The Unbearable Heaviness of Climate Coloniality." *Political Geography*, 102638.

Sultana, F. 2022b. "Critical Climate Justice." *The Geographical Journal* 188 (1): 118–124.

Tanner, K. E., K. A. Moore-O'Leary, I. M. Parker, B. M. Pavlik, S. Haji, and R. R. Hernandez. 2021. "Microhabitats Associated with Solar Energy Development Alter Demography of Two Desert Annuals." *Ecological Applications* 31 (6): e02349.

Taylor, M. 2018. "Climate-Smart Agriculture: What is it Good For?" *The Journal of Peasant Studies* 45 (1): 89–107.

Thomas, K., R. D. Hardy, H. Lazrus, M. Mendez, B. Orlove, I. Rivera-Collazo, and R. Winthrop. 2019. "Explaining Differential Vulnerability to Climate Change: A Social Science Review." *Wiley Interdisciplinary Reviews: Climate Change* 10 (2): e565.

Tsing, A. L., A. S. Mathews, and N. Bubandt. 2019. "Patchy Anthropocene: Landscape Structure, Multispecies History, and the Retooling of Anthropology: An Introduction to Supplement 20." *Current Anthropology* 60 (S20): S186–S197.

van der Ploeg, J. D. 2021. "The Political Economy of Agroecology." *The Journal of Peasant Studies* 48 (2): 274–297.

van der Ploeg, J. D., J. C. Franco, and S. M. Borras. 2015. "Land Concentration and Land Grabbing in Europe: A Preliminary Analysis." *Canadian Journal of Development Studies/Revue Canadienne D'études du Développement* 36 (2): 147–162.

Vergès, F. 2017. "Racial Capitalocene." In *Futures of Black Radicalism*, edited by G. T. Johnson, and A. Lubin, 72–82. London: Verso.

Whitehead, J. 2010. "John Locke and the Governance of India's Landscape: The Category of Wasteland in Colonial Revenue and Forest Legislation." *Economic and Political Weekly* 45 (5): 83–93.

Whyte, K. P. 2018. "Indigenous Science (Fiction) for the Anthropocene: Ancestral Dystopias and Fantasies of Climate Change Crises." *Environment and Planning E: Nature and Space* 1 (1-2): 224–242.

Williams, E. 1944. *Capitalism and Slavery*. Chapel Hill: University of North Carolina Press.

Wolford, W. 2021a. "The Colonial Roots of Agricultural Modernization in Mozambique: The Role of Research from Portugal to ProSavana." *The Journal of Peasant Studies* 48 (2): 254–273.

Wolford, W. 2021b. "The Plantationocene: A Lusotropical Contribution to the Theory." *Annals of the American Association of Geographers* 111 (6): 1622–1639.

Woods, C. 2017. *Development Arrested: The Blues and Plantation Power in the Mississippi Delta.* Brooklyn, NY: Verso Books.

Yenneti, K., and R. Day. 2015. "Procedural (in) Justice in the Implementation of Solar Energy: The Case of Charanaka Solar Park, Gujarat, India." *Energy Policy* 86: 664–673.

Yenneti, K., and R. Day. 2016. "Distributional Justice in Solar Energy Implementation in India: The Case of Charanka Solar Park." *Journal of Rural Studies* 46: 35–46.

Yenneti, K., R. Day, and O. Golubchikov. 2016. "Spatial Justice and the Land Politics of Renewables: Dispossessing Vulnerable Communities Through Solar Energy Mega-Projects." *Geoforum: Journal of Physical, Human, and Regional Geosciences* 76: 90–99.

Yusoff, K. 2018. *A Billion Black Anthropocenes or None*. University of Minnesota Press.

Zalasiewicz, J., C. N. Waters, E. C. Ellis, M. J. Head, D. Vidas, W. Steffen, … J. Zinke. 2021. "The Anthropocene: Comparing its Meaning in Geology (Chronostratigraphy) with Conceptual Approaches Arising in Other Disciplines." *Earth's Future* 9 (3).

Oro blanco: assembling extractivism in the lithium triangle

Daniela Soto Hernandez and Peter Newell

ABSTRACT

As the drive for global electrification proceeds, new pressures are placed on agrarian environments in areas abundant in key minerals for electric batteries. The so-called lithium triangle between Chile, Argentina and Bolivia is one of those places. We develop an account of the 'assemblages of extractivism' at work in this zone that operate at a material, institutional and discursive level. Drawing on fieldwork from the region and conceptualized using different strands of political ecology and political economy, we explore how the construction of a commodity, the materiality of lithium and the role of the state intersect with local understandings and engagements with this latest form of 'renewable extractivism'.

1. Introduction

A transition away from carbon-intensive sources of energy is critical to address the climate crisis, as established by numerous expert bodies and authoritative reports (IPCC 2018; SEI et al. 2021). A prerequisite for the transition to a low-carbon economy lies in technologies for storing energy to bridge gaps that exist between times of peak energy production from renewable sources and times of peak consumption, and for powering vehicles. Given the technological foundations of this move to renewable energy production and storage, the emergent post-fossil-fuel era is characterized by a techno-managerial approach focused on a mixture of socio-technological innovation and pricing mechanisms consistent with neoliberal approaches to energy transition (Newell 2019). Since 2010, the global growth of electricity generation capacity from renewable sources has increased from about 25% in 2001 to 82% in 2020 (IRENA 2021), mostly from the generation of renewable sources such as solar, wind, bioenergy, hydropower and geothermal. This dimension of the proposed solution to the climate crisis aligns with broader capitalist imperatives – permanent technological revolution driven by finance capital (Perez 2002) – albeit in this instance mobilized to provide the same patterns of energy consumption while reducing the production of atmospheric carbon dioxide in pursuit of the elusive decoupling of growth from emissions.

Technological fixes focused on energy storage have followed similar patterns to the technologies of energy production. Lithium-ion batteries, first designed in the 1990s, have become increasingly central to modern lifestyles of consumption and transportation: they are found in numerous items from computers to cell phones, electric cars and prototype electric airplanes. Due to its unequaled lightness as well as its storage capacity (Kazimierski 2019) lithium, often dubbed 'white gold', is a crucial component of current energy storage solutions. Although lithium is relatively abundant on Earth, current technology means that it can only be extracted profitably from certain resources, such as brines. The largest brine reserves are in the so-called 'lithium triangle', a rural territory that stretches over southern Bolivia and northern Argentina and Chile.

While some claim that lithium extraction constitutes 'sustainable mining' given the cost-effectiveness and energy-efficiency of its extraction as it mainly relies on the energy of the sun (Reporte Sostenible 2019; SQM 2021; Tarantini 2020; Voskoboynik and Andreucci 2021), there is mounting evidence that the chemicals and water used for processing lithium cause large-scale environmental disruption to the salt flats and surrounding environment where it is extracted (Babidge et al. 2019; Liu, Agusdinata, and Myint 2019; OCMAL 2018).

The fact that solutions to address energy transition build on and at the same time reinforce older patterns of extractivism has given rise to concerns and rejuvenated broader debates about the role and sustainability of extractive industries in local and international development (Bebbington et al. 2008). Critics have argued that market-led solutions have largely proven unable to address social and environmental issues or support a socially and economically just transition to sustainability (Dunlap and Fairhead 2014; Machado Aráoz 2013a; Zehner 2012). Moreover, since lithium extraction is being carried out in a region where there is a long and painful history of extractivism since colonial times (of saltpeter and silver in Bolivia and copper in Chile, for example), it constitutes what Watts refers to as hyper-extraction: 'expanded, extended, or enhanced extraction' (2021, 212). As such, along with intensifying environmental pressures, 'hyper-extractivism' appears to strengthen the traditional dependency and primary-exporter role of the Global South to the benefit of geopolitically dominant economic powers in the Global North, while further entrenching modes of ecologically uneven exchange (Jerez, Garcés, and Torres 2021; Roberts and Parks 2009; Rodney 1972). Of the 35 countries most dependent on mining, all but Australia and South Korea are developing countries. But this is not just about the Global North and South. So-called 'green mining' is also being used to resuscitate old mining sites for lithium prospecting in poorer parts of richer countries such as Cornwall in the UK, for example, where once tin and copper were mined (*BBC*, September 17, 2020). What this points to instead is the importance of poverty and inequality and in many contexts their close relationship to class, race and gender as determinants of who are the likely beneficiaries of the new wave of extractivism and who is most likely to inhabit the sacrifice zones that resource it.

When considering the key political economy questions of the new accumulation strategy of the mining industry, which we take to be 'Who benefits, who loses, how, and why?', there is arguably little discernible change in the structural matrix of power relations within and between global elites and even in the sources of state revenues when comparing previous and current modes of extractivism. Therefore, when it comes to technologies of energy transition, historically constituted but dynamic global dimensions are critical, whether in terms of relations between states and geopolitical competition (IRENA 2019) or shifting alignments of state and (transnational) capital as they extend their reach and control over new frontiers in the global economy and incorporate new groups within their circuits of capital (Newell and Mulvaney 2013). The large and increasing presence of Chinese companies in the lithium triangle (and in the lithium-ion battery market) is a significant recent illustration of this trend (Zicari, Fornillo, and Gamba 2019).

To address these complexities, in this article we develop an account of the 'assemblages of extractivism' at work in this zone that operate at a material, institutional and discursive level, drawing on primary fieldwork in the region and using different strands of scholarship from political ecology and political economy to conceptualize them. Taking the case of southern Abya Yala and drawing on year-long ethnographic work in the region conducted with those on the front lines of this new resource frontier, we show how accumulation strategies pursued by the state and capital in Chile under the guise of 'green growth' intensify conflicts around access to and the value of lithium, land and water. In particular, we explore the social and colonial dynamics that characterize this site and assemblage of extractivism, as well as potential openings for more transformative politics. In this regard, we draw out the wider implications for understanding and engaging with the chains of transition and 'assemblages of extractivism' across the socio-spatial sites of the lithium economy and what this suggests about the potential to decolonize transitions so that they are more socially and ecologically just.

The analysis contributes to debates about the global political economy of energy transitions by looking at how social and political power shapes outcomes across sites and scales, while enhancing tentative attempts to decolonize transition debates and providing further empirical grounding and conceptual insights on the practical political economies and political ecologies of extractivism. This helps to bridge an understanding of the social dynamics of agrarian change with perspectives from socio-technical transitions, which continue to be more focused on the Global North and the managerial politics of technology, finance and innovation than the deeper politics of social and economic transformation. By building an account of how energy transitions become enmeshed with local politics and their attendant inequities, we are better placed to account for the everyday politics of transition while locating it within the broader structures and power relations that characterize the global political economy. This speaks to the editors' call to explore 'the way agrarian struggles … connect with the challenge of climate change, linking to and going beyond the already widespread challenges to expropriation and extraction in rural areas' as 'a vital focus for both thinking and action' (Borras et al. 2022, 2). Our empirical analysis offers a case study of how different people – in relation to class and other co-constituted axes of social difference such as gender, race and ethnicity – are affected by responses to climate change in agrarian settings as they are mediated through processes of social differentiation, trajectories of accumulation and agrarian politics (Borras et al. 2022, 17).

In particular, therefore, the paper seeks to make three main contributions. Firstly, we elaborate the different elements of 'assemblages of extractivism': their material, institutional and discursive dimensions and the interacting and mutually reinforcing nature of these dimensions contributing both to scholarship on extractivism and work on assemblages (Anderson et al. 2012; Jordhus-Lier, Camilla, and Tale Hammerø 2022). For example, it adds to the work of other scholars working in the region, such as Hope (2021, 208) who explores the 'assemblage of institutions, discourses, landscapes, and infrastructures that are at once disciplined and held together to materialise and legitimise particular interpretations of sustainable development' in Bolivia and how this assemblage sheds light on the way 'powerful, extractivist development logics' are reworked and sustained. Secondly, based on extensive ethnographic work, we explore and engage with the everyday politics of transition and the lived experience of extractivism in the Salar de Atacama (Atacama Salt Flat) watershed: how communities on the front lines of extractivism navigate its complexities and contradictions. This helps to move transition studies beyond a dominant focus on elite socio-technical configurations without furnishing an account of their consequences. Thirdly, by identifying and explicating the relationality of transitions – the global connections between sites of production and consumption – we contribute to scholarship on the global political economy of energy transitions (Newell 2021; Van de Graaf and Sovacool 2020).

2. Understanding the political economies and political ecologies of energy transition

To build our account of rural extractivism in Latin America's lithium triangle we draw upon scholarship from both political economy and political ecology, a combination of which, we argue, is critical to understanding and engaging with the ways in which 'the dynamics of climate change' both shape and are shaped by the rural world, 'whether through state policies, international governance, corporate influence, or agrarian struggles' (Borras et al. 2022, 1). The chains of explanation that link these sites of power and sustain particular strategies of resource extractivism, labor exploitation and value generation and their ecological impacts require critical attention to the role of the state and its relationship to national and international capital. But they also need to be grounded in particular socio-ecological rural realities with their attendant exclusions and patterns of social differentiation. As the editors note, political ecology sheds light on the way in which 'capitalism and climate change are social and ecological processes that are both produced and experienced at multiple sites and scales' (Borras et al. 2022, 4). Analytically, the challenge is to comprehend the ways in which local sites and struggles are related to and embedded within broader structures of power, without reducing what is historically, socially and culturally unique about those sites to abstract global actors and processes (Newell and Bumpus 2012). This more global and relational approach helps to illuminate the dynamics of accumulation and the circuits of capital that tie specific sites of extraction, dispossession and ecological disequilibrium to particular owners of capital. Meanwhile, a complementary, more anthropological approach to the case in question allows us to understand the specific configurations of racialised, class-based and gendered exclusions that interact with and reinforce dominant strategies of accumulation in the region.

Fortunately, despite the early neglect of questions of power and politics in the study of energy transitions (Meadowcroft 2009), theoretical resources from political economy (Baker, Newell, and Phillips 2014; Powers et al. 2015), global political economy (Newell 2021) and (global) political ecology (Lawhon and Murphy 2012; Newell and Bumpus 2012) are increasingly being applied to understand the procedural, distributional and ecological politics of energy transitions. The organization of global production networks (Baker and Sovacool 2017) and commodity-supply zones, understood as 'spaces of immense fecundity and resource abundance' (Bridge 2001, 2155), reflect the imperatives of the new mining boom. The lithium value chain, governed by national states and transnational capital, reflects many of these 'chains of accumulation' (Robbins 2014).

Firstly, these theoretical resources can be usefully combined with scholarship from critical geography on the notion of 'resourcefulness' (Bridge 2011, 2014): the discursive, institutional and material making of commodities and resources such as land (Murray Li 2007), or what Jasanoff and Kim (2013) call 'sociotechnical imaginaries'. These latter approaches underscore the importance of competing imaginaries and knowledge constructions about the value of resources which necessarily precede and enable their extraction and circulation. The socio-spatial politics of lithium combine territory, land, water and indigeneity where regions are constructed as commodity supply spaces (Bridge 2001). The fact that the Atacama Salt Flat is part of the Atacama Desert, usually highlighted as the driest desert on Earth, has helped to establish the misleading idea that the desert is devoid of biodiversity and human life. As a mechanism to legitimize lithium extraction from an anthropocentric view, these territories are imagined as spaces of resource abundance: inhospitable and remote places with nothing but minerals that await exploitation (Bridge 2001). Nevertheless, as we will show in the next sections, lithium itself is not only invisible in the territory – as it forms part of underground water (brines) in the Atacama Salt Flat – but as a chemical product results from destroying its natural environment (brines) and adding many other components to obtain the final products such as lithium carbonate or lithium hydroxide.

Baglioni and Campling (2017) propose a dialectic of materiality of resources given that they are at once physical entities and social categories. As Bridge puts it, two resource dynamics can be outlined when considering resources as a social category that organize our relationship with the non-human world. Firstly, what qualifies and is valued as a resource varies over time and space: 'resources are a relational understanding of the non-human world' (2009, 1219). Secondly, the classification of something as a resource depends on technology, knowledge, economic organization and, we would add, relations of power to project and enforce dominant understandings. Therefore, 'resources are not: they become' (Bridge 2009, 1219) after a process of social translation, showing how different worldings and conceptions of nature are constructed and experienced, including the question of humans' wider connections with the environment of which they are a part (De la Cadena 2010). In Abya Yala, the positions of several indigenous movements highlight a different cosmology of the indivisibility of nature and society that is inclusive of non-human and more-than-human beings (De la Cadena and Blaser 2018; Kohn 2015; Viveiros de Castro 2004).

Secondly, we propose the value of work on extractivism in appreciating these dynamics: understood as economic-political regimes based on the over-exploitation of natural resources (Machado Aráoz 2013b, 129). Extractivism is often defined as a 'modality of accumulation' involving the appropriation of natural resources 'in large volumes and/or high intensity, and that are essentially intended to be exported as unprocessed raw materials, or with minimal processing' (Gudynas 2015, 13). Considering its planetary scale (Arboleda 2020; Chagnon et al. 2022) and effects, extractivism also embodies a particular way of thinking, highlighted by Durante, Kröger, and LaFleur as an extractive onto-logic: a 'violent logic of taking resources – without reciprocity, without stewardship' (2021, 20), extending and expanding the coloniality of capital (Quijano 2000).

Indeed, in Abaya Yala, extractivism carried out by states (in the form of neo-extractivism Gudynas 2009) and by the market has allowed accumulation by dispossession by national and global elites (Green 2003; Harvey 2004), rights violations and violence, including the 'slow violence' associated with climate change (Nixon 2013). In Chile, Núñez, Benwell, and Aliste describe 'eco-extractivism' as a form of 'green-grabbing undertaken by extractivist investors' (2020, 1). As Harvey (2004) suggests, the expulsion of peasant and indigenous populations by ruling classes and the state was key to the commodification and privatization of land and the exploitation of the subaltern class in favor of elites. Such wealth extraction by elites who are predominantly of European heritage can be understood as the result of the dispossession and expulsion of the region's original inhabitants, a form of internal colonialism. As Newell suggests:

> In many ways, the patterns of exploitation we find within countries mirror the forms assumed by global relations of the colonial era. In parts of Latin America a form of 'colonialism within' is practiced when elites of white European or *mestizo* descent either forcibly remove indigenous Indians from their lands or extract resources coercively for global export. Invoking rationales employed by former colonizers, such controversial practices continue to be validated through reference to the need to modernize backward peoples and their 'unproductive' livelihoods. (2005, 77)

These dynamics have resulted in the production of social and environmental injustices premised on class, race and gendered inequalities, manifested, for example, in the overloading of mining, extractive and production waste in the territory of local communities (Perreault 2012). Thus, a process of territorial alienation and de-territorialization has occurred, as local populations are forced to migrate (Machado Aráoz 2015, 186). The extractive activities that rapidly expanded in Latin America from the 1990s onward (Ocampo 2017) are currently being driven by a 'super-cycle' in commodity prices given the demand generated by China. This has led to a reconfiguring of regional politics, from the political and economic reforms under the 'structural adjustment' of the Washington Consensus, to what has been labeled the 'commodity consensus' (Svampa 2015). This refers to the large-scale exportation of raw materials on which current economic development is based in Latin America and the recent process of 'reprimarization' as national economies deepen their dependency on their role as exporters of primary, raw materials and natural resources.

We combine these insights around the notion of an 'assemblage of extraction'. Assemblage is a concept devoted to understanding processes of becoming, 'of putting together, of arranging and organizing the compound of analytical encounters and relations' (Lancione 2013, 359). Originally articulated but not subsequently developed by Deleuze and Guattari (1988), here we use it in a similar way to Hope (2021) to denote how extractivism is simultaneously materialized, institutionalized and discursively produced and legitimized. Where this moves political economies and political economies of energy transition forward is through attention to what Li calls the 'hard work required to draw heterogeneous elements together, forge connections between them and sustain these connections in the face of tension' (2007, 2). The notion of an assemblage is especially useful for understanding the conduct and nature of this work across sites and scales that have to be aligned and operationalized for extractivism to occur. It affords a socio-spatial account (Anderson et al. 2012) of key aspects of the political economies and ecologies of lithium because its extraction rests on a series of imaginaries that have to be continually remade, about the value of a resource spread across territories which are subject to competing ownership claims, whilst being driven by national policy and influenced by mobile transnational capital. Conflicting imperatives and priorities pull in different directions, requiring the political work of negotiation and accommodation and careful boundary work to mine and exchange the lithium, keep the finance flowing and minimize disruption to the commodity supply chain. This more distributed account of agency, which recognizes that there are vulnerabilities, instabilities and impermanence at the intersections of different elements of an assemblage (Haarstad and Wanvik 2017), is not the same, however, as suggesting that systems of consumption, distribution and production are distantly related and loosely interconnected (Jordhus-Lier, Camilla, and Tale Hammerø 2022, 320). In this regard, exploring the practical work of holding together an extractivist assemblage should not 'obscure the relational production of difference and inequality' (Kinkaid 2020, 465), but rather reveal what makes it possible.

3. The political ecologies of extractivism in the lithium triangle: the case of Chile

Lithium became known as a mineral resource after 1817 when it was 'discovered' by a Swedish chemist. Almost 100 years later, a German company started producing lithium for sale, exploiting small deposits mostly in the United States and Australia to meet the needs of a small market. Along with its many uses (in industries from aluminum and glass to pharmaceuticals), lithium was used for nuclear energy and, later, in the first lithium-ion batteries, launched in 1991 by Sony Corporation (Kohn 2015). However, driven by media, corporate and policy discourses around green growth, with the lithium boom at its center, a new extractive wave renamed this transnational territory a triangle marked by three salt flats: the Uyuni salt flat (Bolivian Andean plateau), the Hombre Muerto salt flat (Puna of Atacama, Argentina) and the Atacama Salt Flat (Antofagasta region, Chile). The 'lithium triangle' became what Bridge (2001) calls a 'commodity-supply zone', a narrative that obscures everything but lithium.

Although the 'lithium triangle' is a recent discursive construction, lithium extraction started around 30 years ago in Chile and Argentina, but involves new developments such as Bolivia's state-led extraction project in the Uyuni Salt Flat (Fornillo 2015; Nacif and Lacabana 2015). As such, lithium's boom comes as a new wave of extractivism, rather than as a new site of extraction. In fact, Chile is the largest producer of copper (also needed for batteries), and the Antofagasta region accounts for 54% of national productivity (Sernageomin 2020); and the second largest producer of lithium, accounting for 22% of global supply (USGS 2022). This makes it an interesting case of how 'green growth' intensifies conflicts around access to and the value of lithium, land and water in ways that repeat previous growth strategies.

At the same time, however, lithium provides an interesting case of corruption and failed state control over 'strategic mineral resources': a category created by the US government's growing interest in lithium's potential for nuclear fusion in the context of the Cold War. Consequently, the Chilean military dictatorship changed legislation in 1982 to preserve the mineral as a strategic resource of national interest such that its extraction and exportation was overseen by the Chilean Commission for Nuclear Energy (CChen). This change was part of broader neoliberal reforms at the time, where almost all state companies and public goods such as land and water were privatized. Indeed, around 1978, the military government gave mining rights to exploit brines from the Atacama Salt Flat to the state economic development agency Corfo (Corporación de Fomento de la Producción). Alongside this, two projects were developed to extract brines as private–public partnerships: the first one was the Sociedad Chilena del Litio Ltd (Chilean Society of Lithium) between Foote (with 55%) and Corfo (45%) that started to produce lithium carbonate in 1984. This project was granted rights validated by CChen to exploit, produce, and sell 200,000 tonnes of lithium. However, five years later, Corfo sold its part, and the project was further developed by the transnational company Cyprus Foote (which then sold it to Rockwood Litio Ltd.) and, since 2015, it has been run by the North American company Albemarle (Comisión Nacional del Litio 2015). The second project began in 1986, where Corfo (25%), Amax (63.75%) and Molymet (11.25%) formed the Sociedad Minera Salar de Atacama Ltda (Minsal – Mining Society of Salar de Atacama). The project was granted a contract for 33 years with an agreed leasehold and was allowed a maximum of 180,100 tons of lithium. However, as with the first case, by 1995 the entire project was owned by a national company, SQM Salar SA, and had expanded to produce potassium salts, potassium chloride, potassium sulfate and boric acid, and lithium chloride from lithium carbonate (Comisión Nacional del Litio 2015).

The boom in electromobility was a boost for both lithium extraction quotas; however, it also served to intensify conflict with the indigenous and non-indigenous communities of the Atacama Salt Flat watershed. The Atacamanian indigenous communities, or lickanantay, have occupied the Salar de Atacama basin for more than 9000 years (Nuñez 2007). Traditionally, they developed agro-pastoralist practices, but have been increasingly absorbed by the mining industry (including lithium, but mainly copper mining) and, more recently, tourism sectors (Molina 2019; Nuñez and Pourrut 1995). During the last 30 years, the different villages and ayllus[1] started to constitute 'indigenous communities' legally recognized by the state (Law No.19253) and formed the Consejo de Pueblos Atacameños (CPA – Council of Atacamanian People). These organizations have been key entities in establishing their ancestral territory as well as in pushing their demands with the state and mining companies. Nonetheless, despite their ancestral use and knowledge of the territory, the Chilean state has largely ignored their demands, favoring its ownership of the mineral resources and the profits derived from mining. As a result, the indigenous communities established direct negotiations with Albemarle and, only recently, resumed formal relationships with SQM, the biggest lithium company in Chile today.

3.1. Material assemblages: constructing a commodity

To better understand how the 'assemblage of extractivism' operates in this instance, we first explore the material assemblage of lithium as a commodity. The drive toward electromobility in the Global North has reshaped these territories as lithium reserves. In this context, we argue that the reduction of complex landscapes to a 'commodity-

---

[1]As defined by Romero and Opazo, ayllus are the ancestral territories of high Andean communities:

> The indigenous ayllu term corresponds to the representation of a place and is used by Andean societies to refer to a landscape synthesis made up of components and interactions of natural and social, physical and metaphysical, material and symbolic origin, human living beings and non-living beings. (2019, 43)

supply zone' (Bridge 2001) in the lithium triangle has enabled the state to grant more exploitation rights to current projects and to bid for more rights to extract lithium in the Chilean salt flats, a process we discuss further below.

The construction of lithium as a commodity in this area started in 1962 when the US mining company Anaconda identified huge lithium deposits in the Atacama Salt Flat, while looking for water resources for their copper mining process in Chuquicamata (Slipak and Urrutia 2019). In 1977, the Chilean National Institute for Geological Research confirmed this potential, which, along with the work of the United States Geological Survey (USGS), has been key to supporting the idea of lithium abundance in Chile, said to be the largest reserves worldwide at 9,200,000 tonnes, a claim possible only because the Bolivian reserves (estimated to be 19 million tonnes) have not been incorporated into the US geological survey (USGS 2022).

Lithium is a fairly abundant mineral on Earth. The narratives concerning the potential scarcity in access to lithium deposits refer rather to the potentiality of the technology and knowledge required to *profitably* extract the mineral. In this regard, geological concepts such as 'reserves' and 'resources' highlight deposits of minerals that are susceptible to appropriation under certain conditions. In fact, USGS warns that:

> Reserves data are dynamic. They may be reduced as ore is mined and (or) the feasibility of extraction diminishes, or more commonly, they may continue to increase as additional deposits (known or recently discovered) are developed, or currently exploited deposits are more thoroughly explored and (or) new technology or economic variables improve their economic feasibility. Reserves may be considered a working inventory of mining companies' supplies of an economically extractable mineral commodity. As such, the magnitude of that inventory is necessarily limited by many considerations, including cost of drilling, taxes, price of the mineral commodity being mined, and the demand for it. Reserves will be developed to the point of business needs and geologic limitations … (2022, 197)

As such, lithium reserves constitute a form of 'produced nature', socially constructed by dominant forms of scientific knowledge, obtainable through technology, economic incentives and social valuation (Bustos-Gallardo, Bridge, and Prieto 2021). Even at a material level, lithium is a material/chemical construction, going from lithium carbonate to hydroxide. Brines must be destroyed to obtain lithium (or other minerals), which means water needs to evaporate and the different minerals and salts must be chemically separated and mixed with other components to build a lithium substance suitable for ion-lithium batteries. The process is divided into two: the mining and the chemical. The mining process starts with brine extraction using a pumping system that has more than 370 productive 30-m-deep wells. The drilled brines (with a regular concentration of lithium at 0.2 ppm in brines) are transferred through channels to solar evaporation ponds, with a total surface of pools of 44 km^2. Lithium is concentrated via a process with several stages that can take around 14–15 months until it reaches a concentration of about 5.8% (Garcés 2017). In this process important quantities of lithium are lost, with a lithium recovery efficiency of only 8.2% as shown by Osses in *Minería Chilena* (Osses 2021).

It is also worth mentioning that until the stage of brine extraction, companies can also extract potassium, boron, magnesium or sulfates contained in brines. Therefore, brine extraction affords a multiplicity of opportunities for the extraction of non-renewable sources, offering economic *flexibility* (Bustos-Gallardo, Bridge, and Prieto 2021), since companies can adapt to changing market demand for these different commodities. In fact,

SQM used to be a producer of fertilizers and products associated with potassium. Given the entry of monopolies, SQM stopped basing its sales on potassium once the 'lithium boom' began.

To turn the 'impure lithium carbonate' into a substance that can be used for commercial purposes (mainly lithium carbonate or lithium hydroxide), the companies then take lithium to their industrial plants in La Negra (Albemarle) and Salar del Carmen (SQM) – both located near Antofagasta city. In these industrial complexes lithium and other products are processed, and it is in these chemical processes where most of the drinkable water is consumed and polluted with substances used to obtain the lithium concentration needed for each product. They are then exported through the ports located in the same region to Asian countries such as South Korea, China and Japan, the main buyers of lithium carbonate (Cochilco 2021). Implied within this economic chain of extraction is a process of ecological exchange, as the lithium process does not end in the Salar de Atacama; rather, it continues to absorb water resources and to pollute areas further from the desert in the Antofagasta region.

3.2. Institutional assemblages: states of transition

The second element of the 'assemblage of extractivism' in this context, beyond negotiating the materiality of lithium and intensifying value from its extraction and processing along the supply chain, concerns the role of the state and its interaction with indigenous communities. The grounded anthropological account of the everyday material and cultural politics of extraction and its economies and ecologies as it is experienced by marginalized groups and subaltern voices can be fruitfully complemented, we argue, by a political economy analysis of the role of the state in enabling, (un)governing and enforcing extractivism.

The general background to this is formed by what Leiva (2019) describes as the way strategies adopted in the mining sector and underpinned by support from the Concertación (1990–2010) and Nueva Mayoría (2014–2018) administrations and epistemic communities in Chile have 'played active, generative, and decisive roles in conceptualizing and operationalizing innovative political technologies that institutionalized the power of transnational capital' (2019, 133). This includes the use of political technologies such as the adoption of tools 'to elicit communities' "social license" to start new extractivist investment projects and to upscale existing ones' (2019, 131–132). The overall aim is to construct 'working alliances with domestic and foreign transnational capital, around shared goals of expanding the boundaries of mining capitalism and establishing the requisite conditions for its continued operation' (2019, 135).

While Svampa (2015) positions states as the decisive actors in driving extractivism, Leiva suggests that

> Chile's new extractivist consensus actively incorporates a panoply of new actors including think tanks, consulting firms, non-governmental organizations, international development agencies, academia, political cadre, as well as corporate mining executives as fundamental protagonists in a process in which national government entities play a subservient, supportive role. (2019, 135)

The case of the National Lithium Committee bears this out. An initiative charged with providing a proposal to build a national policy for managing lithium, it included ministers and state representatives, experts in engineering, mining, law and economy (including international guests), union representatives, international organizations such as CEPAL (Economic Commission for Latin America and the Caribbean), the copper state company (Corfo) and a representative of the lickanantay indigenous people.

As noted in the previous section, the role of the state has been highly controversial, with a legal apparatus built during Chile's military dictatorship and hardly modified since, plus an ongoing complicity with mining companies. For example, despite common knowledge about the water-intensive nature of lithium extraction, Chilean law considers brines mineral resources, not water. Consequently, brines are regulated by the Mining Ministry (following the Mining Code) and are not included in the Water Code (the legal body that regulates water resources). The rights to extract brines are given by the Mining Ministry regardless of the water rights given by the state agency, DGA (General Water Direction – Dirección General de Aguas). This difference has major implications in a context where the different agencies of the state do not work together. Therefore, the hydrological cycle of the basin is treated separately, even though brines are part of the water cycle of the Salar de Atacama watershed. Studies carried out by the DGA cannot comprehend the entire water cycle when excluding brines. Equally, as part of mineral resources, the state has so far relied on the information provided by the companies themselves to understand the amounts of brine that remain available. Only recently (and driven by scandals where SQM was found to misreport the total amount of resources used by the state) has an effort been made by the state to develop reliable research.

At the same time, the Chilean state faces internal contestation as there are three scales that constantly interact: the Chilean state (usually located in the Metropolitan Region – 'the center'); the regional state (periphery) and local government (municipalities). All three have conflictive relationships over resources and authority that can shape and reshape the ways in which policies are executed and enacted in local contexts. As a centralist state, the regions and municipalities are not usually the ones that decide about the territory or the ones receiving the main economic revenues. Many conflicts revolve around the capture of revenues and rents. Local actors are often resentful that the lithium companies (and copper companies) are legally registered in the capital, Santiago, which means that all the taxes they pay are received by the Metropolitan Region. They claim that given the fact that the extraction and the environmental consequences are being experienced by the habitants of the region, the companies should pay their taxes in the region as part of a social contract based on taxation and representation. As such, the governance of this chain of accumulation is deeply contested across an assembly linking multiple scales: the national capital and the regional capital (Santiago–Antofagasta); the regional capital and the rural areas where exploitation is taking place (Antofagasta–San Pedro de Atacama); and, finally, Santiago and San Pedro de Atacama, mainly connected through the companies who usually are formally registered in Santiago.

The same centralizing dynamics characterize the ownership of the mineral resources of the Salar de Atacama. As conflicts over lithium extraction have intensified, the state has insisted on securing the benefits and ownership of the Salar de Atacama, which has been a territory claimed as ancestral by the indigenous communities surrounding the salt flat. However, state action is also seen as part of a larger network of corruption in which the Chilean elite is embroiled. For example, a former judge explained in an interview that he was working for the Antofagasta regional government in the early 1990s, when it was decided that the Salar de Atacama could not be exploited anymore since it was recognized as a very fragile ecosystem. This resulted in intensive pressure being brought to bear to give concessions to SQM, driven by the economic and political elite from Santiago. Despite this lobbying, the government initially managed to stay firm in their position. Nonetheless, in a move that clearly shows the close relationship between companies and state elites in Chile, when he came back from holidays, there was a document signed by 'a state officer' (someone unknown to him) overriding the decision and allowing the exploitation of the salt flat by SQM.

The close relationship between the state and the companies has also been exposed through corruption scandals. Not only did Julio Ponce Lerou became a main shareholder of SQM after buying the state company during the dictatorship of his then father-in-law, Augusto Pinochet; but since 2014 the company has been part of many corruption scandals, such as illegal financing of politicians and stock market manipulation. Equally, in 2016, Corfo sued the company for breach of contract. SQM had illegally inscribed water rights and mining easements under its ownership and was also discovered to be misinforming the state about the real quantity of extracted brines and the lithium that was being sold. In response to the public scandal, Corfo reached an agreement with the company and granted more rights to extract brines until 2030, along with slightly higher revenues for local municipalities (Corfo 2019).

The other consequence of the state's proximity to mining capital is a failure to recognize and protect indigenous rights. The conflicts between the lickanantay people and the state over sovereignty and autonomy have shaped the conflicts regarding mining companies. The Observatorio de Conflictos Mineros de América Latina's (OCMAL) online database shows 48 socio-environmental conflicts involving multinational extractivist corporations operating in Northern, Central and Southern Chile, where indigenous people have been a key part of the resistance.[2] In this regard, even though the state ratified ILO (International Labour Organisation) Convention No. 169, which specifies that indigenous people have the right to sovereignty and autonomy in their ancestral territories, the Chilean law has not been updated in line with this requirement. In fact, the indigenous law (Law No. 12,253) regulating the legal apparatus that indigenous communities need to use to be recognized is seen by those communities as 'non appropriate' and detrimental for their autonomy and traditional way of organizing. The colonial nature of recognition policies has granted certain rights insofar as they do not challenge the dominant political economic model, whilst limiting further cultural and sovereign rights – a situation common in Latin America, that has been called 'multicultural neoliberalism' by Hale (2002). The lickanantay have directly questioned the sovereignty of the Chilean state in this region, claiming that they should have both sovereignty and autonomy over *their* ancestral territory. As one lickanantay man stated: 'first of all, you need to realize that you are in an indigenous territory. "But the [Chilean] state … ", I am not interested in what the state says, because I do not consider myself Chilean'.

In this context, public policies are seen as imposed and 'chilenizadoras', or westernized. Some indigenous interviewees recalled the example of the National Reserve 'Los Flamencos', as a conservation project that restricted their access to their ancestral territory and practices within it. In fact, a few lickanantay interviewees said that they would prefer

[2]Conflictos Mineros en Chile. Available at: https://mapa.conflictosmineros.net/ocmal_db-v2/conflicto/lista/02032300.

that the states acted like before 'when they were not present in the territory and we lived our life', as one woman put it. Therefore, for some indigenous actors the state is seen as overseeing a model of development that is having a detrimental effect on their culture and autonomy as communities. As a result of the lack of support from the state, indigenous communities have started to negotiate directly with mining companies (OLCA 2020), and in 2016 the CPA signed an agreement directly with Albemarle so that the 18 communities that constituted the CPA received 3.5% of the annual sales of the company. Reflecting on this process, one lickanantay man said:

> an agreement was reached, but we also base ourselves on what we have achieved as rights today, because in these cases it is agreement 169, article 15–16, which states that companies that are installed in certain [indigenous] territories, have to apply compensation measures and that is an obligation, that is, it is not the will of the company to say 'Hey no, you know, if you want to give them', no, it is an obligation. Based on that, the Albemarle and well, lithium agreements were made, and other communities that even have an agreement with SQM, with Escondida, with Zaldívar.

In fact, recently, the CPA joined an agreement already signed by Corfo and SQM, to receive approximately $45 million of SQM earnings (Chululo, September 28, 2021). Both agreements have been highly contested within indigenous and non-indigenous local communities, where lickanantay legal communities have been labeled as corrupt and only interested in money. As a result, territorial tensions and conflicts have shaped and continue to reshape lithium extraction.

3.3. Discursive assemblages: power/knowledge and coloniality

A last step in the 'assemblage of extractivism' is the ordering of appropriate knowledge to enable lithium extraction. This is important because, as Núñez, Benwell, and Aliste show in their work on eco-extractivism in Chile,

> land ownership changes hands from colonists or pioneers to neocolonists or eco-colonists, and this transaction is imbued with a discursive rationale tightly associated with the environmental value of the land [and water in this case], and equally strongly with capitalist speculation based on that value. (2020, 12)

Indeed, whose knowledge counts is contested in the Salar de Atacama watershed. When interviewing state officials, companies and experts, there was common reference to the 'lack of knowledge' about the Salar de Atacama (its origins, cycle of irrigation, dynamics of groundwaters and changes over time) amid claims that obtaining reliable data to better understand its dynamics requires decades of stable measurement. However, as data was only recently acquired, there were many uncertainties, rather than definitive answers. As Babidge (2019) identified, the lack of scientific knowledge has 'delayed regulatory action and sustained extractive activity' (96), moving *caution* to the future, discounting costs for future generations and allowing companies to keep expanding their operations.

Equally, elite scientific knowledge has been increasingly contested. A biologist explained her frustration that while the state and the companies said more research was needed, when the results provided a basis to oppose extraction, they were not taken seriously or seen as providing insufficient evidence. Corporate actors agreed that it was not just a question of acquiring 'any scientific knowledge' to understand the effects of brine extraction, but rather hydrogeological knowledge. However, one hydrogeologist insisted the 'lack of data' means uncertainty remains until there is data over many years because of the need to understand a dynamic and changing environment:

> it is a system [salt flat] that is constantly in motion, especially because they are highly influenced by seasonality; that is, because of what happens in summer and because of what happens in winter, then it is so dependent on the rains ... So no ... it is no use to go one day, and measure and take a picture because in the end in 6 more months you will see something else ... So, for that reason, it is relevant to have data over time. That is monitoring. ... So now, and at this moment the Dirección General de Aguas [General Directorate of Water] has monitoring in Chile. But their data has very low density. And the ones that do monitor a lot are the mining companies themselves. That is, Albemarle in this case and Soquimich. But they are under no obligation to share their data.

What this points to is the fact the disparity of data between the state and the companies does not allow the state to properly audit the companies, nor to make decisions about the quota taking into account reliable data. The knowledge inequalities are pronounced when considering that in Chile, most hydrogeologists work for mining companies whilst the state mainly audits the companies' reports. This imbalance creates power hierarchies that benefit the companies, preventing the state from making independent and informed decisions. It is also, of course, the case that in order to raise capital to finance new mining ventures, companies undertake extensive surveys and (pre)feasibility studies when prospecting for mining wealth. For example, Albemarle and SQM both trade on the New York Stock Exchange which imposes regulatory requirements upon them. To take one example, a Pre-Feasibility Study running to over 270 pages in length was published in 2022 for the Securities and Exchange Commission in the US by Albemarle on Salar de Atacama Región II, Chile, covering data on mineral resources, value and environmental impacts.[3] Hence, the data exists; it is just not evenly shared, particularly with those who might oppose mining.

Despite the uncertainty claimed by the state, companies and experts, several indigenous interviewees repeatedly said that for them it was evident that the Salar de Atacama was being damaged – although they lack what state officials and mining companies would consider relevant 'data'. A lickanantay woman said that she was raised near the salt flat, and she and her family were struck by the changes: not only were some lagoons and freshwater spring pools less abundant and there was less grass for their livestock, but there were notably fewer flamingos coming to the lagoons of the *Salar*. Equally, she said that it was clear for her the salt flat was damaged as it was a yellow color, instead of the white that it used to be and that was a sign of health. Another lickanantay man highlighted the spiritual understanding of this damage:

> The question is what is the state of health of the Salar de Atacama now? We don't know, we don't know, nobody knows! I mean, what we do know, and I am clear about it in a more spiritual way, we know that changes are happening. We know that the Salar de Atacama is being destroyed ... It's like taking the blood out of the human body and bleeding it out little by little. Therefore, you know that it will die at some point.

[3]See https://investors.albemarle.com/node/23741/html#exhibit9631231202110-k.htm.

The parallels with blood and water were often present when interviewing lickanantay people about Puri (water in the lickanantay ancient language *ckunza*) and the *Salar*. For lickanantay, both are considered more-than-human entities that are part of what Boelens describes as the hydro-cosmological cycle: the interconnection of 'cyclical dynamics of hydrology, agro-ecology, human lifetime and cosmology' (2014, 234). It expresses the lickanantay view of '*Puri's* walk' in the territory and its relationship with the salt flat and its brines. Therefore, lithium extraction is understood as water mining, with the effect of destroying the salt flat and with it, the entire territory, including both human and non-human beings where Puri is the blood of the territory and the *Salar*, 'the heart' of the hydro-cosmological cycle in the basin. An elder lickanantay woman explained it like this:

> I have always said that my people have a trunk, its main trunk is the – the spiritual world and the Salar [salt flat] is like where the roots of a great tree of life are, because that is where life begins in the Salar. In other words, now with the scientific vision that we have of the Salar, with the extremophiles ... there is a beginning of life.

Although this is a vision also supported by scientific research (Cubillos et al. 2018; Dorador et al. 2009; Farías and Contreras 2018), companies and state officials interviewed referred to indigenous knowledge dismissively as 'perceptions' that needed to be evaluated against facts coming from data and scientific knowledge. In contrast, several lickanantay people interviewed understood the need and benefit of using both ancestral and scientific knowledge, as long as there is mutual respect. However, as long as the state considers ancestral knowledge to be merely based on 'perceptions', not only do indigenous people not have tools to combat the environmental damage, but their own knowledge is denigrated and lost. As such, the colonial understanding of what knowledge means and what types of knowledge are valid plays an important enabling role when it comes to expanding mining industries.

4. Decolonizing transitions

Whether through violent forms of land acquisition and dispossession and epistemicide of the knowledge claims of others (De Sousa Santos 2014) or, moving outward, the uneven patterns of resource extraction and exchange within and between countries and within and between rural/urban landscapes, the 'colonial shadow of green electromobility' (Jerez, Garcés, and Torres 2021) is cast over this case. In this sense, the case we have presented here lends weight to calls from others to decolonize energy (Ghosh et al. 2021) and for the study of transitions and to attend more seriously to the role of race (as well as other key social characteristics) in the politics of energy transition (Newell 2021a).

Neo-colonialism is manifest in the way knowledge is mobilized and denied, resources flow, land is acquired, chains of accumulation are built and projects are implemented. This assemblage of extractivism reflects, reveals and embodies multi-scalar inequalities operating within the current and historical geopolitical order. Current energy transition pathways indicate how global elites (and their national counterparts) continue to subdue the Global South, and particular social groups within the majority world, reinforcing a long trajectory of resource regionalism and economic dependence (Humphreys Bebbington and Bebbington 2010, 147). Behind the talk of partnership and green growth we find the appropriation of territory, whereby dominant forms of capitalism preclude possibilities for traditional ways of making use of territory, such as living from livestock and small-scale agriculture. These industries and imaginaries perpetuate a dependency circle: while mining consumes more water and does not allow other economic activities to be developed (such as agriculture), people are forced to sell their labor to mining companies in order to live according to 'modern standards' of development (premised on providing a formal education to their children, living in certain types of houses, having cars, and so on).

However, mining only increases the destruction of the ecosystem that supports peoples' livelihood, which in turn only makes them more dependent on it so they are left without options when minerals for mining are exhausted, and they lose their jobs. Here, lithium operates as both promise and peril: socially constructed as 'oro blanco' (white gold) and the basis of a new mineral boom, but achievable only in ways and through means that repeat historical patterns of dispossession, exclusion and uneven but dependent development: a situation with antecedents in Spanish colonialists' violent search for the sacred gold of the Incas in the fifteenth century.

On the other hand, while lithium extraction has been depicted as environmentally damaging by non-governmental organizations (NGOs), researchers and indigenous communities (Jerez 2018; Liu, Agusdinata, and Myint 2019) and conflicts remain between companies and indigenous communities, past research in Chile (Argento and Puente 2019; Babidge and Bolados 2018; Jerez, Garcés, and Torres 2021; Kohn 2015; Nacif and Lacabana 2015) has overlooked the relevance of money and the job opportunities that communities have received from companies, along with scholarships to help children to study and finance for indigenous initiatives. Even if lithium companies do not offer many job opportunities (Slipak and Urrutia 2019), they do help to fulfill the needs of some rural communities. As opportunities to study and paid work concentrate in urban areas, the past decades have seen a massive emigration from the rural desert and the salt flat to the cities. The prospect of work in lithium mining means some might stay. As one indigenous lithium-worker and former indigenous leader stated: 'This village was dying, everyone left to Calama city to work in mining companies. But with lithium, everyone came back'. This is a familiar development dilemma: better 'take the benefits we can, while it lasts', as this worker stated. This dilemma was echoed by another former leader:

> it is a complex scenario, because we have talked with our brothers, and we have analysed it and mining is directly a mirage. It is directly a mirage because what is going to happen tomorrow when it is over, when the companies have to leave? What will remain? Well apart from the environmental hole that they are going to leave, then what will they be left with if they [atacameños] don't learn to do other things? Because we also have the conflict, especially in the youth, that many young people no longer know how to 'throw a shovel', or how to irrigate, nothing. So, they practically depend 100% on the work of the company, and if they are not in the company, they simply do not know how to do anything else.

As capitalism expands and national development remains tied to ideas of urban progress and modernity, paid work constitutes the only option to access the formal education system, healthcare services, housing, food, technology and so on. In this regard, many interviewees reflected on the future and the possibility to stay on their ancestral lands. If lithium extraction ends, the villages/ayllus 'will die', but if lithium companies remain, the salt flat will die. Therefore, the more dependent the local communities are upon paid labor and the more traditional agriculture and livestock are abandoned and impacted by climate change through drought in particular, the harder it will be to remain in the Atacama Desert. Those who defend 'traditional' ways of living continue to push for alternative futures to the salt flat as they see it as 'death sentence' for them and their ancestral territory. In this context, lithium extraction constitutes not only a depletion of resources, but also the advance of a modernity/coloniality (Quijano 2000) that implies the destruction of other worlds and ways-of-being-in-the-world.

To decolonize the transition in this setting would require a challenge to the dominant but narrow focus on energy 'transitions' centered around elite, largely Northern, framings of 'socio-technical transitions'. This is underpinned by a 'plug and play' mentality of adding new energy sources or technologies to the mix without shifting the means of provision or disturbing the power relations that keep unequal access and uneven exposure to harm in place. It stands in opposition to a more transformative agenda assembled around diverse knowledges and pluriverses (Escobar 2018) and organized institutionally and materially around a regenerative rather than an extractivist economy.

5. Conclusion

The account we have provided here of the assemblages of extractivism organized around the lithium triangle in Latin America links global political economies and ecologies within and beyond the region. These operate internationally between core and periphery in the global political economy of energy transitions, but are closely related to patterns of internal colonization when, as with this case, mestizos continue to appropriate the resources and labor of indigenous groups. As well as emphasizing material appropriation and power and resources, our account also drew attention to the institutional and discursive elements of an 'assemblage of extractivism' required to manage and extract value from the materiality of lithium. Following Bridge (2001), we showed how certain regions get constructed as commodity-supply spaces, which install epistemologies that legitimize discourses of growth, development and construct the role of each territory as a source of comparative advantage in an increasingly competitive global economy.

The analysis contributes to debates about the global political economy of energy transitions and their uneven consequences and distributional outcomes and deepens an understanding of the everyday organization of political economies and political ecologies of extractivism. Our analysis articulates a more relational and transformative account of how responses to climate change can invoke new injustices if they are not attentive to existing social inequalities (Sultana 2022). In rural settings, many of the inequalities revolve around the central dimensions of agrarian political economy – property, labor, income and consumption/reproduction (Borras et al. 2022) – which currently replicate 'extractivist assemblages'. These fundamental issues need to be brought to the fore in the study of 'more than sociotechnical' transitions. In this regard, historical materialist accounts of assemblages can usefully foreground the 'relationship to nature and to each other through the wage relation and systems of social reproduction' (Jordhus-Lier, Camilla, and Tale Hammerø 2022, 319). Politically and strategically, our analysis helps to illuminate what is at stake in attempts to decolonize transition debates: the need to disrupt and displace the discursive, institutional and material assemblages that normalize and reproduce extractivism. As well as challenging these historically constituted and contemporary expressions of class-based and racialised inequalities in particular, moving beyond assemblages of extractivism also means addressing issues of demand and over-consumption that drive extractivism, as well as the narrow models of development that value only what is profitable for investors. It requires us to go beyond a 'plug and play' model of transition which merely adds technologies and energy sources to the mix rather than substituting them and reorganizing systems of provision (Newell and Martin 2020). 'Ecologising' transitions also requires us to account for the ecological flows associated with transitions: the materiality and life course of minerals and the technologies they are used in, the circulation of pollutants and the energy embodied in the circulation of goods from production to exchange to consumption and waste management. But also, we need to see them as part of broader human and nonhuman ecologies in the ways suggested by the lickanantay people in Chile. Finally, democratizing transitions implies challenging questions of ownership and power distribution (Van Veelen and van der Horst 2018), given the problems associated with the concentration of state power in a state–corporate nexus that we have documented here.

Drawing attention to these dimensions does not do away with the need to engage with (rather than just critically reflect upon) the live and real dilemmas about energy pathways, because in the end there is no such thing as clean energy or green mining. There is just clean*er* energy and green*er* mining. Minerals are mined for most everyday goods – titanium for bikes and surgical equipment, copper for cables or minerals for toothpaste – not just for energy; and not just rare earth minerals, though these are often the focus of debate. Therefore, some mining in the near term is inevitable, just as ancient lickanantay engaged in mining activities (Nuñez 2007); but the intensive and extensive nature of hyper-extraction needs to be countered. This means addressing questions around how to reduce the demand for minerals and how to minimize the impacts of mining where it does occur: in this case, improving the efficiency of lithium recovery seems an urgent first step to reduce the extraction of brines. For those on the front lines of this mode of extractivism, it suggests the need for global solidarities to resist communities being played off against one another. Low-carbon, renewable energy and mitigation projects are as conflictive as fossil fuel projects, but out of 649 cases of resistance movements to both fossil fuel and low carbon energy documented by the Environmental Justice Organisations, Liabilities and Trade (EJOLT) atlas, over a quarter of projects encountering social resistance have been canceled, suspended or delayed (Temper et al. 2020). In this context, the discussion about minerals should be an entry point for re-thinking destructive economies. We need, then, to move beyond a narrow conversation about transition – which replaces one technology with another – to think about deeper transformations that are required to assemble knowledge, economies and institutions that are more equitable and sustainable.

Disclosure statement

No potential conflict of interest was reported by the authors.

Acknowledgments

The first author would like to acknowledge the financial support from the Agencia Nacional de Inves tigación y Desarrollo (ANID) in Chile and from the Royal Society of Geography (with IBG) under the Dudley Stamp Memorial Award.

ORCID

Peter Newell http://orcid.org/0000-0002-5371-7668

References

Anderson, Ben, Matthew Kearnes, Colin McFarlane, and Dan Swanton. 2012. "On Assemblages and Geography." *Dialogues in Human Geography* 2 (2): 171–189.

Arboleda, Martín. 2020. *Planetary Mine: Territories of Extraction Under Late Capitalism*. Brooklyn: Verso Books.

Argento, Melisa, and Florencia Puente. 2019. "Entre el boom del litio y la defensa de la vida. Salares, agua, territorios y comunidades en la región atacameña." In *Litio en Sudamérica: geopolítica, energía y territorios*, edited by El Colectivo Bruno Fornillo, 173–220. Colección Chico Mendes. Buenos Aires: Editorial El Colectivo.

Babidge, Sally. 2019. "Sustaining Ignorance: The Uncertainties of Groundwater and Its Extraction in the Salar de Atacama, Northern Chile." *Journal of the Royal Anthropological Institute* 25 (1): 83–102.

Babidge, Sally, and Paola Bolados. 2018. "Neoextractivism and Indigenous Water Ritual in Salar de Atacama, Chile." *Latin American Perspectives* 45 (5): 170–185.

Babidge, Sally, Fernanda Kalazich, Manuel Prieto, and Karina Yager. 2019. "'That's the Problem with That Lake; It Changes Sides': Mapping Extraction and Ecological Exhaustion in the Atacama." *Journal of Political Ecology* 26: 738–760.

Baglioni, Elena, and Liam Campling. 2017. "Natural Resource Industries as Global Value Chains: Frontiers, Fetishism, Labour and the State." *Environment and Planning A: Economy and Space* 49 (11): 2437–2456.

Baker, Lucy, Peter Newell, and Jon Phillips. 2014. "The Political Economy of Energy Transitions: The Case of South Africa." *New Political Economy* 19 (6): 791–818.

Baker, Lucy, and Benjamin K. Sovacool. 2017. "The Political Economy of Technological Capabilities and Global Production Networks in South Africa's Wind and Solar Photovoltaic (PV) Industries." *Political Geography* 60 (September): 1–12.

BBC. 2020. "Cornwall Lithium Deposits 'Globally Significant,'" September 17, sec. England, Local News, Regions, Cornwall. https://www.bbc.com/news/uk-england-cornwall-54188071.

Bebbington, Anthony, Denise Humphreys Bebbington, Jeffrey Bury, Jeannet Lingan, Juan Pablo Muñoz, and Martin Scurrah. 2008. "Mining and Social Movements: Struggles Over Livelihood and Rural Territorial Development in the Andes." *World Development* 36 (12): 2888–2905.

Boelens, Rutgerd. 2014. "Cultural Politics and the Hydrosocial Cycle: Water, Power and Identity in the Andean Highlands." *Geoforum; Journal of Physical, Human, and Regional Geosciences* 57 (November): 234–247.

Borras Saturnino, Ian Scoones, Amita Baviskar, Marc Edelman, Nancy Lee Peluso, and Wendy Wolford. 2022. "Climate Change and Agrarian Struggles: An Invitation to Contribute to a JPS Forum." *Journal of Peasant Studies* 49 (1): 1–28. doi:10.1080/03066150.2021.1956473.

Bridge, Gavin. 2001. "Resource Triumphalism: Postindustrial Narratives of Primary Commodity Production." *Environment and Planning A: Economy and Space* 33 (12): 2149–2173. doi:10.1068/a33190.

Bridge, Gavin. 2009. "Material Worlds: Natural Resources, Resource Geography and the Material Economy." *Geography Compass* 3 (3): 1217–1244.
Bridge, Gavin. 2011. "Resource Geographies 1: Making Carbon Economies, old and new." *Progress in Human Geography* 35 (6): 820–834.
Bridge, Gavin. 2014. "Resource Geographies II: The Resource-State Nexus." *Progress in Human Geography* 38 (1): 118–130.
Bustos-Gallardo, Beatriz, Gavin Bridge, and Manuel Prieto. 2021. "Harvesting Lithium: Water, Brine and the Industrial Dynamics of Production in the Salar de Atacama." *Geoforum; Journal of Physical, Human, and Regional Geosciences* 119 (February): 177–189.
Chagnon, Christopher W., Francesco Durante, Barry K. Gills, Sophia E. Hagolani-Albov, Saana Hokkanen, Sohvi M. J. Kangasluoma, Heidi Konttinen, et al. 2022. "From Extractivism to Global Extractivism: The Evolution of an Organizing Concept." *The Journal of Peasant Studies* May, 1–33.
Chululo. 2021. "CORFO reparte 45 millones," September 28, sec. Reportajes. https://www.chululo.cl/pages/reportajes2.php?id = 28092021_193255.
Comisión Chilena del Cobre (Cochilco). 2021. "Anuario de Estadísticas Del Cobre y Otros Minerales." https://www.cochilco.cl/Lists/Anuario/Attachments/24/Ae2021b.pdf.
Comisión Nacional del Litio. 2015. "Informe Final. Litio: Una Fuente de Energía, Una Oportunidad Para Chile." Ministerio de Minería. https://ciperchile.cl/pdfs/2015/06/sqm/INFORME_COMISION_LITIO_FINAL.pdf.
Corfo. 2019. "Corfo, SQM y Municipalidades Firman Convenios Para Concretar Primera Entrega de Recursos Acordados En Contrato de Salar de Atacama." State. *Corfo.Cl.* November 30. https://corfo.cl/sites/cpp/sala_de_prensa/nacional/convenio_recurso_atacama;jsessionid = JXTAlskeKGiGzl2p4eEldWo1kSCCKK3RyI-D65F0CiQ31vo_sz-z!-2090959747!-288122015.
Cubillos, Carolina F., Pablo Aguilar, Mario Grágeda, and Cristina Dorador. 2018. "Microbial Communities from the World's Largest Lithium Reserve, Salar de Atacama, Chile: Life at High LiCl Concentrations." *Journal of Geophysical Research: Biogeosciences* 123 (12): 3668–3681.
De la Cadena, Marisol. 2010. "INDIGENOUS COSMOPOLITICS IN THE ANDES: Conceptual Reflections Beyond 'Politics.'." *Cultural Anthropology* 25 (2): 334–370.
De la Cadena, Marisol, and Mario Blaser, eds. 2018. *A World of Many Worlds*. Durham: Duke University Press.
Deleuze, Gilles, and Félix Guattari. 1988. *A Thousand Plateaus: Capitalism and Schizophrenia*. London: Athlone Press.
De Sousa Santos, Boaventura. 2014. *Epistemologies of the South: Justice Against Epistemicide*. London New York: Routledge.
Dorador, Cristina, Daniela Meneses, Viviana Urtuvia, Cecilia Demergasso, Irma Vila, Karl-Paul Witzel, and Johannes F. Imhoff. 2009. "Diversity of *Bacteroidetes* in High-Altitude Saline Evaporitic Basins in Northern Chile: HIGH-ALTITUDE *BACTEROIDETES* DIVERSITY." *Journal of Geophysical Research: Biogeosciences* 114 (G2): G00D05.
Dunlap, Alexander, and James Fairhead. 2014. "The Militarisation and Marketisation of Nature: An Alternative Lens to 'Climate-Conflict'." *Geopolitics* 19 (4): 937–961.
Durante, Franceso, Markus Kröger, and William LaFleur. 2021. "Extraction and Extractivisms. Definitions and Concepts." In *Our Extractive Age: Expressions of Violence and Resistance*, edited by Judith Shapiro and John-Andrew McNeish, 19–30. Routledge Studies of the Extractive Industries. New York, NY: Routledge.
Escobar, Arturo. 2018. *Designs for the Pluriverse: Radical Interdependence, Autonomy, and the Making of Worlds*. Duke: Duke University Press.
Farías, María, and Manuel Contreras. 2018. *Extremofilos y Origen de La Vida En Atacama*. SEREMI Chile. Santiago de Chile.
Fornillo, Bruno, ed. 2015. *Geopolítica del Litio: Industria, Ciencia y Energía en Argentina*. Colección Chico Mendes. Buenos Aires: Editorial El Colectivo.
Garcés, Ingrid. 2017. *Litio: Industria, Recursos y Perspectivas Chilenas*. Santiago de Chile.
Ghosh, Bipashyee, Mónica Ramos-Mejía, Rafael Carvalho Machado, Suci Lestari Yuana, and Katharina Schiller. 2021. "Decolonising Transitions in the Global South: Towards More Epistemic Diversity in Transitions Research." *Environmental Innovation and Societal Transitions* 41 (December): 106–109.

Green, Duncan. 2003. *Silent Revolution: The Rise and Crisis of Market Economics in Latin America.* New York: Monthly Press.

Gudynas, Eduardo. 2009. "Diez Tesis Urgentes Sobre El Nuevo Extractivismo: Contextos y Demandas Bajo El Progresismo Sudamericano Actual." *Extractivismo, Política y Sociedad*, November, 187–225.

Gudynas, Eduardo. 2015. *Extractivismos: Ecología, Economía y Política de Un Modo de Entender El Desarrollo y La Naturaleza.* Primera edición. Cochabamba, Bolivia: CEDIB, Centro de Documentación e Información Bolivia.

Haarstad, Hårvard, and Tarje I. Wanvik. 2017. "Carbonscapes and Beyond: Conceptualizing the Instability of oil Landscapes." *Progress in Human Geography* 41 (4): 432–450.

Hale, Charles R. 2002. "Does Multiculturalism Menace? Governance, Cultural Rights and the Politics of Identity in Guatemala." *Journal of Latin American Studies* 34 (3): 485–524.

Harvey, David. 2004. "The 'New' Imperialism: Accumulation by Dispossession." *Socialist Register* 40. https://socialistregister.com/index.php/srv/article/view/5811.

Hope, Jessica. 2021. "The Anti-Politics of Sustainable Development: Environmental Critique from Assemblage Thinking in Bolivia." *Transactions of the Institute of British Geographers* 46 (1): 208–222.

Humphreys Bebbington, Denise, and Anthony Bebbington. 2010. "Anatomy of a Regional Conflict: Tarija and Resource Grievances in Morales's Bolivia." *Latin American Perspectives* 37 (4): 140–160. doi:10.1177/0094582X10372503.

IPCC. 2018. Global Warming of 1.5°C: An IPCC Special Report on the Impacts of Global Warming of 1.5°C above Pre-industrial Levels and Related Global Greenhouse Gas Emission Pathways. IPCC.

IRENA. 2019. *A New World: The Geopolitics of the Energy Transformation.* Abu Dhabi: IRENA.

IRENA. 2021. "Renewable Capacity Highlights." IRENA. https://www.irena.org/-/media/Files/IRENA/Agency/Publication/2021/Apr/IRENA_-RE_Capacity_Highlights_2021.pdf?la = en&hash = 1E1336 89564BC40C2392E85026F71A0D7A9C0B91.

Jasanoff, Sheila, and Sang-Hyun Kim. 2013. "Sociotechnical Imaginaries and National Energy Policies." *Science as Culture* 22 (2): 189–196.

Jerez, Bárbara. 2018. *Impacto Sicioambiental de la extraccion de litio en las cuencas de los salares altoandinos del cono sur.* Santiago de Chile: OCMAL.

Jerez, Bárbara, Ingrid Garcés, and Robinson Torres. 2021. "Lithium Extractivism and Water Injustices in the Salar de Atacama, Chile: The Colonial Shadow of Green Electromobility." *Political Geography* 87 (May): 102382.

Jordhus-Lier, David, Houeland C Camilla, and Ellingvåg Tale Hammerø. 2022. "Alienating Assemblages: Working the Carbonscape in Times of Transformation." *Progress in Human Geography* 46 (2): 319–338.

Kazimierski, Martín. 2019. "Transición Energética, Principios y Retos: La Necesidad de Almacenar Energía y El Potencial de La Batería Ion-Litio." In *Litio En Sudamérica: Geopolítica, Energía y Territorios*, edited by Bruno Fornillo, 25–50. Colección Chico Mendes. Buenos Aires: Editorial El Colectivo.

Kinkaid, Eden. 2020. "Can Assemblage Think Difference? A Feminist Critique of Assemblage Geographies." *Progress in Human Geography* 44 (3): 457–472. doi:10.1177/0309132519836162.

Kohn, Eduardo. 2015. "Anthropology of Ontologies." *Annual Review of Anthropology* 44 (1): 311–327.

Lancione, Michele. 2013. "Homeless People and the City of Abstract Machines: Assemblage Thinking and the Performative Approach to Homelessness: Homeless People and the City of Abstract Machines." *Area* 45 (3): 358–364.

Lawhon, Mary, and James T. Murphy. 2012. "Socio-Technical Regimes and Sustainability Transitions: Insights from Political Ecology." *Progress in Human Geography* 36 (3): 354–378.

Leiva, Fernando. 2019. "Economic Elites and New Strategies for Extractivism in Chile." *European Review of Latin American and Caribbean Studies* 108: 131.

Liu, Wenjuan, Datu B. Agusdinata, and Soe W. Myint. 2019. "Spatiotemporal Patterns of Lithium Mining and Environmental Degradation in the Atacama Salt Flat, Chile." *International Journal of Applied Earth Observation and Geoinformation* 80 (August): 145–156.

Machado Aráoz, Horacio. 2013a. "Crisis Ecológica, Conflictos Socioambientales y Orden Neocolonial: Las Paradojas de NuestrAmérica En Las Fronteras Del Extractivismo." *REBELA* 3 (1): 118–155.

Machado Aráoz, Horacio. 2013b. "Orden Neocolonial, Extractivismo y Ecología Política de Las Emociones." *Revista Brasileira de Sociologia Da Emoção* 12 (34): 11–43.
Machado Aráoz, Horacio. 2015. "El Territorio Moderno y La Geografía (Colonial) Del Capital. Una Arqueología Mínima." *Memoria y Sociedad* 19 (39): 174.
Meadowcroft, James. 2009. "What About the Politics? Sustainable Development, Transition Management, and Long Term Energy Transitions." *Policy Sciences* 42 (4): 323–340.
Molina, Raul. 2019. "Nostalgias, Conversiones y Desbordes En San Pedro de Atacama." *Revista Antropologías Del Sur* 6 (12): 261–281.
Murray Li, Tania. 2007. "Practices of Assemblage and Community Forest Management." *Economy and Society* 36 (2): 263–293.
Nacif, Federico, and Miguel Lacabana, eds. 2015. *ABC Del Litio Sudamericano: Soberanía, Ambiente, Tecnología e Industria*. Colección Economía y Sociedad. Buenos Aires, Argentina: Bernal, Buenos Aires, Argentinal: Centro Cultural de la Cooperación Floreal Gorini Universidad Nacional de Quilmes.
Newell, Peter. 2005. "Race, Class and the Global Politics of Environmental Inequality." *Global Environmental Politics* 5 (3): 70–94.
Newell, Peter. 2019. "Trasformismo or Transformation? The Global Political Economy of Energy Transitions." *Review of International Political Economy* 26 (1): 25–48.
Newell, Peter. 2021a. "Race and the Politics of Energy Transitions." *Energy Research & Social Science* 17: 101839.
Newell, Peter. 2021. *Power Shift: The Global Political Economy of Energy Transitions*. Cambridge, UK: Cambridge University Press.
Newell, Peter, and Adam Bumpus. 2012. "The Global Political Ecology of the Clean Development Mechanism." *Global Environmental Politics* 12 (4): 49–67.
Newell, Peter, and Abigail Martin. 2020. *The Role of the State in the Politics of Disruption & Acceleration*. London: Climate KIC.
Newell, Peter, and Dustin Mulvaney. 2013. "The Political Economy of the 'Just Transition': 1The Political Economy of the 'Just Transition'." *The Geographical Journal* 179 (2): 132–140.
Nixon, Rob. 2013. *Slow Violence and the Environmentalism of the Poor*. First Harvard University Press paperback edition. Cambridge, Massachusetts London, England: Harvard University Press.
Nuñez, Lautaro. 2007. *Vida y Cultura En El Oasis de San Pedro de Atacama*. Imagen de Chile. Santiago de Chile: Editorial Universitaria.
Núñez, Andrés, Matthew C. Benwell, and Enrique Aliste. 2020. "Interrogating Green Discourses in Patagonia-Aysén (Chile): Green Grabbing and eco-Extractivism as a new Strategy of Capitalism?" *Geographical Review* (August): 1–19.
Nuñez, Lautaro, and Pierre Pourrut, eds. 1995. *La Ocupación del Espacio y Economía Campesina en la Región Atacameña: Aspectos Dinámicos*. Antofagasta: Universidad Catolica del Norte, UCN: Institut Français de Recherche Scientifique pour le Developpement en Cooperation.
Observatorio de Conflictos Mineros de América Latina (OCMAL). 2018. "Impacto Socioambiental de La Extracción de Litio En Las Cuencas de Los Salares Altoandinos Del Cono Sur (OCMAL)." Santiago de Chile. https://www.ocmal.org/wp-content/uploads/2018/08/Impacto-Sociambiental-Litio.pdf.
Observatorio Latinoamericano de Conflictos Ambientales (OLCA). 2020. *Institucionalidad Del Dialogo Territorial. La Privatización Del Diálogo*. Quimantú: Santiago de Chile.
Ocampo, José Antonio. 2017. "Commodity-Led Development in Latin America." In *Alternative Pathways to Sustainable Development: Lessons from Latin America*, edited by Gilles Carbonnier, Humberto Campodónico, and Sergio Tezanos Vázquez, 51–76. Boston: Brill | Nijhoff.
Osses, Leonidas. 2021. "Chile 2022: el litio y la era de un nuevo Código de Minería". *Minería Chilena, January 20*. https://www.mch.cl/columnas/chile-2022-el-litio-y-la-era-de-un-nuevo-codigo-de-mineria/#
Perez, Carlota. 2002. *Technological Revolutions and Financial Capital: The Dynamics of Bubbles and Golden Ages*. Cheltenham, UK: E. Elgar Pub.

Perreault, Tom. 2012. "Dispossession by Accumulation? Mining, Water and the Nature of Enclosure on the Bolivian Altiplano." *Antipode*, December, no-no.

Powers, Martha, Poune Saberi, Richard Pepino, Emily Strupp, Eva Bugos, and Carolyn C. Cannuscio. 2015. "Popular Epidemiology and 'Fracking': Citizens' Concerns Regarding the Economic, Environmental, Health and Social Impacts of Unconventional Natural Gas Drilling Operations." *Journal of Community Health* 40 (3): 534–541.

Quijano, Aníbal. 2000. "Colonialidad Del Poder y Clasificación Social." *Journal of World-Systems Research* XI (2): 342–386.

Reporte Sostenible. 2019. "Producción de Litio Verde: El Aporte de SQM al Desarrollo Sostenible." *Reportesostenible.Cl.* November 9. http://reportesostenible.cl/produccion-de-litio-verde-el-aporte-de-sqm-al-desarrollo-sostenible/.

Robbins, Paul. 2014. "Cries Along the Chain of Accumulation." *Geoforum; Journal of Physical, Human, and Regional Geosciences* 54 (July): 233–235.

Roberts, J. Timmons, and Bradley C. Parks. 2009. "Ecologically Unequal Exchange, Ecological Debt, and Climate Justice: The History and Implications of Three Related Ideas for a New Social Movement." *International Journal of Comparative Sociology* 50 (3–4): 385–409.

Rodney, W. 1972. *How Europe Underdeveloped Africa*. New York: Verso Trade.

SEI, IISD, ODI, E3G, and UNEP. 2021. *The Production Gap Report* 2021. https://productiongap.org/2021report/.

Servicio Nacional de Geología y Minería (Sernageomin). 2020. "Anuario de La Minería de Chile." Santiago de Chile: Servicio Nacional de Geología y Minería. https://www.sernageomin.cl/pdf/anuario_de_%20la%20Mineria_de_Chile_2020_290621.pdf.

Slipak, Ariel, and Santiago Urrutia. 2019. "Historias de La Extracción, Dinámicas Jurídico-Tributarias y El Litio En Los Modelos de Desarrollo de Argentina, Bolivia y Chile." In *Litio En Sudamérica: Geopolítica, Energía y Territorios*, edited by Bruno Fornillo, 83–132. Buenos Aires: Editorial El Colectivo.

SQM. 2021. "SQM Sustainable 2020." Sustainability report. Chile: SQM. https://www.sqm.com/en/sustentabilidad/reporte-de-sustentabilidad.

Sultana, Farhana. 2022. "Critical Climate Justice." *The Geographical Journal* 188: 118–124.

Svampa, M. 2015. "Commodities Consensus: Neoextractivism and Enclosure of the Commons in Latin America." *South Atlantic Quarterly* 114 (1): 65–82.

Tarantini, Pablo. 2020. El litio y la oportunidad para el desarrollo sostenible. *Inter Press Service* (*IPS*). October 23. https://ipsnoticias.net/2020/10/litio-la-oportunidad-desarrollo-sostenible/.

Temper, Leah, Sofia Avila, Daniela Del Bene, Jennifer Gobby, Nicolas Kosoy, Philippe Le Billon, Joan Martinez-Alier, Patricia Perkins, Brototi Roy, and Arnim Scheidel. 2020. "Movements Shaping Climate Futures: A Systematic Mapping of Protests Against Fossil Fuel and Low-Carbon Energy Projects." *Environmental Research Letters* 15: 123004.

U.S. Geological Survey. 2022. "Mineral Commodity Summaries." U.S. Geological Survey. https://pubs.usgs.gov/periodicals/mcs2022/mcs2022.pdf.

Van de Graaf, Thijs, and Benjamin K. Sovacool. 2020. *Global Energy Politics*. Cambridge, UK; Medford, MA: Polity.

Van Veelen, B., and D. van der Horst. 2018. "What Is Energy Democracy? Connecting Social Science Energy Research and Political Theory." *Energy Research & Social Science* 46: 19–28.

Viveiros de Castro, Eduardo. 2004. "Perspectival Anthropology and the Method of Controlled Equivocation." *Tipití: Journal of the Society for the Anthropology of Lowland South America* 2 (1): 3–22.

Voskoboynik, Daniel Macmillen, and Diego Andreucci. 2021. "Greening Extractivism: Environmental Discourses and Resource Governance in the 'Lithium Triangle'." *Environment and Planning E: Nature and Space* 0: 1–23.

Watts, Michael John. 2021. "Hyper-Extractivism and the Global Oil Assemblage." In *Our Extractive Age: Expressions of Violence and Resistance*, edited by Judith Shapiro and John-Andrew McNeish, 207–248. Routledge Studies of the Extractive Industries. New York, NY: Routledge.

Zehner, Ozzie. 2012. *Green Illusions: The Dirty Secrets of Clean Energy and the Future of Environmentalism*. Our Sustainable Future. Lincoln: University of Nebraska Press.

Zicari, Julián, Bruno Fornillo, and Martina Gamba. 2019. "El Mercado Mundial Del Litio y El Eje Asiático. Dinámicas Comerciales, Industriales y Tecnológicas." In *Litio En Sudamérica: Geopolítica, Energía y Territorios*, edited by Bruno Fornillo, 51–80. Buenos Aires: Editorial El Colectivo.

OPEN ACCESS

Adapting to climate change among transitioning Maasai pastoralists in southern Kenya: an intersectional analysis of differentiated abilities to benefit from diversification processes

Edwige Marty, Renee Bullock, Matthew Cashmore, Todd Crane and Siri Eriksen

ABSTRACT

With increasingly fragmented rangelands, restricted mobility and climatic stress, diversification has accelerated among East African pastoralists. Diversification is also promoted as a climate change adaptation strategy to reduce climatic exposure. Through a study of a Maasai communal land in southern Kenya, we analyze how pastoralists navigate changing access to key productive resources that are linked to diversification processes, social differentiation, and the reshaping of livelihood practices. By integrating an intersectional approach in access theory, we unpack a deeper level of context specific patterns of inclusion and exclusion embedded within evolving production relations.

Introduction

> 'That time when the world started changing, people started looking for casual jobs, work they could do so that they could take care of their children because cows had finished.'
>
> *Elder Maasai woman, Olkiramatian, 2021.*

In recent decades, pastoralists in East Africa have been diversifying their livelihoods at a scale and speed not previously observed (Lenaiyasa et al. 2020; Little 2021; McCabe, Leslie, and DeLuca 2010; Karmebäck et al. 2015; Caravani 2019). Diversification, now essential to many households' livelihood strategies (Homewood, Kristjanson, and Chenevix Trench 2009; Pollini and Galaty 2021; Nkedianye et al. 2020), includes changing herd composition and livestock production patterns, often to be more intensive and commercially-oriented (Herrero et al. 2016; Rutten 1992). Diversification can also involve investing time and resources away from livestock keeping into ventures such as crop farming, business, or wage employment (Pollini and Galaty 2021). This trend has multiple drivers, both environmental and political economic.

With changing climatic conditions, including increasingly unpredictable rainfall patterns and an increased likelihood of extreme events, diversification is increasingly framed as a cornerstone of climate resilient households, communities and systems in discourses and policies (Cochrane and Cafer 2017). In pastoral areas, the emphasis on diversification comes from recognition of the high vulnerability of the sector to climatic stressors (Herrero et al. 2016). Estimates of climate change's impacts on pastoral production systems predict higher livestock mortality rates and weaker reproductive capacities, with associated declines in milk and meat production (Herrero et al. 2016; Thornton et al. 2009). Climate change is also predicted to affect water availability as well as fodder composition and quality, which could increase competition for grazing grounds and water resources in pastoral areas (Herrero et al. 2016; Thornton et al. 2009).

Diversification, however, is a particularly contentious issue in the pastoralism literature because it also responds to political economic conditions – notably decades of sedentarization and privatization policies imposed on historically marginalized pastoral communities which have led to fragmented rangelands, creating impetus for diversification (Lenaiyasa et al. 2020; Little 2021; Scoones 2021; Nkedianye et al. 2020; Homewood, Kristjanson, and Chenevix Trench 2009). In the Kenyan context, pastoralists were often framed in policies as unproductive land users and pushed towards alternative activities that were more intensive and market-oriented (Nkedianye et al. 2020; Odhiambo 2014; Boles et al. 2019). Nevertheless, diversification is not always a forced response to hardship, but also occurs through important pull factors, such as changing cultural values and socio-economic aspirations (McCabe, Leslie, and DeLuca 2010; Lenaiyasa et al. 2020). Academic debates on pastoral diversification examine the conditions under which it might or might not be considered successful, emphasizing trade-offs between specialization and risk mitigation and hinting at on-going processes of wealth differentiation and class formation (Nkedianye et al. 2020; Little 2021; Crane 2010; Caravani 2019; Galaty and Bonte 2018). Regardless of its drivers, diversification is predicated on shifting the use and access to key productive resources in pastoral landscapes, reshaping long-standing adaptive strategies and cultural institutions (Pollini and Galaty 2021).

Applying Ribot and Peluso's (2003) theory of access to Maasai pastoralists in southern Kenya, we analyse socially differentiated access strategies to resources necessary to diversify production and how these shape people's abilities to benefit from diversification processes in a changing climate. Social differentiation within agrarian societies is (re)produced through differential resource access, labor roles and decision-making power among social groups sharing specific characteristics, such as their age or gender (Ripoll et al. 2017). While pastoralists are often misrepresented as a homogenous group, new production patterns are reshaping resource use, labor roles and means of social reproduction in pastoral areas (Caravani 2019; Rao 2019; Wangui 2008). Similarly, the critical adaptation scholarship has long stressed the need to consider how historical inequalities and social structures differentiate vulnerability to climate change, as well as differentiated abilities to adapt (Ribot 2014). Adaptive processes themselves can challenge or reinforce existing inequalities (Eriksen, Nightingale, and Eakin 2015; Brisebois, Eriksen, and Crane 2022). Our analysis focuses on the embedded mechanisms of social inclusion and exclusion through interrogating how intersectional social positions regulate people's ability to benefit from diversification in the face of climate change.

This article contributes to critical agrarian studies and adaptation literature through an intersectional analysis of how pastoralists navigate the combined pressures of climate change, accelerated capitalist expansion, and changing agrarian politics affecting resource use. The objective is to understand how changing resource access and production relations linked to diversification processes (re)shape pastoralists' resource use in the face of climate change, following three interrelated questions: 1. How do pastoralists access the resources necessary to diversify livelihoods? 2. Which embedded mechanisms of inclusion and exclusion are (re)produced through these strategies of access and changing production relations? and 3. How do these affect the ability of individuals to access potential benefits derived from diversification and community resilience in a changing climate? We start by elaborating a novel analytical framework which analyses how the intersection of power, through gender, age and education, shapes access in a transitioning pastoral area. Understanding access through an intersectional lens directs critical attention to the (re)production of power relations through differentiated access strategies and changing production relations employed by various pastoralists to navigate unfolding diversification processes.

Through an empirical study, we engage with some of the questions raised in the introductory essay of this Forum on the social differentiation dynamics that are associated with climate change adaptation processes in agrarian settings (Borras et al. 2022, 17). We propose a case – one that looks at adaptation through diversifying production practices among Maasai pastoralists – which allows us to reflect on how differentiated people and societies engage with and are affected by adaptation processes responding to both increasing climatic stressors and changing agrarian political economies. This analytical approach also responds to recent calls by scholars to go beyond the classic pastoralist literature readily framing change as a peril to pastoralists and to borrow from critical agrarian studies perspectives to analyse the actual ramifications of changing trajectories of production, accumulation, and agrarian politics on diverse local resource users in order to understand the dynamics changing pastoral practices and associated shifting socio-cultural relations (Scoones 2021, 30; Greiner 2021).

Theoretical framework: accessing, intersectionality and pastoral adaptation

We approach diversification as 'an active social process' in which smallholders increasingly involve themselves with different activities (Ellis 1998, 5). Expected benefits from diversification are not only income-based (Ellis 1998). Instead, for pastoral adaptation, benefits can be understood in terms of resilience to increased climatic stressors through the spreading of risk and improving food security. Furthermore, we do not assume that diversification brings only benefits. Instead, we analyze people's abilities to access benefits from diversification as a way of understanding social differentiation in adaptation in a transitioning pastoral context. Changing resource access is an appropriate empirical entry point for studying diversification processes because pastoral livelihoods are closely linked to flexible and seasonal access to key socio-environmental resources to cope with high level of climatic uncertainty and variability while livelihood diversification implies new patterns of resource access as well as of social organization – both spatially and temporally which affects the resilience of the system (Scoones 2021).

Ribot and Peluso's *Theory of Access* (2003) construes access as the ability to reap benefit, building on earlier works that emphasized that access to resources needs to be considered as 'people's ability to control and use resources effectively' (Berry 1989, 41). The ability to benefit from resources is affected by rights-based, structural and relational mechanisms (Ribot and Peluso 2003). While legal rights are not irrelevant, they are not necessarily a straightforward indication of the ability to benefit. Instead, analysis of structural and relational mechanisms – such as access through authority, knowledge or social identity – serve to interrogate the disconnection between 'formal law and diverse practices' (Peluso and Ribot 2020, 302). Exploring the 'grey zone' between rights and access is particularly crucial in post-colonial contexts, where multiple co-existing institutions exercise normative and legal power (Sikor and Lund 2009, 2).

Recent contributions to access theory elaborate how power relations are embedded within and shape changing patterns of resource use. In particular, they explore authority relations between those controlling and those gaining or maintaining access (Milgroom and Ribot 2019). They also highlight conceptualizations of legitimacy and entanglement to show how changes in the power of customary institutions influence access for different social groups (Lau et al. 2020). Several scholars turn access theory around, to examine the mechanisms that exclude or prevent people from benefiting from resources (Pichler, Schmid, and Gingrich 2021; Hall, Hirsch, and Li 2011). We argue that the mechanisms that include or exclude people are intimately linked and should be further examined together. Another key contribution to access theory emphasizes the need to go beyond analyzing existing rules and norms of access to analyzing actual social practices of access, which the authors conceptualize as 'accessing' (Milgroom, Giller, and Leeuwis 2014). Analyzing practices of accessing resources highlights the relationality of people's agency and power in gaining access, and thus enables a clearer view of how accessing varies by social position. This aligns with the critical adaptation scholarship that understands vulnerability as socio-politically produced (Eriksen, Nightingale, and Eakin 2015; Taylor 2013), as well as with recent feminist works in the climate change field emphasizing the (re)production of social relations at different scales and at the intersection of inequalities (for example, Thompson-Hall, Carr, and Pascual 2016; Brisebois, Eriksen, and Crane 2022; Tavenner and Crane 2022).

Socially locating strategies for accessing resources shows how intersectional social positions differentially affect people's abilities to navigate change as well as the socio-political factors facilitating or hindering the ability to benefit from diversification. Within the domain of climate change adaptation, a previous focus on capacities to adapt was noted to be misleading by conceptualizing capacities as 'innate characteristics of those at risk' (Ribot 2014, 682). Critiquing techno-managerial approaches to climate risk reduction, recent literature emphasizes the (re)production of inequalities and the unintended effects of adaptive processes which can deepen the vulnerabilities of the most marginalized (Atteridge and Remling 2018; Eriksen et al. 2021). As an adaptation process, diversification brings new opportunities for some people, but can also displace risks and bring new exposures for others, acting as 'a socially stratifying capitalist fix providing new avenues for accumulation and market penetration', benefiting a small elite (Mikulewicz 2021, 424). Emphasizing adaptation as occurring within 'relational nature of power' situates individuals and social groups at the intersections of both 'power and oppression', requiring deeper analysis of on-going processes between and within households (Djoudi et al. 2016, 249).

We use an intersectional approach to probe the interplay of different social positions as they shape resource access, ownership, labor roles and decision-making patterns in agrarian contexts (Djoudi et al. 2016; Thompson-Hall, Carr, and Pascual 2016). Intersectionality refers to the ways multiple identify factors, such as age and gender, combine to shape one's experience of the world (Crenshaw et al. 1991). Aligning with recent calls to 'demonstrate how power and agency operate in complex, place-specific, and sometimes contradictory ways to affect lived experiences with adaptation' (Garcia and Tschakert 2022, 2), we focus our analysis on how the interplay of three axes of social differences – age, gender and formal education levels[1] – shapes people's ability to benefit from diversification. While our research was designed to analyze gendered and generational dynamics, we had not initially focused on education as a variable but its importance as an additional marker of social differentiation emerged through the fieldwork and data analysis. Focusing on education as an additional characteristic that intersects with gender and age helped to refine our understanding of the power dynamics. Other salient axes of social differentiation, such as wealth, ethnicity, or location, also influence abilities to benefit from diversification and these are mentioned when relevant but not covered in depth. This approach is a useful part of an intersectional analysis, because it allows us to inductively capture relevant aspects (Kaijser and Kronsell 2014), and grounds our understanding of how inequalities are (re)produced through diversification processes, highlighting both readily considered dimensions of inequality, such as gender, but also context-specific dimensions often not considered in adaptation research, such as education levels and language (Garcia and Tschakert 2022).

Figure 1 presents both rights-based and structural and relational mechanisms considered for our study. Similar to Milgroom, Giller, and Leeuwis (2014), we add another category as a distinct rights-based mechanism. They add 'institutions' to refer to rights secured through informal rules as opposed to rights attributed by law (2014, 201). However, we prefer to use the term 'locally-recognized' to refer to the rights operationalized locally through various forms of informal and formal governance processes, as we find a strict distinction between formal and informal rules difficult to maintain. Several land policies changes, as well as new cross-scale hybrid governance arrangements, have led to an assemblage of institutions coexisting and regulating resource use and management in the southern Kenyan pastoral rangelands (Brehony 2020). Moreover, we add moral economy (Agyei, Hansen, and Acheampong 2020) to the structural and relational mechanisms because it captures the strong cultural norms and values shaping resource governance among Maasai (Pollini and Galaty 2021; Archambault 2016). Agyei, Hansen, and Acheampong (2020, 239) define moral economy following Thompson (1971) as the ways 'economic activities are influenced and structured by moral dispositions, values and norms – a normative behavior that emerges from lived experience and people's intuitive sense of justice'.

[1]References to education in this manuscript imply having attended formal education. However, this does not mean we do not recognize Maasai's own customary educational systems and the transmission of indigenous knowledge which is often unacknowledged and undervalued.

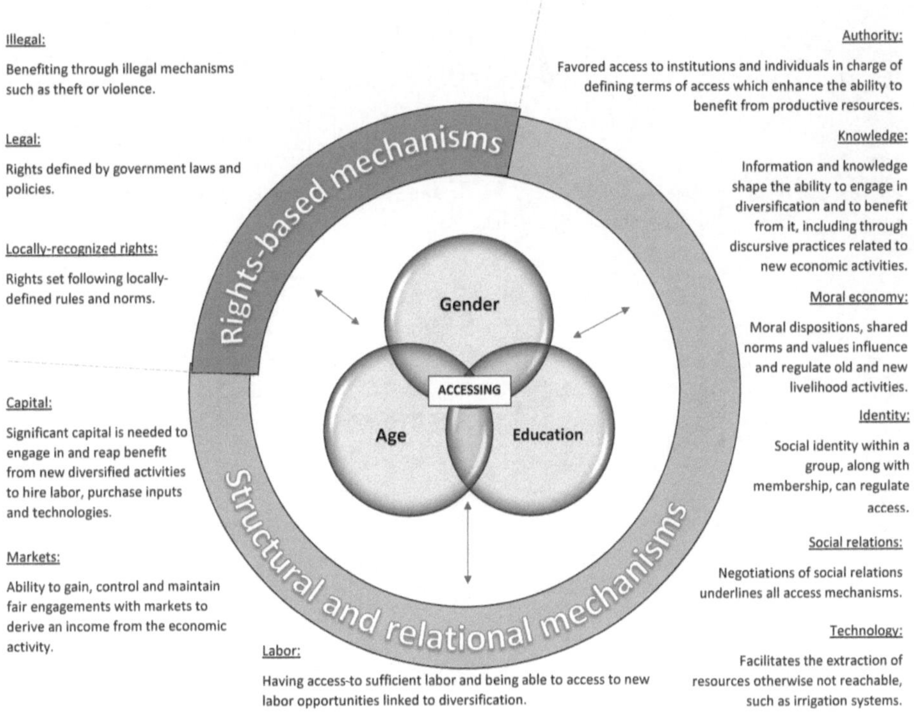

Figure 1. Intersectional approach to analyze access linked to diversification processes – access mechanisms' definitions drawn from Ribot and Peluso (2003); Milgroom, Giller, and Leeuwis (2014, 210); Agyei, Hansen, and Acheampong (2020, 239).

Research design

The setting[2]

The case of Olkiramatian

This study was conducted in the communally owned and managed land of Olkiramatian, Kajiado County, southern Kenya (Figure 2). This is an arid environment with bi-modal rainy seasons, receiving an average annual rainfall of less than <500 mm (Bobadoye et al. 2014). In recent years, higher rainfall variability and unpredictability have been noted within Kajiado County and this variability is predicted to increase with climate change, along with temperatures (Matsaba et al. 2021). These patterns, notably temperatures increases, fit with the observed historical climatic changes and with future climate change predictions for Kenya, but actual changes in rainfall remain hard to pinpoint, as they will vary greatly by location due to the complex topography (Gebrechorkos, Hülsmann, and Bernhofer 2018). While droughts remain the main concern, flooding events have increasingly affected Olkiramatian, as the wetter areas used for dry season grazing have recently experienced floods that destroyed crops, herds, and habitations.

Prior to independence, the British colonial government pushed Maasai into 'reserves', established through the 1904 and 1911 Maasai agreements, with the final southern

[2]While this section references recent research conducted in the area, the background of the study area explained here is also based on the collected research data.

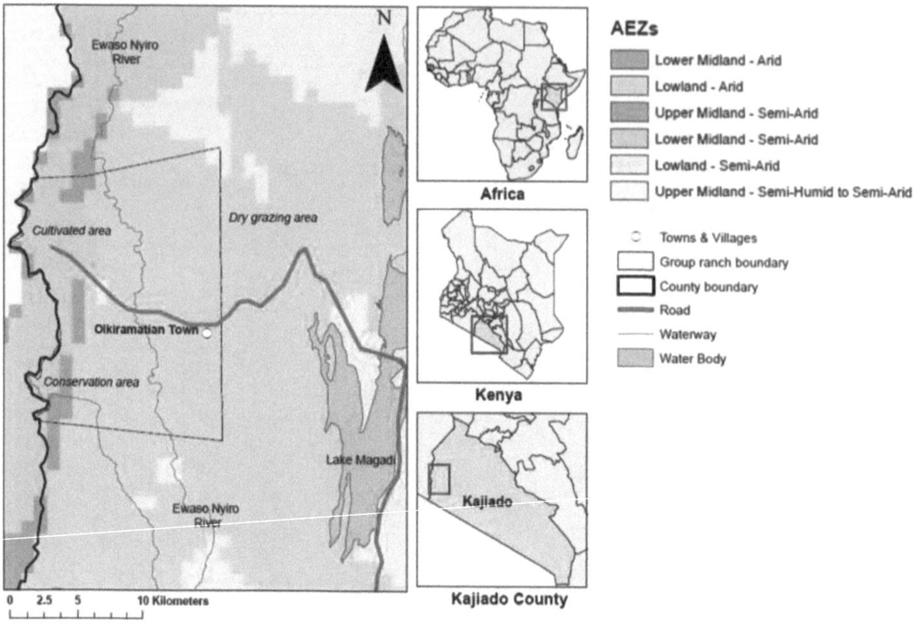

Figure 2. Olkiramatian. Scale: 1:210,537. Nairobi, Kenya: M. W. Graham, 5 April, 2022. ArcMap v. 10.6. ESRI Software, USA, 1995–2022.

reserve's borders mirroring today's Narok and Kajiado Counties (Hughes 2002). For Olkiramatian, the 1911 agreement is contentious for permitting the Crown to regulate mining and lease part of the land to the Magadi Ash Soda Company (Hughes 2008).[3] Following independence and the 1968 Land Adjudication Act (Republic of Kenya 1968), communally owned and managed pastoral areas in Kenya were administered as 'group ranches'. This legal status was applied at different times across Kajiado County, giving locally-elected management committees powers to regulate land uses (Mwangi 2003). This form of communal land title was thought a good compromise to reduce mobility and increase productivity, while conserving customary resource governance arrangements (German et al. 2017). Olkiramatian, with 24,000 hectares of land, became a group ranch in 1978, successfully appointing, and later electing, several management committees to govern alongside an assemblage of customary Maasai institutions, notably age sets, clan, and location-based leaders. Across Kajiado, most communal group ranches established in the 1970s were subdivided into privatized plots in the following decades, stimulating complaints about unfair subdivision and sales of land, as well as general lack of accountability of elected leaders (Kimani and Pickard 1998; German et al. 2017; Rutten 1992; Mwangi 2003).

Olkiramatian is one of the few group ranches within Kajiado County that remained mostly undivided. Yet, significant land use decisions were made in the following decades by the group ranch management committees. The boundaries of the group

[3]The Magadi concession is now run by Tata Chemicals which still legally controls half of Olkiramatian's land through a lease agreement debated in court – this broadly corresponds to the dry grazing area on Figure 2. Producing a map with clear boundaries for Olkiramatian is challenged by the on-going land disputes, not only with Tata Chemicals, but also around the Musenge area, and around the border area with Narok County.

ranch were redefined in 1993, with the subdivision and sale of land north of the group ranch, known as Musenge, to several influential leaders in an attempt to formally exclude the Ilpurko Maasai section, viewed as outsiders by the Illoodokilani majority since their displacement from Laikipia, and counter their settling within Olkiramatian (Brehony 2020). Starting from the 1990s, a partition of the group ranch into two areas was decided upon, with the wetter area of Olkiramatian – referred to as Phase 1, reserved for crop agriculture. Only registered group ranch members are eligible to be allocated a plot for farming. The delimited agricultural area, Nkuruman, is suitable for crop farming due to two rivers, Oloibortoto and Entasopia, coming down the escarpment and providing water for irrigation. Batemi agriculturists from the nearby Kenya/Tanzania border area, as well as Kikuyus and Kambas from the Kenyan highlands, have migrated to farms in Olkiramatian, seeking plot rental agreements with the registered members. Phase 2, the drier area, remained for communal use and pastoral livestock keeping. In 2003, the group ranch committee also delineated a conservation area which serves as a communal dry season reserve grass bank and a site for wildlife tourism (Brehony 2020).

Diversification processes and some of the opportunities that are present in Olkiramatian, notably for cash crop farming in Phase 1, are linked to the unique agro-ecology of the land but also to the better relative road connectivity compared to neighboring group ranches. This context makes Olkiramatian a unique case for analyzing social differentiation dynamics linked to diversification processes in a 'transitioning' pastoral system, one that is still communal but experiencing accelerated land use and production shifts.

Land reforms in the Kenyan drylands

In 2016, the Community Land Act (CLA) was introduced in Kenya with new regulations affecting the few remaining unsubdivided group ranches in Kajiado County, which had to transition towards being legally registered as community lands (Government of Kenya 2016). The status change was envisioned to increase inclusivity and accountability in governance by requiring the registration of all inhabitants of the former group ranches as community land members[4], while asserting indigenous communal land rights. Membership registration patterns within group ranches were key recurring nodes of tension across Kajiado and other Kenyan drylands (Mwangi 2003; Rutten 1992; German et al. 2017). There remains considerable ambiguity under the CLA as to who is and can be recognized as a member within a given community land, thus creating ambiguity about who can claim resource access (Muok et al. 2021). The transition occurs in the context of devolution following the 2010 constitution, with tensions across different scales of governance (Cheeseman, Lynch, and Willis 2016; Lind 2018). Notably, both county and national governmental bodies have land offices and the definition of the powers accorded to different land governance institutions is a very contested issue (Di Matteo 2022; Achiba and Lengoiboni 2020). The imposed transition initiated significant changes within Kajiado County, precipitating renewed pushes for subdivision among some of the unsubdivided group ranches. In Olkiramatian, lingering discussions around the subdivision of the agricultural area into individual land titles create significant

[4]The Community Land Act defines a community vaguely as "a consciously distinct and organized group of users of community land who are citizens of Kenya and share any of the following attributes— (a) common ancestry; (b) similar culture or unique mode of livelihood; (c) socio-economic or other similar common interest; (d) geographical space; (e) ecological space; or (f) ethnicity" (Government of Kenya 2016, 5).

Table 1. Overview of interview participants in the study.

Interview types	Men		Women
Focus group discussions (# of participants in each)			
Adults – wet area	1 (6)		1 (6)
Youth <35 years old[6] – wet area	1 (6)		1 (6)
Adults – dry area	1 (6)		1 (6)
Youth <35 years old – drier area	1 (7)		1 (6)
Total number of FGDs (# of participants)	= 4 (25)		= 4 (24)
Local leaders in resource management committees	7		3
Local resource users			
Adults	12		17
Youth <35 years old	7		7
County government employees			
Individual interview	3		1
Group interview (# of participants)	1 (1)		(1)
Extension officers for Magadi ward			
Group interview (# of participants)	1 (2)		(1)
NGOs /CBOs employees	4		1
Total number of individual interviews (# of participants)		63(67)	

socio-political turmoil that is still ongoing. At the time of writing, Olkiramatian has started to transition into a registered community land but with reservations to subdivide Phase 1 following the pre-existing membership list.

Methodology

This research followed a qualitative case study approach which centered on a small land area as it allows us to analyze in depth the context-specific inclusion and exclusion dynamics associated with changing resource access among diversifying pastoralists. Our aim was to explore processes of social differentiation associated with diversification processes. Our research design did not aim to generate representative data but to delve deeper into the complexities of differentiated strategies of access. We sampled from different demographic groups to gather differentiated individual and group perspectives on resource changes and access patterns (Table 1). Going beyond taking households as the unit of analysis is critical in a Maasai pastoral context, where production is negotiated through several social units with differentiated labor roles and responsibilities (McCabe, Leslie, and DeLuca 2010). The *enkaji* is the unit made of a wife and her children while the *olmarei* consists of the husband, his wives, and children, finally, the *enkang* is made of several *olmarei* living together managing the livestock (Coast 2001).

Data were collected through focus group discussions (FGDs) and semi-structured interviews. Eight FGDs were held with small groups disaggregated by sex, age and location/livelihood to provide insights into the resources most valued and used, as well as perceived changes over time. The FGDs also included a participatory ranking activity that was used to develop an understanding of changing normative ideals. Participants were asked to discuss the factors that made someone best-off or worst-off in the past 10

[6]This demarcation follows the Government of Kenya's definition of youth (Government of Kenya 2019).

years as compared to present. The FGDs broadly followed the GENNOVATE methodology (Petesch and Prain 2018) by adapting the Ladder of Life activity.

Following the FGDs, individual interviews were conducted through a dual sampling approach. First, local leaders active in resource management committees were interviewed. Additionally, a snowball sampling approach was used to select less visible local resource users, often including those less involved in decision-making. Most respondents were Olkiramatian residents, but several residents from the Musenge area were also interviewed, as this was instrumental in questioning further differences between rights based, structural and relational access. The interview guides focused on key resources for the community in different locations and associated access, control and management rules and probed livelihood activities and seasonal changes. Interviews were mainly held at the participants' homesteads, providing an opportunity for direct observation of main livelihood activities and accessed resources. Further interviews with county government officials focused on climate change plans for Kajiado County, as well as natural resources governance arrangements and cross-scales linkages.

Data were collected in several stages between November 2020 and October 2021, with the COVID19 pandemic occasionally disrupting fieldwork activities. The first author conducted the interviews together with research assistants translating the discussion simultaneously in Maa. Notes were taken during interviews, and they were recorded, except in 3 cases where respondents asked for the interview not to be recorded. Transcripts were then translated in English.

The transcripts were analyzed using the qualitative data analysis software NVivo12. Data analysis entailed four steps and was guided by the research questions and the operationalization of our theoretical framework. First, five overarching domain summaries were used to categorize the data, namely resource governance rules and regulations, livelihood-based diversification, livestock-based diversification, pastoral livestock keeping practices, and perceptions of changing livelihoods. Second, we used the mechanisms of access shown in Figure 1 to guide the analysis. Rights-based mechanisms to key productive resources were coded under the governance category as we operationalize it to refer to all locally-defined access rules. The structural and relational mechanisms of access were used as coding markers under each of the livelihood and livestock-based diversification types identified in Olkiramatian (Figure 3). The process also involved inductive coding to capture key themes and sub-themes relevant to our research objectives. Third, we compared different mechanisms of access against what they were used to access, such as productive resources or to the ability to use resources over time to diversify, across the diversification activities. The fourth and final step involved interrogating the access references from an intersectional point of view, by both looking at the social position of the respondents mentioning differentiated strategies of access and the social differences mentioned in the interviews' contents.

Results

The results are presented in four subsections which together examine differential abilities to realize benefits from diversification in the face of a changing climate and at the intersection of age, gender, and education levels. We start by analyzing the changing rights-based access mechanisms to key productive resources linked to diversification, and their

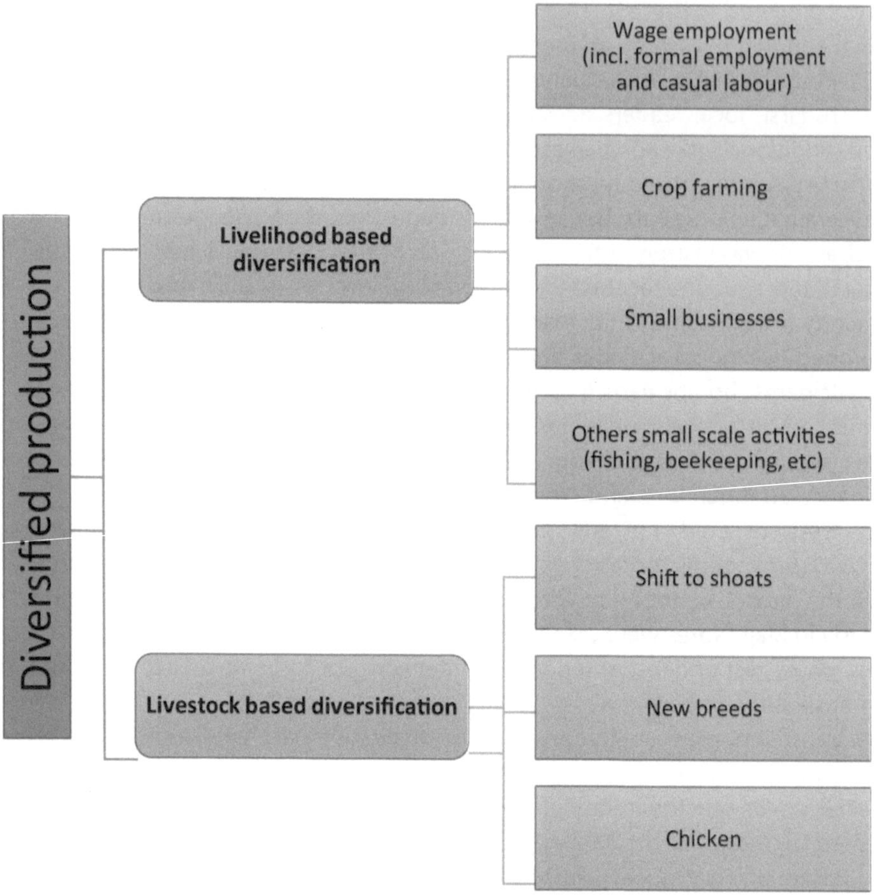

Figure 3. Types of diversification in Olkiramatian – developed from the initial data analysis.

implications for inclusion and exclusion. The second section presents the intersectional analysis of accessing by analyzing access strategies based on changing authority and knowledge relations, as well as the moral economy. The third section focuses on people who are often side-lined from direct access and examines how they negotiate access to key resources through their social networks. Finally, we analyze differential abilities to maintain access to realize long-term climate change adaptation benefits from diversification.

1. Changing rules and norms of access

This section highlights how increased diversification of production activities has led to a reorganization of the landscape to accommodate different resource uses, notably through formalizing access to farmlands, irrigation water, and urban plots. However, these newly defined rights-based mechanisms are negotiated through multiple institutions, with patriarchal customary authority patterns adapting from pastoralism to increased cash cropping and other commercial activities, and with a push towards the individualization of resource rights.

From its onset, membership registration for Olkiramatian group ranch was administered by successive elected management committees, who were also responsible for allocation of farmlands and urban plots. However, registration followed customary patriarchal authority patterns, registering household heads, primarily elder Maasai men. In Olkiramatian, this included male elders from the Illoodokilani section and a few non-Maasai household heads who had settled in Phase 1. Non-members' inclusion was necessary to reach the minimum number for group ranch registration (Ontiri and Robinson 2017). At the time, some Maasai household heads preferred not to register, viewing group ranches as externally-imposed westernized governance that they did not want to legitimize. Maasai women from Illoodokilani section were not generally included in the register, but exceptions were made for widows with no-cowives and some younger unmarried women recognized as household heads. The decision to register new members thereafter fell to the group ranch management committee, which initially allowed for registration of members' children, primarily eldest sons, once transitioned to adulthood.

With crop farming mostly viewed negatively among Maasai in the area, most Maasai residents initially used Phase 1 for late dry season grazing. However, cultivation was progressively perceived as lucrative, leading to more permanent settlements. The government's financial and technical support for farming through extension officers also played a role with the introduction in the 1990s of a cash crop market for export of Asian vegetables and seed distribution. Following the register, and seemingly at a time when Maasai's uptake of crop farming first increased, a first allocation of farmlands into equal size plots took place to formalize and equalize individual farming plots:

> 'The men said in 1993, we want Phase 1 to be subdivided and issued with title deeds. So, we then subdivided Phase 1 into equal portions. But before that, people used to book *[to claim possession of]* land in different sizes according to their wish.' *Elder man in Phase 1.*

As more Maasai asked to be allocated plots in Nkuruman, the group ranch committee decided to close the membership register and forbid new registrations from 2003 onwards. This decision was ostensibly based on the wish to prevent further partition of the farmlands into smaller areas. Young men who came of age after this date could not formally be registered as members and thus could not be allocated a share of land in Phase 1. Less than one third of the current Olkiramatian population can claim direct access to farmlands through being a registered member.

Crop farming was also supported by development of irrigation systems in Nkuruman, namely irrigation pipes and dug canals. However water availability remained limited, notably during dry seasons, and as more farms emerged, irrigation access was also formalized with water being allocated to each farm in rotation. Water governance rules are now decided on by the group ranch committee and subcommittees. Local water committees, composed of appointed village elders, usually settle local resource users' conflicts. Complaints can also be lodged locally through governmental channels – independent of group ranch governance structures – but the costs are higher. Since 2018, another cash crop market for fruits to be sold to Nairobi, mainly papayas, is further incentivizing crop farming, including the clearing of more land for cultivation and investments in water pumping systems to irrigate larger areas.

Similar dynamics towards increased interests in town plots and formalization of access rights are notable around emerging towns. Most of the Maasai were initially uninterested in permanent settlements in towns. However, many non-Maasai settled, notably in Entasopia in Phase 1, which grew exponentially following crop farming. With some villages now in the process of being registered as towns, plot access is formalizing because the county government is responsible for issuing title deeds and receiving payments. This also applies to Olkiramatian town in Phase 2, location of the weekly livestock market. Shifting from group ranch administration to public administration means that non-members of Olkiramatian can purchase town plots. However, in most emerging towns, such as Olkiramatian town, a local committee for plot allocation was also appointed by the elected county ward official to further define rules of access.

The above changing rules and norms of access in Olkiramatian affect patterns of access at the intersection of one's social position. We now turn towards how accessing takes place in practice and shapes the ability of individuals to access benefits from diversification.

2. Accessing through authority and knowledge

Intersecting education, gender and age relations are (re)shaping authority over land, with land registration processes exacerbating intersectional differentiation due to the increased importance of education to navigate complex land policy changes but with the continued gendered patterns of exclusion. Educated younger Maasai men – notably those from age sets more widely formally educated and thus literate – are now leading most governance processes in Olkiramatian. Across Kajiado, the group ranch set-up solidified a shift towards a smaller elected group of educated younger men leading and making decisions for the wider community (Rutten 1992). Accordingly, Olkiramatian elders noted that recognition of who is seen as knowledgeable has shifted towards those with formal education level, which is instrumental to accessing leadership positions and desirable jobs in governmental and non-governmental organizations. In the past, younger people were not recognized as knowledgeable due to their lack of life experience relative to elders.

This shift has important implications for accessing resources for diversification. Some respondents explained the importance of education for communicating with outsiders, such as government agents and even researchers. Most importantly, educated younger men can better understand national and local development processes and secure land for the community and/or for themselves. Even the membership list used for allocation is now digitalized and edited on laptops. Furthermore, the language of policymaking is English, or occasionally Kiswahili, which marginalizes non-speakers. During fieldwork, respondents sometimes asked us for translations of land policies changes, which are written in English, underscoring linguistic barriers as mechanisms of exclusion felt by many Maasai.

Educated Maasai in positions of authority are also better able to influence and redefine patterns of resource access for elite capture (Rutten 1992). In Olkiramatian, while communal rules specify that equal allocation of farmlands shares is a right for all registered members, the actual allocation of farmlands has become skewed with claims that well-connected individuals, often in positions of leadership, used their authority to allocate additional farmlands shares to themselves and others. Some land size differences can be explained by the time of their allocation – people who had settled earlier in Phase 1 have bigger farmlands than latecomers. Yet, some respondents indicated that well-positioned individuals within the group ranch leadership committees have also acquired extra shares of farmlands through different means, for instance through registering additional family members or wealthy outsiders in the group ranch register, in contradiction to locally-defined rights. There were also complaints that the placement of an irrigation project – funded by government through a grant from a multilateral development institution – was heavily influenced by individuals in leadership positions, who decided in advance that part of the irrigation pipe would come closer to their own newly allocated farms.

Structural power dynamics surrounding allocation of farmlands and other agricultural resources are also at the heart of current tensions regarding subdivision and transition to community land. Some interviewees, notably young men seeking land allocation, stressed outsiders' influence on the process, with the illegal registration of non-residents with significant political or financial weight at higher scales of governance or from parts of Kajiado without access to farmland. The registration of outsiders is facilitated by well-positioned individuals in order to secure political back-up for a favorable subdivision of Phase 1, while also responding to a patronage system, common to land allocation processes (Boone 2012). Longtime residents who are unregistered, including non-Maasai, are also seeking to secure farmlands, with some buying or being promised shares ahead of subdivision. One young Maasai man reflected that powerful individuals were spreading misleading information on changing land policies, thus feeding into historical fears of further land losses, and pushing people to seek to secure their rights quickly through official land titles. An employee of a locally active NGO noted that a significant part of their work had shifted towards hiring a lawyer and informing people on land policies.

Yet, the moral economy continues to shape structural access by counterbalancing recent developments towards individual elite accumulations of resources linked to diversification. Maasai governance norms about resource sharing and adaptive decision-making – which should include different clans, age sets, and location representatives – nonetheless continue to be practiced and guide perceptions of what is socially acceptable. Skewed allocation patterns linked to new economic activities are also not acceptable to the wider community and are actively protested, notably by excluded young men. Debates on subdivision have led to the organization of *barazas*, community-wide consultations, in contrast to a small political elite trying to settle matters in closed spaces with different political actors.

The moral economy also deepens exclusionary access patterns along other axes, notably along gendered and ethnic lines. Maasai from the Illoodokilani section often frame longtime non-Maasai residents as outsiders who do not have legitimate claims on membership and ownership, citing their not respecting Maasai rules and norms and being quick to fence and punish livestock harshly for trespassing in Phase 1. Maasai from the IlPurko section are likewise not considered as community members and were never registered. The continued exclusion of most Maasai women from membership and wider community decision making processes – such as consultations to discuss subdivision of Phase 1, as well as from formal land ownership and inheritance through the group ranch constitution – is still widely regarded as acceptable and infrequently contested. This is due to gendered norms relegating women to decision-making at the *enkaji*'s level and, when it comes to resources falling within their management responsibilities such as water for human consumption or milk. Land use issues are perceived as elder men's responsibility which should not be discussed by women (Mwangi 2003).

Women, generally discouraged from engaging in wider consultations, have thus been excluded from land discussions and are mostly unaware of broader land policy changes or notable developments. One middle-aged Maasai woman in Nkuruman reported discovering the new irrigation pipe the day it was installed on her husband's allocated farm. Women's ability to engage in wage employment or casual labor opportunities is also constrained by patriarchal patterns of authority, with important age and gender dynamics:

> 'Women are usually never involved and not recommended for casual jobs. For instance, there was work to build the roads, but the men do not want [*women*] doing it. They would say they are the head of household, and you should just stay home and take care of the livestock. [...] Younger women are not allowed to take part into such ventures, because their husbands do not want them out, to be seen. If you are older as a woman, you can take part if you get the permission from your husband.' *Middle-aged woman, Phase 2.*

Accordingly, the Community Land Act's rule to register everyone as members, as well as the clause to have women representatives in management committees, is generally seen as a top-down bureaucratic imposition. Stakeholders at other governance scales also seem reluctant to address entrenched inequalities, emphasizing their respect for local governance institutions even when they contradict national laws, including the 2010 constitution. County government actors noted that communal lands were easier places to implement projects, including irrigation, as community representatives often willingly give a parcel of land for the proposed project, whereas planning must involve multiple landowners in the subdivided areas across Kajiado County. As such, customary patriarchal authority structures which benefit household heads are being carried over into emerging formal authority structures that shape access to key productive resources linked to new livelihoods and livestock production patterns, even as the new legal structures and recognized forms of knowledge, such as formal education, increasingly benefit younger men. While customary institutions through elders can be key in safeguarding collective user rights, there are inherent exclusions embedded within these decision-making processes, notably on the basis of gender, which can be exacerbated under conditions of increased land stress and declining customary authority (Greiner 2017). Those constrained by limited education and insufficient financial and political capital thus see their access to key productive resources become increasingly restricted.

3. Accessing through old and new social relations

With direct access to key productive resources limited, accessing diversification opportunities is often enacted through social networks and built upon intersecting social relations that have long been crucial to pastoral livestock keeping. However, engagement in increasingly commercialized activities is also leading to renegotiations of some production relations and intra-household patterns of benefit sharing.

Most women and young men, as well as non-members, negotiate access to key productive resources through their social relations, thus reproducing customary Maasai reciprocity and care relations among extended family members, clanmates, friends, and neighbors. Most Maasai women engaging in crop farming reported accessing land through a male relative, often working on the farm of their husband or eldest son. It is also not unusual for women and young men to shift between working on different farmlands, depending on agreements made with family members or other social relations. Young women, once married, are expected to contribute to their husbands' households, and access land and other resources through their husbands. Women nonetheless cited sons and brothers as important for accessing capital to start small business ventures; for instance, through purchasing chickens or improved small-stock breeds. Some respondents in the dry side reported accessing newer animal breeds through social connections, such as borrowing a bull from a neighbor to crossbreed their own herd. Yet, women's accessing new breeds also appears to be negotiated through male relations, as a middle-aged woman indicated that the request for the bull was made by her son to his friend's father. Similarly, with irrigation schedules being quite restrictive, a male crop farmer from Nkuruman reported making arrangements to share irrigation times with his neighbor, lending his four hours of irrigation on his allocated day, while the neighbor shares four hours of irrigation water on his allocated day. This provided both parties with steadier access to water throughout the week. Women's self-help groups were noted as important for saving money to access necessary capital and goods.

Other negotiated accessing strategies are centered around rental arrangements between officially allocated plot owners and renters, both for farmlands and for town plots. Rental agreements are often verbal and can be negotiated for several years or on a seasonal basis. They are seen as a remunerative investment for landlords, providing a buffer against livelihood shocks. Plot owners often choose to rent out because their household members' time is occupied by livestock keeping in the drier part of the landscape or by education. Tenancy relations are nonetheless complex and the ability to benefit is not straightforward. For farmland rental, plot owners often look for skilled farmers, such non-Maasai agriculturalists or Maasai with agricultural experience. Rental plot payment is mostly in kind – for instance through giving the landlord a proportion of the produce – but can also be based on labor agreements. One middle aged Maasai man experienced in farming explained that he was both paying the owner for access to the farmland and hiring him as a casual laborer. Several Maasai respondents reflected that because the potential land productivity is not always evident to Maasai, who are unused to farming, they may be easily misled. This leads to many feeling that those who have the knowledge and skills to farm themselves benefit more from land rentals, especially for cash crops.

> 'They might agree that the owner takes Ksh 20,000 in a year, but when you look at the person who rents, they will get this Ksh 20,000 from papaya every week. […] But Maasai acquired that knowledge. They came to realize that it was a great decision because it was greater than an agreement of renting; […] at the end of it the person who rents it will benefit from this produce and you found yourself having nothing.' *Elder woman, Phase 1.*

Yet, these access strategies are often negotiated on short-term bases, meaning they are unstable. Several respondents reflected that this constrains decision-making and future investments:

> 'The farms that we are cultivating don't belong to us. They're for our parents. So, when you're given a small portion of land you won't complain. You just take the little that was given to

you. Some fail to get it. For instance, in a case where your father has 10 sons, it is difficult for the father to start subdividing among his sons, so some end up borrowing elsewhere.' *Young man, Phase 1.*

'You have to borrow from relatives to farm or rent from someone and agree on terms. You cannot be allocated.' *Young woman, Phase 1.*

Moreover, engagement in new livelihood activities is reshaping intra-household labor roles and patterns of benefits sharing, which adapt to new activities, with the household head typically deciding on labor allocation. Some household heads with multiple wives establish one *enkaji* in Phase 1 to farm, while the rest of the *olmarei* can attend to livestock keeping in the drier side. Alternatively, the *olmarei* is sometimes split across several locations, depending on children's enrollment in schools. In contrast to livestock sales, which are controlled by household heads, some women respondents in Phase 1 reported dividing the allocated farmland into smaller shares between co-wives and often helping on all in addition to the one belonging to the husband, but being able to freely sell produce from their own share. Chickens, which Maasai do not view as livestock, are typically owned and sold by both older and younger women, but also by young men. In some instances, higher education levels among younger women also create new opportunities linked to diversification. Being conversant in English and/or Kiswahili enables younger women residing near town centers to engage in business, such as sourcing household goods from Nairobi to sell to other women in the area. In these changing production environments, pastoralist women's resilience thus needs to be understood in light of continued reliance on social ties and increased labor participation taking diverse shapes (Archambault 2016; Wangui 2008).

4. Realizing long-term benefits from diversification for pastoral adaptation

The analysis above has shown that increased diversification leads to changing patterns of resource access and renegotiated social relations of production, redistributing risks and benefits in the face of climate change. In this section, we note some of the long-term implications for differentiated abilities to realize any benefits for pastoral adaptation, for instance through improved drought resilience, income diversification and/or food security. Such social differentiation also indicate how diversification processes may contribute to emerging class formation processes. The ability to realize benefits from diversification and to strengthen one's livelihood and socio-political position is affected by differentiated abilities to mobilize sufficient labor and capital to engage in diverse livelihood activities across seasons. The increased labor burdens associated with diversification suggest that it can make some households' position more precarious in the long term. Most respondents described difficulties to mobilize sufficient labor to sustain a diversified production over time and to find capital to hire help, especially during dry seasons. Women particularly emphasized the increased labor burden emanating from the need to balance childcare and other household reproduction duties with other livelihood activities. Elder men and women respondents also noted that educated young men are sometimes reluctant to engage in pastoral livestock keeping or are engaged in other activities, reducing the available workforce. Many herdsmen are either children of poorer households who cannot afford school fees or Maasai who have migrated from parts of Tanzania with lower school enrollment rates. During intense droughts, hired herdsmen sometimes leave to help their own household, which necessitates that family members take up herding practices such as traveling with herds to suitable pastures, seeking distant water points, or going to collect acacia pods and cut tree branches. Younger Maasai men increasingly express a desire to keep their herd smaller than their fathers', emphasizing the high financial and labor costs of maintaining it during dry seasons. Household heads decide on labor allocation and prioritize the available capital for maintaining herds which makes it difficult to support other economic activities.

> 'During the drought season, you would try to do your best to protect your cows from dying. So, you would spend most of the time looking after them and finding pastures for them. The men would end up getting stressed out, especially during extreme droughts[5] and we the Maasai are depending on that livestock.' *Elder man, Phase 1.*

> 'It changes because during the drought season, I have to always move and follow the livestock. This slowed down my business. My husband used to tell me to stay and look after the livestock because it was a drought season and for the Maasai women you're not supposed to go against your husband, so I stayed behind.' *Middle-aged woman, Phase 2.*

Several women in Phase 1, both old and young, also reported the high labor burden of watching the livestock as well as taking care of newer breeds that are sometimes less heat resistant, and unable to go far from the homestead. This makes it women's responsibility to cut grass and bring it back to the homestead.

> 'When there is a drought, the children go to school. [...] We take our cows to graze in the people's farms and there are a lot of planted farms around them, and the cows would be scattered all over the place. We have to be there to watch them grazing so that they will not get into people's farms.' *Young woman, Phase 1.*

While increased market engagement, such as through the sales of livestock or crop products, can generate valuable revenues in the short term, the ability to strengthen livelihoods is highly differentiated as the dependency upon unstable market dynamics places people in precarious positions. Cash crop cultivation, for example, is associated with the growth of local brokers linking producers with buyers and transporters from Nairobi. These arrangements reduce farmers' ability to set prices while providing precarious opportunities for unemployed young men and boys, sometimes attracted by quick money, and seeking greater financial independence.

> 'There are young men in this village who are broking paw-paws and getting Ksh10 per kilogram. Every week they can manage to sell over six tons of paw-paws and are able to make Ksh60,000 (USD600).' *Elder men during a focus group discussion, Phase 1.*

Growing cash crops reduces staple crops cultivation, namely maize used for household consumption. Purchasing staple crops, rather than growing them, increases financial pressures on households, and especially on women, whose gendered responsibilities include sourcing household food. A few women respondents noted that while women can sometimes control profits from farming and small business, it can also mean that husbands are no longer expected to provide money for food, resulting in increased financial

[5]In Maa, a dry season is referred to as *Olameyu* while a drought is often referred to *Olameyu sapuk oleng*, namely a very bad / intense dry season, which explains why the terms are sometimes used interchangeably in English transcriptions.

pressure on women. Markets are also often affected by seasonality and extreme events, with droughts reducing livestock prices progressively across intense dry seasons leading to losses for traders. Women who sell milk often do not have enough milk to sell during droughts. The dry season leads to vegetables and flour being sold at higher prices, which can provide some benefits for more established women food sellers in town centers, but creates difficulties for households that purchase.

> 'In farming, it is not profitable at the moment because it is a low season, and we are not planting different varieties of crops. We are growing bananas and papayas. We see it as a good farming activity, but we can't cook the papayas. We sell them and get money to buy food.'
> *Elder woman, Phase 1.*

Finally, diversification also affects long term collective resilience in diverse and contradictory ways. The increased clearing of land for cultivation, settlements, and other activities reduces access to communal late dry season grazing space and water points, leading to a loss of key dry grazing refuge areas and the associated collective loss of flexibility for pastoralists (Nyangena 2018). Yet, in Olkiramatian, it seems that some pastoralists are able to respond to reduced grazing areas in Phase 1 through market-based exchanges. Maize stalks in Phase 1 can for instance be used to feed livestock and are used by farmland owners or sold. Farm owners with few livestock can also rent out their parcel for grazing. Fees vary depending on the size and demand: reported prices ranged from Ksh3000ksh (USD30) for an acre to rising steeply to Ksh8000 (USD80) for a quarter of an acre during a bad drought. The agreement is often valid until there is no more stubble on the rented field which means that people with larger herds rent grazing access from several farms. Respondents in dry areas sometime reported making early arrangements to secure a farm when seeing people planting maize or through their own social network. These changing patterns show that the overall long-term adaptation may be further differentiated by individuals' abilities to mobilize sufficient capital to maintain the herd or to negotiate access to private land. The above highlights how diversification is linked to more individualized and market-based adaptation strategies. These increase the importance of capital relations, thus resulting in benefits and risks being unevenly distributed at the intersections of different social positions. However, it is also affecting long-term collective resilience. Because uneven land access encourages class formation (Galaty and Bonte 2018), current subdivision plans facilitate a landowning agro-pastoral elite as well as an educated urban business class while some pastoralists' socio-economic resilience is likely to worsen with poorer members selling their land shares to support short term needs as has happened across Kajiado. Non-members with sufficient capital can buy lands but those financially constrained are likely to see their possibilities for adaptation drastically reduce and be further obligated to engage in casual labor as communal resources access reduces.

Concluding discussion

Using an original analytical framework that foregrounds an intersectional approach in access theory, this research provides new insights into the relationships between diversification processes, access to resources and the (re)production of social inequalities in adaptation processes in transitioning pastoral areas. Our analysis illustrates that current struggles within the group ranch should be understood not only in the context of Maasai pastoralists' efforts to continue accessing grazing pastures and water; they are also connected to ongoing processes of diversification of pastoral livelihoods wherein accessing a different array of socio-environmental resources is gaining importance in light of climate change. Our intersectional approach to analyzing mechanisms of accessing deepens our understanding of the ways in which gender, age and education influence resource access as a 'dynamic and constantly re-negotiated process' (Milgroom, Giller, and Leeuwis 2014, 201), allowing us to draw some inferences on the ramifications of these changes for both individuals and the pastoral community.

As diversification increases in Olkiramatian, the rules and norms of access to key productive resources are renegotiated while some patterns of exclusion persist, with consequences for social distribution of positive and negative effects of diversification. Complex patterns of social differentiation are emerging in the context of new national land use policies, competing institutions, and rapidly changing commercial environment. Intersectional characteristics differentiate people's accessing of key productive resources for diversification, in part, through the dynamics of (re)negotiating authority and knowledge relations. Younger educated Maasai men are better placed to access information on and navigate changing rights-based access mechanisms as well as to secure desirable wage employment opportunities. This is illustrative of an important socio-cultural shift away from customary governance practices, which favored elder men, to one in which formal education and knowledge (and the resultant authority) take precedence. Nevertheless, the local moral economy continues to play a strong normative role in defining who should be included or not in governance processes based on their social identity, which advantages a local 'patriarchal elite' (Caravani 2019; Scoones 2021). Notably, the prevalent moral economy can help to legitimize Illoodokilani Maasai young men's claims to land allocation, while normalizing the exclusion of most women and others not belonging to the majority group. Excluded women, young men not yet registered in the group ranch register and other non-members often then pursue access to resources linked to diversification by (re)negotiating access through social networks. As Berry (1989) highlighted, access via social identity fosters more involvement in institutions to access – in this research, in order to secure long term access to resources – even as knowledge of 'how to work the system' becomes primordial (Jeppesen and Hassan 2022, 102). Our results also align with recent research evidencing the increased importance of capital relations for grazing access in the context of changing land use across Kajiado (Jeppesen and Hassan 2022), which is likely to further accentuate processes of social differentiation and associated class formation dynamics.

Moreover, our approach to analysing accessing as a strategy shaped by the intersection of multiple axes of social differences provides valuable insights on the differentiated abilities to realize benefits from diversification processes in transitioning systems and the associated implications for pastoral adaptation strategies. The adoption of new livelihood activities in response to both climate and socio-economic changes engenders complex intra-household renegotiation of labor arrangements and benefits sharing. Previous studies across Kajiado have looked primarily at young Maasai men (Mwangi 2003), but analyzing the interplays of age, gender and education shows that young Maasai women are also part of a generational shift in practices and are pursuing their own forms of diversification. Both young and older women engage in forms of diversification that have less potential to yield stable long-term benefits because their involvement is contingent on patriarchal intra-household agreements to access resources and on (re)negotiated household responsibilities. Women, especially, emphasized labor constraints, especially during the dry season, mirroring concerns with trade-offs between diversification and specialization in the literature and the fragmentation of labor (Cochrane and Cafer 2017; Mikulewicz 2021).. These engagement patterns often result in indirect and short-term seasonal access without sufficient stability to deliver economically transformative results in most cases. Similarly, younger landless men who are unable to find off-farm employment face weak long-term diversification prospects because they engage in activities such as day laboring and brokering which are dependent on unstable market dynamics.

Through focusing first on changing rights-based mechanisms of access and then on analyzing accessing in practice, this research foregrounds how the (re)production of power relations is linked to livelihood diversification processes, with tangible consequences for the well-being of individuals and communities. Many applications of access theory have omitted an in-depth analysis of power relations (Myers and Hansen 2019). Our theoretical framing integrates the ways that both power and agency – enacted through access strategies pursued at the intersection of different social positions – shape differentiated engagements with and patterns of benefits from diversification processes. As agrarian transformations in pastoral areas are spatially and temporally complex and play out in very context-specific ways (Scoones 2021), our study illustrates the continued relevance of small-scale case studies and the benefits of extending beyond households as a unit of analysis within adaptation research and existing policies on diversification. By pointing out changing accumulation trajectories in a pastoral context characterized by polygamous relations, extended kinship ties, and shifting communal resource management arrangements, we underscore the importance of social differentiated analysis of livelihood change in transitioning pastoral systems. Taking such an approach underlines the importance of understanding resilience as a multidimensional and relationally negotiated condition. It also raises critical questions about how resilience is locally understood and enacted, as well as what this means for interventions seeking to support such systems, including ones targeting 'gender equity' (Tavenner and Crane 2019, 2022).

Finally, our analysis contributes to building evidence on the diverse ways that adaptation processes are part and parcel of agrarian struggles by highlighting the need to understand diversification processes in pastoral areas both in terms of, firstly, the effects of shifting patterns of production and consequences for accumulation and social differentiation and, secondly, the trade-offs linked to short term benefits versus longer term strengthening of livelihoods and socio-political position. Our findings suggest that diversification tends to promote more individualized and market-based adaptation strategies, but that the drivers and ramifications of increased integration into capitalist production systems and renegotiation of production relations are complex and dynamic. Differentiated engagements with diversification in pastoral areas are not only related to changing material conditions, but also linked to 'intangible' dimensions, such as changing norms and values. New social differentiations emerge through the increased emphasis placed on formal education and how knowledge influences one's position within the community and beyond (e.g. the relation to state or non-governmental actors). At the same time, other entrenched markers of differentiation persist and are crystalized through exclusionary decision-making processes and established roles, perhaps most notably gendered discriminations. The research findings thus underscore the need for climate change adaptation planning in agrarian environments to extend beyond the dominant technical focus (Eriksen, Nightingale, and Eakin 2015), by showing how adaptation processes in pastoral environments are closely intertwined within rapidly evolving socio-political and economic transformations. This requires research and policy initiatives to give greater attention to the ways that in the midst of agrarian change, complex material and immaterial shifts alter the socio-political positions of pastoralists and the resilience of differentiated individuals and the communities they comprise. Future studies on agrarian struggles and climate change should further integrate into intersectional approaches the emerging class dynamics in transitioning pastoral systems as well as investigate more the impacts of diversification processes on the ecological resilience in historically pastoral systems.

Acknowledgements

The work of Edwige Marty, Renee Bullock and Todd Crane is funded through the Program for Climate Smart Livestock (PCSL) which has received financial support from the German Federal Ministry for Economic Cooperation and Development (BMZ) issued through the Deutsche Gesellschaft für Internationale Zusammenarbeit (GIZ), grant number: 81231239. We thank Peadar Brehony for his insightful feedback and comments on an earlier draft of this manuscript as well as Lucy Njuguna and Laura Cramer for helping to think through the presentation of the figures. Thank you as well to the South Rift Association of Land Owners (SORALO) and Olkiramatian leadership for their support during the data collection as well as Winnie Lipina Koinet, Agnes Kamanga and Wilson Rianto for excellent research assistance. The views and perspectives presented in this publication, however, remain solely the responsibility of the authors.

Disclosure statement

No potential conflict of interest was reported by the author(s).

Funding

This work was supported within the framework of a project financed by the GIZ commissioned by the Government of the Federal Republic of Germany.

ORCID

Edwige Marty http://orcid.org/0000-0002-0885-8971
Renee Bullock http://orcid.org/0000-0001-9561-0394
Matthew Cashmore http://orcid.org/0000-0002-1614-3204
Todd Crane http://orcid.org/0000-0002-4395-7545
Siri Eriksen http://orcid.org/0000-0002-6594-2758

References

Achiba, Gargule A., and Monica N. Lengoiboni. 2020. "Devolution and the Politics of Communal Tenure Reform in Kenya." *African Affairs* 119 (476): 338–369. doi:10.1093/afraf/adaa010.

Agyei, Frank Kwaku, Christian Pilegaard Hansen, and Emmanuel Acheampong. 2020. "Access Along Ghana's Charcoal Commodity Chain." *Society & Natural Resources* 33 (2): 224–243. doi:10.1080/08941920.2019.1623358.

Archambault, Caroline S. 2016. "Re-creating the Commons and re-Configuring Maasai Women's Roles on Therangelands in the Face of Fragmentation." *International Journal of the Commons* 10 (2): 728–746. doi:10.18352/ijc.685.

Atteridge, Aaron, and Elise Remling. 2018. "Is Adaptation Reducing Vulnerability or Redistributing It?" *WIRES Climate Change* 9 (1): 1–16. doi:10.1002/wcc.500.

Berry, Sara. 1989. "Social Institutions and Access to Resources." *Africa: Journal of the International African Institute* 59 (1): 41–55. doi:10.2307/1160762.

Bobadoye, A. O., W. O. Ogara, G. O. Ouma, and J. O. Onono. 2014. "Comparative Analysis of Rainfall Trends in Different Sub Counties in Kajiado County Kenya." *International Journal of Innovative Research and Studies* 3 (12): 179–195.

Boles, Oliver J.C., Anna Shoemaker, Colin J. Courtney Mustaphi, Nik Petek, Anneli Ekblom, and Paul J. Lane. 2019. "Historical Ecologies of Pastoralist Overgrazing in Kenya: Long-Term Perspectives on Cause and Effect." *Human Ecology* 47 (3): 419–434. doi:10.1007/s10745-019-0072-9.

Boone, Catherine. 2012. "Land Conflict and Distributive Politics in Kenya." *African Studies Review* 55 (1): 75–103.

Borras, Saturnino M., Ian Scoones, Amita Baviskar, Marc Edelman, Nancy Lee Peluso, and Wendy Wolford. 2022. "Climate Change and Agrarian Struggles: An Invitation to Contribute to a JPS Forum." *The Journal of Peasant Studies* 49 (1): 1–28. doi:10.1080/03066150.2021.1956473.

Brehony, Peadar. 2020. *Conservation and Resilience to Drought: A Social-Ecological Perspective on Conservation Areas in Kenya's Southern Maasailand*. University of Cambridge.

Brisebois, Anouk, Siri Hallstrøm Eriksen, and Todd Andrew Crane. 2022. "The Politics of Governing Resilience: Gendered Dimensions of Climate-Smart Agriculture in Kenya." *Frontiers in Climate* 4 (June), doi:10.3389/fclim.2022.864292.

Caravani, Matteo. 2019. "'De-Pastoralisation' in Uganda's Northeast: From Livelihoods Diversification to Social Differentiation." *The Journal of Peasant Studies* 46 (7): 1323–1346. doi:10.1080/03066150.2018.1517118.

Cheeseman, Nic, Gabrielle Lynch, and Justin Willis. 2016. "Decentralisation in Kenya: The Governance of Governors." *The Journal of Modern African Studies* 54 (1): 1–35. doi:10.1017/S0022278X1500097X.

Coast, E. 2001. Maasai Demography." London School of Economics.

Cochrane, Logan, and Anne Cafer. 2017. "Does Diversification Enhance Community Resilience? A Critical Perspective." *Resilience* 3293 (January): 1–15. doi:10.1080/21693293.2017.1406849.

Crane, Todd A. 2010. "Of Models and Meanings: Cultural Resilience in Social–Ecological Systems." *Ecology and Society* 15 (4), doi:10.5751/es-03683-150419.

Crenshaw, Kimberle. 1991. "Mapping the Margins: Intersectionality, Identity Politics, and Violence Against Women of Color." *Stanford Law Review* 43 (6): 1241–1299.

Djoudi, Houria, Bruno Locatelli, Chloe Vaast, Kiran Asher, Maria Brockhaus, and Bimbika Basnett Sijapati. 2016. "Beyond Dichotomies: Gender and Intersecting Inequalities in Climate Change Studies." *Ambio* 45: 248–262. doi:10.1007/s13280-016-0825-2.

Ellis, Frank. 1998. "Household Strategies and Rural Livelihood Diversification." *Journal of Development Studies* 35 (1): 1–38. doi:10.1080/00220389808422553.

Eriksen, Siri, E. Lisa, F. Schipper, Morgan Scoville-Simonds, Katharine Vincent, Hans Nicolai Adam, Nick Brooks, and Brian Harding, et al. 2021. "Adaptation Interventions and Their Effect on Vulnerability in Developing Countries: Help, Hindrance or Irrelevance?" *World Development* 141: 105383. doi:10.1016/j.worlddev.2020.105383.

Eriksen, Siri Hallstrom, Andrea J Nightingale, and Hallie Eakin. 2015. "Reframing Adaptation: The Political Nature of Climate Change Adaptation." *Global Environmental Change* 35: 523–533. doi:10.1016/j.gloenvcha.2015.09.014.

Galaty, J. G., and P. Bonte. 2018. "The Current Realities of African Pastoralists." In *Herders, Warriors and Traders*, edited by J. G. Galaty, and P. Bonte. New York: Taylor & Francis.

Garcia, Alicea, and Petra Tschakert. 2022. "Intersectional Subjectivities and Climate Change Adaptation: An Attentive Analytical Approach for Examining Power, Emancipatory Processes, and Transformation." *Transactions of the Institute of British Geographers* 47: 651–665. doi:10.1111/tran.12529.

Gebrechorkos, Solomon H., Stephan Hülsmann, and Christian Bernhofer. 2018. "Changes in Temperature and Precipitation Extremes in Ethiopia, Kenya, and Tanzania." *International Journal of Climatology* 39 (1): 18–30. doi:10.1002/joc.5777.

German, Laura, Elizabeth King, Ryan Unks, and Naiputari Paul Wachira. 2017. "This Side of Subdivision: Individualization and Collectivization Dynamics in a Pastoralist Group Ranch Held Under Collective Title." *Journal of Arid Environments* 144: 139–155. doi:10.1016/j.jaridenv.2017.04.009.

Government of Kenya. 1968. *The Land Adjudication Act*. Kenya.

Government of Kenya. 2016. *The Community Land Act* 2016. kenyalaw.org/.

Greiner, Clemens. 2017. "Pastoralism and Land-Tenure Change in Kenya: The Failure of Customary Institutions." *Development and Change* 48 (1): 78–97. doi:10.1111/dech.12284.

Greiner, Clemens. 2021. "African Pastoralism: Plus Ça Change? From Constant Herders to Social Differentiation." *African Futures*, doi:10.1163/9789004471641.

Hall, Derek, P. Hirsch, and Tania Murray Li. 2011. *Powers of Exclusion: Land Dilemmas in Southeast Asia*. Singapore: National University of Singapore Press.

Herrero, M., J. Addison, C. Bedelian, E. Carabine, P. Havlik, B. Henderson, J. Van De Steeg, and P. K. Thornton. 2016. "Climate Change and Pastoralism: Impacts, Consequences and Adaptation." *Revue Scientifique et Technique de l'OIE* 35 (2): 417–433. doi:10.20506/rst.35.2.2533.

Homewood, Katherine M., Patti Kristjanson, and Pippa Chenevix Trench. 2009. *Staying Maasai? Livelihoods, Conservation and Development in East African Rangelands*. doi:10.1186/2041-7136-2-19.

Hughes, Lotte. 2002. *Moving the Maasai : A Colonial Misadventure*. University of Oxford.

Hughes, Lotte. 2008. "Mining the Maasai Reserve: The Story of Magadi." *Journal of Eastern African Studies* 2 (1): 134–164. doi:10.1080/17531050701847326.

Jeppesen, Maja Dahl, and Rahma Hassan. 2022. "Private Property and Social Capital: Dynamics of Exclusion and Sharing in the Subdivided Pastoral Rangelands of Kajiado, Kenya." *Society & Natural Resources* 35 (1): 92–109. doi:10.1080/08941920.2022.2026542.

Kaijser, Anna, and Annica Kronsell. 2014. "Climate Change Through the Lens of Intersectionality." *Environmental Politics* 23 (3): 417–433. doi:10.1080/09644016.2013.835203.

Karmebäck, Vera N, John Ndung, Magnus Jirström, and Gert Nyberg. 2015. "Assessing Gender Roles in a Changing Landscape: Diversified Agro-Pastoralism in Drylands of West Pokot, Kenya." *Pastoralism* 5 (1) doi:10.1186/s13570-015-0039-4.

Kimani, Kamau, and John Pickard. 1998. "Recent Trends and Implications of Group Ranch Sub-Division and Fragmentation in Kajiado District, Kenya." *The Geographical Journal* 164 (2): 202. doi:10.2307/3060370.

Lau, Jacqueline D., Joshua E. Cinner, Michael Fabinyi, Georgina G. Gurney, and Christina C. Hicks. 2020. "Access to Marine Ecosystem Services: Examining Entanglement and Legitimacy in Customary Institutions." *World Development* 126: 104730. doi:10.1016/j.worlddev.2019.104730.

Lenaiyasa, Michael Lugard, Brett L. Bruyere, Jonathan Salerno, and Tomas Pickering. 2020. "Pastoralists' Use of Income Diversification as a Strategy for Adapting to Social-Ecological Change in Samburu, Kenya." *Regional Environmental Change* 20 (1), doi:10.1007/s10113-020-01612-x.

Lind, Jeremy. 2018. "Devolution, Shifting Centre-Periphery Relationships and Conflict in Northern Kenya." *Political Geography* 63: 135–147. doi:10.1016/j.polgeo.2017.06.004.

Little, Peter D. 2021. "Does Livelihood and Asset Diversification Contribute to Pastoralist Resilience?: The Case of Il Chamus, Baringo County, Kenya, 1980-2018." *Nomadic Peoples* 25 (2): 181–205. doi:10.3197/np.2021.250202.

Matsaba, Emmanuel Ochola, Merlijn van Selm, John Mwibanda Wesonga, Hasse Goosen, Ingrid Coninx, and Irene Koomen. 2021. "Agricultural Climate Atlas for Kajiado and Kiambu Counties, Kenya." In *Handbook of Climate Change Management*, edited by W Leal Filho, J Luetz, and D Ayal, 3–23. Springer. doi:10.1007/978-3-030-57281-5_227.

Matteo, Francesca Di. 2022. "Imagining the Kenyan Commons: The Stakes of State Control Over Land in the Formulation of the Community Land Act (2011-2016)." *The Journal of Development Studies* 0 (0): 1–19. doi:10.1080/00220388.2022.2043279.

McCabe, J. Terrence, Paul W. Leslie, and Laura DeLuca. 2010. "Adopting Cultivation to Remain Pastoralists: The Diversification of Maasai Livelihoods in Northern Tanzania." *Human Ecology* 38 (3): 321–334. doi:10.1007/s10745-010-9312-8.

Mikulewicz, Michael. 2021. "Disintegrating Labour Relations and Depoliticised Adaptation to Climate Change in Rural São Tomé and Príncipe." *Area* 53 (3): 422–430. doi:10.1111/area.12630.

Milgroom, Jessica, Ken E. Giller, and Cees Leeuwis. 2014. "Three Interwoven Dimensions of Natural Resource Use: Quantity, Quality and Access in the Great Limpopo Transfrontier Conservation Area." *Human Ecology* 42 (2): 199–215. doi:10.1007/s10745-013-9635-3.

Milgroom, Jessica, and Jesse Ribot. 2019. "Children of Another Land: Social Disarticulation, Access to Natural Resources and the Reconfiguration of Authority in Post Resettlement." *Society & Natural Resources* 0 (0): 1–21. doi:10.1080/08941920.2019.1590668.

Muok, Benard Oula, Marianne Mosberg, Siri Ellen Hallstrøm Eriksen, and Dennis Onyango Ong'ech. 2021. "The Politics of Forest Governance in a Changing Climate: Political Reforms, Conflict and Socio-Environmental Changes in Laikipia, Kenya." *Forest Policy and Economics* 132 (October 2020), doi:10.1016/j.forpol.2021.102590.

Mwangi, Esther N. 2003. "Institutional Change and Politics: The Transformation of Property Rights in Kenya's Maasailand." Indiana University.

Myers, Rodd, and Christian Pilegaard Hansen. 2019. "Revisiting A Theory of Access: A Review." *Society & Natural Resources* 0 (0): 1–21. doi:10.1080/08941920.2018.1560522.

Nkedianye, David K., Joseph O. Ogutu, Mohammed Y. Said, Shem C. Kifugo, Jan de Leeuw, Paul Van Gardingen, and Robin S. Reid. 2020. "Comparative Social Demography, Livelihood Diversification and Land Tenure among the Maasai of Kenya and Tanzania." *Pastoralism* 10 (1), doi:10.1186/s13570-020-00165-2.

Nyangena, John. 2018. *Drought, Land Use Change and Livelihood Diversification among Pastoral Communities in Oltiasika. Kajiado County, Kenya*: Kenyatta University.

Odhiambo, Michael Ochieng. 2014. "The Unrelenting Persistence of Certain Narratives: An Analysis of Changing Policy Narrative about ASALS in Kenya." London.

Ontiri, Enoch, and Lance W Robinson. 2017. "Community-Based Rangeland Management in Shompole and Olkiramatian Group Ranches, Kenya Taking Successes in Land Restoration to Scale Project." https://cgspace.cgiar.org/bitstream/handle/10568/92963/ranches_kenya_report.pdf?sequence=5.

Peluso, Nancy Lee, and Jesse Ribot. 2020. "Postscript: A Theory of Access Revisited." *Society & Natural Resources* 33 (2): 300–306. doi:10.1080/08941920.2019.1709929.

Petesch, Patti, and Gordon Prain. 2018. "Gender Norms, Agency, and Innovation in Agriculture and Natural Resource Management: The GENNOVATE Methodology." Mexico, D. F.

Pichler, Melanie, Martin Schmid, and Simone Gingrich. 2021. "Mechanisms to Exclude Local People from Forests: Shifting Power Relations in Forest Transitions." *Ambio*, doi:10.1007/s13280-021-01613-y.

Pollini, Jacques, and John G. Galaty. 2021. "Resilience Through Adaptation: Innovations in Maasai Livelihood Strategies." *Nomadic Peoples* 25 (2): 278–311. doi:10.3197/np.2021.250206.

Rao, Nitya. 2019. "From Abandonment to Autonomy: Gendered Strategies for Coping with Climate Change, Isiolo County, Kenya." *Geoforum; Journal of Physical, Human, and Regional Geosciences* 102 (March): 27–37. doi:10.1016/j.geoforum.2019.03.017.

Ribot, Jesse. 2014. "Cause and Response: Vulnerability and Climate in the Anthropocene." *The Journal of Peasant Studies* 41 (5): 667–705. doi:10.1080/03066150.2014.894911.

Ribot, Jesse, and Nancy Lee Peluso. 2003. "A Theory of Access*." *Rural Sociology* 68 (2): 153–181. doi:10.1111/j.1549-0831.2003.tb00133.x.

Ripoll, Santiago, Jens Andersson, Lone Badstue, Marion Büttner, Jordan Chamberlin, Olaf Erenstein, and James Sumberg. 2017. "Rural Transformation, Cereals and Youth in Africa: What Role for International Agricultural Research?" *Outlook on Agriculture* 46 (3): 168–177. doi:10.1177/0030727017724669.

Rutten, Marinus Mattheus Eduard Maria. 1992. *Selling Wealth to Buy Poverty: The Individualization of Landownership among the Maasai Pastoralists of Kajiado District, Kenya, 1890-1990*. Catholic Unviersity of Nijmegen.

Scoones, Ian. 2021. "Pastoralists and Peasants: Perspectives on Agrarian Change." *The Journal of Peasant Studies*, doi:10.1080/03066150.2020.1802249.

Sikor, Thomas, and Christian Lund. 2009. "Access and Property: A Question of Power and Authority." *Development and Change* 40 (1): 1–22. doi:10.1111/j.1467-7660.2009.01503.x.

Tavenner, Katie, and Todd A. Crane. 2019. "Implementing 'Gender Equity' in Livestock Interventions." In *Gender, Agriculture and Agrarian Transformations: Changing Relations in Africa, Latin America and Asia*, edited by Car Sachs. Routledge.

Tavenner, Katie, and Todd A. Crane. 2022. "Hitting the Target and Missing the Point? On the Risks of Measuring Women's Empowerment in Agricultural Development." *Agriculture and Human Values*, doi:10.1007/s10460-021-10290-2.

Taylor, Marcus. 2013. "Climate Change, Relational Vulnerability and Human Security: Rethinking Sustainable Adaptation in Agrarian Environments." *Climate and Development* 5 (4): 318–327. doi:10.1080/17565529.2013.830954.

Thompson, E. P. 1971. "The Moral Economy of the English Crowd in the Eithgteenth." *Past and Present* 50 (1): 76–136. doi:10.1215/00382876-2146476.

Thompson-Hall, Mary, Edward R. Carr, and Unai Pascual. 2016. "Enhancing and Expanding Intersectional Research for Climate Change Adaptation in Agrarian Settings." *Ambio* 45 (s3): 373–382. doi:10.1007/s13280-016-0827-0.

Thornton, P. K., J. van de Steeg, A. Notenbaert, and M. Herrero. 2009. "The Impacts of Climate Change on Livestock and Livestock Systems in Developing Countries: A Review of What We Know and What We Need to Know." *Agricultural Systems* 101 (3): 113–127. doi:10.1016/j.agsy.2009.05.002.

Wangui, Elizabeth Edna. 2008. "Development Interventions, Changing Livelihoods, and the Making of Female Maasai Pastoralists." *Agriculture and Human Values* 25 (3): 365–378. doi:10.1007/s10460-007-9111-z.

Advocating afforestation, betting on BECCS: land-based negative emissions technologies (NETs) and agrarian livelihoods in the global South

Pamela McElwee

ABSTRACT

Negative emissions technologies (NETs) for carbon dioxide removal (CDR) are increasingly important responses to achieve global climate change targets, but to date, there has been insufficient attention to land-based NETs (including afforestation, biochar, and other measures) as an agrarian challenge for the global South. This paper explores the implications of different NETs for land, labor, capital, and politics in rural spaces and contributes to articulating agrarian climate justice by demonstrating the potentially unjust implications of many NETs. The paper concludes with how these measures might be designed to be less negative for rural peoples in future implementation.

Introduction

Ongoing failures to achieve sufficient reductions in greenhouse gas emissions have led to increasing discussions of the potential need for negative emissions technologies (NETs) to remove existing CO_2 from the atmosphere (Smith et al. 2015). These carbon dioxide removal (CDR) strategies include direct air capture as well as land-based actions like afforestation or bioenergy with carbon capture and storage (BECCS), often grouped together with other 'geoengineering' solutions to the climate crisis (NAS 2015). NETs are an important element within models that demonstrate how the world could achieve the Paris agreement target of holding global temperature to well below 2°C (Dooley and Kartha 2018; Rogelj et al. 2018).[1] While there is an increasingly large literature discussing NETs, it remains mostly focused on modeling pathways and technical feasibility, with fewer critical discussions of social impacts of deployment, particularly for rural peoples (Williamson 2016; Sovacool 2021a).

The potential environmental impacts of large-scale use of NETs include water scarcity and biodiversity loss, which will also have socioeconomic ramifications (Dooley and Kartha 2018; Yamagata et al. 2018; Dooley, Harrould-Kolieb, and Talberg 2021). A small number of studies have engaged social scientists to identify other key concerns, such

[1]For example, the IPCC 1.5 report has only one modeled pathway (P1) with no use of BECCS: this pathway requires very low energy demand (LED), rapid phase-out of fossil fuels and/or rapid shifts to sustainable food consumption freeing up land for afforestation (IPCC 2018, 122). Other modeled pathways all rely on NETs of some kind.

as public understanding and acceptability as barriers to deployment, or trade-offs with sustainable development goals (Smith et al. 2019; Honegger, Michaelowa, and Roy 2020; McElwee et al. 2020). Several recent reviews have highlighted other key questions around complexity, uncertainty, ethics, and justice (Gough et al. 2018; Forster et al. 2020; Waller et al. 2020), while stronger critiques have also emerged, including concerns that NETs are being used as a smokescreen to avoid difficult fossil fuel emissions reductions (Cox et al. 2018; Carton 2019; Stuart, Gunderson, and Petersen 2020).

There continues to be a strong need for social scientists to engage with discussions regarding the feasibility and desirability of NETs (Markusson et al. 2020). In particular, there has been insufficient attention to land-based NETs as a uniquely *rural* or *agrarian* challenge for the global South, particularly given that climate change is already disproportionately burdening these communities (Sultana 2022). To address this gap, this article aims to: (1) assess the range of technical studies on NETs to determine what rural social challenges are currently considered in modeling studies and which excluded; and (2) examine lessons learned from other agrarian literatures on interventions such as tree planting, biofuels, and reduced emissions from degradation and deforestation (REDD+). Such comparisons are sorely needed, as even those scholars who are critical of the overall techno-optimism in NETs tend to see land-based options as more benign or even beneficial (Stuart, Gunderson, and Petersen 2020).[2]

This paper addresses in part how NETs potentially threaten the concept of 'agrarian climate justice' – that is, a just and equitable response to the impacts of climate change that addresses historical injustices and presents a more progressive vision for the future of agrarian spaces (Borras and Franco 2018). As the call for this special forum has noted, the implications of climate change for rural places and peoples are profound, from the expansion of techno-fixes and extension of neoliberal capital to the need to understand increasingly reactionary agrarian politics (Borras et al. 2021). Notably, without addressing the preexisting power structures that have created climate vulnerabilities in the first place, many potential 'solutions' like NETs run the risk of deepening these problems (Morrow et al. 2020; Newell 2022).

As a growing body of literature on climate justice has noted, there are multiple dimensions to how unfairness and inequities might be created or exacerbated by climate change and responses to these impacts (Newell et al. 2021; Newell 2022; Sultana 2022). These include questions around *procedural justice*, which concerns involvement in decision-making to ensure fairness, transparency, and accountability (Newell et al. 2021). This is a particular concern given the lack of public awareness of issues and involvement for many geoengineering approaches (McLaren 2012; Gough and Mander 2019; Spence, Cox, and Pidgeon 2021). *Recognition* justice relates to what peoples are affected and how, particularly cultural or social differences that need to be recognized (Sovacool 2021b; Sultana 2022); the impact that NETs may have on Indigenous peoples and their lands, for example, is yet to be fully understood (Burns 2016). *Distributional justice* relates to who benefits and who bears the costs from both climate impacts and climate responses like adaptation (Eriksen et al. 2021); in this case, NET interventions

[2]For example, land-based NETs are often presented as providing co-benefits (e.g. forest ecosystem services, increased agricultural productivity, or electricity) with fewer negative trade-offs (Smith et al. 2019). A 2015 National Academy of Sciences report stated that land-based NETs raise fewer ethical issues as compared to others like ocean fertilization or solar radiation management (NAS 2015).

will likely involve difficult trade-offs that may result in exacerbation of already uneven access to land (Smith et al. 2015). *Intergenerational* justice relates to how decisions made today may impact those in the future (Newell et al. 2021); for example, the far-off nature of deployment of some NETs has raised concerns that they may slow urgency for mitigation actions today (Lenzi 2018; Batres et al. 2021; McLaren et al. 2021).

Examining NETs across axes of agrarian change provides a useful way to pre-assess the possible impacts of different options, both in terms of climate justice as well as other priorities like environmental protection. However, there remain large gaps in the literature on rural social impacts from NETs in particular (Robledo-Abad et al. 2017). By explicitly analyzing the implications of different NETs for land, labor, capital, and politics in rural spaces, this paper confirms that existing discourses and plans for NETs continue to follow mostly technocratic and capitalist models. In other words, NETs are seen as technologies, rather than practices that involve and impact people (Buck 2018). Such approaches are bolstered by use of modeling that often fails to include social factors in analysis of the feasibility of NETs (Schweizer et al. 2020). In contrast, agrarian studies scholarship that foregrounds the experiences and expectations of rural peoples can help identify potential risks and roadblocks before NETs are deployed, as well as designing strategies and investments that are more beneficial to rural peoples, including through attention to multiple forms of justice (Morrow et al. 2020; Batres et al. 2021; Healey et al. 2021).

Methods

Literature searches were conducted in Web of Science for a range of land-based NETs, with a particular focus on results in social science journals. For example, the term BECCS alone received 242 hits, of which the most relevant papers were examined for agrarian or social science questions. Because many of the NET options are future-oriented (e.g. BECCS is not yet fully operational), these literatures often rely on modeling of different scenarios for the scale and scope of NETs to the end of the twenty-first century. A total of 81 papers on the technical details of NETs and 105 papers on governance and social implications of NETs were reviewed and examined for discussions of justice-related issues across key categories of land, labor, capital, and rural politics. In addition, an additional 89 studies of existing land-based carbon emissions projects, particularly afforestation for carbon benefits and REDD+, were collected in a conventional 'snowball' fashion and examined for challenges that might hold lessons for future NETs, particularly around achieving co-benefits and avoiding trade-offs (Wittman and Caron 2009; Corbera and Brown 2010; Leach and Scoones 2013; Lund et al. 2017). Additionally, the pledged Nationally Determined Contributions for all countries that are parties to the Paris Agreement were searched for reference to use of NETs in current or future plans through use of Google search engine strings on the UN Framework Convention on Climate Change website.

Examining land-based NETs

NETs aim to remove CO_2 already emitted, providing an opportunity to counterbalance future emissions, particularly 'residual' ones which will be impossible to eliminate entirely

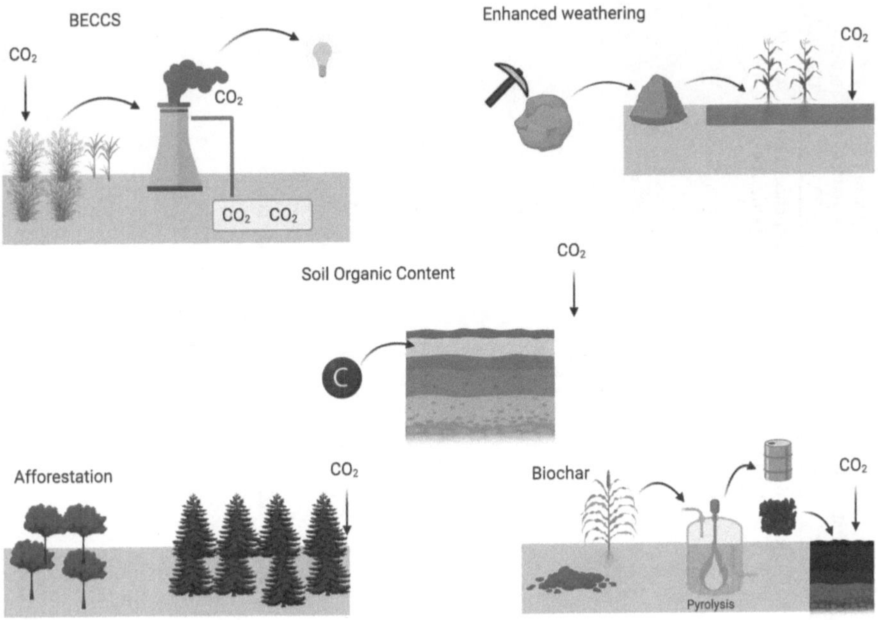

Figure 1. Major land-based NETs.

through existing mitigation measures. There are a range of possible NETs, including carbon capture and storage from power plants and direct air capture, as well as ocean fertilization. These remain technologically far-off, meaning that NETs featuring most prominently in current modeling efforts to explore future emissions pathways tend to be land-based, including bioenergy with carbon capture and storage (BECCS) and afforestation/reforestation. Other land-based NETs include enhanced weathering of minerals, carbon storage in soils, and use of biochar (Figure 1).[3]

Afforestation involves the conversion to forest of land that historically has not contained forests, while *reforestation* occurs on land that has previously contained forests: both can contribute to negative emissions given trees' carbon sequestration potential (see Table 1). Where trees are planted is crucial, as afforestation in boreal areas contributes to the albedo effect, effectively prioritizing afforestation in the tropics instead (Fuss et al. 2018).[4] Both *soil carbon sequestration* and *biochar* relate to the preservation or amendment of soils. Practices to conserve carbon within soils include low or no-tillage, cover crops, improved nutrient and water management and other practices, while biochar is created by the combustion of biological material under pyrolysis which locks up some carbon and which can be added as an amendment to soil (Smith et al. 2019).

[3]Biofuels are not considered a NET, as they are primarily a substitute for fossil fuels and thus a mitigation strategy. REDD+ occupies a more ambiguous position as both a mitigation strategy to reduce forest emissions from deforestation (avoided emissions), as well as a possible NET for negative emissions if forest cover expands (increased carbon sink). In general, existing natural forest sinks should not be counted as NETs, because they are already calculated in global carbon balance estimates (Nolan, Field, and Mach 2021).

[4]Albedo effects result from planting (darker) trees on (lighter) lands in northern regions, which contributes to increased solar radiation absorption and localized warming, and thus offsets the benefits of tree planting in terms of global temperature (IPCC 2018).

Table 1. Land-based NETs Potential Deployment, Impact and Costs

Type of NET	Description	Technical potential for CO_2 removal (flux) (Smith et al. 2020)	'Safe' deployment maximum potential (NAS 2019)*	Current scale of use	Potential scale of deployment	Creates competition for land?	Land required per ton of sequestered CO_2	Cost estimates US$/t$CO_2$
BECCS	BECCS is the combination of bioenergy technology (e.g. production of electricity or fuels from biological sources, ranging from crops to trees) in which generated CO_2 is captured and stored on-site. In theory , BECCS both draws down atmospheric CO_2 concentrations through biological growth and generates low-carbon energy. Rates of CO_2 removal depend on type of feedstock and scale at which bioenergy is produced, but BECCS has not yet been deployed beyond pilots	0.4–11.3 GtCO_2 yr^{-1}	0.5–5 GtCO_2 yr^{-1}	Only handful of demonstration facilities	Wide variation in estimates – 360–2400 Mha	Yes	Dependent on feedstock	$45–250
Biochar	Biochar is a product of pyrolysis, which heats plant matter in absence of oxygen to 'lock-in' carbon and resist microbial decomposition. Anthropogenic soils with biochar are well-known from the Amazon (*terra prietas*). Biochar added	0.03–6.6 GtCO_2e yr^{-1}	0.5–2 GtCO_2e yr	In US, 39,000–77,000 t/y biochar are produced and used	40–260 Mha	Some, for production of feedstocks	<1 ha	$0–185

(*Continued*)

Table 1. Continued.

Type of NET	Description	Technical potential for CO_2 removal (flux) (Smith et al. 2020)	'Safe' deployment maximum potential (NAS 2019)*	Current scale of use	Potential scale of deployment	Creates competition for land?	Land required per ton of sequestered CO_2	Cost estimates US$/t$CO_2$
	to soil can persist for thousands of years, but total carbon removals require life-cycle analysis of production. Feedstocks can include a number of biological materials, from waste to purposively grown. Biochar has potential co-benefits for soil as it can increase water absorptive and nutrient status for crops and may reduce N_2O emissions as well							
Enhanced weathering and mineralization of CO_2	Mineralization occurs when silicate rocks are naturally weathered, and the CO_2 in the air converted to form carbonates like calcite ($CaCO_3$) (NAS 2019). Use of mineralization as a NET has been suggested through various processes, including accelerated weathering of basaltic rocks, which would entail mining rocks, grinding them, and spreading them out on land to expose more surface area	0.5–4 $GtCO_2\ yr^{-1}$	2–4 $GtCO_2\ yr^{-1}$	Only natural weathering in current use and some test sites	Up to 680 Mha in tropics	No	~1t per t of rock	$20–1000
	Afforestation is the conversion to forest of	1.5–17 Gt $CO_2e\ yr^{-1}$	0.5–4 Gt $CO_2e\ yr^{-1}$	Widespread. Estimated that in	Technical land potential up	Yes	Depends on type and age	$0–100

Afforestation/ Reforestation (AF)	land that historically has not contained forests, while reforestation is the conversion to forest of land that has previously contained forests but that has been converted to some other use			2000–2010 23.6 Mha was A/F; 2011–2019 added 3.1 Mha	to 2800 Mha, but more realistically 320–500 Mha		of tree; ranges from <1hato 40 ha	
Soil carbon sequestration	Organic carbon in the soil holds CO_2 and serves as a sink. This carbon content can be increased through land management practices (e.g. from annual cropping to perennial, or agriculture to forest); improved agricultural practices (cover crops, no-till); amending and improving the soil (adding manure, using different crops with deeper roots); and other means	0.4–8.6 Gt $CO_2e\ yr^{-1}$	Up to 5 Gt $CO_2e\ yr^{-1}$	Significant use of some land management techniques already; e.g. 10 Mha enrolled in US Conservation Reserve Program	Potentially up to several thousand Mha of existing land use, but realistically less	No	1–33 ha	<0–$100

Sources: (Moosdorf, Renforth, and Hartmann 2014; Smith 2016; Beerling 2017; Griscom et al. 2017; Bernal, Murray, and Pearson 2018; Minx et al. 2018; NAS 2019; Fuss et al. 2020; Smith et al. 2020; Roe et al. 2021).

*'Safe' maximum rate of CO_2 removal as defined by the National Academy of Sciences means that 'the deployment would not cause large potential adverse societal, economic, and environmental impacts' (NAS 2019). See also Smith (2016).

While soil carbon sequestration, biochar, and afforestation are all existing options, *BECCS* and *enhanced mineralization* are more speculative. BECCS requires the production of bioenergy feedstocks (generally fast-growing species like miscanthus or switchgrass) which must be transported to where they will be converted to steam/heat, liquid fuels, or charcoals. These fuel products are then used for energy generation and the emitted CO_2 is captured either pre- or post-combustion and stored (NAS 2019). For enhanced mineralization, certain basaltic rocks can form carbonites by reacting with CO_2 in the air; mining these rocks, grinding them, and spreading them across land surfaces could increase CO_2 sequestration significantly enough to be a potential NET, although this would require application on large amounts of land (Williamson 2016; Fuss et al. 2018). It is likely that the tropics would be particularly targeted due to the need for warmer climates to enhance the weathering process (Beerling 2017).

NETs and IAMs

NETs have been bolstered by their use in Integrated Assessment Models (IAMs), an important element in recent IPCC reports. IAMs bring together both biophysical climate system models with socioeconomic parameters like population or economic output (often expressed through the concept of Shared Socio-Economic Pathways (SSPs)) to model possible climate futures (McElwee 2021). IAMs can be used to understand how a future end point (like limiting temperature increase to less than 2°) will require specific actions over time, and increasingly, the only way IAMs show the world reaching ambitious targets like 1.5° is to include NETs. How much NETs may need to be used depends on other assumptions, like world population and affluence; for example, in SSP5, a scenario where economic growth is strong and fossil fuel use remains high, trying to go back to 2° or less would require massive use of NETs by 2100 (Popp et al. 2017).

However, only BECCS and afforestation have been modeled in IAMs used in recent IPCC reports, while the other NETs are not included due to technical challenges; this has led to some concerns that BECCS is being overemphasized simply because it can be modeled (Fuhrman et al. 2019). IAMs also show very clearly that the required use of land-based NETs to achieve climate targets will come with trade-offs (Dooley, Christoff, and Nicholas 2018). Both afforestation and BECCS require land conversion (that is, shifting from food production or other uses to growing trees and bioenergy feedstocks), and in different scenarios, this expansion of afforestation and BECCS happens at the expense of natural forests, or agricultural and grazing lands (Popp et al. 2017). However, the *social* impacts of land-based NETs can only be assessed in either general terms or in hypotheticals (Smith et al. 2019). IAMs can only include some development goals, like poverty rates, food security, economic growth, or water use in very general terms. Other potential impacts from land use change like gender differentiation, rising inequality, displacement, or land degradation cannot be forecasted in most IAMs, and thus alternative ways to understand these outcomes are needed.

Examining NETs through an agrarian lens

NETs are usually discussed at the global landscape level, with little attention to the specific places in which they will be used, which makes detailed examinations difficult

(Buck 2018). Further, many NETs remain mostly speculative, with few demonstration sites or models. Thus, it is useful to compare what we do not know about NETs with what do we know about agrarian impacts of existing land-based climate mitigation measures (including REDD+, carbon forestry, and biofuels).[5] In the sections below, key agrarian issues of land, labor, capital, and rural politics are examined, both from the perspective of NETs technical literature regarding what can be modeled and where gaps exist, and from the known outcomes of other carbon and climate-focused projects that have happened in the rural South in recent decades.

Land and NETs

Where and how NETs are likely to induce land use change, and the implications of issues such as land tenure, dispossession, or consolidation, is one of the most important impacts of NETs. Land-based NETs can be divided into two major types: those that will create land competition, because they must be the primary land use where they are deployed (BECCS and afforestation), and those that can be deployed alongside agriculture or other land uses (such as soil carbon, enhanced weathering, or biochar). For land-competing NETs, key issues concern the scale of their potential use and what other land-based activities will be displaced as a result, with knock-on effects for food prices or biodiversity loss (NAS 2019). For example, BECCS used at small scales would not require land use change if fueled solely by residues and wastes from existing agriculture and industrial processes (Fajardy, Chiquier, and Dowell 2018). However, more ambitious emissions reductions would require significant land use change; for example, if BECCS was deployed to remove between 3 and 12 $GtCO_2$ per year, this would likely require between 380 and 700 Mha of land by 2100 (that is, up to 10% of existing total world land use) (Smith et al. 2015).[6] The potential for declines in food production as farmers switch lands to produce trees or feedstocks is clear in most IAMs, some of which suggest that up to 1 billion people could be impacted by large-scale deployment (Fuss et al. 2018).

Where would land use conversion for NETs likely take place? Most IAMs rely on indicators of land suitability and cost to predict where BECCS or afforestation would be most likely to occur (Riahi et al. 2017; Cronin et al. 2020), leading to the assumption that BECCS is most feasible in areas of 'high biomass yields and relatively low carbon stocks (that is, abandoned lands and typically warmer temperate and subtropical areas)' (Hanssen et al. 2020): in other words, predominantly in the global South. BECCS is also more likely to be centered in places with access to ports and export markets for feedstocks. NETs like afforestation are often assumed to be produced on degraded and marginal lands not suitable for agriculture, thereby avoiding competition with food production (Bastin et al. 2019), and Africa and Latin America are projected to experience the most pressure for afforestation, with one model estimating a need for 630 and 600 Mha of land respectively (Kreidenweis et al. 2016). Outside of BECCS and afforestation, the other NETs that do not compete for land are somewhat less constrained in where they can be deployed. Studies of land suitability have noted that poor carbon- or water-holding soils

[5]For example, existing literature has examined community acceptance or opposition of existing energy and climate projects as a proxy for possible reactions to NETs (Buck 2018).

[6]As of 2017, total world land use for agriculture and forestry was 7130 Mha (2429 Mha for forests, 1426 Mha for agriculture, and 3275 Mha of grasslands (for livestock)) (NAS 2019).

tend to be in tropical countries, particularly in Africa, where positive yield effects of inputs like biochar will tend to be greater (Robb et al. 2020).

However, there remain unanswered questions regarding the use of land for NETs in existing models, for which critical agrarian literatures provide useful comparisons, particularly around 'marginal' lands definitions, the potential impact of land grabbing, and the role of colonial histories and global elites in shaping land use. For example, there is no clear definition of what 'marginal lands' means for NETs, with varying definitions used in different models and contexts. Some models assume land is essentially 'abandoned' simply because it is not being used for high-productivity crops (Strengers, Minnen, and Eickhout 2008). There are wide estimates of potential availability of these so-called marginal lands: one study proposed 1300 Mha (including lands where one-third of the world's population is currently farming) as marginal but noted that 'only a fraction would be available for afforestation/reforestation and BECCS' due to these existing land uses (NAS 2019, 118). Critical work on existing biofuels and forest carbon expansion has shown that classifications of marginal lands are often subjective, based on narrative rather than evidence-driven assessments (Unruh 2008; Hajdu and Fischer 2016). What is defined as marginal land in many national contexts is often a political calculation deployed to expropriate existing land users (Baka 2014; Scheidel and Work 2018). Most IAMs also assume that market prices are the most important factor that would drive farmers to stop food production and use their land for investment in trees or feedstocks. Yet the evidence for crop-switching reveals a more complicated picture than simple economic calculations (Li 2014; Borras et al. 2015). Models also usually assume that the lowest productivity farmlands would be abandoned first, but these are often the only places where the poor, women, or marginalized are able to farm (McElwee 2009; Quisumbing and Pandolfelli 2010).

However, despite these concerns, literature searches for BECCS and 'land tenure' as a topic turned up zero studies, as did a search for biochar and 'land tenure.' This is likely because IAMs currently take existing land use as evidence of secure tenure (Cronin et al. 2020), and have no indicators (other than price of land or crops) to inform risks of dispossession. Given the numerous ongoing land tenure conflicts, including a marked rise in land grabs because of biofuel policies in particular (Baka 2014; Hufe and Heuermann 2017), strong concerns have been expressed about potential land grabs driven by NETs (Leach, Fairhead, and Fraser 2012).[7] Experience from existing land grabs shows considerable risks, including food insecurity and poverty increases, particularly from large-scale land acquisitions (Schoneveld, German, and Nutakor 2011; Yengoh and Armah 2015; Müller et al. 2021). Much of the land rush has been speculative, but has nonetheless had significant effects on the ground (Franco and Borras 2019; Hansson et al. 2019), and similar results might be expected with some NETs (Richards and Lyons 2016).

In recent land grabs, dispossession has been more common where farmers had insecure land tenure (e.g. farming on state lands) (Bleyer et al. 2015; Fisher et al. 2018), but even clear property rights do not confer security or the ability to shape land deals

[7]Databases of current and proposed land deals indicate that 3.6 million hectares have been acquired for biofuels, mostly in Africa and Latin America (with no landgrabs in Europe or North America), while large-scale land deals for all types of forestry have affected 31.7 million ha, with many in Eastern Europe/Russia (LandPortal data).

(Vermeulen and Cotula 2010; Franco and Borras 2021). For example, evidence suggests that even where farmers have not themselves lost land but fear the risk of expropriation, they are less likely to invest in on-farm management (Aha and Ayitey 2017). There are also significant cases of local dispossession driven by smallholders themselves (Osborne 2011; Chen 2013; Cavanagh and Benjaminsen 2014; Bleyer et al. 2015; Olwig et al. 2015; Scheidel and Work 2018). Overall, the literature suggests that concentrated ownership of tree plantations has been more associated with dispossession (Kröger 2014; Malkamäki et al. 2018) (an outcome likely to be relevant for BECCS), while smallholder models have been more broadly positive in terms of food security and biodiversity (Jindal, Swallow, and Kerr 2008; Eijck et al. 2014) (more likely for afforestation NETs). Smallholder models have been most successful when local concepts of distributive justice have been recognized (Fisher et al. 2018), although there have been reported problems with benefit sharing due to lack of clarity for carbon rights and contingencies of contracts (Unruh 2008; Corbera et al. 2011; Tienhaara 2012). Further, farmers with smaller landholdings who have engaged in forest carbon projects have often experienced more negative impacts on food security, indicating that the distributions of costs and benefits are often uneven (Aggarwal and Brockington 2020).

Finally, agrarian studies scholars have demonstrated how important historical perspectives on land use change are, as path dependencies from colonial control of land continue to influence current-day trajectories (Chomba et al. 2016; Davis and Robbins 2018). For example, legacies of racialized dispossession have shaped political subjects and forest legibility in REDD+ projects in Guyana and Suriname (Collins 2019), while elsewhere land tenure histories have resulted in unequal benefit distribution systems from REDD+ investments (Kashwan 2015). Further, there is an important role for global elites in shaping land projects, which has privileged small numbers of actors over rural communities whose livelihoods are often discounted (Asiyanbi 2016; Ece, Murombedzi, and Ribot 2017; Hook 2020). Simplistic narratives of crises and degradation that ignore corporate or capitalist drivers are often used to cast blame on local practices for their (perceived negative) climate implications as well as constrain options to those proposed by global financiers (Franco and Borras 2021; Hjort 2020).

Capital and NETs

How financial arrangements for land-based NETs will operate remains an open question. There is likely to be a significant role for private capital, as some corporations are already pledging not just carbon neutrality in their operations but scaling up their investments in NETs specifically.[8] Key economic issues in the literature on NETs primarily focus on what the total potential costs for each technology are (Fajardy and Dowell 2018; Fuss et al. 2018) and how that money might be raised and through what sources. The amount of funding available to NETs will also depend in part on the costs of other emissions reduction options, such as renewables.

The most prominent mechanism within IAMs for incentivizing investments is a carbon tax; models then predict how high a carbon tax would need to be to spur various NET investments. Estimates range from as low as \$6 per ton of CO_2 for afforestation, while

[8]See https://stripe.com/blog/first-negative-emissions-purchases.

BECCS is likely to only be feasible at $100–250 per ton (Strengers, Minnen, and Eickhout 2008; Humpenöder et al. 2014). Carbon prices higher than $55 a ton could stimulate biochar use (Robb et al. 2020), but whether or not farmers benefit from such prices depends on how carbon markets are structured (e.g. if biochar producers rather than individual farmers benefit from incentives). It is currently unclear what form of pricing would be needed to encourage enhanced weathering, given large upfront costs of mining (Edwards et al. 2017). Of the NETs considered here, only soil carbon sequestration shows positive synergies between carbon prices and food security in models (Frank et al. 2017). Achieving 100 $GtCO_2$ of negative emissions from land-based solutions, given a range of prices of $10–100 USD per ton of CO_2 would require a total funding flow on the order of 1 trillion USD (Nolan, Field, and Mach 2021).

Global accounting rules on carbon credits (such as those that will be facilitated by Article 6 of the Paris Agreement) will be crucial in understanding how markets for NETs will work (Zakkour, Kemper, and Dixon 2014; Coffman and Lockley 2017; Lockley and Coffman 2018). Existing studies of voluntary carbon markets have indicated that numerous questions around measurement, verification, and value have challenged the easy uptake of existing carbon offsets (Lohmann 2008; Leach and Scoones 2013). Modeling economic outcomes is also often overly simplistic; for example, IAMs mostly rely on a globally uniform carbon price (unlikely to be realistic), yet variable regional or national prices cannot be modeled well (Fridahl and Lehtveer 2018). Additionally, unlike modeling of costs that fall over time for other technologies, BECCS will likely increase in cost at higher scales and more distant time periods due to less land being available in the future (Honegger and Reiner 2017). Modeling of costs also tends to be limited to initial establishment; for afforestation, IAMs usually assume no costs of maintenance over time, even though labor around fire prevention, thinning, and other practices are often needed. Further, IAMs usually assume zero risk for most NETs (that is, that no enterprises will fail given a consistent level of carbon price), which is an overly optimistic assumption. For example, existing biofuels contracts show many failures because such investments do not price risks well (Tienhaara 2012).

Other questions of interest to agrarian studies scholars, such as what types of investors will be involved, how they will be regulated, and what the equity impacts of north–south investment flows will be, have been less examined for NETs. Funding is very likely to be through multinational actors, as biofuel investments made by complex conglomerates have shown (Borras, McMichael, and Scoones 2011), and is likely to attract start-ups from previously uninvolved actors in land investment (Leach, Fairhead, and Fraser 2012). How such actors and their investments are structured is key to understanding their possible impacts (McCarthy 2010). Evidence from the existing wood pellet industry, which provides a glimpse into what BECCS facilities and supply chains might look like, suggests that multiple investors are required, from family-owned forests to wood brokers to pellet mills to the eventual biomass energy facility across trans-Atlantic chains, with regulatory apparatuses across these different investments often minimal (Ramos 2022).

How NETs use contracts vs. ownership models for investing in supply of feedstocks (e.g. for BECCS in particular) will also result in different outcomes, as seen in evidence from plantation forestry projects. For example, ownership of land by companies can dispossess small farmers, while supply contracts are associated with wealth accruing to shareholders but not local laborers (Richards and Lyons 2016). International investment contracts for forest carbon have often disadvantaged local communities due to lack of transparency and involvement (Tienhaara 2012), while biofuels contracts have been plagued by inflexibility and lack of oversight (German, Schoneveld, and Pacheco 2011). For other NETs that are used in on-farm production by smallholders (e.g. soil carbon and biochar), cost effectiveness is likely to be linked to the ability to enhance agricultural production. Biochar has presumed lower abatement costs in developing countries because it can increase crop yields while decreasing the need for chemical fertilizer amendments or irrigation (Smith et al. 2019). However, farmers do not experience net abatement costs, but rather upfront costs (such as to purchase and deploy biochar), and thus equity issues are still likely to persist; for example, negative debt cycles for poor smallholders have been common in biofuels production (McCarthy 2010).

Labor and NETs

Many of the NETs will require labor for deployment, but currently there is little attention to this topic other than an assumption of low labor opportunity costs in rural areas of the global South (Eijck et al. 2014). For example, most IAMs do not include labor costs explicitly, and assume that with high enough carbon prices (such as $50/ton or more), labor will be available. Whether NETs will generate labor conditions that would either positively or negative impact rural workers is unknown: for example, production of different feedstocks for BECCS might require either smallholders or waged labor (Eijck et al. 2014; Schirmer and Bull 2014). In the latter case, there could be potential risks to investors from not applying appropriate oversight to labor needs, such as violations of child labor laws, safety issues, or non-payment of wages (Eijck et al. 2014).

The experience of existing biofuel plantations and forest carbon projects provide useful comparisons. Low-wage and unsteady labor has been common in many forest carbon projects (Smith and Scherr 2003; Greenleaf 2019), particularly for mega-plantations. For biofuels, the employment generated has varied depending on the feedstock, with jatropha generating more jobs than corn, largely due to mechanization of the latter (Hunsberger, German, and Goetz 2017). In some cases, biofuels like palm oil have generated less jobs than land uses that were displaced by these plantations (Li 2011): estimates indicate that land acquisitions for biofuels in Africa have resulted in the loss of jobs at the farm level, including as high as four people displaced for every 1 ha of land acquired (Renzaho, Kamara, and Toole 2017). Similarly low levels of employment have been reported for forest plantations (Gerber 2011). More employment seems to be generated when smallholders work their own land in out-grower schemes and when processing of goods is done locally (Malkamäki et al. 2018), but mechanization in later phases of production tends to reduce employment opportunities (Deininger 2011). Biofuels projects have also varied in terms of part-time versus stable long-term employment (Hunsberger, German, and Goetz 2017; Pirard, Petit, and Baral 2017).

There are also questions about whether NETs might drive labor migration; some experience from tree planting projects shows migrants may be preferred because of their willingness to work hard jobs for low pay, but this also makes them vulnerable to exploitation (Malkamäki et al. 2018). Similarly, examples from the biofuels literature show that many investors make use of skilled outside labor (Richardson 2010) or migrants for low skill work (Li 2011), with some reports of debt peonage in biofuel plantations or processing in Brazil (Hunsberger, German, and Goetz 2017). There is also little attention to gender within labor practices required for NETs; evidence from other previous afforestation projects indicates that women are often involved as labor but not as owners or beneficiaries of tree planting (McElwee 2009; Gerber 2011). Physically demanding labor may be assumed to require men only, thereby increasing gender gaps and uneven resource access.

Rural politics and NETs

Understanding how rural populations may support or oppose NETs has not been part of the literature in any depth. There is an acknowledgement that NETs will need a social license to operate (Fuss et al. 2020), and community support or opposition is likely to be a function of existing values, framings of risk, and the way benefits are understood or shared (Pidgeon and Spence 2017; Cox et al. 2018). However, most surveys of the acceptability of NETs have been carried out in the global North and have been framed more by concerns about 'tampering with nature' rather than direct livelihood impacts (McLaren et al. 2016; Wolske et al. 2019).

Existing literature on biofuels or carbon tree planting – both less speculative technologies than some NETs – reveals that social licenses do not confer acceptability if other aspects are not considered (Smith and Scherr 2003). One key lesson is that procedural and recognition justice processes have been important considerations in conferring acceptability (Suiseeya and Caplow 2013). Understandably, many rural communities have declined to participate in afforestation projects that do not account for local conceptions of justice and equity (Hendrickson and Corbera 2015), or where corporate interests outweigh local land users (Gerber 2011). Production of feedstocks for BECCS that occur in one location but where the energy generated may flow elsewhere are thus likely to be a sensitive issue that could fuel a sense of injustice (Buck 2018).

Different NETs are likely to unevenly impact households across class, gender, race and ethnicity and other forms of difference (Borras et al. 2021), particularly where NETs generate changes in property rights and labor regimes as noted above. How these axes of difference then become sites of conflict, as well as opportunities for organizing across alliances, has not been considered by the NETs literature, but lessons from other examples are useful here. Forest carbon projects have increased intra-community conflicts between richer and poorer households or those with power and access and those without; between men and women; between generations; and between different ethnic groups (Baynes et al. 2015; Benjaminsen and Kaarhus 2018; Kemerink-Seyoum et al. 2018). There have also been risks of conflict within wider landscapes (Schmid 2022), with negative impacts on overall democratic decision-making among forest-dependent communities (Chomba 2017; Ece, Murombedzi, and Ribot 2017).

Yet in other cases, stronger community organizing has been an outcome of externally driven investments and dispossessions. Indigenous peoples' organizations have led successful efforts to frame REDD+ and other forest projects as a threat to identities and livelihoods if not designed with their rights in mind (Wallbott and Recio 2019; Marín-Herrera, Correa-Correa, and Blanco-Wells 2021), while transnational framing and mobilization against land grabs and biofuels as a form of dispossession have successfully stopped some land appropriation attempts (Franco et al. 2010; Temper 2018). The combination of outside civil society/NGO support and the actions of cohesive user groups threatened by destructive development, such as for plantations or energy production, have proven decisive in many cases (Veuthey and Gerber 2012; Temper et al. 2020). Indigenous women in particular have been strong leaders in framing forest carbon projects as threats to household and community livelihoods, rights, and knowledges (Westholm and Arora-Jonsson 2018; Löw 2020). Food security and sovereignty angles around NETs have potential to be key pivot points for such organizing, particularly for land-competing NETs like afforestation or BECCS where increased food prices are predicted (Kreidenweis et al. 2016). High levels of uncertainty on how NETs will impact food access are thus likely to increase the sense of risk for many local communities (Table 2).

Discussion: challenges and options for anticipating NETs

The IPCC's 1.5° report has warned that 'The impacts of carbon dioxide removal (CDR) options on the Sustainable Development Goals (SDGs) depend on the type of options and the scale of deployment ... Context-relevant design and implementation requires considering people's needs, biodiversity, and other sustainable development dimensions' (IPCC 2018, 21). This argues for reframing the use of NETs as a potentially high-risk gamble with serious justice implications, given uncertainties around deployment (Anderson and Peters 2016).

Modeling projections suggest that the longer it takes to deploy NETs and the higher greenhouse gas emissions rise, the more NETs will be needed at a future point to avert serious climate damage (Skea et al. 2022). If overshoot of 1.5° or 2° targets appears inevitable, the push to use multiple NETs is likely to accelerate (Muratori et al. 2020). Particularly for land-based NETs, many commentators see these as more benign and thus their deployment may be more likely, particularly as support for NET technologies is correlated with their perceived 'naturalness' (Buck 2019; Markusson 2022). This suggests more support for afforestation and less for BECCS, while in fact both strategies compete for land and might have similar agrarian consequences.

Yet for all the increasing discussions of NETs, they are not yet included in many country-level strategies for emissions reductions, including in the Nationally Determined Contributions (NDCs) required under the Paris Agreement. Afforestation and forest restoration is by far the most common NET in existing pledges; no country is currently on record as planning to engage in BECCS, and only a handful have noted their interest in other forms of CDR (see Table 3). This leads to a conundrum: IAM projections of pathways to keep emissions in line with 1.5° and 2° targets often include the use of NETs, while country strategies represented in NDCs rarely do so, outside of afforestation plans. The fact that models require NETs to reach temperature targets, while countries have few explicit plans to deploy them, runs the risk of a situation where there is a last-minute rush to expand these technologies, rather than carefully considering pros, cons, and research needs ahead of time (Moe and Røttereng 2018). It also has neglected careful attention to 'who benefits' from use of NETs versus other approaches. In other words, 'who gets to define what are legitimate mitigation and adaptation measures, involving which and whose natural resources, how, why and with what socio-economic and

Table 2. Summary of key agrarian issues across NETs.

Issue	Key research gaps in NETs literature	Problems identified in agrarian studies literature	Are these problems modeled in IAMs?	Main NETs affected	Potential improvements suggested by literature
Land	- Unclear definition of marginal lands for models - Impacts of displacement of existing land uses - Regional locations of NETs deployment - Risks of uncertainty around land tenure	- Land tenure conflicts - Land grabbing - Elite capture of land resources - Colonial histories and path dependencies	- Total area of deployment only. No inclusion of tenure, land conflicts, or risks of land grabbing.	BECCS, afforestation	- Improved contracts (more transparency, flexibility and oversight) - Inclusion of food production alongside NET feedstock production (e.g. agroforestry)
Capital	- Unclear carbon prices and their incentivization of NETs - Sources of funding (private vs public) - Overall costs of action over time - Risk of failures - Role of govt subsidies	- Role of financial speculation - Complications of measurement and verification - Transparency of contracting - Smallholder rights and costs	- Indirectly – carbon prices included, but not sources of funding	BECCS, weathering, soil carbon, afforestation, biochar	- Transparency in sourcing of investments - Improved contracts (more transparency, flexibility and oversight) - Investor safeguards on procedural and distributional justice - Mandated benefits, quotas, or price floors
Labor	- Types and quality of labor - Length of contracts - Gender issues - Risks to investors of labor violations	- Lower demands for labor due to mechanization - Stability of employment - Demands for migrant labor	- No. Assumed adequate labor costs within carbon price	BECCS, afforestation, biochar	- Quotas for purchases from smallholders in out-grower models - Attention to gender concerns - Safeguards on labor rights - Expanded social safety nets
Rural politics	- Equity in benefit sharing - Perceptions of risks - Impacts on food security	- Procedural and recognition justice - Uneven benefit distribution - Conflicts over benefits - Rural coalitions and organization building in opposition	- Food price changes can be modeled, but not responses to these (e.g. food access and concerns over social license to operate).	BECCS, afforestation	- Participatory planning processes and attention to procedural equity - Access to energy production and other shared benefits - Expanded social safety nets for those at risk of food insecurity

political implications?' (Borras, Franco, and Nam 2020, 8) are not questions that most NDCs have asked.

Addressing the challenge of IAMs in normalizing NETs

As noted previously, nearly all IAM projections in IPCC reports to achieve a climate stabilization target well-below 2° require the inclusion of NETs of some kind (IPCC 2018), and there is increasing recognition that how policy options are framed in these models has an influence on decision-making (Rogelj et al. 2019). Yet the opaque nature of modeling has led to misunderstandings of the role of NETs in these projections and in policy based on them (Fuhrman et al. 2019; Sovacool et al. 2019), with some scholars labeling modelers a 'geoclique' (Cox et al. 2018) or having 'an exclusive character' (Carton et al. 2020). To many, IAMs often operate like a black box where assumptions are unclear or unknown, and which have normalized speculative options like BECCS unnecessarily (Low and Schäfer 2020).

There is increasing interest in ranking NETs across factors like feasibility, effectiveness, and side-effects to determine which portfolios to prioritize in coming decades (Rueda et al. 2021). The empirical evidence from previous carbon forestry and biofuels projects proves that land-based NETs can have detrimental consequences, yet many of these outcomes are not included in IAMs, which makes comparisons among NETs and with other policy options problematic (Rueda et al. 2021). For example, IAMs cannot model many known challenges like ethics and governance issues (Forster et al. 2020): they cannot answer where local populations are likely to be skeptical of NETs due to previous past poor performance of other rural schemes (Montefrio, Sonnenfeld, and Luzadis 2015); and they do not yet consider issues like land tenure or biodiversity well. As a result, IAMs only model what they can – pricing, population, or land quality – and are silent on what they cannot. The modeling community has acknowledged that they need to do a better job incorporating 'implementation limits and obstacles' (Kriegler et al. 2014), as well as considering the 'impacts that NETs will have on sustainable development goals and equity issues' (Fuhrman et al. 2019). Surveys of IAM experts note that they see constraints around resource competition and political feasibility for most NETs, yet most still believe that land-based NETs are important to include in policy portfolios (Fridahl and Lehtveer 2018; Rickels et al. 2019; Low and Schäfer 2020).

Inputs from agrarian studies scholars' work could help introduce some improved indicators and constraints into IAMs. For example, conflicts over tenure rights and existing land grabs could be included by using maps of land acquisition risks in some spatially explicit IAMs. Indicators related to employment and labor that would be required by different NETs as well as recognizing that these costs cannot be captured by carbon prices alone could also improve understanding. Recognizing the potential risks of NETs (e.g. introducing variables around the possibility of lower-than-expected carbon capture or assuming that some percentage of NETs projects will fail) can provide a more realistic understanding of options as well. These improvements to IAMs could potentially reduce the problem of mitigation deterrence by being clearer that NETs are not a panacea and will often involve serious trade-offs among sectors, regions and communities, many of which are yet to be captured in modeling (Grant et al. 2021).

Table 3. NETs discussed in country NDCs.

Afforestation	BECCS	Soil Carbon Sequestration	Enhanced weathering/ mineralization	Biochar	General mention of idea of NETs
Afghanistan	None	Zambia	Iceland	Belize	Canada
Albania		Malawi		Namibia	Fiji
Algeria		UAE		Myanmar	
Angola		Liberia			
Armenia		China			
Bahrain					
Bangladesh					
Belize					
Benin					
Brunei					
Cabo Verde					
Cambodia					
China					
Comoros					
DRC					
EU					
Eritrea					
Ethiopia					
Gambia					
Georgia					
Ghana					
Guinea					
Iceland					
India					
Indonesia					
Jordan					
Kenya					
Kyrgyzstan					
Lebanon					
Lesotho					
Liberia					
Malawi					
Mexico					
Moldova					
Morocco					
Myanmar					
Namibia					
Niger					
Nepal					
North Korea					
Pakistan					
Palestine					
Papua New Guinea					
Rwanda					
Samoa					
Sierra Leone Sri Lanka					
Somalia					
South Sudan					
St. Vincent and Grenadines					
Sudan					
Tajikistan					
Tanzania					
Turkey					
Uruguay					
Uzbekistan					
Uganda					
Vietnam					
Zambia					

Source: Search of NDCs submitted to UNFCCC.

Making NETs less negative through just approaches

As noted, most of the considerations of feasibility of NETs have focused on technical rather than social or justice elements (Morrow et al. 2020). Agrarian studies scholars can highlight problematic assumptions used in these approaches, such as unclear marginal lands definitions that have influenced the deployment of biofuels (German, Schoneveld, and Pacheco 2011). This work can also help temper the enthusiasm for NETs like afforestation as a low-hanging fruit of climate policy by highlighting the slow nature of policy change, given that design of carbon forestry to achieve co-benefits has taken decades and been very complex (Von Hedemann et al. 2020). Indeed, the challenges faced in the past by voluntary forest carbon projects are likely to be even more significant for NETs, such as monitoring and accounting rules across multiple spatial scales (Brander et al. 2021).

NETs will need to have mechanisms for transparency, accountability, responsiveness, and legitimacy in order to be accepted and to reduce their impacts on rural communities. Attention to justice implications will mean that rural peoples need to be part of any discussions and alliances in sites of NET deployment: for example, procedural justice around siting of carbon capture facilities has been key to getting agreement from affected communities (McLaren 2012). Different mechanisms for procedural justice, such as access to information and consent process around land acquisitions, have been used elsewhere (O'Beirne et al. 2020). For example, the engagement of local communities in design and implementation processes has been an important link between local actors and national goals in REDD+ projects (Schroeder and McDermott 2014), which is a particular risk for BECCS given that feedstocks are likely to be produced in one place while energy generated elsewhere in long supply chains (Buck 2019). There is strong skepticism that voluntary codes of conduct or self-regulating mechanisms, such as those that have emerged around palm oil or soy biofuels, are likely to satisfy the need for procedural justice, arguing for stricter regulatory frameworks (Borras and Franco 2010; Blaber-Wegg, Hodbod, and Tomei 2015). This may include the acknowledgement of uneven risks by 'excluding some greenhouse gas removal options from certain regions, areas, or environments' (Smith et al. 2019, 277).

Distributional justice issues are also likely to play a key role in understanding who benefits from and who opposes NET deployment, including how benefit-sharing schemes are organized based on values or opportunity costs (Fisher, Bavinck, and Amsalu 2018). To combat this problem, some biofuels policies have built in specific quotas or rights for smallholders, including 'direct funding, low-interest loans, technical support, a guaranteed minimum price for biofuel feedstock, and a requirement that processors purchase part of their feedstock from smallholders' (Hunsberger, German, and Goetz 2017). NETs projects that ensure producers or users receive mandated benefits like clean electricity could also contribute to acceptability, as could expanded social safety nets as compensation if food prices rise (Fujimori et al. 2018). Additional specific improvements across land, capital, labor, and politics to help minimize the impacts of NETs on rural communities that come from previous experiences are noted in the far right column in Table 2. These can include improving contracts for feedstock production or inclusion of food production alongside feedstocks; improving transparency in financing and contracting; instituting planning processes that ensure inclusion of affected communities; and explicit benefit-sharing mechanisms. All these possibilities have been tested in previous forest carbon or biofuels policies with varying degrees of success and will need to be contextually appropriate for the wide range of rural communities and local land uses affected by future NETs.

Conclusions

Recent reviews have noted that NETs are the latest in long line of 'fads' around land management, conservation, and climate change (Carton et al. 2020), and many are not likely to pan out given the lack of demonstration projects or ability to scale up at needed levels. Nonetheless, they are likely to remain important, both within IAMs and in the real world, and thus critical scholars need to engage with these concepts and projects (Beck and Mahony 2018; Brack and King 2021). The more that countries act now on climate mitigation, the less they must rely on uncertain NETs later on (Lenzi 2018; Hilaire et al. 2019), and research on the implications of NETs is an important part of decision-making in balancing these trade-offs. Concerns about 'mitigation deterrence' of NETs creating a moral hazard in diminishing the urgency of fossil fuel emissions reduction now would particularly benefit from explicit acknowledgement of NETs trade-offs (Fuss et al. 2020; McLaren 2020). Ultimately, the key question is if the negative impacts of geoengineering or stringent mitigation policies are likely to outweigh the also very negative impact of a more than 1.5° world (Hasegawa et al. 2018; Robinson and Shine 2018), and how the uneven burdens of both scenarios might fall on more marginalized peoples (Sultana 2022). There is an additional ethical burden of asking future generations to resolve these thorny issues (Hansen et al. 2017), e.g. to decide between loss of coral reefs and other impacts versus the problematic future deployment of NETs (Vakilifard et al. 2021).

As shown in this review, many NETs that are land-based have potential to raise serious consequences for rural populations, including land tenure conflicts and dispossession, food security risks, gender impacts and poor working conditions, and inadequate benefit-sharing and uneven procedural justice. Many of these impacts are likely to fall hardest on poorer and more marginalized rural farmers and workers, and the land-competing NETs (BECCS and afforestation) are likely to be the most impactful. Additionally, rural populations that are not included, consulted, and given rights of refusal in NETs deployment are likely to influence whether these projects are able to be implemented, as rural protests and rejections of NETs may follow the paths of other climate interventions like forest carbon and biofuels, which have faced disapproval and even sabotage at local levels when faced with inadequate considerate of local values, rights, and benefits (Corbera, Hunsberger, and Vaddhanaphuti 2017).

Given this review's focus on the gaps in knowledge around NETs, and strong evidence that questions surrounding land, capital, labor, and rural politics have been inadequately considered in existing discussions, there is a need for new transdisciplinary research agendas on NETs to assess their technical, biophysical, financial, and societal uncertainties (Fuss et al. 2014, 2016; Minx et al. 2018). Critical social science literature is necessary to contextualize the difficulties that NETs are likely to face in deployment, whether top-down and large-scale or more community-based (Carton et al. 2020), including the ways in which rural peoples may resist or acquiesce to such interventions. Future research agendas are also needed around the use of IAMs in guiding decision-making, such as opening modeling to more participatory approaches and the recognition of alternative world views within them (Forster et al. 2020). Deliberative public engagement on research for these technologies can also help bring more democratic processes in and lead to better consideration of justice in NETs trajectories for the future (Low and Buck 2020).

Acknowledgements

I would like to thank colleagues who worked with me on the IPCC Special Report on Climate Change and Land for stimulating some of the thoughts in this paper.

Disclosure statement

No potential conflict of interest was reported by the author(s).

Funding

The work for this paper was funded in part by a National Science Foundation grant #1853759 on Understanding the Use of Ecosystem Services Concepts in Policy.

ORCID

Pamela McElwee http://orcid.org/0000-0003-3525-9285

References

Aggarwal, A., and D. Brockington. 2020. "Reducing or Creating Poverty? Analyzing Livelihood Impacts of Forest Carbon Projects with Evidence from India." *Land Use Policy* 95: 104608.

Aha, B., and J. Z. Ayitey. 2017. "Biofuels and the Hazards of Land Grabbing: Tenure (In)security and Indigenous Farmers' Investment Decisions in Ghana." *Land Use Policy* 60: 48–59.

Anderson, K., and G. Peters. 2016. "The Trouble with Negative Emissions." *Science* 354 (6309): 182–183.

Asiyanbi, A. P. 2016. "A Political Ecology of REDD+: Property Rights, Militarised Protectionism, and Carbonised Exclusion in Cross River." *Geoforum; Journal of Physical, Human, and Regional Geosciences* 77 (C): 146–156.

Baka, J. 2014. "What Wastelands? A Critique of Biofuel Policy Discourse in South India." *Geoforum; Journal of Physical, Human, and Regional Geosciences* 54: 315–323.

Bastin, J. F., Y. Finegold, C. Garcia, D. Mollicone, M. Rezende, D. Routh, C. M. Zohner, and T. W. Crowther. 2019. "The Global Tree Restoration Potential." *Science* 365 (6448): 76–79.

Batres, M., F. M. Wang, H. Buck, R. Kapila, U. Kosar, R. Licker, D. Nagabhushan, E. Rekhelman, and V. Suarez. 2021. "Environmental and Climate Justice and Technological Carbon Removal." *The Electricity Journal* 34 (7): 107002.

Baynes, J., J. Herbohn, C. Smith, R. Fisher, and D. Bray. 2015. "Key Factors Which Influence the Success of Community Forestry in Developing Countries." *Global Environmental Change* 35: 226–238.

Beck, S., and M. Mahony. 2018. "The Politics of Anticipation: The IPCC and the Negative Emissions Technologies Experience." *Global Sustainability* 1: e8.

Beerling, D. J. 2017. "Enhanced Rock Weathering: Biological Climate Change Mitigation with Co-benefits for Food Security?" *Biology Letters* 13 (4): 20170149.

Benjaminsen, G., and R. Kaarhus. 2018. "Commodification of Forest Carbon: REDD+ and Socially Embedded Forest Practices in Zanzibar." *Geoforum; Journal of Physical, Human, and Regional Geosciences* 93: 48–56.

Bernal, B., L. T. Murray, and T. R. H. Pearson. 2018. "Global Carbon Dioxide Removal Rates from Forest Landscape Restoration Activities." *Carbon Balance and Management* 13 (1): 22.

Blaber-Wegg, T., J. Hodbod, and J. Tomei. 2015. "Incorporating Equity Into Sustainability Assessments of Biofuels." *Current Opinion in Environmental Sustainability* 14: 180–186.

Bleyer, M., M. Kniivilä, P. Horne, A. Sitoe, and M. P. Falcão. 2015. "Socio-economic Impacts of Private Land Use Investment on Rural Communities: Industrial Forest Plantations in Niassa, Mozambique." *Land Use Policy* 51: 281–289.

Borras, S. M., and J. C. Franco. 2010. "From Threat to Opportunity? Problems with the Idea of a 'Code of Conduct' for Land-Grabbing." *Yale Human Rights and Development Law Journal* 13 (1): 507–523.

Borras, S. M., and J. C. Franco. 2018. "The Challenge of Locating Land-Based Climate Change Mitigation and Adaptation Politics Within a Social Justice Perspective: Towards an Idea of Agrarian Climate Justice." *Third World Quarterly* 39 (7): 1–18.

Borras, S. M., J. C. Franco, S. R. Isakson, L. Levidow, and P. Vervest. 2015. "The Rise of Flex Crops and Commodities: Implications for Research." *Journal of Peasant Studies* 43 (1): 93–115.

Borras, S. M., J. C. Franco, and Z. Nam. 2020. "Climate Change and Land: Insights from Myanmar." *World Development* 129: 104864.

Borras, S. M., P. McMichael, and I. Scoones. 2011. "The Politics of Biofuels, Land and Agrarian Change: Editors' Introduction." *Journal of Peasant Studies* 37 (4): 575–592.

Borras, S. M., I. Scoones, A. Baviskar, M. Edelman, N. L. Peluso, and W. Wolford. 2021. "Climate Change and Agrarian Struggles: An Invitation to Contribute to a JPS Forum." *Journal of Peasant Studies* 49 (1): 1–28.

Brack, D., and R. King. 2021. "Managing Land-Based CDR: BECCS, Forests and Carbon Sequestration." *Global Policy* 12 (S1): 45–56.

Brander, M., F. Ascui, V. Scott, and S. Tett. 2021. "Carbon Accounting for Negative Emissions Technologies." *Climate Policy* 21 (5): 1–19.

Buck, H. J. 2018. "The Politics of Negative Emissions Technologies and Decarbonization in Rural Communities." *Global Sustainability* 1: e2.

Buck, H. J. 2019. "Challenges and Opportunities of Bioenergy with Carbon Capture and Storage (BECCS) for Communities." *Current Sustainable/Renewable Energy Reports* 6 (4): 124–130.

Burns, W. 2016. "Human Rights Dimensions of Bioenergy with Carbon Capture and Storage: A Framework for Climate Justice in the Realm of Climate Geoengineering." In *Climate Justice: Case Studies in Global and Regional Governance Challenges*, edited by R. Abate, 149–170. Washington, DC: Environmental Law Institute.

Carton, W. 2019. ""Fixing" Climate Change by Mortgaging the Future: Negative Emissions, Spatiotemporal Fixes, and the Political Economy of Delay." *Antipode* 51 (3): 750–769.

Carton, W., A. Asiyanbi, S. Beck, H. J. Buck, and J. F. Lund. 2020. "Negative Emissions and the Long History of Carbon Removal." *Wiley Interdisciplinary Reviews: Climate Change* 11 (6): e671.

Cavanagh, C., and T. A. Benjaminsen. 2014. "Virtual Nature, Violent Accumulation: The 'Spectacular Failure' of Carbon Offsetting at a Ugandan National Park." *Geoforum; Journal of Physical, Human, and Regional Geosciences* 56: 55–65.

Chen, J. C. 2013. "Sustainable Territories: Rural Dispossession, Land Enclosures and the Construction of Environmental Resources in China." *Human Geography* 6 (1): 102–125.

Chomba, S. 2017. "Choices Have Consequences: REDD+ and Local Democracy in Kenya." *Conservation and Society* 15 (4): 400–413.

Chomba, S., J. Kariuki, J. F. Lund, and F. Sinclair. 2016. "Roots of Inequity: How the Implementation of REDD Plus Reinforces Past Injustices." *Land Use Policy* 50: 202–213.

Coffman, D., and A. Lockley. 2017. "Carbon Dioxide Removal and the Futures Market." *Environmental Research Letters* 12 (1): 015003.

Collins, Y. A. 2019. "Colonial Residue: REDD+, Territorialisation and the Racialized Subject in Guyana and Suriname." *Geoforum; Journal of Physical, Human, and Regional Geosciences* 106: 38–47.

Corbera, E., and K. Brown. 2010. "Offsetting Benefits? Analyzing Access to Forest Carbon." *Environment and Planning A* 42 (7): 1739–1761.

Corbera, E., M. Estrada, P. May, G. Navarro, and P. Pacheco. 2011. "Rights to Land, Forests and Carbon in REDD+: Insights from Mexico, Brazil and Costa Rica." *Forests* 2 (1): 301–342.

Corbera, E., C. Hunsberger, and C. Vaddhanaphuti. 2017. "Climate Change Policies, Land Grabbing and Conflict: Perspectives from Southeast Asia." *Canadian Journal of Development Studies / Revue Canadienne D'études du Développement* 38 (3): 297–304.

Cox, E. M., N. Pidgeon, E. Spence, and G. Thomas. 2018. "Blurred Lines: The Ethics and Policy of Greenhouse Gas Removal at Scale." *Frontiers in Environmental Science* 6: 38.

Cronin, J., F. Zabel, O. Dessens, and G. Anandarajah. 2020. "Land Suitability for Energy Crops Under Scenarios of Climate Change and Land-Use." *GCB Bioenergy* 12 (8): 648–665.

Davis, D. K., and P. Robbins. 2018. "Ecologies of the Colonial Present: Pathological Forestry from the taux de boisement to Civilized Plantations." *Environment and Planning E: Nature and Space* 1 (4): 447–469.

Deininger, K. 2011. "Challenges Posed by the New Wave of Farmland Investment." *Journal of Peasant Studies* 38 (2): 217–247.

Dooley, K., P. Christoff, and K. A. Nicholas. 2018. "Co-producing Climate Policy and Negative Emissions: Trade-Offs for Sustainable Land-Use." *Global Sustainability* 1: e3.

Dooley, K., E. Harrould-Kolieb, and A. Talberg. 2021. "Carbon-Dioxide Removal and Biodiversity: A Threat Identification Framework." *Global Policy* 12 (S1): 34–44.

Dooley, K., and S. Kartha. 2018. "Land-based Negative Emissions: Risks for Climate Mitigation and Impacts on Sustainable Development." *International Environmental Agreements: Politics, Law and Economics* 18 (1): 79–98.

Ece, M., J. Murombedzi, and J. Ribot. 2017. "Disempowering Democracy: Local Representation in Community and Carbon Forestry in Africa." *Conservation and Society* 15 (4): 357–370.

Edwards, D. P., F. Lim, R. H. James, C. R. Pearce, J. Scholes, R. P. Freckleton, and D. J. Beerling. 2017. "Climate Change Mitigation: Potential Benefits and Pitfalls of Enhanced Rock Weathering in Tropical Agriculture." *Biology Letters* 13 (4): 20160715.

Eijck, J. V., H. Romijn, E. Smeets, R. Bailis, M. Rooijakkers, N. Hooijkaas, P. Verweij, and A. Faaij. 2014. "Comparative Analysis of Key Socio-Economic and Environmental Impacts of Smallholder and Plantation Based Jatropha Biofuel Production Systems in Tanzania." *Biomass and Bioenergy* 61: 25–45.

Eriksen, S., E. L. F. Schipper, M. Scoville-Simonds, K. Vincent, H. N. Adam, N. Brooks, B. Harding, et al. 2021. "Adaptation Interventions and Their Effect on Vulnerability in Developing Countries: Help, Hindrance or Irrelevance?" *World Development* 141: 105383.

Fajardy, M., S. Chiquier, and N. M. Dowell. 2018. "Investigating the BECCS Resource Nexus: Delivering Sustainable Negative Emissions." *Energy & Environmental Science* 11 (12): 3408–3430.

Fajardy, M., and N. M. Dowell. 2018. "The Energy Return on Investment of BECCS: Is BECCS a Threat to Energy Security?" *Energy & Environmental Science* 11 (6): 1581–1594.

Fisher, E., M. Bavinck, and A. Amsalu. 2018. "Transforming Asymmetrical Conflicts Over Natural Resources in the Global South." *Ecology and Society* 23 (4): 28.

Fisher, J. A., C. J. Cavanagh, T. Sikor, and D. M. Mwayafu. 2018. "Linking Notions of Justice and Project Outcomes in Carbon Offset Forestry Projects: Insights from a Comparative Study in Uganda." *Land Use Policy* 73: 259–268.

Forster, J., N. E. Vaughan, C. Gough, I. Lorenzoni, and J. Chilvers. 2020. "Mapping Feasibilities of Greenhouse Gas Removal: Key Issues, Gaps and Opening up Assessments." *Global Environmental Change* 63: 102073.

Franco, J. C., and S. M. Borras. 2019. "Grey Areas in Green Grabbing: Subtle and Indirect Interconnections Between Climate Change Politics and Land Grabs and Their Implications for Research." *Land Use Policy* 84: 192–199.

Franco, J. C., and S. M. Borras. 2021. "The Global Climate of Land Politics." *Globalizations* 18 (7): 1277–1297.

Franco, J., L. Levidow, D. Fig, L. Goldfarb, M. Hönicke, and M. L. Mendonça. 2010. "Assumptions in the European Union Biofuels Policy: Frictions with Experiences in Germany, Brazil and Mozambique." *Journal of Peasant Studies* 37 (4): 661–698.

Frank, S., P. Havlík, J.-F. Soussana, A. Levesque, H. Valin, E. Wollenberg, U. Kleinwechter, et al. 2017. "Reducing Greenhouse Gas Emissions in Agriculture Without Compromising Food Security?" *Environmental Research Letters* 12 (10): 105004.

Fridahl, M., and M. Lehtveer. 2018. "Bioenergy with Carbon Capture and Storage (BECCS): Global Potential, Investment Preferences, and Deployment Barriers." *Energy Research & Social Science* 42: 155–165.

Fuhrman, J., H. McJeon, S. C. Doney, W. Shobe, and A. F. Clarens. 2019. "From Zero to Hero? Why Integrated Assessment Modeling of Negative Emissions Technologies is Hard and How We Can Do Better." *Frontiers in Climate* 1: 11.

Fujimori, S., T. Hasegawa, J. Rogelj, X. Su, P. Havlik, V. Krey, K. Takahashi, and K. Riahi. 2018. "Inclusive Climate Change Mitigation and Food Security Policy Under 1.5°C Climate Goal." *Environmental Research Letters* 13 (7): 074033–11.

Fuss, S., J. G. Canadell, P. Ciais, R. B. Jackson, C. D. Jones, A. Lyngfelt, G. P. Peters, and D. P. V. Vuuren. 2020. "Moving Toward Net-Zero Emissions Requires New Alliances for Carbon Dioxide Removal." *One Earth* 3 (2): 145–149.

Fuss, S., J. G. Canadell, G. P. Peters, M. Tavoni, R. M. Andrew, P. Ciais, R. B. Jackson, et al. 2014. "Betting on Negative Emissions." *Nature Climate Change* 4 (10): 850–853.

Fuss, S., C. D. Jones, F. Kraxner, G. P. Peters, P. Smith, M. Tavoni, D. P. van Vuuren, et al. 2016. "Research Priorities for Negative Emissions." *Environmental Research Letters* 11 (11): 115007.

Fuss, S., W. F. Lamb, M. W. Callaghan, J. Hilaire, F. Creutzig, T. Amann, T. Beringer, et al. 2018. "Negative Emissions – Part 2: Costs, Potentials and Side Effects." *Environmental Research Letters* 13 (6): 063002.

Gerber, J. F. 2011. "Conflicts Over Industrial Tree Plantations in the South: Who, How and Why?" *Global Environmental Change* 21 (1): 165–176.

German, L., G. C. Schoneveld, and P. Pacheco. 2011. "Local Social and Environmental Impacts of Biofuels: Global Comparative Assessment and Implications for Governance." *Ecology and Society* 16 (4): 29.

Gough, C., S. Garcia-Freites, C. Jones, S. Mander, B. Moore, C. Pereira, M. Röder, N. Vaughan, and A. Welfle. 2018. "Challenges to the Use of BECCS as a Keystone Technology in Pursuit of 1.5^0C." *Global Sustainability* 1: e5.

Gough, C., and S. Mander. 2019. "Beyond Social Acceptability: Applying Lessons from CCS Social Science to Support Deployment of BECCS." *Current Sustainable/Renewable Energy Reports* 6 (4): 116–123.

Grant, N., A. Hawkes, S. Mittal, and A. Gambhir. 2021. "Confronting Mitigation Deterrence in Low-Carbon Scenarios." *Environmental Research Letters* 16 (6): 064099.

Greenleaf, M. 2019. "The Value of the Untenured Forest: Land Rights, Green Labor, and Forest Carbon in the Brazilian Amazon." *Journal of Peasant Studies* 47 (2): 1–20.

Griscom, B. W., J. Adams, P. W. Ellis, R. A. Houghton, G. Lomax, D. A. Miteva, W. H. Schlesinger, et al. 2017. "Natural Climate Solutions." *Proceedings of the National Academy of Sciences of the United States of America* 114 (44): 11645–11650.

Hajdu, F., and K. Fischer. 2016. "Problems, Causes and Solutions in the Forest Carbon Discourse: A Framework for Analysing Degradation Narratives." *Climate and Development* 9 (6): 1–11.

Hansen, J., M. Sato, P. Kharecha, K. von Schuckmann, D. J. Beerling, J. Cao, S. Marcott, et al. 2017. "Young People's Burden: Requirement of Negative CO_2 Emissions." *Earth System Dynamics* 8 (3): 577–616.

Hanssen, S. V., V. Daioglou, Z. J. N. Steinmann, J. C. Doelman, D. P. V. Vuuren, and M. A. J. Huijbregts. 2020. "The Climate Change Mitigation Potential of Bioenergy with Carbon Capture and Storage." *Nature Climate Change* 10 (11): 1023–1029.

Hansson, A., M. Fridahl, S. Haikola, P. Yanda, N. Pauline, and E. Mabhuye. 2019. "Preconditions for Bioenergy with Carbon Capture and Storage (BECCS) in sub-Saharan Africa: The Case of Tanzania." *Environment, Development and Sustainability* 22 (7): 6851–6875.

Hasegawa, T., S. Fujimori, P. Havlik, H. Valin, B. L. Bodirsky, J. C. Doelman, T. Fellmann, et al. 2018. "Risk of Increased Food Insecurity Under Stringent Global Climate Change Mitigation Policy." *Nature Climate Change* 8 (8): 1–7.

Healey, P., R. Scholes, P. Lefale, and P. Yanda. 2021. "Governing Net Zero Carbon Removals to Avoid Entrenching Inequities." *Frontiers in Climate* 3: 672357.

Hendrickson, C. Y., and E. Corbera. 2015. "Participation Dynamics and Institutional Change in the Scolel Té Carbon Forestry Project, Chiapas, Mexico." *Geoforum; Journal of Physical, Human, and Regional Geosciences* 59 (C): 63–72.

Hilaire, J., J. C. Minx, M. W. Callaghan, J. Edmonds, G. Luderer, G. F. Nemet, J. Rogelj, M. Del, and M. Zamora. 2019. "Negative Emissions and International Climate Goals – Learning from and about Mitigation Scenarios." *Climatic Change* 157 (2): 189–219.

Hjort, M. 2020. "Who Should Be Governed to Reduce Deforestation and How? Multiple Governmentalities at the REDD+ Negotiations." *Environment and Planning C: Politics and Space* 38 (1): 134–152.

Honegger, M., A. Michaelowa, and J. Roy. 2020. "Potential Implications of Carbon Dioxide Removal for the Sustainable Development Goals." *Climate Policy* 21 (5): 1–21.

Honegger, M., and D. Reiner. 2017. "The Political Economy of Negative Emissions Technologies: Consequences for International Policy Design." *Climate Policy* 18 (3): 1–16.

Hook, A. 2020. "Following REDD+: Elite Agendas, Political Temporalities, and the Politics of Environmental Policy Failure in Guyana." *Environment and Planning E: Nature and Space* 3 (4): 999–1029.

Hufe, P., and D. F. Heuermann. 2017. "The Local Impacts of Large-Scale Land Acquisitions: A Review of Case Study Evidence from Sub-Saharan Africa." *Journal of Contemporary African Studies* 35 (2): 168–189.

Humpenöder, F., A. Popp, Jan Philip Dietrich, D. Klein, H. Lotze-Campen, M. Bonsch, B. L. Bodirsky, I. Weindl, M. Stevanovic, and C. Müller. 2014. "Investigating Afforestation and Bioenergy CCS as Climate Change Mitigation Strategies." *Environmental Research Letters* 9 (6): 064029.

Hunsberger, C., L. German, and A. Goetz. 2017. "'Unbundling' the Biofuel Promise: Querying the Ability of Liquid Biofuels to Deliver on Socio-Economic Policy Expectations." *Energy Policy* 108: 791–805.

IPCC. 2018. "Summary for Policymakers." In: *Global Warming of 1.5°C. An IPCC Special Report on the Impacts of Global Warming of 1.5°C Above Pre-industrial Levels and Related Global Greenhouse Gas Emission Pathways, in the Context of Strengthening the Global Response to the Threat of Climate Change, Sustainable Development, and Efforts to Eradicate Poverty*. Geneva: Intergovernmental Panel on Climate Change.

Jindal, R., B. M. Swallow, and J. M. Kerr. 2008. "Forestry-Based Carbon Sequestration Projects in Africa: Potential Benefits and Challenges." *Natural Resources Forum* 32 (2): 116–130.

Kashwan, P. 2015. "Forest Policy, Institutions, and REDD+ in India, Tanzania, and Mexico." *Global Environmental Politics* 15 (3): 95–117.

Kemerink-Seyoum, J. S., T. M. Tadesse, W. K. Mersha, A. E. C. Duker, and C. D. Fraiture. 2018. "Sharing Benefits or Fueling Conflicts? The Elusive Quest for Organizational Blue-Prints in Climate Financed Forestry Projects in Ethiopia." *Global Environmental Change* 53: 265–272.

Kreidenweis, U., F. Humpenöder, M. Stevanović, B. L. Bodirsky, E. Kriegler, H. Lotze-Campen, and A. Popp. 2016. "Afforestation to Mitigate Climate Change: Impacts on Food Prices Under Consideration of Albedo Effects." *Environmental Research Letters* 11 (8): 085001.

Kriegler, E., J. Edmonds, S. Hallegatte, K. L. Ebi, T. Kram, K. Riahi, H. Winkler, and D. P. van Vuuren. 2014. "A New Scenario Framework for Climate Change Research: The Concept of Shared Climate Policy Assumptions." *Climatic Change* 122 (3): 401–414.

Kröger, M. 2014. "The Political Economy of Global Tree Plantation Expansion: A Review." *Journal of Peasant Studies* 41 (2): 235–261.

Leach, M., J. A. Fairhead, and J. Fraser. 2012. "Green Grabs and Biochar: Revaluing African Soils and Farming in the new Carbon Economy." *Journal of Peasant Studies* 39 (2): 285–307.

Leach, M., and I. Scoones. 2013. "Carbon Forestry in West Africa: The Politics of Models, Measures and Verification Processes." *Global Environmental Change* 23 (5): 957–967.

Lenzi, D. 2018. "The Ethics of Negative Emissions." *Global Sustainability* 1: e7.

Li, T. M. 2011. "Centering Labor in the Land Grab Debate." *Journal of Peasant Studies* 38 (2): 281–298.
Li, T. M. 2014. "Can There be Food Sovereignty Here?" *Journal of Peasant Studies* 42 (1): 205–211.
Lockley, A., and D. Coffman. 2018. "Carbon Dioxide Removal and Tradeable Put Options at Scale." *Environmental Research Letters* 13 (5): 054034.
Lohmann, L. 2008. "Carbon Trading, Climate Justice and the Production of Ignorance: Ten Examples." *Development* 51 (3): 359–365.
Löw, C. 2020. "Gender and Indigenous Concepts of Climate Protection: A Critical Revision of REDD+ Projects." *Current Opinion in Environmental Sustainability* 43: 91–98.
Low, S., and H. J. Buck. 2020. "The Practice of Responsible Research and Innovation in 'Climate Engineering'." *Wiley Interdisciplinary Reviews: Climate Change* 11 (3): e644.
Low, S., and S. Schäfer. 2020. "Is Bio-Energy Carbon Capture and Storage (BECCS) Feasible? The Contested Authority of Integrated Assessment Modeling." *Energy Research & Social Science* 60: 101326.
Lund, J. F., E. Sungusia, M. B. Mabele, and A. Scheba. 2017. "Promising Change, Delivering Continuity: REDD+ as Conservation Fad." *World Development* 89: 124–139.
Malkamäki, A., D. D'Amato, N. J. Hogarth, M. Kanninen, R. Pirard, A. Toppinen, and W. Zhou. 2018. "A Systematic Review of the Socio-Economic Impacts of Large-Scale Tree Plantations Worldwide." *Global Environmental Change* 53: 90–103.
Marín-Herrera, M., H. Correa-Correa, and G. Blanco-Wells. 2021. "Territorialization of the REDD+ Strategy in the Bribri Indigenous People, Talamanca, Costa Rica." *Ambiente & Sociedade* 24: e02312.
Markusson, N. 2022. "Natural Carbon Removal as Technology." *Wiley Interdisciplinary Reviews: Climate Change* 13 (2): e767.
Markusson, N., N. Balta-Ozkan, J. Chilvers, P. Healey, D. Reiner, and D. McLaren. 2020. "Social Science Sequestered." *Frontiers in Climate* 2: 2.
Mccarthy, J. F. 2010. "Processes of Inclusion and Adverse Incorporation: Oil Palm and Agrarian Change in Sumatra, Indonesia." *Journal of Peasant Studies* 37 (4): 821–850.
McElwee, P. 2009. "Reforesting 'Bare Hills' in Vietnam: Social and Environmental Consequences of the 5 Million Hectare Reforestation Program." *AMBIO: A Journal of the Human Environment* 38 (6): 325–333.
McElwee, P. 2021. "Anthropological Engagements with Integrated Assessment Modeling." *Economic Anthropology* 8 (1): 168–171.
McElwee, P., K. Calvin, D. Campbell, F. Cherubini, G. Grassi, V. Korotkov, A. L. Hoang, et al. 2020. "The Impact of Interventions in the Global Land and Agri-Food Sectors on Nature's Contributions to People and the UN Sustainable Development Goals." *Global Change Biology* 26 (9): 4691–4721.
McLaren, D. P. 2012. "Procedural Justice in Carbon Capture and Storage." *Energy & Environment* 23 (2–3): 345–365.
McLaren, D. 2020. "Quantifying the Potential Scale of Mitigation Deterrence from Greenhouse Gas Removal Techniques." *Climatic Change* 162: 2411–2428.
McLaren, D., K. A. Parkhill, A. Corner, N. E. Vaughan, and N. F. Pidgeon. 2016. "Public Conceptions of Justice in Climate Engineering: Evidence from Secondary Analysis of Public Deliberation." *Global Environmental Change* 41: 64–73.
McLaren, D., R. Willis, B. Szerszynski, D. Tyfield, and N. Markusson. 2021. "Attractions of Delay: Using Deliberative Engagement to Investigate the Political and Strategic Impacts of Greenhouse Gas Removal Technologies." *Environment and Planning E: Nature and Space* 251484862110662.
Minx, J. C., W. F. Lamb, M. W. Callaghan, S. Fuss, J. Hilaire, F. Creutzig, T. Amann, et al. 2018. "Negative Emissions: Part 1 – Research Landscape and Synthesis." *Environmental Research Letters* 13 (6): 063001.
Moe, E., and J.K. S. Røttereng. 2018. "The Post-Carbon Society: Rethinking the International Governance of Negative Emissions." *Energy Research & Social Science* 44: 199–208.
Montefrio, M. J. F., D. A. Sonnenfeld, and V. A. Luzadis. 2015. "Social Construction of the Environment and Smallholder Farmers' Participation in 'Low-Carbon', Agro-Industrial Crop Production Contracts in the Philippines." *Ecological Economics* 116: 70–77.

Moosdorf, N., P. Renforth, and J. Hartmann. 2014. "Carbon Dioxide Efficiency of Terrestrial Enhanced Weathering." *Environmental Science and Technology* 48: 4809–4816.

Morrow, D. R., M. S. Thompson, A. Anderson, M. Batres, H. J. Buck, K. Dooley, O. Geden, et al. 2020. "Principles for Thinking About Carbon Dioxide Removal in Just Climate Policy." *One Earth* 3 (2): 150–153.

Müller, M. F., G. Penny, M. T. Niles, V. Ricciardi, D. D. Chiarelli, K. F. Davis, J. Dell'Angelo, et al. 2021. "Impact of Transnational Land Acquisitions on Local Food Security and Dietary Diversity." *Proceedings of the National Academy of Sciences* 118 (4): e2020535118.

Muratori, M., N. Bauer, S. K. Rose, M. Wise, V. Daioglou, Y. Cui, E. Kato, et al. 2020. "EMF-33 Insights on Bioenergy with Carbon Capture and Storage (BECCS)." *Climatic Change* 163: 1–17.

NAS. 2015. *Climate Intervention: Carbon Dioxide Removal and Reliable Sequestration*. Washington, DC: National Academy of Sciences.

NAS. 2019. *Negative Emissions Technologies and Reliable Sequestration*. Washington, DC: National Academy of Sciences.

Newell, P. 2022. "Climate Justice." *Journal of Peasant Studies* 49 (5): 915–923.

Newell, P., S. Srivastava, L. O. Naess, G. A. T. Contreras, and R. Price. 2021. "Toward Transformative Climate Justice: An Emerging Research Agenda." *Wiley Interdisciplinary Reviews: Climate Change* 12 (6): e733.

Nolan, C. J., C. B. Field, and K. J. Mach. 2021. "Constraints and Enablers for Increasing Carbon Storage in the Terrestrial Biosphere." *Nature Reviews Earth & Environment* 2 (6): 436–446.

O'Beirne, P., F. Battersby, A. Mallett, M. Aczel, K. Makuch, M. Workman, and R. Heap. 2020. "The UK Net-Zero Target: Insights Into Procedural Justice for Greenhouse Gas Removal." *Environmental Science & Policy* 112: 264–274.

Olwig, M. F., C. Noe, R. Kangalawe, and E. Luoga. 2015. "Inverting the Moral Economy: The Case of Land Acquisitions for Forest Plantations in Tanzania." *Third World Quarterly* 36 (12): 2316–2336.

Osborne, T. M. 2011. "Carbon Forestry and Agrarian Change: Access and Land Control in a Mexican Rainforest." *Journal of Peasant Studies* 38 (4): 859–883.

Pidgeon, N. F., and E. Spence. 2017. "Perceptions of Enhanced Weathering as a Biological Negative Emissions Option." *Biology Letters* 13 (4): 20170024.

Pirard, R., H. Petit, and H. Baral. 2017. "Local Impacts of Industrial Tree Plantations: An Empirical Analysis in Indonesia Across Plantation Types." *Land Use Policy* 60: 242–253.

Popp, A., K. Calvin, S. Fujimori, P. Havlik, F. Humpenöder, E. Stehfest, B. L. Bodirsky, et al. 2017. "Landuse Futures in the Shared Socio-Economic Pathways." *Global Environmental Change* 42: 331–345.

Quisumbing, A. R., and L. Pandolfelli. 2010. "Promising Approaches to Address the Needs of Poor Female Farmers: Resources, Constraints, and Interventions." *World Development* 38 (4): 581–592.

Ramos, S. J. 2022. "Biomass Logistics: Mythistory and Sociotechnical Imaginary in Trans-Atlantic Wood Pellet Assemblage." *Environment and Planning E: Nature and Space* 5 (1): 318–339.

Renzaho, A. M. N., J. K. Kamara, and M. Toole. 2017. "Biofuel Production and Its Impact on Food Security in Low and Middle Income Countries: Implications for the Post-2015 Sustainable Development Goals." *Renewable and Sustainable Energy Reviews* 78: 503–516.

Riahi, K., D. P. van Vuuren, E. Kriegler, J. Edmonds, B. C. O'Neill, S. Fujimori, N. Bauer, et al. 2017. "The Shared Socioeconomic Pathways and Their Energy, Land Use, and Greenhouse gas Emissions Implications: An Overview." *Global Environmental Change* 42: 153–168.

Richardson, B. 2010. "Big Sugar in Southern Africa: Rural Development and the Perverted Potential of Sugar/Ethanol Exports." *Journal of Peasant Studies* 37 (4): 917–938.

Richards, C., and K. Lyons. 2016. "The New Corporate Enclosures: Plantation Forestry, Carbon Markets and the Limits of Financialized Solutions to the Climate Crisis." *Land Use Policy* 56: 209–216.

Rickels, W., C. Merk, F. Reith, D. P. Keller, and A. Oschlies. 2019. "(Mis)conceptions About Modeling of Negative Emissions Technologies." *Environmental Research Letters* 14 (10): 104004.

Robb, S., S. Joseph, A. A. Aziz, P. Dargusch, and C. Tisdell. 2020. "Biochar's Cost Constraints Are Overcome in Small-Scale Farming on Tropical Soils in Lower-Income Countries." *Land Degradation & Development* 31: 1713–1726.

Robinson, M., and T. Shine. 2018. "Achieving a Climate Justice Pathway to 1.5°C." *Nature Climate Change* 8 (7): 564–569.

Robledo-Abad, C., H. Althaus, G. Berndes, S. Bolwig, E. Corbera, F. Creutzig, J. Garcia-Ulloa, et al. 2017. "Bioenergy Production and Sustainable Development: Science Base for Policymaking Remains Limited." *GCB Bioenergy* 9 (3): 541–556.

Roe, S., C. Streck, R. Beach, J. Busch, M. Chapman, V. Daioglou, A. Deppermann, et al. 2021. "Land-Based Measures to Mitigate Climate Change: Potential and Feasibility by Country." *Global Change Biology* 27 (23): 6025–6058.

Rogelj, J., D. Huppmann, V. Krey, K. Riahi, L. Clarke, M. Gidden, Z. Nicholls, and M. Meinshausen. 2019. "A New Scenario Logic for the Paris Agreement Long-Term Temperature Goal." *Nature* 573 (7774): 357–363.

Rogelj, J., A. Popp, K. V. Calvin, G. Luderer, J. Emmerling, D. Gernaat, S. Fujimori, et al. 2018. "Scenarios Towards Limiting Global Mean Temperature Increase Below 1.5°C." *Nature Climate Change* 8 (4): 325–332.

Rueda, O., J. M. Mogollón, A. Tukker, and L. Scherer. 2021. "Negative-Emissions Technology Portfolios to Meet the 1.5°C Target." *Global Environmental Change* 67: 102238.

Scheidel, A., and C. Work. 2018. "Forest Plantations and Climate Change Discourses: New Powers of 'Green' Grabbing in Cambodia." *Land Use Policy* 77: 9–18.

Schirmer, J., and L. Bull. 2014. "Assessing the Likelihood of Widespread Landholder Adoption of Afforestation and Reforestation Projects." *Global Environmental Change* 24: 306–320.

Schmid, D. V. 2022. "Are Forest Carbon Projects in Africa Green but Mean? A Mixed-Method Analysis." *Climate and Development*, 1–15.

Schoneveld, G., L. German, and E. Nutakor. 2011. "Land-based Investments for Rural Development? A Grounded Analysis of the Local Impacts of Biofuel Feedstock Plantations in Ghana." *Ecology and Society* 16 (4): 10.

Schroeder, H., and C. L. McDermott. 2014. "Beyond Carbon: Enabling Justice and Equity in REDD+ Across Levels of Governance." *Ecology and Society* 19 (1): art31.

Schweizer, V. J., K. L. Ebi, D. P. van Vuuren, H. D. Jacoby, K. Riahi, J. Strefler, K. Takahashi, B. J. van Ruijven, and J. P. Weyant. 2020. "Integrated Climate-Change Assessment Scenarios and Carbon Dioxide Removal." *One Earth* 3 (2): 166–172.

Skea, J., P. Shukla, A. Reisinger, and R. Slade. 2022. *Working Group 3 Contribution to the IPCC Sixth Assessment Report: Summary for Policymakers*. Geneva: Intergovernmental Panel on Climate Change.

Smith, P., J. Adams, D. J. Beerling, T. Beringer, K. V. Calvin, S. Fuss, B. Griscom, et al. 2019. "Impacts of Land-Based Greenhouse Gas Removal Options on Ecosystem Services and the United Nations Sustainable Development Goals." *Annual Review of Environment and Resources* 44 (1): 1–32.

Smith, P., S. J. Davis, F. Creutzig, S. Fuss, J. Minx, B. Gabrielle, E. Kato, et al. 2015. "Biophysical and Economic Limits to Negative CO_2 Emissions." *Nature Climate Change* 6 (1): 42–50.

Smith, J., and S. J. Scherr. 2003. "Capturing the Value of Forest Carbon for Local Livelihoods." *World Development* 31 (12): 2143–2160.

Smith, P. 2016. "Soil Carbon Sequestration and Biochar as Negative Emission Technologies." *Global Change Biology* 22 (3): 1315–1324.

Smith, P., K. Calvin, J. Nkem, D. Campbell, F. Cherubini, G. Grassi, V. Korotkov, et al. 2020. "Which Practices Co-Deliver Food Security, Climate Change Mitigation and Adaptation, and Combat Land Degradation and Desertification?" *Global Change Biology* 26 (3): 1532–1575.

Sovacool, B. K. 2021a. "Reckless or Righteous? Reviewing the Sociotechnical Benefits and Risks of Climate Change Geoengineering." *Energy Strategy Reviews* 35: 100656.

Sovacool, B. K. 2021b. "Who Are the Victims of Low-Carbon Transitions? Towards a Political Ecology of Climate Change Mitigation." *Energy Research & Social Science* 73: 101916.

Sovacool, B. K., L. Baker, M. Martiskainen, and A. Hook. 2019. "Processes of Elite Power and Low-Carbon Pathways: Experimentation, Financialisation, and Dispossession." *Global Environmental Change* 59: 101985.

Spence, E., E. Cox, and N. Pidgeon. 2021. "Exploring Cross-National Public Support for the Use of Enhanced Weathering as a Land-Based Carbon Dioxide Removal Strategy." *Climatic Change* 165 (1–2): 23.

Strengers, B. J., J. G. V. Minnen, and B. Eickhout. 2008. "The Role of Carbon Plantations in Mitigating Climate Change: Potentials and Costs." *Climatic Change* 88 (3–4): 343–366.

Stuart, D., R. Gunderson, and B. Petersen. 2020. "Carbon Geoengineering and the Metabolic Rift: Solution or Social Reproduction?" *Critical Sociology* 46 (7–8): 1233–1249.

Suiseeya, K. R. M., and S. Caplow. 2013. "In Pursuit of Procedural Justice: Lessons from an Analysis of 56 Forest Carbon Project Designs." *Global Environmental Change* 23 (5): 968–979.

Sultana, F. 2022. "Critical Climate Justice." *The Geographical Journal* 188 (1): 118–124.

Temper, L. 2018. "From Boomerangs to Minefields and Catapults: Dynamics of Trans-Local Resistance to Land-Grabs." *Journal of Peasant Studies* 46 (1): 1–29.

Temper, L., S. Avila, D. D. Bene, J. Gobby, N. Kosoy, P. L. Billon, J. Martinez-Alier, et al. 2020. "Movements Shaping Climate Futures: A Systematic Mapping of Protests Against Fossil Fuel and Low-Carbon Energy Projects." *Environmental Research Letters* 15 (12): 123004.

Tienhaara, K. 2012. "The Potential Perils of Forest Carbon Contracts for Developing Countries: Cases from Africa." *Journal of Peasant Studies* 39 (2): 551–572.

Unruh, J. D. 2008. "Carbon Sequestration in Africa: The Land Tenure Problem." *Global Environmental Change Part A* 18 (4): 700–707.

Vakilifard, N., E. P. Kantzas, N. R. Edwards, P. B. Holden, and D. J. Beerling. 2021. "The Role of Enhanced Rock Weathering Deployment with Agriculture in Limiting Future Warming and Protecting Coral Reefs." *Environmental Research Letters* 16 (9): 094005.

Vermeulen, S., and L. Cotula. 2010. "Over the Heads of Local People: Consultation, Consent, and Recompense in Large-Scale Land Deals for Biofuels Projects in Africa." *Journal of Peasant Studies* 37 (4): 899–916.

Veuthey, S., and J.F. Gerber. 2012. "Accumulation by Dispossession in Coastal Ecuador: Shrimp Farming, Local Resistance and the Gender Structure of Mobilizations." *Global Environmental Change* 22 (3): 611–622.

Von Hedemann, N., Z. Wurtzebach, T. J. Timberlake, E. Sinkular, and C. A. Schultz. 2020. "Forest Policy and Management Approaches for Carbon Dioxide Removal." *Interface Focus* 10 (5): 20200001.

Wallbott, L., and E. Recio. 2019. "Practicing Human Rights Across Scale: Indigenous Peoples' Affectedness and Recognition in REDD+ Governance." *Third World Thematics: A TWQ Journal* 3 (5–6): 1–22.

Waller, L., T. Rayner, J. Chilvers, C. A. Gough, I. Lorenzoni, A. Jordan, and N. Vaughan. 2020. "Contested Framings of Greenhouse Gas Removal and Its Feasibility: Social and Political Dimensions." *Wiley Interdisciplinary Reviews: Climate Change* 11 (4): e649.

Westholm, L., and S. Arora-Jonsson. 2018. "What Room for Politics and Change in Global Climate Governance? Addressing Gender in Co-benefits and Safeguards." *Environmental Politics* 27 (5): 1–22.

Williamson, P. 2016. "Scrutinize CO_2 Removal Methods." *Nature* 530: 153–155.

Wittman, H. K., and C. Caron. 2009. "Carbon Offsets and Inequality: Social Costs and Co-benefits in Guatemala and Sri Lanka." *Society and Natural Resources* 22 (8): 710–726.

Wolske, K. S., K. T. Raimi, V. Campbell-Arvai, and P. S. Hart. 2019. "Public Support for Carbon Dioxide Removal Strategies: The Role of Tampering with Nature Perceptions." *Climatic Change* 152 (3–4): 345–361.

Yamagata, Y., N. Hanasaki, A. Ito, T. Kinoshita, D. Murakami, and Q. Zhou. 2018. "Estimating Water–Food–Ecosystem Trade-Offs for the Global Negative Emission Scenario (IPCC-RCP2.6)." *Sustainability Science* 13 (2): 301–313.

Yengoh, G. T., and F. A. Armah. 2015. "Effects of Large-Scale Acquisition on Food Insecurity in Sierra Leone." *Sustainability* 7 (7): 9505–9539.

Zakkour, P., J. Kemper, and T. Dixon. 2014. "Incentivising and Accounting for Negative Emission Technologies." *Energy Procedia* 63: 6824–6833.

Food, famine and the free trade fallacy: the dangers of market fundamentalism in an era of climate emergency

Matias E. Margulis, Kristen Hopewell and Edi Qereshniku

ABSTRACT
The market fundamentalist logic underpinning WTO agricultural trade rules poses a profound threat to global food security in the context of climate change. The growing frequency of extreme weather events raises the prospect of large-scale disruption of agricultural production and distribution. A trading system oriented towards maximizing economic efficiency and reducing redundancy risks leaving the world population acutely vulnerable to food supply disruptions. Ensuring the resilience of global food systems will require considerably more, rather than less, state intervention in agricultural and food production, including cultivating agricultural surpluses to protect against supply shocks.

Critical agrarian studies has long drawn attention to the flaws in the market rationality of the World Trade Organization (WTO), and its attempts to organize global agricultural production along neoliberal lines. In this article, we argue that climate change adds new urgency to that critique: the market fundamentalist logic underpinning the WTO poses a profound threat to global food security in the context of climate change.

The neoliberal logic on which the WTO rests has prioritized maximizing economic efficiency and reducing redundancy as the core goals of the established global trading order. For nearly its entire history, the central problématique driving the governance of agricultural trade has been how to manage and reduce the agricultural surpluses that have been the unintended consequence of government policies to support their farm sectors. Agricultural surpluses – such as the EU's infamous 'milk lakes' and 'butter mountains' – have been seen as a prime illustration of the perverse outcomes of government intervention, a symbol of the irrationality of government interference in markets. The rules of the multilateral trading system – restricting the use of policy tools such as tariffs, subsidies, supply management, marketing boards, price supports, and so forth – are all about rooting out such inefficiency, by organizing global food production based primarily on market principles, and deeply constraining the scope for state intervention in agriculture markets. Fetishizing the 'rationality' of free markets, market-based approaches to global food systems promise to 'rationalize' agricultural production by eliminating waste and inefficiency, and fostering specialization based on comparative advantage to maximize economic gains.

However, as we will show, what may appear rational from the narrow logic of economic efficiency is deeply irrational amid a growing global climate emergency. Climate change represents a profound threat to global food security. Extreme weather events (such as drought, flooding, extreme heat, wildfires and hurricanes) will only grow more frequent, raising the prospect of large-scale disruption of agricultural production and distribution, across entire regions and even multiple regions simultaneously. The possibility of widespread crop failure raises the specter of mass famine, once thought confined to the dustbins of history. A global food system built along neoliberal lines is not only entirely unprepared to deal with this prospective reality, but key features of this system – including the market logic of 'just-enough' production (the agricultural equivalent of 'just-in-time' production in the industrial realm) designed to minimize waste and inefficiency, along with the heavy concentration of agricultural production in just a few major growing regions driven by the logics of specialization and comparative advantage – leave the world population acutely vulnerable to large-scale food supply disruptions.

Recent shocks, including both the Covid pandemic and the war in Ukraine, have painfully highlighted the vulnerability of global food systems. Responding to the threat posed by climate change, we argue, will require a radical overhaul of the rules and institutions governing global agricultural trade. While, until now, the entire system has been oriented towards reducing agricultural surpluses, as the product of waste and inefficiency, going forward, precisely the opposite will be needed: redundancy – building extra capacity into systems to act as a safeguard or insurance mechanism – will be essential to ensuring the resilience of global food systems, and reducing risk and vulnerability. This will require considerably more, rather than less, state intervention in agricultural and food production, including the state playing an important role in intentionally creating, and appropriately managing, agricultural surpluses to protect against supply shocks caused by climate change.

To deal with the current and future risks posed by climate change, states will need to invest in measures such as diversifying supply, supporting producers, and building food stocks. In other words, states will need to spend money – that is, provide various forms of agricultural subsidies – along with employing a host of other policy measures. But the WTO's agricultural trade rules regulate, and severely circumscribe, virtually every aspect of agriculture policy that would be relevant to these goals, including limiting the amount of money that states can spend and how they can spent it.

The climate crisis has laid bare capitalism's logic of annihilation, with the relentless pursuit of profit, built upon the commodification of both people and nature, now threatening the very foundations of human existence. In keeping with this *Journal of Peasant Studies* (JPS) Forum's focus on the politics of agrarian struggle for the era of climate change, and the different strategic logics of anti-capitalist struggle presented in the introduction (Borras et al. 2021), the analysis presented here accords most closely with the strategy of a gradual dismantling or taming of capitalism. Our approach is informed by an emphasis on the role of government in mitigating the harmful effects of global capitalism. In contrast to the dominant neoliberal logic that has informed global institutions like the WTO and efforts to reduce the scope for state intervention in agriculture markets, we argue that activist government policy has a critical role to play in fostering more resilient food systems to respond to the growing threat posed by climate change.

The article is organized in three main sections. The first sets out the logic and objectives behind the WTO's rules and principles governing agriculture. The second examines the ways in which climate change is expected to disrupt global agricultural production and distribution, potentially endangering both the availability of, and access to, food, particularly for the world's poorest and most vulnerable populations. The third draws together the preceding two sections to analyze why and how WTO agricultural trade rules threaten to hamper efforts by governments to build greater resilience in global food systems in order to protect and promote global food security.

The ethos of WTO agricultural trade rules

As a core pillar in the project of neoliberal globalization, the WTO has been a key driver of the neoliberalization of agriculture worldwide. In large part due to the major farm programs run by dominant powers such as the US and EU, agriculture was effectively excluded from the General Agreement on Tariffs and Trade, which governed trade since the end of the Second World War. Agriculture was not brought under the authority of the multilateral trading system until the Uruguay Round of trade negotiations (1986–1994), which resulted in the creation of the WTO and its Agreement on Agriculture (AoA) in 1995. The AoA was, as one World Bank analyst put it, 'a historic breakthrough' (Ingco 1995, 1); for the first time, it established international rules and disciplines to govern agricultural trade.

The establishment of the WTO's rules governing agricultural trade was fundamentally driven by the desire to reduce structural surpluses created by subsidies. This was the key 'problem' policymakers were seeking to solve with AoA. By the 1980s, subsidies had fueled an overproduction crisis, and states sought to tackle the buildup of surpluses and the escalation of subsidies through the creation of WTO rules on agriculture. There was widespread condemnation of the 'waste,' 'expense,' 'failures,' and 'perversities' of agricultural subsidies (Josling, Sanderson, and Warley 1990, 46). Then, as now, critics across the political spectrum argued that agricultural subsidies disproportionately benefitted wealthy land owners, large-scale farmers and agribusiness, rather than small farmers (Green and Griffith 2002; Schechinger 2021), decrying such subsidies as 'corporate welfare' (Riedl 2007) and a 'notorious' instance of 'government waste and abuse' (Lincicome 2020). Economists called for eliminating 'distortions' in global agricultural markets caused by subsidies and other government policies, arguing that subsidies are economically inefficient, 'stimulate excess production' and 'distort international competition' (Josling, Sanderson, and Warley 1990, 441).

Negotiated during the Uruguay Round, the principal purpose of the AoA was to end a growing agricultural trade war between the US and EU. This 'farm war' was caused by a structural surplus in agricultural production in OECD countries, fueled by subsidies (Wolfe 1998). High levels of agricultural support in the developed world – particularly the US and EU – led to a significant increase in agricultural production, resulting in large surpluses and a collapse in prices, which in turn set off intense competition for global market share through further subsidies (Josling, Sanderson, and Warley 1990; Wolfe 1998). The farm subsidy war was extremely costly for the OECD countries themselves, and led to the dumping of cheap subsidized grain on international markets, with devasting consequences for farmers elsewhere, especially in the Global South.

Given the dominance of the US, EU and other advanced-industrialized states in the Uruguay Round negotiations, an emerging new political consensus within the OECD on the need for agricultural reform played a major role in shaping the AoA negotiations and the resulting rules (Wolfe 1998). According to the 1987 OECD Ministerial Communiqué, these states were concerned about 'serious imbalances' in global agricultural markets:

> Boosted by policies which have prevented an adequate transmission of market signals to farmers, supply substantially exceeds effective demand. The cost of agricultural policies is considerable, for government budgets, for consumers and for the economy as a whole. Moreover, excessive support policies entail an increasing distortion of competition on world markets; [and] run counter to the principle of comparative advantage which is at the root of international trade ...

At the time, policymakers were facing the twin problems of excess supply and surplus disposal, which deeply shaped their ideas about what rules were needed to govern international trade. At the forefront was the 'pressing need' to 'implement measures which, by reducing guaranteed prices and other types of production incentives, by imposing quantitative production restrictions, or by other means, will prevent an increase in excess supply' (OECD 1987). The OECD states called for 'a concerted reform of agricultural policies' via the AoA, with the guiding principles being to 'minimize possible economic distortions' and 'permit better functioning of market mechanisms' through 'a progressive reduction of assistance to and protection of agriculture' (OECD 1987). The long-term objective, they argued, was to:

> allow market signals to influence by way of a progressive and concerted reduction of agricultural support, as well as by all other appropriate means, the orientation of agricultural production; this will bring about a better allocation of resources which will benefit consumers and the economy in general. (OECD 1987)

Government intervention was thus seen as the problem and freeing markets the solution.

The creation of the AoA marked a profound shift. For the first time, governments – who had previously been largely free to design and pursue their own agricultural policies as they saw fit – were bound by international laws constraining their use of a wide range of policy instruments. Moreover, WTO rules have the status of 'hard law,' enforceable through the WTO's powerful dispute settlement system, which makes its rules legally binding on states. The AoA thus represented the most important attempt 'ever made to fundamentally reshape the relationship between national governments and the farmers they represent and to have market forces determine agricultural trade' (Josling, Sanderson, and Warley 1990, 459). The AoA sought to reduce levels of agricultural support and protection, constraining the scope for interventionist state policy and driving a process of reform of national agricultural policies to make them more market-oriented. Countries were required to lower their tariffs and other non-tariff barriers to imports of agricultural and food products, curb their use of export subsidies, and reduce and modify their farm support programs. The agreement sought both to reduce government support to agriculture as well as to restrict the forms such support may take.

The creation of the WTO's agricultural trade rules occurred at the height of the Washington Consensus in the 1980s and 1990s and was part of a larger set of forces pushing towards the neoliberal restructuring of global agriculture. Under IMF and World Bank structural adjustment programs, developing countries came under heavy pressure to open, liberalize, and deregulate their markets, including reducing tariffs and other import restrictions, dismantling agricultural cooperatives and marketing boards, and reducing or eliminating production and input subsidies (Clapp and Moseley 2020). Developing countries were pressed to restructure and redirect their agriculture sectors towards export-oriented agriculture based on comparative advantage, often focused on a single commodity like cotton, rather than producing food staples for domestic consumption. The prevailing neoliberal orthodoxy promised that growing trade and integration into global markets would provide the solution to poverty, hunger and underdevelopment (McMichael 2020) – a promise that for most developing countries proved to be an abysmal failure, resulting in economic stagnation; rising inequality, poverty, and food insecurity; dispossession, growing land concentration, and forced migration (Bayliss, Fine, and van Waeyenberge 2011; Saad-Filho 2014). A proliferation of bilateral and regional free trade agreements also forced open agricultural markets, particularly in the Global South – often with dire consequences for local farmers. As a result of the North American Free Trade Agreement (NAFTA), for example, Mexico was flooded with cheap imports of corn from the US, resulting in the dispossession of nearly two million peasant farmers (Carlsen 2003). In addition, under the sway of neoliberal ideology, many governments around the world also undertook unilateral reforms to liberalize and deregulate their agriculture sectors, such as a the privatization of the Australian Wheat Board in 1999 and dismantling of the Canadian Wheat Board in 2012.

Importantly, the AoA was intended to 'provide a framework for the long-term reform of agricultural trade and domestic policies over the years to come' (WTO 2022). As such, the agreement contained a 'continuation clause' committing states to begin new negotiations almost immediately to further deepen and expand the rules and continue to drive forward the process of liberalization. This became part of the Doha Round of trade negotiations, which began in 2001. Although the Doha negotiations collapsed in 2011 due to intense inter-state conflict, the WTO agriculture negotiations have continued. As a result of these negotiations, states reached an agreement on agricultural export competition, which prohibits export subsidies and established stricter disciplines on export credit, international food aid, and state trading enterprises, at the Nairobi Ministerial in 2015 (Margulis 2018a). Such measures are all directly tied to the issue of surplus disposal. States are currently engaged in negotiations seeking to reach a standalone WTO agreement on agriculture, which would continue the process of trade liberalization begun with the Uruguay Round AoA, by tightening restrictions limiting the scope for state intervention across a range of areas, including subsidies, tariffs and other market access barriers, and export restrictions.

Remarkably, within the ongoing WTO agriculture negotiations, there has been virtually no mention of climate change – despite overwhelming scientific evidence that climate change will upend global food systems and pose a significant threat to global food security. The negotiations are proceeding as though the climate emergency does not exist, or has no bearing for agriculture and food. As the following sections will show, this risks putting the world on a dangerous collision course.

The impact of climate change on agriculture and food security

Climate change will have profoundly destabilizing effects on global agricultural production and distribution. A world population of 10 billion people by 2050 is expected to increase the need for food by as much as 50%–60% above current levels (Bailey et al. 2015; Quiggin et al. 2021),[1] but since agriculture is highly weather dependent, it is among the most exposed and vulnerable sectors to climate change. Climate change will negatively affect food security in a multitude of complex ways, including through impacts on food availability, access, utilization and stability (IPCC 2022, 519, 442).[2] Here, in this paper, we focus primarily on the issue of *shocks* to food supply and distribution. According to the Intergovernmental Panel on Climate Change (IPCC):

> Climate change will increasingly put pressure on food production and access, especially in vulnerable regions, undermining food security and nutrition (*high confidence*). Increases in frequency, intensity and severity of droughts, floods and heatwaves, and continued sea level rise will increase risks to food security (*high confidence*). (IPCC 2022, 14)

This disruption is expected to be greatest in some of the world's most impoverished regions, particularly in the countries that sit around the equator (Murphy and Lilliston 2017). Moreover, multiple climate hazards will occur simultaneously and interact, compounding overall risk and causing risks to cascade across regions (IPCC 2022, 18–19). The IPCC indicates that climate change is already negatively impacting agricultural production and distribution, and these trends will continue.

Recent years have witnessed a growing number of record-breaking climate extremes, such as unprecedented rainfall or heatwaves, that have shattered previous records by large margins. Climate change is expected to lead to more frequent harvest failures, as key food producing regions will fall under growing stress from floods, droughts, extreme heat and other manifestations of climate volatility. It is estimated that 40% of global cropland will be exposed to severe drought for three or more months each year by 2050 (compared to 9% between 1981 and 2010) (Quiggin et al. 2021). The risk of a 1-in-100 year production shock is expected to increase to 1-in-30 or more by 2040 (Bailey et al. 2015).

Global vulnerability to climate-induced production shocks is exacerbated by heavy dependence on a small number of crops and growing regions for much of the world's food supply. Four staple crops – wheat, maize, rice, and soybeans – provide two-thirds of the world's food calories (Kim, Iizumi, and Nishimori 2019). Global production of these staple crops is concentrated in a handful of 'breadbasket' regions: at least 72% of global production of each of those four staple crops occurs in just five countries, specifically China, the US, India, Brazil, and Argentina (Caparas et al. 2021). The geographical concentration of global food production leads to 'inherent systemic risk' (Bailey et al. 2015). Given the high degree of reliance on a handful of major agricultural regions for a large portion of the world's food supply, the global food system 'is not robust to a shock in one or more major production regions' (Bailey et al. 2015).

---

[1]Critics challenge this projection, arguing that it could be mitigated by reversing the meatification of diets (Weis 2015).

[2]While climate change may increase yields in some high-latitude regions, the overall effect on global crop yields is expected to be negative, with an average of 17% decline in yields of the four main crops accounting for 70% of global harvested crop area by 2050 (FAO 2015, xi).

The growing risk of large-scale crop failures due to climate change (i.e. failure within a breadbasket or across multiple breadbaskets) poses a profound threat to global food security. As extreme weather events increase in intensity and frequency due to climate change, breadbasket failures – a decline in yields of 10% of more – will become more likely (Caparas et al. 2021). Global breadbasket failures are expected to be nearly 5 times more likely by 2030 and up to 25 times more likely by 2050 (Caparas et al. 2021). Rice and maize failures – events that are 'extremely unlikely' today in the global breadbasket regions – will occur at least every other year by 2050. For soybeans, the outlook is even bleaker, with an 81% chance of failure in any given year by mid-century. Wheat is the most vulnerable of the four staple crops: with 65% produced in regions with high water scarcity – a problem that will only worsen with climate change – wheat breadbasket failures could be seen every year by 2050.

What is more, climate change increases the likelihood of global 'synchronized' production shocks – multiple major staple food producing and exporting countries facing simultaneous crop shortfalls simultaneously (Tigchelaar et al. 2018). For example, just four countries – the US, China, Brazil and Argentina – account for 87% of the world's maize exports. It is forecasted that by mid-century the probability of a greater than 10% yield loss in any given year will be between 40% and 70% in each of these countries (Tigchelaar et al. 2018). The probability of a synchronous, greater than 10% crop failure across all countries (which is currently near zero) will rise to nearly 50%.

Climate change is already having adverse impacts on food production and reducing food security (IPCC 2022, 10–11). In recent years, regional droughts and heatwaves have resulted in 20%–50% losses in crop harvests (Quiggin et al. 2021). Examples abound. In Australia, severe drought caused wheat harvests to collapse by 50% two years in a row (2006 and 2007). The 2018 European heatwave led to multiple crop failures and yield losses of up to 50% in Central and Northern Europe. At that time, Central Europe experienced severe drought across 52% of its cropland area. In 2015–2016, Central America (specifically Guatemala, Honduras and El Salvador) experienced prolonged drought that resulted in losses of 50%–90% of crop harvest (Holleman et al. 2020). In that same year, Southern Africa experienced its worst drought in 35 years, resulting in an 8 million tonne regional cereal deficit. In Western Canada, multiple climate disasters – including extreme heat, drought and forest fires, followed swiftly by unprecedented rainfall, landslides and flooding – wreaked havoc on food production in 2021: wheat production plummeted by 35% and canola by 14%, 1.3 million farm animals died, and 80% of commercial shellfish stocks were wiped out in a massive die-off (Edmiston 2021; Nickel 2021). India's 2022 heat wave resulted in up to a 15%–25% decrease in wheat yields (Bal, Prasad, and Singh 2022).

Infrastructure and distribution

In addition to affecting food supply through production shocks, climate change will also impact critical infrastructure that is essential to the transportation and distribution of agricultural products, as well as inputs like fertilizer. The IPCC (2022, 19) indicates that climate impacts on key infrastructure leading to losses and damages to food systems are already occurring and will increase dramatically in future. A large portion of the world's grain and fertilizer supply passes through a relatively small number of export and transit hubs, or 'chokepoints.' Three primary types of chokepoint have been identified as critical to global food security: (1) maritime corridors such as straits and canals; (2) coastal infrastructure in major crop-exporting regions; and (3) inland transport infrastructure in major crop-exporting regions (Bailey and Wellesley 2017). All of these chokepoints are vulnerable to weather hazards, including storms or floods, which will increase with climate change, and a serious interruption at one or more of these chokepoints could lead to supply shortfalls and price spikes (Bailey and Wellesley 2017).

As one Chatham House study stated, 'supply chains are only as reliable as their weakest links' (Bailey and Wellesley 2017, 5). And the risks are many: hurricanes present a significant threat to ports; flooding can close shipping channels, and runoff from intense rainfall can increase silt and debris buildup, leading to shallower and less accessible channels; drought can disrupt inland waterways by reducing water levels and making channels impassible; road and rail transportation corridors can be damaged by flooding, wildfires, and landslides (EPA 2021). In the US, for example, inland waterways carry about 60% of American exports of the four staple crops (which account for 13% of worldwide exports), primarily to ports on the Gulf Coast, which is vulnerable to hurricanes (EPA 2021). Similarly, in Brazil, four ports on the southeastern coastline are responsible for nearly a quarter of global soybean exports; these ports in turn depend on the roads linking them to the farms in the country's interior, but these roads are often rendered impassable by extreme rainfall, as occurred in 2013 for instance (Bailey and Wellesley 2017).

In addition to moving food and feed, most countries also depend on global transportation networks to supply critical inputs such as fertilizer. Like agricultural production, fertilizer production is also highly concentrated, with just six countries – Canada, China, India, Russia, Belarus, and the US – accounting for 70% of global fertilizer production and over 50% of exports (Bailey and Wellesley 2017). The concentration of supply is even higher for specific types of fertilizer: Canada, Russia and Belarus produce over 90% of the global supply of potassium chloride (muriate of potash), for example (Bailey and Wellesley 2017).

By increasing the frequency and severity of extreme weather events, climate change will also lead to more frequent closures of chokepoints. It also increases the risk of concurrent disruptions at different locations:

> Examples might include distant chokepoints being simultaneously disrupted by different weather systems, or a major chokepoint in one part of the world being closed during a harvest failure in a crop-growing region elsewhere. In such circumstances, market impacts are compounded. (Bailey and Wellesley 2017)

A concurrent disruption of production and distribution occurred recently in the South American soy belt, due to a prolonged drought affecting Brazil, Argentina, Uruguay and Paraguay since 2019. Amid record heat and the worst drought in almost a century, southern states in Brazil, for example, saw as much as 90% of the soybean harvest wiped out in some areas (Harris and Pulice 2021; Vara and Mano 2022). At the same time, the extreme drought lowered water levels on the Parana River, the main transport route for grain exports from Argentina – the world's largest exporter of soymeal and third largest corn exporter. At the port of Rosario, where 80% of Argentina's grain exports are loaded, water levels fell to just 0.06 meters (compared to a normal height of nearly 3 meters), disrupting shipping capacity (Bronstein 2021).

Cascading and unequal effects

Climate change will have thus profound effects on agricultural production and distribution. The impacts on global food supply will be compounded, leading to food insecurity, loss of livelihoods and instability. Extreme weather events have spillover effects across national boundaries through supply chains, markets and resource flows (IPCC 2022, 19). Past experience has shown that market responses often amplify the effects of even a small supply disruption – such that even a relatively modest drop in supply can be highly destabilizing and lead to a large increase in prices – and cause those effects to reverberate through the system. (Bailey et al. 2015). The global food crises in 2007–2008 and 2010–2011, for example, were spurred by relatively modest climate impacts (drought in Australia and regional crop failures), which interacted with other factors – such as biofuel policies diverting grain to ethanol production, rising oil prices, low stocks, and speculative trading – to create a run on grain markets, leading to a dramatic spike in global food prices, a decline in availability and access, and a sharp increase in hunger and food insecurity (Quiggin et al. 2021). The food crises drove tens of millions of people into poverty and triggered food riots around the world. Many countries responded by implementing food export bans that further exacerbated the spike in prices.

It is clear that the impact of climate change on agricultural production and distribution will be significant, but as Borras et al. (2021) argue in the introduction to this *JPS* Forum, these effects will not be felt equally. It is the poorest and most vulnerable populations that will be hit the hardest. As the IPCC (2022, 10–11) details, climate shocks have already increased acute food insecurity and malnutrition 'in many communities, especially for Indigenous Peoples, small-scale food producers and low-income households, with children, elderly people and pregnant women particularly impacted.' Climate variability and extremes negatively impact food access through three channels:

> (a) loss of food production for own consumption; (b) loss of income for people whose livelihoods depend on agriculture and natural resources, reducing their ability to purchase food; and (c) spikes and volatility in food prices following climate shocks, which reduce purchasing power of people dependent on markets to purchase food. (Holleman et al. 2020)

As consumers, the poor and other vulnerable groups are most vulnerable when food supply shocks lead to price spikes (FAO 2015). Mirroring larger imbalances in the global distribution of power, those with the least resources will inevitably loose out in a market-based competition to access food under conditions of increasing scarcity. In addition, the negative impacts of climate change on agriculture production also directly affect food access for people that depend on agriculture for both food and livelihoods; those most affected are 'smallholder rural agriculture households, especially the poorer households, who have limited options to cope with climate shocks' (Holleman et al. 2020). Such small-scale farmers produce 63% of food in Kenya, for example, 69% in Tanzania, 70% in Nepal and 85% in Bolivia (Rapsomanikis 2015). As a result, it is estimated that the livelihoods of approximately 70% of the population in Sub-Saharan Africa, for instance, are exposed to rainfall and climate risk (Hansen et al. 2011). Climate-related events such as droughts have already had a devastating impact on millions of smallholder subsistence farmers who rely on crop production for their food and livelihoods (Holleman et al. 2020).

The dangers of market-rule in the era of climate change

Climate change will mean increasing ecological shocks and crises, posing a significant and growing threat to global food security. For much of recent history, food insecurity has been caused not by lack of supply – which has been abundant and more than sufficient to meet the needs of the world population – but by unequal distribution and a lack of access among poor and vulnerable groups. The problem, in other words, has been one not of scarcity but of 'hunger amidst abundance' (Araghi 1999). That could change, however, in the era of climate change. Global breadbasket failure or disruption of critical infrastructure chokepoints could lead to supply shortages or the inability to move food to where it is needed. The threat of such shocks is magnified by the dominant logic currently governing the global trading system. There is a profound contradiction between the market-oriented rules of the WTO, which have increasingly constrained the policy space available to governments, and the growing need for state action occasioned by the climate crisis. Strikingly, however, the issue of climate change has been entirely absent from the WTO's ongoing negotiations on agriculture, where there has been no acknowledgement or recognition of the relevance of climate change for the global rules governing national agricultural policies. The fact that a threat of such magnitude is being ignored within the current WTO agriculture negotiations is itself a signal of the outdated paradigm that continues to actively drive the institution's governance of agricultural trade.

Markets and magical thinking

As the editors detail in the introduction to this *JPS* Forum, one of the painful ironies of the climate crisis is the way in which 'corporate-driven, technological narratives' often portray 'capitalism as a self-correcting system' that can effectively manage climate change mitigation and adaptation through technological innovation. The fact that capitalism has been a primary driver of climate change is considered merely 'accidental,' while its boosters insist that the climate crisis can be resolved 'though an open market-place with the right commitments and incentives' (Borras et al. 2021, 9). Central to this narrative is a supreme faith that the 'magic of the market' can be relied on to deliver the solution – the profit-motive will spur corporations and other private sector actors to innovate, develop new technologies, and find solutions to avert catastrophe. Championing market mechanisms to solve the crisis, this is not just about *technological fixes* but above all *market fixes*. When it comes to agriculture and food production, this kind of magical thinking is nothing short of dangerous.

As Clapp (2017) demonstrates, the dominant narrative promotes trade liberalization as a key solution to ensuring sustainable and resilient food systems in the context of climate change. The rationale focuses on the efficiency gains from trade. According to Pascal Lamy, former Director-General of the WTO, by fostering greater competition, trade allows food production to shift to countries where it can be done most efficiently (e.g. countries with abundant supplies of fertile land and water): 'with the climate crisis … it will become imperative that we produce food in the right places, and not where we would be wasting scarce water or other natural resources' (Lamy 2012; see also Baldos and Hertel 2015). A global food system structured around free trade and comparative advantage is thus considered not only the most economically efficient but also the most environmentally efficient. This leads proponents to call for more trade liberalization – a deepening and expansion of the WTO's rules, to further reduce subsidies, tariffs and other barriers – as the means to promote food security in the context of climate change (Lamy 2012, 63; UNEP and WTO 2009).

The architects of the multilateral trading system, actors such as the US government that played a dominant role in the design of its rules, have long argued that 'food security – the ability to acquire the food you need when you need it – is best provided through a smooth-functioning world market' (USDA quoted in Ritchie 1993, 29). According to this view, vulnerability comes from insufficient integration into global markets and distortions that prevent those markets from functioning smoothly (Lamy 2011). The answer therefore lies in introducing greater competition into global agricultural markets through trade liberalization, in order to allow trade to operate as a 'transmission belt between supply and demand' with 'as little friction as possible' (Lamy 2011). The problem, however, as repeated crises have demonstrated, is that the market has never been 'smooth-functioning' and will be even less so as climate change causes the frequency and severity of shocks to multiply. The neoliberal orthodoxy insists that equilibrium will be automatically restored through the market's own self-regulating function. In a situation of supply shortages, where there is insufficient supply to meet demand, this will lead to an increase in prices, which will in turn prompt farmers to increase production, thereby eliminating the mismatch between supply and demand (Lamy 2012). In the long run, economists insist, markets are self-correcting. But as John Maynard Keyes famously quipped, 'In the long run, we are all dead.' Nowhere is this more true than agriculture. Food is not simply a widget or a commodity like any other. It is essential for human survival. In an instance of supply shortage, starving people cannot wait for price signals to trickle through global markets and incentivize increased production. In addition, agriculture is also far less flexible than other sectors: given the inherent time lags involved in growing food, planting decisions must be made long before food can be harvested. Unlike manufacturing, agricultural production cannot simply be stopped and started in response to the immediate demands of the market.

According to the free market orthodoxy currently underpinning the trading system, only the market can effectively transmit information about how much food is needed and therefore what quantity of food farmers should produce. However, as the COVID pandemic has starkly demonstrated, leaving decisions about the supply of critical goods to the market can lead to severe failures. When the pandemic struck, for example, the market had failed to produce a sufficient supply of personal protective equipment (PPE) and vital medical equipment like ventilators to meet the surging need (a demand shock) and could not ramp up production fast enough to match demand, resulting in acute shortages that greatly exacerbated the loss of life (Hopewell and Tafel 2020). Moreover, while manufacturers have greater flexibility to reorient production to address a supply shortage – a distillery retooling to produce hand sanitizer, for example – such quick pivots are simply not possible with food production.[3] In a crisis involving supply

[3]Given the large volume of grain used to make biofuel and feed animals, there could, in theory, be scope for substitution, by shifting out of biofuels or reducing meat consumption to increase food supply in a time of shortage. However, practical and political impediments make this extremely challenging.

or demand shocks, relying on the market will result in an under-supply of critical goods. And when it comes to the goods on which human life depends – whether food or, in a pandemic, PPE and ventilators – the consequences of a shortfall can be catastrophic. This is why it is necessary to prepare for potential future shocks and build resilience into global food systems, and why markets cannot be relied upon to guarantee food security.

In the existing food regime, agricultural production in excess of demand (i.e. surplus) has been defined as wasteful. The WTO's intended purpose of getting the state out of agriculture is that markets will calibrate supply to match demand, thereby eliminating waste and maximizing efficiency. However, there is an inherent trade-off between efficiency and resilience (Gölgeci, Yildiz, and Andersson 2020). In the manufacturing sector, for instance, market forces led to the widespread adaption of 'just-in-time' or 'lean' manufacturing – where manufacturers maintain low inventories and acquire inputs and produce goods only as immediately needed – to minimize inefficiency and reduce waste. Yet, while it may be more efficient, as the COVID pandemic demonstrated, the danger of just-in-time production is that it leaves manufacturing acutely vulnerable to supply chain disruption. Similarly, the equivalent in the agriculture sector is what we might call 'just-enough' production – that is, producing just enough to meet current demand. However, in agriculture, the danger of 'just-enough' production is that it leaves no buffer if supply falls short due to extreme weather events that disrupt food production or distribution. As extreme weather events become both more frequent and more severe, building in such a buffer – intentionally producing more than current needs, in order to store the surplus for future consumption in case of emergency – will be essential to ensuring food security.

It is thus essential to recognize the inherent risks, dangers and limitations of capitalist modes of organizing food production, and relying on free trade and the private sector to coordinate world food supply and prices. Food security, in short, is too important to be simply left to the market.

Pushing back against the neoliberal retrenchment of the state

Amid the ascent of neoliberalism, many traditional forms of government policy intervention in agriculture were suppressed and delegitimized. Today, for example, as Graddy-Lovelace and Diamond (2017, 70) detail, there is 'near universal disdain' for agricultural subsidies, which are framed as 'waste and corruption.' Meanwhile, policy measures like supply management and price supports have become taboo topics, 'systematically blacklisted in policy and political circles for decades.' But it is important to recollect why farm policy measures such as subsidies, supply management and price supports were created in the first place.

Historically, many agricultural support programs emerged from the need to ensure adequate incomes for farmers and ensure the stability of the food supply – precisely the same problems that the world will increasingly confront moving forward in the era of climate change. In North America, for example, many farm support policies originated in response to the agricultural crisis of the 1930s, when extreme drought and land degradation led to severe dust storms in the prairies. Both an ecological and economic crisis, the 'Dust Bowl' triggered a wave of farm bankruptcies, forcing many impoverished farmers to abandon their farms, which threatened to decimate food production capacity. The US, as part of the New Deal, introduced a range of policy measures, including price floors, production controls, stockholding and import restrictions, intended to bring greater stability to agriculture markets, ensuring both that farm incomes were high enough to enable farmers to cover the costs of production and that consumers had an adequate and affordable supply of food. Similarly, Canada introduced agricultural support policies to maintain the livelihoods of farmers in order to keep farmers on the land and producing a crop. One such measure to create greater economic security for farmers was the creation of the Canadian Wheat Board in 1935, as a state-trading monopsony or marketing board established by the government at the request of farmers seeking to obtain stable, viable farmgate prices for their crop. As Robert Wolfe puts it, during the depression, faced with a profound crisis that threatened the food supply, 'farm policy became social policy' in many countries (Wolfe 1998, 147). While not launched until later, in 1962, the EU's Common Agricultural Policy (CAP) was spurred by similar motives. Coming out of the widespread experience of hunger across Europe during and after WWII, the CAP was prompted by the desire to provide farmers with a decent standard of living to keep them engaged in farming and ensure a stable food supply, amid the pressures of widespread urbanization.

What history shows is that governments can play an important role in boosting farm incomes and production, buffering farmers and consumers from deleterious market forces, and ensuring the stability of food supplies. Of course, not all government interventions have been successful, and certain policies have had significant defects and shortcomings. The EU CAP, for example, proved highly successful in stabilizing prices and raising both agricultural production and farmers' incomes; yet ultimately it was *too* successful, creating an overproduction crisis (which, in turn, led EU policymakers to resort to export subsidies to reduce surpluses). But it is important to recognize the flaws, contradictions and limitations of specific historical policy interventions without falling into the trap of condemning all forms of government intervention. In other words, we must not throw the baby out with the bathwater. Much depends on the design of specific policies, attention to their ecological and distributional consequences, and safeguards and international cooperation to prevent potentially adverse outcomes (such as agricultural dumping). Many of the policy tools that were purportedly discredited in the era of neoliberalism will need to be revisited, retooled and redirected to help adapt to growing climate volatility. Moreover, given the growing threat to global food security posed by climate change, it is once again time to see farm policy as social policy.

The clash between WTO rules and food system resilience

The broad consensus governing the neoliberal food regime has been that governments should not interfere in the market. This is what underpins the existing rules of the WTO and what has been behind the ongoing effort to further deepen and expand those rules to ever more effectively restrict the scope for state intervention in agriculture. But the climate crisis demands a sea change in thinking regarding the role of the state in global food systems, and a corresponding rejection of the neoliberal imperative to 'get the state out of the market.' Adaptation will require effective public policies to help stabilize global food systems in a context of growing climate risk and ensure food availability and access. Climate scientists have stressed that with global breadbasket failures 'expected to occur at least every other year by mid-century,' governments and international organizations 'need to prepare for global breadbasket failures now, and in such a way that centers on the people who will be most adversely impacted,' including smallholder farmers, low-income consumers and other vulnerable groups (Caparas et al. 2021). Confronted with the threats posed by climate change, governments need to take a more active role in supporting farmers and ensuring the stability of global food supplies.

The food crisis caused by the Russian invasion of Ukraine offers insight into the potential effects of a climate-induced breadbasket failure for global food security and underscores the dangers of supply concentration. The war has disrupted exports from two countries that together account for over a quarter of the world's wheat supply, causing wheat prices to skyrocket by over 50% and fueling a global food crisis (Demirjian 2022; Terazono and Pooler 2022). This has in turn prompted over 20 countries to impose restrictions on food and/or fertilizer exports, exacerbating the crisis and causing prices to soar further; such restrictions could impact 17% of globally traded calories in 2022, while restrictions on fertilizer exports limit the ability of other countries to maintain food production amid global supply shortfalls (Reuters 2022). Coming on the back of Covid – which caused the number of people without regular access to food to double – UN Secretary General António Guterres has warned that the world faces 'the spectre of a global food shortage' that could last for years (The Economist 2022). High food prices have raised the number of food insecure by 440 million to 1.6 billion, nearly 250 million people are on the brink of famine, and hundreds of millions more are at risk of being driven into poverty (The Economist 2022).

A fundamental conflict exists between the types of policies that will be needed to address the impacts of climate change on agriculture and the current rules governing the trading system. Trade based on comparative advantage has led to a high degree of concentration in global agriculture markets and dependence on just a handful of exporters and crops for a large share of globally traded food staples. Building greater resilience in global food systems will require efforts to diversify supply, including both investing in domestic production to reduce dependence on imports, as well as supporting the promotion of a greater diversity of crops and production techniques to increase resilience. Policies to bolster resilience are especially critical for import dependent developing countries with high numbers of poor food consumers and producers. Two-thirds of countries are net food importers (Lamy 2012) – most of which are developing and least developed countries with limited resources to compete for food on international markets amid climate-induced supply shortages and price shocks. Many countries that are heavily dependent on imports for food staples would like to reduce their risk by boosting domestic production. Diversifying supply to reduce the risk of shocks will, however, require that states have the ability to make strategic use of policies such as tariffs, import restrictions, and other infant industry protections, in combination with production-enhancing subsidies. Yet all such measures are currently heavily restricted or outright prohibited under existing WTO rules. In addition to diversifying supply, strengthening global food systems and building greater resilience will also require supporting rural producers and building food stocks, all of which in turn will require that states have access to a broad range of agricultural policy tools, including tariffs, import quotas, subsidies, supply management, marketing boards, and price supports. However, the principal purpose of the WTO's trade rules has been to increasingly restrict and eliminate such measures. Today, the policy space available to governments – their ability to use these types of instruments – is highly circumscribed by the rules of the global trading system.

The WTO's agricultural trade rules were created in the midst of an overproduction crisis. The central problem the WTO AoA was designed to address was a crisis of excess supply fueled by subsidies, and the resulting problem of surplus disposal. But the era of abundant supply leading to 'cheap food' appears to have come to an end (Marsden 2013).[4] The challenge going forward may instead be quite different in nature – one of supply shortages due to climate-induced production and distribution shocks, in a context of rising food demand fueled by population growth, alongside declining yields caused by climate change and other sources of ecological degradation.

Subsidies are one example of an important policy tool that can be used to increase resilience, by, for example, diversifying supply, supporting producers, and building food stocks. But the WTO's rules tightly restrict how governments can use subsidies to support their agriculture sectors. They are designed to engineer a shift away from subsidy measures that are linked to production and thereby stimulate output towards subsidies that are 'production neutral' – meaning subsidies that are decoupled from (i.e. not tied to) output levels and hence do not incentivize increased production. Under the AoA, subsidies are therefore classified according to their effects on production and trade. Subsidies that incentivize increased production – and thus distort trade – are classified as 'Amber Box' and subject to strict limits. Measures that have no or minimal effect on production and trade are classified as 'Green Box' subsidies, which are permitted and exempt from any limits or reduction commitments. The rules are intended to prompt states to shift from providing trade-distorting support (Amber Box) to providing less trade-distorting forms of support (Green Box) by redesigning their farm support programs to eliminate production-stimulating measures.

The core principle underpinning WTO subsidy rules is that actions should be avoided that stimulate production and lead to excess production, agricultural surpluses (i.e. supply outpacing demand), and the accumulation of stocks. The whole purpose of the WTO subsidy regime is to move states away from forms of farm support that incentivize increased production and towards measures that are production neutral. But going forward stimulating agricultural output by incentivizing increased production (in order to, for example, build emergency food reserves or diversify supply) maybe exactly what will be needed. Building stronger and more resilient food systems may require intentionally building redundancy into the system as an insurance mechanism. This would include intentionally cultivating surpluses – production levels in excess of current consumption needs – in order to build stocks as a safeguard against future shocks. This is akin to collectively 'saving for a rainy day' (… or drought, flood, wildfire, etc.). If a government wants to create a sustained surplus that can be used as insurance against future shocks, it may seek to employ subsidies that contain incentives to increase production. But this is what WTO rules are specifically designed to prevent. Subsidies that are linked to production

[4]Although, of course, seemingly 'cheap food' always masked the high environmental and social costs of food production in the corporate food regime.

and stimulate output are trade-distorting and therefore considered undesirable and restricted by WTO subsidy caps.

Similarly, for an import-dependent country seeking to diversify its supply by boosting domestic production, production-enhancing subsidies may be a desirable policy tool. Measures such as price supports and input subsidies are considered the fastest and most effective way to boost production and relatively easy to operate compared with other (less trade-distorting) instruments (Bellmann 2019, 7). However, these are Amber box subsidies, and the WTO sets strict limits on such subsidies that cannot be exceeded. As part of their efforts to foster adaptation and mitigation, governments may want to provide sector-specific support for things like irrigation, transportation infrastructure, or storage facilities, yet these too can be considered trade distorting and subject to WTO subsidy limits. Finally, subsidies can play a critical role in supporting the livelihoods of small farmers (Bellmann 2019, 9), which will be particularly important amid the growing volatility caused by climate change, but the types of subsidies governments can provide are strictly limited by their WTO commitments. The same is true of any other measure that provides farmers with a higher price than they would receive if exposed to foreign competition, such as tariffs or import restrictions (Bellmann 2019). WTO rules have thus significantly constrained the scope for states to use subsidies – as well as a host of other trade policy tools – to foster greater food systems resilience in the face of climate change.

Illustrative example: public food stockholding vs. WTO subsidy rules

As the example of public food stockholding illustrates, the provisions of the WTO Agreement on Agriculture are increasingly outdated and inadequate to address emerging food security challenges in the context of climate change. Public food reserves – whether operated at the local, national, regional or global level – could provide a vital means of responding to climate shocks, by providing a buffer against shortages and stabilizing prices. Many countries historically used food reserves to protect against drought, manage food supply and stabilize prices (Lilliston and Ranallo 2012). These included emergency food reserves, where food stocks are acquired and then released during major disruptions in food supply, such as when there is a catastrophic domestic crop failure and/or shortages on the world market. It also included price stabilization food reserves, where the principal objective is to maintain domestic food prices within a certain range by regulating fluctuations in supply. Food reserves can accomplish this objective by smoothing out supply through the strategic acquisition and release of stocks. However, many countries (both developed and developing) reduced or eliminated their public stockholding programs during the 1980s and 1990s as a result of structural adjustment programs and neoliberal agriculture reforms (FAO 2021; Margulis 2018b). The climate crisis creates an urgent need to revisit these and other agricultural policy tools, but global trade rules constrain the scope for states to use such measures.

Existing WTO rules exert real and significant constraints on national policy space, including the ability of governments to generate and manage food stocks. For example, while government purchases and stockholding are permitted under WTO rules, as are subsidies for consumers, a potential conflict arises if there is a subsidy provided to farmers as part of a public stockholding program (Margulis 2018b). Under existing WTO rules, if government purchases to build stocks and reserves for food security take place at 'administered' prices (i.e. fixed or minimum prices set by the government in advance), they are considered trade distorting and therefore subject to WTO subsidy limits. WTO rules stipulate that only minimal amounts of trade-distorting subsidies are allowed: trade-distorting subsidies to farmers cannot exceed 10% of the value of agricultural production for developing countries and 5% for developed countries (known as *de minimis* levels). (A small number of countries also have additional room to provide trade-distorting subsidies under their Aggregate Measure of Support (AMS) limits, however these are primarily developed countries and represent less than 20% of WTO member states.) Moreover, negotiations are currently underway at the WTO to further reduce these subsidy caps.

There are many reasons why a government may want to use administered prices to create a guaranteed price floor for farmers. As the FAO (2021, 9) indicates, 'Even if purchases are done at prices that are in line with world market prices, by providing a guaranteed market outlet and predictable prices, public procurement can support farm incomes by reducing price risks, particularly for smallholder farmers.' Providing a stable market and less variable income for farmers can also be important to encourage on-farm investment (FAO 2021, 10). In addition, if a country wants to increase domestic agricultural production, government purchases at above-market prices can be a highly effective means to stimulate output, with higher prices incentivizing farmers to expand production.

However, for countries seeking to operate public food stockholding programs, WTO rules are highly constraining. Importantly, under the WTO Agreement on Agriculture, the value of the subsidy provided under a public stockholding program is *not* assessed based on actual budgetary expenditure. Instead, it is calculated as the difference between the administered price and a fixed external reference price, multiplied by the quantity of eligible production (see Figure 1). As a result, WTO rules now create significant difficulties for countries seeking to engage in public stockholding.

First, for most countries, the fixed external reference price (FERP, which is used as a proxy for the global market price) is based on a 3-year average price between the years 1986–1988. In other words, for a government operating a stockholding program using administered prices, the value of the subsidy is calculated based not on market prices today but on market prices in the 1980s. The WTO's rules were created in a time of relatively low and stable prices, which remained largely flat during the first decade of the Agreement on Agriculture (IISD 2021). But there has been a steep rise in world food prices since then, significantly reducing the policy space for countries to engage in public stockholding (FAO 2021, 10). Food price inflation has dramatically increased the gap between current prices and the reference price. Between 2004 and 2020, for

$$\text{Subsidy} = (\text{Administered Price} - \text{Fixed External Reference Price}) \times \text{Eligible Production}$$

Figure 1. WTO formula for calculating subsidy provided via public stockholding.

example, world market prices for key food staples such as wheat and rice were 50%–100% higher than the 1986–1988 reference price (IISD 2021), and prices have spiked again amid the current food crisis. What this means is that even if a government procures food at an administered price equivalent to (or even potentially below) the current market price, it would still be considered a subsidy (FAO 2021, 10).

Second, governments are further constrained by the fact that, under WTO rules, the value of the support provided under public stockholding programs is calculated based on all eligible production – that is, the quantity of production that is 'eligible' to receive the benefit of the price support provided through the administered price (FAO 2021, 42). It is calculated based on the total volume of production in the country (e.g. the total domestic production of wheat) – *not* the volume actually purchased by the government. Even if a government purchases only a relatively small fraction of total domestic output for its public stockholding programs, the entire volume of domestic production in that country is used for counting the value of the subsidy.

As a result, the WTO grossly overestimates the value of support provided to farmers under public stockholding programs. In keeping with the objectives of the AoA, this was intentional, as 'an inbuilt mechanism to ensure continuous reduction in domestic support and thus encourage governments to shift towards less trade-distorting forms of support' (Bellmann 2019, 13). But due to the design of WTO subsidy rules, today most governments seeking to operate public food stockholding programs using administered prices would be in danger of breaching WTO subsidy limits – even if their administered prices are set at a level in line with, or below, market prices (FAO 2021; IISD 2021). To provide an illustration: in 2017, the world market price for rice was US$405 per tonne, but the Fixed External Reference Price (FERP) (i.e. the average price of rice from 1986 to 1988) is $226 (IISD 2021). For a government purchasing rice for public stockholding, even if the administered price was in line with the current market price of $405, it would be considered a $179 subsidy ($405–226) for each tonne of rice produced in that country (*not* the total amount actually purchased by the government for public stockholding). Assume the country produced 10 million tonnes of rice in 2017, and the government purchased 1 million tonnes of rice for stockholding. Under WTO rules, the value of the subsidy would be considered $1.79 billion ($179 × 10 million). In reality, however, the government spent only $179 million to procure rice for its stockpile and it did so at an administered price equivalent to the market price.

The constraints of WTO rules are compounded by the fact that a country's WTO subsidy cap is unknowable in advance. In agriculture, production levels and prices are highly variable and unpredictable. But since *de minimis* levels are based on the current value of production, a government cannot know in advance what that limit will be in any given year – the maximum value of subsidies it is allowed to provide can be only determined *ex post* (FAO 2021, 41). This inherent uncertainty in WTO subsidy caps further reduces the scope for state intervention in agriculture via public stockholding as well as other forms of subsidization.

Finally, public food stockholding programs may require complementary trade policy measures such as import tariffs or quotas, to maintain minimum procurement prices at the stated level (FAO 2021, 9, 11). If the administered price is set above the world market price, private actors could import the commodity to sell to the government at a higher price. It may therefore be necessary for the government to implement some form of import protection – such as tariffs to raise the price of imports to the level of the procurement price – to prevent arbitrage and ensure the public procurement program can contain costs and meets its objective of supporting domestic producers. However, in addition to its constraints on subsidies, WTO rules similarly restrict the ability of states to use tariffs and quotas to control imports. Each country has an individual tariff schedule setting out the maximum tariff rates it is allowed to apply, constraining the ability of governments to raise tariffs in support of their public food stockholding programs or other agricultural policy goals.

Existing WTO rules have created problems for many countries seeking to create or expand public food stockholding programs. In 2013, the Indian government, for instance, dramatically expanded is system of public food stockpiling and distribution, with the establishment of a landmark new National Food Security Act that made approximately two-thirds of its population (or over 800 million people) eligible to received subsidized food grains (India 2013). India's public stockholding program is intended to provide food security for consumers, subsidize farmers, regulate domestic supply and ensure price stability (ICTSD 2016). However, since India's food stockholding program involves the government subsidizing farmers by purchasing food at a guaranteed 'minimum support price' (i.e. a price floor), and the associated subsidies (the difference between the support price and the FERP) have breached the 10% subsidy limit, India has found itself in violation of WTO rules and in danger of legal challenge (WTO 2020).

The Indian government has pushed to exempt public food stockholding by developing countries from WTO subsidy limits. However, this initiative has faced fierce opposition from the US and other agricultural exporters. The Indian government, which in recent years has emerged as one of the most powerful states at the WTO, secured a temporary solution – an interim due restraint mechanism, or 'peace clause' – to exempt its public food stockholding program from legal challenge at the WTO (Hopewell 2018, 2022; Margulis 2018a). In the 2013 Ministerial Decision on Public Food Stockholding for Food Security Purposes, states agreed to refrain from challenging public food stockholding programs operated by developing countries under the WTO's dispute settlement system (Margulis 2018b; Wilkinson, Hannah, and Scott 2014). Ultimately, however, this exemption is extremely limited: it is only a temporary peace clause, applies only to developing countries, and only covers programs that were already established at the time of the Decision (December 2013) (WTO 2013). As a result, the agreement only applies to a handful of countries – primarily larger emerging economies, such as India, Thailand and Indonesia – that already had existing food stockholding programs in place (Bekele 2014). The vast majority of developing countries, who may want to create similar programs to India's in future, remain prevented from doing so. As this example illustrates, the WTO's rules may therefore significantly constrain efforts by governments to support food security by building more resilient food systems.

Socializing risk and supporting rural producers

Strengthening the resilience of global food systems to cope with climate change goes hand in hand with supporting rural producers. As the introduction to this *JPS* Forum highlights, industrial capitalism 'has often treated nature as inexhaustible and rural inhabitants as disposable,' while climate change 'exacerbates the uncertainty and amplifies the risks attached to capitalist agriculture, thereby increasing the vulnerability of rural populations' (Borras et al. 2021, 5). In a capitalist market economy, farmers are forced to fully bear the biophysical risks inherent in farming, subject to the vagaries of weather, pests and highly volatile markets (Goodman and Redclift 1991). But ensuring an adequate and stable food supply is a public good. Government policy can therefore play an important role in helping to socialize, and thereby reduce, the risks of farming. This is all the more necessary given that the inherent volatility of farming will only increase with climate change. Indeed, 'protection for those who provide sustenance for all has never been more needed' (Graddy-Lovelace and Diamond 2017).

A critical part of climate adaptation and ensuring the stability of food systems will require supporting the livelihoods of farmers to help socialize the growing risks of farming in an era of climate change. This includes the need to expand income support and insurance programs to carry rural producers through droughts and other climate hazards, in order to enable them to continue producing food in increasingly volatile and unpredictable circumstances. However, governments seeking to operate such programs are constrained by the WTO Agreement on Agriculture. Government-supported crop insurance and other price and income support programs can be considered trade-distorting support – by insulating producers from market forces and thereby encouraging 'excess' production – and are tightly restricted under WTO subsidy rules (Glauber 2015; Schnepf 2019). Depending on their structure, design and operation, countries seeking to use income insurance and safety-net programs are vulnerable to legal challenge at the WTO (Glauber 2015).

The need for insurance and income support is particularly acute for smallholder farmers, who are most vulnerable to climate disruption. Extreme weather events could force many smallholder farmers, who are least equipped to weather such shocks, out of agriculture altogether. If an extreme weather event, such as a drought, wipes out a crop, this could destroy a smallholder's entire livelihood, forcing them to abandon farming and leave the countryside in search of more stable income. Crucially, smallholders are responsible for a significant portion of the world's food supply. It is estimated that 500 million smallholder farms in the developing world support nearly 2 billion people, and in Asia and sub-Saharan Africa those small farms provide 80% of the food supply (FAO 2015, ix).

A significant part of climate adaptation therefore needs to be focused on ensuring the resilience of rural producers themselves – that is, ensuring their ability to sustain themselves with a descent livelihood, remain on the land and continue engaging in food production over the long-term, even in the face of growing climate volatility and the disruption caused by increasingly frequent extreme weather events (FAO 2015). Conceptualized in this way, climate adaptation is directly tied to struggles by rural producers for better incomes and wages, as well as greater certainty and stability of income. Indeed, ensuring the livelihoods of farmers and the stability of agricultural production is precisely why agricultural policies like subsidies, price supports and supply management emerged in the first place.

Designing effective mechanisms to support peasants, pastoralists, fisherfolk, rural workers and others who are on the front lines of climate change will require being guided by producers themselves in determining what forms of support are needed. Rather than a top-down, technocratically-driven approach, this will require greater deference to these actors and respect for their local perspectives, experience and knowledge.

Developing sound agricultural policies in response to climate change therefore requires the active participation of agrarian movements in policymaking. Solutions need to be attuned to concerns of equality and justice, including 'related to knowledge (whose knowledge counts?), procedure (who is involved in deciding?), distribution (who gets which benefits and who suffers what costs/risks?), and correctives (how are past wrongs addressed?)' (Borras et al. 2021, 12).

Of course, policy measures to support rural producers and enhance the resilience of global food systems will be costly. But far from being 'wasteful' as free-market advocates allege, such costs are a necessary insurance against mass hunger and food insecurity. Importantly, though, it is the poorest countries that are likely to be hit hardest by climate change and least equipped to bear these costs. Ensuring the stability of global food systems is a shared responsibility, and the costs must therefore be shared globally, with international cooperation and redistributive measures to ensure that all countries, even the poorest, have the resources needed to support their farmers and agriculture sectors. In addition, poorer countries with limited resources to provide subsidies may need to rely more heavily on other trade policy tools, such as tariffs, and require comparatively greater flexibility under WTO rules to do so. International cooperation could also be used to improve the effectiveness of agricultural policies by fostering greater coordination across countries, such as through cooperative efforts to diversify supply and counter the current concentration in global agricultural production, or efforts to coordinate the management of public food reserves (Murphy and Lilliston 2017). In addition, to avoid repeating the mistakes of the past – where subsidized surplus production from rich countries was dumped on international markets to the detriment of the world's poorest and most vulnerable farmers – international cooperation will also be needed to create safeguards to hold countries accountable for the external consequences of their domestic policies and ensure that one country's agriculture policies do not undermine farmers and food production in other countries; here WTO rules could play an important role.

Conclusion

As this analysis has shown, the market rationality underpinning the rules and principles of the WTO – which are designed to circumscribe the scope for state action and grow progressively tighter with each round of successive negotiations – poses a profound threat to global food security in the context of climate change. As climate-induced shocks and disruptions to agricultural production and distribution become more frequent, there is a looming collision between agricultural trade rules, on one hand, and the growing need for more activist government policies to support food security, on the other.

The current trading order created under the WTO, oriented towards maximizing economic efficiency and reducing redundancy, leaves global food systems acutely vulnerable to supply shocks, which will become increasingly frequent and severe due to climate change. Building more resilient food systems to cope with the climate crisis will require that states have the capacity to utilize a broad range of agricultural policy tools to support farmers, diversify supply, and cultivate and manage food stocks as a buffer against supply shortfalls. However, the scope for states to use such measures has been increasingly constrained by a trade regime build on the premise of getting the state out of agriculture.

Given the profound susceptibility of agriculture to climate change, the climate crisis is also inherently an agrarian crisis. The core assumptions underpinning WTO rules governing agricultural trade are fundamentally unfit for this emerging crisis. Responding to climate change will require a major paradigm shift and a profound rethinking of the goals and principles of the trading system. In response to the growing climate emergency, struggles and solutions will both have to be multi-scalar. What we have presented here is only one piece of this larger struggle. Rejecting neoliberalism's antipathy towards the state and fetishization of market-based solutions to both climate change and food security, governments – at local and national levels as well as through regional and international cooperation – have a central role to play in ushering in a new food regime amidst a growing climate crisis, and they must have the freedom of action needed to do so effectively.

Acknowledgements

We would like to thank the three anonymous reviewers for their very helpful comments on the paper.

Disclosure statement

No potential conflict of interest was reported by the author(s).

Funding

This research was supported by a Hampton Grant, the Social Science and Humanities Research Council [grant number: #435-2021-0192], and the Canada Research Chairs program .

References

Araghi, Farshad. 1999. "The Great Global Enclosure of Our Times: Peasants and the Agrarian Question at the End of the Twentieth Century." In *Hungry for Profit: The Agribusiness Threat to Farmers, Food and the Environment*, edited by Fred Magdoff, John Bellamy Foster, and Frederick H. Buttel, 145–160. New York: Monthly Review Press.

Bailey, Rob, et al. 2015. *Extreme Weather and Resilience of the Global Food System*. Final Project Report from the UK-US Taskforce on Extreme Weather and Global Food System Resilience.

Bailey, Rob, and Laura Wellesley. 2017. *Chokepoints and Vulnerabilities in Global Food Trade*. London: Chatham House.

Bal, Santanu Kumar, J. V. N. S. Prasad, and Vinod Kumar Singh. 2022. *Heat Wave 2022: Causes, Impacts and Way Forward for Indian Agriculture*. Hyderabad: ICAR-Central Research Institute for Dryland Agriculture.

Baldos, Uris Lantz C., and Thomas W. Hertel. 2015. "The Role of International Trade in Managing Food Security Risks from Climate Change." *Food Security* 7 (2): 275–290.

Bayliss, K., B. Fine, and E. van Waeyenberge. 2011. *The Political Economy of Development*. London: Pluto.

Bekele, Martha Getachew. 2014. "Offer of a Truce: The Peace Clause Agreement on Food Stockholding in Bali." CUTS International, Briefing Paper No. 7.

Bellmann, Christophe. 2019. *Subsidies and Sustainable Agriculture*. London: Chatham House.

Borras, Saturnino M., Ian Scoones, Amita Baviskar, Marc Edelman, Nancy Lee Peluso, and Wendy Wolford. 2021. "Climate Change and Agrarian Struggles: An Invitation to Contribute to a JPS Forum." *The Journal of Peasant Studies* 49 (1): 1–28.

Bronstein, Hugh. 2021. "'Once in 100 Years' Drought Seen Affecting Argentine Grains Exports Into Next Year." *Reuters*, August 12.

Caparas, Monica, Zachary Zobel, Andrea D. A. Castanho, and Christopher R. Schwalm. 2021. "Increasing Risks of Crop Failure and Water Scarcity in Global Breadbaskets by 2030." *Environmental Research Letters* 16 (10): 104013.

Carlsen, Laura. 2003. *The Mexican Farmers' Movement: Exposing the Myths of Free Trade*. Americas Program Policy Report. Silver City, NM: Interhemispheric Resource Center.

Clapp, Jennifer. 2017. "The Trade-ification of the Food Sustainability Agenda." *The Journal of Peasant Studies* 44 (2): 335–353.

Clapp, Jennifer, and William G. Moseley. 2020. "This Food Crisis is Different: COVID-19 and the Fragility of the Neoliberal Food Security Order." *The Journal of Peasant Studies* 47 (7): 1393–1417.

Demirjian, Karoun. 2022. "Rapid Escalation of Ukraine Crisis Fuels Fear of Confrontation Between Russia and the West." *Washington Post*, March 1.

The Economist. 2022. "The Coming Food Catastrophe." May 19.

Edmiston, Jake. 2021. "Extreme Heat and Drought in Western Canada Wreak Havoc on Food System." *Financial Post*, July 16.

EPA. 2021. *Climate Impacts on Transportation*. Washington, DC: US Environmental Protection Agency.

FAO. 2015. *Climate Change and Food Security*. Rome: UN Food and Agriculture Organization.

FAO. 2021. *Public Food Stockholding – A Review of Policies and Practices*. Rome: UN Food and Agriculture Organization.

Glauber, Joseph. 2015. "Agricultural Insurance and the World Trade Organization." IFPRI Discussion Paper 01473.

Gölgeci, Ismail, Harun Emre Yildiz, and Ulf Andersson. 2020. "The Rising Tensions Between Efficiency and Resilience in Global Value Chains in the Post-COVID-19 World." *Transnational Corporations* 27 (2): 127–141.

Goodman, D., and M. Redclift. 1991. *Refashioning Nature*. Abingdon: Routledge.

Graddy-Lovelace, Garrett, and Adam Diamond. 2017. "From Supply Management to Agricultural Subsidies—And Back Again? The U.S. Farm Bill & Agrarian (In)viability." *Journal of Rural Studies* 50: 70–83.

Green, Duncan, and Matthew Griffith. 2002. *Dumping on the Poort*. London: CAFOD Trade Justice Campaign.

Hansen, J. W., S. J. Mason, L. Sun, and A. Tall. 2011. "Review of Seasonal Climate Forecasting for Agriculture in Sub-Saharan Africa." *Experimental Agriculture* 47 (2): 205–240.

Harris, Bryan, and Carolina Pulice. 2021. "Worst Drought in a Century Hits Brazil as it Fights to Overcome Covid." *Financial Times*, June 19.

Holleman, C., F. Rembold, O. Crespo, and V. Conti. 2020. *The Impact of Climate Variability and Extremes on Agriculture and Food Security*. Agricultural Development Economics Technical Study. Rome: FAO.

Hopewell, Kristen. 2018. "Recalcitrant Spoiler? Contesting Dominant Accounts of India's Role in Global Trade Governance." *Third World Quarterly* 39 (3): 577–593.

Hopewell, Kristen. 2022. "Heroes of the Developing World? Emerging Powers in WTO Agriculture Negotiations and Dispute Settlement." *The Journal of Peasant Studies* 49 (3): 561–584.

Hopewell, Kristen, and Joshua Tafel. 2020. "Why Trade Restrictions Must Be Eliminated During COVID-19's Second Wave." *The Conversation*, October 6.

ICTSD. 2016. *Public Stockholding for Food Security Purposes*. Geneva: ICTSD.

IISD. 2021. *Procuring Food Stocks Under World Trade Organization Farm Subsidy Rules*. Geneva: IISD.

India. 2013. "The National Food Security Act." In *The Gazette of India*. New Delhi: Government of India.

Ingco, Merlinda D. 1995. *Agricultural Trade Liberalization in the Uruguay Round*. Policy Research Working Paper. Washington, DC: World Bank.

IPCC. 2022. *Climate Change 2022: Impacts, Adaptation and Vulnerability*. Intergovernmental Panel on Climate Change.

Josling, Timothy E., Fred H. Sanderson, and T. K. Warley. 1990. "The Future of International Agricultural Relations: Issues in the GATT Negotiations." In *Agricultural Protectionism in the Industrialized World*, edited by Fred H. Sanderson, 433–464. London: Routledge.

Kim, Wonsik, Toshichika Iizumi, and Motoki Nishimori. 2019. "Global Patterns of Crop Production Losses Associated with Droughts from 1983 to 2009." *Journal of Applied Meteorology and Climatology* 58 (6): 1233–1244.

Lamy, Pascal. 2011. "Trade is Vital for Food Security, Lamy Tells Agricultural Economists." Speech by WTO Director-General, August 30.

Lamy, Pascal. 2012. "Pascal Lamy Speaks on the Challenge of Feeding 9 Billion People." Speech by WTO Director-General, February 8.

Lilliston, Ben, and Andrew Ranallo, eds. 2012. *Grain Reserves and the Food Price Crisis*. Minneapolis: Institute for Agriculture and Trade Policy.

Lincicome, Scott. 2020. *Examining America's Farm Subsidy Problem*. CATO Institute. https://www.cato.org/commentary/examining-americas-farm-subsidy-problem#.

Margulis, Matias E. 2018a. "Negotiating from the Margins: How the UN Shapes the Rules of the WTO." *Review of International Political Economy* 25 (3): 364–391.

Margulis, Matias E. 2018b. "The World Trade Organization Between Law and Politics: Negotiating a Solution for Public Stockholding for Food Security Purposes." *Transnational Legal Theory* 9 (3-4): 343–360.

Marsden, Terry. 2013. "From Post-Productionism to Reflexive Governance: Contested Transitions in Securing More Sustainable Food Futures." *Journal of Rural Studies* 29: 123–134.

McMichael, Philip. 2020. "Does China's 'Going Out' Strategy Prefigure a New Food Regime?" *The Journal of Peasant Studies* 47 (1): 116–154.

Murphy, Sophia, and Ben Lilliston. 2017. *Food Reserves, Climate Adaptation and the World Trade Organization*. Minneapolis: Institute for Agriculture and Trade Policy.

Nickel, Rod. 2021. "Drought shrinks Canada's Wheat Crop to 14-Year Low, Shrivels Canola Harvest." *Reuters*, August 30.

OECD. 1987. *Ministerial Communiqué*. Paris: OECD.

Quiggin, Daniel, Kris De Meyer, Lucy Hubble-Rose, and Antony Froggatt. 2021. *Climate Change Risk Assessment*. London: Chatham House.

Rapsomanikis, G. 2015. *The Economic Lives of Smallholder Farmers*. Rome: FAO.

Reuters. 2022. "The War in Ukraine is Fuelling a Global Food Crisis." May 30.

Riedl, Brian. 2007. *How Farm Subsidies Harm Taxpayers, Consumers, and Farmers, Too*. Washington: Heritage Foundation.

Ritchie, Mark. 1993. *Breaking the Deadlock*. Minneapolis: IATP.

Saad-Filho, Alfredo. 2014. "The 'Rise of the South': Global Convergence at Last?" *New Political Economy* 19 (4): 578–600.

Schechinger, Anne. 2021. *Under Trump, Farm Subsidies Soared and the Rich Got Richer*. Environmental Working Group. https://www.ewg.org/interactive-maps/2021-farm-subsidies-ballooned-under-trump/.

Schnepf, Randy. 2019. *U.S. Farm Support: Compliance with WTO Commitments*. Washington: Congressional Research Service.

Terazono, Emiko, and Michael Pooler. 2022. "Wheat Prices Hit Record Highs as War Halts Exports from Ukraine and Russia." *Financial Times*, March 4.

Tigchelaar, Michelle, David S. Battisti, Rosamond L. Naylor, and Deepak K. Ray. 2018. "Future Warming Increases Probability of Globally Synchronized Maize Production Shocks." *Proceedings of the National Academy of Sciences* 115 (26): 6644–6649.

UNEP and WTO. 2009. *Trade and Climate Change*. Geneva: UN Environment Program and World Trade Organization.

Vara, Diego, and Ana Mano. 2022. "Brazil Farmers Struggle as Drought Batters Southern Soy Crop." *Reuters*, January 12.

Weis, Tony. 2015. "Meatification and the Madness of the Doubling Narrative." *Canadian Food Studies* 2 (2): 296–303.

Wilkinson, Rorden, Erin Hannah, and James Scott. 2014. "The WTO in Bali: What MC9 Means for the Doha Development Agenda and Why It Matters." *Third World Quarterly* 35 (6): 1032–1050.
Wolfe, R. 1998. *Farm Wars*. Cambridge: Cambridge University Press.
WTO. 2013. "Ministerial Decision: Public Stockholding for Food Security Purposes." Bali Ministerial Conference, WT/MIN(13)/38, WT/L/913.
WTO. 2020. "Domestic Support Notification from India." WTO Committee on Agriculture, March 31, G/AG/N/IND/1.
WTO. 2022. *Legal Texts of the WTO Agreements: A Summary of the Final Act of the Uruguay Round*. Geneva: World Trade Organization.

OPEN ACCESS

Uneven resilience and everyday adaptation: making Rwanda's green revolution 'climate smart'

Nathan Clay

ABSTRACT
Regimes of agricultural modernization and climate change adaptation have converged in Rwanda under the banner of 'climate smart agriculture'. Findings from a study with four agrarian communities show how external agendas of climate smartness can undermine locally rooted strategies for navigating social and environmental uncertainties. Through a focus on two crops (maize and sweet potato), this paper illustrates how climate resilience can be viewed as an uneven and incomplete process situated in peasants' struggles for viability, autonomy, and wellbeing. I suggest that attention to *everyday adaptations* can help researchers and practitioners think beyond the technical adjustments that currently dominate institutionalized responses to climate change.

The adaptation imperative

In May 2016, as Rwanda faced its 'worst drought in 60 years' (Ntirenganya 2016), the World Economic Forum on Africa held its yearly meetings in Kigali. During a 'Tackling Climate Change' event, panelists listed alarming figures (increased temperatures, decreased rainfall, millions more people food insecure) before turning to the investment priorities that could avert disaster: climate information services, crop insurance, high-yielding seeds, stronger markets, and water infrastructure. A press release for the WEF Forum succinctly captures this framing of climate change adaptation:

> Small-scale producers do not have the resources or ability to mitigate or protect themselves from the effects of climate change. Adaptation strategies need to be implemented. The effects of global warming can be managed by optimizing inputs – i.e. fertilizer application according to soil analysis and good-quality seeds with high germination potential. There needs to be a shift from traditionally grown staple crops ... to cash crops – niche products with higher yields and margin. (WEF 2016)

Having recently arrived in Kigali following two years of fieldwork in rural Rwanda, I was struck by how this technocratic narrative mirrored the agricultural modernization program that had been the focus of my research. Since 2007, Rwanda's government and development partners have pursued policies intended to rapidly shift away from polyculture, semi-subsistence family farming towards scaled-up commercial systems reliant on monoculture fields of high-yielding grain crops intended for export. Rwanda is a flagship member in the Alliance for a Green Revolution in Africa (AGRA[1]), whose stated mission is to install commercial farming throughout the continent. When I set out in 2012 to study how this dream of agricultural modernization impacted Rwandan farming families' abilities to cope with climatic uncertainties, I encountered little formal discussion about climate change. By 2016, agricultural modernization programs in Rwanda were commonly branded as 'climate smart,' despite few evident adjustments to underlying practices.

This paper examines efforts to integrate climate change anxieties into agricultural development agendas and the implications for agrarian livelihoods. I consider the consequences of framing climate change adaptation as commensurate with technology-led agricultural intensification – what Marcus Taylor has termed the 'adaptation-modernization nexus' (2014, 99). My analysis centers on four Rwandan communities' experiences with overlapping programs of agricultural intensification and climate change adaptation, including Rwanda's Green Growth and Climate Resilience Strategy (GGCRS), Crop Intensification Program (CIP), Land Use Consolidation program, and associated agricultural improvement and climate adaptation efforts operated by various development agencies.[2] These initiatives have found fertile meeting ground under the umbrella of 'climate smart agriculture' (CSA).

Developed in 2010,[3] CSA has become a prominent framework to mainstream climate change concerns into development practice. Its proponents claim that CSA offers a 'triple win' of climate change adaptation, greenhouse gas mitigation, and increased agricultural productivity – all of which are viewed as crucial to meeting food security goals and economic development targets during an era of global climate change (FAO 2019; World Bank 2016). Skeptics have painted CSA as an effort to renew a paradigm of input-led agricultural intensification by attaching to it a narrative of climate emergency (Borras and Franco 2018; Clapp, Newell, and Brent 2018; Newell and Taylor 2018). They have argued that CSA neglects power relations, allocates responsibility for climate change mitigation to marginalized people, and can be readily co-opted by dominant policy discourses (Chandra, McNamara, and Dargusch 2018; Karlsson et al. 2018; Taylor 2018). Little research, however, has examined on-the-ground realities of CSA (though see Cavanagh et al. 2017; Taylor and Bhasme 2021).

This paper considers CSA's place in Rwanda's 'Long Green Revolution' (Patel 2013). My intention is to examine how CSA emerges in historical-geographical contexts, including how it shapes – and is shaped by – broader arcs of agrarian change. I focus on Rwanda's Land Use Consolidation (LUC) program (Bizoza 2021), which strongly incentivizes individual producers to consolidate land toward the cultivation of crops deemed

[1]AGRA was founded by the Rockefeller and Bill and Melinda Gates Foundations in 2006 with the purpose of moving the continent away from subsistence farming and towards commercial, intensive farming.

[2]These initiatives hew closely to a vision of climate change adaptation informed by a conviction in the transformative power of agricultural modernization. This adaptation-as-modernization framing is not unique to Rwanda. It has become paradigmatic in development financing and practice throughout Africa (Mikulewicz and Taylor 2020).

[3]CSA was developed by the World Bank, the Consultative Group on International Agricultural Research (CGIAR), and the United Nations' Food and Agriculture Organization (FAO).

by the state to be economically viable. LUC is at the center of efforts to increase crop productivity and enable agricultural-led economic growth through scaling up and commercializing the production of six priority crops: maize, wheat, rice, potato, bean, and cassava. LUC has become closely aligned with CSA mechanisms of crop insurance and marshland intensification. Findings from fieldwork in southwest Rwanda show how these conjoined efforts of agricultural modernization and climate change adaptation create an uneven landscape of resilience by privileging wealthier men and undermining the livelihoods and everyday forms of climate adaptation practiced by semi-subsistence producers, particularly women in farming households.

Through this case study of Rwanda, I advance an argument for approaching resilience as an uneven and incomplete process that is situated in peasants' daily struggles for autonomy, security, and wellbeing. I show how climatic stressors entwine with unpredictable market dynamics, resource access constraints, gender relations, and other forces that impinge on agrarian societies. I highlight the value of considering *everyday adaptation*: incremental changes to land use and livelihood practices by which people address these myriad and intersecting challenges. I document the everyday nature and inherent unevenness of resilience through the stories of two plants: maize and sweet potato. While agrarian studies scholarship as tended to center humans and their institutions (Galvin 2018), a focus on human-plant relations enables us to see how climate resilience resides not in individuals or societies but in assemblages of people, crops, livestock, soil, technology, and climate. I show how some plants (maize in this case) become enrolled in assemblages that allow wealthier male farmers to mitigate climate risk while other plants (in this case sweet potatoes) form alliances with poorer households and women, enabling them to avert hunger following climatic shocks. This suggests the importance of examining interrelations among climate, crop ecology, and social dynamics in research and practice on climate change and development.

The resilience of development

Over the past decade, climate adaptation and greenhouse gas mitigation have become explicit goals of development policy and practice. These efforts to 'mainstream' climate change have drawn sustained criticism from political ecologists. Many have taken issue with the decontextualized nature of these climate-development programs, which tend to depict vulnerability and adaptation as 'internal' processes that are challenged by 'external' stressors, hiding how political-economic structures and cross-scale social processes give rise to climate vulnerabilities (e.g. Brown 2014; Ribot 2014; Watts 2015). Such a framing often leads to adaptation solutions of a technical and managerial nature (Taylor 2014; Nightingale et al. 2020) and to greenhouse gas mitigation policies that exacerbate land conflicts and support land grabbing in the name of the environment (Fairhead, Leach, and Scoones 2012; Corbera, Hunsberger, and Vaddhanaphuti 2017). Many have also pointed out that policy interpretations of climate 'resilience' tend to evade attention to the concept's political nature, often referring to stability and self-sacrifice while overlooking how power dynamics influence people's abilities to persevere or not (Mikulewicz and Taylor 2020; Holt-Giménez, Shattuck, and Van Lammeren 2021). This partiality towards the colloquial meaning of resilience allows climate change to be invoked in ways that secure development's existing biases rather than catalyze long-needed reforms to development theory and practice (Carr 2019).

Together with this focus on shortcomings the shortcomings of climate-development research, political ecologists have advanced calls for understanding vulnerability, adaptation, and resilience as *relational processes* (Turner 2016). A relational approach views climate adaptation as embedded in social-environmental dynamics rather than a set of technical adjustments that societies employ in response to biophysical change (Eriksen, Nightingale, and Eakin 2015). It allows us to see how adaptation or resilience for one group of people at one scale often implies vulnerability for another group at another scale (Taylor 2014; Goldman, Turner, and Daly 2018). It encourages us to focus on the negotiations about which adaptations are to be prioritized (Matin, Forrester, and Ensor 2018), how the meanings of 'resilience' and 'adaptation' are contested, and how these debates expose potentially contrasting values about future agrarian worlds (Borras et al. 2022). For instance, Burnham and Ma (2018) show how drought adaptation strategies that emphasize building producers' capacities for irrigated agriculture implicitly value commercial farming futures over subsistence futures. Within these negotiations lie opportunities for reworking power differentials or maintaining them (Forsyth and Evans 2013). A relational lens has thus been foundational in political ecology research on the cross-scalar social relations that co-produce climate injustices (Sultana 2022).

In this article, I combine this political ecology of resilience lens with emerging scholarship on human-plant relations. The latter illustrates how agrarian worlds are composed of assemblages linking plants and humans together with technologies, knowledge, infrastructure, capital, and finance (Galvin 2018). James Scott's (2017) book *Against the Grain* provides a useful starting point. Scott calls for attention to how crops' biophysical characteristics and life cycles become dynamically linked with human social relations in ways that can enable and constrain certain groups of people. He argues that grains such as wheat became quintessential *political crops* for early states in part because their biophysical properties make them innately governable. Grains mature above ground and ripen in unison, facilitating legible rural landscapes. Once harvested, grains can be dried, stored, and transported with high value relative to their volume and weight. Grains' legibility thereby accompanies them from farms to cities. By contrast, Scott depicts tubers such as cassava as *escape crops* because they demand little labor once planted, ripen underground, and their roots remain edible for two years after maturation. These traits entice subsistence producers who prioritize flexibility, autonomy, and a secure localized food source.

While Scott (2017) developed this concept through historical assessments of early states, scholars have recently applied it to examine class and labor dynamics in contemporary agri-food systems. For instance, Sinha (2022) adapts the political crops lens to account for the diffuse power that characterizes agrarian capitalism in India, where capital and labor circulate within and beyond state boundaries. She demonstrates how the material properties of the plants shape production relations and markets downstream of production, which in turn can impinge on producers' livelihoods and risk management. With case studies from Malawi and India, Jakobsen and Westengen (2021) show maize can help consolidate agribusiness power over food systems while also supporting assemblages of food sovereignty – that is, where the decisions about what and how to produce and consume food reside with subsistence producers. Similarly, Roman and

Westengen (2022) document how cassava was central to Brazil's colonization and to resistance in communities of former escaped slaves because the tuber's material properties resist standardization, instead fostering collective identity and strengthening community bonds. A red thread in this nascent literature is that political outcomes of control or resistance do not result solely from plants' biophysical characteristics or from pure human intention, but instead emerge from social-material relations linking humans and plants in a given place and time.

Merging a political ecology approach of relational resilience with this thinking on human-plant relations helps me peel apart interlocking processes of agrarian change and climate change amid Africa's Green Revolution. As I demonstrate below, some plant-human assemblages distribute climate risk onto the most marginalized people while others support efforts to forge more equitable adaptations in the face of climatic uncertainty and top-down agricultural policies.

Making Rwanda's green revolution 'climate smart'

Rwanda's agricultural intensification policies have earned the country praise from development agencies while consistently drawing criticism from academics. What both groups would likely agree upon is that the policies are ambitious. Since 2007, Rwandan households have been compelled to settle in villages, consolidate land in monoculture, and cultivate market-oriented crops (potato, bean, rice, maize, and wheat). A top-down governance regime compels local and regional authorities to ensure that land is converted to 'modern' techniques (Heinen 2022). What this means in practice is that traditional crops and polyculture land use systems are strongly discouraged. In accordance with national targets, regional agronomists and government-certified agrodealers select the crops to grow each season. The strictness of land use regulations varies within and between rural communities depending on where the government has invested in agronomic infrastructure. In drained marshlands and terraced fields, producers have a little leeway. They can be fined or have land confiscated by the state if they fail to plant the crop selected by authorities. Penalties are less severe for other fields, particularly those not visible from a principal road, where administrators tend to look the other way.

Development scholars tend to portray Rwanda's Green Revolution as an imposed vision of rural modernization (Dawson, Martin, and Sikor 2016). Criticism has often centered on the authoritarian nature of the government's approach and how it poorly fits with local realities. For instance, accounts emerged of how agricultural modernization enabled the state and private sector to appropriate common property (Huggins 2014). Studies have found that these policies have exacerbated food insecurity for the rural poor (Dawson, Martin, and Sikor 2016; Clay 2018), with particularly negative consequences for women (Bigler et al. 2019). In their various ways, these and other studies find that Rwanda's agricultural policies have reproduced inequalities by inadequately addressing (or even aggravating) their underlying causes. The present paper builds on this work to consider how a mindset of agricultural modernization infuses the climate adaptation regime that has emerged in Rwanda over the past decade.

Since Rwanda's green revolution policies began in 2008, numerous initiatives of agricultural intensification have since been rebranded as 'climate smart'. A key example is the

Land Husbandry, Water Harvesting, and Hillside Irrigation project (LWH), a 135 million USD effort operated by the World Bank and Government of Rwanda from 2009 to 2018. LWH began with a focus on constructing terraces and draining marshlands for intensive, commercial agriculture. By 2015 the aims were described as 'enhancing hillside agriculture resilience to climate change and variability through increased productivity and income' (World Bank 2015, 10). At the project's close, the Bank claimed that it had created 'climate-smart productive landscapes' by 'increasing productivity and livelihoods and reducing climate vulnerability'. This construction of climate resilience as synonymous with modernization is similarly articulated in Rwanda's 2011 Green Growth and Climate Resilience Strategy and is further fleshed out in a 2015 report, *Climate Smart Agriculture in Rwanda* (World Bank 2015). This report is articulated through the language of agricultural efficiency. For example, it attributes low yields to a combination of 'the predominance of small-scale subsistence farming' and climate change and states that 'producers in drought-prone areas lack the knowledge, skills, and the adequate infrastructure to cope with such harsh conditions' (p.4). In turn, CSA is framed as a set of technologies and practices that maintain or increase productivity while facilitating climate adaptation or greenhouse gas mitigation. Various crops and associated cultivation practices (e.g. 'crop rotation' and 'recycling of crop residues' for maize) are given a numeric ranking in terms of their 'climate smartness.' As I illustrate below, this framing of climate change and development falters because climate risk management is not reducible to discrete technologies or practices, but is situated within complex social-environmental dynamics that comprise human-plant relations in a given place and time.

CSA in Rwanda combines what Borras and co-authors (2022) refer to as a 'climate-emergency narrative' with a 'corporate-driven, technological narrative'. Smallholder producers are cast as victims because they lack technologies and knowledge to adapt. This positions states and corporations as natural protagonists because they offer the education and technology to 'optimize inputs.' Another example of this reductionist framing of CSA can be seen in the One Acre Fund (OAF), an American company that has partnered with the Rwandan government since 2011 to distribute hybrid seeds together with fertilizer and agricultural training. Through their extensive networks which reach hundreds of thousands of households (One Acre Fund 2016), OAF has been a critical cog in Rwanda's Green Revolution ambitions. Despite little change to their business model, OAF recently began claiming that their market-based input dissemination service is 'the ideal delivery system for our growing suite of climate smart products and services' (2020). To validate this system as a climate smart solution, OAF deploys a climate change framing of smallholder producers as simultaneously vulnerable to the impacts of climate change and responsible for climate change:

> At One Acre Fund we understand that smallholders are on the front lines of tackling climate change around the globe. But, inevitably, as part of the agricultural ecosystem, they are also responsible for contributing to that change. For example, in Sub-Saharan Africa, producers without sustainable avenues for increasing their yields are often compelled to "harvest" their natural environments through deforestation, converting natural land to farmland, and participating in other unsustainable agricultural practices such as monoculture which can lead to issues of soil degradation. This scenario creates a cycle of land breakdown and poor yields that, over time, accelerates the effects of climate change. (One Acre Fund 2020)

In Summary, CSA bolsters the Rwandan government's schema of agricultural modernization in three ways. First, by depicting rural populations as innately vulnerable to climate change, CSA narratives paint smallholders as passively impacted by external climatic shocks – and therefore in need of external technical assistance. Second, by presenting smallholders as a homogenous group, these narratives gloss over differences in climate vulnerability and depoliticize the processes of agrarian change and social differentiation that give rise to it. Third, CSA narratives rely on the modernist tactic of separating nature and society – in this case, crops from humans. While this binary creates the preconditions for the agricultural modernization project, below I demonstrate how it also underpins CSA's failure to support equitable climate resilience.

Study site and methodology

This research was conducted from 2014 to 2016 in four *umudugudu* (rural communities) in Southwest Rwanda's Nyamagabe district (Figure 1). Nyamagabe has long counted among Rwanda's poorest regions. Its underperformance on economic growth and food security indicators is often attributed to the steep slopes, narrow valleys, and infertile, rocky, acidic soils that characterize the region. Yet, for hundreds of years, Nyamagabe has supported high population densities through intensive agricultural systems that centered on the integration of crops and livestock, with pastures maintained through controlled burning (Olson 1994). Farming households manage soil fertility through locally-viable practices including intercropping, agroforestry, leasing additional fields, caring for others' animals in exchange for manure, and collecting grasses for green manure.

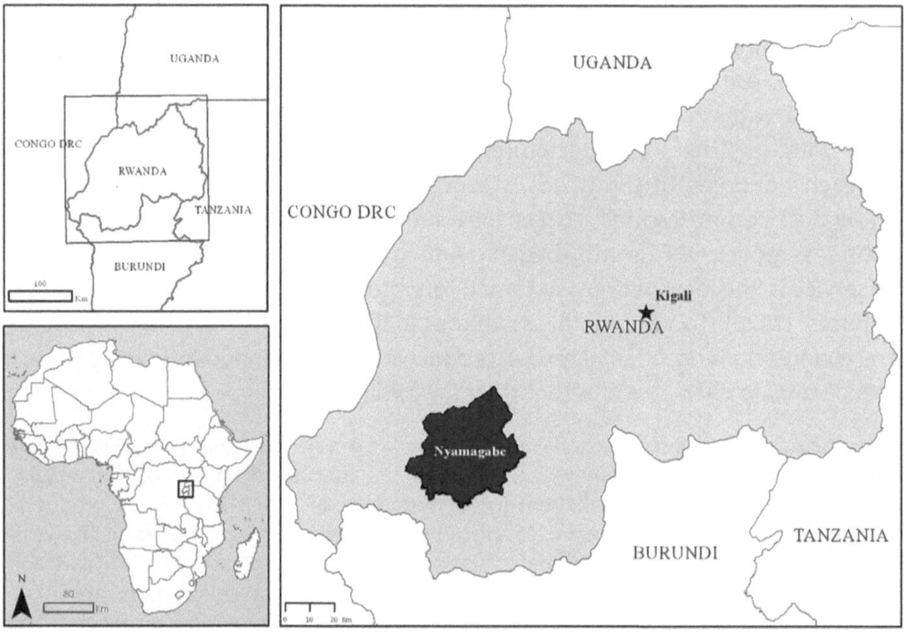

Figure 1. Map of Nyamagabe (study region) in Rwanda.

Although these labor-intensive systems evolved to counter Nyamagabe's environmental limits, these agroecologies also expose deeply-rooted social disparities. As Jennifer Olson (1994) shows with a detailed account of the region's political ecology, inequities in land and animals are at the heart of rural Nyamagabe's continued poverty. In precolonial Nyamagabe, land use was organized through a feudal monarchy, where ruling *Tutsi* pastoralists allotted land and animals to *Hutu* agriculturalists in exchange for portions of their crop production (Maquet and Naigiziki 1957). Through manure, animals transferred nutrients from distant grasslands to intensively cultivated fields near households (de Lame 2005). Inequities in resource access were exacerbated during the colonial period, leading to further intensification as land grew scarcer. When famines resulted, administrators brushed away demands for land redistribution by attributing low crop yields to climatic fluctuations and soil erosion caused by inappropriate farming techniques (Harroy 1944). These inequitable though finely-tuned agro-ecological practices have persisted (albeit with important evolutions whose precise nature is beyond the scope of this paper) into the present day. As Olson (1994) notes and my findings uphold, uneven access to these resources has meant that the soil fertility gap has continuously widened on class and gender lines.

Contemporary agriculture in Nyamagabe and throughout the country (at least prior to the influence of Rwanda's LUC Program) is characterized by diverse polyculture systems. Soil moisture and fertility vary greatly over short distances and depending on the season. A typical household cultivates a wide range of crops each season across a patchwork of small fields at various elevations. Figure 2 (from Ford 1990) shows a simplified rendition of this land use pattern. *Urutoke* and *Imirima* (permanent fields of perennial crops) surround the household, which until Rwanda's villagization program was typically located somewhere near the middle of the hill. *Indare* (rainy season fields of staple tubers and grains) are planted above or below the household, and *Imigende* (dry season fields) are located in low-lying valleys or marshland, which retain more moisture throughout the year and therefore serve an important purpose of enabling continued food supply during the lean months (Ford 1990).

This system takes advantage of the diverse microclimates offered by the mountainous landscape. It enables producers to optimize a combination of low-risk food security crops with crops that are typically sold. Cultivating fields in diverse locations allows farm families to distribute risk of either insufficient rainfall or torrential rains – both of which have become increasingly common in Nyamagabe. Nearly all households in the study site could be classified as semi-subsistence producers. The preference being to consume food produced by the household (mainly bean, sweet potato, peas, and potato), with surplus crops and cash crops (banana, tea, sorghum, potato, cassava, and sweet potato) sold to meet household expenses. With two distinct rainy and dry seasons each year, the labor and social calendar has been structured around key periods of agricultural labor. This system also ensures a nearly continuous supply of food throughout the year.

Nyamagabe makes for an important case study of Rwanda's efforts to improve agricultural productivity in part because of the innate challenges to commercial agriculture in the region. The Crop Intensification Program is strongly represented in the region, with

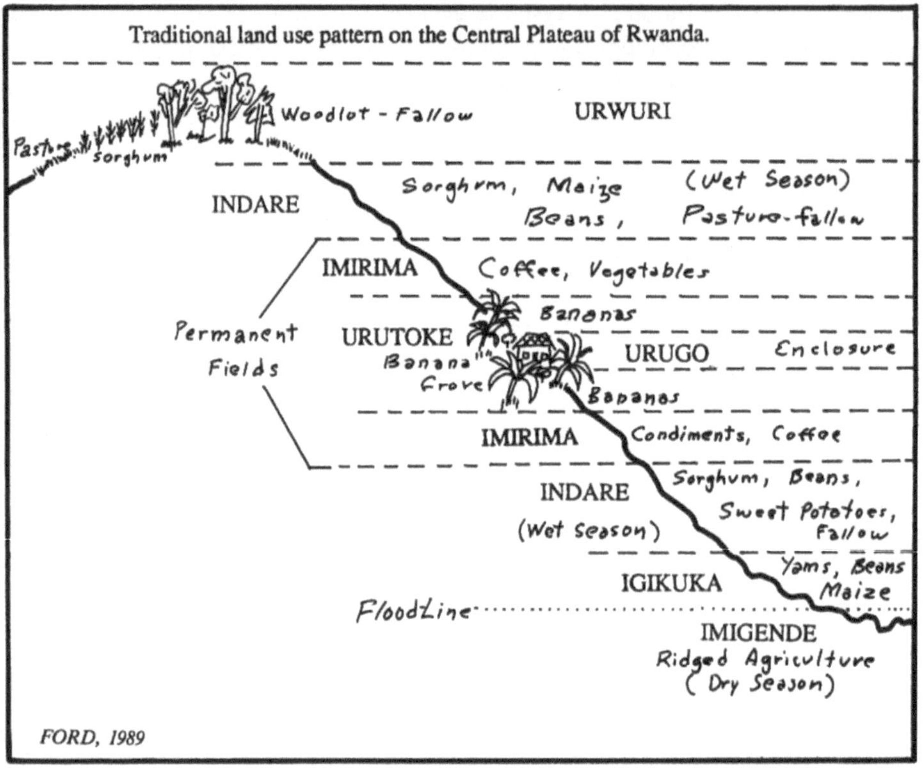

Figure 2. Rendition of a spatially and temporally heterogeneous agroecology in twentieth century Rwanda. Figure from Ford (1990).

90% of the study participants in the four communities operating land that came under the jurisdiction of the program. Within each of these communities, there are strong discrepancies in income, livestock, and access to land and other productive resources (see Table 1). The wealthiest 10% of households owns around eight times as much land as the poorest

Table 1. Relative asset classes and correlating characteristics of survey respondent households. Each grouping of household was found to be significantly different ($p < .01$) from the others in terms of the factors in the leftmost column. Note that these data are based on participant responses to survey questions.

	Household asset class relative to other study participants				
	Lowest 25 percent	Below average	Above average	Highest 10 percent	Sig
Number of households	117	137	128	49	0.000
Female headed households (%)	45.8%	29.2%	20.8%	4.2%	0.000
Mean landholding (ha)	0.12	0.30	0.65	1.54	0.000
Median landholding (ha)	0.1	0.25	0.56	1.46	0.000
Mean number of cows owned	0.09	0.31	0.64	1.29	0.000
Mean pigs, goats, and sheep owned	0.52	0.8	1.24	2.1	0.000
Total annual income (USD)	$171	$170	$278	$536	0.000
Percent of income from farm labor	94%	75%	59%	42%	0.000
Mean age of household head (years)	44	49	50	50	0.010
Mean number of people in household	3.85	4.68	5.22	6.55	0.000

60% of households and five times as many livestock. Day in and day out, these class divides translate into starkly different experiences with Rwanda's agricultural policies. As I detail elsewhere (Clay 2018; Clay and King 2019), wealthier households are able to generate surplus crops and income from renting out their land while poorer households are forced to work in others' fields to earn money for inputs prior to cultivating their own land, and they care for others' animals in exchange for the manure, generally without benefiting from the profit of animal sale. These class divides mean that wealthier households have significant power in shaping decisions about what crops are grown in a particular area. Because landowners work with agronomists to select the crops grown in consolidated areas, they play a role in controlling what their tenants plant.

This study is based on two years of fieldwork conducted from 2014 to 2016 in four *umudugudu* (rural communities) in Nyamagabe together with a Rwandan research team. The findings reported here draw from qualitative and quantitative methods. We first conducted six open-ended focus group interviews to gain a sense of issues at the interface of climate change and agriculture, (each with between 7 and 16 participants). We built on this understanding through three months of ethnographic work (daily informal conversations and participant observation within the four communities), which helped us apprehend how agricultural practices and climate risk management are interrelated and shifting amid the ongoing Green Revolution campaign. This qualitative understanding allowed us to design a structured questionnaire, which (following pretests with 30 households to calibrate the questions) was administered to all households in the four *umudugudu* who were available and willing to participate. Surveying all households reduced selection bias as our 428 respondents represent more than 90% of households.

Following initial analysis of the above materials, we designed and conducted a parcel-level survey in which these 428 households' 3017 parcels of land were visited, with detailed measurements of the crops grown, yields produced, and land use strategies employed. While research assistants were administering the second survey, I visited 40 households' fields, asking follow-up questions that probed about why they made certain decisions. Together, the two surveys gave us a robust understanding of how livelihoods, resource access, food security, biophysical aspects of fields, and crop yields intersect in the context of Rwanda's Green Revolution policies and climatic shocks (two drought seasons and several flooding events that occurred from 2013 to 2016) to shape differential experiences. With the help of a research assistant, I then conducted 72 semi-structured interviews with 36 male and 36 female residents across the four study *umudugudu* (villages). Respondents were selected randomly after grouping survey respondents according to assets and income. Approximately half of respondents classified themselves as poorer than average while 25% classified themselves as wealthier than average and another 25% as among the poorest. Interviews averaged 1 hour and 22 minutes and focused on changing land use and labor dynamics, climate change experience, and gender roles in the context of Rwanda's Green Revolution policies. Most included visits to agricultural parcels, where further unstructured questions were asked. Finally, I conducted 23 semi-structured interviews with local and regional administrators (mayors, umudugudu leaders, agronomists, cooperative presidents, input distribution companies, and security officials, among others), also with a research assistant.

Shifting human-plant-climate dynamics

Maize: uneven resilience

On a dry day in September, around 80 community members gathered on the newly-constructed terraces in Nyamagabe to learn 'modern maize planting techniques'. They watched as an OAF field manager – a young Rwandan man who hailed from an urban center – used a tape measure to space out holes in the soil and insert neon pink hybrid maize seeds. Producers, mostly men, only half-watched this performance. They had seen such demonstrations before and were more interested in the hundreds of sacs of fertilizer and seed that awaited distribution by OAF. These transactions of expert knowledge and modern agricultural inputs have become common across Rwanda. Maize (*Zea mays*) is a centerpiece of efforts to transform rural areas through agricultural modernization. By extension, it has also become key to efforts to make Rwanda's green revolution climate smart. To understand the challenges of assuring climate resilience through what is widely seen as a climatically risky crop, we must look to the human-plant relations surrounding maize.

Across East and Southern Africa, colonial governments compelled producers to adopt maize in the early twentieth century and the crop became a staple in most countries (Jakobsen and Westengen 2021). The story of colonial maize diverged significantly in Rwanda. Belgian administrators had little success encouraging its adoption in food or farming systems. Producers maintained a preference for sorghum and finger millet, which relied on complex land use practices that were embedded in systems of social reciprocity and collective labor. In an effort to evict these native grains, colonial administrators implemented land-use regulations. Citing fears of land degradation, they outlawed the burning of grasses, a vital practice for millet cultivation (Uwizeyimana 1991). Still, maize never became a staple in rural Rwanda. Instead, sweet potato filled the caloric niche left vacant by the millet that could no longer be cultivated (Olson 1994).

Over the past 15 years, however, maize has been the star of Rwanda's Green Revolution. Countrywide, the area planted annually in maize rose from 100,000 hectares in 2007–260,000 hectares in 2017 (Ngango and Hong 2021). This surge is attributed to Rwanda's LUC program and to OAF, which work synergistically to compel planting of maize and supply the required inputs. Rwanda's emerging maize regime thus ties smallholder producers to numerous external actors, including seed and fertilizer companies, private-sector index-based insurance providers, government agronomists, and consumers in urban marketplaces. Several biophysical properties of maize attract the attention of these diverse entities. The straight, tall, golden maize stalks are visible from afar, ensuring legible landscapes of uniform plants. Maize goes hand-in-hand with terrace construction and marshland draining, both of which function to render land legible by the state and investible in its vision of modern agriculture. The biophysical nature of hybrid maize further reinforces the need for large contiguous areas planted with the same seeds. This is because hybrid maize is open-pollinating, meaning it can cross-pollinate with nearby maize plants, whether they are hybrid or locally-adapted varieties. To grow hybrid maize, Rwandan producers, therefore, need to purchase new seeds every season. This contrasts to the archetypal hybrid rice varieties of Asia's Green Revolution, which producers needed only to purchase once because rice is self-pollinating (Patel 2013). The fact that maize is open-pollinating incentivizes the state to ensure that producers consolidate land in maize planting areas. If they do not, then more expensive high-yielding varieties may cross-pollinate with local varieties, resulting in irregularly-shaped seeds that are worth less in commercial markets.

The need to purchase seed each season also ties Rwandan producers into agreements with seed distributors, enabling global circuits of capital and knowledge to reach into rural landscapes and households. OAF distributes the majority of maize seed in 'technology bundles' that include synthetic fertilizer and crop insurance. OAF allocates these bundles on credit, with a high-interest loan (in 2015 the interest rate was 19%) to be repaid at the end of the growing season. OAF is powerful in Rwanda's maize assemblage, acting as a vehicle for ideals, capital, and seeds. Through OAF, maize has finally made inroads in Rwanda after years of failed colonial efforts. And in maize, OAF has found an ideal conduit for its brand of corporate philanthropy that relies on continuously scaling up through recruiting more producers in order to generate enough revenue to pay investors back and keep expanding. As OAF's Investments Director put it, the company aspires to become 'the Amazon for rural producers' (Parrucci 2018).

Rwanda's maize assemblage enrolls smallholder producers in ways that reinforce gendered labor practices and class divides. Respondents described how men had all but abandoned agriculture in Nyamagabe until the promise of commercialized, modern agriculture pulled them back. Men described maize as a potent symbol of the country's progress and criticized systems of intercropping as ignorant – adhering to the state's narrative. Many women pointed out that men contribute relatively little to the household agricultural labor burden but that they increasingly pressure their wives to plant maize. Women also condemned hybrid maize as risky due to its need for high soil fertility and its intolerance of dry spells relative to locally-adapted varieties. As such some women hid locally-adapted maize seeds in nondescript places (e.g. broken earthenware containers), which they planted in LUC areas so that administrators and husbands would see that they complied with the policy, yet without losing money to purchased inputs that they saw as likely to fail in the event of drought. Maize earned a reputation as environmentally risky for good reason. In 2014 and 2016, droughts decimated nearly all of the hybrid maize planted in LUC areas on hillsides. And in 2014 and 2015 floods completely destroyed maize planted in LUC areas in marshlands. Many female respondents described maize as a gamble that is not worth the wager of household food security.

By 2021, OAF supplies more than one million customers, claiming to support the poorest of the poor. However, in Nyamagabe, respondents described how OAF is only viable for wealthy producers who can succeed with maize because they have more fertile land, more access to inputs, and more assets that enable them to bear the risk of not being able to repay a high-interest loan in the event of a drought or heavy rain that destroys their crops. These wealthier male respondents often see maize as an important step out of subsistence agricultural livelihoods. Through fertilizer inputs for maize and the increased availability of people who are willing to farm via sharecropping arrangements, wealthier landowners have found a pathway to enhance the fertility of their fields. On the flip side, many poorer producers described their frustrations at laboring in rented fields for years and applying copious manure only to have the owner discontinue the agreement once the land was fertile enough. In this sense, maize enables control and appropriation of rents not only by the state and corporations, but also by local elites. Most problematically, the poorest households (who are arguably the least prepared to succeed with agricultural modernization) were forced to plant a higher proportion of their land (45%) in government-selected crops than wealthier households, who only planted 27% of their land.

Recognizing the substantial climate risk of maize in Rwanda, OAF has sought options for producers to insure its maize. OAF contacted Agriculture and Climate Risk Enterprise (ACRE), a crop insurance initiative founded by the seed and pesticide conglomerate Syngenta and with ties to AGRA. Syngenta Foundation conducted a feasibility study of ten crop value chains in Rwanda to guide provision of crop weather insurance for smallholders. The study identified maize as the most viable crop because it would garner enough insurance premiums after three years to make it financially viable for private sector investment (Syngenta Foundation 2012). ACRE's insurance is 'index-based,' using satellite imagery to determine aberrations in rainfall in a given season. The smallest pixel that can be viewed is 4 km, meaning that microclimates cannot be assessed, a crucial oversight in a mountainous country such as Rwanda. My study participants reported that this has resulted in them not being compensated for numerous climatic shocks that have decimated maize harvests. Nevertheless, OAF now requires customers to purchase crop insurance for maize as part of its bundle of services. Through this financialization of climate risk, smallholder producers have thus become tied into global capital circuits. While maize is a cornerstone of efforts to make Rwanda's green revolution climate smart, it offers producers a costly illusion of climate resilience and a disproportionate burden of climatic risk.

Despite the hopes for maize to generate economic growth, the crop has continuously failed to deliver countrywide. Even assessments by Rwanda's Ministry of Agriculture find that maize had among the lowest revenues per hectare from 2014 to 2017 – providing only a third the revenue of sweet potato (GoR 2018). The lackluster performance of maize is frequently attributed to 'sub-optimal agricultural practices,' with 'agricultural inputs and best management practices' prescribed as solutions to overcome persistent 'yield gaps' (Bucagu et al. 2020, 1269). However, even in years when maize yields are relatively high, respondents noted that the price of maize bottoms out due to oversupply. While in theory producers could wait to sell their harvests until demand increases, those without adequate storage risk contamination by aflatoxins (a pathogenic fungi that thrives in hot, humid conditions). For example, in 2017 Africa Improved Foods (a large maize processor in Kigali established by Rwanda's government with support from AGRA) rejected 90% of maize due to the presence of aflatoxins (AIF 2021). Such market dynamics and value chain stages are seldom considered within CSA, where the overriding focus is on the production phase. In part through the illusion of climate-smart development, maize has pulled semi-subsistence producer deeper into uneven risk allocation and strengthened control by government and corporate actors.

Sweet potato: everyday adaptation

Above I explored the paradox of enrolling a climatically risky crop in a 'climate-smart' program of agricultural modernization. Here I focus on sweet potato to consider the inverse: the ramifications of limiting cultivation of a crop that has long been a fixture of climate risk mitigation. A staple throughout rural Rwanda, sweet potato (*Ipomoea batatas*) has been the country's most important source of calories since at least the 1980s (Tardif-Douglin 1991). At 88 kg per person, Rwanda maintains the highest per capita production of sweet potatoes in the world (Theisen 2020). In southwest Rwanda, acidic, low-phosphorus soils have made sweet potatoes particularly important due to the challenges of growing grain crops in the absence of manure and synthetic fertilizer (Olson 1994, and discussed above about maize). As my study participants repeatedly exclaimed: 'sweet potato is the food of this place.'

This was not always the case. Nearly 100 years ago, producers resisted colonial administrators' demands that they grow sweet potatoes. A 1924 law sought to prevent famines by compelling producers to maintain 0.15 hectare of sweet potato and cassava per adult (Ministere des Colonies 1954).[4] Yet, Rwandans preferred sorghum and millet to sweet potatoes, often uprooting the tuber once colonial authorities were no longer in sight (Everaerts 1939). Nevertheless, as land became increasingly scarce and millet production untenable due to laws against burning grasses, sweet potatoes took on a greater importance. The tuber offered dependable, calorie-rich (if relatively low nutrient) yields. By the 1980s, government food security planning centered on sweet potatoes (Tardif-Douglin 1991). In the chaos that followed the 1994 war and genocide, sweet potatoes became particularly important as a low-risk crop.

How did sweet potatoes shift from unwanted colonial imposition to a fixture of risk-averse semi-subsistence farming in Rwanda? To answer this question, we can look to human-plant relations. Sweet potatoes' malleability has enabled them to co-adapt with complex farming systems that developed amid land scarcity, inequality, irregular rainfall, and gendered household responsibilities. First, sweet potato life cycles are well-suited to Rwanda's bimodal rainy seasons. If planted during one of the two rainy seasons, they typically retain sufficient soil moisture through dry seasons, enabling continuous cultivation throughout the year. Second, because sweet potatoes develop below ground, they invite intercropping with above-ground beans, banana, sorghum, or maize. Intercropping has a dual purpose of ensuring a continuous supply of sweet potatoes and reducing soil erosion by maintaining ground cover.[5] Sweet potatoes further optimize subsistence systems because they require relatively little labor and no purchased inputs. While synthetic and organic fertilizers are essential for maize, sweet potatoes require only 'green manure' cut from trees. And the tubers are propagated by planting vines, which are typically sourced from a recently harvested field or from neighbors and family.

Sweet potatoes are also malleable in terms of consumption. They are typically harvested immediately before consumption or sale and can remain in the ground for two years after ripening, where they will continue to grow. Women frequently likened this to 'a bank account.' Sweet potatoes and their vines can also be diverted to feed livestock, playing an important role in mixed-crop livestock systems. Their reliability has given sweet potatoes a central role in feeding the household (particularly children) and its

---

[4]Famines were attributed to irregular rainfall and these crops were selected because they resisted drought, did not need to be harvested at the same time as other crops, and could be rapidly transported in case of regional food shortages (Harroy 1944).

[5]When sweet potatoes are intercropped, the goal is often to reserve their vines as seedstock for a subsequent season. Other times, sweet potato is planted in an already-established field of sorghum, maize, or bean, with the plan to continue cultivating sweet potato after the principal crop is harvested.

livestock. As such, sweet potatoes are a strongly gendered crop in Rwanda, associated exclusively with women given their roles in ensuring food for children and livestock. Underscoring this deep gendered nature, respondents laughed at the idea of a man harvesting sweet potatoes. As important as they are for food security, sweet potatoes are also a vital cash crop for women, who sell the tubers at night markets to pay for clothing, food, or children's school fees. Sweet potatoes' unique ability to operate as both a food security crop and a market crop has long provided small holders the flexibility vital to supporting household resilience in the face of uncertainty.

In short, sweet potatoes help households – through the labor of women – to diffuse climatic risk and space out labor across the year. However, the same features that make sweet potato emblematic of complex semi-subsistence farming systems have made them the target of government efforts to abolish subsistence farming. Respondents discussed the rampant uprooting of sweet potatoes by local leaders who were enforcing the government's LUC program. This limitation on the undisputed staple food had drastic effects in Nyamagabe. It led to increased hunger for 39% of all households surveyed and 49% of female-headed households. The effects were most stringent for poorer women, who explained how their food security, sovereignty, and wellbeing had eroded due to government policies. The places where the poorest households had once planted sweet potatoes – hillsides and marshlands – are the most strictly controlled. These are the government's priority areas of agricultural modernization and sites of 'climate smart' initiatives that constructed terraces and drained marshlands. Prior to this, poorer households planted sweet potatoes in low-lying areas and marshlands as a way to buffer against drought. The substantial labor required to prepare marshlands meant that only those most in need of a safety net would cultivate there. Marshlands became especially important in low-rainfall years. After consolidation, small plots of marshland cost 3000 to rent for a three-month period. Producers were obligated to cultivate maize or horticulture crops for export. This made marshlands off-limits to poorer households and undesirable to all but the wealthiest who could pay for inputs, including labor. During drought seasons in 2014 and 2016, drastic reduction of sweet potato acreage had cascading effects for poor households and women. When drought devastated maize harvests, they worked in the fields of wealthier landholders for money to buy food. Yet, the vastly reduced supply of sweet potatoes caused market prices to quadruple. And the shortage of planting material led to selling vines, a once-unheard of practice.

However, the materiality of sweet potatoes also enabled everyday adaptation. Following the 2014 drought, producers gradually returned to cultivating sweet potatoes, although in different ways than before. Women explained that 'it is like stealing.' They discussed how people did in fact continue to plant sweet potatoes, but only in 'hidden places' in *akabande* (the deep valleys that mark the border between communities). While the authorities uprooted sweet potatoes planted on hillsides, they looked the other way when the crop was planted in valleys. This was because valleys are hidden from view and because it is difficult to attribute valley land to a particular household as the plots are often a 20-minute walk or more from the homestead. Valley land thus became crucial for the mitigation of climatic shocks.

The practice of planting sweet potatoes in valley land can be considered a mode of everyday adaptation to coercive land-use governance as well as droughts and floods that routinely devastated fields planted in the LUC program. Yet, access to hidden areas in the *akabande* is uneven. Only 37% of female-headed households had access compared to 56% of all households. The importance of sweet potato amid drought, coupled with the fact that the crop could only be grown in valleys, caused valley land to skyrocket in value. Wealthier households, who own the majority of this land, have increased land rental fees. Before long, many sweet potato plots in hidden places were operated on a sharecropping basis, where tenants generally owe half of the resulting crop production to landowners (although this can be negotiated). Even while this is recognized as an exploitative institution, respondents noted that it has become increasingly common due to the need for valley land for sweet potatoes. Moreover, while 31% of female-headed households no longer sell any sweet potatoes, wealthier households have accumulated more valley land and increased sales of sweet potatoes, capitalizing on their scarcity. Additionally, where households of modest means own some valley land, it is typically located far away, creating a substantial labor burden for women who need to walk daily to the fields to harvest sweet potatoes for their families to eat. In summary, the inability of agricultural policies and CSA programs to acknowledge the keystone role of sweet potatoes in climate risk management spurred transformations that have deepened coercive land-labor dynamics. This allows us to see how agrarian change, climate change, and agricultural development are linked processes that cannot be managed in isolation.

Reconceptualizing resilience as everyday struggle for wellbeing

To address the seemingly boundless uncertainties of global climate change, development agencies increasingly promote resilience as a vital aspect of their programs and policies. This paper has examined CSA, a prominent form of 'climate resilient development' that is currently being unrolled across Africa and throughout the global South. My case study of Rwanda demonstrates how CSA can have complex and uneven implications for the wellbeing of Africa's smallholder food producers. It showed how CSA has taken shape through large-scale, top-down initiatives that are embedded in a pre-existing paradigm of agricultural-led economic growth. In an effort to make Africa's green revolution climate smart, resilience is framed as the mitigation of external threats to increasing agricultural yields. Adaptation is viewed as a series of technological and managerial solutions intended to plan for and control climatic uncertainty. These solutions (crop insurance, water management infrastructure, climate information services, and improved seeds) all help to streamline a Green Revolution technology package that prioritizes input-led intensification to install economies of scale. By focusing on rural communities' experiences with climate-development initiatives in Rwanda, I have identified crucial limits to these framings of resilience and adaptation. Simply put, the technocratic vision of resilience advanced by CSA is out of step with the lived experiences of climate and agrarian change among smallholder food producers, for whom resilience is inseparable from efforts to procure security, wellbeing, and autonomy.

More specifically, my findings show how programs that reduce climate change adaptation to a set of technologies can undermine the intricate social-ecological dynamics at the heart of smallholders' climate risk management strategies. This suggests that equating climate resilience with agricultural efficiency can have negative implications for those already on the margins, for whom land use strategies that may appear inefficient can be carefully orchestrated efforts to mitigate climate variability, shocks, and change. The technological, finance, infrastructure, and market innovations put forward as climate solutions (as in the case above of 'climate smart' maize) exacerbate agrarian inequities because they favor wealthier male producers who are already integrated in cash economies – the same groups typically favored by Green Revolution technologies. Perhaps unsurprisingly then, Nyamagabe residents unable to benefit from agricultural modernization described CSA technologies as indistinguishable from other top-down management efforts that limit their sovereignty over land use. This shows how a Green Revolution mindset can all too easily permeate efforts to align agricultural development with climate adaptation. It underscores the need to actively reimagine agricultural development paradigms during a time of global climate change. It also suggests the inseparability of climate justice and agrarian justice (Borras and Franco 2018) and the importance of articulating climate resilience through attention to peasant rights over land and decision-making processes (Walsh-Dilley, Wolford, and McCarthy 2016).

More generally, this study exposes a harsh irony of climate-development initiatives. As shown in the WEF quote that opened this article, development programming frequently depicts subsistence producers as innately vulnerable and unable to adapt to climate change. The technologies put forth as climate smart solutions (crop insurance, high-yielding seeds, climate information services, irrigation infrastructure, and terraces) further consolidate agency in the hands of experts, solidifying state and corporate control over rural communities. This further strips away agency from the vulnerable, debilitating locally rooted systems of managing social, economic, and environmental uncertainties. Of course, as others have pointed out, appropriating the cause of the vulnerable may be a deliberate adaptation by states and corporations, many of which stand to benefit from advancing climate solutions that rely on market, financial, and technical innovations (Barnett 2020; Paprocki 2021). I am certainly not the first to consider the stranglehold that a technocratic mindset has on institutional responses to climate change. Abundant research has documented how efforts to build climate resilience through targeted adaptations can reproduce exclusionary power dynamics (for a summary see Nightingale et al. 2020). This has in turn inspired calls for explicitly decolonial approaches to climate adaptation research and practice (Borges-Méndez and Caron 2019; Haverkamp 2021; Santiago-Vera et al. 2021) and for climate-development initiatives that open up space for democratic deliberation about adaptation priorities (Mikulewicz 2018). In support of these more inclusive climate-development initiatives, I advance three interrelated arguments.

First, I corroborate calls (Allen et al. 2019; Chaigneau et al. 2022) for reframing resilience not as an idealized outcome of stability that automatically confers wellbeing but as a contested and contingent process: a struggle for viability, security, wellbeing, and autonomy – or what I call *uneven resilience*. By ignoring or obfuscating these struggles, programs that claim to offer climate resilient development risk effacing people's multifarious efforts to secure wellbeing amidst adverse social and environmental conditions. Conceptualizing resilience in this way throws into stark relief some of the conceits of CSA. However, I believe that approaching resilience as inherently uneven has applicability beyond assessing development programs that conflate agricultural modernization and climate change adaptation. It affords a simple reminder to bring resilience back down to Earth, to see it not as a balm but as a messy and contested process that is inseparable from broader agrarian change.

Second, and relatedly, I suggest the need for greater attention to how vulnerable populations creatively employ their knowledge and skills throughwhat I term *everyday adaptations*: land use and livelihood adjustments that are woven into the fabric of daily life.

There is no need to reinvent the wheel here. Political ecologists and agrarian political economists have long focused on how everyday 'local' practices are dynamically linked to extra-local processes. As one example, Frances Cleaver (2012) offers the useful metaphor of development as *bricolage*, where institutions of resource management are shaped through a continuous process of negotiation. Yet, such an analytic has not commonly featured in research on the interface of climate and agrarian change. Even while it is increasingly well known that global climate change has uneven and inequitable impacts on society, efforts to catalog these impacts remain abstracted from the experiences of people struggling with climate change alongside myriad other social-ecological challenges in their daily lives (Turner 2016; Goldman, Turner, and Daly 2018; Sultana 2022). There can be a fine line between climate adaptations and broader agrarian struggles (Borras et al. 2022). An empirical focus on the diverse, everyday adaptations that encompass yet are not solely provoked by climate change can help to subvert the predominant interpretation of climate as an external threat that can be managed by merely 'getting the institutions right.' If applied as more of a mindset than a framework, it may also help to open research, policy, and practice to subaltern knowledge and processes of adaptation, or what Haverkamp (2021) calls 'adaptation otherwise'.

Third, I suggest that attention to human-plant relations may prove fruitful for further work on climate change and agrarian struggles. This is one way (though certainly not the only way) to consider everyday adaptation because it helps ground interview and survey questions in context-specific practices, values, and challenges. Indeed, my analytical narrative centers on human-plant relations with maize and sweet potato because my interlocutors presented their experiences with Green Revolution and CSA programs through discussion of everyday activities that bound them to these plants in various ways. They explained how maize facilitates the further enrichment of those who can tolerate risk in accordance with a new regime of financialized resilience that is upheld through CSA. They also explained how sweet potatoes facilitate adaptation to climate change and the subtle resistance of state efforts to modernize rural landscapes. The plant's biophysical properties – together with the necessity of feeding families – encouraged women to defy top-down directives of agricultural modernization through creative yet labor-intensive practices. Yet, access to land to cultivate sweet potatoes is extremely uneven and requires significant labor by women to reach the far-away valley parcels. Thus, a focus on human-plant relations helps discern how resilience is uneven and partial. It is not intrinsic to plants, agronomic technologies, or social institutions but rather an emergent result of human-plant relations in a specific place and time. Analysis of shifting plant-human relations helps to develop a more embedded and relational understanding about how climate change adaptation and agrarian change are materially, historically and geographically situated.

Conclusions

The findings presented here speak to the need to revise CSA and related efforts to incorporate climate change concerns into agricultural development. By ascribing quantitative values of 'climate smartness' to crops, technologies, and management practices, CSA makes climate change governable by a constellation of powerful corporate and state actors. Empowering these actors now – through a discourse that simultaneously disempowers 'climate-vulnerable peasant producers' – will guarantee that external experts continue to govern climate change for years to come. Indeed, since its inception CSA has continued to tighten its grip on institutionalized responses to climate change in agrarian settings. As Scoones and Stirling (2020) discuss regarding sustainability more broadly, such a foreclosure of alternative futures represents a significant agrarian injustice.

This article has reflected on how we might build more equitable and democratic responses to climate change by recognizing climate resilience as embedded in broader agrarian struggles for rural viability, autonomy, and wellbeing. It echoes calls (Walsh-Dilley, Wolford, and McCarthy 2016) for a rights-based interpretation of resilience to guide development operations in a time of global climate change. As the authors of the Working Group II of the Intergovernmental Panel on Climate Change (IPCC) sixth assessment report (2022) suggest, climate resilient development should not be expected to manifest through a single decisive action or policy, but through the concerted work of diverse actors who strive to transform the values, ideologies, and social structures that underpin existing institutions. A focus on the everyday practices that constitute adaptation in agrarian environments could help seed such efforts.

Acknowledgements

I am grateful for insightful feedback from four anonymous referees and the JPS editorial team. Numerous colleagues have helped me develop these arguments, including Alfred Bizoza, Dan Clay, Elissa Dickson, Brian King, and Karl Zimmerer. Thousands of people in Rwanda, who I will not name here, made this study possible. Any errors are my responsibility alone.

Disclosure statement

No potential conflict of interest was reported by the author(s).

Funding

The research informing this study was supported by the Fulbright U.S. Student Program and a Borlaug Fellowship in Global Food Security (supported by the United States Agency for International Development–USAID). The paper was written while my position was funded by the Formas - Swedish Research Council for Environment, Agricultural Sciences and Spatial Planning (grant number 2021-02063).

References

AIF. 2021. *The Fight Against Aflatoxin*. African Improved Foods. Accessed 22 March 2022. https://africaimprovedfoods.com/the-fight-against-aflatoxin/.

Allen, Craig R., David G. Angeler, Brian C. Chaffin, Dirac Twidwell, and Ahjond Garmestani. 2019. "Resilience Reconciled." *Nature Sustainability* 2: 898–900.

Barnett, Jon. 2020. "Global Environmental Change II: Political Economies of Vulnerability to Climate Change." *Progress in Human Geography* 44 (6): 1172–1184.

Bigler, Christine, Michèle Amacker, Chantal Ingabire, and Eliud Birachi. 2019. "A View of the Transformation of Rwanda's Highland Through the Lens of Gender: A Mixed-Method Study About Unequal Dependents on a Mountain System and Their Well-Being." *Journal of Rural Studies* 69: 145–155.

Bizoza, Alfred R. 2021. "Investigating the Effectiveness of Land use Consolidation– a Component of the Crop Intensification Programme in Rwanda." *Journal of Rural Studies* 87: 213–225.

Borges-Méndez, Ramón, and Cynthia Caron. 2019. "Decolonizing Resilience: The Case of Reconstructing the Coffee Region of Puerto Rico After Hurricanes Irma and Maria." *Journal of Extreme Events* 6 (1). doi:10.1142/S2345737619400013.

Borras, Saturnino M. Jr., and Jennifer C. Franco. 2018. "The Challenge of Locating Land-Based Climate Change Mitigation and Adaptation Politics Within a Social Justice Perspective: Towards an Idea of Agrarian Climate Justice." *Third World Quarterly* 39 (7): 1308–1325.

Borras, Saturnino M. Jr., Ian Scoones, Amita Baviskar, Marc Edelman, Nancy Lee Peluso, and Wendy Wolford. 2022. "Climate Change and Agrarian Struggles: An Invitation to Contribute to a JPS Forum." *The Journal of Peasant Studies* 49 (1): 1–28.

Brown, Katrina. 2014. "Global Environmental Change I: A Social Turn for Resilience?" *Progress in Human Geography* 38 (1): 107–117.

Bucagu, Charles, Alain Ndoli, Athanase R. Cyamweshi, Leon N. Nabahungu, Athanase Mukuralinda, and Philip Smethurst. 2020. "Determining and Managing Maize Yield Gaps in Rwanda." *Food Security* 12: 1269–1282.

Burnham, Morey, and Zhao Ma. 2018. "Multi-Scalar Pathways to Smallholder Adaptation." *World Development* 108: 249–262.

Carr, Edward. 2019. "Properties and Projects: Reconciling Resilience and Transformation for Adaptation and Development." *World Development* 122: 70–84.

Cavanagh, Connor Joseph, Anthony Kibet Chemarum, Paul Olav Vedeld, and Jon Geir Petursson. 2017. "Old Wine, New Bottles? Investigating the Differential Adoption of 'Climate-Smart' Agricultural Practices in Western Kenya." *Journal of Rural Studies* 56: 114–123.

Chaigneau, Tomas, Sarah Coulthard, Tim M. Daw, Lucy Szaboova, Laura Camfield, F. Stuart Chapin III, Des Gasper, et al. 2022. "Reconciling Well-Being and Resilience for Sustainable Development." *Nature Sustainability* 5: 287–293.

Chandra, Alvin, Karen E. McNamara, and Paul Dargusch. 2018. "Climate Smart Agriculture: Perspectives and Framings." *Climate Policy* 18 (4): 526–541.

Clapp, Jennifer, Peter Newell, and Zoe W. Brent. 2018. "The Global Political Economy of Climate Change, Agriculture and Food Systems." *Journal of Peasant Studies* 45 (1): 80–88.

Clay, Nathan. 2018. "Seeking Justice in Green Revolutions: Synergies and Trade-Offs Between Large-Scale and Smallholder Agricultural Intensification in Rwanda." *Geoforum; Journal of Physical, Human, and Regional Geosciences* 97: 352–362.

Clay, Nathan, and Brian King. 2019. "Smallholders' Uneven Capacities to Adapt to Climate Change Amid Africa's 'Green Revolution': Case Study of Rwanda's Crop Intensification Program." *World Development* 116: 1–14.

Cleaver, Frances. 2012. *Development Through Bricolage: Rethinking Institutions for Natural Resource Management*, 240. London: Routledge.

Corbera, Esteve, Carol Hunsberger, and Chayan Vaddhanaphuti. 2017. "Climate Change Policies, Land Grabbing and Conflict: Perspectives from Southeast Asia." *Canadian Journal of Development Studies* 38 (3): 297–304.

Dawson, Neil, Adrian Martin, and Thomas Sikor. 2016. "Green Revolution in Sub-Saharan Africa: Implications of Imposed Innovation for the Wellbeing of Rural Smallholders." *World Development* 78: 204–218.

de Lame, Danielle. 2005. *A Hill Among a Thousand: Transformations and Ruptures in Rural Rwanda*. Madison: University of Wisconsin Press.

Eriksen, S. H., A. J. Nightingale, and H. Eakin. 2015. "Reframing Adaptation: The Political Nature of Climate Change Adaptation." *Global Environmental Change* 35: 523–533.

Everaerts, E. 1939. "Monographie Agricole du Ruanda-Urundi." *Bulletin Agricole du Congo Belge* 30 (3): 343–396.

Fairhead, James, Melissa Leach, and Ian Scoones. 2012. "Green Grabbing: A New Appropriation of Nature?" *The Journal of Peasant Studies* 39 (2): 237–261.

FAO. 2019. *Agriculture and Climate Change – Challenges and Opportunities at the Global and Local Level – Collaboration on Climate-Smart Agriculture*. Rome. 52 pp. https://www.fao.org/3/CA3204EN/ca3204en.pdf.

Ford, Robert E. 1990. "The Dynamics of Human-Environment Interactions in the Tropical Montane Agrosystems of Rwanda: Implications for Economic Development and Environmental Stability." *Mountain Research and Development* 10 (1): 43–63.

Forsyth, T., and N. Evans. 2013. "What is Autonomous Adaptation? Resource Scarcity and Smallholder Agency in Thailand." *World Development* 43: 56–66.

Galvin, Shaila Seshia. 2018. "Interspecies Relations and Agrarian Worlds." *Annual Review of Anthropology* 47 (1): 233–249.

Goldman, Mara J., Matthew D. Turner, and Meaghan Daly. 2018. "A Critical Political Ecology of Human Dimensions of Climate Change: Epistemology, Ontology, and Ethics." *WIREs Climate Change* 9(4). doi:10.1002/wcc.526.

GoR. 2018. *Strategic Plan for Agriculture Transformation 2018-2024: Planning for Wealth*. Government of Rwanda.

Harroy, Jean-Paul. 1944. *Afrique, Terre Qui Meurt: La Degradation De Sols Africains Sous l'Influence De La Colonisation*. Brussels: Academie Royale de Belgique.

Haverkamp, Jamie. 2021. "Collaborative Survival and the Politics of Livability: Towards Adaptation Otherwise." *World Development* 137. doi:10.1016/j.worlddev.2020.105152.

Heinen, Sebastian. 2022. "Rwanda's Agricultural Transformation Revisited: Stagnating Food Production, Systematic Overestimation, and a Flawed Performance Contract System." *The Journal of Development Studies*, doi:10.1080/00220388.2022.2069494.

Holt-Giménez, Eric, Annie Shattuck, and Ilja Van Lammeren. 2021. "Thresholds of Resistance: Agroecology, Resilience and the Agrarian Question." *The Journal of Peasant Studies* 48 (4): 715–733.

Huggins, C. 2014. "'Control Grabbing' and Small-Scale Agricultural Intensification: Emerging Patterns of State-Facilitated 'Agricultural Investment' in Rwanda." *The Journal of Peasant Studies* 41 (3): 365–384.

Jakobsen, Jostein, and Ola T. Westengen. 2021. "The Imperial Maize Assemblage: Maize Dialectics in Malawi and India." *The Journal of Peasant Studies* 49 (3): 536–560.

Karlsson, L., L. O. Naess, A. Nightingale, and J. Thompson. 2018. "'Triple Wins' or 'Triple Faults'? Analyzing the Equity Implications of Policy Discourses on Climate-Smart Agriculture (CSA)." *Journal of Peasant Studies* 45 (1): 150–174.

Maquet, Jacques J., and S. Naigiziki 1957. "Les droits fonciers dans le Ruanda ancien." *Revue Congolaise* 11 (4): 339–359.

Matin, Nilufar, John Forrester, and Jonathan Ensor. 2018. "What is Equitable Resilience?" *World Development* 109: 197–205.

Mikulewicz, Michael. 2018. "Politicizing Vulnerability and Adaptation: On the Need to Democratize Local Responses to Climate Impacts in Developing Countries." *Climate and Development* 10 (1): 18–34.

Mikulewicz, Michael, and Marcus Taylor. 2020. "Getting the Resilience Right: Climate Change and Development Policy in the 'African Age'." *New Political Economy* 25 (4): 626–641.

Ministere des Colonies. 1954. *L'Agriculture au Congo Belge et au Ruanda-Urundi de 1948 a 1952*. Brusselles: Royaume de Belgique, Ministere des Colonie, Service de l'Agriculture.

Newell, Peter, and Olivia Taylor. 2018. "Contested Landscapes: The Global Political Economy of Climate-Smart Agriculture." *Journal of Peasant Studies* 45 (1): 108–129.

Ngango, Jules, and Seungjee Hong. 2021. "Impacts of Land Tenure Security on Yield and Technical Efficiency of Maize Farmers in Rwanda." *Land Use Policy* 107. doi:10.1016/j.landusepol.2021.105488.

Nightingale, Andrea J., Siri Eriksen, Marcus Taylor, Timothy Forsyth, Mark Pelling, Andrew Newsham, Emily Boyd, et al. 2020. "Beyond Technical Fixes: Climate Solutions and the Great Derangement." *Climate and Development* 12 (4): 343–352.

Ntirenganya, Emmanuel. 2016. Rwanda's Longest Drought in Six Decades. *New Times*. 16 September 2016. Accessed 10 February 2018. http://www.newtimes.co.rw/section/read/203577.

Olson, Jennifer. 1994. *Producer Responses to Land Degradation in Kikongoro, Rwanda*. East Lansing: Michigan State University.

One Acre Fund. 2016. *Press Release: Tubura Announces Open Enrollment for Producers in Rwanda*. May 24. Accessed 20 March 2022. https://oneacrefund.org/blog/press-release-tubura-announces-open-enrollment-producers-rwanda/.

One Acre Fund. 2020. *Building Resilient Climate Smart Agricultural Systems for Smallholder Producers*. September 17. Accessed 20 March 2022. https://impakter.com/one-acre-fund-building-resilient-climate-smart-agricultural-systems-for-smallholder-producers/.

Paprocki, Kasia. 2021. *Threatening Dystopias: The Global Politics of Climate Change Adaptation in Bangladesh*. Ithaca: Cornell University Press.

Parrucci, Andrew. 2018. *A New Loan for Sustainable Agriculture: One Acre Fund*. August 6. Accessed 30 March 2022. https://calvertimpactcapital.org/resources/a-new-loan-for-sustainable-agriculture-one-acre-fund.

Patel, Raj. 2013. "The Long Green Revolution." *The Journal of Peasant Studies* 40 (1): 1–63.

Ribot, Jesse. 2014. "Cause and Response: Vulnerability and Climate in the Anthropocene." *The Journal of Peasant Studies* 41 (5): 667–705.

Roman, Gabriel G, and Ola T. Westengen. 2022. "Taking Measure of an Escape Crop: Cassava Relationality in a Contemporary Quilombo-Remnant Community." *Geoforum; Journal of Physical, Human, and Regional Geosciences* 130: 136-145.

Santiago-Vera, T., P. M. Rosset, A. Saldivar, B. G. Ferguson, and V. E. Méndez. 2021. "Re-conceptualizing and Decolonizing Resilience from a Peasant Perspective." *Agroecology and Sustainable Food Systems* 45 (10): 1422–1440.

Scoones, Ian, and Andy Stirling. 2020. *The Politics of Uncertainty: Challenges and Transformation*. London: Routledge.

Scott, James C. 2017. *Against the Grain: A Deep History of the Earliest States*. New Haven: Yale University Press.

Sinha, Shreya. 2022. "From Cotton to Paddy: Political Crops in the India Punjab." *Geoforum; Journal of Physical, Human, and Regional Geosciences* 130: 146–154.

Sultana, Farhana. 2022. "Critical Climate Justice." *The Geographical Journal* 188 (1): 118–124.

Syngenta Foundation. 2012. *Crop and Livestock Insurance Feasibility Study: Final Report*.

Tardif-Douglin, David G. 1991. *The Role of Sweet Potato in Rwanda's Food System: The Transiton from Subsistence Orientation to Market Orientation*. Ithaca: Cornell University.

Taylor, Marcus. 2014. *The Political Ecology of Climate Change Adaptation: Livelihoods, Agrarian Change, and the Conflicts of Development*. London: Routledge.

Taylor, Marcus. 2018. "Climate-smart Agriculture: What is it Good for?" *Journal of Peasant Studies* 45 (1): 89–107.

Taylor, Marcus, and Suhas Bhasme. 2021. "Between Deficit Rains and Surplus Populations: The Political Ecology of a Climate-Resilient Village in South India." *Geoforum; Journal of Physical, Human, and Regional Geosciences* 126: 431–440.

Theisen, Kelly. 2020. *Africa: Sweetpotato Production*.

Turner, Matthew D. 2016. "Climate Vulnerability as a Relational Concept." *Geoforum; Journal of Physical, Human, and Regional Geosciences* 68: 29–38.

Uwizeyimana, Laurien. 1991. *Croissance demographique et production agricole au Rwanda: Impossible adequation? Cahiers du CIDEP 8*. Louvain-la-Neuve: CIDEP.

Walsh-Dilley, Marygold, Wendy Wolford, and James McCarthy. 2016. "Rights for Resilience: Food Sovereignty, Power, and Resilience in Development Practice." *Ecology and Society* 21 (1): 11.

Watts, Michael.2015. "Now and Then: The Origins of Political Ecology and the Rebirth of Adaptation as a form of Thought." In *The Routledge Handbook of Political Ecology*, edited by Tom Perreault, Gavin Bridge, and James McCarthy, 19–50. London: Routledge..

WEF. 2016. *4 Factors Holding Back Africa's Small-Scale Producers*. Accessed 20 March 2022. https://www.weforum.org/agenda/2016/05/4-factors-holding-back-african-producers/.

World Bank. 2016. *World Bank Group Climate Change Action Plan*. Accessed 20 March 2022. https://openknowledge.worldbank.org/handle/10986/24451.

World Bank; CIAT. 2015. *Climate-Smart Agriculture in Rwanda*. Washington D.C.: The World Bank Group.

Rethinking 'just transitions' from coal: the dynamics of land and labour in anti-coal struggles

Amod Shah

ABSTRACT
Communities resisting large coal mining projects navigate the significant tensions between imperatives of urgent climate action and economic growth in complex and contingent ways. Drawing on empirical research in a mining region of Central-Eastern India, this paper examines how the changing 'agrarian' context of rural livelihoods and household reproduction within mining-affected communities shapes the motivations of local anti-coal struggles, and the articulation of climate-change related concerns within them. It argues that such a conceptualization of political contestations over coal extraction points to crucial possibilities for building broader counter-hegemonic movements for more inclusive 'just transitions' away from coal.

1. Introduction

Coal is a key driver of climate change and the single largest contributor of global CO_2 emissions (IEA 2019; Olivier and Peters 2020). With the window to avoid impending climate catastrophe steadily narrowing, it is not surprising that 'end of coal as emerged as a crucial rallying call for climate movements and global climate policy (Rose-warne, Goodman, and Pearse 2014; UNFCCC 2021). Such national and global-level anti-coal mobilizations driven predominantly by the need for urgent climate action, in combination with the numerous local struggles against proposed and existing coal mines across the world, have been instrumental in reducing overall dependence on coal. Global coal demand has not yet peaked, but its share of the energy mix remains on a slow decline (IEA 2019, 16). Recent commitments to limit and eventually phase out coal-based power generation during the Glasgow COP26 climate conference in November 2021 will likely accelerate this shift, though at a much slower pace than needed (Abnett and Piper 2021; de Hoog and Kirk 2021).

These efforts stand in contrast to the continuing importance of coal-based economies in many parts of the world. Global coal power generation capacity has doubled since 2000 with rapid growth in China, India and other industrializing countries (Carbon Brief 2020; Goodman et al. 2020, 1). More than 70% of India s electricity use depends on coal, and domestic production for thermal power and industry has risen by 40% in the last decade (Carbon Brief

2019; Permual 2021). The state-owned Coal India Limited expects to increase yearly production to 1 billion tonnes by 2024, a 44% increase from 2019 levels (PIB 2020). These estimates exclude private sector coal mining, which is also rapidly expanding with a massive ongoing push to liberalize the sector as part of the *Aatmanirbhar Bharat* (self-reliant India) programme announced during the Covid-19 pandemic (Aggarwal 2021a; Jagannath 2020).

Coal thus remains central to economic growth strategies of India and much of the industrializing world. These countries have strongly resisted a rapid phase out of coal use, pointing to their relatively low per capita emissions and the need to ensure that global climate action efforts do not unfairly hinder their development trajectories (Sethi 2021). India's Ministry of Coal recently declared that '[B]eing an affordable source of energy with substantial reserve, coal is going to stay as major source of energy in the foreseeable future' and that, notwithstanding its commitment to achieve net zero emissions by 2070 at the Glasgow conference, 'the pace of transition to cleaner energy sources in India is to be viewed in the light of national circumstances, and principle of common but differentiated responsibilities and respective capabilities, the transfer of climate finance and low cost climate technologies' (Ministry of Coal 2022).

Crucially, actual policy efforts to facilitate an eventual shift away from fossil fuel-driven growth remain narrowly defined and limited in scope. In March 2022, for instance, the Indian government announced a 'just transition' project in select districts of the country with declining coal production (Aggarwal 2021b; Srivastava 2022). Over an eight-year period, the Central coal ministry aims to spend about $1 billion on the socio-economic rehabilitation of communities in the vicinity of closed or abandoned coal mines, and those likely to cease production in the next few years.[1] Such efforts are a long overdue recognition of the significant challenges and uncertainties of post-coal futures for communities that are heavily dependent on mining-related employment. However, vast regions of the country where coal mining is more recent, and likely to rapidly expand going forward, have been left out of their ambit. More generally too, narratives of a 'just transition' away from coal tend to emphasize the labour and employment dimensions of coal mine closures (Roy, Kuruvilla, and Bhardwaj 2019, 285–6; World Bank 2021)

The fragmented global consensus and widespread uncertainties underlying the 'end of coal' have crucial implications for how we understand the politics of anti-coal struggles and just transitions, particularly in India and other industrializing countries. If calls for stronger climate action are not translating into rapid declines in coal extraction in these settings, can they provide a sustained and effective basis for mobilizing against coal mining? Indeed, local resistance to coal mining projects has increased in scale and prominence, but climate-related concerns often play a limited role in the framing and demands of such struggles (Goodman et al. 2020, 67–8; Paprocki 2021). Conde (2017, 84–5) also argues that motivations and narratives underlying opposition to mining may be strategically combined with more 'global' discourses on indigenous rights and environmental and climate justice, but cannot be reduced to the latter.

This paper argues that a closer focus on the changing 'agrarian' dynamics within mining affected communities are useful for understanding why local anti-coal struggles coalesce

[1]Such plans also point to the Increased global funding being made available for 'just transitions' away from coal as part of climate action efforts (see, for instance, CIF 2021). India's recently announced project is similarly contingent on financial support from the World Bank.

around their particular demands, and how environmental and climate concerns are articulated within them. Through an examination of political contestations over a large coal mining project in the Tamnar region of India, it highlights the significant impacts of mining-related dispossession on processes of agrarian production, dependence on agricultural and non-agricultural wage labour, and practices of daily and inter- generational social reproduction. Local communities have mobilized strongly in response to these impacts, combining demands for agrarian and climate justice in meaningful and instructive ways. The paper ends with brief reflections on the possible implications of such a broadened conceptualization of anti-coal struggles for counter-hegemonic movements seeking truly 'just transitions' away from coal.

2. Conceptualizing the politics (and political possibilities) of anti-coal struggles

The varied political character of local struggles against resource extraction has been an important focus of existing research. Studies point to the crucial role of project or sector-specific characteristics, political opportunity structures and cross/multi-scalar alliances in shaping strategies, discourses and outcomes of resistance to large-scale mining projects (Conde and Le Billon 2017; Dietz and Engels 2017; Prause and Le Billon 2021; Temper et al. 2020). Others emphasize the distinctive dynamics of resource governance and politics that tend to emerge within sub-national 'extractive regimes' and 'political settlements' linked to mining (Adhikari and Chhotray 2020; Bebbington et al. 2018, 12). Such interventions provide a more nuanced perspective on how the particular contexts of extractive projects are implicated in the emergence and character of political contestations over them. At the same time, understanding if, and how, diverse place-based struggles over resource extraction can come together within wider social movements requires a complementary focus on their linkages to the broader structural dynamics of contemporary capitalism and climate change.

In an influential recent essay, Nancy Fraser (2021) points to the possibility of building an explicitly anti-capitalist politics around the urgent task of confronting the climate crisis and dominant responses to it. Her arguments, which draw extensively on long-running debates in ecological Marxism and feminist critiques of social reproduction, centre around capitalism's inherent tendency to perpetuate multiple ecological and social-reproductive crises. The deeply interlinked nature of these crises in capitalist societies also implies that any such progressive eco-politics must necessarily incorporate both environmental and non-environmental objectives. Fraser's call for counter-hegemonic climate movements to transcend the 'merely environmental' (Fraser 2021, 96) extends – to a systemic level – grassroots conceptions of environmental and climate justice that encompass concerns of well-being and livelihoods, claims for recognition of alternative ways of life, and challenges to dominant understandings of relationships between humans and nonhuman natures (Schlosberg and Collins 2014, 360–1). Such perspectives see environmental change and social justice as closely interlinked, calling attention to how affected communities have mobilized strongly against forms of ecological degradation that threaten their material interests and value systems (Bebbington et al. 2008, 2891; Guha and Martinez-Alier 1997; Martinez-Alier 2002). Indeed, more radical demands for climate justice in coal mining contexts have also pushed for an explicit recognition that anti-coal struggles cannot be limited to reducing CO2 emissions, but must challenge long-running systems of capitalist exploitation and oppression (Still Burning 2021, 9–12; see also Borras et al. 2021, 12–3 for a broader discussion on radical climate justice approaches).

The systemic nature of the climate crisis thus offers a common material basis for bringing together diverse movements, both 'environmental' and 'non-environmental'. However, struggles over the 'climate' are not uniform, but rather must be understood as the outcome of the context-specific and historically situated unfolding of climate change politics in particular settings (Borras et al. 2021, 5–8). In discussing contemporary agrarian struggles and the possibilities they offer for building a counter-hegemonic climate politics, the authors of the forum framing paper draws on Wright's (2019) typology of multiple strategic logics of anti-capitalism. They suggest that even if movements understand and engage with climate change politics in diverse ways, their combined actions can still contribute to the ultimate goal of 'eroding capitalism' (Borras et al. 2021, 14–16). We can therefore think of a plurality of objectives of local responses confronting climate change impacts and dominant responses to them, many of which may nevertheless be consistent with a broadly anti-capitalist orientation.

Finally, such an expanded understanding of the motivations and composition of climate-related movements calls attention to the enabling conditions for broader coalition building. Reflecting on the emergence of struggles for 'agrarian climate justice' that combine demands for agrarian and climate justice (Franco and Borras 2019) in Myanmar, Sekine (2021) argues that a changing political context of rural democratization and expanding political opportunity structures, especially at the national level, have played an important role in complementing and supporting the efforts of diverse local struggles. Calmon, Jacovetti, and Koné (2021) similarly explore the possibilities for building alliances between peasant and environmental movements in Mali. They suggest that agrarian climate justice has emerged as a progressive mobilizing frame to counter dominant narratives of climate security, in large part through the implementation of localized initiatives like village land commissions. Both these interventions also highlight the crucial, but difficult, task of scaling up and broadening agrarian climate justice struggles in a contemporary global context where agrarian and land politics on the one hand, and climate politics on the other, are increasingly closely intertwined.

The above discussions speak to key considerations for conceptualizing the politics and political possibilities of contemporary anti-coal struggles. First, despite clear systemic links between fossil fuel-driven capitalist growth and the climate crisis, coal mining-affected communities often do not explicitly articulate their opposition in relation to climate change. The mobilization of climate-centred narratives is thus best understood as a contingent outcome of local resistance to coal extraction and related coalition-building processes playing out in specific political-economic contexts. Second, an emphasis on the diverse non-environmental concerns that shape motivations and strategies of anti-coal struggles does not imply that the 'environment' or 'climate' are not important considerations for them. Rather, these reflect varied ways of understanding and engaging with contemporary capitalism and climate change politics, and the need for more expansive notions of climate justice and 'just transitions' which meaningfully incorporate them.

Discussions within critical agrarian studies provide important insights for unpacking these complex political dynamics of local anti-coal struggles and their incorporation of climate change-related concerns. More specifically, they draw attention of the changing context of livelihoods and social reproduction within the predominantly rural settings of coal mining projects, and its influence on the motivations and political character of collective opposition by affected communities. In this sense, political contestations over coal extraction can be understood as simultaneously 'climate' struggles confronting a key driver of the climate crisis and 'agrarian' struggles around the transformation of land and labour relations in mining-affected areas.

3. Agrarian dynamics of anti-coal struggles

Critical agrarian studies scholarship has focused closely on how processes linked to contemporary capitalist development shape and reconfigure the dynamics of agricultural production, wage labour and social reproduction, and its implications for rural politics (Akram-Lodhi and Kay 2009; Bernstein 1996, 39, 2006). Of particular relevance to the present analysis is the argument that a key feature of neoliberal globalization is the intensified fragmentation of the bases of reproduction of a large and growing share of the working population (Bernstein 2006, 454–5). Within agriculture, for instance, it is increasingly rare to come across landless rural workers depending solely on agricultural wage labour, or small holding farmers completely reliant on agricultural petty commodity production for their survival (Bernstein 2006, 454). Rather, a large proportion of rural households now reproduce through diverse combinations of waged and self-employment, both within and outside agriculture and across multiple spatial locations. They represent heterogenous 'classes of labour' that 'depend – directly and indirectly – on the sale of their labour power for their own daily reproduction' (Panitch and Leys 2001: ix, cited in Bernstein 2006, 455).[2]

Two assumptions underlying Bernstein's conceptualization of classes of labour require closer attention. First, rural classes of labour are highly diverse in how they combine land and labour to resolve their particular crises of reproduction. Nevertheless, they all remain, directly or indirectly, net sellers of labour power (see also Pattenden 2016, 23 and Lerche 2010, 65–66 for an elaboration of Bernstein's arguments). Focusing on the political implications of such an understanding, Pattenden (2018, 1042) argues that, heterogeneous classes of labour nevertheless share a common (and primary) interest in improving the levels and terms of their access to wage employment.[3] Second, in its insistence that classes of labour do not reproduce outside of conditions of generalized commodity production and, in particular, the sale of their labour power, such a conceptualization tends to de-emphasize the role of non-commodified forms of land and labour within households' social reproduction strategies. Both assumptions have been challenged by other recent contributions.

Jacobs (2018) and Zhan and Scully (2018), for instance, both position land as a key basis of political struggle. They argue that a large proportion of the working population in South Africa and China are increasingly engaged in insecure and poorly compensated wage

[2]There are important similarities to Shivji's (2017) conceptualization of 'working people', which argues that under neoliberalism, capital expropriates not just surplus value from wage labour, but also part of the necessary consumption of small producers. While such an understanding is broadly consistent with Bernstein's understanding of the growing challenges to simple reproduction of classes of labour, Shivji emphasizes the functional role of self-exploitation of the working population for contemporary capital accumulation.

[3]Bernstein (2006, 456–457) himself takes a more ambivalent stance, arguing that contemporary political struggles of classes of labour – many of which centre around demands for land – are best understood as disparate and context-specific responses to their crises of reproduction, rather than earlier systematic class struggles.

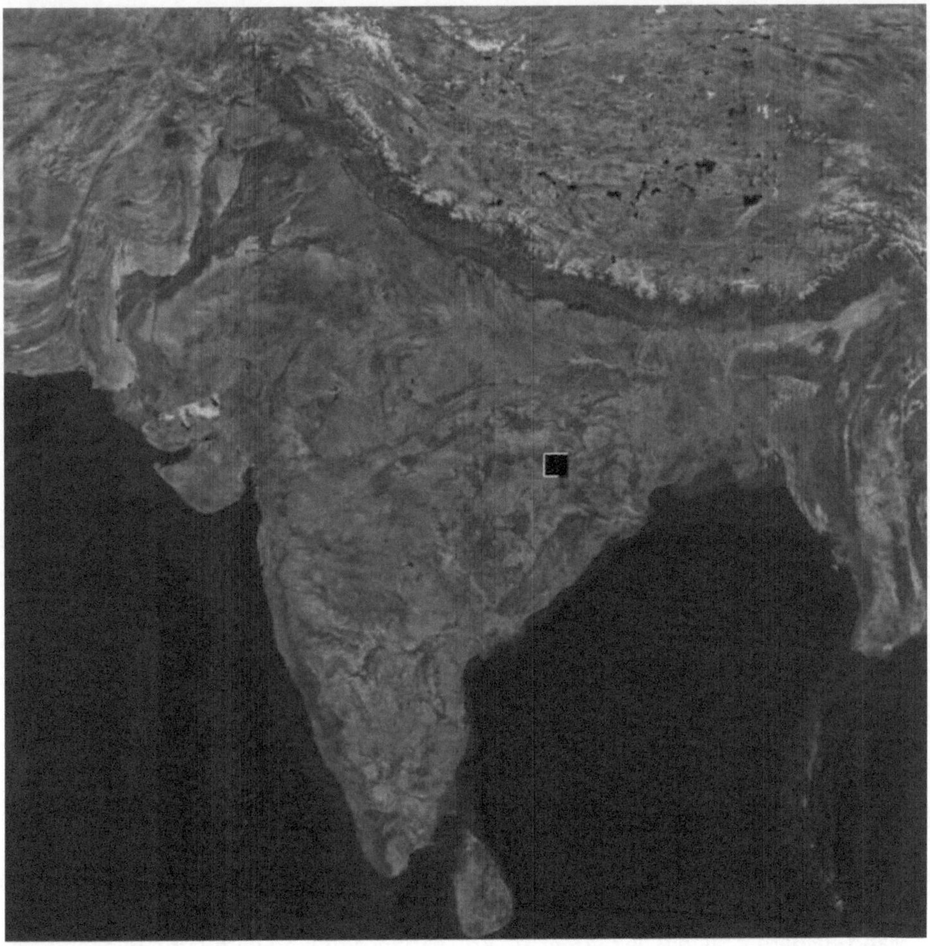

Figure 1. Geographical location of Tamnar and Mand-Raigarh coal field. Source: Coal Mine Surveillance & Management System, National Center of Geo-Informatics (NCoG), https://ncog.gov.in/CMSS/guest/guestGisDashboard.

labour which does not guarantee their survival. Smallholding agriculture and other land-based occupations are thus essential for their livelihood security and simple reproduction. In both countries, (semi)proletarian households have strongly resisted efforts to expropriate and dispossess them from their landholdings (Jacobs 2018, 895; Zhan and Scully 2018, 1031–2). Such resistance calls attention to the continuing importance of land and agrarian production within the otherwise fragmented reproduction strategies of many households engaged in wage labour, and the proliferation of anti-dispossession movements motivated by communities' 'defence of the semiproletarian condition' (Levien, Watts, and Hairong 2018, 869).[4] It suggests that land and labour may both serve as crucial and complementary loci of rural political mobilization. The focus then shifts to the diverse ways in which they

[4]Both authors do, however, caution that such resistance rarely indicates a conscious effort to withdraw from or reduce workers' dependence on wage labour, nor does it highlight a broader trend towards 're-peasantization'. See, for instance, Moyo and Yeros (2005) for this latter perspective.

are actually combined in particular local contexts, and on how this can influence the motivations and character of the collective struggles that emerge.

Such perspectives also highlight the importance of land beyond its role in alleviating wage labouring households' crises of simple reproduction, the predominant focus of Bernstein's classes of labour. Dependence on land represents more than just a survival strategy brought about by low wages and insecure wage employment. Jacobs (2018, 893) points out that deep historical ties to agriculture and consciousness of land rights have been a key motivation for residents of Zabalaza in South Africa to occupy urban land for farming and raising livestock. For many wage working populations, land continues to provide the primary basis for reproducing the wider social relations and identities that underpin production processes (Borras et al. 2022; Cousins et al. 2018). This broader understanding of social reproduction processes also suggests that households depend on 'a range of access to a range of land' to facilitate their engagement in (agricultural) production and wage labour (Borras et al. 2022).

Key focus areas for unpacking the 'agrarian' context underlying political contestations over coal mining thus include the dependence on diverse combinations of agricultural production *and* wage labour within diversifying and fragmented rural livelihoods, and the incorporation of commodified *and* non-commodified labour and land within household social reproduction strategies. The following sections examine a specific case of local political contestations over coal mining in the Tamnar region of India to elaborate on how such an approach can be operationalized, and what it can tell us about the intermingling of 'agrarian' and 'climate' concerns in anti-coal struggles.

4. Introducing the case: anti-coal struggles in Tamnar

The empirical focus of this paper is on local opposition to coal mining in Tamnar, an administrative block of Raigarh district in Chhattisgarh, India (see Figure 1). Tamnar lies within the vast Mand-Raigarh coalfield, a 3445.77 sq. km. expanse of land with large mineable coal reserves. By one 2014 estimate, only about 1.1% of the coalfield's total land area has already been directly affected by mining activity (CMPDI 2014). However, Tamnar and other relatively new regions of coal extraction have become increasingly important to India's coal economy, particularly as attention shifts away from more established and heavily mined areas (GSI 2021).

The Mand-Raigarh coalfield is further sub-divided into 72 'coal blocks' earmarked for development as independent mining projects, of which mining activity has so far commenced for 11 projects (CMPDI 2022). The present paper relies primarily on extensive field research conducted in two villages situated adjacent to one such currently operational coal mine. This included 70 qualitative semi-structured interviews with village residents and detailed notes from 10 local meetings during the period from September 2019 to March 2020. These findings have been supplemented by data from an earlier land dispossession study conducted in the same villages in February 2012, and from interviews with local movement participants and other actors conducted during shorter visits to the region, in July 2016 and April 2019.

The two study villages are among eight surrounding villages whose lands have been partially dispossessed by the same mining project. This mine was originally allocated by the government to a private thermal power producer, which commenced mining operations here in 2006. Prior to its establishment, village residents received very limited information about the project, so that the arrival of the mine and initial land acquisition processes were a surprise for many. This period saw limited and mostly individualized resistance from land-losing households. However, organized community opposition in the two study villages has emerged gradually and strengthened, as the mine's cumulative impacts on local livelihoods and the environment have amplified.

By the time of a large subsequent round of land acquisition in 2010, there were sustained protests against the substantial irregularities in land acquisition procedures by the mining company, including the falsification of mandatory consent for the project from village *gram sabhas* (local governance institutions), illegal transfers of indigenous *Adivasi* land, and coercion and intimidation of landholders.[5] Residents also began to mobilize against the mine's negative impacts on the local environment, in particular high levels of air and water pollution, declining groundwater levels, damage to village habitations due to mine blasting, and loss of access to village commons and forests. These impacts were particularly severe in the two study villages, which saw the expansion of mining activities closer to inhabited areas of the villages.[6] Remediation and compensation for these impacts has remained a key and consistent demand of local protests in subsequent years. Such efforts also received an unexpected boost in 2014, when the Indian Supreme Court ordered the cancellation of 214 coal blocks nationwide due to illegalities in their allocation processes. Since then, the mine has been operating under the temporary custodianship of a state-owned coal mining corporation, which has limited further expansion of mining and land acquisition.

Demands for greater employment in the project have been another crucial focus of local struggles. State-level rehabilitation policies in Chhattisgarh mandate the provision of at least one permanent job for each family whose land is acquired but, in practice, these job opportunities were almost exclusively provided to large landholders. In 2016, after a sustained blockade of the mine, there was a large increase in the number of jobs made available to affected households. At least one member of most land-losing households in the two study villages currently has an informal 'contract' job in the mine. These jobs are poorly paid and highly insecure, but provide an important support for household incomes. The risk that renewed protests will stall production has also been a strong motivation for the mine operator to ensure the continued availability of informal jobs.

Such on-ground mobilizations have been complemented with an increasing reliance on the legal system. Direct legal challenges to mining-related land acquisition processes are difficult, since India's eminent domain laws allow for the compulsory acquisition of land for any loosely defined 'public purpose'. In response, residents of the study villages have increasingly opposed the expansion of the mine through more progressive environmental protection laws. In 2013, they approached the National Green Tribunal (NGT) which has legal jurisdiction over cases related to environmental protection. This first complaint demanded action against the company for illegal diversion of forest land for mining and mining-related pollution in the study villages. A second NGT case, filed in 2018, called for action against environmental violations and widespread pollution due of poorly regulated

[5]Shrivastava, Gupta Bhaya, and Worsdel (2020) highlight that there have been similarly widespread violations and subversion of consent provisions for large-scale land acquisitions across India.

[6]This was an outcome of the private mine operator's plans to not relocate inhabited areas of the affected villages, but to limit open-cast mining to agricultural and common lands and forests.

coal mining and industrial activity in the Tamnar region. In both cases, the court has passed favourable interim rulings requiring the state to actively intervene and redress these violations. They represent important legal victories for affected communities' efforts to limit unchecked land acquisition for coal extraction and its substantial socio-environmental costs.

Within the evolving and multi-faceted political dynamics of organized resistance to coal mining in the study villages, preventing the further loss of land for mining activities, expanding opportunities for wage work in the mine, and protecting the local environment thus serve as crucial – and seemingly contradictory – motivations for community mobilization. Local anti-coal struggles were unsuccessful in preventing the establishment of the mine, but they have been instrumental for the uneasy present status quo in which the project operates with substantial restrictions on its continued expansion and a growing recognition of the need to mitigate its negative impacts. At the same time, there remain significant challenges for a complete closure of the mine, not just at the level of law and state policy, but also in terms of the contraction of local wage employment opportunities this will likely cause. To better understand how these goals and demands have emerged, and how they link to a broadened conceptualization of climate justice struggles, the following section examines the changing dynamics of land, labour and social reproduction in the two study villages.

5. Coal mining and agrarian change: land, labour and social reproduction in mining-affected communities

Land dispossession for mining has functioned as a large external shock to existing, predominantly agrarian livelihoods in the study villages. A comparison of village-level population Census data from 2001 (prior to establishment of the mine) and 2011 (soon after two large rounds of land acquisition) highlights the scale of this disruption (see Table 1). Over this period, the share of the working population in the study villages reported to be engaged in agricultural cultivation declined from 75.7% to 47.3%, while that of agricultural labourers rose from 14.4% to 41%. Total village households also declined by almost a tenth due to (partial or complete) migration of land-losing households.

Census findings are instructive but present a limited picture of the changing livelihood dynamics in the study villages. They neglect the gradual pace of mining-related land dispossession for many households. Formally acquired land has often remained unutilized for many years, during which time land losers are able to continue cultivating it. As of December 2016, only about half of the mine's total area had already been excavated or utilized for project-related infrastructure. The large reported increase in proportion of agricultural labourers also almost entirely comprises the Census sub-category of 'marginal workers',

Table 1. Change in sources of work in study villages, 2001 and 2011.

	Share of total workers			
Year	Cultivators	Agricultural labourers	Household industry workers	Other workers
2001	75.7%	14.4%	5.3%	4.6%
2011	47.3%	41.0%	0.0%	11.7%

Source: Census of India 2001 and 2011.

suggesting a high likelihood of their engagement in other non-agricultural work. Perhaps most crucially, household livelihoods and reproduction strategies have reoriented and stabilized over time, as they have responded to mining-related land dispossession and collectively mobilized against its impacts. For instance, many have utilized the compensation from land acquisition to purchase agricultural land in nearby villages, cultivating it alongside their remaining landholdings in the study villages. This has facilitated a continued engagement in agricultural cultivation despite the loss of land to the mine.

It is extremely important to recognize that, as with rural dispossession for mining and industry in other settings, the impacts of the coal mine in the study villages have been diverse and unequal, based on multiple overlapping and intersecting axes of social differentiation. Nevertheless, organized resistance to the project has gradually coalesced around particular demands for land, wage labour and environmental protection. This evolving political consensus among affected households reflects specificities in how mining-related dispossession has transformed the local agrarian context. The discussion that follows presents wider research findings and vignettes summarizing the experiences of a few households to elaborate on these differentiated but patterned impacts of mining.

5.1. Agrarian petty commodity production and accumulation

Agriculture was central to household livelihoods and reproduction in the study villages prior to the establishment of the mine, though the nature of this dependence varied widely. Among indigenous *Adivasi* households, which comprised more than 70% of the study village population, a small proportion were large landholding households engaged in petty commodity production, predominantly of rice. Most *Adivasi* households however had small to medium landholdings, and generally combined cultivation with the seasonal hiring out of their labour. There were relatively few landless or marginal landholders that relied exclusively on agricultural wage work. Dominant caste farming communities, officially designated as Other Backward Classes (OBCs), were the other major population group in the study villages. The large majority of these households were petty commodity producers with medium to large landholdings. *Dalits* and so-called 'upper caste' communities made up a very small share of the population. Mining- related land dispossession has not led to major shifts in the social composition of the study villages. However, differences in initial landholdings are an important determinant of how households have navigated its impacts.

Most large landholding households – both *Adivasis* and OBCs – were able to use compensation from land acquisition for the mine to buy additional agricultural land in nearby villages, often ones where they have existing familial ties. Some were also co-opted through higher compensation levels, or benefited from rent-seeking opportunities as brokers and aggregators during the land acquisition process. As a result, most of them have not experienced a substantial reduction in overall landholdings. Nevertheless, the fragmentation of landholdings has led many households to either split across different villages (often one adult son moving with their immediate family) or to lease out part of their land instead of cultivating it themselves. Many also report significantly lower agricultural productivity on their remaining land in the study villages, linked to high levels of air and water pollution from mining activity. These households thus face growing constraints on their ability to accumulate and expand through agrarian petty commodity production.

> The household continues to cultivate about 7–8 acres of land in the village and another 9 acres of land bought in another village 10 kms away. They rely mainly on family labour, with some additional labour hired in from the village during sowing and harvesting periods. However, output from remaining land in the study village is about half of the other landholdings. In the previous year they earned about Rs. 4,00,000 (USD 5300) in total, after accounting for all expenses. (Interview 9, 15 October 2019 and Interview 41, 1 November 2019)

The vast majority of *Adivasi* households with small- and medium-sized landholdings have experienced very significant disruptions to their existing agrarian livelihoods. Most were able to purchase some agricultural land in other villages but given the low levels of compensation, their total landholdings have generally fallen. Earlier strategies of household reproduction focused on predominantly subsistence-oriented production of their own landholdings and limited hiring out labour to larger landholders have also become increasingly untenable given the decreasing landholding sizes in the study villages. Responses vary widely within this broad strata of affected households, but tend to involve a combination of more intensive cultivation of their remaining landholdings to produce a marketable surplus, and a growing reliance on non-agricultural wage employment.

> The household cultivates about 4 acres of land in the village, split between rice paddy and lentils and oilseeds (they also own 2.5 acres of land in a nearby village). They rely on family labour for cultivation. A tractor has to be hired in for ploughing rice fields, seeds and pesticides are purchased from nearby private shops, while fertilizer can be procured on loan from state-owned providers. They are able to sell a part of their produce, but also depend on income from their son's job in the mine. (Interview 6, 13 October 2019)

These evolving dynamics of agricultural production due to coal mining in the study villages are reinforced by broader structural processes of agrarian change. In recent years, Chhattisgarh and its neighbouring states have emerged as priority areas for a 'Second Green Revolution in Eastern India' (Bhatt et al. 2016), where relatively low levels of agricultural productivity and favourable agro-ecological conditions are seen to provide vital opportunities for agricultural development. Towards this end, there has been significant expansion of state support for developing irrigation infrastructure, provisioning improved high-yielding seeds and subsidised fertilizers, and improving agrarian extension services and marketing

Table 2. Indicators of agricultural cropping intensity, input use and credit access for Raigarh district, 2006–07 and 2016–17.

Year	Share of net cropped area cultivated more than once each year	Share of gross cropped area treated with fertilizers	Share of operational holdings with institutional credit
2006–07	5.2%	40.2%	3.4%
2016–17	13.9%	93.9%	44.9%

Source: Tables 1B, 2A and 2B, Input Survey, Agricultural Census, https://inputsurvey.dacnet.nic.in/districttables.aspx.

infrastructure. Crucially, public procurement of rice at state-mandated minimum support prices has also steadily increased (Varma 2018).

Such state interventions have led to a significant acceleration in the processes of agricultural commercialization and intensification. Chhattisgarh's production of rice paddy rose by 60% between 2005–06 and 2016–17 (Varma 2018). Agricultural Census data for Raigarh district highlights how rural producers have rapidly moved away from traditional rice varieties to cultivate input-intensive and high-yielding varieties which are procured through the public system. Levels of agricultural mechanization and reliance on institutional credit have also risen (see Table 2).

Agrarian livelihoods in the study villages are thus characterized by the accelerating development of productive forces and market linkages. These processes have facilitated a deepening of petty commodity relations, including for households previously engaged in subsistence-oriented agriculture. However, losses of cultivable land due to mining and the fragmented nature of household landholdings mean that avenues for agrarian accumulation and further social differentiation remain highly limited. In this sense, mining-affected households in the study villages share a strong common interest in retaining and cultivating their remaining landholdings.

5.2. Wage labour in rural livelihoods

While a continued engagement in agriculture provides vital support for household incomes in the study villages, very few households can rely solely on the cultivation of their remaining landholdings for their reproduction. Most large landholders were able to secure a 'permanent' job in the mining company for one male household member along with monetary compensation for their acquired land. Other family members remain substantively engaged in the cultivation of their remaining landholdings. The combination of better paid, regular wage employment in the mine and agricultural petty commodity production is able to ensure sufficient means for the reproduction of these households. They do remain dependent on the hiring in of some agricultural labour, but the requirement for wage labour is limited by the reduced size of landholdings in the study villages.

> The household retains 7.5 acres of land in the one of the study villages. They cultivate rice paddy on their landholdings, relying predominantly on labour from immediate and extended family members. During peak periods of sowing and harvesting (approximately one month in the year), they also have to hire in labour. (Interview 36, 29 October 2019)

The vast majority of land losing households, however, have secured wage employment in the mining project in the form of informal 'contract' jobs. About 60 men from the study villages are currently employed for tasks like pump operation, blasting, fire control, surveying and sweeping. Each worker only gets 16–18 days of work per month, and labour contractors that employ them are frequently changed. The mining company has also subcontracted actual coal excavation and transportation activities that require more skilled workers like machinery operators and drivers to smaller companies, which rarely hire local residents. Most small to medium landholding households therefore combine poorly paid, insecure wage employment in the mine with the intensified cultivation of their remaining land and, in some instances, land hired in from larger landholders.

> The household is presently able to cultivate about 2.5 acres of their remaining landholdings. They cultivate the land themselves, without the need to hire in labour. They are able to sell a part of their produce, but also need to rely on income from the younger son's job as a contract worker in the mine. (Interview 3,12 October 2019)

The loss of agricultural land for mining in the study villages substantially limits the demand for agricultural labour, especially outside of seasonal periods of peak cultivation, as most households are able to cultivate their limited landholdings with family labour or kinship-based arrangements. The small proportion[7] of landless and marginal landholding *Adivasi* households in the study villages have borne the brunt of this contraction. Very few of them have been able to secure even informal jobs in the mining project, limiting their ability to diversify livelihoods away from agricultural wage work. These households therefore continue to rely primarily on hiring out their labour to large cultivators in nearby villages, along with other occupations like basket weaving and iron working, and income from rural works programmes.

> The household is landless, and depends on income from making iron farm implements and hiring out agricultural labour. They earn about Rs. 120 (USD 1.5) for a day's agricultural work. However, since people have less land now, such work is harder to come by in the village. (Interview 68, 15 November 2019)

Mining-related land dispossession has thus led to a growing dependence on (non-agricultural) wage labour, even as sources of wage employment and the extent of households' dependence on them vary widely. However, with few exceptions, wage labour is best understood as an important but far from sufficient component of affected households' increasingly fragmented livelihood and reproduction strategies, which tend to involve diverse combinations of the cultivation of remaining landholdings, agricultural wage labour, and insecure wage employment in the mining project.

5.3. Non-commodified bases of social reproduction

Key aspects of rural households' social reproduction are directly linked to or otherwise enabled by their ability to access different types of land, including cultivable agricultural land, village commons and pastures, and forests (Cousins, Forthcoming, 5). As a result, land losses for the mining project, besides altering local dynamics of agricultural production and wage labour, have significantly impacted existing social reproduction strategies in the study villages.

Households report a large reduction in the number of cattle they raise because there is little land available nearby to graze them. The destruction of surrounding forests due to mining activities also means that villagers have to commute much longer distances to

[7]It is possible that some landless and marginal households migrated out from the village post-land acquisition and were not included in the present analysis.

collect forest products. Such shifts have a direct impact on the levels and diversity of domestic consumption, as well as on household incomes. For instance, the collection and sale of *tendu* leaves at government mandated prices previously provided a crucial source of cash income, especially for landless and small and medium landholding *Adivasi* households, but has become increasingly scarce with the expansion of the mine.

> The household was earlier able to collect large amounts of produce from dense forests nearby – particularly *tendu* leaves and *char-chironji* seeds – and from *mahua* and mango trees growing on their own land. These were sold to local traders and government agencies, providing significantly higher monetary income than rice cultivation. The loss of forests and trees, coupled with higher pollution, means that this income is very limited now. (Interview 16, 18 October 2019)

Households must therefore increasingly rely on income from agricultural production and wage labour for their consumption needs and survival. *Adivasi* and OBC households with large landholdings and stable employment have been better able to manage this shift. For most other households, however, challenges from limited access to agricultural land and wage labour are intensified by the growing integration of their social reproduction within commodity relations. Women in these households also increasingly engage in income-generating agricultural production and wage labour, alongside unpaid care and reproductive work, to supplement male family members' incomes from informal employment in the mining project.

Nevertheless, substantial elements of household social reproduction in the study villages have remained outside the purview of commodified land and labour relations. With very few exceptions, land-losing households have been able to continue cultivating small amounts of rice and other staple cereals, pulses, and vegetables for their own consumption. This, coupled with the support provided by a relatively well-functioning Public Distribution System (PDS) for rice and a few essential commodities at highly subsidised prices, limits the necessity of securing crucial consumption and nutritional needs through the market. Rural works undertaken under the Mahatma Gandhi National Rural Employment Guarantee Act (MGNREGA) also provide crucial supplementary income, particularly for the landless and marginal landholders during lean agricultural periods. In addition, Corporate Social Responsibility (CSR) funds from the mining project have been mobilized for some limited public services – mainly common water supply and lighting – and small contributions to village activities. Daily and intergenerational social reproduction in the study villages thus retains a substantial non-commodified component, the preservation of which is a shared interest of mining-affected households.

5.4. Reconciling demands for 'agrarian' and climate' justice in anti-coal struggles

The above discussion highlights the key concerns that motivate political contestations over coal mining in the study villages. Mining-related dispossession has significantly restricted the extent to which agriculture – whether through direct cultivation or hiring out of labour – can function as a viable basis for household subsistence and reproduction. However, insecure and poorly paid wage employment in the mine also provides only limited support to household incomes. As a result, most households face varying degrees of a reproduction squeeze, which has to be resolved by combining the intensive cultivation of limited agricultural landholdings, engagement in agricultural and non-agricultural wage labour, and non-commodified forms of land and labour in diverse ways.

Anti-coal struggles in the study villages reflect the evolution of a broad-based political consensus on how to collectively navigate this shifting and increasingly challenging 'agrarian' context of household livelihoods and social reproduction. This includes a recognition of the urgent need of prevent continued losses of agricultural and non-agricultural land, while also pushing for the expansion of wage labour opportunities in the mining project, even if these are insecure and poorly paid. Such ongoing struggles for land and labour may lack the emblematic appeal of anti-dispossession movements grounded in the 'defence of land', but they reflect the reality of many mining projects where strong and sustained local opposition extends well beyond the initial point of land enclosure.

Struggles for socially just and equitable outcomes from coal mining also align closely with more radical understandings of climate justice, even if climate change-related concerns are often not explicitly articulated in their demands and strategies. Residents of the study villages have consistently mobilized against widespread degradation of the local environment due to mining activity, including by partnering with civil society organizations to conduct multiple impact studies that document the extremely high levels of air, water and soil pollution, and related health complaints among affected communities. These studies have also been presented as evidence in their ongoing legal cases before the NGT. Besides imposing multiple financial penalties on the mining company and local government departments for violating environmental norms, the tribunal has set up an independent expert committee to monitor a range of remedial measures. These include the appropriate disposal of fly ash from nearby thermal power plants and industries, restrictions on road transportation of coal by trucks, ensuring drinking water and healthcare facilities for residents of mining-affected villages, and stringent monitoring of air and water pollution levels.

Such efforts have effectively prevented further expansion of the coal mine's production capacity and significantly raised the costs of environmental compliance and remediation. The project thus remains active, but with substantial constraints on its continued operation and economic viability. Such outcomes also become particularly relevant in terms of the possibilities they offer to undermine the rapid ongoing commercialization of coal mining in India, predicated on high private profits through the low cost exploitation of a seemingly abundant resource. There are thus important linkages between the diverse motivations of local anti-coal struggles and calls within radical climate justice movements to confront more systemic dynamics of capitalist accumulation and exploitation that underpin the climate crisis.

6. Mobilizing for 'just transitions' in coal mining regions

Residents of the two study villages that form the empirical basis of this paper are part of a large population spread across numerous coal-rich regions of India. The diverse political economies of coal extraction that characterize these different regions of coal extraction play a crucial role in shaping the impacts of mining-related dispossession and political contestations over them. However, the present analysis also points to areas where the varied rural contexts of anti-coal struggles converge, and the possibilities these may offer for building broader counter-hegemonic movements that effectively counter dominant understandings of 'just transitions' for mining-affected communities.

There are differences in the extent of reliance on wage labour in coal mining projects between regions like Tamnar and older, more established regions where substantial coal extraction and a significant transition away from agrarian livelihoods have already occurred. In these latter regions, affected households predominantly reproduce through a range of direct and indirect forms of wage employment in mining (Dsouza and Singhal 2021; Montrone, Ohlendorf, and Chandra 2021). However, processes of subcontracting and informalization of the workforce have been a widespread and persistent feature of the post-1990s liberalization of India's coal mining sector (Nayak 2022; Roy 2003). These have led to the proliferation of informal and highly precarious wage work within contemporary coal economies, similar to the experience in the study villages. This points to the need for wider mobilizations for 'just transitions' from coal to incorporate a coherent critique of popular claims that such extractive projects ensure decent employment for affected communities. Crucially, such a critique also has to be linked to the systemic dynamics of coal mining, so that demands for employment *outside* mining – rather than the expansion of work within the sector – emerge as a central focus of political struggle, particularly in newly emerging mining regions.

A second important area of convergence relates to political contestations over the impacts of mining on the local environment. Anti-coal struggles in the study villages demonstrate how a combination of sustained on-ground protests and legal mobilizations focused on progressive environmental protection laws can limit mining activity and the economic viability of projects. Cumulatively, such efforts offer the possibility to significant undermine the dynamics of profitmaking and accumulation that are central to coal extraction and its detrimental climate impacts. Integrating more localized environmental concerns which often motivate such mobilizations – for instance, around air and water pollution, destruction of forests and commons, decreasing groundwater levels, and rising disease burdens – into broader discussions of how the climate crisis and its related politics impact rural areas is a crucial task for counter-hegemonic movements against coal extraction. A related, and more practical, consideration concerns the substantial legal and technical expertise that local struggles need to successfully monitor and challenge violations of environmental laws and regulations by mining projects (Oskarsson and Bedi 2018).

Finally, the broader agrarian context underlying political contestations over coal extraction also points to possible linkages between contemporary mining and agrarian struggles in India. In the study villages, expanding state support for agriculture has been crucial for enabling a vital role for petty commodity production within mining-affected households' increasingly fragmented livelihood strategies. Similar dynamics of state-driven agricultural intensification and commercialization can be seen across many coal mining regions of Central-Eastern and Eastern India. While discussions on recent protests against the proposed liberalization and corporatization of India's agricultural sector have often focused on the role of farmers' movements from former Green Revolution regions like Punjab and Western Uttar Pradesh, the experience of mining-affected communities in Tamnar suggests that demands for sustained and reliable state support in agriculture may provide a common basis for mobilizing across the country's diverse agrarian contexts. At the same time, broadening the scope of such agrarian movements to engage more directly with the demands of landless workers and smallholding producers is a shared concern, and challenge, both for struggles seeking progressive 'just transitions' from coal extraction and radical agrarian justice struggles in non-mining regions.

Acknowledgements

I am grateful to the forum editors, participants of the 3rd Annual Writeshop in Critical Agrarian Studies and Scholar-Activism, and two anonymous reviewers for their patient support and detailed comments on the paper. Thanks especially to Jun Borras, Julien-François Gerber, Amrita Chhachhi, Shapan Adnan, Nadya Karimasari, Atakan Buke, Suravee Nayak, Babur Hussain, Bruna Figueiredo Gonçalves, Vikas Dubey, Bhoopendra Kumar, Yukari Sekine and Daniela Calmon who provided extremely helpful feedback on earlier drafts. This paper would not have been possible without the guidance and support of many friends in Tamnar who remain anonymous in the text.

Disclosure statement

No potential conflict of interest was reported by the author(s).

ORCID

Amod Shah http://orcid.org/0000-0002-2127-7071

References

Abnett, K., and E. Piper. 2021. "COP26 Coal Pledge Falls Short on Support as Emissions Surge." *Reuters,* November 4. https://www.reuters.com/business/cop/coal-crosshairs-un-climate-summit-stirs-clean-energy-pledges-2021-11-04/.

Adhikari, A, and V. Chhotray. 2020. "The Political Construction of Extractive Regimes in Two Newly Created Indian States: A Comparative Analysis of Jharkhand and Chhattisgarh." *Development and Change* 51 (3): 843–873.

Aggarwal, M. 2021a. "India's Mining Reforms Juggernaut Continues Despite Concerns." *Mongabay,* February 23. https://india.mongabay.com/2021/02/indias-mining-reforms-juggernaut-continues-despite-concerns/.

Aggarwal, M. 2021b. "India to Seek Billion-dollar Assistance from the World Bank for Just Transition Programme." *Mongabay,* November 3. https://india.mongabay.com/2021/11/india-to-seek-billion-dollar-assistance-from-the-world-bank-for-just-transition-programme/.

Akram-Lodhi, A. H., and C. Kay. 2009. *Peasants and Globalization: Political Economy, Rural Transformation and the Agrarian Question.* London; New York: Routledge.

Bebbington, A., A. Abdulai, D. H. Bebbington, M. Hinfelaar, and C. A. Sanborn. 2018. *Governing Extractive Industries: Politics, Histories, Ideas.* Oxford: Oxford University Press.

Bebbington, A., D. H. Bebbington, J. Bury, J. Lingan, J. P. Munoz, and M. Scurrah. 2008. "Mining and Social Movements: Struggles Over Livelihood and Rural Territorial Development in the Andes." *World Development* 36 (12): 2888–2905.

Bernstein, H. 1996. "Agrarian Questions Then and Now." *The Journal of Peasant Studies* 24 (1–2): 2259.

Bernstein, H. 2006. "Is There an Agrarian Question in the 21st Century?" *Canadian Journal of Development Studies / Revue Canadienne D'études du Développement* 27 (4): 449–460.

Bhatt, B. P., J. S. Mishra, A. Dey, A. K. Singh, and S. Kumar. 2016. *Second Green Revolution in Eastern India: Issues and Initiatives.* Patna: ICAR Research Complex for Eastern Region.
Borras, S. M., Jr., J. C. Franco, D. Ra, T. Kramer, M. Kamoon, P. Phyu, K. K. Ju, et al. 2022. "Rurally Rooted Cross-Border Migrant Workers from Myanmar, Covid-19, and Agrarian Movements." *Agriculture and Human Values* 39: 315–338.
Borras Jr., S. M., I. Scoones, A. Baviskar, M. Edelman, N. L. Peluso, and W. Wolford. 2021. "Climate Change and Agrarian Struggles: An Invitation to Contribute to a JPS Forum." *The Journal of Peasant Studies* 49 (1): 1–28.
Calmon, D., C. Jacovetti, and M. Koné. 2021. "Agrarian Climate Justice as a Progressive Alternative to Climate Security: Mali at the Intersection of Natural Resource Conflicts." *Third World Quarterly* 42 (12): 2785–2803.
Carbon Brief. 2019. "The Carbon Brief Profile: India." *Carbon Brief,* March 14. https://www.carbonbrief.org/the-carbon-brief-profile-india.
Carbon Brief. 2020. "Mapped: The World's Coal Power Plants." *Carbon Brief,* March 26. https://www.carbonbrief.org/mapped-worlds-coal-power-plants.
CIF. 2021. "CIF Begins Historic $2.5B Coal Transition Pilot in Four Developing Countries." *Climate Investment Funds Press Release,* November 4. Accessed 29 April 2022. https://www.climateinvestmentfunds.org/news/cif-begins-historic-25b-coal-transition-pilot-four-developing-countries.
Conde, M. 2017. "Resistance to Mining. A Review." *Ecological Economics* 132: 80–90.
Conde, M., and P. Le Billon. 2017. "Why do Some Communities Resist Mining Projects While Others do Not?" *The Extractive Industries and Society* 4: 681–697.
Cousins, B., A. Dubb, D. Hornby, and F. Mtero. 2018. "Social Reproduction of 'Classes of Labour' in the Rural Areas of South Africa: Contradictions and Contestations." *The Journal of Peasant Studies* 45 (5-6): 1060–1085.
Cousins, B. Forthcoming. "Land, Social Reproduction and Agrarian Change." In *Oxford Handbook of Land,* edited by S. M. Borras Jr., and J. C. Franco. Oxford: Oxford University Press.
CMPDI. 2014. *Land Use / Vegetation Cover Mapping of Mand Raigarh Coalfield Based on Satellite Data for the Year 2014.* December 2014, Remote Sensing Cell Geomatics Division CMPDI (HQ), Ranchi.
CMPDI. 2022. "Online Coal Block Information System." *CMPDI.* Accessed April 29, 2022. https://www.cmpdi.co.in/OCBIS/dashboard.php.
de Hoog, N., and A. Kirk. 2021. "Current Coal Phaseout Pledges 'Absolutely Not Enough', Warn Experts." *The Guardian,* December 23. https://www.theguardian.com/environment/ng- interactive/2021/dec/23/why-cop26-coal-power-pledges-dont-go-far-enough-visualised.
Dietz, K., and B. Engels. 2017. "Contested Extractivism, Society and the State: An Introduction." In *Contested Extractivism, Society and the State: Struggles Over Mining and Land,* edited by B. Engels, and K. Dietz, 1–20. London: Palgrave Macmillan.
Dsouza, S., and K. Singhal. 2021. *Socio-economic Impacts of Coal Transitions in India: Bottom-up Analysis of Jobs in Coal and Coal-Consuming Industries.* New Delhi: National Foundation of India.
Franco, J. C, and S. M. Borras Jr. 2019. "Grey Areas in Green Grabbing: Subtle and Indirect Interconnections Between Climate Change Politics and Land Grabs and Their Implications for Research." *Land Use Policy* 84: 192–199.
Fraser, N. 2021. "Climates of Capital: For a Trans-Environmental Eco-Socialism." *New Left Review* 127: 94–127.
Goodman, J., L. Connor, D. Ghosh, K. Kohli, J. P. Marshall, M. Menon, K. Mueller, et al. 2020. *Beyond the Coal Rush: A Turning Point for Global Energy and Climate Policy?* Cambridge: Cambridge University Press.
GSI. 2021. Inventory of Coal Resources of India (as on 1st April 2021). *Geological Survey of India.* Accessed August 25, 2022.
Guha, R, and J. Martinez-Alier. 1997. *Varieties of Environmentalism: Essays North and South.* London: Earthscan.
IEA. 2019. *Global Energy & CO2 Status Report 2019.* International Energy Agency. Accessed April 29, 2022. https://www.iea.org/reports/global-energy-co2-status-report-2019.

Jacobs, R. 2018. "An Urban Proletariat with Peasant Characteristics: Land Occupations and Livestock Raising in South Africa." *The Journal of Peasant Studies* 45 (5-6): 884–903.

Jagannath, J. 2020. "Coal Sector Reforms Will Make Eastern and Central India Pillars of Development: PM Modi." *Mint,* June 18. https://www.livemint.com/news/india/coal-sector-reforms-will-make-eastern-and-central-india-pillars-of-development-pm-modi-11592482554129.html.

Lerche, J. 2010. "From 'Rural Labour' to 'Classes of Labour': Class Fragmentation, Caste and Class Struggle at the Bottom of the Indian Labour Hierarchy." In *The Comparative Political Economy of Development Africa and South Asia,* edited by Barbara Harriss-White, and Judith Heyer, 6687. London: Routledge.

Levien, M, M. Watts, and Y. Hairong. 2018. "Agrarian Marxism." *The Journal of Peasant Studies* 45 (56): 853–883.

Martinez-Alier, J. 2002. *The Environmentalism of the Poor: A Study of Ecological Conflicts and Valuation.* Cheltenham, UK: Edward Elgar.

Ministry of Coal. 2022. *Coal Production.* Lok Sabha Starred Question No. 490, Answered on: 06.04.2022. Ministry of Coal, Government of India.

Montrone, L, N. Ohlendorf, and R. Chandra. 2021. "The Political Economy of Coal in India – Evidence from Expert Interviews." *Energy for Sustainable Development* 61: 230–240.

Moyo, S., and P. Yeros. 2005. *Reclaiming the Land: The Resurgence of Rural Social Movements in Africa, Asia and Latin America.* London: Zed Books.

Nayak, S. 2022. "Migrant Workers in the Coal Mines of India: Precarity, Resilience and the Pandemic." *Social Change* 52 (2): 203–222.

Olivier, J. G. J, and J. A. H. W. Peters. 2020. *Trends in Global CO2 and Total Greenhouse Gas Emissions: 2020 report.* PBL Netherlands Environmental Assessment Agency, The Hague.

Oskarsson, P, and H. P. Bedi. 2018. "Extracting Environmental Justice: Countering Technical Renditions of Pollution in India's Coal Industry." *The Extractive Industries and Society* 5 (3): 340–347.

Panitch, L,, and C. Leys. 2001. "Preface." In *The Socialist Register 2001,* edited by L. Panitch, and C. Leys, vii–xi. London: Merlin Press.

Paprocki, K. 2021. "Primacy and Absence of Climate Change." *Himal Southasian.* https://www.himalmag.com/primacy-and-absence-of-climate-change-bangladesh-threatening-dystopias-excerpt-2021/.

Pattenden, J. 2018. "The Politics of Classes of Labour: Fragmentation, Reproduction Zones and Collective Action in Karnataka, India." *The Journal of Peasant Studies* 45 (5-6): 1039–1059.

Pattenden, J. 2016. *Labour, State and Society in Rural India: A Class-Relational Approach.* Manchester: Manchester University Press.

Permual, J. P. 2021. "Explained. Why is India's Coal Usage Under Scrutiny?" *The Hindu,* November 21. https://www.thehindu.com/sci-tech/energy-and-environment/explained-why-is-indias-coal-usage-under-scrutiny/article37604012.ece.

PIB. 2020. "Year End Review 2020 - Ministry of Coal." *Press Information Bureau, Government of India,* December 31. Accessed August 25, 2022. https://pib.gov.in/Pressreleaseshare.aspx7PRID = 1685058.

Prause, L., and P. Le Billon. 2021. "Struggles for Land: Comparing Resistance Movements Against Agro-Industrial and Mining Investment Projects." *The Journal of Peasant Studies* 48 (5): 1100–1123.

Rosewarne, S., J. Goodman, and R. Pearse. 2014. *Climate Action Upsurge. The Ethnography of Climate Movement Politics.* London: Routledge.

Roy, A. K. 2003. "Disinvestment and Outsourcing in Coal." *Economic & Political Weekly* 38 (49): 5153–5154.

Roy, A., B. Kuruvilla, and A. Bhardwaj. 2019. "Energy and Climate Change: A Just Transition for Indian Labour." In *India in a Warming World,* edited by Navroz K. Dubash, 284–300. New Delhi: Oxford University Press.

Schlosberg, J., and L. B. Collins. 2014. "From Environmental to Climate Justice: Climate Change and the Discourse of Environmental Justice." *WIRES Climate Change* 5 (3): 359–374.

Sekine, Y. 2021. "Emerging 'Agrarian Climate Justice' Struggles in Myanmar." *The Journal of Peasant Studies* 48 (3): 517–540.
Sethi, N. 2021. "India Saved Itself from Dubious Demands at the Glasgow Summit." *Mint,* November 17. https://www.livemint.com/opinion/online-views/india-saved-itself-from-dubious-demands-at-the-glasgow-summit-11637166857304.html.
Shivji, I. G. 2017. "The Concept of 'Working People'." *Agrarian South: Journal of Political Economy* 6 (1): 1–13.
Shrivastava, K. S., S. Gupta Bhaya, and T. Worsdel. 2020. "Subversion of the Due Process for Seeking Consent of Communities in Land Acquisition and the Resultant Land Conflicts." Paper prepared for presentation at the '2020 World Bank Conference on Land and Poverty', Washington DC, March 16–20.
Srivastava, R. 2022. "India's First Social Plan for Closed Coal Hubs Aims for 'Honourable' Lives." *Thomson Reuters Foundation,* March 7. https://news.trust.org/item/20220307103908-a5v9s/.
Still Burning. 2021. "Coal, Colonialism & Resistance: Centring Voices of Affected Communities and Warning of False Solutions." *Still Burning.* Accessed April 29, 2022. https://stillburning.net/wp-content/uploads/2021/03/Still_Burning_2021.pdf.
Temper, L„ S. Avila, D. Del Bene, J. Gobby, N. Kosoy, P. Le Billon, J. Martinez-Alier, et al. 2020. "Movements Shaping Climate Futures: A Systematic Mapping of Protests Against Fossil Fuel and Low-Carbon Energy Projects." *Environmental Research Letters* 15 (12): 123004.
UNFCCC. 2021. "End of Coal in Sight at COP26." *External Press Release,* November 4. Accessed April 29, 2022. https://unfccc.int/news/end-of-coal-in-sight-at-cop26.
Varma, S. 2018. "Why Are MP, Chhattisgarh Farmers Angry Despite High Production?" *Newsclick,* November 10. https://www.newsclick.in/why-are-mp-chhattisgarh-farmers-angry-despite-high-production.
World Bank. 2021. "For a Just Transition Away from Coal, People Must Be at the Center." *Feature Story,* November 3. Accessed October 24, 2022. https://www.worldbank.org/en/news/feature/2021/11/03/for-a-just-transition-away-from-coal-people-must-be-at-the-center.
Wright, E. O. 2019. *How to be an Anti-Capitalist in the Twenty-First Century.* New York: Verso Books.
Zhan, S., and B. Scully. 2018. "From South Africa to China: Land, Migrant Labor and the Semi Proletarian Thesis Revisited." *The Journal of Peasant Studies* 45 (5-6): 1018–1038.

OPEN ACCESS

Rescaling the land rush? Global political ecologies of land use and cover change in key scenario archetypes for achieving the 1.5°C Paris agreement target

Jevgeniy Bluwstein and Connor Cavanagh

ABSTRACT

Advancing future-oriented perspectives in political ecology and critical agrarian studies, this paper examines projected land use and land cover change (LULCC) dynamics in four 'archetypal' scenarios foregrounded by the IPCC for limiting global warming to 1.5°C by 2100. Focusing on the Global South, we explore how these archetypes project a radical reversal of historical LULCC and rural population trends, potentially implying a considerable rescaling of contemporary land rush dynamics. Taken together, land-based climate mitigation futures highlight risks related to the (re)production of relative surplus populations through processes of rural enclosure and accumulation by dispossession in the Global South.

1. Introduction

Readers of this journal will be familiar with the land rush that swept through many parts of the globe, and particularly the Global South, throughout the first two decades of this century. During the apex of the global land rush (2007–2014), an annual mean of around 8 million hectares (Mha) of land deals was recorded by the Land Matrix database, with nearly 80 percent of these transactions occurring in the Global South. In short, these and similar land rush dynamics have been the subject of a generative, and still-growing, literature at the intersection of political ecology and critical agrarian studies (Borras et al. 2011; Fairhead, Leach, and Scoones 2012; White et al. 2012; Dell'Angelo et al. 2017). Important questions remain, however, about how both historical and contemporary land rush processes will articulate with the drivers and impacts of global climate change in the future (Franco and Borras 2021; Liao et al. 2021), as well as with attempts to mitigate climate change in the agriculture, forestry, and land use (AFOLU) sector (Davis, Rulli, and D'Odorico 2015).

In this respect, it is notable that key scenario archetypes in the IPCC's (2018) influential *Special Report on Global Warming of 1.5 Degrees* (hereafter SR1.5) project land use and land cover change (LULCC) for cropland and pasture contraction – as well as forest cover and bioenergy cropland expansion – at rates of implementation similar to those observed within the aforementioned 'peak period' of the global land rush (2007–2014), albeit sustained throughout the remainder of the twenty-first century.[1] The sheer scale and rapidity of these projected changes in pasture, cropland, and forest land cover may thus provide us with a useful lens or prism through which to explore potential interrelations between climate change mitigation, land rush dynamics, and the risk of associated socio-environmental injustices across a range of future scenarios. Indeed, a growing body of literature demonstrates how land-based climate mitigation initiatives can entail adverse side effects or significant trade-offs between different 'sustainable development' objectives (Hasegawa et al. 2018; Doelman et al. 2019), and particularly so in low(er)-income countries with a high agricultural share of GDP or total employment (Hurlbert et al. 2019, 675). Especially severe trade-offs can occur, for instance, if land area requirements for afforestation or bioenergy carbon capture and storage (BECCS) initiatives are dramatically increased in lieu of aggressive near-term reductions in fossil fuel-related emissions, amounting to several million km^2 above 2020 reference levels in some scenarios (Doelman et al. 2020).

Exploring the implications of these 1.5 °C scenario archetypes – and in response to the *Journal of Peasant Studies'* call for critical research examining interrelations between climate and agrarian change – this paper seeks to deepen our understanding, in particular, of 'how climate change and the rural world intersect' (Borras et al. 2022a, 5) in the context of emergent land-based mitigation responses. Departing somewhat from the laudably incisive focus on contemporary and *historical* dynamics in both political ecology and critical agrarian studies, we explore how differential *future* intersections between 'climate change and the rural world' may vary across a range of archetypal mitigation scenarios. Although there is a significant knowledge base in sustainability science, Earth system science, and related fields appraising trade-offs between climate change mitigation and livelihood or development objectives across a range of future scenarios (e.g. Doelman et al. 2019, 2020; Henry et al. 2022), studies of the interface between climate and agrarian change have in some ways only just begun to emerge in earnest across the wider critical literature (see, for instance, Franco and Borras 2019; Borras, Franco, and Nam 2020; Franco and Borras 2021). Whilst there has now been more than a decade of empirical and historical studies of the asymmetrical impacts of land and resource acquisitions or 'green grabbing' for conservation, climate change mitigation, and other ecological restoration schemes (e.g. Kelly 2011; Fairhead, Leach, and Scoones 2012; Leach and Scoones 2015), only rarely have political ecologists and critical agrarian studies scholars examined the empirical substance of *future* climate change mitigation scenarios or projections in detail (for emerging exceptions, see *inter alia* Buck 2019;

[1]Land-based climate mitigation includes a mix of different mitigation strategies or 'integrated response options' (IPCC 2020) across the agriculture, forestry, and land use (AFOLU) sector. Climate change mitigation in this sector can be realized both by reducing or avoiding emissions from agriculture, as well as from catalysing forms of land use or cover change that are deemed to be 'desirable' from a mitigation perspective. Accordingly, relevant 'integrated response options' in the AFOLU sector often entail vast scales of projected reforestation or afforestation, as well as the widespread establishment of bioenergy carbon capture and storage (BECCS) plantations (see also Carton 2019; Carton et al. 2020).

Carton 2019). In other words, we suggest that there is significant potential for political ecologists and other critical scholars to generate more granular theorizations of how Blaikie and Brookfield's (1987, 17) famously 'shifting dialectic between society and land-based resources' may resolve differentially across a range of possible mitigation futures, and with what implications.

Empirically, this paper seeks to encourage further discussion and debate on these issues by drawing upon four scenario archetypes recently foregrounded by the IPCC for limiting global warming to 1.5 °C above pre-industrial averages (see IPCC 2018). Specifically, these archetypes include: i) the Low Energy Demand (LED) scenario (Grubler et al. 2018), ii) Shared Socioeconomic Pathway 1 (SSP1-19), iii) Shared Socioeconomic Pathway 2 (SSP2-19), and iv) Shared Socioeconomic Pathway 5 (SSP5-19).[2] Particularly the latter three pathways are 'archetypal' in the sense that they are distilled from a broader suite of approximately 90 scenarios for meeting the 1.5 °C target with limited or no overshoot, thus illustrating the 'variety of underlying assumptions and characteristics' that underpin the full range of mitigation trajectories in the IPCC's SR1.5 assessment (Rogelj et al. 2018b, 110). Importantly, each archetype envisions differential future pathways for urbanization, economic growth, socioeconomic inequality, international cooperation, and institutional or technological innovation. For instance, the LED scenario projects a 40 percent reduction of final global energy demand below 2020 levels by 2050, including a downscaling of industrial activity in the Global North by 42 percent (Grubler et al. 2018, 520). At the other end of the spectrum, SSP5-19 entails 'accelerated globalization' (O'Neill et al. 2017, 174), an increase in final global energy demand that exceeds 2010 reference levels by a factor of at least 1.4 (Rogelj et al. 2018a, 328), and the maintenance of significant rates of global compounding economic growth (Kriegler et al. 2017).

Methodologically, we assess the political-ecological implications of the four scenario archetypes by contextualizing projected land use and cover change (LULCC) vis-à-vis the observed history of LULCC in the second half of the twentieth century (1960–2000) (Lay et al. 2021; Winkler et al. 2021), as well as the scale and rate of transactions documented throughout the 2007–2014 'peak period' of the global land rush (Rulli, Saviori, and D'Odorico 2013; Liao et al. 2021). To meaningfully assess changes in available pasture, cropland, and forest land cover in the past and in the future, however, we also account for parallel 'megatrends' in population, economic development, and human settlement dynamics. In doing so, we extract related conceptual implications from our analysis for the present *Journal of Peasant Studies* forum, highlighting how these future projections threaten to articulate with concerns related to rural enclosures (White et al. 2012), primitive accumulation or 'accumulation by dispossession' (Kelly 2011; Fairhead, Leach, and Scoones 2012), and the production of relative surplus populations (Benanav 2014; Li 2017) across the four scenario archetypes.

The contributions of this approach are several. As Borras et al. (2022a, 16) emphasize in their introduction to this forum, the 'processes of enclosure and extraction that neoliberalism accelerated' throughout the last several decades 'have changed agrarian class dynamics [...] Today, there is a staggering rise in the number of people who originated from rural areas but are now partly or fully separated from their means of production

[2]For an overview of the Shared Socioeconomic Pathways (SSPs) and their corresponding 'narratives', see the associated special issue of *Global Environmental Change* (e.g. O'Neill et al. 2017).

and social reproduction'. Indeed, analyses of enclosure and subsequent partial or total dispossession in the land rush literature have deepened our understanding of the often formally-unstated or disavowed social, political, and economic consequences of large-scale land and resource acquisitions (see, especially, Wily 2012; Cotula 2013; Dell'Angelo et al. 2017). Crucially, our analysis furthers the existing scholarship on these themes by highlighting how 1.5 °C scenario archetypes project future LULCC to unfold in a radically different demographic context relative to the recent historical record, and particularly so in the Global South. Whereas cropland and pasture expansion unfolded over the last 60 years against a demographic background of a growing rural population in the Global South, the conversion of cropland and pastures to forests and bioenergy plantations in 1.5 °C scenario archetypes is projected to occur in the context of a rapidly shrinking rural population. Not least, then, we illuminate how this anticipated reversal of historical trends raises critical questions for political ecology and critical agrarian studies about, *inter alia*: 1) the conditions under which rural areas depopulate in diverse future scenarios; 2) the agrarian political economy of transforming rural smallholdings to both 'sustainably intensified' sites of agricultural production and forests or other plantation-based carbon sinks; and 3) the implications of transitioning rural areas from their historically latent function of 'warehousing' relative surplus populations (Li 2010; Benanav 2014) to a context in which cities and other urban areas are presumed to fulfil this role (see also Davis 2006; Harvey 2012).

In unpacking these contributions, the article proceeds as follows. First, we revisit seminal works in political ecology and critical agrarian studies on global land rush dynamics, highlighting how a new generation of empirical inquiries into emergent land use and cover change trajectories may both extend and enrich the significance of these contributions. Secondly, we present empirical findings from our analysis of the above four scenario archetypes, disaggregating the spatial extent and rate of projected LULCC across several world regions, and highlighting related socio-environmental justice considerations. Thirdly, we situate findings on the magnitude of future LULCC and rural population change projected within each archetype with two key 'analogues': i) observed LULCC trends throughout the mid-to-late twentieth century (1960–2000), as well as ii) Land Matrix data on the scale and rate of acquisitions over the first two decades of the twenty-first century (2000–2020). We conclude with an overture to this JPS forum – as well as to broader communities of practice in political ecology and critical agrarian studies – inviting further inquiries into the potential for emergent climate change mitigation trajectories to entail a 'rescaling of the land rush', as it were.

2. Political ecology, sustainability transformations, and emergent trajectories of global climate change mitigation

In political ecology, critical agrarian studies, and related fields, recent scholarship has productively examined the multi-scalar politics of various 'transformations to sustainability' on a planetary scale (Scoones et al. 2020). Importantly, this literature highlights contestations over which pathways to sustainability are most suitable, feasible, or desirable – and for whom (Leach, Raworth, and Rockström 2013; Cavanagh and Benjaminsen 2017; Newell, Paterson, and Craig 2021). Scholars have recently highlighted, for instance, the emerging 'decarbonization divide' between Global North and South, implicating Northern transport and energytransitions in the production of 'toxic ecologies' or 'sacrifice zones' with deleterious implications for human health, environmental pollution, and rural displacement in the Global South (Sovacool et al. 2019, 2020; see also Dunlap and Laratte 2022). As critical agrarian studies scholars have been at pains to highlight, in particular, the emerging impacts of these sustainability transformations will inevitably articulate with a baseline context already characterized by large and growing inequalities within and across world regions, potentially thus accelerating ongoing processes of socioeconomic differentiation or class formation in both rural and urban areas (Edelman and Wolford 2017; see also Akram-Lodhi et al. 2021).

Indeed, recent studies of sustainability transformation across multiple socio-ecological domains build upon decades of critical scholarship examining interventions to foster climate change mitigation in the agriculture, forestry, and land use (AFOLU) sector. Here, numerous scholars have illuminated how these interventions are often predicated on a false equivalence between 'luxury emissions' from industry or transportation in the Global North and 'survival emissions' from agriculture or biomass-related energy production in the Global South (e.g. Agarwal and Narain 1991; see also Bumpus 2011; Carton, Lund, and Dooley 2021). Particularly when related interventions incentivize, or depend upon, the displacement of rural land users, such initiatives raise well-documented concerns about multifaceted social and environmental injustices. Amongst others, these risks include the 'moral hazard' that offsetting or cross-sectoral mitigation schemes may result in emissions reduction deterrence (Anderson and Peters 2016; Carton 2019), as well as human rights concerns regarding displacement, dispossession, or restricted access to lands and resources (see, *inter alia,* Beymer-Farris and Bassett 2012; Cavanagh and Benjaminsen 2014; Fisher et al. 2018; Cavanagh et al. 2021). Indeed, both political ecologists and critical agrarian studies scholars have repeatedly highlighted important trade-offs and associated environmental (in)justice risks associated with the implementation of climate mitigation policies across different regions, scales, and sectors (Akram-Lodhi et al. 2021; Newell 2022).

To date, however, much of the recent literature has focused on the implications of contemporary or *historical* interventions for conservation or climate change mitigation, rather than on the scale and potential consequences of projected *future* interventions. Laudably, a small number of recent studies have begun to redress this knowledge gap, for instance by producing critical analyses of fossil fuel-intensive mitigation pathways and their reliance on bioenergy carbon capture and storage (BECCS) or other speculative carbon capture and storage technologies (Dooley and Kartha 2018; Buck 2019; Carton 2019; Carton et al. 2020). In addition, a small cluster of studies has generated more focused analyses of shifting relations between climate change impacts, climate change mitigation initiatives, and their implications for agrarian politics (Borras and Franco 2018, 2019; Borras, Franco, and Nam 2020, 2011; Franco and Borras 2021). However, engagement between political ecology, critical agrarian studies, and the wider spectrum of projected mitigation pathways that have arisen from a range of different integrated assessment and other modelling frameworks remains quantitatively limited. Indeed, as Figure 1 illustrates, relatively few analyses in the critical literature overtly engage integral aspects of the mitigation scenario scholarship, as reflected by a paucity of studies exploring – for instance – both political ecology and the Shared Socioeconomic Pathways (SSPs) or Representative Concentration Pathways (RCPs).

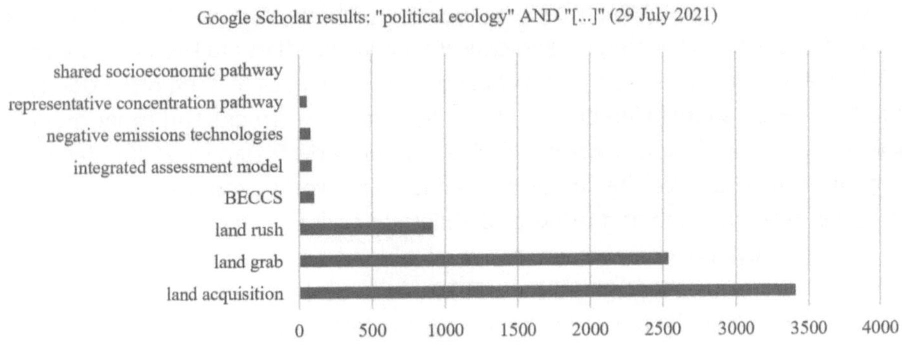

Figure 1. Google scholar results ('political ecology' AND ' … '). Data: Google scholar (29 July 2021).

These and related lacunae point to a growing need for studies situating emergent trajectories of climate change mitigation vis-à-vis diverse critical engagements with a perceived 'global land rush' emerging in the aftermath of the 2007–8 financial crisis. As will be well-known to readers of this journal, historically significant spikes in global food and energy prices followed the financial crisis, prompting a range of investors and other actors to 'rediscover' investment opportunities in agriculture and other key natural resource sectors (Borras et al. 2011; Dell'Angelo et al. 2017). Rulli, Saviori, and D'Odorico (2013), for instance, suggest that these price dynamics catalysed a measurable uptick in large-scale land acquisitions involving transnational investors, many – but not all – of which targeted lands and resources in the rural Global South. Importantly, such investments were often underpinned or legitimized by narratives of unused or idle land (Geisler 2012), as well as by arguments concerning the potential for efficiency gains to be realized via agricultural intensification and the associated closure of 'yield gaps' in ostensibly undercapitalized forms of smallholder agriculture (Li 2014).

In turn, this land rush sparked a 'literature rush' of academic publications in political ecology, critical agrarian studies, and related fields (Scoones et al. 2013). Here, key foci include the role of corporate actors (White et al. 2012); the involvement of states and state agencies (Wolford et al. 2013); and the various ways in which land or resource acquisitions elicit diverse 'responses from below' from both individuals and communities (Hall et al. 2015). Increasingly, critical scholars have also addressed the implications of these processes for ecosystems and natural resource management beyond 'land' as such, including biodiversity conservation, landscape-based environmental change mitigation (including REDD+ and other forms of forest or ecosystem restoration), and the acquisition of water resources, fisheries, or other marine ecosystems (Benjaminsen and Bryceson 2012; Fairhead, Leach, and Scoones 2012; Dell'Angelo, Rulli, and D'Odorico 2018). Again, however, much of this literature engages land and resource acquisitions in historical perspective, illuminating how contemporary dynamics map onto the often-unresolved legacies and injustices of colonial, authoritarian, or other contested forms of land management and environmental governance (Fairhead, Leach, and Scoones 2012; Geisler 2012; Wily 2012). Whilst critical research on low-carbon pathways and the associated 'processes and structures of inequality, exclusion and injustice' is growing (Sovacool et al. 2019, 1), these tensions and contradictions have largely not yet been examined across the full range of projected land-based mitigation pathways in detail.

On one hand, such relatively limited engagement with the scenario analysis literature is understandable. Political ecologists and critical agrarian scholars are keenly aware of how LULCC processes are governed by political-economic dynamics that are often poorly represented – or even simply illegible – within integrated assessment and other modelling frameworks (e.g. Leach and Scoones 2015; Asiyanbi and Massarella 2020). Behind the abstract net LULCC figures that these modelling frameworks produce lies the complex ground-level reality of land and resource conflicts, as well as patterns of capital accumulation, commodification, and the concentration of land or resource ownership (Dell'Angelo et al. 2017; Borras, Franco, and Nam 2020). As critical scholars know all too well, in 'reality' the drivers of land use or cover change are anything but straightforward. Indeed, attempts to reconfigure the ownership of lands and resources via transnational investment often fail, as the case study literature illustrates in particular (Cavanagh and Benjaminsen 2014; Borras et al. 2022b; see also Li 2014). Considering the gravity of these and related assumptions embedded within the 'black box' of specific IAM methodologies, critical scholars might thus be tempted to dismiss these scenario archetypes altogether as being simply 'unrealistic' or irreparably blind to certain political-economic realities. Others might demur that – in light of observed GHG emissions trajectories and the potential activation of malign 'tipping points' or planetary feedback loops (e.g. Kemp et al. 2022) – the 1.5 °C mitigation target is simply increasingly unattainable, undermining the utility of a detailed critical analysis of associated mitigation scenarios.

Epistemologically, however, it is somewhat problematic to conclude that a projected *simulation* – or more specifically, an *optimization* or 'optimized simulation' – is '(un)realistic' in this sense. In other words, IAM frameworks have been 'optimized' or otherwise parameterized in such a way that they are effectively forced to produce scenarios that result in the achievement of the 1.5 °C Paris Agreement target or other specified objectives. In these scenarios, desired objectives are thus in fact 'achieved' via the specified mechanisms and pathways. Figuratively speaking, the metaphorical 'dominos' in these scenarios are compelled to fall in ways that are essentially necessitated by the achievement of the specified objectives or parameters – such as 1.5 °C of warming above pre-industrial averages, or radiative forcing of 1.9 watts/m^2 by 2100 – albeit via distinct pathways across the LED and SSPx archetypes. As Keppo et al. (2021: 5) note, the production of these scenarios unavoidably requires IAMs to deliberately simplify complex system dynamics, denoting that they are 'generally not meant to be normative, nor provide a blueprint for policy makers'. Differently put – as Rogelj et al. (2018a, 331) write – 'because models are stylized, imperfect representations of the world, feasible dynamics in a model might be infeasible in the real world, while vice versa infeasibility in a model might not mean that an outcome is infeasible in reality'.

Seen this way, one could suggest that the value of IAM outputs does not lie in their analytical precision or implied feasibility, but in their generation of a range of future 'archetypes' or illustrative scenarios in relation to which both historical and emergent empirical trajectories can be contextualized. In this article, we thus offer a critical analysis of four such scenario archetypes, seeking to broker an initial conversation and to encourage further research on these themes in political ecology and critical agrarian studies. To this end, in what follows we reconstruct LULCC implications within and across four mitigation archetypes: LED, SSP1-19, SSP2-19, and SSP5-19. Subsequently, we distil key insights or 'takeaways' from this analysis vis-à-vis the political ecology and critical agrarian studies literature on the global land rush, highlighting considerable potential for further engagement in these fields.

3. Context and methodological approach

Methodologically, we draw upon an analysis of quantitative output from IIASA's SSP and 1.5 °C scenario explorer database (see Huppmann et al. 2018), which facilitates access to the empirical substance of scenarios associated with the four archetypal mitigation pathways foregrounded by the IPCC (2018) *Special Report on Global Warming of 1.5 °C*. Utilizing IIASA's scenario explorer, we obtained output for projected LULCC by 2100 relative to 2010 reference levels. Specifically, this describes land cover changes in – *inter alia* – pasture, cropland with and without second generation bioenergy crops, and forests (both natural and managed) over time, measured in millions of hectares (Mha).

Output is available for three shared socioeconomic pathways (SSP1, SSP2, SSP5) which have been fed into six IAM frameworks to simulate future land cover changes for a mitigation target of 1.5 °C above pre-industrial averages.[3] This mitigation target (1.5 °C) broadly corresponds to Representative Concentration Pathway (RCP) 1.9, which entails radiative forcing of 1.9 watts/m^2 by 2100. Hence, our analysis is based on the following three SSPx-RCP1.9 archetypes: SSP1-RCP1.9, SSP2-RCP1.9, and SSP5-RCP1.9. In addition, we also examined the LED scenario, which is based in part upon SSP2-RCP1.9, albeit with adjustments specifically intended to model a divergent 'low energy demand' scenario that would avoid reliance on land-intensive negative emissions technologies, specifically BECCS (Grubler et al. 2018). Whereas multiple model 'runs' of every individual SSP were completed across different IAMs – each of which entails somewhat varying methodologies and assumptions, resulting in significant differences in output across modelling frameworks even for the same SSPx scenario (see also Skeie et al. 2021) – the LED scenario was quantified through the MESSAGEix-GLOBIOM 1.0 framework alone (Grubler et al. 2018). Although our approach relies on mean LULCC values distilled from multiple different 'runs' of the same broad SSPx archetype – rather than median values and ranges for LULCC output across individual scenarios within each archetype – this approach nonetheless reflects important general tendencies across the full range of the four illustrative mitigation pathways, and is thus practicable for encouraging an initial wave of critical discussion on these issues.

Indeed, in aggregate the four scenario archetypes present us with an indicatively broad range of mitigation futures. This is particularly so insofar as each archetype is infused with a distinct storyline or 'narrative' about future socio-ecological trajectories (O'Neill et al. 2017). These trajectories reflect divergent outcomes across key indicators including human population growth, international cooperation (or the lack thereof), urbanization, and economic development. Given our focus on LULCC from pastures and cropland to forests and biofuel cultivation, we are particularly interested in implications for shifting rural population and human settlement dynamics. To this end, Table 1 and Figure 2 highlight projected population and urbanization figures for both the three SSPx scenarios and

[3]These IAM frameworks include AIM, IMAGE, MESSAGE-GLOBIOM, REMIND, and GCAM. For the purposes of our analysis, WITCH was excluded due to incomplete datasets on land use.

Table 1. Future population and urbanization dynamics and underlying assumptions. (Based on Jiang and O'Neill 2017; Riahi et al. 2017; Grubler et al. 2018). Data for urbanization rate: Jiang and O'Neill 2017. Population data: FAO-historical, IIASA-projected.

	SSP1 and SSP5	SSP2 and LED
Global population, first half of twenty-first century	Moderate population growth	High population growth
Global population, second half of twenty-first century	High overshoot (at 8.5 billion) and shrinking population (to 7 billion)	Low overshoot (9.5 billion) and stabilization at 9 billion
Global population, 2010 vs 2100	Almost identical (around 7 billion)	Growth from 7 to 9 billion
Urbanization rate (World)	High urbanization rate (from 60% in 2020 to 92% in 2100)	Moderate urbanization rate (from 56% in 2020 to 79% in 2100)
Urbanization rate (Global South)	High (from 49/84/53% in 2020 to 90/95/91% in 2100 –Africa/Latin America/Asia)	Moderate (from 46/82/48% in 2020–73/91/78% in 2100 –Africa/Latin America/Asia)
Modelled assumptions	Increased migration and fast urbanization due to medium to high economic growth, high income growth, relatively low income inequality, 'environmentally friendly living arrangements' (SSP1); high economic growth and technological change, increasing agricultural productivity, growing wealth (SSP5)	Moderate income and economic growth, and moderate socio-economic inequality lead to moderate urbanization rate, in line with historical development. LED scenario with much more aggressive assumptions concerning reductions in consumption, energy conservation or efficiency, industrial activity, and therefore final global energy demand

LED to illuminate how rural population is projected to change from 2020 until 2100 (see also Jiang and O'Neill 2017).[4]

Several key trends are notable. First, rural population has almost doubled in the last 60 years in the Global South, while it has stagnated in the North (Figure 2). Second, all four scenario archetypes project a significant and absolute depopulation of rural areas across the world (Table 1 and Figure 2). Here, SSP1 and SSP5 futures project that around 500 out of 600 million rural people will live in the Global South. By contrast, SSP2 and LED futures project that around 1.7 out of 1.8 billion rural people will live in the Global South. These reductions converge with unprecedented urbanization processes. Here, SSP1 and SSP5 reflect the highest rates of urbanization (92% globally), whereas SSP2 and LED evince a slightly more 'moderate' degree of urbanization, both simulating an 80% urban share of world population by 2100 (Riahi et al. 2017, 158). Importantly, this is relative to an approximately 57% urban share of world population as of 2021 (World Bank 2022).

Although most population and economic development assumptions in LED are based upon the 'middle of the road' SSP2 scenario, LED also includes a unique 'food security' constraint. This is intended to ensure that 'increased populations in the global South are not worse off in terms of animal and vegetal calorie intake as a result of climate mitigation efforts based on land use (for example, the expansion of bioenergy crops)' (Grubler et al. 2018, 525). Although this specific constraint is unique to LED, SSP1 also includes somewhat optimistic socioeconomic assumptions. van Vuuren et al. (2017: 241), for instance, highlight how assumptions of 'significant gains in access to food' underpin IMAGE-based SSP1 projections. This denotes that in SSP1 'policies to reduce poverty

[4]We follow the IIASA regional level definition, which includes 'countries from the reforming economies of Eastern Europe and the Former Soviet Union' as part of Global North. See IIASA (2021), https://tntcat.iiasa.ac.at/SspWorkDb/dsd?Action=htmlpage&page=about (accessed 15 April 2021).

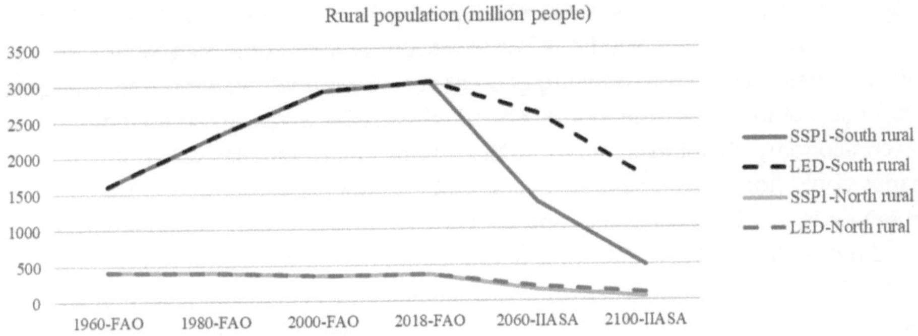

Figure 2. Rural population across global south and global north according to different population projections. 1960–2018 historical data (FAO). 2060–2100 projection (IIASA). SSP5 data not included, given a strong similarity with SSP1. SSP2 data not included, as it is identical with LED. SSP1 is based on IMAGE model, LED is based on MESSAGEix-GLOBIOM model.

and hunger in combination with increased welfare lead to an increase in per capita consumption of food' (van Vuuren et al. 2017: 243). Similarly, Popp et al. (2017, 340–341) emphasize that mitigation response options largely do not 'influence food prices [in SSP1] due to a general "food first" policy, which can restrict agricultural expansion to avoid deforestation, but further only allows bio-energy on areas not needed for food and feed production.' Moreover, Riahi et al. (2017, 158) show how the three SSPs compatible with a 1.5 °C climate future are underpinned by significant reductions in income inequality.

Given the above parameters and assumptions, we reiterate that these scenario archetypes are perhaps best conceptualized as *optimizations*, rather than as forecasts – much less as 'predictions' – in the conventional sense. That is to say, even the most sophisticated integrated assessment frameworks necessarily entail simplifications of the empirical dynamics that characterize all complex social, economic, and biophysical systems in practice (Gambhir et al. 2019; Peng et al. 2021). This underscores how SSP1, SSP2, and SSP5-based archetypes are inherently 'optimized' – and therefore, remain somewhat 'optimistic' – in the sense that all mitigation-relevant choices, decisions, and transactions take place in relatively well-governed, technologically advanced, and administratively expedient worlds. With respect to land use and cover change, the most prominent of these assumptions include: i) demand changes (e.g. dietary changes that shape agricultural and livestock demand and supply in desirable ways); ii) innovations that enable efficiency gains (e.g. sustainable intensification in agriculture, or energy conservation measures that reduce final global energy demand); iii) efficient markets and price discovery mechanisms that successfully 'internalize externalities' (e.g. by successfully pricing carbon emissions to improve the affordability or attractiveness of alternative energy sources and land uses); and iv) globally expedient governance and policy implementation (allowing for rapid and uniformly competent rollout of large-scale response options, such as afforestation or BECCS).

Crucially, these 'optimistic' assumptions are not prevalent to the same extent in all five SSPs. 'Resurgent nationalism' and regional conflicts dominate SSP3, for example, denoting that economic development is slow and inequalities persist or worsen, limiting the kind of mitigation effectiveness that is a necessary precondition for meeting the 1.5 °C target (O'Neill et al. 2017). Similarly, SSP4 is characterized by increasing inequality across and within countries, leading to a growing gap between cosmopolitan elites and 'a fragmented collection of lower-income, poorly educated societies that work in a labor intensive, low-tech economy' (Riahi et al. 2017, 157). As Rogelj et al. (2018a, 325) note, '1.9 W/m^2 scenarios could not be achieved in several models under SSPs with strong inequalities'. In other words, 1.5 °C compatible scenarios are seemingly difficult – if not simply impossible – to model successfully in a world that is characterized by growing geopolitical tensions (SSP3) and/or inequalities (SSP4). This is significant, as it also suggests that 1.5 °C *can* be achieved if the world 'follows a path in which social, economic, and technological trends do not shift markedly from [recent] historical patterns' (Riahi et al. 2017, 157). Such is the definition of SSP2, upon which LED is based.

In the following section, we turn to an analysis of historical patterns to illustrate the significance of LULCC projected in these future scenario archetypes. Firstly, we examine projected LULCC across the four archetypes, disaggregating results to illuminate divergent implications across world regions. Secondly, we contextualize scenario projections in relation to broader patterns of historical land use and cover change across the twentieth and early twenty-first centuries. Thirdly, we situate the scale and rate of projected LULCC vis-a-vis the recent 'global land rush' – as documented by the Land Matrix database (2022) – highlighting the contribution of global land rush dynamics to global LULCC patterns in the early twenty-first century.

4. Emergent land use and cover change trajectories in historical perspective

4.1 Land use and land cover change in 1.5 °C futures: differential impacts across scenarios, world regions, and modelling frameworks

Across the four scenario archetypes, the magnitude of associated land use and cover change is quite striking both in absolute terms, as well as when considered relative to 'baseline' scenarios for each SSP (Figure 3). Here, baselines reflect future trajectories resembling business as usual trends, with no additional climate change mitigation policies beyond those in place at a given reference year. A comparable future baseline is not available for the LED scenario; however, its shared premises with SSP2 allow for a degree of useful contextualization vis-à-vis the SSP2 baseline.

Averaging quantitative output across available IAM modelling frameworks (Figure 4, left panel), all four scenario archetypes reflect a vast reduction in pasture land cover globally, ranging from 847 to 683 Mha between 2010 and 2100, and amounting to a mean annual pasture reduction of 7.6–9.4 Mha. This is accompanied by a significant absolute reduction in cropland, ranging from 480 Mha in the SSP1-19 'sustainability' archetype, to 69 Mha in the LED archetype (mean annual reduction of 0.8–5.3 Mha). In general, pasture land cover reductions are necessary to enable the implementation of land-based mitigation response options on a vast scale across the four archetypes. Land area requirements for bioenergy cropland and BECCS, for instance, range from 759 Mha in SSP5-19 to 158 Mha in LED, amounting to a mean annual growth of 1.8–8.4 Mha (Figure 4, left panel). This is accompanied by a similarly vast land area

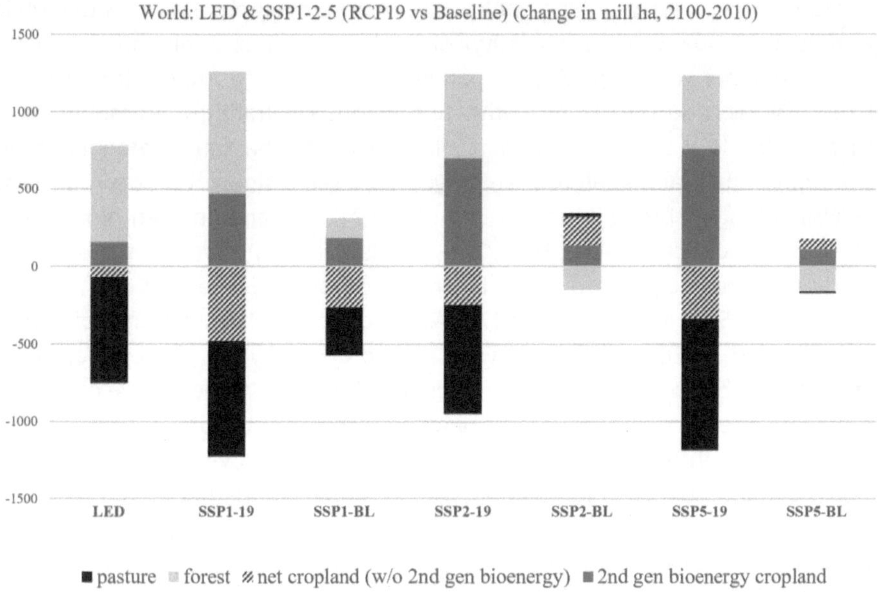

Figure 3. World land use/cover change (in millions of hectares in 2100, relative to 2010 reference levels) in key emissions reduction pathways under baseline (no additional mitigation policies) and mitigation policy conditions for limiting global warming to 1.5 °C above pre-industrial averages (RCP1.9). Data: © IAMC 1.5 °C scenario explorer hosted by IIASA https://data.ene.iiasa.ac.at/iamc-1.5c-explorer (see Huppmann et al. 2018).

requirement for afforestation, ranging from 786 Mha in SSP1-19 to 475 Mha in SSP5-19, or an annual mean growth of 5.2–8.7 Mha (Figure 4, left panel). Even the LED scenario, which is notable in its aversion to allocating bioenergy cropland for BECCS, compensates for this avoidance by projecting afforestation on a scale that exceeds expansions of forest land cover in both SSP2-19 and SSP5-19, amounting to approximately 620 Mha of afforestation by 2100, or roughly 7 Mha per year on average (Figure 4, left panel).

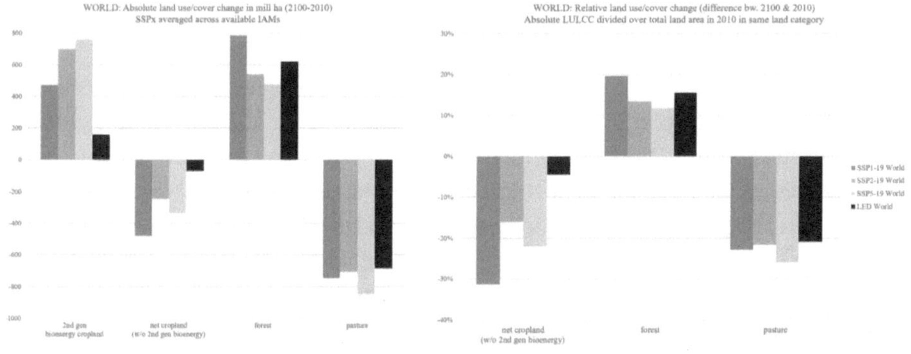

Figure 4. World land use/cover change in absolute (left panel) and relative terms (right panel), in key emissions reduction pathways for limiting global warming to 1.5 °C above pre-industrial averages. Each SSP archetype represents averaged outputs across all available IAMs. LED archetype is based on a single IAM model. Data: © IAMC 1.5 °C scenario explorer hosted by IIASA https://data.ene.iiasa.ac.at/iamc-1.5c-explorer (see Huppmann et al. 2018).

In absolute terms, comparing and contrasting these projections allows us to highlight average rates of annual change (intensity) of land-based climate mitigation initiatives, expressed in Mha per year (Figure 4, left panel). Another way to explore the magnitude of change is by plotting LULCC in relative terms, thus illustrating the amount of land affected by 2100 as a share of totally available land of the same category in 2010 (Figure 4, right panel). Across all four archetypes, we see a similar reduction in pasture (ca. 21–26%), a bigger spread in afforestation (ca. 12–20%), and an even bigger spread in cropland reduction (ca. 5–31%). Here, LED stands out from the other three scenarios by projecting 'merely' a 5% reduction in cropland cover. Unlike LED, the 'food first policy' scenario in SSP1 (Popp et al. 2017, 340) leads to a 31% contraction of cropland in 2100 compared to 2010 levels. Likewise, LED projects by far the least amount of additional land for bioenergy cultivation (Figure 4, left panel) – although we cannot illustrate it in relative terms given the lack of bioenergy cropland in 2010.

In short, both the intensity and the scale of projected land cover change raise serious questions regarding how rural populations will be impacted, particularly in the case of wholly or partially subsistence-oriented pastoralists, smallholder farmers, and other populations whose livelihoods are closely tied to the natural resource base (see also Doelman et al. 2019, 2020). To further examine these dynamics, we disaggregate the magnitude of projected LULCC across world regions to highlight divergent consequences across the Global North and South (Figure 5). Comparing projected LULCC (2100-2010) relative to the available land in 2010 in the same category and region, we see significant differences between Global North and South,[5] and between different scenario archetypes (see Figure 5, right panel). Here, SSPx reductions in cropland (without bioenergy) amount to around 23–36% of available cropland in the North, and 9–28% in the South. By contrast, LED offers a less interventionist scenario, projecting 4% reductions in cropland in the North, and 8% in the South. With regard to changes for pasture, similar projections apply for LED, but somewhat reversed for non-LED outputs. SSPx reductions in pastures amount to around 19–23% of available pasture in the North and 23–27% in the South, whereas LED projects reductions amounting to 16% in the North and 24% in the South. Finally, looking at forest cover change, the Global North is projected to gain 6–13% of forest cover, while the South is to experience afforestation between 18 and 26% for SSPx projections. Again, LED follows the trend by projecting 9% growth in forest cover in the North and 24% in the South.

Overall, relative to 2010 land area size, the comparison between North and South highlights a higher loss of cropland in the North (although not for LED, where it is reversed), more afforestation in the South and more losses of pasture in the South (Figure 5, right panel). As above, LED projects the least losses in cropland and the least increases in bioenergy cultivation (Figure 5). However, in absolute terms (Figure 5, left panel), all four scenarios project a disproportionally affected Global South compared to the North, across all four land cover change categories (with one exception of cropland in the SSP2 archetype, where the Global North is losing more cropland than the South, in absolute terms).

[5]In this context, the term 'Global South' refers to an aggregation of the R5MAF, R5LAM, and R5ASIA regions. For country-level definitions, see IIASA's SSPs database: https://tntcat.iiasa.ac.at/SspWorkDb/dsd?Action=htmlpage&page=about (accessed 26.07.2021)

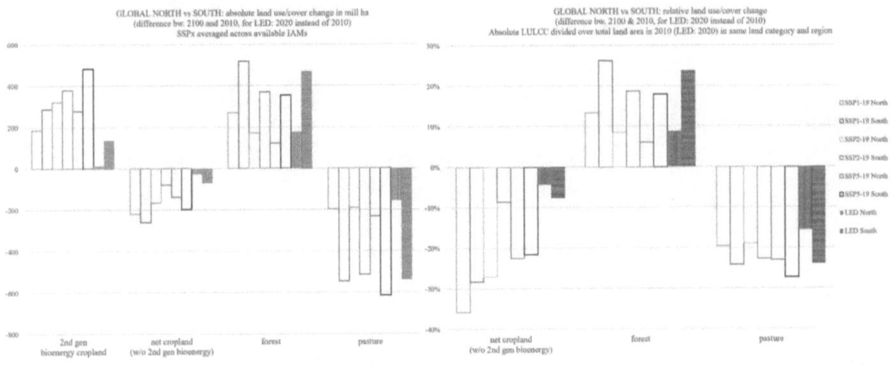

Figure 5. Global north vs Global south - absolute (left panel) and relative (right panel) land use/cover change, in key emissions reduction pathways for limiting global warming to 1.5 °C above pre-industrial averages. Data: © IAMC 1.5 °C scenario explorer hosted by IIASA https://data.ene.iiasa.ac.at/iamc-1.5c-explorer (see Huppmann et al. 2018).

Disaggregating land cover change in spatial terms even further, Figure 6 illustrates differences within the three regions that make up the Global South according to IIASA's regional definitions: Middle East and Africa (MAF), Latin America (LAM), and Asia (ASIA).[6] In the MAF region, for instance, land cover for pasture declines dramatically across the SSPx archetypes, ranging from 304 Mha in SSP-19 to 208 Mha in SSP2-19, or a mean annual reduction between 2.3 and 3.4 Mha. Mirroring global dynamics, this is accompanied by the vast implementation of response options entailing expanded land cover for bioenergy crops and forests. Bioenergy cropland requirements range from 166 Mha in SSP5-19 to 73 Mha in SSP1-19, for example, whereas forest expansion ranges from 140 Mha in SSP1-19 to 93 Mha in SSP2-19. Annually, this amounts to a mean growth of cropland for bioenergy at the scale of 0.8–1.8 Mha, and of forest land at the scale of 1–1.6 Mha. One positive aspect is the moderate cropland reduction in the MAF region, with SSP2 even projecting growth in cropland (Figure 6).

A comparison between the three regional blocks within the Global South suggests that in absolute terms ASIA is projected to lose the least amount of pasture, while losing the most cropland and experiencing the highest growth in bioenergy cropland (Figure 6, left panel). Relative to 2010 reference levels (Figure 6, right panel), ASIA is still projected to lose the most cropland, while LAM is projected to lose the most pasture land cover. All three regions are projected to experience comparable levels of afforestation, most of which takes place in the SSP1 archetype.

Taken together, the compounding losses in cropland and pasture – particularly in the Global South – suggest an overarching scramble for land to satisfy mitigation demands via either afforestation or BECCS. Further unpacking the significance of these land area requirements, in what follows we contextualize the scale and intensity of projected future LULCC for climate mitigation vis-à-vis historical analogues from the second half of the twentieth century (section 4.2) and the recent land rush from the early twenty-first century (section 4.3).

[6]We do not include LED projections at this scale, due to lack of disaggregated regional output (LED output only includes figures at the World and Global North-Global South scales).

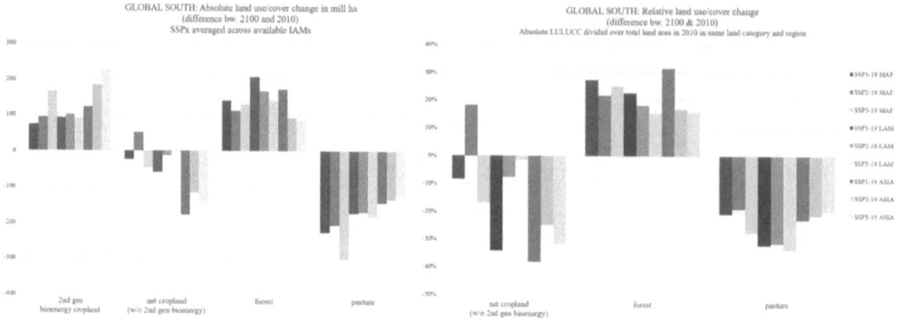

Figure 6. Global south - regional land use/cover change in absolute (left panel) and relative terms (right panel) in key emissions reduction pathways for limiting global warming to 1.5 °C above pre-industrial averages. MAF: Middle East & Africa; LAM: Latin America; ASIA: most Asian countries with exception of Middle East, Japan and former Soviet Union states. Data: © IAMC 1.5 °C scenario explorer hosted by IIASA https://data.ene.iiasa.ac.at/iamc-1.5c-explorer (see Huppmann et al. 2018).

4.2 Land use and land cover change dynamics throughout the second half of the twentieth century

To unpack the significance of 1.5 °C land use and cover change trajectories, we situate projected LULCC in the four scenario archetypes vis-à-vis the historical record compiled by the FAO. This allows us to contextualize projected scenarios in relation to historical LULCC data, which reflects how a land class (such as cropland) expands or shrinks over time (e.g. annually) at a given scale. Figure 7 illustrates how – unlike in the Global North (grey) – net cropland and pasture land cover have steadily increased between 1961–2000 across the Global South (black), while forest land cover has decreased between 1990–2000 (FAO data). To assess the significance of these historical data, we situate these trends vis-à-vis projected (2010–2100) global LULCC trajectories (for clarity, plotting SSP1 and LED data only).[7] In short, Figure 7 indicates a radical reversal of historical LULCC dynamics in the Global South. Here, historical pasture and cropland expansion are to be transformed into an absolute reduction, while historical deforestation trends are to be reversed via large-scale afforestation (in varying combinations of both 'natural' and 'managed' plantation forests).

In Figure 8, we compare the intensity of historical and future LULCC for cropland, pasture, and forest, spanning the periods 1961–2000 (FAO),[8] and 2010/2020-2100 (IIASA). Several aspects are noteworthy. First, there is a significant difference between the *direction* and *intensity* of annual LULCC in the Global North and Global South throughout the second half of the twentieth century. With regard to the *direction* of change, pasture and cropland have increased in the Global South, unlike in the Global North. As to the *intensity*, the Global South has experienced significant rates of LULCC across the three land classes, again, unlike in the North.

[7]With the exception of LED, the IIASA database does not provide a North-vs-South breakdown of global LULCC trajectories.

[8]However, for lack of pre-1990 data for forest LUC in Global South and Global North we only include two datapoints for forest based on FAO: 1990 and 2000.

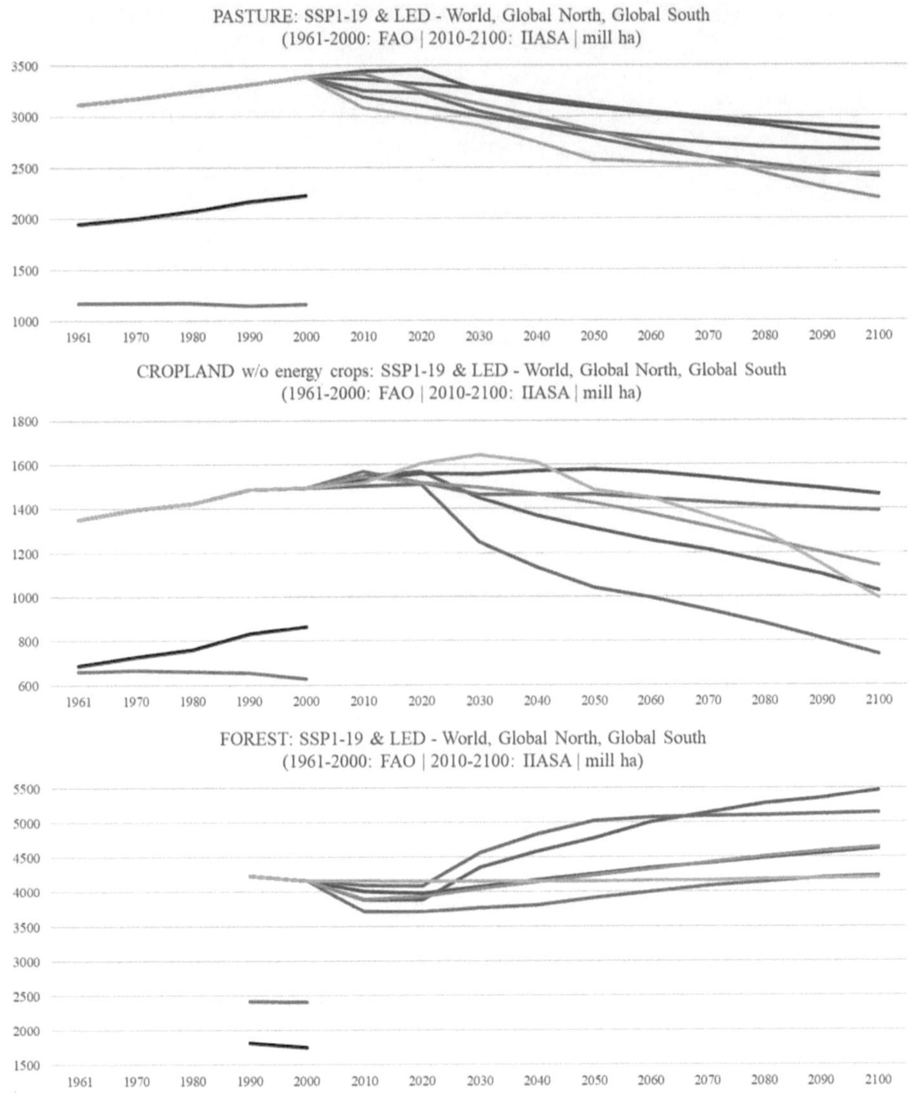

Figure 7. Historical and future land use/cover, *globally* (in different shades of blue). Historical land use/cover in Global South (black) and Global North (grey). Historical data (1961–2000) from FAO, future data (2010–2100) from IIASA.

Second, these differences in *intensity* between Global North and Global South are projected to continue in 1.5 °C climate futures, although this time LULCC assumes a similar *direction* across North and South. The annual rate of projected change in the Global North is situated between −2.3 and +2.2 Mha, hence somewhat higher than in the past (−0.2 to −1.1 mill ha p.a.). In the Global South, the annual rate of net LULCC evinces a much higher rate than in the North, amounting to −6.8 to +5.9 mill ha p.a. in the future and −6.2 to +6.9 mill ha p.a. in the past. In other words, the Global South is not only expected to halt its historically intense rates of cropland and pasture *expansion* (4.4–6.9 mill ha), but to undergo similarly intense rates of cropland and pasture *reduction*

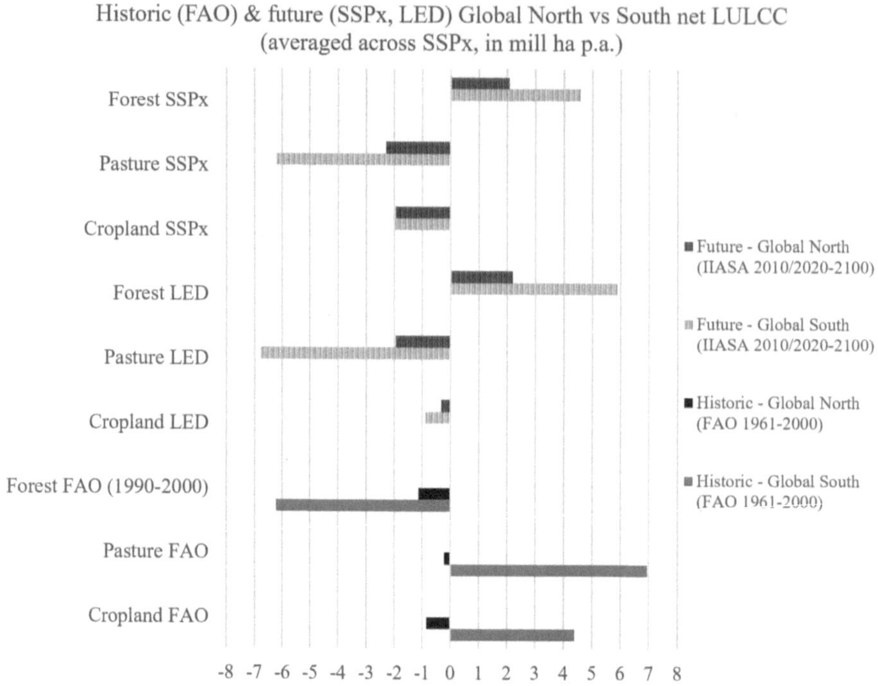

Figure 8. Rate of mean annual net LULCC: Future (LED and SSPx archetypes) and historic (FAO 1961-2000). SSPx = averaged across SSP1, 2, and 5. Data: IIASA/Huppmann et al. (2018), FAO.

(6.2–6.8 mill ha), therefore dwarfing both historical and projected LULCC dynamics in the Global North.

Importantly, the twentieth-century history of pasture expansion took place in the context of a growing rural population in the Global South. By contrast, future projections of pasture contraction are underpinned by rural depopulation. As discussed in relation to Figure 2 above (see Section 3), the Global South is projected to lose almost half of its current rural population by 2100 in SSP2 and LED futures, amounting to around 1.7 billion rural people in 2100. SSP1 and SSP5 futures are characterized by even more depopulation in the rural South, amounting to roughly half a billion people living in rural areas by 2100. The reconfiguration of LULCC dynamics in predominantly rural areas to mitigate climate change is thus projected to take place in spaces that are much less densely populated in the future.

Taking into account future population projections, in Figure 9 we compare the rate of cropland and pasture availability per rural person in 1960 (FAO), 2020, and 2100 (IIASA). Here, we draw attention to the Global South in particular, given that rural population is already at low levels in the Global North today. As Figure 9 illustrates, the rate of cropland and pasture availability per rural person is less pronounced in LED and SSP2 futures compared to SSP1 and SSP5, given that the former are projected to experience more modest urbanization rates than the latter. Taken together, Figures 7–9 hint at significant transformations of LULCC dynamics and rural demographics in the Global South. In the following Section 4.3, we draw upon an additional analogue in the form of the recent global

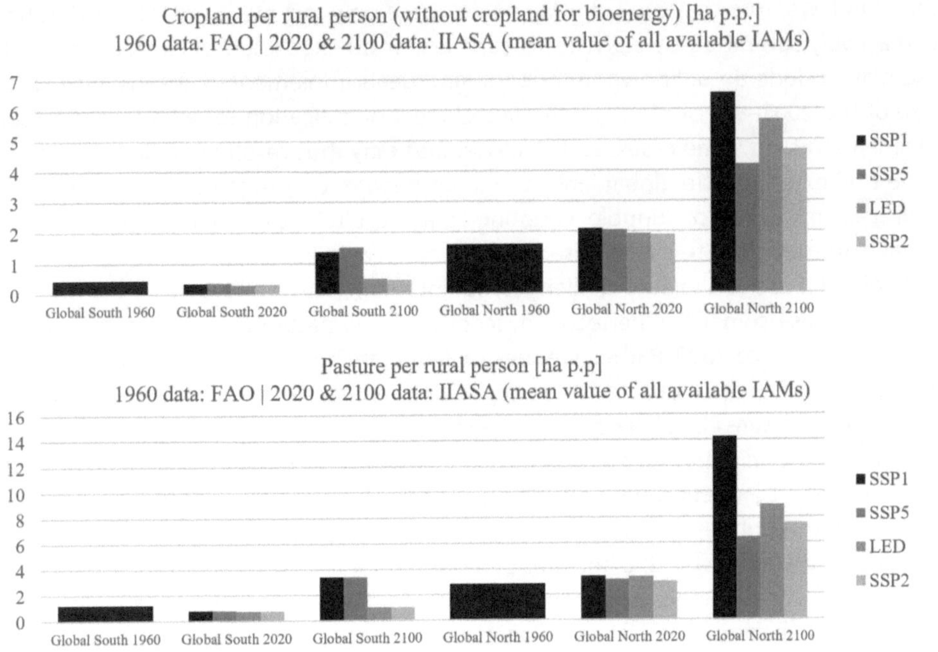

Figure 9. Global North vs Global South - historic and future land availability per rural person (top panel: Cropland, bottom panel: Pasture).

land rush to further contextualize the significance of both the scale and rate of projected LULCC in the above four scenario archetypes.

4.3 The land rush as spatial analogue for projected 1.5 °C land use and cover change futures

Important concerns have been raised in recent literature regarding the theme of figuratively 'messy hectares' (Edelman 2013) in the global land rush debate, or the challenge of reliably quantifying land acquisition dynamics since the turn of the twenty-first century (see also Scoones et al. 2013; Borras et al. 2022b). In this context, some scholars have suggested that quantitative analyses risk unduly asserting 'false precision' (Oya 2013) regarding the spatial extent of such acquisitions. Conversely, others have noted that all quantitative datasets typically evince limitations with respect to their reliability, and that – if these limitations are accounted for – the kinds of 'imperfect data' (Rulli and D'Odorico 2013, 907) generated by initiatives like the Land Matrix database still potentially facilitate useful insights into relevant empirical dynamics.

Simply put, we share many of the methodological concerns raised by critical scholars regarding the reliability of quantitative land rush data. Indeed, we do not seek to detract from critiques that have rightfully highlighted the implicit 'politics of evidence' (Scoones et al. 2013) within differing calculations of the spatial extent of global land and resource acquisitions (see, especially, Edelman 2013; Oya 2013). Such concerns are perhaps especially acute in relation to the possible 'double-counting' of overlapping or sequential transactions involving the same lands or properties (e.g. Scoones et al. 2013, 475). On the other hand, we note that the Land Matrix database only includes transactions that have been actively submitted by activists, researchers, or other scholars. As a result, the database may exclude an unknown number of land deals implemented, for instance, in the wake of the 2007–8 crisis, but which have eluded identification for a variety of reasons (Edelman 2013). In some cases, such uncertainties may thus result in an *under*estimation of the true extent of the global land rush in aggregate (Borras et al. 2022b), even while small-scale instances of 'double counting' may result in the over-estimation of the spatial extent of discrete land transactions at the local scale.

In light of these and related challenges, our intention here is not to imply that the Land Matrix database somehow 'perfectly' describes the full nature or spatial extent of land rush dynamics since 2000. Rather, we seek to illuminate how these data may nonetheless provide us with a useful qualitative *analogue* or reference point in relation to which the significance of future LULCC trajectories may be contextualized, rather than 'compared' in the formal sense. In doing so, we follow the Land Matrix (2022) database in defining a land acquisition as a transaction that involves: i) a 'transfer of rights to use, control, or ownership of land through sale, lease, or concession'; which ii) has been 'initiated since the year 2000'; iii) covers 'an area of 200 hectares or more'; and iv) implies 'the potential conversion of land from smallholder production, local community use, or important ecosystem service provision to commercial use'. Overall, the database tracks a total of approximately 93.2 Mha of land deals (2000–2020) globally (around 4.4 Mha annually) and 73 Mha in the Global South (around 3.5 Mha annually), across all acquisition types and implementation statuses.

Importantly, acquisitions logged in the Land Matrix database are not distributed evenly across the first two decades of the twenty-first century. Rather, available data reflect a clear spike of acquisitions between 2007 and 2014, and particularly so in the Global South (Figure 10, top panel). This 'peak' period of acquisitions correlates with two parallel spikes in food and energy prices that followed the global financial crisis of 2007–2008 (Figure 10, bottom panel). Considering these food and energy price dynamics, it is notable that land acquisitions in the Global South in the year 2011 alone (12.4 Mha) account for nearly 17% of all Global South acquisitions logged in the Land Matrix database from 2000 to 2020. Moreover, such correlations are perhaps especially ominous in light of recent macroeconomic developments. Whilst energy prices have fallen somewhat in recent years (2015–2020), both food and energy prices are once again rising sharply in the tumultuous context of the global COVID-19 pandemic, associated supply chain disruptions, and the geopolitical fallout from Russia's war of aggression in Ukraine. An important, empirically-open question thus remains concerning whether – or to what extent – an observable spike in land acquisitions will once again accompany resurgent food and energy prices.

More immediately than in the past, however, contemporary responses to the food and energy crisis must also navigate emissions reductions pledges associated with (*inter alia*) the 1.5 °C Paris Agreement target, which according to the IPCC (2018) would entail a decline in global net anthropogenic CO_2 emissions of approximately 45% below 2010 levels by 2030. Incentives for both mitigation and non-mitigation-associated land acquisitions are in some ways thus even more pressing than in the aftermath of the 2007–8 financial crisis. Methodologically, however, one cannot directly 'compare' the magnitude or spatial extent of historical land acquisitions (for instance, as documented within the

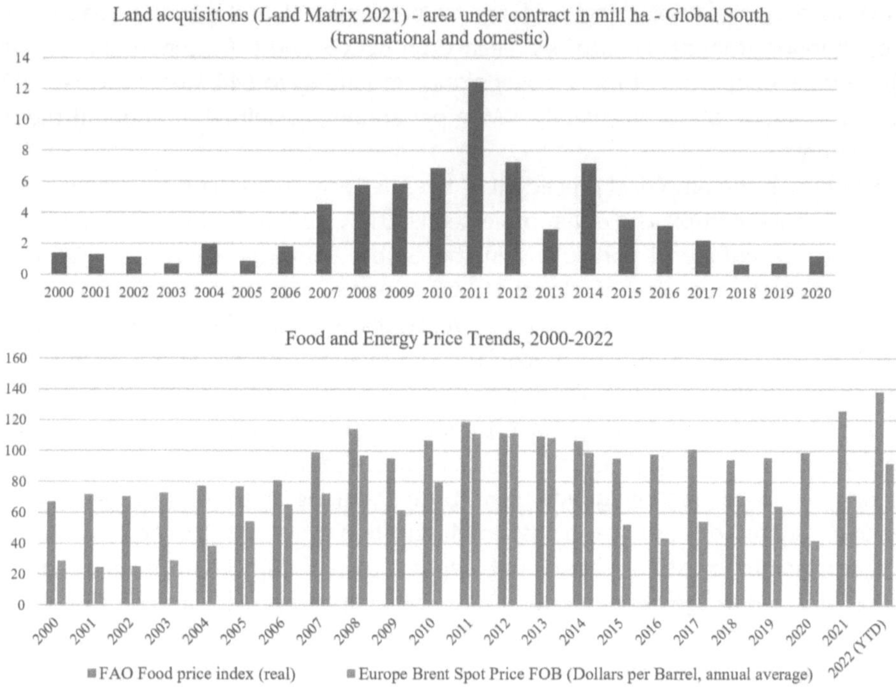

Figure 10. Land acquisitions per year, Global South, 2000–2020 (top panel) and FAO food price index and europe brent spot price (annual average, 2000–2022) (bottom panel). Data: Land Matrix Database, FAO and US EIA (2022).

Land Matrix database) with future LULCC projections for meeting the 1.5 °C target across the above four scenario archetypes. Amongst other constraints, individual Land Matrix transactions may or may not entail either land use or land cover *changes* per se (e.g. from forest cover to cropland or vice versa). Moreover, even if such changes are purportedly desired by investors, the speculative nature of many land deals implies that corresponding alterations of land use or land cover may not actually materialize on the ground (Borras et al. 2022b).

Assuming that a significant proportion of projected LULCC for climate change mitigation will materialize through single change events in the future, however, it is nonetheless salient that average rates of projected LULCC (in Mha per annum) resemble rates of acquisition throughout the 'peak period' documented in the Land Matrix database. Indeed, across the three SSPx and LED archetypes, mean future LULCC in the Global South is projected at approximately 6 Mha p.a. of pasture reductions and 4.5–6 Mha p.a. of forest cover expansion (Figure 7). Temporally, this is reminiscent of the rate of land acquisitions during the 'peak period' (2007–14) of the global land rush – during which the Land Matrix database tracked approximately 6.6 Mha p.a. of land deals in the Global South – albeit, importantly, sustained throughout the remainder of the twenty-first century. This is significant given that – whilst the critical literature increasingly takes issue with the reliance on afforestation and BECCS in scenarios that envision the slowest rates of transition away from fossil fuels (e.g. SSP5-19; see Carton 2019; Hickel 2019) – rates of projected LULCC even in the Low Energy Demand (LED) archetype still remain considerable relative to both the historical record and the 'peak period' of the global land rush (Cavanagh 2021). Indeed, although it laudably eschews reliance on BECCS, the LED scenario nonetheless entails a global expansion of forest cover alone amounting to 646 Mha between 2020-2100, or an average of approximately 8 Mha per annum, sustained throughout the rest of the century.

Drawing upon the empirical precedent of twentieth-century LULCC (as highlighted in Section 4.2), it is not implausible that projected LULCC would unfold through single mitigation-related events, and particularly in the Global South. Winkler et al. (2021), for instance, estimate that 38% of global LULCC (pasture, cropland and forestry) in the period 1960–2019 involved single change events, the majority of which occurred in the Global South. Potapov et al. (2022) echo this observation, highlighting how in the last two decades, 79, 61, and 39% of cropland expansion in Africa, South-East Asia, and South America, respectively, took place through the conversion of natural vegetation: in other words, through single LULCC events. Moreover, achieving the 1.5 °C Paris Agreement target implies that mitigation-related interventions at the landscape scale will largely necessitate both direct and sustained transition from – for example – pasture to forests, cropland to forests, or pasture to bioenergy cropland. That said, even if projected LULCC is achieved as the *net* outcome of multiple change events -- and on top of a significant volume of other, more conventional land acquisitions -- our reference to the historical analogue of the global land rush simply becomes more conservative. In the latter case, projected net LULCC futures (4.5–6 Mha p.a.) will amount to gross LULCC above 6 Mha p.a., likely exceeding what the Land Matrix database has tracked for the 'peak period' of the global land rush, albeit sustained throughout the remainder of the century. Given these considerations, we suggest that the Land Matrix offers a reasonable – potentially even a somewhat conservative – qualitative analogue that allows us to make sense of the projected scale and rate of mitigation-related transformations at the landscape scale.

Differently put, discrete instances of land use or cover change for climate change mitigation can often entail similar socio-economic consequences as conventional land or resource acquisitions. This is particularly so insofar as these changes imply conversion to non-agricultural or non-productive land uses, as well as corresponding opportunity costs or restrictions of access to natural resources for rural populations. Already, a substantial case study literature engaging dynamics of 'green grabbing' (Fairhead, Leach, and Scoones 2012) illustrates how such restrictions may catalyse outright dispossession (e.g. Beymer-Farris and Bassett 2012; Cavanagh and Benjaminsen 2014), as well as more subtle forms of 'control grabbing' (Hall et al. 2015) with negative implications for rural livelihoods (see also Fisher et al. 2018; Cavanagh et al. 2021). Given both the magnitude and rate of projected LULCC across the four scenario archetypes, we thus infer that considerable potential exists for the latter dynamics to entail a significant 'rescaling of the land rush'. This is particularly so in light of the role of the global land rush as a significant contributor to both GHG emissions (Liao et al. 2021) and broader LULCC dynamics in recent history (Winkler et al. 2021), as well as the significant reconfiguration of prevailing historical patterns implied within future LULCC projections (Figures 7–9). Yet the question remains: what lessons, exactly, should critical scholars extract from historical dynamics of land and resource acquisition to guide the analysis of emergent trajectories of land-based climate change mitigation? Seeking to encourage further discussion and debate surrounding these concerns, the ensuing section sketches out an initial series of analytical 'contours' or emerging conceptual foci that may warrant further attention in political ecology, critical agrarian studies, and related fields.

5. Global political ecologies of 1.5 °C land use and cover change futures

To situate key insights from our analysis, we begin by highlighting the likely enduring significance of six key trends that White et al. (2012) argue have underpinned the global land rush in the first two decades of the twenty-first century. These include: 1) 'the global anticipation of food insecurity', which led to a new wave of corporate investment in the agricultural sector; 2) 'new forms of resource extraction for fuel security', incentivizing biofuel production and the acquisition of land for alternative energy projects; 3) 'new environmental imperatives and tools', precipitating resource acquisitions in the name of conservation or environmental change mitigation; 4) 'extensive infrastructure corridors and Special Economic Zones' that link extractive frontiers to metropolitan areas and foreign markets; 5) the 'creation of new financial instruments', which led to speculative and risk hedging investments; and finally, 6) a new 'set of rules, regulations and incentives provided by the international community' that promote land or resource acquisitions institutionally, legislatively, and financially. Taken together, these trends highlight the 'dispossession of land, water, forests and other common property resources; their concentration, privatization and transaction as corporate (owned or leased) property; and [...] the transformation of agrarian labour regimes' within the recent land rush (White et al. 2012). In other words, through the shifting politics and political ecologies of land and resource control, some individuals, communities and/or land users have become gainfully incorporated into land deals. Simultaneously, others have either been excluded outright, or incorporated in accordance with terms and conditionalities that they perceive to be detrimental (Hall et al. 2015).

As Borras et al. (2022a) highlight in their introduction to the present *Journal of Peasant Studies* forum, these dynamics of dispossession, (adverse) inclusion, and outright exclusion increasingly already play out vis-à-vis the uneven impacts of both climate change and attempts at its mitigation. Importantly, these malign outcomes can manifest via processes of 'accumulation [or dispossession] from above' as well as through dynamics of 'accumulation [or dispossession] from below' (Cousins 2013; see also Amanor 2012). That is to say, simultaneous accumulation (by some) and dispossession (of others) may result from large-scale land or resource acquisitions for climate change mitigation, ecosystem restoration, or agribusiness enterprises that are imposed 'from above', for instance via transnational investment flows or donor-driven schemes facilitated at the national and regional scales by state agencies or other intermediaries (Wolford et al. 2013). Yet these intertwined phenomena of simultaneous accumulation and dispossession can also occur at much 'smaller' scales, for instance when the imposition of ostensibly 'climate smart' or 'sustainably intensified' agricultural practices result in the expansion of already well-capitalized small or medium scale enterprises at the expense of less well-capitalized 'competitors' (Franco and Borras 2021; see also Cavanagh et al. 2021). As Borras, Franco, and Nam (2020) remind us, ostensibly 'non-corporate' or smaller-scale land transactions can contribute to land rush dynamics in ways that either *exacerbate* climate change (by heightening emissions from intensified agricultural production – see Liao et al. 2021) or contribute to its mitigation (i.e. when medium-scale enterprises respond to new economic incentives for emissions reductions, land use change, or land cover change, displacing already marginal households or agricultural producers in the process).

Ultimately, these emerging concerns once again underscore White et al.'s (2012, 620) observation that recent rush dynamics have threatened to precipitate 'a truly wide-ranging "land reform"', albeit a largely regressive one in which 'governments take land from the poor and give (or sell or lease) it to the rich'. Reflecting on the recent turn toward mainstreaming climate change adaptation and mitigation concerns in land governance frameworks, Borras and Franco (2018, 1314) thus rightfully highlight how tendencies to favour 'the landed classes and elite actors engaged in capital accumulation, while marginalising social justice land policies of redistribution, recognition and restitution' threaten to become entrenched in prevailing climate change responses as well. As Section 4 has illustrated above, projected LULCC for climate change mitigation can certainly be conceptualized as a wide-ranging land reform at the global scale. What remains less clear, however, is the extent to which land-based climate mitigation initiatives will further amplify these *regressive* patterns of land reform, in practice amounting to a 'rescaling of the land rush', as it were.

Three key insights from the above analysis of our four scenario archetypes suggest that projected LULCC futures may indeed indicate an incipient rescaling of the land rush in this regard, amounting to what Borras, Franco, and Nam (2020, 2) term 'regressive climate change politics' (also see Franco and Borras 2021). Firstly, it is notable that none of the above four scenario archetypes overtly explore post-capitalist or post-growth pathways. The SSP2 ('middle of the road') scenario, for instance, assumes that future development trajectories will 'not shift markedly from historical patterns' (Riahi et al. 2017). Likewise, SSP5 ('fossil fuelled development') echoes a return to the spirit of the post-World War II era, assuming the continuation of 'resource and energy intensive lifestyles' through rapid economic growth, albeit in ways that are offset through the deployment of large-scale 'techno-fixes', such as afforestation or BECCS (Riahi et al. 2017). By contrast, the SSP1 ('sustainability') archetype anticipates decreasing inequality between and within countries, underpinned by wider transitions to a broadly-defined 'green economy' (O'Neill et al. 2017). More radically, the LED ('low energy demand') scenario envisions the pursuit of sustainable development via considerable reduction in energy demand and material throughput, particularly in the Global North (Grubler et al. 2018) – albeit in ways that largely remain ambivalent about possibilities to maintain modest levels of compounding economic growth. Differently put, the four archetypes continue relying on economic growth – albeit in progressively more 'green' iterations over the coming decades – and in which processes of compounding GDP growth are absolutely decoupled from greenhouse gas emissions over time.

Such an aversion to explicitly post-growth scenarios is concerning, given that a rescaled land rush to mitigate climate change in a broadly 'green' capitalist scenario risks perpetuating most – if not all – of the six trends that White et al. (2012) have observed in the recent 'peak' land rush period, and particularly so in the Global South. Indeed, past correlations of heightened land acquisitions, food prices, and energy prices (see Figure 10, bottom panel) are perhaps especially concerning in light of emerging geopolitical developments, the implications of which will reverberate for years – if not the next decade and more – to come. Whilst energy prices declined somewhat in recent years (2015–2020), both food prices and energy prices are once again rising sharply in the tumultuous context of both the global COVID-19 pandemic and the ongoing war in Ukraine. In particular, the second-order implications of the latter are expected to significantly disrupt global markets for energy, agricultural commodities, and – importantly – fertilizers. Indeed, following the outbreak of the war in Ukraine, the food price index reached an all-time high of 159.7 in March 2022 – almost 21% above the 2011 peak, and nearly 16% higher than the previous historical maximum, recorded during the OPEC oil crisis in 1974. Similarly, Brent spot prices have rallied in early 2022 to an average of almost 108 USD over the year to date. This represents a 63% increase in the food price index and a 157% increase in average Brent spot prices from 2020 levels, respectively. In short, this confluence of trends raises concerns about the extent to which 'the global anticipation of food insecurity' (White et al. 2012) will once again precipitate a global rush for land and resources, albeit this time mapping onto the context of emergent mitigation-related LULCC via afforestation and biofuel plantations at the expense of cropland and pasture on a vast scale.

Secondly, projected LULCC in the four scenario archetypes is envisioned to take place under radically different demographic conditions relative to the recent historical record. Over the course of the late twentieth and early twenty-first centuries, land use and cover change dynamics in the Global South were underpinned by growing rural populations and an increasing – or at least not stagnant – land base for cropland and pasture (Figure 7). By contrast, all four scenario archetypes project a massive loss of pasture and cropland, dovetailing with a rapidly shrinking rural population. In other words, this radical reversal of historical LULCC trends in terms of land availability and rural demography implies an acceleration of historical patterns of de-agrarianization and de-peasantization (Bryceson, Kay, and Mooij 2000), realized through ostensibly 'sustainable' forms of agricultural intensification and associated practices of 'land sparing' to facilitate growth in forest land cover or other mitigation-related land uses.

In this respect, contemporary land rush dynamics may once again provide us with a glimpse into the implications of anticipated LULCC trajectories to meet the 1.5 °C Paris agreement target. As Peters (2013, 538) cautions with respect to the future 'upscaling' of large-scale agriculture on the African continent, in particular:

> If the currently influential view that large-scale agriculture is the only and proper way to produce foods and other agricultural products maintains its dominance, the fate of people who live on and from that land is to be rendered 'surplus' to perceived development needs. At best, the agro-industrial vision of the future marginalizes small-medium scale farming into enclaves or as appendages to large-scale, industrial agriculture; at worst, it is erased.

Differently put, if land-based climate mitigation futures unfold as projected through the combination of a rescaled land rush, exponentially increasing carbon prices, and rural depopulation (see Figure 8), rural spaces across the Global South will become increasingly economically valuable as carbon sinks and 'sustainably intensified' agricultural zones through a new wave of enclosures. Conversely, large segments of (formerly) rural populations will at least initially be rendered surplus to capital's requirements, pending their ability to transition to non-agricultural livelihoods in cities or other urban areas. In short, there is a significant risk that dispossession from rural landscapes through yet another wave of new enclosures will underpin land-based climate mitigation responses across the four scenario archetypes.

Thirdly – and perhaps most importantly – the above dynamics of simultaneous rural depopulation and enclosure once again highlight the need for a corresponding analysis of shifting human settlement patterns vis-à-vis the production of what Marx (1990 [1867], 782–802) once termed 'relative surplus populations', or populations that are 'superfluous to capital's average requirements for its own valorization'. Such dynamics are important to our analysis of projected LULCC futures, given that – in the past – cropland and pastures have expanded in the Global South in lockstep with growing rural and urban populations (see Figures 2 and 7). Broadly, this denotes that rural landscapes have often acted as a 'temporary sink for excess urban workers' (Benanav 2014, 110). At times, this has facilitated practices of 'managing dispossession' (Li 2010) that allow 'latent' surplus populations to assume agrarian livelihoods as peasants or smallholders. In stark contrast, the future mitigation scenarios we have examined in this article largely do not offer such an option across broad swathes of the rural sphere in the Global South, implying that latent surplus populations are expected to become 'manifest' via processes of urbanization and rural depopulation (Benanav 2014, 110, 178).

Global urbanization and rural depopulation of this kind risk prefiguring futures characterized by insecure, informal, and precarious urban labour, with little or no possibility to return to either subsistence-oriented or modestly market-oriented production in rural areas. Yet integrated assessment model scenarios largely cannot directly capture or represent such informal phenomena, given their inevitable simplifications of complex system dynamics (Peng et al. 2021). In this sense, IAMs may not be strictly 'wrong' to claim that certain 1.5 °C-compatible scenarios are underpinned by reduced inequality, continued economic growth, and relatively more 'sustainable' forms of development (e.g. Riahi et al. 2017). That is to say, a reduction of *formal* inequalities and an increase in precarious, *informal* surplus populations are not necessarily mutually exclusive outcomes within the context of these scenario projections.

In sum, the above three insights present us with a series of pressing questions that may warrant further exploration and debate in political ecology, critical agrarian studies, and related fields. For instance – even in futures characterized by the effective limitation of global warming to 1.5 °C above pre-industrial averages by 2100 – how 'successful' will these simultaneous processes of expansive LULCC and widespread urbanization be in unleashing a new phase of capital accumulation through ostensibly 'green' growth? To what extent will such growth in fact gainfully integrate newly landless populations of former smallholders into formal wage relations within the context of rapidly urbanizing human settlements? In other words: how confident should we be that land-based climate mitigation responses will in fact *achieve* the simultaneous ambitions of securing both a spatial and a socio-ecological fix for capital in the 'age of stagnation' (Copley 2022)?

In Jason Moore's (2010, 395) terms, such a 'success' would effectively mark the end of the long economic downturn since the 1970s, which has all-too-often been understood as a 'developmental crisis' of capitalism to be overcome 'through new forms of productivity and plunder' (see also Harvey 2003). On one hand, liberal analysts might contend that there is perhaps nothing *inherently* undesirable about such an outcome, particularly if it is compatible with SSP1-esque ambitions to reduce formal inequalities both between and within nations. Indeed, if Moore (2010) is correct that our present conjuncture has led to the unravelling of the 'cheap' global food regime – as evidenced by the 2007–8 crisis and its aftermath in the form of the global land rush (see also Figure 10) – then a 'rescaled' land rush driven by the implementation of climate mitigation initiatives could conceivably result in such a combined spatial and socio-ecological fix. Conversely, the following observation from Davis (2006, 16) provides two key reasons for concern in this regard:

> Rather than the classical stereotype of the labor-intensive countryside and the capital-intensive industrial metropolis, the Third World now contains many examples of capital-intensive countrysides and labor-intensive deindustrialized cities. 'Overurbanization,' in other words, is driven by the reproduction of poverty, not by the supply of jobs. This is one of the unexpected tracks down which a neoliberal world order is shunting the future.

Differently put, a critical analysis of scenario projections for a 1.5 °C future resonates with Davis' anticipation of incipient 'capital-intensive countrysides' and 'labor-intensive deindustrialized cities'. In the context of Global South, the former eerily resembles newly 'green' projections regarding the rapid decline of croplands and pastures in hitherto undercapitalized rural areas. Indeed, across the projected LULCC futures examined above, these areas are envisioned to rapidly transform into capital-intensive carbon sinks and 'sustainably intensified' agricultural enterprises. By contrast, the latter will increasingly manifest as growing urban spaces to absorb the future landless poor, that 'floating' part of relative surplus population that is slated for displacement from rural areas (Marx 1990 [1867]).

In this respect, we emphasize that our analysis offers a complementary take on the question of 'green labour' in the unevenly emergent green economy. Neimark et al. (2020), for instance, have recently highlighted the rise of what they call the 'eco-precariat' in the context of environmental interventions in rural areas of the Global South. Whilst these authors focus on people labouring precariously *for* environmental interventions, however, here we highlight the simultaneous production of populations who are rendered surplus *through* or as a result of environmental interventions, such that their labor is often not 'needed' except perhaps in a variety of emergent informal or precarious settings (see also Kelly 2011; Fairhead, Leach, and Scoones 2012). Here, Neimark et al. (2020, 8) suggest that – through accumulation by dispossession – land-based rural livelihoods are transformed into a 'working class of proletariat or wage workers' that can be subsequently enrolled in environmental interventions and projects as labourers, leading to the production of what they call the 'eco-precariat' and the 'hyper-eco-precariat'. Whereas this seems to imply a largely 'complete' process of primitive accumulation or accumulation by dispossession, we note that – in practice – the actual extent of proletarianization often remains highly uneven or 'incomplete' (Bluwstein et al. 2018). Differently put, a portion of the rural population previously engaged in land-based rural livelihoods will often not be absorbed into the waged working class – however precarious its status – but rather risks being rendered as an expandable 'surplus' altogether. Simply put, such a predicament constitutes a highly fraught foundation for agrarian politics in future scenarios characterized by both the ostensibly 'successful' and 'sustainable' limitation of global warming to 1.5 °C above pre-industrial levels.

6. Conclusion

Engaging recent debates on multi-sector sustainability transformations in political ecology and related fields, this article has explored the land use and land cover change (LULCC) implications of four scenario archetypes recently foregrounded by the IPCC for limiting global warming to 1.5 °C above pre-industrial averages. Empirically, we have disaggregated LULCC projections across the four scenario archetypes, examining their divergent impacts across several world regions and highlighting a suite of associated socio-environmental justice implications. Overall, we underscore how all four mitigation archetypes may imply a considerable 'rescaling of the land rush' with respect to the spatial extent of projected LULCC, as well as simulated rates of implementation, which evince few empirical analogues in the twentieth and early twenty-first centuries. Differently put, prevailing 1.5 °C scenario archetypes imply unprecedented both rural and urban transformations, and particularly so insofar as they entail a radical reversal of historical LULCC dynamics in the Global South. We note, for instance, that the observed history of twentieth-century pasture and cropland expansion is broadly simulated to rapidly transform into an absolute reduction across the four archetypes, while historical deforestation trends are projected to be reversed via large-scale reforestation and afforestation (involving varying combinations of both 'natural' and 'managed' plantation forests). In other words, the Global South is not only projected to rapidly halt its historically intense rates of cropland and pasture *expansion* (4.4–6.9 Mha p.a.), but to undergo similarly intense rates of cropland and pasture *reduction* (6.2–6.8 Mha p.a.) throughout the remainder of the twenty-first century, thereby dwarfing both historically observed trends and simultaneously projected LULCC dynamics in the Global North.

To contextualize the significance of this reversal of historical LULCC trends, we have drawn upon the recent land rush as an additional spatial 'analogue' or reference point that will be familiar to many political ecologists and critical agrarian studies scholars. Methodologically, we have not sought to formally 'compare' the spatial magnitude of historical land acquisitions with future LULCC projections across the above four scenario archetypes. Indeed, as emphasized above, individual transactions logged in the Land Matrix database may or may not entail either land use or land cover *changes* per se (e.g. from forest cover to cropland or vice versa). Nonetheless, it remains notable that average rates of projected LULCC in the four scenario archetypes resemble rates of acquisition throughout the 'peak period' (2007–2014) of the global land rush, albeit sustained throughout the remainder of the century. Importantly, the database tracks a total of approximately 73 Mha of land deals in the Global South (2000-2020), amounting to an average of roughly 3.5 Mha of acquisitions per year, or 6.6 Mha per year throughout the 'peak period' of the global land rush. This is notable, given that – across the SSPx and LED archetypes – mean future LULCC in the Global South is projected at approximately 6 Mha p.a. of pasture contraction and 4.5–6 Mha p.a. of forest cover expansion. Even the Low Energy Demand (LED) scenario – promoted by critical scholars for its avoidance of certain land-intensive mitigation initiatives, such as BECCS (e.g. Hickel 2019) – projects an increase in forest land cover alone amounting to approximately 646 Mha globally and 470 Mha in the Global South between 2020 and 2100, or around 8 Mha per year on average (5.9 Mha p.a. in the Global South). Simply put, the sheer scale and rate of these LULCC projections even in low(er) energy demand scenarios raise important questions about their articulation with both contemporary and historical land rush dynamics, as well as their corresponding potential to be accompanied by related socio-environmental injustices.

As a result of these and similar risks, our analysis has sought to illuminate associated conceptual implications for the present *Journal of Peasant Studies* forum, highlighting concerns related to rural enclosures, primitive accumulation or 'accumulation by dispossession', and the implied production of relative surplus populations in future scenario archetypes. Here, we conclude by emphasizing that – despite these risks – none of the 1.5 °C scenario archetypes examined above would be likely to amount 'in practice' to what one might term a 'conclusively' stable political-ecological formation (see Wainwright and Mann 2013). Indeed, each of the four archetypes are potentially characterized by their own implicit contradictions or latent political-ecological tensions, some of which may precipitate the emergence of what Franco and Borras (2019) call 'agrarian climate justice' movements. That is to say, mitigation futures characterized by 'green' growth, large-scale land sparing dynamics through LULCC, and widespread urbanization may only temporarily yield spatial or socio-ecological 'fixes' for capital. Conversely, these dynamics will inevitably also foment contradictory political-ecological tendencies of their own, which will almost certainly precipitate further contestation, mobilization, and conflict. Many rural people in the Global South, for instance, will simply not volunteer to be dispossessed and urbanized into informal and precarious wage relations without the prospect of a suitable and/or viable alternative livelihood. Similarly, both the urban poor and middle classes will undoubtedly oppose the erosion of their standard of living via food or energy price inflation, and particularly so if 'demand reduction' initiatives are haphazardly implemented, resulting in demand *destruction* via sustained price increases instead. Already, a growing segment of the urban – or 'newly urbanizing' – poor is increasingly engaging in struggles demanding basic rights and respect for human dignity, at times escalating into strikes, riots, and other forms of organized or semi-organized resistance (Davis 2006; Harvey 2012; Clover 2019). Ultimately, this predicament echoes Tania Li's (2011, 281) observation that 'any program that robs rural people of their foothold on the land must be firmly rejected [...] unless vast numbers of jobs are created, or a global basic income grant is devised'. Land-based climate mitigation under predominantly capitalist political-ecological conditions (perhaps approximating the 'climate leviathan' scenario described by Wainwright and Mann 2013) may represent such a programme, with a rescaled land rush and the production of an unprecedented eco-surplus population as its outcome.

Acknowledgments

Both authors contributed equally to this article. Order of authorship is alphabetical. We would like to thank the editors of the *Journal of Peasant Studies* Forum on Climate Change and Critical Agrarian Studies, as well as the three anonymous reviewers, for helpful comments on past versions of the manuscript.

Disclosure statement

No potential conflict of interest was reported by the author(s).

ORCID

Jevgeniy Bluwstein http://orcid.org/0000-0002-1162-5028
Connor Cavanagh http://orcid.org/0000-0001-8373-2124

References

Agarwal, A., and S. Narain. 1991. *Global Warming in an Unequal World*. New Delhi: Centre for Science and Environment.

Akram-Lodhi, A. H., K. Dietz, B. Engels, and B. M. McKay. 2021. "An Introduction to the Handbook of Critical Agrarian Studies." In *Handbook of Critical Agrarian Studies*, edited by A. H. Akram-Lodhi, K. Dietz, B. Engels, and B. M. McKay, 1–7. Cheltenham: Edward Elgar.

Amanor, K. S. 2012. "Global Resource Grabs, Agribusiness Concentration and the Smallholder: Two West African Case Studies." *The Journal of Peasant Studies* 39 (3-4): 731–749. doi:10.1080/03066150.2012.676543

Anderson, K., and G. Peters. 2016. "The Trouble with Negative Emissions." *Science* 354 (6309): 182–183. doi:10.1126/science.aah4567

Asiyanbi, A., and K. Massarella. 2020. "Transformation is What you Expect, Models are What you get: REDD+ and Models in Conservation and Development." *Journal of Political Ecology* 27 (1): 476–495.

Benanav, A. S. 2014. *A Global History of Unemployment: Surplus Populations in the World Economy, 1949-2010*. Berkeley: University of California Press.

Benjaminsen, T. A., and I. Bryceson. 2012. "Conservation, Green/Blue Grabbing and Accumulation by Dispossession in Tanzania." *Journal of Peasant Studies* 39 (2): 335–355.

Beymer-Farris, B. A., and T. J. Bassett. 2012. "The REDD Menace: Resurgent Protectionism in Tanzania's Mangrove Forests." *Global Environmental Change* 22 (2): 332–341. doi:10.1016/j.gloenvcha.2011.11.006

Blaikie, P., and H. Brookfield. 1987. *Land Degradation and Society*. London: Methuen.

Bluwstein, J., J. F. Lund, K. Askew, H. Stein, C. Noe, R. Odgaard, F. Maganga, and L. Engström. 2018. "Between Dependence and Deprivation: The Interlocking Nature of Land Alienation in Tanzania." *Journal of Agrarian Change* 18: 806–830. doi:10.1111/joac.12271

Borras Jr., S. M., and J. C. Franco. 2018. "The Challenge of Locating Land-Based Climate Change Mitigation and Adaptation Politics Within a Social Justice Perspective: Towards an Idea of Agrarian Climate Justice." *Third World Quarterly* 39 (7): 1308–1325. doi:10.1080/01436597.2018.1460592

Borras Jr., S. M., J. C. Franco, T. Moreda, Y. Xu, N. Bruna, and B. A. Demena. 2022b. "The Value of so-Called 'Failed' Large-Scale Land Acquisitions." *Land Use Policy* 119: 106199. doi:10.1016/j.landusepol.2022.106199

Borras Jr., S. M., J. C. Franco, and Z. Nam. 2020. "Climate Change and Land: Insights from Myanmar." *World Development* 129: 104864. doi:10.1016/j.worlddev.2019.104864

Borras Jr., S. M., R. Hall, I. Scoones, B. White, and W. Wolford. 2011. "Towards a Better Understanding of Global Land Grabbing: An Editorial Introduction." *Journal of Peasant Studies* 38 (2): 209–216. doi:10.1080/03066150.2011.559005

Borras Jr., S. M., I. Scoones, A. Baviskar, M. Edelman, N. L. Peluso, and W. Wolford. 2022a. "Climate Change and Agrarian Struggles: An Invitation to Contribute to a JPS Forum." *The Journal of Peasant Studies* 49 (1): 1–28. doi:10.1080/03066150.2021.1956473

Bryceson, D. F., C. Kay, and J. Mooij. 2000. *Disappearing Peasantries? Rural Labor in Africa, Asia and Latin America*. London, UK: Intermediate Technology Publications.

Buck, H. 2019. *After Geoengineering: Climate Tragedy, Repair, and Restoration*. New York: Verso.

Bumpus, A. G. 2011. "The Matter of Carbon: Understanding the Materiality of tCO2e in Carbon Offsets." *Antipode* 43 (3): 612–638. doi:10.1111/j.1467-8330.2011.00879.x

Carton, W. 2019. ""Fixing" Climate Change by Mortgaging the Future: Negative Emissions, Spatiotemporal Fixes, and the Political Economy of Delay." *Antipode* 51 (3): 750–769. doi:10.1111/anti.12532

Carton, W., A. Asiyanbi, S. Beck, H. J. Buck, and J. F. Lund. 2020. "Negative Emissions and the Long History of Carbon Removal." *WIRES Climate Change* 11 (6): e671. doi:10.1002/wcc.671

Carton, W., J. F. Lund, and K. Dooley. 2021. "Undoing Equivalence: Rethinking Carbon Accounting for Just Carbon Removal." *Frontiers in Climate* 3: 664130. doi:10.3389/fclim.2021.664130

Cavanagh, C. J. 2021. "Limits to (de)Growth: Theorizing 'the Dialectics of Hatchet and Seed' in Emergent Socio-Ecological Transformations." *Political Geography* 90: 102479. doi:10.1016/j.polgeo.2021.102479

Cavanagh, C., and T. A. Benjaminsen. 2014. "Virtual Nature, Violent Accumulation: The 'Spectacular Failure' of Carbon Offsetting at a Ugandan National Park." *Geoforum; Journal of Physical, Human, and Regional Geosciences* 56: 55–65. doi:10.1016/j.geoforum.2014.06.013

Cavanagh, C. J., and T. A. Benjaminsen. 2017. "Political Ecology, Variegated Green Economies, and the Foreclosure of Alternative Sustainabilities." *Journal of Political Ecology* 24: 200–216. doi:10.2458/v24i1.20800

Cavanagh, C. J., P. O. Vedeld, J. G. Petursson, and A. K. Chemarum. 2021. "Agency, Inequality, and Additionality: Contested Assemblages of Agricultural Carbon Finance in Western Kenya." *The Journal of Peasant Studies* 48 (6): 1207–1227. doi:10.1080/03066150.2019.1707812

Clover, J. 2019. *Riot. Strike. Riot: The new era of Uprisings*. New York and London: Verso Books.

Copley, J. 2022. "Decarbonizing the Downturn: Addressing Climate Change in an age of Stagnation." *Competition & Change*. doi:10.1177/10245294221120986

Cotula, L. 2013. "The New Enclosures? Polanyi, International Investment law and the Global Land Rush." *Third World Quarterly* 34 (9): 1605–1629. doi:10.1080/01436597.2013.843847

Cousins, B. 2013. "Smallholder Irrigation Schemes, Agrarian Reform and 'Accumulation from Above and from Below' in South Africa." *Journal of Agrarian Change* 13 (1): 116–139. doi:10.1111/joac.12000

Davis, M. 2006. *Planet of Slums*. New York: Verso.

Davis, K. F., M. C. Rulli, and P. D'Odorico. 2015. "The Global Land Rush and Climate Change." *Earth's Future* 3 (8): 298–311. doi:10.1002/2014EF000281

Dell'Angelo, J., P. D'Odorico, M. C. Rulli, and P. Marchand. 2017. "The Tragedy of the Grabbed Commons: Coercion and Dispossession in the Global Land Rush." *World Development* 92: 1–12. doi:10.1016/j.worlddev.2016.11.005

Dell'Angelo, J., M. C. Rulli, and P. D'Odorico. 2018. "The Global Water Grabbing Syndrome." *Ecological Economics* 143: 276–285. doi:10.1016/j.ecolecon.2017.06.033

Doelman, J. C., E. Stehfest, A. Tabeau, and H. van Meijl. 2019. "Making the Paris Agreement Climate Targets Consistent with Food Security Objectives." *Global Food Security* 23: 93–103. doi:10.1016/j.gfs.2019.04.003

Doelman, J. C., E. Stehfest, D. P. van Vuuren, A. Tabeau, A. F. Hof, M. C. Braakhekke, … H. van Meijl. 2020. "Afforestation for Climate Change Mitigation: Potentials, Risks and Trade-Offs." *Global Change Biology* 26 (3): 1576–1591. doi:10.1111/gcb.14887

Dooley, K., and S. Kartha. 2018. "Land-based Negative Emissions: Risks for Climate Mitigation and Impacts on Sustainable Development." *International Environmental Agreements: Politics, Law and Economics* 18 (1): 79–98. doi:10.1007/s10784-017-9382-9

Dunlap, A., and L. Laratte. 2022. "European Green Deal Necropolitics: Exploring 'Green' Energy Transition, Degrowth & Infrastructural Colonization." *Political Geography* 97: 102640. doi:10.1016/j.polgeo.2022.102640

Edelman, M. 2013. "Messy Hectares: Questions About the Epistemology of Land Grabbing Data." *Journal of Peasant Studies* 40 (3): 485–501. doi:10.1080/03066150.2013.801340

Edelman, M., and W. Wolford. 2017. "Introduction: Critical Agrarian Studies in Theory and Practice." *Antipode* 49 (4): 959–976. doi:10.1111/anti.12326

Fairhead, J., M. Leach, and I. Scoones. 2012. "Green Grabbing: A new Appropriation of Nature?" *Journal of Peasant Studies* 39 (2): 237–261. doi:10.1080/03066150.2012.671770

Fisher, J. A., C. J. Cavanagh, T. Sikor, and D. M. Mwayafu. 2018. "Linking Notions of Justice and Project Outcomes in Carbon Offset Forestry Projects: Insights from a Comparative Study in Uganda." *Land Use Policy* 73: 259–268. doi:10.1016/j.landusepol.2017.12.055

Franco, J. C., and S. M. BorrasJr. 2019. "Grey Areas in Green Grabbing: Subtle and Indirect Interconnections Between Climate Change Politics and Land Grabs and Their Implications for Research." *Land Use Policy* 84: 192–199. doi:10.1016/j.landusepol.2019.03.013

Franco, J. C., and S. M. BorrasJr. 2021. "The Global Climate of Land Politics." *Globalizations* 18 (7): 1277–1297. doi:10.1080/14747731.2021.1979717

Gambhir, A., Butnar, I., Li, P. H., Smith, P., and Strachan, N. (2019). A review of criticisms of integrated assessment models and proposed approaches to address these, through the lens of BECCS. *Energies*, 12 (9), 1747.

Geisler, C. 2012. "New Terra Nullius Narratives and the Gentrification of Africa's "Empty Lands"." *Journal of World-Systems Research* 18 (1): 15–29. doi:10.5195/jwsr.2012.484

Grubler, A., C. Wilson, N. Bento, B. Boza-Kiss, V. Krey, D. L. McCollum, … J. Cullen. 2018. "A low Energy Demand Scenario for Meeting the 1.5 °C Target and Sustainable Development Goals Without Negative Emission Technologies." *Nature Energy* 3 (6): 515–527. doi:10.1038/s41560-018-0172-6

Hall, R., M. Edelman, S. M. Borras, I. Scoones, B. White, and W. Wolford. 2015. "Resistance, Acquiescence or Incorporation? An Introduction to Land Grabbing and Political Reactions 'from Below'." *The Journal of Peasant Studies* 42 (3-4): 467–488. doi:10.1080/03066150.2015.1036746

Harvey, D. 2003. *The new Imperialism*. Oxford: Oxford UP.

Harvey, D. 2012. *Rebel Cities: From the Right to the City to the Urban Revolution*. New York: Verso books.

Hasegawa, T., S. Fujimori, P. Havlík, H. Valin, B. L. Bodirsky, J. C. Doelman, … D. Mason-D'Croz. 2018. "Risk of Increased Food Insecurity Under Stringent Global Climate Change Mitigation Policy." *Nature Climate Change* 8 (8): 699–703. doi:10.1038/s41558-018-0230-x

Henry, R. C., A. Arneth, M. Jung, S. S. Rabin, M. D. Rounsevell, F. Warren, and P. Alexander. 2022. "Global and Regional Health and Food Security Under Strict Conservation Scenarios." *Nature Sustainability* 5 (4): 303–310.

Hickel, J. 2019. "Degrowth: A Theory of Radical Abundance." *Real-World Economics Review* 87 (87): 54–68.

Huppmann, D., E. Kriegler, V. Krey, K. Riahi, J. Rogelj, S. K. Rose, J. Weyant, et al. 2018. *IAMC 1.5°C Scenario Explorer and Data Hosted by IIASA*. Vienna: Integrated Assessment Modeling Consortium & International Institute for Applied Systems Analysis. http://doi.org/10.22022/SR15/08-2018.15429 (accessed 22.10.2020).

Hurlbert, M., J. Krishnaswamy, E. Davin, F. X. Johnson, C. F. Mena, J. Morton, … Z. Zommers. 2019. "Risk Management and Decision Making in Relation to Sustainable Development." In *Climate Change and Land: IPCC Special Report*, 673–800. Geneva: IPCC.

Intergovernmental Panel on Climate Change (IPCC). 2018. *Special Report on Global Warming of 1.5 °C*. Geneva: IPCC.

Intergovernmental Panel on Climate Change (IPCC). 2020. *Climate Change and Land: IPCC Special Report. Revised 2020 Edition*. Geneva: IPCC.

Jiang, L., and B. C. O'Neill. 2017. "Global Urbanization Projections for the Shared Socioeconomic Pathways." *Global Environmental Change* 42: 193–199. doi:10.1016/j.gloenvcha.2015.03.008

Kelly, A. B. 2011. "Conservation Practice as Primitive Accumulation." *Journal of Peasant Studies* 38 (4): 683–701. doi:10.1080/03066150.2011.607695

Kemp, L., C. Xu, J. Depledge, K. L. Ebi, G. Gibbins, T. A. Kohler, … T. M. Lenton. 2022. "Climate Endgame: Exploring Catastrophic Climate Change Scenarios." *Proceedings of the National Academy of Sciences* 119 (34): e2108146119. doi:10.1073/pnas.2108146119

Keppo, I., I. Butnar, N. Bauer, M. Caspani, O. Edelenbosch, J. Emmerling, ... F. Wagner. 2021. "Exploring the Possibility Space: Taking Stock of the Diverse Capabilities and Gaps in Integrated Assessment Models." *Environmental Research Letters* 16 (5): 053006.

Kriegler, E., N. Bauer, A. Popp, F. Humpenöder, M. Leimbach, J. Strefler, … I. Mouratiadou. 2017. "Fossil-fueled Development (SSP5): An Energy and Resource Intensive Scenario for the 21st Century." *Global Environmental Change* 42: 297–315. doi:10.1016/j.gloenvcha.2016.05.015

Land Matrix. 2022. Frequently Asked Questions (FAQ). International Land Coalition (ILC), Centre de Coopération Internationale en Recherche Agronomique pour le Développement (CIRAD), Centre for Development and Environment (CDE), German Institute of Global and Area Studies (GIGA) and Deutsche Gesellschaft für Internationale Zusammenarbeit (GIZ). https://landmatrix.org/faq/ (accessed 08 September 2020).

Lay, J., W. Anseeuw, S. Eckert, I. Flachsbarth, C. Kubitza, K. Nolte, and M. Giger. 2021. *Taking Stock of The Global Land Rush: Few Development Benefits, Many Human and Environmental Risks*. Analytical Report III. Hamburg: German Institute for Global and Area Studies.

Leach, M., K. Raworth, and J. Rockström. 2013. "Between Social and Planetary Boundaries: Navigating Pathways in the Safe and Just Space for Humanity." In *World Social Science Report 2013: Changing Global Environments*, edited by H. Hackmann, 84–89. Paris: UNESCO.

Leach, M., and I. Scoones. 2015. "Political Ecologies of Carbon in Africa." In *Carbon Conflicts and Forest Landscapes in Africa*, edited by M. Leach and I. Scoones, 1–42. London and New York: Routledge.

Li, T. M. 2010. "Indigeneity, Capitalism, and the Management of Dispossession." *Current Anthropology* 51 (3): 385–414. doi:10.1086/651942

Li, T. M. 2011. "Centering Labor in the Land Grab Debate." *Journal of Peasant Studies* 38 (2): 281–298. doi:10.1080/03066150.2011.559009

Li, T. M. 2014. "What is Land? Assembling a Resource for Global Investment." *Transactions of the Institute of British Geographers* 39 (4): 589–602. doi:10.1111/tran.12065

Li, T. M. 2017. "After Development: Surplus Population and the Politics of Entitlement." *Development and Change* 48 (6): 1247–1261. doi:10.1111/dech.12344

Liao, C., K. Nolte, J. A. Sullivan, D. G. Brown, J. Lay, C. Althoff, and A. Agrawal. 2021. "Carbon Emissions from the Global Land Rush and Potential Mitigation." *Nature Food* 2 (1): 15–18. doi:10.1038/s43016-020-00215-3

Marx, K. 1990 [1867]. *Capital: A Critique of Political Economy* (Vol. 1). Moscow: Progress Publishers.

Moore, J. W. 2010. "The End of the Road? Agricultural Revolutions in the Capitalist World-Ecology, 1450-2010." *Journal of Agrarian Change* 10 (3): 389–413. doi:10.1111/j.1471-0366.2010.00276.x

Neimark, B., S. Mahanty, W. Dressler, and C. Hicks. 2020. "NotJustParticipation: The Rise of the Eco-Precariat in the Green Economy." *Antipode* 52 (2): 496–521. doi:10.1111/anti.12593

Newell, P. 2022. "Climate Justice." *Journal of Peasant Studies*, 1–9. doi:10.1080/03066150.2022.2080062

Newell, P., M. Paterson, and M. Craig. 2021. "The Politics of Green Transformations: An Introduction to the Special Section." *New Political Economy* 26 (6): 903–906. doi:10.1080/13563467.2020.1810215

O'Neill, B. C., E. Kriegler, K. L. Ebi, E. Kemp-Benedict, K. Riahi, D. S. Rothman, … M. Levy. 2017. "The Roads Ahead: Narratives for Shared Socioeconomic Pathways Describing World Futures in the 21st Century." *Global Environmental Change* 42: 169–180. doi:10.1016/j.gloenvcha.2015.01.004

Oya, C. 2013. "Methodological Reflections on 'Land Grab' Databases and the 'Land Grab' Literature 'Rush'." *Journal of Peasant Studies* 40 (3): 503–520. doi:10.1080/03066150.2013.799465

Peng, W., G. Iyer, V. Bosetti, V. Chaturvedi, J. Edmonds, A. A. Fawcett, … J. Weyant. 2021. "Climate Policy Models Need to get Real About People — Here's how." *Nature* 594 (7862): 174–176. doi:10.1038/d41586-021-01500-2

Peters, P. E. 2013. "Land Appropriation, Surplus People and a Battle Over Visions of Agrarian Futures in Africa." *Journal of Peasant Studies* 40 (3): 537–562. doi:10.1080/03066150.2013.803070

Popp, A., K. Calvin, S. Fujimori, P. Havlik, F. Humpenöder, E. Stehfest, … T. Hasegawa. 2017. "Land-use Futures in the Shared Socio-Economic Pathways." *Global Environmental Change* 42: 331–345. doi:10.1016/j.gloenvcha.2016.10.002

Potapov, P., S. Turubanova, M. C. Hansen, A. Tyukavina, V. Zalles, A. Khan, … J. Cortez. 2022. "Global Maps of Cropland Extent and Change Show Accelerated Cropland Expansion in the Twenty-First Century." *Nature Food* 3 (1): 19–28. doi:10.1038/s43016-021-00429-z

Riahi, K., D. P. Van Vuuren, E. Kriegler, J. Edmonds, B. C. O'neill, S. Fujimori, … W. Lutz. 2017. "The Shared Socioeconomic Pathways and Their Energy, Land use, and Greenhouse gas Emissions Implications: An Overview." *Global Environmental Change* 42: 153–168. doi:10.1016/j.gloenvcha.2016.05.009

Rogelj, J., A. Popp, K. V. Calvin, G. Luderer, J. Emmerling, D. Gernaat, … M. Tavoni. 2018a. "Scenarios Towards Limiting Global Mean Temperature Increase Below 1.5 °C." *Nature Climate Change* 8 (4): 325–332. doi:10.1038/s41558-018-0091-3

Rogelj, J., D. Shindell, K. Jiang, S. Fifita, P. Forster, V. Ginzburg, … M. V. Vilariño. 2018b. "Mitigation Pathways Compatible with 1.5°C in the Context of Sustainable Development." In *Global Warming of 1.5°C: Special Report*, 93–174. Cambridge: Cambridge University Press.

Rulli, M. C., and P. D'Odorico. 2013. "The Science of Evidence: The Value of Global Studies on Land Rush." *Journal of Peasant Studies* 40 (5): 907–909. doi:10.1080/03066150.2013.853044

Rulli, M. C., A. Saviori, and P. D'Odorico. 2013. "Global Land and Water Grabbing." *Proceedings of the National Academy of Sciences* 110 (3): 892–897. doi:10.1073/pnas.1213163110

Scoones, I., R. Hall, S. M. Borras, B. White, and W. Wolford. 2013. "The Politics of Evidence: Methodologies for Understanding the Global Land Rush." *Journal of Peasant Studies* 40 (3): 469–483. doi:10.1080/03066150.2013.801341

Scoones, I., A. Stirling, D. Abrol, J. Atela, L. Charli-Joseph, H. Eakin, … P. van Zwanenberg. 2020. "Transformations to Sustainability: Combining Structural, Systemic and Enabling Approaches." *Current Opinion in Environmental Sustainability* 42: 65–75. doi:10.1016/j.cosust.2019.12.004

Skeie, R. B., G. P. Peters, J. Fuglestvedt, and R. Andrew. 2021. "A Future Perspective of Historical Contributions to Climate Change." *Climatic Change* 164 (1): 1–13.

Sovacool, B. K., L. Baker, M. Martiskainen, and A. Hook. 2019. "Processes of Elite Power and low-Carbon Pathways: Experimentation, Financialisation, and Dispossession." *Global Environmental Change* 59: 101985. doi:10.1016/j.gloenvcha.2019.101985

Sovacool, B. K., A. Hook, M. Martiskainen, A. Brock, and B. Turnheim. 2020. "The Decarbonisation Divide: Contextualizing Landscapes of low-Carbon Exploitation and Toxicity in Africa." *Global Environmental Change* 60: 102028. doi:10.1016/j.gloenvcha.2019.102028

van Vuuren, D. P., E. Stehfest, D. E. Gernaat, J. C. Doelman, M. Van den Berg, M. Harmsen, … B. Girod. 2017. "Energy, Land-use and Greenhouse gas Emissions Trajectories Under a Green Growth Paradigm." *Global Environmental Change* 42: 237–250. doi:10.1016/j.gloenvcha.2016.05.008

Wainwright, J., and G. Mann. 2013. "Climate Leviathan." *Antipode* 45 (1): 1–22. doi:10.1111/j.1467-8330.2012.01018.x

White, B., S. M. BorrasJr, R. Hall, I. Scoones, and W. Wolford. 2012. "The new Enclosures: Critical Perspectives on Corporate Land Deals." *The Journal of Peasant Studies* 39 (3-4): 619–647. doi:10.1080/03066150.2012.691879

Wily, L. A. 2012. "Looking Back to see Forward: The Legal Niceties of Land Theft in Land Rushes." *The Journal of Peasant Studies* 39 (3-4): 751–775. doi:10.1080/03066150.2012.674033

Winkler, K., R. Fuchs, M. Rounsevell, and M. Herold. 2021. "Anomalous Collapses of Nares Strait ice Arches Leads to Enhanced Export of Arctic sea ice." *Nature Communications* 12 (1): 1–10. doi:10.1038/s41467-020-20314-w

Wolford, W., S. M. BorrasJr, R. Hall, I. Scoones, and B. White. 2013. "Governing Global Land Deals: The Role of the State in the Rush for Land." *Development and Change* 44 (2): 189–210. doi:10.1111/dech.12017

World Bank. 2022. *Urban Population (% of total population)*. https://data.worldbank.org/indicator/SP.URB.TOTL.IN.ZS

Producing nature-based solutions: infrastructural nature and agrarian change in San Martín, Peru

Will Lock

ABSTRACT

The concept of 'nature-based solutions' has brought together a broad range of actors with seemingly diverse goals. This article, contributing to the forum on climate change and agrarian struggles, argues that these nature-based solutions follow a growing trend to frame the natural world as infrastructure, providing an ethnographic case study of agroforestry and reforestation projects in San Martín, Peru. It shows that as infrastructural nature claims ever more of the metaphorical space of discourse and the material space of landscapes, it becomes not just a new way of producing nature but central to the ideologies and ecologies of green capitalism.

Introduction

The recent rise in interest in nature-based solutions has seen carbon markets become increasingly integrated into rural politics and development strategies. In turn, carbon has become central to agrarian change in a number of contexts. As will be explored in this article, Peru is a paradigmatic case. A national drive to promote carbon-based projects has been met with regional enthusiasm to expand reforestation schemes and integrate carbon credits into export commodity production systems through agroforestry. A broad consensus of policymakers, development NGOs and private sector actors have buttressed this trend, creating a powerful development narrative in the forest frontiers of Peru's Amazonian regions, which is producing distinct landscapes and livelihoods.

This article explores the role of infrastructural nature in these emerging carbon offsetting landscapes, contributing to the forum on climate change and agrarian struggles. Responding to Borras Jr. et al.'s (2022, 5) call to 'situate "climate" within a wider set of environmental struggles in agrarian settings', it analyses the narratives and strategies that frame carbon offsetting, but considers too the kind of socionatures it produces – or constrains. Engaging with an emerging literature on the anthropology of infrastructure (Star 1999; Dourish and Bell 2007; Venkatesan et al. 2018) and the production of nature

(Ekers and Loftus 2013; Boyd and Prudham 2017; Carton and Andersson 2017), the article will show how nature-based solutions are driving a view of the natural world that ontologically and epistemologically conflates it with that of the built environment.

The logic behind nature-based solutions follows a growing trend to frame the natural world as infrastructure, reflected in both its use in practitioner and policy circles, alongside increasing academic interest (Nelson and Bigger 2022). Studies of 'infrastructural nature' have explored the historical tendency to treat nature as infrastructure (Swyngedouw 2007; Carse 2012) and the framing of infrastructural nature in modern iterations of 'ecosystem management' (Wakefield 2020). This article seeks to push these ideas further by grounding the discourse in a tangible ethnographic case study of nature-based solutions in the region of San Martín, Peru. In doing so, the article firstly explores how we might operationalise infrastructure as an analytical tool for political ecology and critical agrarian studies; expanding on the production of nature that it entails and the risks that it incurs. Secondly, it analyses the ideology that underpins infrastructural nature and its relation to green capitalism; highlighting how narratives of control and stability enrol a greater range of actors into support for nature-based solutions.

In exploring the new landscapes of nature-based solutions, the article shows how viewing the natural world as a distinct type of infrastructure attributes to it the same sense of predictability and quantifiable value that we expect of the built environment. In the region of San Martín, where landscapes are increasingly dominated by supposedly climate-friendly plantations – be they of commodity crops, agroforestry models or timber – infrastructural nature risks underestimating the unruly ecology and economic uncertainty that can arise from treating the natural world in this manner. The tendency towards simplification, technification and repetition of ecologies of control calls into question not just the well-publicised tree planting targets and net zero claims accelerated in the post-Paris governance agenda, but also the very logic that underpins nature-as-infrastructure.

The article begins with a brief summary of the concept of infrastructural nature and its connection to political ecological thought. It will then move on to describe the pursuit of carbon projects in San Martín, drawing on ethnographic fieldwork conducted in a suite of conservation and reforestation projects in the region between 2017 and 2019, to show how it has manifested in tangible policies and landscapes of production. Drawing on participant observation and 68 interviews conducted with foresters, local and international NGOs, regional government workers and cacao, coffee and cattle farmers, it will highlight how certain voices and approaches were prioritised in the region, ultimately shaping the conflicts over socionatural relations and how nature and space are produced. This is followed by a discussion of the potential consequences of this approach for local populations, the regional politics of conservation and global climate change goals.

By introducing the example of San Martín's production of carbon to the forum on climate change and agrarian struggles, this paper situates carbon landscapes as a central axis between global climate governance and local agrarian struggles. As carbon offsetting brings together surprising coalitions of actors under new narratives of rural development, it claims ever more of the metaphorical space of discourse and the material space of landscapes. This paper thus argues that in infrastructural nature we are not just seeing a new way of producing the natural world but dominant new ideologies and ecologies of green capitalism and agrarian change.

The evolution of nature as infrastructure

The conceptual framing of the natural world as a form of infrastructure has a long history in policy, practice and academic work. Swyngedouw (2007), for example, discusses the megaprojects of Franco's Spain in terms of their infrastructural framing of water systems, while Wakefield (2020) highlights the longer genealogy of using green spaces in urban planning and Carse (2012) discusses the ongoing integration of the natural world into the infrastructure of the Panama Canal.

The relevance to modern environmental governance has been perhaps most comprehensively elaborated by Nelson and Bigger (2022, 2), who define infrastructural nature as the 'policy approaches, scientific practices, discourses, and investment strategies that make ecosystems legible, governable, and investable as systems of critical functions that sustain and secure (certain forms of) human life'. While Nelson and Bigger trace this ideology to the emergence of systems ecology in the 1960s and 1970s, similar logics could be traced in numerous projects to bend nature to the will of humans, to the technical and simplified structure of plantations or even to biblical tenets to master the natural world (Merchant 1982).

While such presentations and simplifications of the natural world have long been a focus of political ecology and critical agrarian studies, the language has spread beyond the world of ecology to politicians, non-governmental organisations (NGOs), activists, practitioners and more. Calls to manage the 'earth system' or even to marshal 'spaceship earth'[1] have proliferated in recent years in response to the threat of climate change and form the basis of the nature-based solutions promoted by figures as diverse as Greta Thunberg and Donald Trump, George Monbiot and the executives of Shell.

The growing use of the language and logics of natural infrastructure reflects not just the growth in an interest in 'rebuilding' the natural world but how a whole semantic field connecting architecture and construction is being integrated into environmental governance. The Executive Secretary of the Convention on Biological Diversity, for example, has noted how 'biodiversity is the "infrastructure" that keeps our planet going' (Hance, 2018), while the Nature Conservancy argues that we should view nature as 'green infrastructure' to leverage investment (Tercek 2017).

The idea of Green Infrastructure – as opposed to the blue infrastructure of water or more general natural infrastructure or ecological infrastructure – has perhaps pushed this logic furthest in policy circles. While the term has its roots in urban studies, often relating to green spaces within cities or that join the urban with the rural, definitions have been expanded by conservation practitioners and policy makers. The European Union's Green Infrastructure Strategy (EC 2013, 7), for example, refers to it as 'a strategically planned network of natural and semi-natural areas with other environmental features designed and managed to deliver a wide range of ecosystem services', clearly connecting the emerging policy focus on ecosystem services to the discursive logic of infrastructural nature.

While to date the concept that has mostly been applied to trees, parks and other providers of 'green services' in urban settings or managing water systems, this article centres

[1]A term coined by architect Richard Buckminster Fuller. Fuller notably linked his vision of spaceship earth to a time when we wouldn't need the 'distraction' of politics (Murphy 2015).

its role in the growth of nature-based solutions in rural contexts. Green infrastructure in particular has been seen as synonymous with nature-based solutions (Nesshöver et al. 2017), with a common focus on the creation, control and optimisation of ecosystems. Some proponents frame this as a move away from previous extractive approaches with nature 'no longer solely seen as a source of commodities' but inspiring 'more systemic economic solutions' as ecosystems 'evolve to diverse but locally optimal equilibriums between productivity, adaptability, and resilience' (Maes and Jacobs 2017, 122).

Rather than synonymous, infrastructural nature can be seen as an ideological underpinning of nature-based solutions as they are emerging in practice. As noted by Nelson and Bigger (2022), nature is not simply treated as infrastructure and allowed to 'evolve' but *made* to function as such. Wakefield (2020, 776), for example, shows how oysters are discursively produced as infrastructure in New York as a 'risk management solution'. Here, the biological processes of oysters are 'tamed', so as to be used as a plannable, predictable and scalable 'solution' to coastal management and disaster mitigation. Nature, in the words of Carse (2012, 540), must 'be built, invested in, made functional, and managed' in the service of human needs. This requires both a view of human mastery over nature and a view of nature itself as a stable, predictable system to build upon.

While I concur with Wakefield's (2020, 763) assertion that this 'question of how living beings are made to be infrastructure has not been explored' on a biopolitical level, I argue that we must go further than analysing how actors make 'nature do what it does' (776), to analyse how it *produces* very specific types of socionatures. Carse (2012, 551), for example, examines the processes by which 'watershed forests became infrastructure through the purposeful work that went into linking them with the existing water management system'. Here too, further nuance could be developed. Rather than looking at how 'a landscape becomes infrastructure for one system of production' (Carse 2012, 540), approaching this question from a critical agrarian perspective allows us to analyse the overlapping narratives and infrastructures that exist in these landscapes. The contestations that occur, and the processes of negotiation and negation that alter the dynamics of production.

The production of carbon for nature-based solutions reflects this complexity and deserves greater focus as infrastructural nature. A range of literature has built on Henri Lefebvre's ([1974] 1991) concept of the production of space and Neil Smith's (1990) theory of the production of nature to analyse the rise of carbon and nature-based solutions in recent years (Ekers 2015; Boyd and Prudham 2017; Carton and Andersson 2017). The production of nature thesis, in particular, draws our attention to the specific ways that economic relationships can shape socionatural systems, which can be usefully applied to these new markets. As noted by Neil Smith (1990, 86): 'It is not merely that different production processes have different "space requirements"; rather, in the process of building productive forces into the environment, space is produced according to the spatial properties of this set of productive forces'.

While criticisms of Smith's work point to the potentially prescriptive or reductive framing of a 'singular' production of nature (Andueza 2021), a more nuanced reading will explore the different productions of nature that co-exist – or could exist. Following Ekers and Loftus (2013, 237) 'the key is to historicize the specific forms that the making of natures takes, and to be able to do this in geographically situated ways'. As will be shown below, the production of nature for carbon markets in San Martín is neither as totalising as an 'historical nature' (Moore 2015) nor as clear as state territorialisation (Ballvé

2012), but a reflection of the narrative politics of different groups co-constituting and spreading *specific* productive regimes.

As this article will go on to show, the infrastructural logic and narratives behind interventions in the Peruvian Amazon are more than marketing rhetoric used by diverse groups but a critical reframing of the natural world in need of investigation and analysis, particularly as carbon markets and nature-based solutions are scaled up. I argue that such dynamics call for a combination of the insights of economic geography with those of ethnography, and specifically the ethnography of infrastructure to interrogate the (re)structuring of socionatural relations in agrarian settings. Viewing the way power operates through the production of nature-based solutions in San Martín highlights a novel ideological approach that is embraced by politicians, practitioners and activists across the spectrum, and one that requires greater focus in academic work.

Infrastructural nature from global ideal to local policy

What are nature-based solutions? What does it mean to produce nature *as a* solution? And how does it change how we relate to the natural world when it *is a* solution to our – or society's – problems? The following sections draw on research conducted over the space of three trips between 2017 and 2019, working and living amongst communities involved in a suite of for-profit nature-based solutions projects run by a French carbon vendor in the Western provinces of Mariscal Cáceres and Huallaga in San Martín (see Figure 1). While much of this time involved working with local farmers to understand how international schemes and regional goals were translated into policies and demands, analysis also focused on project and policy documentation, as well as government geodata, to assess how landscape planning had evolved over time, and how nature-

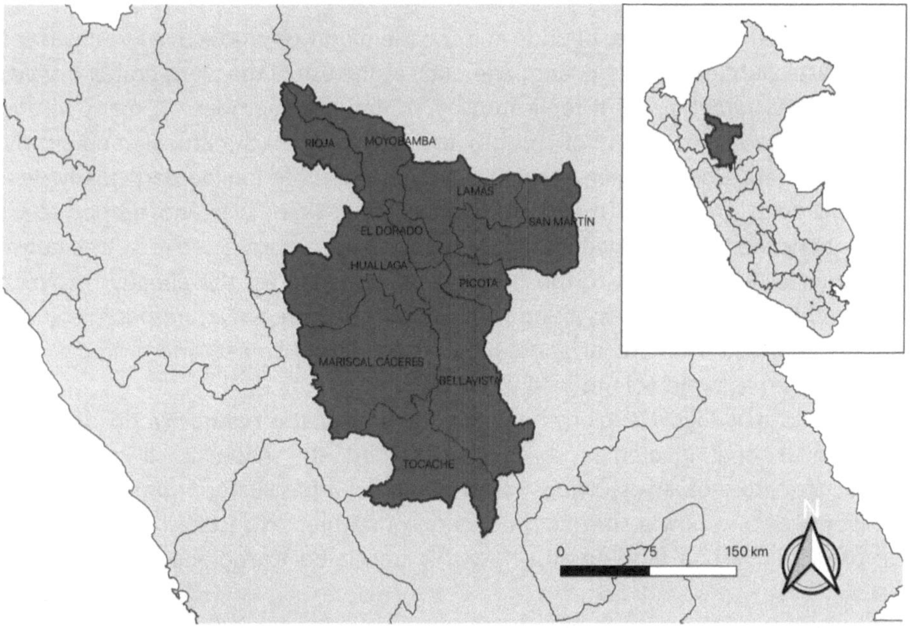

Figure 1. San Martín with provincial borders in the context of Peru.

based solutions were now being integrated into it. Interviews were conducted across these spaces in local villages and regional towns, alongside numerous informal conversations with farmers, foresters, and planners.

On my first visit to one village involved in a nature-based solutions project in San Martín run by a collaboration between an international for-profit company and a local NGO set up to manage a number of projects, I was given a tour by one of the members of the local cacao association, Eduardo. Our walk around the project area took in various sites: areas of cacao plantations, a (failed) fish farm and a plantation of *capirona,* a rapid-growth tree species favoured by farmers in most reforestation projects in this valley. On requesting to see some of the forest ('*bosque*') I was taken to another vast plantation of *capirona*, with very few other species of tree and limited undergrowth.

While I grew used to the eerie quietness of the cacao plots and plantations, the sheer lack of life in these new forests was, on first impressions, striking. The trees were planted in uniform rows and the long bare trunks of the *capirona* provided no habitat for local birdlife, which was normally so abundant in the forests of this valley. The soil was dry and tough to break by hand, with little sign of insect life above or below it.

My tour of the village concluded with another cacao farmer, Julián, taking me to his newly cleared plot on the side of a large hill, ready for planting a new *capirona* plantation. As we climbed up the land, with all vegetation cut back and then burned to clear the early growth and prepare the soil, Julián explained that in a few short years the area would be a profitable timber plantation funded by carbon credits. As we sat overlooking the village and the wider valley, he pointed to one of the overgrown areas to the bottom of his parcel of land and explained the process of clearing such land, as he had the very patch on which we were sitting. In contrast to his great pride in the reforestation plots, the weedy growth of returning forest was deemed ugly and worthless.

The *plantations* were a source of pride and the seemingly unproductive *forest* was seen simply as waste. Subtle changes in language, such as the use of the word timber (*madera*) rather than trees (*arboles*), were replicated by community members in many villages. When I highlighted the diversity of shrubby areas growing back naturally and slowly, I was told they were 'poor' – when I enquired about areas that looked particularly dense with undergrowth, I was told they were worthless, as there were no timber species present[2] – or needed to be turned into plantations. In a local protected area, one conservation worker even suggested to me that they should reforest the already recovering jungle, to help speed the process along with rapid-growth trees. Nature-based solutions had not just changed the material landscape but the local discourse, and produced landscapes were seen as preferable to wild.

The sections that follow explore how this framing of the natural world developed and the ideas and ideologies that have helped to shape it. It shows how narratives of nature-based solutions tend towards infrastructural framings of the natural world, tracing how a specific production of nature has resulted from a national pursuit of global carbon funding and how this translated to regional goals and local practices.

---

[2]Literally, 'This forest is poor, there's no timber' ('*es pobre este bosque, no hay madera*').

Capturing the carbon market

The Government of Peru has actively pursued opportunities to expand nature-based solutions in the country, designing a range of laws and rules to facilitate the abstraction and legal ownership of carbon. The country's carbon strategy is specifically geared towards the growth of a carbon offset economy through 'improving' agricultural and forestry productivity, REDD+ projects, and new land management systems (Government of Peru 2016).

The strategy of the National Forest and Wildlife Service (*Servicio Nacional Forestal y de Fauna Silvestre*, SERFOR), reflects this focus on the development of forests for new green economies through the 'promotion of the sustainable use of forests' and the 'incentivising of forest plantations for recovering degraded areas and increasing forested areas' (SERFOR 2016). This fits with a wider approach to conservation in Peru, which places it within sustainable development plans and prioritises its profitable aspects (Shanee et al. 2020).

The national strategy for climate change and the work of establishing the national carbon market is the responsibility of two key bodies – the Ministry of the Environment (MINAM) and the National Environment Fund (FONAM). FONAM specifically manages the 'National Carbon Portfolio' (which includes registered projects in voluntary markets, the Clean Development Mechanism and the development of state-based REDD+ projects) and promotes environmental investments in Peru to state and non-state actors globally. This includes forestry management and reforestation, which at a local level have additional supervision from the Ministry of Agriculture, SERFOR and from Regional Environmental Authorities, as well as the National Park Service for those that fall in state protected areas.

The scale of ambition for carbon credit-funded projects in the country can also be seen in the marketing of protected areas or reforestation land as a means of investment throughout the country, by both private groups and government bodies. On an international level, FONAM takes these proposals to carbon trade fairs and conferences to engage with investors around the world, producing investor booklets which boast of the potential for forestry production in the Amazon and advertise vast areas of land 'available' for reforestation. Private companies are often also in attendance at Amazonian investment events, producing similar booklets advertising reforestation models of intensively designed plantations that promise spectacular returns from the combination of carbon credits and sustainable logging.

On the subnational level, the expansion of the carbon portfolio has been aided by ongoing processes of decentralisation in Peru. Regional governments are able to register plantations for reforestation or circumvent lengthy application systems for protected areas by establishing regional conservation areas or 'conservation concessions' and REDD+ working groups operate on a regional level to promote opportunities and share best practice, with San Martín a leader in these processes.

The creation of privately protected areas[3] has also been key to the growth of REDD+ projects in Peru, as the designation of legal ownership is the basis of creating a saleable asset in carbon and an essential element to listing with certification groups such as Verra.

[3]Private conservation in Peru is largely facilitated through three types of official park designation – private conservation areas, ecotourism concessions and conservation concessions.

In total, privately protected areas now total almost 2 million hectares of the country (Shanee et al. 2020) alongside an ever-expanding number of regional conservation areas. Responsibility for these areas falls to regional bodies, rather than the more bureaucratic national groups (Piu and Menton 2014), allowing carbon investment opportunities to be identified and applied for through regional or local governments.

Peru's enthusiasm for REDD+ has made it a leader in both 'REDD readiness' projects for the nascent compliance markets and voluntary market projects. As of 2017, the National Carbon Portfolio included 50 REDD+ projects covering more than 6 million hectares (6,412,082), of which over 4 million fall outside of national parks (FONAM 2017, plus author's analysis) and 41 afforestation/reforestation projects covering just 195,441 hectares but with the potential to generate almost 60 MtCO2e[4] in 20 years (FONAM 2017).

The country's private sector projects have been even more dynamic, driven in particular by the boom in interest in nature-based solutions from 2016. Analysis from Ecosystem Marketplace (2019) suggests that between 2016 and 2018 over half (57%) of the overall global increase in voluntary market credits came from Peru – accounting for 86% of the overall 22.8 MtCO2e increase in volume from Latin America. Without Peru, global REDD+ volume would have been virtually unchanged in the 2016–2018 period (Ecosystem Marketplace 2019) and while other countries have made progress since, nature-based solutions continue to supercharge the market (Ecosystem Marketplace 2021).

While the rapid growth of projects in Peru in this era reflects a wider rise in interest in nature-based solutions, it was also a result of the proactive attempts to translate this global interest into national strategies and regional policies, as well as the integration of green and natural infrastructure into Peruvian law (Tomateo 2021). The region of San Martín was at the forefront of this trend. Reforestation, REDD+ and climate smart agriculture – and their promotion alongside increasing ties to global markets – have combined to create a powerful new narrative of agrarian development that appeals to diverse groups. These new initiatives, and the ideologies on which they rely, reflect the top-down solutions to climate change, deforestation and biodiversity loss promoted by global bodies (and markets) but are also inscribed in local landscape planning through processes of 'zonification', which segment areas of land for their greatest delivery of goods and services.

Optimising the landscape

The 'ecological and economic zonification' ('*Zonificación Ecológica y Económica*', ZEE) of areas at a macro, meso and micro scale has been pursued by successive national governments in Peru and has been supported by and developed in tandem with regional governments as well as through bilateral aid and international NGOs (Augusto 2018). San Martín is a national leader in this regard, having pioneered a strategy to organising its territory from around 2008. In this strategy – termed 'Production, Protection and Inclusion'[5] – areas are designated for (intensive) agriculture, strict exclusion zones or areas of low intensity use or agroforestry production.

[4]59,826,569.68 tCO2e.
[5]Also referred to in international reports as the 'production protection compact'.

The core of the argument behind zonification in San Martín is to optimise the landscape to meet the needs of both agricultural production and biodiversity protection. In doing so, it is assumed that local farming communities will have the tools and support to produce sufficient crops on their current land, without needing to expand or claim new areas to farm from the region's forests, thus stemming deforestation. Put simply by a representative of the regional government in an interview for this research: 'we realised that we can't protect, without producing. We can't ensure the protection [of forests], without ensuring the production [of crops]'.

The strategy of using increased commodity production to avoid the expansion of the agricultural frontier into forested areas is, in itself, not new. Land-sparing 'proposes that increased productivity will satisfy crop demands without extending production area, allowing other areas to be set aside for "strong" conservation, largely protected from all human use' (Oliveira and Hecht 2016, 269). The concept of land sparing is commonplace to the point of received wisdom for development and environmental practitioners (Angelsen and Kaimowitz 2000) and Peru is no exception. The current approach in San Martín, however, is not only rooted in conservation thinking, but inextricably tied up with the work of private organisations and new opportunities to profit from international offsets and nature-based solutions. This brings together a broader coalition of actors, from cacao exporters to conservation NGOs and carbon groups to politicians, under the narrative of zonification, further integrating the logic and funding of sustainable development into conservation projects (see also Chambers et al. 2020).

Before moving on to discuss the types of nature produced, it is worth detailing the success of the scheme in terms of headline figures of production and protection – and why it is so attractive to policymakers and development funders. Land sparing has resulted in an enormous growth of productive activity in San Martín, with agricultural production soaring while the regional government pursued its conservation agenda in the early 2000s. Between 2001 and 2009 the agricultural sector in San Martín grew by 80%, while from just two official protected areas in 2001, protected areas now cover a total of 1.95 million hectares in the region (almost 40% of the total area), not including the numerous buffer zones (Shanee et al. 2020).

Reforestation has also boomed with multiple projects around the region. Vast areas have been designated 'suitable for reforestation' and millions of trees planted. From 2013 to 2018, for example, 441 certificates for forest plantations were authorised covering an area of 1758 hectares.[6] Forestry is a key focus of private involvement with the state with many buffer zones and wider areas planning large plantations. Agroforestry has also been widely promoted, with the sustainable intensification of commodities such as cacao, coffee and cattle combined with replanting schemes to generate carbon credits while attempting to regenerate ecosystems.

The power of this logic – promising protected areas with increased production – has had profound outcomes on power relations in the region, bringing together an eclectic range of small and large actors under a broadly shared goal of landscape management. In officially demarcating the land, new rules and regulations were standardised in a way that makes land more investable for the private sector (rendering it legible, see Scott

[6]These figures were quoted in a presentation by a Regional Government representative on 01/06/2018 which also highlighted the potential profits of these schemes.

[1998]), something highlighted by the aggressive marketing of reforestation in regional trade fairs. This provided fertile ground for groups looking to establish projects through voluntary carbon markets but also produced specific natures that could provide the seeming stability and predictability of this planning process.

Producing infrastructural nature

The zonification process in San Martín made the region a national leader in nature-based solutions, it also resulted in a specific production of nature pursued by state, private and non-governmental actors hoping to achieve goals of social development, conservation and agricultural production, previously seen as contradictory. While these goals were formalised from 2008, the introduction of carbon credits and nature-based solutions became the catalyst that helped to spread the model across the region. Three core aspects of the infrastructural nature of the landscapes created through the combination of zonification and nature-based solutions will be explored in this section: the simplification of socionatural systems, the intensification of production and the expansion of specific productive regimes across the region.

Simplification

In seeking to optimise and rationalise the landscape, the zonification process creates areas of specialisation, simplifying potentially diverse landscapes to prioritise ways in which they can achieve their optimal value. Detailed maps designate zones of rice production, cattle farming, maize or broader designations for agroforestry systems such as coffee or cacao. Agricultural extension activities and incentives can then be targeted to encourage local farmers to convert to certain crops or follow best practice methods – in turn strengthening local supply chains and infrastructure for delivering these crops to market.

While the zonification process seeks to optimise and rationalise the landscape, the simplification of ecosystems emerges as much out of the models prescribed and local practices. Plans are not simply imposed on areas (nothing is to stop a farmer planting avocados in a maize zone, for example), but the incentives and subtle processes of rationalisation, promotion and celebration tend to reinforce certain crops. For example, in certain areas of San Martín, farmers who do not have secure land tenure have been offered titles for a 40-year lease on their land by the regional government if it is entered into micro-zonification programmes. These '*cesión en uso*' contracts encourage farmers to impose strict separation of crops on their land in a bid to promote the intensification of certain crops alongside areas of conservation or reforestation at the most local level.

With cacao, preferred varieties are promoted (often the globally favoured CCN51) and agroforestry plantation designs are shown to farmers in events and workshops across the region. One NGO even proudly displays these standards on their office walls by a model plantation, with farmers asked to sign the poster to declare that they will adhere to specific modes of production. The preferred methods promoted means that a single species and plantation style dominates the local landscape.

The gradual reduction in diversity in favour of ecosystem simplification is most clearly seen, though, in the timber species in reforestation and agroforestry plantations around

San Martín. Typically, plantations are optimised to meet minimum forest definitions under Peruvian and UNFCCC accords,[7] the plantations can be harvested sustainably in a rotational system to allow for the sale of timber, while also accruing carbon credits. In accordance with these rules and the demands of the markets, plantations are designed to contain a combination of trees of short, mid and long-term growth to ensure the greatest returns in carbon and timber in the space provided.

In reality, however, many of the farmers in San Martín prefer the rapid growth species of tree, which can be matured, cut and sold in 5–7 years. While original designs might show semi-diverse grids of different tree species, most plantations are largely – or entirely – composed of one of these species, such as *capirona* and *bolaina*. These two trees alone account for 84% (54% and 30% respectively) of all trees planted in regional government schemes, with the remaining 16% representing just four other species (GORESAM 2022).

The processes of simplification were reiterated to me by a local reforestation group which predominantly integrates timber into coffee agroforestry. From 2010 to 2015 work had focused on a *minimum* of eight species of timber in any plot but had since reduced to a *maximum* of two or three. As noted by the head of reforestation, Jorge, this was for 'reasons of business, of economics, logistics' but often meant a focus on a single species which 'could be *capirona*, it could be *bolaina*, or it could be another species, purely eucalyptus'. Jorge explained that while diverse plots were better for environmental reasons, the number of species was reduced because for farmers the ultimate goal is the extraction of as much timber as possible. Another reforestation project manager noted that even where they tried to promote more diverse plots and insisted on three species per parcel minimum, farmers continued to almost exclusively plant *capirona* or *bolaina*. The actual diversity of species in plantations thus rarely goes further than some fast-growth trees and, perhaps, plantain, leaving vast areas of smallholder agroforestry in virtual monocultures.

Intensification

Plantations are not only simplified, but being asked to deliver more through intensification. The intensification of production is promoted by a number of groups – and notably across a consensus of national and regional politicians, international development funds (such as USAID), exporters and private companies, and a range of local and international conservation NGOs and carbon groups pushing for the 'sustainable intensification' of production to support the land-sparing strategies. The drive for intensification is manifested locally in the call for technical 'improvements' and the ubiquitous concept of '*aprovechamiento*',[8] through which local NGOs and regional officials criticise local farming practices that do not produce commodities for export or take advantage of technology such as strimmers, increased fertiliser use and high-yielding seed clones.

The push for intensification extends to the production of numerous crops across the region. State incentives such as the aforementioned '*cesion en uso*' contracts encourage

[7]A minimum area of 0.5 hectares; A minimum tree crown cover of 30%; And a minimum tree height of 5 meters.

[8]There is a notable similarity between the term *aprovechar* (literally, to make good use of or to take advantage of) as it is being used in San Martín and the discourses of improvement in the English Enclosures, wherein: 'making land productive – that is, *improving* it – was becoming the basis for property rights; and more particularly, the *failure* to improve could mean *forfeiting* the rights of property' (Wood 2017, 157).

practices to rapidly increase the yield and productivity of key crops, such as coffee and cacao. The intensification of livestock in San Martín is equally instructive in its ambition. Current plans for support for cattle-farming communities located within a restricted conservation concession area, for example, were described to me by local residents and NGO workers as intensifying production to reduce land use – to have ten cows per hectare where farmers currently have two. One long-time conservation worker in the region highlighted the paradoxical nature of expanding the number of cattle within conservation areas and the inherent contradictions of 'sustainable cattle', even as they explained their own organisation's shift in recent years towards 'productive' conservation that focused on livelihood opportunities and agricultural support.

The intensification is notable too in the approach to agroforestry and reforestation projects. Within plantations, despite being planted at 'optimum' distances as stated in project plans, trees are still densely planted. Clones are produced in nurseries around San Martín to grow straighter, higher and faster. Promotional materials for reforestation groups promise ever-taller trees, grown in ever-denser blocks, with increased carbon sequestration and quicker turnaround for sustainable logging cycles. The trees themselves in these models are assumed to be consistent, stable and machine-like in their production and are being planted in vast numbers by private and public groups alike. The emerging productive regime thus seeks to deliver ever more yield through pushing nature to work harder, particularly at the margins and frontiers.

Massification

As production is intensified to deliver ever more goods and services, it also leads to its expansion across geographies. Multiple groups, including within the regional government, the local business community and local farmers, spoke of plans to expand the production and intensification of strategic commodities in the region. Amongst these productive goods that are being promoted, few are discussed with as much enthusiasm as tree-planting, a potential value-add to previously unproductive land, exhausted crop plantations and providing economic diversification for farmers across the region.

One international reforestation organisation has even formed a new company to deal with the sale of timber from its numerous small-scale plantations. The aim of the company, as expressed by one project leader was to allow farmers in the valley to have access to 'the best technology to take advantage of the most timber possible'. The company – and its promises of scaling to ever more areas – began to dominate the topics of many meetings and discussions with reforestation workers in 2018, and even more so in 2019, as the timber industry became the focus of ever greater claims of intensification and profitability.

One regional government head of *conservation* even considered a functioning timber trade to be their 'dream' for the future of the area and NGO reports talk of plans for the planting of millions of trees across the San Martin region to turn it into a *'forest power'*. In 2019, this dream took a step towards reality with the first REDD+ -funded reforestation plantations obtaining certification by the Forest Stewardship Council. Conservation and government workers enthused about this possibility, placing it at the core of multiple projects and the regional productive strategy of one valley. Project designs were shared, plantation layouts replicated and models of 'success' promoted.

At a regional level, many groups have a stake in the success of the reforestation and agroforestry projects, from the organisations directly involved in the project to local companies and regional politicians keen to show that the model of agriculture and carbon can deliver benefits for communities across San Martín. While this was something reproduced and reiterated to me by a wide variety of actors in the region, events were an especially instructive example of how (and whose) success is communicated to wider audiences. In both 2017 and 2018, the San Martín Regional Government held large events celebrating the food, farming and future of the forests of the region – with both years receiving visits from the incumbent President of Peru. These events provided an opportunity to see how the projects and plantations were made infrastructural, woven into wider ideals of development for the region, and marketed as investment opportunities.

In presentations, agroforestry and reforestation were presented as simple models of neat lines of trees, with carbon sequestration and potential profits highlighted. Having worked amongst these largely monocultured plantations, the gap between rhetoric and reality at the events was notable. The hype created however, made competing visions of socionatural relations seem risky, uncertain and even wasteful, as they are outperformed in the metrics collected and targeted by the regional government and development organisations. A leader of one local project regenerating soils through gradual processes of biochar creation and production for self-sufficiency told me that funding or support from the regional government was increasingly difficult as their approach did not measure carbon or provide opportunities to scale up in a profitable manner.

As I will go on to argue, what works and spreads reflects not just a drive for profit, however, but a specific ideology of the natural world that assumes the *possibility* of optimisation, predictability and stability. While there has not been the top-down drive for monoculture production systems of old, the optimisation of the natural world for its delivery of ecosystem services has led to specialised zones of production and decreasing diversity. Thus, despite proposing an alternative green economy, landscapes of small-scale producers tend to specialise in a single crop, in some cases a single variety, with nature-based solutions beginning to resemble the agroindustrial models they claim to supplant. This is hard to ignore when travelling anywhere in the region, with areas dedicated to cacao production often spreading across vast quantities of fields, where a combination of success-stories and state-based incentives see an increasing number of small-scale producers turn to cacao production. The new assemblages that make up infrastructural nature, however, overestimate nature's capacity, underestimate its instability and narrow the development pathways for local communities.

Understanding infrastructural nature

The production of nature-based solutions in San Martín has resulted in distinct landscapes and emerging socionatural relations. But what exactly is produced? What are these ecologies of green capitalism? Perhaps more pertinently for this forum on climate change and agrarian struggles, how do they differ from previous top-down models and plantation systems? Connecting the literature on infrastructural nature with nature-based solutions highlights how 'forested landscapes' are not just 'assigned an infrastructural function' (Carse 2012, 552), nor simply made to function as such (cf. Nelson and Bigger 2022); certain aspects of nature are emphasised discursively and promoted materially. This section will elaborate on these aspects, and the potential issues with them, through the infrastructural framing of nature as stable, as repeatable and as a carbon sequestering solution to climate change.

Nature-as-stable

The framing of nature-based solutions makes a familiar assumption about the natural world: that its processes are manageable and controllable. Infrastructural nature pushes this further however, assuming stability and predictability akin to the inert materials that make up the grey infrastructure of cities. On regional planning maps, plantation models and statistical summaries, trees, plants and other lifeforms are treated as building blocks of human-designed ecosystems.

The assumption of stability is a particular issue for a region that is subject to environmental stresses (such as the constant erosion of waterways and conversion of forest to wetland) and shocks (such as frequent floods and droughts), not to mention political change and the shifting population dynamics driven by local population growth and migration. The patchwork of small farms of densely packed trees and specific commodity crops has left producers increasingly vulnerable to market swings and crop failure, but also created the perfect breeding ground for pests and plagues.

In the cacao sector, for example, disease has particularly come in the form of the cacao pod borer, *mazorquero*, and, the fungus, frosty pod rot, both of which have decimated harvests for some farmers in recent years. Moreover, as the model of production creates ever more problems, more time, money and expertise are invested in it, with numerous projects seeking to prop up ailing commodity crops such as cacao, often in the form of Western expertise that prescribe the same solution of further intensification to make up for shortfalls. The density of cacao plots in itself is a challenge for these farmers. While clearing a plot of diseases and pests may be possible, various farmers complained of them quickly spreading back from neighbouring plots. The simplification and intensification of plantations has thus created new geographies of risk in San Martín as they have been confronted by the 'inherent unruliness in ecologies, trees and therefore carbon' (Leach and Scoones 2015, 14).

Infrastructure provides a lens to understand not only the unruliness that causes projects to fail but, critically, how this is ignored in plans, maps and market strategies. In San Martín, the continuous struggles with pests and diseases threaten outright ruin for farmers, yet these threats fail to register in the regional planning documents that assume stability. This provides some insights into the 'insistence on "technical fixes" that do not work' in the region (Paredes and Kaulard 2022, 5).

Where Osborne (2015, 67) has noted 'the uncooperative nature of carbon commodification' and Prudham (2015) points to the always contested nature of commodification, infrastructural nature can focus our analysis on the narratives that *frame* unruly nature as stable, manageable and commodifiable, and how failures and crises in turn are compensated for by success stories and expansion.

Nature-as-repeatable

The infrastructural approach treats the natural world as not just stable, but repeatable. This leads to unique – and at times stark – landscapes of production. The mapping and titling of land, for example, is seen as both a prerequisite for investment and as critical to producing commodities in sufficient quantities to tap into high value global markets, whether that be carbon and timber markets or through niche superfoods such as the locally produced nut *sacha inchi*. This logic is pervasive in San Martín and leads to specific commodities, seeds and clones being incentivised, spreading ever further across the region.

Locally, the repetition of certain crops leads to the geographic centralisation of production (Smith 1990), in many cases locking in landscape uses. While a *campesino* in San Martín may be able to secure a form of land tenure thanks to agroforestry projects (and this is no minor accomplishment), communities may be tied into specific crops promoted by regional support and infrastructure or limited in expansion or rotational cropping practices by strictly delineated areas of use or protection. As favoured models and varieties spread ever further across the region, entire valleys of small-scale producers begin to resemble industrial plantations. Over time, this reinforces a logic that favours intensively grown export crops at the expense of diverse production systems and crops for self-sufficiency. While not without its positive aspects, the landscape planning and zonification in San Martín therefore has the potential to disadvantage alternative development pathways (as detailed in this forum by Paredes and Kaulard 2022).

Here, an important contrast with traditional plantation systems is emerging in the practices of green infrastructure. Where plantations are typically 'large-scale agricultural operations characterized by their intensive use of capital investments as well as the exploitation of wage labour' (Perfecto, Jiménez-Soto, and Vandermeer 2019, S236), infrastructural nature involves the *integration* of smallholders into similar systems of production. Such dynamics have been noted in corn production in Myanmar by Franco and Borras Jr. (2019, 195), in which they observe 'the ongoing transformation of much of Northern Shan State's once-biodiverse and variegated land uses into an increasingly uniform small farm-based corn monoculture subordinated to merchant capital'. Equally, Pye (2019) has pointed to processes of 'accumulation by inclusion' in certified palm oil production, as increasing numbers of small-scale farmers in Southeast Asia switch to industrial methods of palm oil production.

Infrastructural nature provides a lens to understand these new waves of accumulation occurring under the banner of ecosystem services and nature-based solutions, in which the machine-like structure of the plantation becomes the infrastructure-like structure of produced landscapes. In nature-based solutions, we can observe and analyse this process as not simply driven by large operations of capital accumulation, but by the emerging narratives, models and goals of climate change mitigation and how diverse groups coalesce around them.

Nature-as-solution

Nature-based solutions make an explicit connection between goals of climate change mitigation and landscape management. Applying the logic of infrastructural nature to this dynamic can highlight not only the ways that ecosystems are being optimised to offset carbon emissions and deliver ever-greater benefits, but the wider implications of valuing specific socionatural systems.

The reality of plantations of nature-based solutions are far removed from the marketing and original plans of many projects, with ideals of forested landscapes being replaced with select fast growth species in semi-monocultures. Even in contexts where agroforestry may have been promoted as a way to boost biodiversity within agricultural systems and diversify production, it has resulted in a few profitable species of timber, chosen for their rapid growth to ensure quicker turnaround times for logging and carbon sequestration. In San Martín, this reflects a growing conflation between development and conservation goals under the auspices of regulated and 'greened' production (Chambers et al. 2020).

While agroforestry systems promise to increase production in conditions that 'more closely mimic natural forest' (Clough, Faust, and Tscharntke 2009, 197), the resulting 'reforested' plots are not what many would recognise as a forest or jungle. Trees may well be planted to match official definitions of forest in height, crown coverage and area, but rapid-growth species planted in neat rows with trimmed undergrowth provide an eerie silence in place of the normal competing sounds of the wildlife of the Peruvian Amazon. From a purely sensory perspective they seem devoid of any sort of life beyond the hulking trees which rapidly stretch up to the light.

In being framed as a solution to environmental and developmental ills, infrastructural nature not only works as a technology of land management, but as noted in the literature on the production of carbon, is '(re)made to work harder, faster and better' (Boyd and Prudham 2017, 877). There is a move from the *formal* subsumption of nature, the expansion of commodity production based on the commodification of existing carbon, to the *real* subsumption of nature, trees planted for their specific ability to capture carbon and ease of control (Smith 2007; Carton and Andersson 2017). Here, it is no longer enough to hem nature in and tame it, but ecosystems must be *made* and optimised. Infrastructural nature can take our understanding of this real subsumption of nature further to probe at the ideological underpinnings that it represents and how it draws new actors into practices of accumulation.

While the production of nature as stable, as repeatable and as a solution reflects longer processes of landscape management and control in San Martín (Augusto 2018), nature-based solutions have provided a narrative that internalises previous critiques. The changing language to describe the wealth of plantations and the poverty of the natural forests or overgrown land, or the subtle use of the term 'timber' to discuss trees, reflects a changing narrative politics being driven by climate change. This allows for expanded (assumed) control over the forest frontiers and leads to novel forms of inclusion and exclusion under the banner of the green economy.

The socionatural poverty of nature as infrastructure

This article has expanded on the concept of infrastructural nature and highlighted some of the intended and unintended consequences of this logic when framed as a response to climate change. Responding to this forum's call to interrogate the 'combinations of narratives and strategies [that] frame climate change and the institutionalised responses to it in agrarian settings' (Borras Jr. et al. 2022, 17), it has shown how groups in San Martín have coalesced under the idea of nature-based solutions to extend ongoing processes of rural commodification and control. The evidence presented here builds on papers in this forum analysing the narratives and appropriations associated with carbon offsets (Schwartzman 2022) and the emerging assemblages of actors in agrarian contexts responding to climate change goals (Hernandez and Newell 2022), to interrogate too the types of nature being produced and the ways in which they reflect infrastructural ideologies.

The narrative of infrastructural nature holds a seductive logic for many and has helped to create a coalition of actors under a stated mission of optimisation and land-sparing, rather than simply privatisation and profit. In doing so, it has managed to enrol actors that may well disavow neoliberalism – including local conservation groups, 'ethical' brands and progressive policymakers – into an approach that ultimately still enables expanded capital accumulation. It can be summed up with a response to the role of conservation by one informant from the regional government, 'conservation for me is a way of protecting resources and at the same time exploiting them'. Infrastructural nature is, as such, both a core narrative in ongoing agrarian transitions in response to climate change and a critical pillar in an emerging ideology of green capitalism.

Infrastructural nature is not only reflected in changing discourses and desires at the global level, but in tangible changes to landscapes, lifestyles and livelihoods, such as in San Martín. Where previous utopian projects to produce artificial natures 'attempted to come to terms with the chaos of the natural world, by defining it as something apart from human reality, something completely "outside"' (Murphy 2015, 210), current attempts seek to remake it as a repeatable, stable, and service providing system, which humans *must* manage to avert climate disaster.

Sullivan (2009, 23) has highlighted the 'cultural poverty' in Payments for Ecosystems Services, in which 'the non-human world in all its diversity and mystery becomes the provider of services'. I argue that in applying the logic of infrastructural nature to nature-based solutions we can see too the 'socionatural poverty' of the landscapes produced. Strolling through the densely packed trees of carbon-credit funded plantations, devoid of bird life and biodiversity, one can reflect on a new nature being produced in the name of sustainable development. This is the natural world through the lens of infrastructure: a discreet, simplified, technified, manageable unit designed to deliver optimal services.

Pointing out the failures of infrastructural nature and the emerging risks from 'unruly nature' in this article is not to call for a hands-off approach to the natural world that would reinforce similarly stark dichotomies between nature and society, but to highlight the specific production of nature it entails and the alternative approaches it obscures. Cloaked in a language of optimisation and stability, the green infrastructure paradigm reinforces specific spatial relations and structures of power in San Martín.

While some progressive actors may seek to reclaim an infrastructural view of nature, or even claim that it could have decolonial dimensions (Tomateo 2021), the ontological view of the natural world that it relies on fails to account for the uncertainty or humility that these complex, contextual and changing systems require. By failing to engage with the holistic, but messy, political work of what truly diversified landscapes may look like, natural infrastructure can but only continue to reproduce the inequalities and failings of extractivist policies elsewhere.

Disclosure statement

No potential conflict of interest was reported by the author(s).

Funding

This work was supported by Economic and Social Research Council [grant number ES/J500173/1].

ORCID

Will Lock http://orcid.org/0000-0003-2782-1602

References

Andueza, Luis. 2021. "Value, (Use) Values, and the Ecologies of Capital: On Social Form, Meaning, and the Contested Production of Nature." *Progress in Human Geography* 45 (5): 1105–1125. doi:10.1177/0309132520947473.

Angelsen, A., and D. Kaimowitz. 2000. "When Does Technological Change in Agriculture Promote Deforestation?" In *Tradeoffs or Synergies? Agricultural Intensification, Economic Development and the Environment*, edited by David R. Lee, and Christopher B. Barrett, 89–114. Wallingford, UK: CABI.

Augusto, María Claudia. 2018. "When State and Society Meet: The Land-Use Planning Policy in San Martin (2002-2018)." *Debates En Sociología* 47: 9–39. doi:10.18800/debatesensociologia.201802.001.

Ballvé, Teo. 2012. "Everyday State Formation: Territory, Decentralization, and the Narco Landgrab in Colombia." *Environment and Planning D: Society and Space* 30 (4): 603–622. doi:10.1068/d4611.

Borras Jr., M. Saturnino, Ian Scoones, Amita Baviskar, Marc Edelman, Nancy Lee Peluso, and Wendy Wolford. 2022. "Climate Change and Agrarian Struggles: An Invitation to Contribute to a *JPS* Forum." *The Journal of Peasant Studies* 49 (1): 1–28. doi:10.1080/03066150.2021.1956473.

Boyd, W., and S. Prudham. 2017. "On the Themed Collection, 'The Formal and Real Subsumption of Nature.'" *Society & Natural Resources* 30 (7): 877–884. doi:10.1080/08941920.2017.1304600.

Carse, Ashley. 2012. "Nature as Infrastructure: Making and Managing the Panama Canal Watershed." *Social Studies of Science* 42 (4): 539–563. doi:10.1177/0306312712440166.

Carton, W., and E. Andersson. 2017. "Where Forest Carbon Meets Its Maker: Forestry-Based Offsetting as the Subsumption of Nature." *Society & Natural Resources* 30 (7): 829–843. doi:10.1080/08941920.2017.1284291.

Chambers, J., M. D. Aguila Mejía, R. Ramírez Reátegui, and C. Sandbrook. 2020. "Why Joint Conservation and Development Projects Often Fail: An in-Depth Examination in the Peruvian Amazon." *Environment and Planning E: Nature and Space* 3 (2): 365–398. doi:10.1177/2514848619873910.

Clough, Y., H. Faust, and T. Tscharntke. 2009. "Cacao Boom and Bust: Sustainability of Agroforests and Opportunities for Biodiversity Conservation." *Conservation Letters* 2 (5): 197–205. doi:10.1111/j.1755-263X.2009.00072.x.

Dourish, P., and G. Bell. 2007. "The Infrastructure of Experience and the Experience of Infrastructure: Meaning and Structure in Everyday Encounters with Space." *Environment and Planning B: Planning and Design* 34 (3): 414–430. doi:10.1068/b32035t.

EC (European Commission). 2013. *Building a Green Infrastructure for Europe*. Luxembourg: Publishing Office of the European Union.

Ecosystem Marketplace (Forest Trends' Ecosystem Marketplace). 2019. *Financing Emissions Reductions for the Future: State of the Voluntary Carbon Markets 2019*. Washington DC: Forest Trends.

Ecosystem Marketplace (Forest Trends' Ecosystem Marketplace). 2021. *'Market in Motion', State of Voluntary Carbon Markets 2021*. Washington, DC: Forest Trends.

Ekers, Michael. 2015. "A Fix in the Forests: Relief Labor and the Production of Reforestation Infrastructure in Depression-Era Canada." *Environment and Planning A: Economy and Space* 47 (12): 2537–2554. doi:10.1177/0308518X15609211.

Ekers, Michael, and Alex Loftus. 2013. "Revitalizing the Production of Nature Thesis: A Gramscian Turn?" *Progress in Human Geography* 37 (2): 234–252. doi:10.1177/0309132512448831.

FONAM (Fondo Nacional del Ambiente). 2017. Peru Low Carbon Economy 2017 Carbon Projects, Voluntary Markets, REDD+, NAMAs & NDCs. http://fonamperu.org.pe/wp-content/uploads/2017/06/Peru-low-carbon-economy-2017.pdf.

Franco, J. C., and S. M. Borras Jr. 2019. "Grey Areas in Green Grabbing: Subtle and Indirect Interconnections between Climate Change Politics and Land Grabs and Their Implications for Research." *Land Use Policy* 84: 192–199. doi:10.1016/j.landusepol.2019.03.013.

GORESAM (Regional Government of San Martín). 2022. "Dirección Regional de Agricultura: Proyecto Cacao." Accessed August 15, 2022. https://portal.regionsanmartin.gob.pe/portal/apps/MapJournal/index.html?appid=6984d24b63794be28aa12a5153e097ff.

Government of Peru. 2016. *Estrategia Nacional de Bosques y Cambio Climático*. https://www.unredd.net/documents/un-redd-partner-countries-181/national-redd-strategies-1025/15861-peru-national-strategy-on-forests-a-climate-change/file.html.

Hance, J. 2018, June 28. "Biodiversity is the 'infrastructure that supports all ife'." *The Guardian, UK.* https://www.theguardian.com/environment/radical-conservation/2018/jun/28/biodiversity-is-the-infrastructure-that-supports-all-life.

Hernandez, Daniela Soto, and Peter Newell. 2022. "Oro Blanco: Assembling Extractivism in the Lithium Triangle." *The Journal of Peasant Studies* 49 (5): 945–968. doi:10.1080/03066150.2022.2080061.

Leach, Melissa, and Ian Scoones. 2015. *Carbon Conflicts and Forest Landscapes in Africa*. London: Routledge.

Lefebvre, Henri. [1974] 1991. *The Production of Space*. Oxford: Blackwell.

Maes, J., and S. Jacobs. 2017. "Nature-Based Solutions for Europe's Sustainable Development." *Conservation Letters* 10 (1): 121–124. doi:10.1111/conl.12216.

Merchant, C. 1982. *The Death of Nature: Women, Ecology, and the Scientific Revolution*. London: Wildwood House.

Moore, Jason W. 2015. *Capitalism in the Web of Life: Ecology and the Accumulation of Capital*. London: Verso Books.

Murphy, D. 2015. *Last Futures Nature, Technology, and the End of Architecture*. London: Verso Books.

Nelson, Sara H., and Patrick Bigger. 2022. "Infrastructural Nature." *Progress in Human Geography* 46 (1): 86–107. doi:10.1177/0309132521993916.

Nesshöver, C., T. Assmuth, K. N. Irvine, G. M. Rusch, K. A. Waylen, B. Delbaere, D. Haase, et al. 2017. "The Science, Policy and Practice of Nature-Based Solutions: An Interdisciplinary Perspective." *Science of the Total Environment* 579: 1215–1227. doi:10.1016/j.scitotenv.2016.11.106.

Oliveira, G., and S. Hecht. 2016. "Sacred Groves, Sacrifice Zones and Soy Production: Globalization, Intensification and Neo-Nature in South America." *The Journal of Peasant Studies* 43 (2): 251–285. doi:10.1080/03066150.2016.1146705.

Osborne, T. 2015. "Tradeoffs in Carbon Commodification: A Political Ecology of Common Property Forest Governance." *Geoforum; Journal of Physical, Human, and Regional Geosciences* 67: 64–77. doi:10.1016/j.geoforum.2015.10.007.

Paredes, Maritza, and Anke Kaulard. 2022. "Forest as 'Nature' or Forest as Territory? Knowledge, Power, and Climate Change Conservation in the Peruvian Amazon." *The Journal of Peasant Studies*, 1–22. doi:10.1080/03066150.2022.2134010.

Perfecto, Ivette, M. Estelí Jiménez-Soto, and John Vandermeer. 2019. "Coffee Landscapes Shaping the Anthropocene: Forced Simplification on a Complex Agroecological Landscape." *Current Anthropology* 60 (S20): S236–S250. doi:10.1086/703413.

Piu, H. C., and M. Menton. 2014. *The Context of REDD+ in Peru: Drivers, Agents and Institutions*. Occasional Paper 106. Bogor, Indonesia: CIFOR.

Prudham, Scott. 2015. "Property and Commodification." In *The Routledge Handbook of Political Ecology*, edited by Tom Perreault, Gavin Bridge, and James McCarthy, 430–445. Oxford, UK: Routledge.

Pye, Oliver. 2019. "Commodifying Sustainability: Development, Nature and Politics in the Palm Oil Industry." *World Development* 121: 218–228. doi:10.1016/j.worlddev.2018.02.014.

Schwartzman, Gabe. 2022. "Climate Rentierism after Coal: Forests, Carbon Offsets, and Post-Coal Politics in the Appalachian Coalfields." *The Journal of Peasant Studies* 49 (5): 924–944. doi:10.1080/03066150.2022.2078710.

Scott, James C. 1998. *Seeing Like a State: How Certain Schemes to Improve the Human Condition Have Failed*. New Haven: Yale University Press.

SERFOR (Servicio Nacional Forestal y de Fauna Silvestre). 2016. *Primer Informe Parcial del Inventario Nacional Forestal y de Fauna Silvestre*. Lima: SERFOR.

Shanee, Sam, Noga Shanee, Will Lock, and Maria Espejo-Uribe. 2020. "The Development and Growth of Non-Governmental Conservation in Peru: Privately and Communally Protected Areas." *Human Ecology* 48 (6): 681–693. doi:10.1007/s10745-020-00188-8.

Smith, Neil. 1990. *Uneven Development: Nature, Capital and the Production of Space*. 2nd ed. Oxford: Blackwell.

Smith, Neil. 2007. "Nature as Accumulation Strategy." *Socialist Register* 43 (43): 16–36.

Star, S. L. 1999. "The Ethnography of Infrastructure." *American Behavioral Scientist* 43 (3): 377–391. doi:10.1177/00027649921955326.

Sullivan, Sian. 2009. "Green Capitalism, and the Cultural Poverty of Constructing Nature as Service Provider." *Radical Anthropology* 3: 18–27.

Swyngedouw, Erik. 2007. "Technonatural Revolutions: The Scalar Politics of Franco's Hydro-Social Dream for Spain, 1939–1975." *Transactions of the Institute of British Geographers* 32 (1): 9–28. doi:10.1111/j.1475-5661.2007.00233.x.

Tercek, Mark. 2017. "The Best Investment Opportunity Ever: Investing in Nature 101." *Forbes*, US, October 11. https://www.forbes.com/sites/marktercek/2017/10/11/the-best-investment-opportunity-ever-investing-in-nature-101/#648533303f1f.

Tomateo, Claudia. 2021. "Indigenous Land Systems and Emerging of Green Infrastructure Planning in the Peruvian Coastal Desert: Tensions and Opportunities." *Journal of Environmental Policy & Planning* 23 (5): 683–700. doi:10.1080/1523908X.2021.1960806.

Venkatesan, S., L. Bear, P. Harvey, S. Lazar, L. M. Rival, and A. Simone. 2018. "Attention to Infrastructure Offers a Welcome Reconfiguration of Anthropological Approaches to the Political." *Critique of Anthropology* 38 (1): 3–52. doi:10.1177/0308275X16683023.

Wakefield, Stephanie. 2020. "Making Nature into Infrastructure: The Construction of Oysters as a Risk Management Solution in New York City." *Environment and Planning E: Nature and Space* 3 (3): 761–785. doi:10.1177/2514848619887461.

Wood, Ellen Meiksins. 2017. *The Origins of Capitalism: A Longer View*. London: Verso.

Climate refugees or labour migrants? Climate reductive translations of women's migration from coastal Bangladesh

Camelia Dewan

ABSTRACT

Climate reductive translations of migration attract international attention, but result in three problematic misreadings of Bangladesh's socioecological landscape. First, attributing migration to climate change misreads coastal vulnerabilities and the importance of migration as a gendered livelihood strategy to deal with rural precarity and debt- both in the past and present. Second, misreading migration caused by brackish tiger-prawn cultivation, infrastructure-related waterlogging and riverbank erosion as 'climate-induced' hinders a discussion of long-term solutions for rural underemployment, salinisation, siltation and land loss. Lastly, framing climate change as causing 'gendered displacement' ignores the importance of affective kinship relations in shaping single women's migration choices.

Introduction: beyond climate reductive translations of migration

In 2014, I met with Bangladeshi migration researchers in Dhaka who had gained significant funding for a project on 'climate-related migration' following Cyclone Aila in 2009. I asked [in Bangla] how migration was related to climate change. The researchers replied that since their main focus is on migration, they could continue their work by using climate change as a *masala* to attract funding. They further pointed out that shrimp farming was the main cause of outmigration:

> The main reason why the flooding was so devastating during Aila was due to the damages shrimp farms [*ghers*] made on flood-protection embankments. The devastation was caused by a cyclone, but its impact would have been less had it not been for *ghers*.[1]

Yet an increasing number of papers frame such migration as climate-related (Kartiki 2011; Saha 2017; Ahsan 2019) or even climate-induced (M. R. Islam and Shamsuddoha 2017; R. Islam, Schech, and Saikia 2021) while also creating a false binary of internal migration as 'climate-induced' vs. 'non-climate-induced' (Adri and Simon 2018) in ways

[1]Words in italics are in Bangla unless otherwise indicated.

HUFFINGTON POST

Haunting Photos Show Effects Of Climate Change In Bangladesh

The number of climate change refugees in Bangladesh is expected to increase dramatically in the coming decades.

Stefanos Nikitas HuffPost Greece

01/28/2016 04:32 pm ET

Probal Rashid

Figure 1. Huffington Post article on 'Climate Change Refugees'. Source: Nikitas (2016). Photo courtesy Probal Rashid.

that removes from view political problems of land use that make rural livelihoods even more precarious (Figure 1).

Bangladesh is often portrayed as a 'climate hotspot' where its coastal migrants are described as victims of climate change who are 'forced' to migrate due to climate change, or else risk drowning in rising sea levels or in cyclonic events (Collectif Argos 2010; Stellina Jolly and Ahmad 2019; McDonnell 2019). Such recasting of migrants in Bangladesh and other regions of the world as 'climate refugees' ignores how migration was, and continues to be, a central part of agrarian lives (Farbotko and Lazrus 2012) where causes of migration are multifaceted and contradictory, as are the motivations of those who move (Amrith 2014).

Striking images of Bangladeshis as 'climate refugees' can be seen as an attempt to advocate for global climate justice, but they are also anchored in a longer history of Bangladesh being a key destination of international aid. As donors are increasingly allocating funds to climate change, development brokers engage in 'climate reductive translations', that is they create a causal narrative that links their particular activities to the policy theory of climate change, in order to continue with their ongoing activities (Dewan 2021b; 2020).[2] 'Climate reductionism' refers to 'the increasing trend to ascribe all changes in environment

[2]By translation I refer to the processes by which development brokers produce coherence [make projects real] by stabilising representations so as to match causal events to the prevailing policy theory (Mosse and Lewis 2006, 13; Mosse 2005, 9) This draws on Latour's use of translation 'as a relation that does not transport causality but induces two mediators into coexisting [with each other]' (Latour 2005, 108).

and society to climate change' (Hulme 2011, 255–256) – a climate reductive translation of migration represents migration as *caused by* climate change. While contemporary capitalism fuels the climate crisis (Borras Jr. et al. 2022), capitalist actors increasingly draw on climate reductive translations to reframe flood-protection embankments, brackish aquaculture and capital-intensive high-yield agriculture as *climate adaptation solutions* and misread the Bengal delta (Dewan 2021b). Climate reductive translations of migration may attract international attention for Bangladesh's plight, but it misreads coastal vulnerabilities while normalising a victimising discourse where there is no space to discuss *political solutions* that could alleviate agrarian livelihoods such as rural employment schemes, silt management and freshwater agriculture.

The current framing of migration as caused by climate change deflects attention away from how coastal vulnerabilities are fundamentally entwined with political structures of social and economic inequalities. Similar to how policymakers have misread Kissidougou's forest landscape by reading history backwards and wrongly blaming local villagers for deforestation and environmental destruction in the present (Fairhead and Leach 1996, 3), attributing migration from Bangladesh's coastal zone to climate change alone misreads the socio-environmental landscape and the multi-faceted drivers of migration. As the 2022 floods in Pakistan highlight, devastation is often accepted as largely unavoidable and unpredictable when blamed on climatic change rather than poor governance (Kelman 2022). Misreadings of environmentally related migration (e.g. salinisation, floods and riverbank erosion) as 'climate-induced' removes existing structural socio-economic push factors of migration and potential ways to address agrarian struggles from public debate. Even so, 'push' dimensions of migration do not help to fully explain why some women in similar socio-economic positions are unable to migrate, while others can. Climate reductive translations of migration not only misread socio-environmental and gendered vulnerabilities, but also ignore the historical importance of migration for agrarian livelihoods while removing women's agency to choose to migrate.

Women in the Global South are constructed as more 'vulnerable' to climate change (Arora-Jonsson 2011) in ways that tends to obfuscate their agency (Cuomo 2011, 695). Bangladeshi women are assumed to be more likely to suffer in climate-related events due to being restricted by patriarchal rules of female seclusion (Cannon 2009), that they do not migrate on their own (Poncelet et al. 2010) or that they are extra vulnerable to trafficking (Stellina Jolly and Ahmad 2019). This stereotyping of gendered vulnerabilities homogenises Bangladeshi women in a way that ignores how class, education, urban-rural lifestyle, and kinship networks shape different migration possibilities (Lahiri-Dutt and Samanta 2004; Mehta 2018). Paying attention to the migration decisions of landless single mothers in coastal areas highlights the differential agentive capacity of women migrants and how this is entangled with kinship-based reproductive support.

A historically grounded ethnographic account of migration shows how such processes are embedded in environmental and socio-economic vulnerabilities rooted in political structures of gendered inequality. Through its close attention to the intimate everyday migration decisions of landless single mothers in coastal areas, this ethnography critiques how 'coastal vulnerabilities' conflates the vulnerability of a particular place to climatic risk with the socioeconomic constraints of the people living there (Dewan 2021b, 18–20). It reveals the importance of kinship relations in shaping the differential agentive capacity of women in choosing to migrate and argues that climate reductive translations of migration obfuscate the affective relations that sustain agrarian livelihoods. This historical and ethnographically grounded account of 'climate reductive translations of migration' thus contributes to critical debates on the increasingly politicised knowledge production of climate change (Barnes and Dove 2015; Eriksen, Nightingale, and Eakin 2015) and climate displacement in an agrarian world already on the move (Kelley, Shattuck, and Thomas 2021).

Three problematic misreadings of the socioecological landscape arise from 'climate reductive translations' of migration. First, attributing migration to climate change misreads coastal vulnerabilities and the importance of migration as an agrarian livelihood strategy to deal with rural precarity and debt both in the past and – also for single women (divorced, widows). Second, misreading migration caused by saline tiger-prawn cultivation, infrastructure-related waterlogging and riverbank erosion as 'climate-induced migration' or 'climate displacement' hinders a discussion of long-term solutions that may remediate such damaging anthropogenic floods. Lastly, ideas of climate change causing 'gendered displacement' misread gendered vulnerabilities. Landless women consider various aspects of reproductive kinship support intimate relations and social reputation as they choose whether or not to migrate to brick kilns, Dhaka garments industry and the Gulf for domestic work. Misreading causes of migration based on tropes of Bangladesh as a 'climate change victim' hinders public debate on solutions for rural underemployment, floods, land loss and salinisation by aquacultures. Crucially, it misses how migration is constrained/enabled by affective kinship relations to sustain social reproduction at home.

The article draws on 12 months of multi-sited qualitative research from August 2014 to July 2015. I conducted forty in-depth, unstructured interviews with development professionals in Dhaka and Khulna cities and nine months of ethnographic fieldwork in southwest coastal Bangladesh, mainly in the embanked floodplain of 'Nodi'[3] in Khulna District. This included participant observation of village life in the two unions of Nodi: 'Lonanodi' were brackish tiger-prawn cultivation was dominant, and 'Dhanmarti' where a grassroots movement ended such saline shrimp farming. My main interlocutors were two groups of women doing 'earthwork' – repairing small roads in a donor-funded rural employment scheme targeting landless 'female-headed' households (equated with the 'poorest of the poor' See Kabeer 1991; Lewis 1993). I met the twenty women regularly at their work sites, and they soon invited me home to meet their families. I hired some of them as my field assistants to go door-to-door and carry out a qualitative household survey (*jorib*) in Nodi with what ended up as approximately four hundred households. NGOs starting new projects often carry out a *jorib*, resulting in villagers being accustomed to talking to outsiders about their livelihoods and this helped legitimate my presence. Spending whole days with the women further facilitated participant observation, where I immersed myself into rural life, to 'as far as possible, to think, see, feel and sometimes act as a member of its culture' (Powdermaker 1966, 9).

[3]All names have been changed to protect the anonymity of my interlocutors. I obtained oral informed consent by explaining to each interlocutor the purpose of my work, who I am, what my affiliations were, the nature of my research, that I was not part of a project that they could benefit from and that all information would be anonymised. I asked open-ended questions about environmental change and livelihood strategies and qualitatively analysed the transcripts of these conversations. I obtained ethics approval for this fieldwork from the Department of Social Anthropology, SOAS (University of London).

Misreading coastal vulnerabilities

Rural precarity and agrarian debt

It was payday during one of the last months of the donor-funded rural employment scheme in Nodi. I sat with the landless women labourers outside the local government building in Shobuj town as they waited for their salaries. When they received news that they would not receive the two-year extension to their existing four-year contract, they were visibly upset and disappointed. Riparna, married with two children exclaimed:

> [After two years], we are barely stepping out of the poverty we're in and now we're going to fall straight back down again. If we only got two more years, we would have been able to secure our livelihoods.

The women expressed anxiety over finding a new income that would enable them to repay microcredit debt used to keep their children in school, to save up money for their daughters' marriages, and – for some of the women – their ill husband's healthcare costs.

All these expenditures are internalised as individual private costs in the absence of state-entitlements for Bangladeshi citizens (Dewan 2021b). This reflects another key finding from my survey: the large amount of agrarian debt accumulated through women household members taking four to six microcredit loans from different NGOs (Dewan 2021b; See also Karim 2011; Paprocki 2016; S. B. Banerjee and Jackson 2017). Nodi comprises of Dhanmarti union where a grassroots movement ended tiger-prawn cultivation in 2008 in favour of freshwater crop cultivation, and Lonanodi union where the movement failed. In both unions, households used microcredit to pay for healthcare, education, dowry and fees/bribes to labour brokers, illustrating the importance of debt for agrarian formation (Fairbairn et al. 2014)

Development projects and NGOs, both through the access to microcredit and to various development projects targeting the 'poor', constitute an important element of the political economy in Bangladesh's southwest coastal zone, considered to be the most poor and vulnerable in the country. My *jorib* in Nodi revealed that there were significantly more landless women (without capital or male adult earners in the household) that coveted these short-term rural employment schemes despite low wages and no social or health insurance: they provided an income for a fixed duration. The project-mentality of the development industry serves to perpetuate a system that only provides temporary patches to large-scale problem of structural rural un(der)employment and sustain a form of labour market based on precarious, low-cost labour (Dewan 2021a; See also Qureshi 2014). Coastal vulnerabilities here are fundamentally tied to socio-economic inequalities.

Migration as an agrarian livelihood strategy

High levels of out-of-pocket expenses incentivised people in Nodi to migrate – but not only among those facing the livelihood hardships of living in saline barren deserts of shrimp farms (see Feldman and Geisler 2012; Paprocki 2019b). Local work was mostly available during the planting of monsoon rice in July–August and during its harvest in the Bengali month of *Poush* (mid-December to mid-January). Seasonal labour migration provides an important source of income for Bangladeshi villagers (Afsar 2003; Toufique 2002). Able-bodied male daylabourers migrated for a month during the dry season to harvest *goromer dhan* [dry season rice] in the northern parts of Khulna district and received rice as payment, sustaining a family of four for six months. Other forms of circular migration among men in Nodi include crab fishing in the Sundarbans in the winter months at risk of being held at ransom and beaten up by gangs of *dacoits* [pirates], earth-work in labour markets or rickshaw pulling in Khulna or Dhaka.

Climate reductive translations of migration thus fail to capture the importance of seasonal labour migration as an agrarian livelihood strategy in South Asia – both historically and in the present (Rogaly et al. 2002). With the 1947 partition of India, 'internal' seasonal migration was obstructed through artificial borders; separating the local people of coastal Bengal despite affinal and cultural ties (Van Schendel 2001). Now that global warming is increasingly framed through security and borders (Cons 2018), Indian geopolitical fears of Bangladeshi climate refugees are increasingly used to legitimise further securitisation and tightening of border control (Chaturvedi and Doyle 2010, 207). For my interlocutors, India – *Bharat* – was just across the river and not *bidesh* [abroad]. Working in factories or in regionally located rice mills or brick kilns even across the border were typical forms of season-based migration for landless when local labour opportunities are scarce (Mahbubar Rahman and Van Schendel 2003; Samaddar 1999). Such circularity enables migrants to remain deeply embedded in rural social networks (Mosse et al. 2002), a form of translocal householding (Etzold and Mallick 2016; Jacka 2017; Gidwani and Ramamurthy 2018).

Historical migration among 'female-headed households'

The historical equivalents of my interlocutors, working-class divorced women and young widows ('female-headed households') in Bengal, also migrated as a livelihood strategy (Sen 1999; Engels 1996). Mughal Bengal's famous muslin textile industry was an important employer of women spinners in the eighteenth century – they were often widows from all castes and classes (N. Banerjee 1990). Bengal's textile industries were de facto deindustrialised under East India Company rule (Chattopadhyay 1990; Faraizi 1993). A Dhaka resident described how 'The import of the cotton thread from England into the country at a low price threw the Bengal spinners out of employment. Hapless widows, who subsisted on spinning were the hardest hit' (20th August 1831 in *The Samachar Sarpan*).[4]

For women, cottage-based subsistence crafts remained as an important source of income outside of the rice harvesting season – these were no different from ordinary household chores; any woman of any background could resort to them in circumstances of distress. Examples include animal husbandry, making dairy products like clarified butter [*ghee*], preserving and processing grains and pulses, making puffed rice [*muri*], preparing vegetable oil, as well as collecting, processing and selling forest produce. They also pounded rice, ground flour, produced rock salt, collected and sold fuel and fodder, as well as made leaf plates, baskets, mats, nets and brooms (N. Banerjee 1990, 284). Until the 1880s, 20% of Bangladesh's workforce comprised of women who were primarily engaged in these food processing industries that they could do from home (Sen 1999).

[4]Cited from B.N Banerjee op.cit, Vol II, pp. 336–337 in Chattopadhyay (1990).

In the 1890s, the jute industry in Bengal grew rapidly and saw a mass movement of migrants to Calcutta and its vicinities; by 1912, there were 61 jute factories in this region employing nearly 200,000 workers (Basu 2008). Like today's pattern of circular migration, adult males migrated to urban industries, while periodically returning to their wives and children in the villages (Basu 2008). Women from Bihar and Uttar Pradesh denied support from kin (widows, deserted by husbands) – also migrated to the jute mills (Sangari and Vaid 1990; N. Banerjee 1990; Sen 1999). The colonial expansion of jute exports saw the sector becoming male-dominated, where women were gradually subjected to irregular hiring and informal recruitment (Sen 1999).

By the 1920s, Bengal's jute industry was in turmoil and this was compounded by a major agrarian crisis in Bengal. Rising prices, the commercialisation of the rice trade, changing forms of rent collection, declining agricultural yields, a decrease of arable land due to British-planned public works contributed to rural indebtedness that caused three million peasant families to become functionally landless (Chatterjee 1982, 122; Engels 1996, 198). Furthermore, the introduction of mechanised rice huskers and the opening of male-operated rice mills in the 1920s impacted women workers disproportionately. While 270,000 women registered rice manual husking as their key occupation in 1911, this number dropped to 136,000 in 1931 (Engels 1996).[5] With further decline in cottage-based incomes and the lowered demand for women in jute mills, the 1920s saw landless women leaving their villages to work as domestic servants in the cities; migrating with their families or leaving with single women villagers. The number of women domestics increased by 300% within a decade (Engels 1996, 198).

These historiographies of working-class women in Bengal illustrate how the penetration of capitalism and its restructuring of subsistence economies often increased gender-based disadvantages (Momsen 2003). They also show the importance historically of migration as a key livelihood strategy that enabled women to adapt to changing circumstances: from reduced reliability of the land due to environmental stressors and shocks to the loss of family/kin support, to remediating debt and the lack of rural income opportunities. Both in the past and present, rural indebtedness serves as a push factor for migration. Yet viewing migration as an outcome mainly of uneven capitalist economic change may be limited. In the remaining paper, I discuss the importance of environmental factors for sustained agrarian livelihoods and the complexity of migration decisions made actively by women who dare to work outside the home and the social stigma they may face for doing. Drawing on ethnographic insights, I argue for the importance of kinship support for social reproduction and intimate relations in enabling landless divorced single mother's to migrate.

Misreading 'climate-induced migration'

Shrimp-related migration

Like in the past, most women in Nodi prefer local, cottage-based work that enable them to live at home and take care of their children and still have an income source. After the grassroots movement to end brackish tiger-prawn cultivation, Dhanmarti women were

[5]With the rapid mechanisation of the 1970s Green Revolution, landless women lost their main income source from processing rice (Greeley 1987).

able to yet again engage in subsistence cottage crafts (weaving palm leave mats and making fuel sticks of cow dung). In contrast, women in Lonanodi lamented that saline aquaculture degrades soil fertility and damages water quality in ways that reduces their ability to gain income from cottage-based crafts and to conduct homestead activities that rely on freshwater domestic ponds (Dewan 2021b). Since the 1990s, tiger-prawn cultivation expanded through land grabbing (Adnan 2013) and changed agrarian structures (Ito 2002). It contributed to a 'fake Blue Revolution' (Deb 1998) that reduced local labour opportunities and contributed to outmigration from rural coastal areas (Swain 1996; Guhathakurta 2003; Samaddar 1999).

Rural livelihoods thrive on fertile agricultural soils and the original purpose of flood-protection embankments in this region was to keep out saline intrusion in the rivers during the dry season (Dewan 2021a). Shrimp businessmen drill unauthorised pipes and cut embankments in order to flood these arable lands with brackish waters – rendering them structurally unstable (de Silva 2012). Such forced salinisation of the soil reduces local food production and increases the cost of food and other household items (Paprocki and Cons 2014; Swapan and Gavin 2011), worsening rural livelihood opportunities (Pouliotte, Smit, and Westerhoff 2009). Salinisation negatively affects reproductive activities (cooking, cleaning, housekeeping, crafts) as well as mental and bodily wellbeing (Dewan 2021b) – illustrating the gendered workings of aquaculture extractivism (Ojeda 2021). Based on these socio-environmental effects, the International Organisation of Migration (2010) states that even without the projected impacts of climate change, increasing salinity and population pressure will escalate emigration pressures in Bangladesh's coastal zone. This illustrates how the vulnerability of agrarian livelihoods in this delta is shaped by a complex interplay of land use practices.

Such an ecological vulnerability is compounded by the lack of state provision in an aid-dependent landscape where discourses of climate change obscures the unequal access to water-land resources held by the more powerful (Dewan 2021b). Considering this historical and political context of inequality, climate reductive translations of migration that represent people migrating from tiger-prawn producing villages in the wake of cyclones as 'climate migrants' (Lovatt 2016) are deeply problematic. A reframing of shrimp-induced migration as 'climate migration' could be perceived as part of a grander scheme of depeasantization through 'anticipatory ruination' by an 'adaptation regime' (Paprocki 2019a), where agrarian dispossession fuels migration supporting urban expansion (Paprocki 2018). In addition, a climate reductive translation of migration from tiger-prawn producing areas obfuscates the environmental degradation that exacerbates structural socio-economic push factors of migration. It is therefore important to separate migration caused by environmental degradation and broken embankments resulting from shrimp farming from 'climate-induced' migration.

Flood-related migration

As shrimp-induced migration suggests, floods in Bangladesh are not solely about climatic change. Furthermore, not all kinds of floods must be prevented. Monsoon *borsha* floods have long been integral to agricultural activities when silt-laden river water mixed with rain and inundated these deltaic wetlands, this fertilised the soil and irrigated rice fields while providing essential breeding ground for fish (Zaman 1993). The construction of donor-funded 'flood-protection' embankments in the 1960s obstructed these *borsha* floods and trapped silt in the outside river (Dewan 2021a). The difference in elevation of a raised riverbed and a sunken floodplain resulted in rainwater being unable to drain and caused damaging *jalabaddho* floods [drainage congestion, waterlogging] (Iqbal 2010) that we often see in media portrayals of 'flooded Bangladesh' (Dewan 2021b) (Figure 2).

Embankments contribute to waterlogging and obstruct the delta's ability to raise its land levels through sediment deposits, and therefore pose greater flood risks than rising sea levels (Auerbach et al. 2015). Indeed, the unembanked Sundarbans have gained 1–1.5 m of elevation, while embanked floodplains (like Nodi) are sediment deprived (van Staveren, Warner, and Shah Alam Khan 2017). A reductive translation of migration as caused by climate-related floods removes attention from the fact that 'flood-protection' embankments – through a century of poor infrastructure design by colonial powers and western donors – actually worsen *jalabaddho* floods (Dewan 2021b). Migration due to waterlogging is not 'climate-induced'.

Embankments further contribute to the silting up several important water bodies, rivers and canals (Dewan, Mukherji, and Buisson 2015). This further exacerbates Bangladesh's climatic risks as the monsoon is expected to shift and this poses a grave threat to lives and livelihoods in South Asia (Amrith 2018). Silted waterbodies have less capacity to retain freshwater monsoon rains over the dry season and this hampers their ability to counter the salinity of the yearly dry saline season (Dewan 2020).

This siltation and salinisation is also affected by India's diversion of freshwater through the Farakka barrage that has increased salinity of Ganges flows since the 1970s (Government of Bangladesh 1976; Khan et al. 2015) and the conversion of agricultural land into saltwater tiger-prawn farms that salinize soil and water since the 1980s (Bernzen, Jenkins, and Braun 2019). In order to address flooding as the monsoon rains and dry seasons are expected to shift with climatic change, silt management of the delta and long-term solutions against waterlogging that increase the water-retention capacities of rivers and canals is of urgent importance to help sustain agrarian livelihoods.

Displacement from riverbank erosion

Floods and silt capture the geological feature of the Bengal delta that accretes and erodes. The Ganges River has meandered eastwards for millennia eroding land,

Figure 2. Embankment and raised banks due to siltation. Photo by Author.

riverbanks and now flood-protection embankments as it changes its course while creating new lands (Brammer 2012). This held an important role in shaping land revenue policies from Mughal to colonial times (Dewan 2021b). O'Malley (1908) writes '... the streams are constantly eating away on one bank and depositing on the other, until the channel in which they formerly flowed becomes choked up'. Riverbank erosion swallows up land into the river. This is not a new phenomenon causing vulnerability in the Bengal delta, recognising how land was swallowed up by the rivers. As a senior government official in Bangladesh's water sector stated: 'you cannot correlate climate change [directly] with riverbank erosion ... The delta is in a formation stage, it eroded in the past, it's eroding now, and it will continue to erode in the future as well'[6] (Figure 3).

Riverbank erosion (*nodi bhangon*) is damaging and disruptive; ruining water-sided road cum embankments causing those who live on top, or outside, the embankment to lose their homes and move away. My interlocutors in Nodi perceived riverbank erosion as part of the course of the river – this displacement is one they had seen for generations, resulting in people having to migrate. Such erosion is increasingly cast as being *caused by* climate change where Western media and the aid industry portray these erosion-displaced peoples as 'climate refugees' (McDonnell 2019; Nikitas 2016), their land 'swallowed by the river' (Montu 2020). A short-term environmental shock like riverbank erosion does not itself constitute climatic change (Bernzen, Jenkins, and Braun 2019) nor are riverbank displaced peoples necessarily climate migrants (Lewis 2010). Such a climate reductive translation of migration deflects attention away from the fact that riverbank-induced displacement forms part of a wider political problem of regular land loss in a context of land fragmentation and a growing population. Labelling this as 'climate-induced' makes it difficult to properly discuss the long-term political solutions of equitable and pro-poor land settlement that forms part of contemporary agrarian struggles in Bangladesh.

Misreading gendered vulnerabilities

Reproductive support, intimate relations and migration to brick kilns

Lipi is a divorced single mother living with her elderly mother and adolescent daughter in Lonanodi. She was particularly upset over the roads project ending since migration was not an option for her. Her elderly mother could not take care of her daughter on her own, nor could she rely on her older brothers' wives who lived in the same homestead compound. She also expressed strong fear of migrating:

> I have never migrated. I'm too afraid to go to an unknown place as a lone woman. *Kharap* [bad, spoiled] things can happen to me. I will never do it! What if someone kidnaps me or if I get lost and cannot find my way back again? It's too terrifying, I won't do that, I'd rather die poor, at home.

Thus, she tried to get irregular work in Lonanodi, struggling to ensure that her daughter could continue studying rather than being married off.

---

[6]Bangladeshi researchers I met stated that up until 2015 there had been no conclusive study confirming that climatic change is exacerbating *nodi bhangon*.

Figure 3. Riverbank erosion in Satkhira district January 2015. Photo by Author.

Lipi lives in southwest coastal Bangladesh – a region portrayed by donors and NGOs as a place where migration is 'climate-related' and where women are particularly vulnerable to 'displacement' (World Bank 2009; UNDP 2019). In this section, I critically examine this misreading of gendered vulnerabilities by discussing how Nodi women's ability and motivations to migrate differ based on affective and intimate relations that help them navigate social reproduction in a context of both socio-economic and environmental vulnerabilities.

Furthermore, Lipi referred to brick kilns as places of illicit extramarital relationships and stated that she was above *kharap* behaviour. Her colleague Yasmine, on the other hand, regularly migrated to the brick kilns each year, spending six months there at a time. She was relaxed about the contract ending stating, 'Once the project ends, I will go to the brick kilns. I can earn a lot of money'. It was only after I left Bangladesh that she introduced me to her *bariwallah* [colloq. husband] on the phone. As a divorced single mother, she was now in a new relationship in a society that condemns extra-marital romantic relations. In the brick kilns, she and her partner could live as 'husband and wife' – a practice of 'temporary marriages' also common in the Bengal jute factories of Calcutta more than a century ago, where women would symbolically marry a man as protection from other men (Curjel 1923). Similar to how migration to brick kilns may be romantically motivated in India (Shah 2006), the privacy of remote brick kilns in Bangladesh provide the opportunity for an alternative life. For Yasmine, who was married off when she was twelve and divorced when her daughter was only a few months old, this romantic relationship was one that she chose for herself as an adult woman. It provided an important incentive for migration, while ensuring personal security among strangers in patriarchal spaces – her choice to migrate was affective and thus not 'climate-induced' or due to environmental pressures caused by shrimp farming.

While Lipi had no female relatives to leave her daughter with, and did not want to migrate alone, Yasmine lived with her mother in a female-headed household along with her many sisters and younger brothers' wives that could take care of her now adolescent daughter while she worked with her 'husband'. This illustrates how women's choices to migrate depends on their access to kinship-based reproductive support at home to take care of their children as well as personal contacts (incl. intimate relations) to facilitate migration.

The pull of Dhaka ready-made garments

Both Lipi and Yasmine in Lonanodi lacked educational qualifications and were limited to the informal sector, such as brick kilns. Dhaka's global textile industry, on the other hand, only employs workers above the age of eighteen with a secondary school certificate, incentivising villagers to allow their daughters to study longer. While landless families in Dhanmarti also migrated to brick kilns, villagers in Dhanmarti knew more people who could help them get jobs in Dhaka's Ready-Made Garment's industry. Bangladesh's garments industry received much negative attention after the 2013 Rana Plaza fire, but it remains one of the popular migration destinations in Bangladesh especially for poor women attracted to the higher urban wages (Seeley et al. 2006). They account for nearly two-thirds of the total employment (R. I. Rahman and Islam 2013; Jamaly and Wickramanayake 1996; Kabeer 2000). Dhaka RMG employs more than four million people and accounts for eighty per cent of Bangladesh's exports (Gardner and Lewis 2015, 177). In contrast to migration to the garment industry being a response to capitalist displacement like in the Caribbean (Werner 2016), landless women workers in Nodi, perceived work in Dhaka's RMG industry to hold a higher status compared to seasonal brickwork. Garments work is an employment (*chakri*) that requires full-time work throughout the year, where they can choose to earn more through paid overtime.

In Nodi, labour migration to Dhaka's RMG industries provided a stable, not-too-expensive, higher income opportunity to repay large microcredit debt, especially for families whose small businesses or migration abroad had failed. For example, Sunil is a young boy staying with his maternal grandparents in a village in Dhanmarti while his parents migrated to Dhaka RMG to pay off a microcredit loan for a failed poultry business. Rather than such rural-to-urban migration being climate-induced, or that these women workers are necessarily environmental migrants (McDonnell 2019; Evertsen and van der Geest 2019), this was a much coveted migration destination. Indeed, some of the landless single mothers in Lonanodi asked if I could help their daughters get work in the Dhaka RMG sector.

But these migration choices were not only economic. Salma was one of my closest interlocutors. At the time of my fieldwork, she was a divorced single mother in her mid-twenties living with her young daughter and divorced mother. Salma's father left them at an early age, and she grew up in the lush homestead compound of her maternal grandparents in Dhanmarti. At the age of 16, Salma went with a distant female 'relative' to work in the Dhaka RMG industry. As the earthwork project was ending, she was considering re-migrating to Dhaka RMG, but this time with her new secret partner. Like in Yasmine's case, intimate relations shaped Salma's desire to migrate, while her kinship relations provided her with the reproductive support to raise her child in Nodi while she was working in Dhaka.

Migration to the gulf: social stigma as a key gendered vulnerability

These romantic choices are kept highly secret also due to the wider societal stigmatisation of working-class women without husbands, be they young and single, widows or divorcés/abandoned. In the past and present, unmarried/divorced/widowed women of poorer classes often faced social stigma when working outside side of the home. Women workers in the jute mills were increasingly seen as breaking rules of conduct for women and received negative treatment by their male colleagues (Engels 1996). By the 1920s, these women workers were deemed as social outcasts (Sangari and Vaid 1990) and were negatively described as 'disease-ridden', 'degraded' and engaging in open prostitution (Curjel 1923). Similarly, women migrating to Dhaka or the Middle East today are perceived to be engaging in sexual work/relations. Salma, who met her ex (and father of her child) while working in Dhaka, emphasised that she lived with a family with strict controls, not at girls' hostels:

> When you have six to seven girls living together in a *mess* [hostel], they can go wherever they want. They have too much freedom; they can go to parks and live like bachelors. No one asks what they are doing. They can stay out late at night, meet boys. There are various problems with these things; they can have [extra-marital] affairs.

Whether women migrate to brick kilns, to Dhaka's Ready Made Garment's sector or the Gulf, women who enter new spaces or leave home to work without a formal 'male guardian' are often perceived as *kharap* as they break social conventions of women's purity. Several of my interlocutors, including Lipi, described this type of normative stigma as a reason for why they would not leave Nodi for work.

Yet, Salma told me how three women from their village had already gone to the Middle East for domestic work via a locally based labour broker. The broker approached Salma and offered a two-year work contract in Oman. She knew the families receiving remittances were doing well. Despite her good relations to her maternal uncles and their wives, Salma pointed out: 'It is uncommon that brothers allow their sisters to inherit. We must plan for the day my grandfather is no longer here ... we must save up money to buy our own homestead land we can live on'. Salma's 40-year-old divorced single mother Jhorna, worried about the [sexual] risks and Salma's safety, stating that Salma was too young and beautiful and that she would rather go in her place. The broker agreed and said the fee would only be 30,000 taka. Afterwards, Salma regretted the decision:

> Once we started the process of training, tickets and applying for passports, the cost went up to 50,000 taka. We have spent 80,000 taka in total. We feel cheated. We borrowed 60,000 taka and I used 20,000 taka of my own savings. Now my mother is crying daily. They're making her work in two homes rather than just one.

Jhorna struggled for the first few months and homesickness was a common issue for many of the men and women going to work abroad as they struggled to adapt, without their kin, in a foreign country, surrounded by a foreign language and foreign foods. Salma confided that she will not recommend any woman to work in the Middle East as a domestic worker: 'It's better to be a beggar in Bangladesh, than indebted there'. Salma pleaded with Jhorna to stay another six months so that they could at least repay the loans taken to migrate. If Jhorna broke her contract and returned earlier, the journey to Oman would have been for nothing; they would not have acquired the savings needed to buy land and would be forced to migrate elsewhere (such as the Dhaka RMG sector) to repay their loans.

Among my interlocutors, climate or environmental reasons were never given as a cause of migration to the Gulf. Prior to my fieldwork, I mistakenly assumed that extreme poor women and female-headed households would not have the financial resources to migrate internationally as domestic workers.[7] Yet Salma and her colleagues did consider such options, borrowing from brokers to migrate. For example, Sahira a youthful and attractive divorced single mother living in her brother's homestead compound in Lonanodi village was considering an offer from her cousin's friend in Dhaka to help her get a job in Saudi Arabia after her earthwork project ended: 'I only have to pay 30,000 taka for a two-year contract, they will arrange the rest'. I recalled Salma and her mother's experience and the costs of visas and flights, none of which could be covered by a 30,000-taka fee.

Her colleagues pointed out that she did not know this *dallal* [broker] and began to discuss the risks of sexual exploitation of domestic workers abroad covered in the newspapers. Afrina, the oldest, was the most critical:

> We do not want to send our daughters/girls to Arab countries. They [the brokers] only want young and beautiful girls. We have read in the papers that these returned girls were tortured. Did you know that one of our Ministers said that when Bangladeshi girls are being exploited in Bangladesh, they might as well be exploited in Saudi and earn good money?

Arguably, concerns of sexual exploitation of women migrant workers can easily be coopted where negative representations of women's sexuality as *kharap* collude with patriarchal framings that support the policing of women (see Bakare-Yusuf 2013). However, as out-of-pocket expenses for healthcare, education, labour broker and marriage/dowry costs increase and microcredit debt with it in a context of structural underemployment and inequitable access to land (Dewan 2021b), women must look farther afield. Such options include migration to seasonal brick kilns and rice mills, the Dhaka garment factories, as domestic workers both in urban areas and to the Middle East to provide for themselves and their families where romantic relations and kinship support both motivate and shape migration decisions.

Discussion and conclusion

Countries of the Global South have been subject to colonial powers and external actors (donors, corporations) intervening in their societies, economies – and their environments – for centuries. Development interventions such as embankments, brackish

[7]Research on Bangladeshi domestic workers migrating to the Middle East and South East Asia has already been long established (Gardner 1995; Siddiqui 2001). In recent years, the Government of Bangladesh has implemented incentives to recruit women to work abroad as domestic workers (Migrant Rights 2016).

tiger-prawn cultivation for exports and agrochemical-dependent high-yield agriculture fundamentally change and renders more vulnerable Bangladesh's deltaic wetlands, exacerbating risks to climatic change (Dewan 2021b). Development interventions supposed to help Bangladesh adapt to climatic change may actually worsen the social vulnerability that they are supposed to address by removing focus from how such socio-ecological vulnerabilities are compounded by the lack of state provision in an aid-dependent landscape. This in turn results in high out-of-pocket expenditure for healthcare, education and labour brokering costs – fuelling rural indebtedness.

Combining environmental history, political economy and historiographies of working class women in Bengal with long-term ethnography with landless single mothers in the southwest coastal zone of Bangladesh reveal how longstanding agrarian injustices and gender-based constraints contribute to migration choices. Such historically anchored ethnography yields insights into agrarian political economy which can be leveraged against simplified understandings of contemporary migration being *caused by* climate change. Any discussion on 'climate displacement' in agrarian worlds drastically changing due to environmental degradation and climatic change must also therefore take into account changing political economies and the importance of migration as a rural livelihood strategy. Labour migration continues to be important for Bangladeshis. Contrary to beliefs that Bangladeshi [Muslim] women are more sedentary and extra 'climate vulnerable', working-class women in Bengal have a history of migrating for work, especially divorcées and widows.

Attention to environmental history further sheds light on various forms of misreading of 'climate displacement' in Bangladesh, an iconic symbol of a 'climate change victim' that is portrayed as drowning in rising sea levels. Firstly, the long-term salinisation of arable lands by brackish aquaculture has contributed to rural hardship and outmigration since it was introduced to the coastal zone already in the early 1990s as its damages to embankments have further contributed to cyclone-related migration. Secondly, waterlogging floods are deeply entangled with man-made infrastructure impeding the drainage of monsoon rains. Thirdly, land being 'swallowed by the river' is due to river-bank erosion (rather than floods) and is a geological feature of the Ganges. By portraying river-bank displaced peoples, migrants from tiger-prawn producing villages and people suffering from waterlogging as 'climate refugees'-political solutions are ignored such as land-compensation schemes, stopping shrimp farming, and remediating the flaws of flood-protection embankments through tidal river management and regular canal excavation through rural employment schemes.

A healthy ecology is crucial for cottage-based crafts and rural livelihoods, where women in saline Lonanodi faced greater hardship than those in freshwater Dhanmarti. Yet, due to debt and high costs of living, women were migrating from both these unions. Like rural women in colonial times, lack of rural income motivated them to work outside the home and village. Though the jute industry is no longer as important as it once was, Dhaka's export-oriented garment industries resembles the Mughal textile industries by being one of the main employers of working class women, particularly those without adult male earners. In the 1920s and 1930s, agrarian crisis resulted in many women working as domestic workers in Bengali cities, today they go to the Middle East. But as this ethnography shows, women migrated not only due to environmental degradation, debt or aspirations to increase incomes but to also be with their secret romantic partners. The ability of landless, divorced single mothers to migrate depended on their reproductive kinship support and their social relations that could help them migrate safely. Indeed, some women chose not to migrate due to the social stigmatisation of migrant women, especially those migrating to the Middle East as domestic workers or to Dhaka RMG – both high-income destinations.

To conclude, climate reductive translations of migration misread coastal vulnerabilities and the push factors of migration and ignore the heterogeneity of women and their differential agentive capacities, thereby failing to address policy solutions that remediate rural underemployment, floods, riverbank erosion and salinisation by aquaculture. By highlighting the disconnect between development policy discourses with everyday gendered lived experiences of socio-environmental agrarian vulnerabilities, this article contributes to the growing critical scholarship on climate adaptation, migration and development and 'agrarian worlds on the move'.

Acknowledgements

The research on which this article is based was funded by a Bloomsbury Colleges studentship from the University of London (Birkbeck College and SOAS) and writing up was funded by the Norwegian Research Council [grant number 275204/F10]. I am indebted to my (sister-)interlocutors in Nodi who shared their lives with me. I am grateful for comments on early versions of this article from David Mosse, Penny Vera-Sanso, Katy Gardner, James Fairhead, Caroline Osella and Christopher Davis. Thanks to Nicholas Loubere, Nari Senanayake, Anwesha Dutta and Elisabeth Schober, as well as the participants of the 2022 AAG panel 'Conceptualizing Climate Displacement in an Agrarian World Already on the Move' and Session 11 of the 2022 JPS Climate Change and Agrarian Justice conference for critical and constructive comments. Special thanks to Annie Shattuck, as well as to the four anonymous reviewers for their excellent feedback and suggestions.

Disclosure statement

No potential conflict of interest was reported by the author(s).

Funding

This research was funded by Bloomsbury Colleges through a PhD studentship at Birkbeck College and SOAS (2013–2017); and writing up through the Norwegian Research Council under [grant number 275204/F10]; Norges Forskningsråd.

ORCID

Camelia Dewan http://orcid.org/0000-0003-3377-2413

References

Adnan, Shapan. 2013. "Land Grabs and Primitive Accumulation in Deltaic Bangladesh: Interactions Between Neoliberal Globalization, State Interventions, Power Relations and Peasant Resistance." *The Journal of Peasant Studies* 40 (1): 87–128. doi:10.1080/03066150.2012.753058

Adri, Neelopal, and David Simon. 2018. "A Tale of Two Groups: Focusing on the Differential Vulnerability of "Climate-Induced" and "Non-Climate-Induced" Migrants in Dhaka City." *Climate and Development* 10 (4): 321–336. doi:10.1080/17565529.2017.1291402.

Afsar, Rita. 2003. *Internal Migration and the Development Nexus: The Case of Bangladesh*. Dhaka: Refugee and Migratory Movements Research Unit and UK Department for International Development.

Ahsan, Reazul. 2019. "Climate-Induced Migration: Impacts on Social Structures and Justice in Bangladesh." *South Asia Research* 39 (2): 184–201. doi:10.1177/0262728019842968.

Amrith, Sunil S. 2014. "Currents of Global Migration." *Development and Change* 45 (5): 1134–1154. doi:10.1111/dech.12109

Amrith, Sunil S. 2018. "Risk and the South Asian Monsoon." *Climatic Change* 151 (1): 17–28. doi:10.1007/s10584-016-1629-x.

Arora-Jonsson, Seema. 2011. "Virtue and Vulnerability: Discourses on Women, Gender and Climate Change." *Global Environmental Change* 21 (2): 744–751. doi:10.1016/j.gloenvcha.2011.01.005.

Auerbach, L. W., S. L. Goodbred Jr, D. R. Mondal, C. A. Wilson, K. R. Ahmed, K. Roy, M. S. Steckler, C. Small, J. M. Gilligan, and B. A. Ackerly. 2015. "Flood Risk of Natural and Embanked Landscapes on the Ganges–Brahmaputra Tidal Delta Plain." *Nature Climate Change* 5 (2): 153–157. doi:10.1038/nclimate2472.

Bakare-Yusuf, Bibi. 2013. "Thinking with Pleasure: Danger, Sexuality and Agency." In *Women, Sexuality and the Political Power of Pleasure. Feminisms and Development*, edited by Susie Jolly, Andrea Cornwall, and Kate Hawkins, 28–41. London: Zed Books.

Banerjee, Nirmala. 1990. "Working Women in Colonial Bengal: Modernization and Marginalization." In *Recasting Women: Essays in Indian Colonial History*, edited by Kumkum Sangari and Sudesh Vaid, 269–301. New Jersey: Rutgers University Press.

Banerjee, Subhabrata Bobby, and Laurel Jackson. 2017. "Microfinance and the Business of Poverty Reduction: Critical Perspectives from Rural Bangladesh." *Human Relations* 70 (1): 63–91. doi:10.1177/0018726716640865.

Barnes, Jessica, and Michael Dove. 2015. *Climate Cultures: Anthropological Perspectives on Climate Change. Yale Agrarian Studies Series*. New Haven: Yale University Press.

Basu, Subho. 2008. "The Paradox of Peasant Worker: Re-Conceptualizing Workers' Politics in Bengal 1890-1939." *Modern Asian Studies* 42 (1): 47–74. doi:10.1017/S0026749X0700279X.

Bernzen, Amelie, J. Jenkins, and Boris Braun. 2019. "Climate Change-Induced Migration in Coastal Bangladesh? A Critical Assessment of Migration Drivers in Rural Households Under Economic and Environmental Stress." *Geosciences* 9 (1): 51. doi:10.3390/geosciences9010051.

Borras Jr., Saturnino M., Ian Scoones, Amita Baviskar, Marc Edelman, Nancy Lee Peluso, and Wendy Wolford. 2022. "Climate Change and Agrarian Struggles: An Invitation to Contribute to a JPS Forum." *The Journal of Peasant Studies* 49 (1): 1–28. doi:10.1080/03066150.2021.1956473.

Brammer, Hugh. 2012. *The Physical Geography of Bangladesh*. Dhaka: The University Press.

Cannon, Terry. 2009. "Gender and Climate Hazards in Bangladesh." In *Climate Change and Gender Justice. Working in Gender & Development*, edited by Geraldine Terry, 54–70. Oxford, UK: Practical Action; Oxfam.

Chatterjee, Partha. 1982. "Agrarian Structure in Pre-Partition Bengal." In *Three Studies on the Agrarian Structure in Bengal 1850-1947*, edited by Asok Sen, Partha Chatterjee, and Saugata Mukherji, 113–224. Calcutta: Centre for Studies in Social Sciences.

Chattopadhyay, Dilip Kumar. 1990. *Dynamics of Social Change in Bengal, 1817-1851*. Calcutta: Punthi Pustak.

Chaturvedi, Sanjay, and Timothy Doyle. 2010. "Geopolitics of Fear and the Emergence of 'Climate Refugees': Imaginative Geographies of Climate Change and Displacements in Bangladesh." *Journal of the Indian Ocean Region* 6 (2): 206–222. doi:10.1080/19480881.2010.536665.

Collectif Argos. 2010. *Climate Refugees*. Cambridge, MA: MIT Press.

Cons, Jason. 2018. "Staging Climate Security: Resilience and Heterodystopia in the Bangladesh Borderlands." *Cultural Anthropology* 33 (2): 266–294. doi:10.14506/ca33.2.08.

Cuomo, Chris J. 2011. "Climate Change, Vulnerability, and Responsibility." *Hypatia* 26 (4): 690–714. doi:10.1111/j.1527-2001.2011.01220.x

Curjel, Dagmar F. 1923. *Women's Labour in Bengal Industries. Bulletins of Indian Industries and Labour 31*. Calcutta: Superintendent Government Printing.

Deb, Apurba Krishna. 1998. "Fake Blue Revolution: Environmental and Socio-Economic Impacts of Shrimp Culture in the Coastal Areas of Bangladesh." *Ocean & Coastal Management* 41 (1): 63–88. doi:10.1016/S0964-5691(98)00074-X

Dewan, Camelia. 2020. "'Climate Change as a Spice': Brokering Environmental Knowledge in Bangladesh's Development Industry." *Ethnos: Journal of Anthropology* 87 (3): 538–559. doi:10.1080/00141844.2020.1788109.

Dewan, Camelia. 2021a. "Embanking the Sundarbans: The Obfuscating Discourse of Climate Change." In *The Anthroposcene of Weather and Climate: Ethnographic Contributions to the Climate Change Debate*, edited by Paul Sillitoe, 294–321. Oxford: Berghahn Books.

Dewan, Camelia. 2021b. *Misreading the Bengal Delta: Climate Change, Development, and Livelihoods in Coastal Bangladesh. Culture, Place, and Nature*. Seattle: University of Washington Press.

Dewan, Camelia, Aditi Mukherji, and Marie-Charlotte Buisson. 2015. "Evolution of Water Management in Coastal Bangladesh: From Temporary Earthen Embankments to Depoliticized Community-Managed Polders." *Water International* 40 (3): 401–416. doi:10.1080/02508060.2015.1025196

Engels, Dagmar. 1996. *Beyond Purdah? Women in Bengal 1890-1939. SOAS Studies on South Asia*. Delhi: Oxford University Press.

Eriksen, Siri H., Andrea J. Nightingale, and Hallie Eakin. 2015. "Reframing Adaptation: The Political Nature of Climate Change Adaptation." *Global Environmental Change* 35 (November): 523–533. doi:10.1016/j.gloenvcha.2015.09.014.

Etzold, Benjamin, and Bishawjit Mallick. 2016. "Moving Beyond the Focus on Environmental Migration Towards Recognizing the Normality of Translocal Lives: Insights from Bangladesh." In *Migration, Risk Management and Climate Change: Evidence and Policy Responses. Global Migration Issues*, edited by Andrea Milan, Benjamin Schraven, Koko Warner, and Noemi Cascosne, 105–128. Cham: Springer International Publishing.

Evertsen, Kathinka Fossum, and Kees van der Geest. 2019. "Gender, Environment and Migration in Bangladesh." *Climate and Development* 0 (0): 1–11. doi:10.1080/17565529.2019.1596059.

Fairbairn, Madeleine, Jonathan Fox, S. Ryan Isakson, Michael Levien, Nancy Peluso, Shahra Razavi, Ian Scoones, and K. Sivaramakrishnan. 2014. "Introduction: New Directions in Agrarian Political Economy." *The Journal of Peasant Studies* 41 (5): 653–666. doi:10.1080/03066150.2014.953490.

Fairhead, James, and Melissa Leach. 1996. *Misreading the African Landscape: Society and Ecology in a Forest-Savanna Mosaic*. Cambridge: Cambridge University Press.

Faraizi, Aminul Haque. 1993. *Bangladesh: Peasant Migration and the World Capitalist Economy. South Asian Publications Series, Asian Studies Association of Australia, no. 8*. New Delhi: Sterling.

Farbotko, Carol, and Heather Lazrus. 2012. "The First Climate Refugees? Contesting Global Narratives of Climate Change in Tuvalu." *Global Environmental Change* 22 (2): 382–390. doi:10.1016/j.gloenvcha.2011.11.014

Feldman, Shelley, and Charles Geisler. 2012. "Land Expropriation and Displacement in Bangladesh." *The Journal of Peasant Studies* 39 (3–4): 971–993. doi:10.1080/03066150.2012.661719

Gardner, Katy. 1995. *Global Migrants, Local Lives: Travel and Transformation in Rural Bangladesh. Oxford Studies in Social and Cultural Anthropology*. Oxford: Clarendon Press.

Gardner, Katy, and David Lewis. 2015. *Anthropology and Development: Challenges for the Twenty-First Century*. London: Pluto Press.

Gidwani, Vinay, and Priti Ramamurthy. 2018. "Agrarian Questions of Labor in Urban India: Middle Migrants, Translocal Householding and the Intersectional Politics of Social Reproduction." *The Journal of Peasant Studies* 45 (5–6): 994–1017. doi:10.1080/03066150.2018.1503172.

Government of Bangladesh. 1976. *Deadlock on the Ganges*. Dhaka: Ministry of Water, Government of Bangladesh.

Greeley, Martin. 1987. *Postharvest Losses, Technology, and Employment: The Case of Rice in Bangladesh*. Boulder, CO: Westview Press.

Guhathakurta, Meghna. 2003. "Globalisation, Class and Gender Relations: The Shrimp Industry in Southwestern Bangladesh." In *Globalisation, Environmental Crisis, and Social Change in Bangladesh*, edited by Matiur Rahman, 295–308. Dhaka: The University Press.

Hulme, Mike. 2011. "Reducing the Future to Climate: A Story of Climate Determinism and Reductionism." *Osiris* 26 (1): 245–266. doi:10.1086/661274.

International Organization for Migration. 2010. *Assessing the Evidence: Environment, Climate Change and Migration in Bangladesh*. Dhaka: International Organization for Migration.

Iqbal, Iftekhar. 2010. *The Bengal Delta: Ecology, State and Social Change, 1840-1943. Cambridge Imperial and Post-Colonial Studies Series*. Basingstoke: Palgrave Macmillan.

Islam, Rafiqul, Susanne Schech, and Udoy Saikia. 2021. "Climate Change Events in the Bengali Migration to the Chittagong Hill Tracts (CHT) in Bangladesh." *Climate and Development* 13 (5): 375–385. doi:10.1080/17565529.2020.1780191.

Islam, M Rezaul, and M. Shamsuddoha. 2017. "Socioeconomic Consequences of Climate Induced Human Displacement and Migration in Bangladesh." *International Sociology* 32 (3): 277–298. doi:10.1177/0268580917693173.

Ito, S. 2002. "From Rice to Prawns: Economic Transformation and Agrarian Structure in Rural Bangladesh." *The Journal of Peasant Studies* 29 (2): 47–70. doi:10.1080/714003949.

Jacka, Tamara. 2017. "Translocal Family Reproduction and Agrarian Change in China: A New Analytical Framework." *The Journal of Peasant Studies* 45 (7): 1341–1359. doi:10.1080/03066150.2017.1314267.

Jamaly, R., and E. Wickramanayake. 1996. "Women Workers in the Garment Industry in Dhaka, Bangladesh." *Development in Practice* 6 (2): 156–161.

Jolly, Stellina, and Nafees Ahmad. 2019. "Climate Refugees: The Role of South Asian Judiciaries in Protecting the Climate Refugees." In *Climate Refugees in South Asia: Protection Under International Legal Standards and State Practices in South Asia. International Law and the Global South*, edited by Stellina Jolly and Nafees Ahmad, 203–254. Singapore: Springer Singapore. doi:10.1007/978-981-13-3137-4_6.

Kabeer, Naila. 1991. "Gender Dimensions of Rural Poverty: Analysis from Bangladesh." *Journal of Peasant Studies* 18 (2): 241–262. doi:10.1080/03066159108438451.

Kabeer, Naila. 2000. *The Power to Choose: Bangladeshi Women and Labour Market Decisions in London and Dhaka*. New York: Verso Books.

Karim, Lamia. 2011. *Microfinance and Its Discontents: Women in Debt in Bangladesh*. Minnesota: University of Minnesota Press.

Kartiki, Katha. 2011. "Climate Change and Migration: A Case Study from Rural Bangladesh." *Gender & Development* 19 (1): 23–38. doi:10.1080/13552074.2011.554017.

Kelley, Lisa C., Annie Shattuck, and Kimberley Anh Thomas. 2021. "Cumulative Socionatural Displacements: Reconceptualizing Climate Displacements in a World Already on the Move." *Annals of the American Association of Geographers* 0 (0): 1–10. doi:10.1080/24694452.2021.1960144.

Kelman, Ilan. 2022. "Pakistan's Floods Are a Disaster – But They Didn't Have to Be." *The Conversation* (blog), 20 September 2022. http://theconversation.com/pakistans-floods-are-a-disaster-but-they-didnt-have-to-be-190027.

Khan, Zahirul H., F. A. Kamal, N. A. A. Khan, S. H. Khan, M. M. Rahman, M. S. A. Khan, A. K. M. S. Islam, and Bharat R. Sharma. 2015. "External Drivers of Change: Scenarios and Future Projections of the Surface Water Resources in the Ganges Coastal Zone of Bangladesh." In *Revitalizing the Ganges Coastal Zone: Turning Science Into Policy and Practices Conference Proceedings*, edited by Elizabeth Humphreys, T. P. Tuong, Marie-Charlotte Buisson, I. Pukinskis, and Michael Phillips, 27–38. Colombo: CGIAR Challenge Program on Water and Food (CPWF).

Lahiri-Dutt, K., and G. Samanta. 2004. "Fleeting Land, Fleeting People: Bangladeshi Women in a Charland Environment in Lower Bengal, India." *Asian and Pacific Migration Journal : APMJ* 13 (4): 475–495. doi:10.1177/011719680401300404.

Latour, Bruno. 2005. *Reassembling the Social: An Introduction to Actor-Network-Theory*. Oxford: Oxford University Press.

Lewis, David. 1993. "Going It Alone: Female-Headed Households, Rights and Resources in Rural Bangladesh." *European Journal of Development Research* 5 (2): 23–42. doi:10.1080/09578819308426586.

Lewis, David. 2010. "The Strength of Weak Ideas? Human Security, Policy History, and Climate Change in Bangladesh." *Critical Interventions* 11: 113–129.

Lovatt, Joanna. 2016. "The Bangladesh Shrimp Farmers Facing Life on the Edge." *The Guardian*, 17 February 2016, sec. Global Development Professionals Network. http://www.theguardian.com/global-development-professionals-network/2016/feb/17/the-bangladesh-shrimp-farmers-facing-life-on-the-edge.

McDonnell, Tim. 2019. "Climate Change Creates a New Migration Crisis for Bangladesh." *National Geographic*, 24 January 2019. https://www.nationalgeographic.com/environment/2019/01/climate-change-drives-migration-crisis-in-bangladesh-from-dhaka-sundabans/.

Mehta, Rimple. 2018. *Women, Mobility and Incarceration: Love and Recasting of Self Across the Bangladesh-India Border*. London: Routledge.

Momsen, Janet Henshall. 2003. *Gender and Development. Routledge Perspectives on Development*. London: Routledge.

Montu, Rafiqul Islam. 2020. "'It's over for Us': How Extreme Weather Is Emptying Bangladesh's Villages." *The Guardian*, 16 December 2020, sec. Global Development. https://www.theguardian.com/global-development/2020/dec/16/how-floods-are-emptying-bangladesh-villages.

Mosse, David. 2005. *Cultivating Development: An Ethnography of Aid Policy and Practice*. London: Pluto Press.

Mosse, David, Sanjeev Gupta, Mona Mehta, Vidya Shah, Julia Fnms Rees, and KRIBP Project Team. 2002. "Brokered Livelihoods: Debt, Labour Migration and Development in Tribal Western India." *Journal of Development Studies* 38 (5): 59–88. doi:10.1080/00220380412331322511

Mosse, David, and David Lewis. 2006. "Theoretical Approaches to Brokerage and Translation in Development." In *Development Brokers and Translators: The Ethnography of Aid and Agencies*, edited by David Lewis and David Mosse, 1–26. Bloomfield, CT: Kumarian Press.

Nikitas, Stefanos. 2016. "Haunting Photos Show Effects of Climate Change in Bangladesh." *The Huffington Post*, 28 January 2016. http://www.huffingtonpost.com/entry/bangladesh-climate-change_us_56aa5cd8e4b0d82286d53900.

Ojeda, Diana. 2021. "Social Reproduction, Dispossession, and the Gendered Workings of Agrarian Extractivism in Colombia." In *Agrarian Extractivism in Latin America. Routledge Critical Development Series*, edited by Ben M. McKay, Alberto Alonso Fradejas, and Arturo Ezquerro-Cañete, 85–98. New York, NY: Routledge.

O'Malley, L. S. S. 1908. "Bengal District Gazetteers: Khulna." Calcutta. V/27/62/111. India Office Records, British Library.

Paprocki, Kasia. 2016. "'Selling Our Own Skin': Social Dispossession Through Microcredit in Rural Bangladesh." *Geoforum; Journal of Physical, Human, and Regional Geosciences* 74 (August): 29–38. doi:10.1016/j.geoforum.2016.05.008.

Paprocki, Kasia. 2018. "Threatening Dystopias: Development and Adaptation Regimes in Bangladesh." *Annals of the American Association of Geographers* 108 (4): 955–973. doi:10.1080/24694452.2017.1406330.

Paprocki, Kasia. 2019a. "All That Is Solid Melts Into the Bay: Anticipatory Ruination and Climate Change Adaptation." *Antipode* 51 (1): 295–315. doi:10.1111/anti.12421

Paprocki, Kasia. 2019b. "The Climate Change of Your Desires: Climate Migration and Imaginaries of Urban and Rural Climate Futures." *Environment and Planning D: Society and Space*, December, 026377581989260.

Paprocki, Kasia, and Jason Cons. 2014. "'Life in a Shrimp Zone: Aqua- and Other Cultures of Bangladesh's Coastal Landscape'." *The Journal of Peasant Studies* 41 (6): 1109–1130. doi:10.1080/03066150.2014.937709.

Poncelet, A., F. Gemenne, M. Martiniello, and H. Bousetta. 2010. "A Country Made for Disasters: Environmental Vulnerability and Forced Migration in Bangladesh." In *Environment, Forced Migration and Social Vulnerability*, edited by Tamer Afifi and Jill Jèager, 211–222. Heidelberg: Springer Berlin.

Pouliotte, Jennifer, Barry Smit, and Lisa Westerhoff. 2009. "Adaptation and Development: Livelihoods and Climate Change in Subarnabad, Bangladesh." *Climate and Development* 1 (1): 31–46. doi:10.3763/cdev.2009.0001.

Powdermaker, Hortense. 1966. *Stranger and Friend: The Way of an Anthropologist. Nachdr.* New York: Norton & Company.

Qureshi, Ayaz. 2014. "'Up-Scaling Expectations among Pakistan's HIV Bureaucrats: Entrepreneurs of the Self and Job Precariousness Post-Scale-Up'." *Global Public Health* 9 (1–2): 73–84. doi:10.1080/17441692.2013.870590.

Rahman, Rushidan I., and Rizwanul Islam. 2013. *Female Labour Force Participation in Bangladesh: Trends, Drivers and Barriers*. ILO Asia-Pacific Working Paper Series. New Delhi: International Labour Organization. http://www.ilo.org/wcmsp5/groups/public/—asia/—ro-bangkok/—sro-new_delhi/documents/publication/wcms_250112.pdf.

Rahman, Mahbubar, and Willem Van Schendel. 2003. ""I Am Not a Refugee": Rethinking Partition Migration." *Modern Asian Studies* 37 (3): 551–584. doi:10.1017/S0026749X03003020.

Rights, Migrant. 2016. "Saudi Lifts Ban on Bangladeshi Workers, While Ignoring Plight of Thousands of Stranded Workers," 12 August 2016. https://www.migrant-rights.org/2016/08/saudi-lifts-ban-on-bangladeshi-workers-while-ignoring-plight-of-thousands-of-stranded-workers/.

Rogaly, Ben, Daniel Coppard, Abdur Safique, Kumar Rana, Amrita Sengupta, and Jhuma Biswas. 2002. "Seasonal Migration and Welfare/Illfare in Eastern India: A Social Analysis." *Journal of Development Studies* 38 (5): 89–114. doi:10.1080/00220380412331322521

Saha, Sebak Kumar. 2017. "Cyclone Aila, Livelihood Stress, and Migration: Empirical Evidence from Coastal Bangladesh." *Disasters* 41 (3): 505–526. doi:10.1111/disa.12214

Samaddar, Ranabir. 1999. *The Marginal Nation: Transborder Migration from Bangladesh to West Bengal*. New Delhi: Sage Publications.

Sangari, Kumkum, and Sudesh Vaid. 1990. "Introduction: Recasting Women." In *Recasting Women: Essays in Indian Colonial History*, Edited by Kumkum Sangari and Sudesh Vaid. New Jersey: Rutgers University Press.

Seeley, J., S. Ryan, M. I. Hossain, and I. A. Khan. 2006. "Just Surviving or Finding Space to Thrive? The Complexity of the Internal Migration of Women in Bangladesh." In *Poverty, Gender and Migration*, edited by S. Arya and A. Roy, 171–191. London: SAGE Publications.

Sen, Samita. 1999. *Women and Labour in Late Colonial India: The Bengal Jute Industry*. Cambridge: Cambridge University Press.

Shah, Alpa. 2006. "The Labour of Love: Seasonal Migration from Jharkhand to the Brick Kilns of Other States in India." *Contributions to Indian Sociology* 40 (1): 91–118. doi:10.1177/006996670504000104.

Siddiqui, Tasneem. 2001. *Transcending Boundaries: Labour Migration of Women from Bangladesh*. University Press: Dhaka.

Silva, Sanjiv de. 2012. "Situation Analysis for Polder 3. [Project Report Prepared by IWMI for the CGIAR Challenge Program on Water and Food (CPWF) under the Project "Increasing the Resilience of Agricultural and Aquacultural Systems in the Coastal Areas of the Ganges Delta: Project G3 - Water Governance and Community Based Management"]." Report. International Water Management Institute (IWMI). https://cgspace.cgiar.org/handle/10568/41802.

Staveren, Martijn F. van, Jeroen F. Warner, and M. Shah Alam Khan. 2017. "Bringing in the Tides. From Closing Down to Opening up Delta Polders via Tidal River Management in the Southwest Delta of Bangladesh." *Water Policy* 19 (1): 147–164. doi:10.2166/wp.2016.029.

Swain, Ashok. 1996. "The Environmental Trap: The Ganges River Diversion, Bangladeshi Migration and Conflicts in India." Report, Uppsala University, Department of Peace and Conflict Research, No. 41. Uppsala, Sweden: Dept. of Peace and Conflict Research, Uppsala University.

Swapan, Mohammad Shahidul Hasan, and Michael Gavin. 2011. "A Desert in the Delta: Participatory Assessment of Changing Livelihoods Induced by Commercial Shrimp Farming in Southwest Bangladesh." *Ocean & Coastal Management* 54 (1): 45–54. doi:10.1016/j.ocecoaman.2010.10.011

Toufique, Kazi Ali. 2002. "'Agricultural and Non-Agricultural Livelihoods in Rural Bangladesh: A Relationship in Flux'." In *Hands Not Land: How Livelihoods Are Changing in Rural Bangladesh*, edited by Kazi Ali Toufique and Cathryn Turton, 57–64. Dhaka: Bangladesh Institute of Development Studies.

UNDP. 2019. "CBA Bangladesh: Coping with Climate Risks by Empowering Women in Coastal Areas (GBSS) | UNDP Climate Change Adaptation." https://projects/spa-cba-bangladesh-coping-climate-risks-empowering-women-coastal-areas-gbss.

Van Schendel, Willem. 2001. "Working Through Partition: Making a Living in the Bengal Borderlands." *International Review of Social History* 46 (03): 393–421. doi:10.1017/S0020859001000256.

Werner, Marion. 2016. *Global Displacements: The Making of Uneven Development in the Caribbean. Antipode Book Series*. West Sussex: Wiley-Blackwell.

World Bank. 2009. "Poverty and Climate Change: Reducing the Vulnerability of the Poor through Adaptation." Working Paper 52176. Washington, DC: World Bank. http://documents.worldbank.org/curated/en/534871468155709473/Poverty-and-climate-change-reducing-the-vulnerability-of-the-poor-through-adaptation.

Zaman, M. Q. 1993. "Rivers of Life: Living with Floods in Bangladesh." *Asian Survey* 33 (10): 985–996. doi:10.2307/2645097.

Certificated exclusion: forest carbon sequestration project in Southwest China

Jun He and Jiping Wang

ABSTRACT

This research presents a case study of a high-profile carbon forest sequestration project in Southwest China, which claims to achieve 'triple-win' outcomes for livelihoods, biodiversity and climate change. However, over the last 15 years, each household that participated in the project has only received 10–20 USD (about 0.67–1.33 USD annually), despite the requirement for farmers to convert their agricultural land into tree plantations. We argue that contemporary capitalism engaged in climate change mitigation efforts has effectively recast the rural governance of land from a territorial arrangement to multiple actors exerting control over carbon credit certification, which creates a new form of climate injustice and social exclusion in agrarian change.

1. Introduction

It is widely acknowledged that natural ecosystems can play an important role in reducing the concentration of carbon dioxide (CO_2) in the atmosphere to mitigate the effects of climate change (IPCC 2021). Current research estimates that from 2010 to 2019, 54 percent of global CO_2 emissions were absorbed by terrestrial and ocean ecosystems, accounting for 31 percent and 23 percent, respectively (Friedlingstein et al. 2020). Given current limitations in increasing the carbon storage capacity of the ocean, a great deal of attention has been given to increasing the carbon storage capacity of terrestrial vegetation via improved land use. As such, climate change mitigation efforts are now largely interlinked with land-use strategies (Franco and Borras 2019; IPCC 2019; Borras, Franco, and Nam 2020). Contemporary capitalism has increasingly turned to investment in land-related activities for mitigating climate change impacts,[1] including programs like Reducing Emission from Deforestation and Forest Degradation plus (REDD +) (e.g. Larson et al. 2013; Fletcher et al. 2016), Afforestation and Reforestation as Clean Development Mechanisms (A/R CDM) (e.g. Boyd, Gutierrez, and Chang 2007; Ma et al. 2014) and

[1]This phenomenon has been also broadly categorized under the umbrella of neoliberal conservation, involving the use of market instruments to reduce greenhouse gas emissions (see Igoe and Brockington 2016; Fletcher 2020).

climate-smart agriculture (e.g. Newell and Taylor 2018; Cavanagh et al. 2021). At the same time, climate change and land management intersect with global capitalism in ways inextricably bound to rural society and agrarian studies (Franco and Borras 2021; Borras et al. 2022). While free-market capitalism has been promoted as an alternative to command-and-control approaches to conservation, the ultimate effects of capitalistic engagement in climate change mitigation efforts on rural societies remain a subject of ongoing debate.

There is a growing body of literature documenting the unexpected environmental and socioeconomic outcomes resulting from capitalism engaged in land-use change for climate change mitigation. Notably, Fairhead, Leach, and Scoones (2012) have observed the global land rush phenomenon as interconnected with climate change politics, which they termed 'green grabbing', that indicates the alienation of land for conservation (bio-carbon sequestration in this case).[2] This surge in worldwide 'green grabbing' has implications for social inequality as well as ecological consequences (Fairhead, Leach, and Scoones 2012; Yang and He 2021). Using a case study from Myanmar, Borras, Franco, and Nam (2020) provide a deeper understanding of 'green grabbing' in climate change politics by comparing a legitimating process for land grabbing with a de-legitimating process for local peoples' mitigation and adaptation practices. Apart from focusing on the elimination of local participation in land decisions, other researchers have also observed limited financial return to local people involved with carbon offset projects (e.g. Corbera and Brown 2010; Karlsson et al. 2018; Cavanagh et al. 2021) and social inequality on gender issues (Collins 2018). Ecological outcomes have also varied in effectiveness, such as expanding monoculture planting alongside increasing carbon prices (Johnston and Withey 2017), and the overestimation of carbon sequestration capacities from afforestation projects or the REDD+ program (van Minnen et al. 2008; Alexander et al. 2011; Lederer 2011). In response to those criticisms, scholars have attempted to explore the socio-political structures that form the underlying mechanisms enabling 'green grabbing'. Borras and Franco (2018) were among those who proposed the notion of 'agrarian climate justice' to understand the challenges of land-based climate change, as they are embedded in the fundamental issues of agrarian struggle and anti-capitalistic thought and approaches (Hunsberger et al. 2017; Sekine 2021; Borras et al. 2022).

To deepen understanding of climate change and agrarian struggle, Borras et al. (2022) suggested four contrasting narratives in analysis: (1) a corporate-driven technology narrative; (2) a climate emergency narrative; (3) a climate justice narrative; and (4) a structural transformation narrative. Borras et al. (2022) argue that these narratives overlap, containing multiple strands within each narrative, and are often combined in response to agrarian settings, exerting different impacts on localities. While this framework is useful in identifying interconnection between climate change and agrarian struggle, empirical analysis to enrich understandings of interconnection and develop existing theoretical discussions is urgently needed. Importantly, Borras et al. (2022, 17) highlighted the need to examine the specific local and global impacts resulting from environmental exclusion and inclusion in agrarian settings, particularly in the context of overlapping narratives and

[2]The existing literature documents green grabbing for conservation as occurring in the forms of national park establishment, ecotourism project, afforestation programs for climate change and biodiversity conservation (Fairhead, Leach, and Scoones 2012; Yang and He 2021).

strategies. Thus, there is a call for robust empirical evidence to advance understandings of local dynamics, which would contribute significantly to both theoretical and empirical discussions.

In response to this call, we examined the empirical case of a forest carbon sequestration project in China. Differing from existing studies that have tended to probe the questionable effectiveness and equity of capitalism in mitigating the effects of climate change, our approach aimed at deepening understanding of the ways in which capitalism for mitigating the effects of climate change drives on-site agrarian changes, through examining the emerging mechanisms that enable green grabbing in agrarian climate struggles. Using an in-depth case-study approach, the research presents a high-profile carbon-financing forest initiative in Southwest China, an economically underdeveloped region with rich biodiversity. As such, this project is widely advertised as a 'triple-win' carbon forest project for achieving biodiversity conservation, climate mitigation and community development. In light of the four overlapping narratives outlined by Borras et al. (2022), we examine the exclusionary processes and outcomes using an analytical framework that combines Hall, Hirsch, and Li (2011)'s 'powers of exclusion' and Sikor, He, and Lestrelin (2017)'s 'property rights regime'. This analytical combination allows for a broad discussion regarding the possession of property rights across multi-level actors for a particular resource to help clarify the mechanisms behind environmental exclusion under contrasting framings of the climate narratives.

The new national policy for 'net-zero emission' alongside an increase in the number of forest carbon sequestration projects in China forms the backdrop of our study. In 2020, the Government of China made an ambitious commitment to achieve 'net-zero emissions' by 2060, widely known as carbon neutrality (Liu et al. 2022). China's prominent role in the Global South in utilizing technocentric and large-scale afforestation to cope with climate change has been widely acknowledged and promoted (Qi and Dauvergne 2022). However, the effects of forest carbon sequestration projects conducted in China are poorly understood from an empirically grounded approach. The existing body of research has been largely dominated by natural scientists, who have focused on recording carbon storage in forest ecosystems and for various forest programs (e.g. Liu et al. 2012; Sun and Liu 2020). A few studies have assessed the social aspects of forest carbon sequestration projects, and have observed negative outcomes in the financial and social capital of local people who participated in these projects (Qiu et al. 2018), effects on local traditional culture (Yang et al. 2018) and the lack of villagers' voices in project implementation (Ba et al. 2020). Despite these efforts from social scientists, analysis remains limited regarding the long-term results of negative social outcomes. A deeper understanding of the ways in which capitalism engages with climate change mitigation is therefore required. Robust empirical evidence can meaningfully contribute to current policy development in China and beyond.

The paper is structured into seven sections. After this introduction, we present the analytical framework derived from theories of property rights and agrarian climate change, followed by detailing the research methods and study site. The fourth section presents an overview of the carbon finance project, and the fifth section analyzes the possession of property rights among different stakeholders to paint a holistic picture of exclusion, after which a discussion of observed ecological and socioeconomic outcomes is presented. The paper concludes by discussing the theoretical and empirical implications of the research.

2. Exclusion, property rights and agrarian climate change

The concept of exclusion has been widely used in natural resource management around the world. In writings on political economy, much literature refers to exclusion as a process for restricting stakeholders' decision-making in the governance of a given resource (e.g. Ribot 1995; Adams and Hutton 2007; Lele et al. 2010). Accordingly, exclusion is the social process diametrically opposed to participation. Exclusion is an inherent characteristic of resource management that all too often results in inequality and dispossession (Peluso and Lund 2011). Using a theory of property rights to analyze natural resource management, Schlager and Ostrom (1992) developed an influential framework, defining exclusion as the right that 'determine[s] who will have an access right, and how that right may be transferred' (1992, 251). This definition explicitly draws together the relation between exclusion and access. Superseding the right to access, the right to exclusion can be granted and legitimated by a given authority that is associated with a given resource (Tubtim and Hirsch 2004; Sikor and Lund 2009).

The concept of access was further developed to be broadly understood as the 'ability to benefit from things' (Ribot and Peluso 2003). Hall and colleagues suggested the understanding of exclusion as opposite to access (Hall, Hirsch, and Li 2011), defining the concept of exclusion as 'the ways in which people are prevented from benefitting from things [resources]' (Hall, Hirsch, and Li 2011, 7). This broader definition, with its origins in access and property theory, provides an insightful analytical framework to examine the power relations that underline the social process of exclusion in the management of a given resource, offering possible answers regarding who benefits or loses and why. Emphasizing the importance of embedded power relations, Hall, Hirsch, and Li (2011) suggested four types of power that prevent people from receiving benefits from natural resources: (1) *regulation* as formal and informal rule; (2) *force* as explicit and implicit violence; (3) *market* that builds capitalistic relations; and (4) *legitimation* as justification in supporting exclusion. Those four powers act synergistically, but they are distinctively effective across scales. This framing of powers of exclusion has been widely used in agrarian studies, particularly in understanding changing property regimes along social relations in capitalism; at the same time, Hall, Hirsch, and Li (2011) have suggested that their framework is somewhat incomplete, as it relies upon concepts in property and access theory to generate a complete picture of exclusion (Hall, Hirsch, and Li 2011, 197).

To enrich the framework of Hall, Hirsch, and Li (2011), we propose examining exclusion from a broader perspective of 'bundles of rights'. Instead of focusing on Schlager and Ostrom's (1992) narrow definition of the five types of rights associated with natural resource management, we adopted the broader framework developed by Sikor, He, and Lestrelin (2017) to expand the 'bundle of rights' analysis in property regime. According to Sikor, He, and Lestrelin (2017), the property rights regime is organized into three orders of rights: (1) use rights; (2) control rights; and (3) authoritative rights. *Use rights* are the rights to enjoy a direct benefit derived from a resource or an indirect benefit associated with a resource. *Control rights* are second-order rights that determine the scope of use, which include the rights to management, exclusion, transaction and monitoring. The highest order of property rights is authoritative rights. *Authoritative rights* define control rights, including definition and allocation. The combination of Hall, Hirsch, and Li (2011) and Sikor, He, and Lestrelin (2017) offers an enriched framework of particular use to our project, as empirically there are many more actors involved in resource management than just local communities. Each of those actors exercises various forms of power (i.e. regulation, market, force and legitimacy) to obtain rights or limit others' rights (i.e. use right, control right, authoritative rights) to directly or indirectly benefit from a resource. Thus, understanding what rights are held by whom can help generate a holistic picture of 'exclusion' associated with the governance of a particular resource.

Integrating Sikor, He, and Lestrelin (2017) and Hall, Hirsch, and Li (2011) into a singular framework offers analytical and contextual advantages. First, the analytical advantage stems from resource governance increasing in complexity along with the global spread of neoliberal policies and values. Resource governance has shifted from conventional biophysical control of a resource to what Sikor et al. (2013) have termed a 'flow-centered' arrangement of governance, such as the certification of agriculture, wood products and ecosystem services[3] (also see Liu, Yang, and Li 2016; Hull and Liu 2018; Munroe et al. 2019). The access to and control of resources and the myriad ways in which we benefit from them have largely been determined by multiple actors at different levels. In this sense, the management of a resource increasingly provides access to various benefits outside the immediate context, including indirect benefits from, for example, carbon sequestration or watershed functions. In this case, the carbon trade has involved multiple stakeholders from international, national, provincial and local levels exercising their power to gain benefits or prevent others from benefitting from carbon projects. As such, the contemporary governance of resources requires a broader perspective to combine updated property rights theory and complex power relations, while Sikor, He, and Lestrelin (2017) provide a broader rights-based perspective to understand the powers of exclusion explicated by Hall, Hirsch, and Li (2011).

Second, the contextual advantage for discussing climate change in this study stems from modern sustainability discourses in agrarian studies, forming four contrasting narratives that intersect in rural society, as suggested by Borras et al. (2022). These narratives, i.e. the corporate-driven technology narrative, climate emergency narrative, climate justice narrative and structural transformation narrative, are embedded in land relations across fundamental dimensions of property, labor, income and consumption reproduction (Borras et al. 2022). Contextually, agrarian climate change narratives inter-penetrate climate *regulation*, *force*, *market* and *legitimation*, all of which enable the exclusionary process via the allocation of various property rights (i.e. use rights, control rights and authoritative rights). Thus, understanding the climate narratives proposed by Borras et al. (2022) provides an insight for contextual analysis of agrarian struggle and local dynamics in response to climate mitigation initiatives (a forest carbon sequestration project in this case).

Building upon this theoretical foundation, we suggest certification in climate change as an exclusionary process that prevents certain groups of people from benefitting from climate mitigation efforts. As such, it is what we called 'certificated exclusion'.[4]

---

[3]Sikor et al. (2013) described this shift as from a 'territorial' to a 'flow-centered' governance arrangement to indicate increasing numbers of outside actors who govern contemporary land-use systems. To explain this phenomenon, Liu, Yang, and Li (2016) and Hull and Liu (2018) used the term 'telecoupling', referring to the connections between distant land systems in ecosystem services.

[4]Unlike the typology of exclusion developed by Xu (2019) as an analytical framework, we suggest 'certificated exclusions' as an emerging phenomenon in natural resource governance that describes the exclusionary process through the certification of a given resource.

Differing from licensed exclusion, which refers to exclusionary processes in land titling (Hall, Hirsch, and Li 2011), what we call certificated exclusion includes a wide range of certification processes, such as those responsible for carbon credits, organic food, fair trade, the legality of timber trade, non-timber forest products and more, which provide the illusion and expectation of high economic return through the certification. However, reality is often at odds with expectations. Regarding the forest carbon sequestration project, one clear example of certificated exclusion is the imposition of a sophisticated technocentric procedure to certify carbon credits for trade. The powers of exclusion are exercised in this case via the combination of climate regulation, a state monopoly on force, carbon-financing markets and climate legitimation, as informed by Hall, Hirsch, and Li (2011). To expand the framework of power of exclusion, we examined certificated exclusion by adopting the property rights regime approach developed by Sikor, He, and Lestrelin (2017) in order to identify and understand different rights possessed by different actors across the forest carbon sequestration project. This conceptual combination helps unpack power relations at different levels of rights along with a given product or service – in this case, ecosystem services from climate mitigation. It also provides an insightful analysis to understand the underlying mechanisms of green grabbing. Equipped with the property rights approach, the research paints a convincing holistic picture of exclusion through examining different powers that determine the allocation of rights among local, national and international actors. Moreover, this combined framework is contextualized in current contrasting narratives of agrarian climate change, as informed by Borras et al. (2022), which comprise important contexts for determining which kinds of powers of exclusion are exercisable.

3. Study site and methods

This research was carried out in Tengchong County, Yunnan Province, Southwest China, an area situated along the border of China and Myanmar (Figure 1). Gaoligong Nature Reserve, located in Tengchong, contains the most diverse flora and fauna in China, and even in the world, and it also functions as an important watershed of the Salween and Irrawaddy rivers. The region lies at the intersection of the Eastern Himalayas and the Hengduan Mountains of China, which are recognized as global biodiversity hotspots (Myers et al. 2000). Diverse ethnic groups have inhabited the remote region of Tengchong for centuries, and traditional livelihood strategies have fostered a close reliance on nearby natural resources, leading to vulnerability and historical poverty (Zinda and He 2020). Recently, the region has undergone a massive transition from a subsistence-based economy to a market-based economy, facilitated by the shared cross-border region with Myanmar (Dong and He 2018; Hua, Kono, and Zhang 2021).[5] However, economic marketization has also led to an increase in the number and intensity of threats to local ecosystems and biodiversity and unexpected agrarian changes in the locality (He et al. 2021; Hua, Kono, and Zhang 2021).[6]

[5]Within this transition to modernization, land markets emerged as a result of market liberalizations and the reform to redistribute collective-owned land to individual households, which allowed the capitalistic engagement that changed the land relations (Ye 2015).

[6]Broadly speaking, agrarian change driven by booms in tree crop planting along the borderland of Southwest China has been also examined by Xu (2019) and Zinda and He (2020).

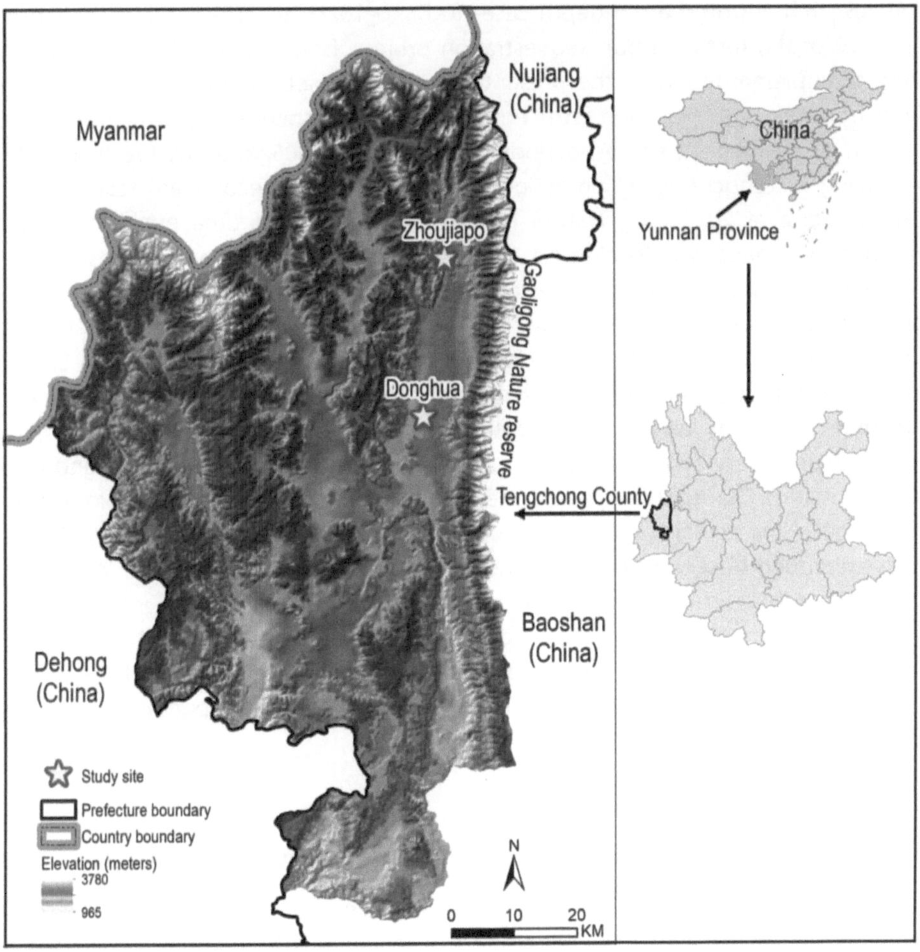

Figure 1. Location of study sites.

To address its degraded ecosystem, biodiversity threats and local poverty, Tengchong County was selected as a site for implementing the forest carbon sequestration project.

The empirical fieldwork focused on two villages, Zhoujiapo Village and Donghua Village, comprising two of the four sites selected in the forest carbon sequestration project with relatively large areas. These villages have traditionally cultivated rice paddies in the lowlands and corn and buckwheat in the highlands for subsistence, while rape seed and tobacco were introduced in the late 1980s as cash crops for generating cash income. As for forestry, both villages feature large forest areas dominated by fir and birch plantations for timber use, along with a substantial amount of walnut cultivation, which is promoted by the County Government as a poverty alleviation strategy. Local people gather a range of non-timber forest products (NTFPs), including diverse edible mushrooms, wild vegetables and firewood from the forest. In both villages, the carbon sequestration forest was planted on agricultural land previously used for growing corn, buckwheat and tobacco.

This research adopted an in-depth case-study approach to obtain a comprehensive perspective of the forest carbon sequestration project, from the initial planning to the process of implementation to the present day. The dataset used in this research was derived from the authors' long-term engagement in Tengchong County since 2005, which can be divided into three periods of time. From 2005 to 2009, the first author was working in Tengchong on an action-research project for ecosystem restoration in the buffer zone of Gaoligong Nature Reserve. This period of time enabled the first author to fully engage with local forest officials and international organizations involved with planning the forest carbon sequestration project in Tengchong. From 2010 to 2012, extensive fieldwork was carried out by the authors who were conducting research into payment for ecosystem services (PES), focusing on a comparative study of the Sloping Land Conversion Program (SLCP) and this current forest carbon sequestration project in Tengchong. This study used various research methods, including participatory observation, in-depth interviews, questionnaire surveys, semi-structured interviews and focus group discussions. During this research, the authors were able to examine the immediate socioeconomic and ecological outcomes of both PES projects. From 2018 to 2022, both authors continued observing the outcomes of this forest carbon sequestration project, focusing on its long-term post-implementation impacts. Semi-structured interviews and participatory observation were applied during this period to understand the project's socio-economic and ecological outcomes alongside peoples' perceptions of the project. These three phases of research have allowed us to assess and compare local dynamics across different time periods and stages of this project.

4. The 'triple-win' forest carbon sequestration project in Tengchong

Given global interest regarding carbon sequestration and the growing carbon market in China, The Nature Conservancy (TNC) and Conservation International (CI) signed an MoU (Memorandum of Understanding) with the China Forestry Administration[7] in 2003. The three parties agreed to jointly promote collaboration in biodiversity conservation, sustainable forest management and carbon sequestration, particularly in the field of climate change mitigation through forestry management (Ma, Tang, and Yang 2019). Under this agreement, the three parties jointly implemented a forest carbon sequestration project, officially called the Forest Restoration for Climate, Community and Biodiversity (FCCB) project, in order to cope with climate change, protect biodiversity and help local community development, comprising the 'triple-win' objectives of the project. The project was also widely known as the Multifunctional Forest Restoration project. The project was financed by 3M (Minnesota Mining and Manufacturing Corporation) with 3 million USD to cover two pilot sites, one in Yunnan Province and the other in Sichuan Province. The funds from 3M were intended to cover operational costs of TNC/CI as technical supporters as well as the certification of carbon emission reduction and transaction costs of carbon trade, while the provincial government provided matching funds to cover all on-site planting activities. For China, this project represented the first

[7]The State Forest Administration was renamed the State Forestry and Rangeland Bureau in 2018. In response, the provincial- and county-level forest bureaus were respectively renamed to the Provincial Forestry and Rangeland Bureau and the County Forestry Rangeland Bureau in 2018.

joint project that involved collaboration with the private sector, non-governmental organizations and government agencies for forest carbon sequestration, which has been regarded as a major structural transformation toward facilitating the entry of international corporations into the authoritarian state for coping with climate change. The project setting clearly links to climate justice and climate emergency narratives, as suggested by Borras et al. (2022), which is reflected in the Global North (3M Corporation) financing the Global South (poor communities in China) for an afforestation project with the goal to produce carbon offsets and facilitate poverty alleviation and biodiversity conservation.

As for project implementation in Yunnan Province, the Yunnan Forest Bureau was assigned by the State Forestry Administration to be the implementor in leading collaboration between TNC and CI for this FCCB project. To meet project objectives, the project team selected Tengchong County as the pilot site because it met all three objectives of the project as a global biodiversity hotspot, a relatively underdeveloped region and an area prime for afforestation (Ma, Tang, and Yang 2019). Since 2007, the project has established a total of 467.7 ha of mixed forestland, among which 37.6 ha of lands are linked directly to and 78.2 ha located on the outskirts of the Gaoligong Nature Reserve. Native tree species have been planted, such as fir (*Taiwania flousiana*), birch (*Betula luminifera*), pine (*Pinus yunnanensis*) and alder (*Alnus nepalensis*). Reforestation activities have involved 1367 villagers in four villages, one state forest farm and one private forest farm,[8] leading to the planting of 750,000 trees across 467.7 ha of land. The planting took place on agricultural or barren land that had not been forested in the last 20 years in order to meet carbon-financing criteria for the purpose of additionality. The project is expected to last from 2007 to 2036, leading to a 152,000-tonne carbon emissions reduction (CER), as documented in the project baseline report.

Operationally, the Tengchong Forest Department as the line agency of the Yunnan Forest Bureau worked together with TNC/CI to mobilize local communities for planting. Afforestation was carried out by villagers and both state and private forest farms, and a contract was also signed between the three parties including TNC/CI, the government and farmers or forest farms. The Tengchong Forest Department provided a subsidy of RMB 3000/ha (about 444 USD/ha) to farmers, which included tree seedling and laborers, while only free seedlings were provided to the state and private forest farms. Planted trees were arranged in a 30 cm × 30 cm matrix, and fertilizer was only added one time to ensure minimal carbon leakage resulting from afforestation. The plantation model was designed to be mixed forest for state forest farms, monoculture for for private forest farms and mixed (both monoculture alder and mixed forest were applied) for communities. The Forest Department transported plastic-potted tree seedlings to farmers, and farmers did the manual planting; afterward, the Forest Department evaluated survival rates at the end of the year. Once farmers met planting requirements and survival rates, the subsidy, calculated based on the area of planting, was provided as the labor fee after deducting seedling costs from the initial RMB 3000/ha subsidy.

As for carbon financing, this project was planned to be a small-scale afforestation and reforestation project and expected to generate income through the sale of Certified Emission Reductions on the global voluntary carbon market through the Clean Development

---

[8]A forest farm (lin chang 林场) is a type of forest enterprise engaged in afforestation and commercial logging for profit.

Mechanism (CDM) of the Kyoto Protocol. With its 'triple-win' objectives, a CCB standard (climate, community and biodiversity) was applied for the project[9] in line with small-scale A/R CDM. This complex methodology took a technocentric approach for climate mitigation, requiring a high degree of involvement from external actors.[10] In 2006, the project invited the third-party agency TUV/SUD[11] for project verification, which awarded the project a Gold Rating under the CCB Alliance, recognizing the superior project design and objectives. The project became the first CCB Gold-Rated project in the world. As the project report stated, the project achieved the following to qualify for CCB Gold certification:

(1) Afforestation/reforestation acting as a buffer zone for the nature reserve to facilitate biodiversity conservation efforts;
(2) Afforestation/reforestation helping to prevent widespread invasive species (*Ageratina adenophora*);
(3) High-quality afforestation and reforestation contributing to climate mitigation;
(4) Carbon financing and involvement in afforestation generating income for the community and alleviating poverty.

The visible profile of the project raised expectations quite high. The project team estimated that total financial returns from the project would reach about 20 million USD, including carbon finance and timber revenues, after 30 years. Accordingly, the Yunnan Foundation of Green and Environment Development was established with seed funding from CI and the Yunnan Provincial Forest Bureau. The foundation was assigned to operate the Tengchong Forest Carbon project and manage the funds generated from the project, which also has the long-term goal to develop more carbon projects in the future. In 2008, the first sale of certified CER occurred, when TNC/CI sold verified emissions reductions of approximately 21,000 tons on the voluntary carbon market at about 10 USD per ton. Around 1,046,600 RMB,[12] excluding the transaction fee, was transferred to the Tengchong Forestry Bureau. Farmers only received the first payment in 2012 after evaluation was carried out by Yunnan Provincial Forest Bureau, amounting to about 1800 RMB per ha for the first five years. However, according to the project design, around 25 percent of income from total carbon sales has gone to transaction fees for TNC/CI, who sell the carbon. Around 65 percent of the carbon sales income, excluding the transaction fee, has been distributed to participating farms, households and companies. The remaining 15 percent income has been used for administrative fees for the County Forest Department. At the time of writing, the first farmer payment is also the last farmer payment; no further transactions have taken place due to the decreasing price of carbon sequestration related to falling demand. The carbon transaction process is largely opaque; local forest officials and farmers do not have access to information and are simply notified when events happen. The project was closed after the first payment was made in 2012, after

[9]CCB Standards can be applied throughout the lifetime of the project to evaluate the social and environmental impacts of land-based carbon projects. The standards are effectively combined with a carbon-accounting standard, including the Carbon Fix Standard (CFS), the CDM and the Voluntary Carbon Standard (VCS).
[10]Borras et al. (2022) also suggested the technocentric approach for climate mitigation as a narratives in agrarian climate struggle.
[11]A German-based company for certification. See https://www.tuvsud.com.
[12]1 USD = 6.77 RMB in 2008.

which the Forest Department took over in hiring forest rangers to oversee forest management.

In sum, this forest carbon sequestration project in Tengchong featured four narratives of climate change as suggested by Borras et al. (2022), forming an agrarian setting at the locality. The global climate emergency calls for action, so Tengchong was selected to meet climate justice for afforestation to sell carbon credits from the poor in the Global South to a corporation in the Global North. Developing the CDM methodology for the carbon trade required not only a technocentric approach but also a structural change to allow multi-stakeholder involvement in resource governance in the context of an authoritarian state. However, these narratives practiced at the locality led to a change in the local governance of the property rights regime and unexpected environmental and socioeconomic outcomes.

5. Analyzing exclusion through the property rights regime lens in the forest carbon sequestration project

5.1 Exclusion in use rights

In this project, farmers hold use rights to all direct benefits derived from the involved land as they possess the land titles. Within the 30-year project duration, timber is prohibited from being harvested. After the project concludes in 2035, the timber will belong to those who planted the trees. The farmers, state forest farm and private forest farm will have sole access to the timber resources. However, the timber harvest will be subject to a restrictive harvest quota,[13] which will be issued by the Tengchong County Forest Department. The harvest quota is thus an area of future uncertainty. Many farmers we interviewed were concerned about the possibility of their carbon forests being classified as ecological forests,[14] as ecological forests are protected from harvesting. Apart from that, harvest activities would also need to follow CDM principles to ensure carbon capture after the project ends. As for the use rights to indirect benefits, farmers shared use rights with TNC/CI and the County Forest Department. As mentioned above, farmers were eligible to receive 60 percent of the carbon revenues after deducting 25 percent as transaction costs to TNC/CI, and 15 percent to fund administrative salaries in the Forest Department. With regards to this arrangement, people raised several concerns. 'If that [classification of carbon forests as ecological forests] happens, no harvest activities can be carried out according to forest law', remarked one local forest official.[15] Another farmer noted that 'harvesting timber requires us to have a quota and permit, which is hardly to apply, as we only have small area and limited timber volume'.[16]

In sum, as shown from an analysis of use rights, the harvest quota system as the regulatory power of exclusion is most likely to prevent farmers' direct benefit from timber. The carbon trade has also limited farmers' use rights to harvest timber before the project ends.

[13]For more about the harvest quota system see He (2016).

[14]Forest is classified in two categories in China: (1) ecological forests, which are used only for protecting ecological functions and where no harvest activities can be carried out; and (2) commercial forests, which are used for economic purposes that allow for harvest to profit.

[15]Interviewed at Tengchong in 10 July 2014.

[16]Interviewed at Tengchong in 15 July 2014.

Moreover, farmers cannot fully enjoy the indirect benefits from carbon revenues because of the exclusionary process of carbon trade (see details below).

5.2 Exclusion in control rights

The County Forest Department and TNC/CI shared most control rights. The county Forest Department and TNC/CI together possessed the *management* and *exclusion rights*. TNC/CI define the management activities to be carried out by local villagers who are hired by the Forest Department and TNC/CI. Management activities were defined to include pest control, fire control, daily inspection activities, etc. Farmers were also required to maintain record-keeping for their daily inspections. In accordance with the FCCB project contract, every five years, thinning for silviculture can be carried out based on approval from the Forest Department and TNC/CI, who will design and manage silvicultural activities to ensure avoiding carbon leakage and the additionality principle. As for the exclusion rights in the project, TNC/CI and the Forest Department could decide where the project should be implemented and who could be involved in accordance with the small-scale A/R CDM. In this sense, they exercise their rights for land selection and zoning to meet the CDM requirements in which land has experienced deforestation and not supported forest growth in the last 20 years. Next, the Forest Department consulted with villagers to gauge willingness to participate in the project. Tree species selection was also done by TNC/CI in collaboration with the Forest Department, who consulted with local villagers. The selection of tree species and mode of planting aimed to prioritize CCB criteria over local preferences, as stated by one TNC staff. As a result, some multi-stakeholder consultation occurred at this stage, but TNC/CI and the Forest Department largely dominated the process. People felt generally unhappy about the planning and implementation, as the following quotations show. 'We would like to have more fir plantation instead of other species. However, to meet the project goal, we have to follow their [TNC/CI and government] command', said one villager.[17] One local forest official said: 'The goal of poverty reduction has rarely [been] considered in species selection, and as TNC/CI want to prioritize … biodiversity conservation, we were expecting good return from carbon market'.[18]*Transaction rights* are held by TNC/CI, who undertake responsibility for the carbon trade in the international market, while the Tengchong Forest Department oversees distributing payment to local project participants. The process is generally murky to local forest officials, who stated, 'We never know about the market and transactions, given lack of knowledge and language barriers'.[19] The County Forest Department, Provincial Forest Department and TNC/CI share *monitoring rights* over tree plantations in the FCCB project. The County Forest Department assesses tree survival and growth, checks compliance with the required silvicultural practices and hires farmers as guards to protect the plantations against theft, fire and other forms of destruction. The Provincial Forest Department audits the disbursement of carbon payments to farmers by the County Forest Department. TNC/CI sends in designated consultants every few years to verify carbon stocks.

[17]Interviewed at Tengchong in 17 February 2021.
[18]Interviewed at Tengchong in 15 July 2014.
[19]Interviewed at Tengchong in 17 February 2020.

In short, as shown from an analysis of control rights, farmers were excluded from decision-making in the planning, management and implementation process, as technological requirements were defined by international organizations in alignment with CCB standards, which reflects the power of regulation and legitimacy in the exclusionary process. Farmers were also unable to participate in the transaction and monitoring process, as international markets feature barriers that require sufficient knowledge and language skills – which also limited the involvement of the local government. The international organizations exercised power regarding the market, carbon trade standards and regulations, which enabled them to obtain the highest degree of control rights and excluded other benefits.

5.3 Exclusion in authoritative rights

Authoritative rights are shared by the Forestry Department and international organizations. *Definition rights* are exercised by the State Forestry Administration by way of restrictions imposed on the management of land designated for forestry and the regulations on the allocation of timber harvest quotas. Additionally, definition rights are also held by the CDM and the CCB Alliance. The CDM delineates the discretionary space available for the exercise of management, transaction and monitoring rights by issuing methodologies for the calculation of carbon stocks, verified emissions reduction and the verification of carbon stocks.[20] The CCB Alliance influences the exercise of control rights through formulating and issuing standards for the protection of local communities and biodiversity, such as the requirement to inform and consult with local people. *Allocation rights* are also shared by the Forest Department and international organizations. The project implementation area is determined by the involved Forestry Departments and TNC/CI. Tengchong Forestry Bureau determines the specific individuals, groups and organizations in charge of forest management planning, financial transactions, exact quotas of timber for logging, harvest permits and forest management monitoring. The Provincial Forest Department has the right, along with TNC/CI, to assign the handling of financial transactions to the Tengchong Forestry Bureau. They determine the budget allocation from carbon credits in a non-transparent operation. Any use of the carbon fund, including disbursement to participants as well as budgeting for daily management and for thinning and silvicultural practices, also requires their approval. Statements from local officials and farmers express their feeling of being excluded from the carbon trade. 'It was said that we sold all the carbon credits, but they [TNC/CI] only provided a one-time payment. We have no idea how the rest of the funds were spent', stated one local forest officials in Tengchong.[21] One villager stated: 'We do not know how much money there is, where is the market, how come the CO_2 is sellable, we can just follow them [TNC/CI]'.[22]

Exclusion in authoritative rights is explicitly reflected in the technology and standards controlled by international organizations, which enable them to limit the involvement of farmers and governments in the process of selling carbon. TNC/CI play a critical role by

---

[20]For example, the Country Forest Department and TNC/CI knew that they had to select for reforestation deforested land that had been barren for at least 20 years.
[21]Interviewed at Tengchong 12 July 2014.
[22]Interviewed at Tengchong 15 July 2014.

bridging the certification body, local government and farmers, so TNC/CI can benefit from the project via their access to knowledge and standards.

To summarize this section, the exclusionary process occurred when stakeholders aimed to ensure the project would be certified as CCB in the carbon market, which was unpacked via governance of the property rights regime in the project. As for local farmers, they are largely excluded from the processes of project planning and implementation. The state exercises control and authoritative rights without participation by local farmers: central state agencies hold authoritative rights and define control rights in a restrictive manner, whereas lower-level state units are confined to control rights. On the other hand, there is involvement of international actors in the exercise of control and authoritative rights. TNC/CI hold control rights together with local state units, and the CDM and CCBA possess authoritative rights alongside China's State Forestry Administration. As the carbon market involves a transaction at the international level, the international organizations play a crucial role as bridges to facilitate the project and transaction. Their role in the process is irreplaceable, as they have a monopoly on the technology and standards, ultimately controlling the management and allocation of carbon funds. Thus, at higher levels, controlling the technology and standards of carbon certification is the mechanism that enables the international organizations to limit the benefit of others in the carbon sequestration project. It is also the factor that allows these international organizations to share control rights and authoritative rights with the government and exercise power in defining use rights at a lower level. However, this process of exclusion through carbon certification has also led to unexpected outcomes, as discussed below.

6. 'Triple-win' outcomes in the forest carbon sequestration project?

6.1 Ecological outcomes

At this point in time, ecological outcomes derived from the project remain limited. While there is evidence of some tree growth, and the survival rate and plantation area both meet baseline requirements, carbon sequestration amounts and the multifunctionality of ecosystem restoration have both been exaggerated in documentation. Established tree plantations are insufficient to restore the ecosystem at the buffer zone of the nature reserve. The mixed-forest mode of planting has involved the selection of only two or three species that comprise only broadleaved (e.g. alder and birch) and conifer (e.g. fir and pine) trees, which are local fast-growing species. Although these trees are all native species, lack of species diversity is crippling ecosystem restoration efforts compared to other forest restoration practices in Gaoligong Nature Reserve (e.g. Lu et al. 2017). Given the lack of diversity in tree species selection, one village-based forest ranger reported the scarcity of small mammals and birds within established plantation lands. Moreover, it has been observed that an invasive species (*Ageratina adenophora*) remains distributed throughout the established forests owing to lack of diverse tree species, while project documents claim that restoration efforts are preventing the spread of invasive species (Figure 2).

Forest structure is also not well established in plantation land. Silviculture activities have not been carried out in five-year increments as planned due to a lack of funding

Figure 2. (A) Fir planted by farmers under the established alder forests for high economic return; (B) invasive species that have spread widely under the plantation; (C) established plantations lack silviculture.

from carbon financing after the first sale in 2008. In Zhoujiapo, the monoculture of broad-leaved alder has been affected by disease and pests. Villagers had to cut down the alders affected by pests for use as firewood. This problem has been raised by both villagers and local forest officials. As one villager stated: 'Alders are more susceptible to pest and disease; 10 years from now, only 60 percent of the trees will remain, and the market for alder is not that good'.[23] Local forest officials have also raised concerns about forest structure, stating that 'Silviculture should be carried out earlier to improve the forest structure. However, we still need funding from the government, as we have not received money from carbon sales'.[24] The remaining alder trees are unfavorably scattered. In Donghua Village, farmers have begun interplanting fir in the alder forest after the alder died of natural causes. Fir has been noted to yield higher economic returns and require less management. According to farmers, fir is performing well, with over 80 percent of the trees remaining after the initial planting.

6.2 Socioeconomic outcomes

The forest carbon sequestration project in Tengchong has generated very limited financial benefits. While income from the sales of carbon should be distributed every five years per contractual requirements, farmers have only received one payment, in 2012. Furthermore, the payment was a small figure that covered neither opportunity costs nor forest management costs. On average, farmers received only about 1800 RMB per ha in 2012. In many cases, farmers have about 0.06–0.13 ha of land involved in the project. The participating

[23]Interviewed at Zhoujiapo in 17 February 2022.
[24]Interviewed at Tengchong in 18 February 2022.

farmer with the most involved land has about 1.3 ha, and received up to 2340 RMB (about 350 USD). This farmer's individual case was prominently advertised in the promotional newsletter of the project. Many farmers we interviewed mentioned that they have only received about 100–200 RMB (about 15–30 USD) since 2007, a figure significantly lower than revenues historically generated from growing tobacco or even buckwheat. At the same time, these farmers have lost the opportunity to participate in a national SLCP program for restoring forests on agriculture land, which provides a far more substantial payment of approximately 3000 RMB per ha per year (about 440 USD/ha/year).[25] In both Zhoujiapo and Donghua, some farmers have begun planting camellia under the remaining alder trees to obtain oil for cash. In addition, fir has been interplanted sparsely in the alder forest to produce timber. Farmers and local forest officials have expressed their views on this project, with some advocating for more fir planting:

> We have been informed that nobody wants to buy the gas [carbon] now, so we cannot receive any payment from the sale of gas [carbon]. However, a carbon forest is just a plantation activity – I'd rather plant fir for good environmental and economic return.[26]

Others were skeptical about the financial operations of the project, as the following quotation demostrates:

> Planting a carbon forest is in fact easy in practice – it doesn't take advanced technology, and it's just like our traditional afforestation practices. We simply do not understand why the transactional cost for international organizations is so high, while it brings such little benefit to our farmers.[27]

In addition, benefits delivered to farmers were also exaggerated. As the project document stated for CCB, farmers can involve themselves by providing manual labor for planting, thus receiving project benefits. However, this involvement has also generated little income for farmers. Farmers estimated that their labor cost works out to roughly 1350 RMB per ha (about 200 USD),[28] and on average farmers received 90–180 RMB (13–26 USD) as a one-time payment. In Donghua, one forest ranger was also involved in forest management for the project, retaining employment from 2007 to 2012. Therefore, the project has delivered underwhelming employment and income-enriching opportunities to farmers.

In sum, outcomes from the carbon sequestration project remain severely limited in both ecological and socioeconomic terms. Poorly optimized tree combinations and weak forest structures have led to scarce ecological benefits via carbon sequestration and biodiversity conservation. The unforeseen decline of the carbon market has led to the disappearance of promised income from carbon financing. Revenue from timber is also uncertain owing to restrictions in the timber quota system and alder pest control measures. The 'triple-win' objectives for climate, biodiversity and community development claimed by the project are seriously questionable in practice.

---

[25]The SLCP is a Central Government-financed project for forest restoration that provides payment for 16 years for farmers to establish forestland in marginal agricultural land (see more details in He 2020).
[26]Interviewed at Zhoujiapo in 21 February 2022.
[27]Interviewed at Tengchong in 14 July 2014.
[28]The labor cost is equivalent to 3000 RMB per ha subsidy minus the seedling cost (1650 RMB per ha).

7. Conclusion: certificated exclusion in climate change mitigation

This research examined a high-profile forest carbon sequestration project through an analytical framework that combines exclusion and property rights. Situating the discussion in the broader context of agrarian climate change and the contrasting narratives described by Borras et al. (2022), theoretical and empirical conclusions drawn from this research represent three aspects.

First, per Hall, Hirsch, and Li (2011), this research suggests 'certificated exclusion' as a phenomenon in carbon-financing forest projects and more broadly in the green grabbing of capitalistic land relations. Differing from common understandings in which certification comprises a type of entitlement, our research reveals certification in forest carbon-financing project as a process that includes certain groups but excludes others. As shown in this case, forest carbon has become a resource that can be marketed only when it is certificated. Carbon certification in forestry involves a sophisticated and technocentric procedure necessary to meet the standards of various certification agencies, such as the calculation of additionality, avoiding leakage, carbon capture, and more. As such, knowledge as a mechanism of exclusion precludes local involvement in this process. Following the contrasting narratives theory of Borras et al. (2022), this study presents a technological narrative combined with the climate emergency and justice narratives to create the legitimation of carbon certificates that allows for the credible trade of carbon in the global marketplace. Powers of exclusion in carbon certification were evident in the carbon market, as well as authoritative force from the state forest agency and regulation of carbon offsetting and legitimation of climate mitigation. Empirically, certificated exclusion manifested via property rights regimes regarding what actors possess which rights. As the case showed, local people were largely excluded from the project, as they held only lower-level property rights, such as use rights, while control and authoritative rights were possessed and shared across forestry departments and international organizations. Because authoritative rights determined control rights, and control rights defined use rights (Sikor, He, and Lestrelin 2017), the cascade effect stripped local people of most power in the carbon-financing project, and they ultimately did not receive many tangible benefits derived from possessing control and authoritative rights, despite their local biophysical possession of forestland. Thus, as also observed by Ribot and Peluso (2003), the biophysical possession of resources does not automatically generate and deliver benefits (e.g. Ribot and Peluso 2003; He 2016). Thus, as in this case, green grabbing is indeed possible without grabbing land or forest, but rather by grabbing carbon certification to ensure a certain group benefits and limits others' benefits derived from land or forest.

Second, this research suggests a new form of governance in land relations that emerges when capitalism engages in climate change mitigation efforts. Recently, in response to increasing attention to the intersection between climate change politics and land (e.g. Franco and Borras 2019), research indicates that new land relations should go beyond biophysical characteristics endemic to terrestrial-centered arrangements. Currently, access and control in land relations have been dismantled from terrestrial-centered arrangements toward legitimation narratives in the wake of technological advance, climate justice awareness and conservation effectiveness (Borras et al. 2022). As such, mechanisms in climate change, such as carbon certification, PES transaction,

REDD+ regulations and the CDM have been mapped onto the ground in practical terms to exclude certain social groups and institutionally hinder or even preclude their capacity to benefit from land use (see also Sikor et al. 2013, 2017). As the case here has shown, guidelines under these agreements or certifications issued by international organizations strongly influence the exercise of authoritative rights over natural resources. Conventional land governance practices have now shifted from access to and control over land to control of standard-setting, certification, technologies and discursive practices, forming a new core at the center of land governance. This new form of land governance makes use of climate-financing, carbon-offsetting and sequestration schemes that transform rural landscapes through multiple forms of enclosure and exclusion (Borras et al. 2022). In these transformed rural landscapes, access to and use of resources are in flux, especially amid the backdrop of coercive climate-smart practices that disregard on-site agrarian dynamics. Agrarian land struggle in the context of climate change is multi-level and multi-scale, with local, regional and global implications strewn across any given rural locality, going beyond terrestrial-centered arrangements.

Third, this research suggests that broader property right analysis can help advance an understanding of agrarian study in the context of climate change. While a great deal of attention has been given to green grabbing (Fairhead, Leach, and Scoones 2012), diversification of property rights with regard to natural resources should be further developed within the theory of property rights (e.g. Sikor, He, and Lestrelin 2017; He et al. 2020). In particular, neoliberal conservation explicitly involves capitalism in the project of conservation, making possible the selling of environmental goods and services that are associated with the newly emerged property rights regime. In this case, diversification of property rights enabled various mechanisms, including legal land titles to the forest, knowledge of carbon calculation and market operation, access to certification agencies, state regulations and more. As such, analyzing the diversification of property rights with regard to a given resource and its associated services helps provide a holistic picture of the power relations between actors across different levels and scales, which has also been observed by Corbera and Brown (2010). As shown in this case, understanding which entities possess use rights, control rights and authoritative rights in this forest carbon sequestration project helped uncover the underlying mechanisms of exclusion and reveal the processes that prevent locals from enjoying benefits related to the forest and forest carbon. While property is widely acknowledged to include numerous kinds of social relations among actors with respect to things (Bromley 1992; Benda-Beckmann, Benda-Beckmann, and Wiber 2006), property rights analysis extended from Schlager and Ostrom (1992) enriches the theory of agrarian studies and struggle, and also provides a novel perspective to unpack broad mechanisms behind exclusion in green grabbing.

Acknowledgements

This research financially benefited from the National Natural Science Foundation of China (Project No. 72063037) and Yunnan Provincial Senior Talent Support Program. The authors thank the editors and Prof. Jun Borras for the invitation to contribute to the *JPS* Forum on Climate Change and Critical Agrarian Studies. The early version of this manuscript benefited from four anonymous reviewers' constructive comments. English editing by Austin G. Smith and mapping by Huafang Chen are also acknowledged.

Disclosure statement

No potential conflict of interest was reported by the authors.

Funding

This research financially benefited from the National Natural Science Foundation of China (Project No. 72063037) and the Yunnan Provincial Senior Talent Support Program.

References

Adams, W. N., and J. Hutton. 2007. "People, Parks and Poverty: Political Ecology and Biodiversity Conservation." *Conservation and Society* 5 (2): 147–183.

Alexander, S., C. R. Nelson, J. Aronson, D. Lamb, A. Cliquet, K. L. Erwin, and C. Murcia. 2011. "Opportunities and Challenges for Ecological Restoration Within REDD+." *Restoration Ecology* 19 (6): 683–689. doi:10.1111/j.1526-100X.2011.00822.x.

Ba, F., J. Liu, T. Zhu, Y. Liu, and J. Zhao. 2020. "CDM Forest Carbon Sequestration Projects in Western China: An Analysis Using Actor-Centered Power Theory." *Forest Policy and Economics* 115: 102–139. doi:10.1016/j.forpol.2020.102139.

Benda-Beckmann, F. V., K. V. Benda-Beckmann, and M. G. Wiber. 2006. "The Properties of Property." In *Changing Properties of Property*, edited by F. V. Benda-Beckmann, K. V. Benda-Beckmann, and M. G. Wiber, 1–39. New York: Berghahn.

Borras, S. M. Jr, and J. C. Franco. 2018. "The Challenge of Locating Land-Based Climate Change Mitigation and Adaptation Politics Within a Social Justice Perspective: Towards an Idea of Agrarian Climate Justice." *Third World Quarterly* 39 (7): 1308–1325. doi:10.1080/01436597.2018.1460592.

Borras, S. M. Jr, J. C. Franco, and Z. Nam. 2020. "Climate Change and Land: Insights from Myanmar." *World Development* 129 (104864): 1–11. doi:10.1016/j.worlddev.2019.104864.

Borras, S. M. Jr, I. Scoones, A. Baviskar, M. Edelman, N. L. Peluso, and W. Wolford. 2022. "Climate Change and Agrarian Struggles: An Invitation to Contribute to a JPS Forum." *The Journal of Peasant Studies* 49 (1): 1–28. doi:10.1080/03066150.2021.1956473.

Boyd, E., M. Gutierrez, and M. Chang. 2007. "Small-scale Forest Carbon Projects: Adapting CDM to low-Income Communities." *Global Environmental Change* 17 (2): 250–259. doi:10.1016/j.gloenvcha.2006.10.001.

Bromley, D. 1992. *Making the Commons Work: Theory, Practice and Policy*. San Francisco: Institute for Contemporary Studies Press.

Cavanagh, C. J., P. O. Vedeld, J. G. Petursson, and A. K. Chemarum. 2021. "Agency, Inequality, and Additionality: Contested Assemblages of Agricultural Carbon Finance in Western Kenya." *The Journal of Peasant Studies* 48 (6): 1207–1227. doi:10.1080/03066150.2019.1707812.

Collins, A. 2018. "Saying all the Right Things? Gendered Discourse in Climate-Smart Agriculture." *The Journal of Peasant Studies* 45 (1): 175–191. doi:10.1080/03066150.2017.1377187.

Corbera, E., and K. Brown. 2010. "Offsetting Benefits? Analyzing Access to Forest Carbon." *Environment and Planning A: Economy and Space* 42 (7): 1739–1761. doi:10.1068/a42437.

Dong, M., and J. He. 2018. "Linking the Past to the Future: A Reality Check on Cross-Border Timber Trade from Myanmar (Burma) to China." *Forest Policy and Economics* 87: 11–19. doi:10.1016/j.forpol.2017.11.002.

Fairhead, J., M. Leach, and I. Scoones. 2012. "Green Grabbing: A new Appropriation of Nature?" *The Journal of Peasant Studies* 39 (2): 237–261. doi:10.1080/03066150.2012.671770.

Fletcher, R. 2020. "Neoliberal Conservation." In *Oxford Research Encyclopedia of Anthropology*. September 28. doi:10.1093/acrefore/9780190854584.013.300.

Fletcher, R., W. Dressler, B. Büscher, and Z. R. Anderson. 2016. "Questioning REDD+ and the Future of Market-Based Conservation." *Conservation Biology* 30 (3): 673–675. doi:10.1111/cobi.12680.

Franco, J. C., and S. M. Borras. 2019. "Grey Areas in Green Grabbing: Subtle and Indirect Interconnections Between Climate Change Politics and Land Grabs and Their Implications for Research." *Land Use Policy* 84: 192–199. doi:10.1016/j.landusepol.2019.03.013.

Franco, J. C., and S. M. Borras Jr. 2021. "The Global Climate of Land Politics." *Globalizations* 18 (7): 1277–1297. doi:10.1080/14747731.2021.1979717.

Friedlingstein, P., M. O'Sullivan, M. W. Jones, R. M. Andrew, J. Hauck, A. Olsen, G. P. Peters, et al. 2020. "Global Carbon Budget 2020." *Earth System Science Data* 12: 3269–3340. doi:10.5194/essd-12-3269-2020.

Hall, D., P. Hirsch, and T. M. Li. 2011. *Powers of Exclusion: Land Dilemmas in Southeast Asia*. Singapore: National University of Singapore.

He, J. 2016. "Rights to Benefit from Forest? A Case Study of the Timber Harvest Quota System in Southwest China." *Society & Natural Resources* 29 (4): 448–461. doi:10.1080/08941920.2015.1062949.

He, J. 2020. "Situated Payments for Ecosystem Services: Local Agencies in the Implementation of the Sloping Land Conversion Programme in Southwest China." *Development and Change* 51 (1): 73–93. doi:10.1111/dech.12539.

He, J., B. Kebede, A. Martin, and N. Gross-Camp. 2020. "Privatization or Communalization: A Multi-Level Analysis of Changes in Forest Property Regimes in China." *Ecological Economics* 174 (106629): 1–11. doi:10.1016/j.ecolecon.2020.106629.

He, J., A. Martin, R. Lang, and N. Gross-Camp. 2021. "Explaining Success on Community Forestry Through a Lens of Environmental Justice: Local Justice Norms and Practices in China." *World Development* 142 (105450): 1–13. doi:10.1016/j.worlddev.2021.105450.

Hua, X., Y. Kono, and L. Zhang. 2021. "Excavating Agrarian Transformation Under 'Secure' Crop Booms: Insights from the China-Myanmar Borderland." *The Journal of Peasant Studies* 1–30. doi:10.1080/03066150.2021.1926993

Hull, V., and J. Liu. 2018. "Telecoupling: A New Frontier for Global Sustainability." *Ecology and Society* 23 (4): 41. doi:10.5751/ES-10494-230441.

Hunsberger, C., E. Corbera, S. M. Borras Jr, J. C. Franco, K. Woods, C. Work, and C. Vaddhanaphuti. 2017. "Climate Change Mitigation, Land Grabbing and Conflict: Towards a Landscape-Based and Collaborative Action Research Agenda." *Canadian Journal of Development Studies / Revue canadienne d'études du développement* 38 (3): 305–324. doi:10.1080/02255189.2016.1250617.

Igoe, J., and D. Brockington. 2016. "Neoliberal Conservation: A Brief Introduction." In *The Environment in Anthropology: A Reader in Ecology, Culture, and Sustainable Living*, edited by N. Haenn, R. R. Wilk, and A. Harnish, 324–331. New York: NYU Press.

IPCC. 2019. *IPCC Special Report on Climate Change, Desertification, Land Degradation, Sustainable Land Management, Food Security, and Greenhouse gas fluxes in Terrestrial Ecosystems*. Approved Draft. 7 August 2019. https://www.ipcc.ch/srccl/.

IPCC. 2021. *Climate Change 2021: The Physical Science Ba- sis. Contribution of Working Group I to the Sixth Assessment Report of the Intergovernmental Panel on Climate Change*. Cambridge: Cambridge Press.

Johnston, C. M., and P. Withey. 2017. "Managing Forests for Carbon and Timber: A Markov Decision Model of Uneven-Aged Forest Management with Risk." *Ecological Economics* 138: 31–39. doi:10.1016/j.ecolecon.2017.03.023.

Karlsson, L., L. O. Naess, A. Nightingale, and J. Thompson. 2018. "'Triple Wins' or 'Triple Faults'? Analysing the Equity Implications of Policy Discourses on Climate-Smart Agriculture (CSA)." *The Journal of Peasant Studies* 45 (1): 150–174. doi:10.1080/03066150.2017.1351433.

Larson, A. M., M. Brockhaus, W. D. Sunderlin, A. Duchelle, A. Babon, T. Dokken, and T. B. Huynh. 2013. "Land Tenure and REDD+: The Good, the Bad and the Ugly." *Global Environmental Change* 23 (3): 678–689. doi:10.1016/j.gloenvcha.2013.02.014.

Lederer, M. 2011. "From CDM to REDD+—What do we Know for Setting up Effective and Legitimate Carbon Governance?" *Ecological Economics* 70 (11): 1900–1907. doi:10.1016/j.ecolecon.2011.02.003.

Lele, S., P. Wilshusen, D. Brockington, R. Seidler, and K. Bawa. 2010. "Beyond Exclusion: Alternative Approaches to Biodiversity Conservation in the Developing Tropics." *Current Opinion in Environmental Sustainability* 2 (1–2): 94–100. doi:10.1016/j.cosust.2010.03.006.

Liu, Z., Z. Deng, G. He, H. Wang, X. Zhang, J. Lin, and X. Liang. 2022. "Challenges and Opportunities for Carbon Neutrality in China." *Nature Reviews Earth & Environment* 3: 141–155. doi:10.1038/s43017-021-00244-x.

Liu, J., W. Yang, and S. Li. 2016. "Framing Ecosystem Services in the Telecoupled Anthropocene." *Frontiers in Ecology and the Environment* 14 (1): 27–36. doi:10.1002/16-0188.1.

Liu, S., T. Zhou, L. Wei, and Y. Shu. 2012. "The Spatial Distribution of Forest Carbon Sinks and Sources in China." *Chinese Science Bulletin* 57 (14): 1699–1707. doi:10.1007/s11434-012-4998-1.

Lu, Y., S. Ranjitkar, R. D. Harrison, J. Xu, X. Ou, X. Ma, and J. He. 2017. "Selection of Native Tree Species for Subtropical Forest Restoration in Southwest China." *PloS one* 12 (1): e0170418. doi:10.1371/journal.pone.0170418.

Ma, M., T. Haapanen, R. B. Singh, and R. Hietala. 2014. "Integrating Ecological Restoration Into CDM Forestry Projects." *Environmental Science & Policy* 38: 143–153. doi:10.1016/j.envsci.2013.11.008.

Ma, J., C. Tang, and B. Yang. 2019. "Restoration Project of Small-Scale Reforestation in Tengchong of Yunnan Province." In *Forest Carbon Practices and Low Carbon Development in China*, edited by Z. Lu, X. Zhang, J. Ma, and C. Tang, 175–208. Beijing: Springer and Peking University Press.

Munroe, D. K., M. Batistella, C. Friis, N. I. Gasparri, E. F. Lambin, J. Liu, and J. O. Nielsen. 2019. "Governing Flows in Telecoupled Land Systems." *Current Opinion in Environmental Sustainability* 38: 53–59. doi:10.1016/j.cosust.2019.05.004.

Myers, N., R. A. Mittermeier, C. G. Mittermeier, G. A. Da Fonseca, and J. Kent. 2000. "Biodiversity Hotspots for Conservation Priorities." *Nature* 403 (6772): 853–858. doi:10.1038/35002501.

Newell, P., and O. Taylor. 2018. "Contested Landscapes: The Global Political Economy of Climate-Smart Agriculture." *The Journal of Peasant Studies* 45 (1): 108–129. doi:10.1080/03066150.2017.1324426.

Peluso, N. L., and C. Lund. 2011. "New Frontiers of Land Control: Introduction." *Journal of Peasant Studies* 38 (4): 667–681. doi:10.1080/03066150.2011.607692.

Qi, J. J., and P. Dauvergne. 2022. "China's Rising Influence on Climate Governance: Forging a Path for the Global South." *Global Environmental Change* 73 (102484): 1–13. doi:10.1016/j.gloenvcha.2022.102484.

Qiu, L. L., F. Yang, K. P. Paudel, M. A. Koondhar, and W. Z. Zeng. 2018. "Influence of Rural Households' Livelihood Capital on Income Derived from Participation in the Forest Carbon Sequestration Project: A Case from the Sichuan and Yunnan Provinces of China." *International Forestry Review* 20 (4): 538–558. doi:10.1505/146554818825240700.

Ribot, J. C. 1995. "From Exclusion to Participation: Turning Senegal's Forestry Policy Around?" *World Development* 23 (9): 1587–1599. doi:10.1016/0305-750X(95)00060-P.

Ribot, J. C., and N. L. Peluso. 2003. "A Theory of Access." *Rural Sociology* 68 (2): 153–181. doi:10.1111/j.1549-0831.2003.tb00133.x.

Schlager, E., and E. Ostrom. 1992. "Property-rights Regimes and Natural Resources: A Conceptual Analysis." *Land Economics* 68 (3): 249–262. doi:10.2307/3146375.

Sekine, Y. 2021. "Emerging 'Agrarian Climate Justice'struggles in Myanmar." *The Journal of Peasant Studies* 48 (3): 517–540. doi:10.1080/03066150.2020.1839054.

Sikor, T., G. Auld, A. J. Bebbington, T. A. Benjaminsen, B. S. Gentry, C. Hunsberger, and C. Upton. 2013. "Global Land Governance: From Territory to Flow?" *Current Opinion in Environmental Sustainability* 5 (5): 522–527. doi:10.1016/j.cosust.2013.06.006.

Sikor, T., J. U. N. He, and G. Lestrelin. 2017. "Property Rights Regimes and Natural Resources: A Conceptual Analysis Revisited." *World Development* 93: 337–349. doi:10.1016/j.worlddev.2016.12.032.

Sikor, T., and C. Lund. 2009. "Access and Property: A Question of Power and Authority." *Development and Change* 40 (1): 1–22. doi:10.1111/j.1467-7660.2009.01503.x.

Sun, W., and X. Liu. 2020. "Review on Carbon Storage Estimation of Forest Ecosystem and Applications in China." *Forest Ecosystems* 7 (1): 1–14. doi:10.1186/s40663-019-0210-2.

Tubtim, N., and P. Hirsch. 2004. "Common Property as Enclosure: A Case Study of Backswamp in Southern Laos." *Society & Natural Resources* 18 (1): 41–60. doi:10.1080/08941920590881925.

van Minnen, J. G., B. J. Strengers, B. Eickhout, R. J. Swart, and R. Leemans. 2008. "Quantifying the Effectiveness of Climate Change Mitigation Through Forest Plantations and Carbon

Sequestration with an Integrated Land-use Model." *Carbon Balance and Management* 3 (1): 1–20. doi:10.1186/1750-0680-3-3.

Xu, Y. 2019. "Politics of Inclusion and Exclusion in the Chinese Industrial Tree Plantation Sector: The Global Resource Rush Seen from Inside China." *The Journal of Peasant Studies* 46 (4): 767–791. doi:10.1080/03066150.2017.1405936

Yang, B., and J. He. 2021. "Global Land Grabbing: A Critical Review of Case Studies Across the World." *Land* 10 (3): 324. doi:10.3390/land10030324.

Yang, F., K. P. Paudel, R. Cheng, L. Qiu, T. Zhuang, and W. Zeng. 2018. "Acculturation of Rural Households Participating in a Clean Development Mechanism Forest Carbon Sequestration Program: A Survey of Yi Ethnic Areas in Liangshan, China." *Journal of Forest Economics* 32: 135–145. doi:10.1016/j.jfe.2018.04.006.

Ye, J. 2015. "Land Transfer and the Pursuit of Agricultural Modernization in China." *Journal of Agrarian Change* 15 (3): 314–337. doi:10.1111/joac.12117

Zinda, J. A., and J. He. 2020. "Ecological Civilization in the Mountains: How Walnuts Boomed and Busted in Southwest China." *The Journal of Peasant Studies* 47 (5): 1052–1076. doi:10.1080/03066150.2019.1638368.

OPEN ACCESS

Resilience and conflict: rethinking climate resilience through Indigenous territorial struggles

Noémi Gonda, Selmira Flores, Jennifer J. Casolo and Andrea J. Nightingale

ABSTRACT

Resilience to climate change demands a transformation in social and political relations, but the literature has largely neglected how these are embedded within legacies of conflict. We explore the roles socioenvironmental conflicts play in the scaling up of transformation amidst ongoing settler colonial projects in Indigenous territories in Nicaragua. Drawing on insights from resilience, climate change, and critical agrarian studies, this article reframes resilience as a process produced within socioenvironmental conflicts, placing contestation and negotiation in the centre frame. By re-signifying the meanings and practices of resilience, Indigenous agrarian struggles contribute to 'eroding capitalism' and its entwinement with climate change.

1. Introduction

> The attacks have been escalating. (…) [A Mayangna Indigenous community in the heart of the biosphere reserve of BOSAWAS] was attacked by 80 heavily armed settlers with the result that 16 homes were burned, 10 people went missing, and four men from the Indigenous community were killed and two wounded, one of whom was left paraplegic as a result of the gunshot wounds he received during the attack. The settlers also slaughtered the Indigenous people's livestock. (100% Noticias 2020; IWGIA 2020)

Across the world, protecting valuable ecosystems and Indigenous land rights through biosphere reserves is taking on new significance as reserves become enrolled in efforts to promote resilience to climate change. Such efforts require clear resource tenure rights along with stable institutional arrangements, both of which are normally packaged under 'good governance' initiatives. Yet the target areas and purported beneficiaries of these projects are often embroiled in multiple socioecological conflicts, forged in the crucible of multi-scaled dynamics of climate and agrarian change. The emphasis on property rights, stability and reducing conflict often sits in stark contrast with empirical realities wherein climate change initiatives have been shown to exacerbate land conflicts, struggles over authority, and perpetuate fraught relations embedded within colonial histories and capitalist expansionism (Nightingale 2017; Sultana 2022; Whyte 2017).

The BOSAWAS biosphere reserve in Nicaragua is no exception. The above violent attacks occurred eleven months before the Central American Bank for Economic Integration (CABEI) approved 84 million US dollars from the Green Climate Fund for Bio-CLIMA; a climate change mitigation project aimed at integrating climate action, reducing deforestation and strengthening resilience in Nicaraguan biosphere reserves, including BOSAWAS (Nicaraguan Ministry of Finance and Public Credit 2019). Since its creation in 1987, the BOSAWAS reserve has become a locus of land/forest use conflicts and struggles over whose vision of livelihood security will provide resilience for communities and ecosystems in the long term. The promise of abundant resources beyond the 'agrarian frontier' – an imagined frontier between agriculture and forested areas (Maldidier 2004) – spur migration by non-Indigenous populations to these Indigenous areas, felling the rainforest and deepening processes of conflict and primitive accumulation (Larson 2010; Nygren 2004).

In this article, we show that Indigenous resilience is not innate. Rather, resilience practices and processes emerge on the terrain of conflict and collaboration tied to territorialisation processes of settler colonialism, understood as 'the ongoing process which maintains European economic systems, political structures, social norms and occupation on Indigenous lands' (Thompson and Ban 2021, p.230), and associated capitalist political economies. By doing so, we challenge the notion that resilience in the face of climate change requires reducing conflict. Instead, we show how complex relations between the state, people and more-than-humans, characterised by both conflict and collusion, are instrumental in creating the openings wherein 'right' resilience can emerge. Rather than an outcome of climate change interventions or a form of essentialized flexibility, we theorise resilience to be a process produced within socioenvironmental conflicts. This process framing draws attention to the messy and often unruly practices, moments, and relations through which people seek livelihood security and assertion of their lifeways (Nightingale 2018). There is no blueprint or predictable outcome for resilience processes. Nevertheless, in a more normative vein, we follow feminist political ecologists to imagine 'right' resilience as a process of building commons, affective, democratic, equitable and just relations with humans and non-humans (Mehta and Harcourt 2021) for long-term livelihood security, while acknowledging that each of these processes is contested (Cote and Nightingale 2012) and based upon divergent definitions (Forsyth 2018). As it is used today, resilience signals an idealised response to the uncertainties and extreme events of climate change that can produce sustainability, itself another slippery boundary object. Indeed, it is precisely because resilience is a boundary object which is increasingly shaping the strategies of development agencies, state policies and even local communities, that inspires us to reclaim it and insist that its ontologically and politically contested nature is foregrounded.

Indigenous peoples, we argue, caught in the crosshairs of climate change /biodiversity conservation initiatives and 'agrarian and resource frontier' expansion, are engaging in class and more-than class politics in ways that do not neatly fit within Wright's (2019) discussion of anticapitalist strategies[1] but nevertheless wear away at agrarian, extractivist and environmental attempts that threaten lands, lives, and livelihoods. In Nicaragua, as climate change policies and effects intensify settler colonialism[2] and associated political economic processes, Mayangna communities with recognised historic land rights (GoN 2003) manoeuvre to protect their territory in ways that both resemble *and* diverge from non-Indigenous struggles for agrarian justice and autonomy.

Rethinking resilience at the conjuncture of Indigenous territorial struggles, climate change migration, projects and policies, and the ongoing dynamics linked to agrarian dynamics of colonial-extractive capitalism[3] help us to make two crucial theory-practice leaps. First, that resilience processes arising out of socioenvironmental conflict can produce a plurality of strategies that can erode current racialised processes of agrarian and extractive capitalism from the inside-out. In particular, we show that ontological and spiritual dimensions of land and territory are not the 'flip side' of agrarian political economy understandings, rather they are reproduced, sustained and sometimes suspended in tension with them. As a result, challenges to the current political economy can occur within capitalist relations of production and exchange, not simply in resistance to them. Conflicts and dissent are the cement in this continuum underpinning the practices through which resilience emerges.

Second, the ways in which Indigenous peoples embrace or refuse capitalist and market relations, attempting to reconfigure them in their own terms, is a resilience strategy that is equally valid as efforts to gain autonomy from the market through (e.g.) agroecology or food security (Santiago Vera et al. 2022). The former strategy is often missed in agrarian studies that focus on class relations and land as a physical terrain to be claimed (Coulthard 2014; Koshy et al. 2022; Tzul Tzul 2018). In addition, understanding resilience through Indigenous territorial struggles, rather than through class, allows us to better draw attention to resilience processes' relationship with the non-human (Whyte 2018) and with efforts for autonomy and self-governance which lie at the heart of counter capitalist initiatives (Gahman 2020). By making these analytical moves, Indigenous territorial struggles at the conjuncture of a biosphere reserve re-purposed for climate change mitigation efforts, and 'agrarian frontier' expansion, help us envision how resilience processes can lead to transformative change. Moreover, how Indigenous peoples operate within and refuse colonial-capitalist ontologies and epistemologies helps make visible the colonial underpinnings and continuities of agrarian capitalism under climate change.

Drawing on insights from feminist political ecology, critical agrarian studies and Indigenous studies, this article thus reframes resilience, placing conflict, contestation, and negotiation in the centre frame. We build from recent work on power in resilience (Garcia et al. 2022) and theoretical insights on authority, knowledge and affect in

[1]Wright's 'strategic logics' that historically animated anti-capitalist struggles are: 'smashing capitalism', 'dismantling capitalism', 'taming capitalism', 'resisting capitalism' and 'escaping capitalism'(2019).

[2]Settler colonialism is based on Eurocentric ideas that see land 'as a physical terrain to be claimed; as the seat of political sovereignty; as the resource of capitalist development; and as the homeland of the settler'. It downplays the role of land in Indigenous rights struggles. It is grounded in physical, epistemological and ontological 'elimination'of Indigenous peoples (Taylor and Lublin 2021, 263).

[3]Extractive capitalism is a form of capitalism that is based on the extraction of profit from humankind and nature often jeopardising agrarian and Indigenous populations' livelihoods and ways of living with non-human nature. It is 'a logic and practice of colonialist domination that involves the extraction of natural resource rents without restoration or care of the territories to which the resources belonged nor of the socio-natural relations that they previously sustained' (Fash 2022, 38). Extensive ranching is one form of contemporary extractivism.

climate change (Eriksen, Nightingale, and Eakin 2015; Nightingale, Gonda, and Eriksen 2022). This work shows how resilience occurs within the operation of power and political economies, rejecting earlier accounts which overly emphasised resilience as an adaptive capacity to biophysical change (Cote and Nightingale 2012; IPCC 2007). We thus reinforce recent calls, – in particular in the Forum on Climate Change and Agrarian Justice (Borras et al. 2021) – for seeing local negotiations, contestations struggles, and conflicts, as reworking, resisting and re-signifying resilience not just as material processes borne out of class struggle, but also as products of the agency and desires of different people (Camargo 2022), and as part of broader agrarian and Indigenous movements for recognition and autonomy (Mills-Novoa et al. 2022). Following Indigenous philosopher Kyle Whyte's (2017) critique of universalist framings of climate change as the singular threat of modernity, while casting Indigenous people as either victims or saviours of it, we argue that these conflicts are a vital part of on-going relations through which resilience emerges, rather than impediments to a normative end point. More pluralistic engagements with the complexities of these emplaced power asymmetries and relations (Nightingale 2018) open space for the co-production and co-existence of multiple climate realities to serve as leverage points for transformative change. Climate change is the most recent ruination (Stoler 2013) in a long history of environmental changes tied to the dynamics of creative-destructive capitalist-colonial dynamics. Understanding resilience through the lens of violent colonial continuities reveals pathways to building 'right' socio-environmental resilient relations (Gram-Hanssen, Schafenacker, and Bentz 2022).

Empirically, we examine the relationships between socioenvironmental conflicts and resilience practices in Indigenous territories that overlap the BOSAWAS biosphere reserve. Colonial nation state imaginaries and projections of the 'agrarian frontier' have fuelled an intensification of land grabbing and illegal selling of Indigenous lands. The dynamics of capitalist accumulation in the biosphere reserve vary: from land poor climate migrants to cattle-ranchers and their hired hands, to timber barons, to large-scale exploration and exploitation of the subsoils. Nevertheless, all take advantage of a repeated narrative of empty or unproductive spaces available for the taking, and all operate through intense pressures due to how climate change adaptation and mitigation schemes rework agrarian and extractive capitalist spatial relations. These processes of land and forest dispossession are also embedded within political splits between and within Indigenous communities that draw them into contradictory relations of production and exchange. Rather than mitigating conflict, climate change interventions fuel increasingly violent clashes between state forces, settlers, Indigenous communities and environmentalist organisations. For Indigenous populations, their need to reduce conflict subsumes their livelihood aspirations and socioenvironments to the priorities of an authoritarian state and a (largely) well-meaning international development community focused on carbon capture rather than on human well-being. It is within these messy dynamics that we find the trajectories of 'resilience' and how local people work within problematic relations to assert their lifeways.

In the next section, we theorise the nexus of resilience, critical agrarian studies and climate change in Indigenous territorial struggles by discussing how 'frontier' imaginaries, contested subjectivities, and visions of autonomy feed into the politics of resilience in Nicaragua. Section 3 presents the process of engaged research in the Mayangna territories through which this paper emerged, while section 4 describes the climate contours and overlapping conflicts in the BOSAWAS reserve. Section 5 shows how Indigenous people maintain cooperation and conflict in continuous balance to survive, anchoring their lifeways in the past to anticipate future uncertainties, and how their climate resilience strategies are articulated in affective relations and emotional practices which also have the potential to erode capitalism. In Section 6 we conclude by elaborating how conflict and contestations productive of resilience require scrutiny of everyday practices, collusions, contestations and struggles; understanding of historical continuities and legacies of resilience practices; uncovering their complexity and uncertain character; and understanding how resilience actions support dissent and the building of affective relations.

2. Theorising the nexus of resilience, critical agrarian studies and climate change in indigenous territorial struggles

In this section, we first highlight the conceptual foci that emerge from our empirical research: namely 'frontier' imaginaries, contested subjectivities, and visions of autonomy. Second, we underline the main ideas from diverse trends of the resilience literature on which we build our analysis. Through bringing these two aspects together, we unpack the nexus of resilience, critical agrarian studies and climate change in Indigenous territorial struggles. Our analysis shows how complex relations of conflict and collusion between the state, capitalism, migration and settlement, people and more-than-humans are instrumental in creating the openings wherein 'right' resilience can emerge. More mainstream resilience literature suggests that conflicts represent a risk that needs to be reduced for building resilience (Sultana et al. 2019). Climate change mitigation programmes specifically list conflict as a risk (CABEI 2019). We theorise that rather than an outcome of climate change interventions or a form of essentialised flexibility, resilience is a process produced through socioenvironmental conflicts, themselves rooted in processes of extractivism, capitalist land use, and the politics of the so-called 'frontier'.

2.1. At the intersection of capitalism and anti-capitalism: 'frontier' imaginaries, contested subjectivities, and visions of autonomy

The tensions we observe in the rural world are underpinned by the contradiction between sanctioned policies and practices based on exploiting so-called 'agrarian and resource frontiers' (Kröger and Nygren 2020) and those to mitigate climate change and protect biodiversity through a different, yet overlapping space: biosphere reserves. 'Agrarian and resource frontiers' have long been political: from the 1950s onwards, peasant migration from the Eastern, non-Indigenous parts of Nicaragua towards the Western, Indigenous territories was encouraged by the Somoza dictatorship, aimed at avoiding political instability while supporting elites involved in export-oriented, large-scale agriculture. By keeping the best quality lands of the Pacific, Central and Northern regions of the country for ruling elites, and pushing smallholder producers towards the East, the colonisation of the 'agrarian frontier' contributed to capitalist expansion while avoiding social conflicts: Indigenous territories were the escape valve. For our discussion, it is immaterial whether the contemporary case is that of an 'agricultural frontier', or a so-called post-frontier where the only land 'free' to be appropriated through colonisation remains within the protected conservation areas. What matters for us is that the lived environments of local forest-dwellers, – the Indigenous people – continue to be framed as unused lands (Kröger and Nygren 2020), rendering the people within them as non-existent.

Understanding the everyday processes of an imaginary, physical, epistemological and ontological erasure brings a new dimension to critical agrarian studies discussions of territory (Giraldo and Rosset 2018). It helps recognise the entanglements between subjectivities and land rights, suggesting common ground between Indigenous peoples and campesinos. 'Frontier' imaginaries work in relation to one another, hardening lines between peasant/small farmer subjects and Indigenous peoples through simplistic binaries: invaders vs. natives; destructive settlers vs. natural stewards (Devine 2018; Larson 2010; Nightingale 2018), conveniently erasing the colonial-capitalist continuities that produce the conditions for conflict. While this shift represents powerful possibilities for uniting struggles across the rural world,[4] it is crucial that what is distinct about Indigenous processes of resilience is visible in the quest to forge unity around anti-capitalist politics.

Indigenous studies and anti-colonial analyses highlight the centrality of land/nature and the historical global connections of capitalism in ways that bring into focus two logics of agrarian expansion and green grabbing (Coulthard 2014; Koshy et al. 2022; Ybarra 2018): (i) the plurality of uses and meanings of territory and; (ii) the elimination of Indigenous territories-bodies. We argue that overlooking the significance of Indigenous territorial struggles runs the risk of contributing to further de-politicising the resilience debate and reinforcing the colonial-capitalist status quo (Ojha et al. 2022). Further, such an omission can normalise violence as well as Human and land rights abuses.

Mayangnas in and around the BOSAWAS reserve, like many Indigenous peoples in the rural world, have protected rights to land and resources as well as to self-government under the International Labour Organisation's (ILO) 169 convention. But Indigenous 'autonomous territories' overlap with the biosphere reserve, which is governed by UNESCO linked institutions in collaboration with the central state. On the ground, attention to the everyday efforts of managing conflicts and re-negotiating power relations of Mayangnas living in the BOSAWAS communities suggests the centrality of autonomy as a normative goal for resilience. Autonomy resides in the collective ancestral practice through which Mayangnas experience, govern, care for, and reproduce their relationship to each other and to more-than-human nature, including their right to define how they engage with market/class relations. As we show below, Mayangna autonomy does not mean becoming autonomous from market relations through food sovereignty and agroforestry practices: rather, Mayangnas' social reproduction is interlinked with both capitalist accumulation and building alternative social relations to dependence, deprivation and marginalisation (Douwe van der Ploeg 2010).

In the subsequent discussion, we highlight seven main ideas from diverse trends of the resilience literature (socioenvironmental resilience, Indigenous studies and settler colonial critiques that expand critical agrarian engagements with resilience) that have inspired our thinking about the everyday politics of resilience.

---

[4]The Vía Campesina social movement has been particularly influential in advancing a plurality of uses and meanings of territory to articulate struggles for access and control of land with struggles for the defence of territories and lives (Giraldo and Rosset 2018) as a way of embracing myriad uses and meanings at work in the rural world.

2.2. Everyday resilience politics

A first aspect of our argument is that the historically produced knowledge, needs and values of marginalised groups cannot be subsumed to concerns over adaptive capacity and planning for the future. We build on ecological resilience research that makes clear that social, political and biophysical changes are deeply entangled, meaning that understanding one in isolation of the other is not only ontologically flawed but also leads to empirically problematic policy outcomes. Socioecological systems thus capture the dynamics of change (Folke et al. 2016; Haider et al. 2021), but too often end up reifying stability and resistance to change in livelihood systems (Carr 2019). Recent efforts at rethinking resilience from the socioecological perspective recommend shifts in practice: from capitals to capacities, from objects to relations, from outcomes to processes, from closed to open systems, from generic interventions to context sensitivity, and from linear to complex causality (Reyers et al. 2022). While we welcome these efforts, we believe that they still too often gloss over how intersectional subjects that are necessarily classed (racialised, gendered, abled, aged) can refuse, albeit in ambiguous ways, the dispossessing and dis/placing effects of climate change.

A second aspect of our argument is that efforts at building resilience may create an *undesirable* resilient system. Resilience in ecological science refers to the ability of a system to withstand shocks and disturbance without flipping into a different phase, such as a forest becoming a grassland from too much disturbance. In socio-ecological systems, this thinking has led to analyses concerned about humanity's 'operating space' in relation to environmental change (Rockström et al. 2009) and the institutional and political arrangements that promote adequate social and human capital to avoid the collapse of current societies and economies (Stone-Jovicich et al. 2018). Many scholars in this tradition are concerned about the unsustainability of today's industrial economy and thus promote shifting into a 'better phase' (Olsson, Folke, and Moore 2022), without also considering that such efforts may indeed create an *undesirable* resilient system.

Third, we want to bring the agrarian question back into the debate on Indigenous resilience but without subsuming the latter to the former. Resilience policies and practice linked to Indigenous territorial struggles in so-called 'agrarian and resource frontiers' speak directly to the still present agrarian question (Watts 2021) posed by Kautsky: 'whether and how capital is seizing hold of agriculture, revolutionizing it, making old forms of production and property untenable and creating the necessity for new ones' (1988 [1899], p.12). Both discussions on the recognition of Indigenous lands without full restitution of the accompanying rights, and resilience studies in the face of climate change, seek to show how individuals and communities navigate market forces or hegemonic state-society relations (Correia 2019; Sekine 2021). Further, resilience studies emerging from critical urban studies make visible the spatial relations of racial capitalism. Grove, Cox, and Barnett (2020, 1627) frame it as both 'a mechanism for securing … racialized economic trajectories' and a 'medium' through which anti-racial capitalist practices can emerge and gain traction. Resilience, in this light, in spaces where agrarian capitalism, extractive capitalism and 'green grabbing' are all at work, can signal the possibilities and persistence of subjects that straddle class relations and non-class relations (Santiago Vera et al. 2022).

In Nicaragua, it draws attention to Indigenous peoples' resistance to the ways that existing forms of production perpetuate ecological destruction. Analytical attention needs to pivot upon what work resilience does, who takes it up and for what purposes. Development agencies and the state promote resilience as an idealised way to support communities and marginalised peoples in the face of uncertainty and disasters, yet critics show that subjects become incorporated into shifting capitalist relations in uneven ways (Gonda 2019). Projects that promote resilience to support communities and Indigenous peoples often demand that they *further* engage with the dynamics of capitalist colonialism. Yet, resilience also signals the ways that Indigenous communities navigate multiple and diverging pressures from settler migrants, public actors and international projects. In these ways, we unsettle the idea that Indigenous resilience automatically requires resistance to capitalism, and focus instead on the dynamics of conflict and collusion to both market dynamics and historical lifeways through which resilience is asserted.

A fourth aspect to our argument is that the ontologies behind materiality and meaning matter. As we show, Indigenous peoples reclaim capitalist processes by alternatively seeking redress *and* engage in land transactions to assure the land/forest continues to anchor their identity, history, and community. The insidious power of capitalist relations of exchange – resting upon historical regimes of accumulation – to define land, property, identity politics and conservation agendas is a core frame for understanding the dynamics of the biosphere reserves. However, it is not the only frame that shapes how processes of resilience unfold, nor is resistance to capitalism the only way through which local people push back against these pressures. Political ecology and critical agrarian studies literature have shown how struggles over the (micro) politics of contestation and reworkings of hegemonic rule/ governmentality vis à vis rural dispossession, reveal political fissures and openings (Devine 2018; Li 2014; Moore 2005; Peet and Watts 2004; Sawyer 2004; Wainwright and Bryan 2009; Watts 2003). Yet, this focus needs re-centering Indigenous epistemologies (in the fashion of Ybarra 2018).

Fifth, to accomplish such re-centering, we draw from work on how socioenvironmental change is embedded within the operation of power in relation to climate change adaptation (Eriksen, Nightingale, and Eakin 2015; Nightingale 2017). Precisely what this means varies from trying to better integrate political economy dynamics into socioecology systems thinking (Van Hecken et al. 2021), to showing how cross scalar dynamics and struggles over knowledge and subject-making shape resilience dynamics (Garcia et al. 2022). Garcia et al. (2022) highlight the inequitable legacies of colonialism that underpin modern exclusions, exploitation, identities and representations through which 'resilience' is negotiated. Their argument invokes earlier feminist and Indigenous attention to the everyday spaces, places and processes that are often ignored as sites of social change (Abu-Lughod 1990; Cumes 2012; Rivera Cusicanqui 2010). Highlighting these situated social and material practices that support life and livelihoods allows us to problematise when and how 'everyday acts of resilience, reworking and resistance' can remake social and material relations (Betteridge and Webber 2019, 944), including capitalism.

Sixth, we are inspired by recent critical literature that has begun to engage the role of resistance and everyday conflicts in building resilience. Here, resilience is shown as complex and uncertain process (Harris, Chu, and Ziervogel 2018), as opposed to some naturalised characteristic of a people (Kaika 2017), or desired outcome of an intervention

(Boyd et al. 2008). Our argument draws on a theorisation of power that captures its ambivalent and multidirectional nature (Butler 1997; Nightingale 2011, 2017; Tuana 2013), side stepping the debate about 'resistance' versus 'resilience' by showing how all acts of resistance include dimensions of collusion and cooperation. Indeed, practices of resilience are diverse, contested, and often contradictory – sometimes individual, collective, or redistributive, while at other times reinforcing existing inequalities (Betteridge and Webber 2019). In Nicaragua, this insight is vital to avoid essentialising how various members of Indigenous communities and settlers engage with fraught land exchange dynamics.

Finally, conflicts are not just about oppositional politics, violence or land use; they have the potential to create affective relations. Yet affective relations always carry with them uncertainty and unpredictability. As outcomes of relational encounters, they are not easy to direct or orchestrate, even if it is these affects that hold the most promise to bridge social and political divides and help generate transformative engagements with each other (González-Hidalgo and Zografos 2019). Efforts to link affect with resilience in the climate change debate are rare (Nightingale, Gonda, and Eriksen 2022, are one exception) but affective relations between humans and non-humans are central in Indigenous thinking and decolonial scholarship (e.g. Simpson 2007; Tuhiwai Smith 2013).

These constrained but not foreclosed possibilities and limits of socio environmental struggles in Indigenous territories/protected areas suggest the need to better understand entanglements between conflicts and resilience to highlight terrains of struggle that can contribute to unsettling capitalist and colonialist structures and relations. It is to these terrains in Nicaragua that we now turn.

3. Engaged research in uncertain times

This paper brings new research (2018-present) in conversation with the insights drawn from long-term engagement (30 + years) with Nicaraguan *mestizo* peasants and agrarian processes, and deepening relations with Indigenous peoples and territories (10 + years). Fieldwork around conflict and climate change began in 2019–2020 with twenty-five open interviews with Indigenous, and *mestizo* professionals and activists involved in defending human, environmental and indigenous territorial rights. Those interviewed (12 women and 13 men), traced their experiences with struggles in and around biosphere reserves in Nicaragua. These interviews led to 6 workshops (2021) with two groups of Mayangna forest guardians and one group of Mayangna women. The security situation in Nicaragua made further community research impossible and posed dilemmas both in relation to what we could research and how we could write. Nevertheless, ongoing communication has continued as we try to deepen non-extractive research praxis and 'deep reciprocity' (Casolo et al. 2022; Simpson 2017) that honours Indigenous knowledges, struggles, and autonomy.

The convergence of three crises: political, public health and climate both reshaped and became part of the research process itself (Gonda et al. 2021, 2022). Together these crises highlighted for us the multiple scales of precarity that differently situated peoples navigate, and revealed the ways in which the politics of emotion penetrate struggles over lands, lives and livelihoods.

4. Climate contours of territorial conflict in the BOSAWAS reserve

> Some institutions and [even] some Mayangna say we, the Mayangna, live in a reserve … [but] we live in a [territory] that is inherited from our ancestors, a place that we are owners, we do not live in a reserve of the State. (Interview 2/06/ 2020).

Making sense of Mayangna resilience strategies requires placing their efforts within the overall political economy of environmental governance in Nicaragua. Governance regimes in relation to the biosphere reserve often come into direct contradiction with Indigenous rights. For example, state institutions manipulate the 'free and informed consent' requirement of projects[5] – co-opting leaders, consulting few people, and exchanging bribes. Rather than safeguarding biodiversity and strengthening the 'resilience' of tropical forest ecosystems, the resulting 'overlapping systems of governance have encouraged rapid ecological destruction and social differentiation as well as corruption and violence' and violate Indigenous peoples' rights (Kaimowitz, Faune, and Mendoza 2003, 6). Between 2011 and 2016, for example, over half million hectares of forests in Nicaraguan biosphere reserves were destroyed by forest fires, hurricanes, and clearance for extraction and other activities (Campos Cubas 2018; Guevara Flores 2017, 22–23). State officials and political elites are complicit in giving mining concessions or rights for timber extraction and cattle ranching in direct violation of the reserve's regulations and the desires of many resident peoples. The police supposed to defend the territorial rights of local people, often are enrolled into such illegal land deals.

Who is living around the biosphere reserve and how they make claims to land is central to our story. 'Settlers' refers to people of usually mixed Spanish-Indigenous ancestry (called *mestizo* in Nicaragua) who obtained access to land on Indigenous territories through a variety of pathways including, buying land titles and resource concessions from Indigenous leaders and state officials who sell them illegally, from impoverished local people who are desperate for cash, and through land traffickers. Land traffickers offer cheap land, operating in pseudo-legality created through connections with local elites and Indigenous territorial leaders in collusion with municipal or regional authorities. Land traffickers, especially on land for artisanal mining and livestock, breed chains of settler families, as one settler family opens the door to their kin in a recurrent process. Once on the land, settlers often engage in aggressive deforestation to install pastures for cattle ranching and sometimes attack Indigenous residents in an attempt to stake claims in the rich forest lands (Figueroa Romero and Pérez 2021; Oakland Institute 2020).

Since the imposition in 1987 of a state vision over an area of 20,000 km^2 overlapping with Mayangna autonomous territories[6] (GoN 2001), areas of forest have been converted to pasture and Mayangna territories are increasingly inhabited by *mestizos*. For example, in the BOSAWAS territory, in 1990 there were only 15 non-Mayangna families, in 2007 there were 121 while in 2013 the number increased to 314. Although there is no recent data, according to Mayangnas this number has been increasing. The underlying problem with *mestizo* settlers is the actions they implement, in particular the

[5]Required by the ILO's Convention 169, Nicaragua being one of its signatories since 2010.

[6]In 1987, the then Sandinista government accorded autonomy to Nicaraguan Indigenous territories to lessen Indigenous support for the US funded counterrevolutionary forces. In parallel, the central state supported agrarian reform policies that were pushing non-Indigenous farmers in search of pastures towards the 'agrarian frontier' and these Indigenous territories.

deforestation of large areas of forests to introduce livestock that change land use and jeopardise Mayangnas' autonomy and lifeways as well as their aspirations and rights to govern the territory and preserve the forest. Gendered and racialised dualistic imaginaries reinforce the frontier ideal of a male cattle rancher with leather boots and cowboy hats, looking over his cattle from the back of his horse, as opposed to an Indigenous woman relying on the forest for food and medicine (Flores Cruz and Torres 2012; Gonda 2021).

Meanwhile, BOSAWAS continues to be conceived from the perspective of the central, colonial capitalist state. For example, the recently cemented roads that connect the capital city of Managua to Bonanza (the closest medium-sized city to BOSAWAS) have been built with the idea of extracting resources (mining, timber, cattle) for development, conceived as national economic growth through extraction. Human Rights defenders, environmentalists and social activists highlight that the progressive invasion into Indigenous territories has been encouraged by the state through licences and mining concessions for private and public-private interests (interview 1, 13 May 2020; interview 2, 21 May 2020; interview 3, 6 June 2020). Alternatives that support Indigenous territories' autonomy, for example possibilities to engage in agro-ecotourism, are not considered.

Mayangna communities most affected by conflicts, find themselves caught between two paths of action. The first: to obtain state protection from an increasingly undemocratic and corrupt political regime (Martí i Puig and Serra 2020), or second: to work in an uneven terrain of state neglect and ultimately accept co-existence with settlers. Environmental activists and some environmentalist organisations defend Indigenous people's rights to their ancestral lands, amplifying the important roles local people play in maintaining the forests. Yet, Matamoros-Chávez (2014, 83) highlights that

> the evidence that the settlers remain untouchable in the Indigenous territories suggests that they have learned to navigate between a network of institutional and economic interests, which means that they can achieve their strategic objective of establishing roots in the Indigenous territory.

In this context, Indigenous Mayangna people have developed their own strategies for reducing conflict and navigating through complex relations of dispossession and exchange.

5. Re-signifying resilience at the interstices of conflict and transformation

In this section, we illustrate how re-signifying resilience at the interstices of conflict and transformation can contribute to 'eroding capitalism'. Through the realities of environmental change and critical agrarian debates we see the ways in which Indigenous communities navigate climate change related conflicts and violence whilst holding onto their territory and lifeways. By doing so, they contribute to eroding capitalism from the inside-out by redefining the terms of engagement in the biosphere reserve.

5.1. *Balancing cooperation and conflict*

Conflicts in the biosphere reserve reflect the ways in which the actors outlined above engage in conflict, cooperation or collusion to further their interests, and sustain their relationships in their territory. Those interviewed identified three actors: Indigenous territorial authorities, non-Indigenous state actors, and *mestizo* migrant occupiers of various stripes. Together they pointed to multiple threads: the significance of Indigenous ways of feeling and knowing (*senti-pensar*) the land/territory itself, awareness of the relationship between agrarian capitalist expansion and land speculation and deforestation (what they call 'invasion'), intensification of climate change, and extractivism. Despite recognising complexity of these dynamics, they focus on how Indigenous territorial authorities fail to uphold the interests of those they represent, while state actors are absent, or worse, actively participate in furthering the interests of settlers and those seeking mining concessions or access to forests for extraction or Bio-CLIMA perks. Indigenous territorial authorities admit that they find themselves caught between the conflicting demands of the Indigenous people they are supposed to represent; their duties to prosecute settlers; and the designs of the state to open the forest for economic growth and development (mining licenses, pine resin and timber extraction, cattle ranching). Yet, they omit their role in the crafting of the Bio-CLIMA project and its conflict minimisation logic.

A flash point for these conflicts is between territorial authorities attending to agrarian and environmental pressures, and forest guardian groups tasked with safeguarding Mayangna autonomous territory from settlers engaged in deforestation. Forest guardians constitute the first line of protection by routinely convening a group of men from different communities to monitor territorial boundaries, restore boundary markers – clearing paths, checking 'papers' (documents of settlers they hold to prove their 'rights' on the land), and apprehend and present for prosecution any would-be invaders. If caught within the territory, persons must show a deed or leasing agreement. Recently, however, collusion between settlers, Indigenous territorial authorities, the state, and agrarian and extractive capitalist processes linked climate/cattle/mineral dynamics have become more frequent, undermining the actions of the forest guardians who attempt to exercise Mayangna territorial rights.

While practices of coercion through corruption are widespread in relation to peasant and peasant-indigenous struggles (Nuijten 2004), when Indigenous territories are involved, more than life and livelihoods are at stake. The ability to autonomously protect the deep roots of lifeways bound together with more-than-human nature is under threat. Bribes and threats from extractive enterprises, cattle ranchers, loggers, and state officials, corrupt Indigenous territorial authorities, leading them to rent and sell land without consultation undermine ancestral values (interview 4, man, June 2, 2020). Community Indigenous leaders are targeted by politicians and government officials who invite them to travel to the regional or national capital to stay in hotels, offering 'free food', alcohol and even paying for prostitutes in return for signatures or support (Interview 1, 13 May 2020). According to community members and forest guardians, 'territorial authorities are more interested in augmenting their salaries than investing in or defending our territory' (Mayangna man, personal communication, 12 April 2021). Given this collusion with Indigenous territorial authorities, when certain forest guardians attempt to gather evidence (papers, pictures or even witnesses) to prove that territorial authorities are part of the problem, they enter into conflict with other Indigenous authorities: those recognised hierarchically by the colonial state due to their support for extractive businesses linked to national economic growth.

For example, in June 2021, forest guardians captured and evicted a group of 12 *mestizo* men from a given territory, who subsequently produced papers signed by the president of the Indigenous territorial government giving them permission to settle and use communal land. The forest guardians' protests that they had not given such permission to the territorial authorities were met with silence. They then sought state support, calling on local police to defend Indigenous territorial rights; but again the police who should protect them let the settlers go free. Ultimately, in the face of the betrayal of their own authorities, forest guardians had to coordinate with the *mestiz,* colonial state, with the territorial government, the local police, the Ecological Battalion of the Nicaraguan Army and the attorney general's office (personal communication, 6 June 2021) to evict the settlers. They sacrificed some autonomy in order to save more immediate practices.

Discussions with forest guardians and interviews with women have shown that opinions about these actions are divided on generational and political (level of authority) grounds, sowing discord in communities and families based on the intertwined values of Indigenous autonomy and a just relationship with nature. For example: 'young Mayangna men think maybe … I must sell seven hectares out of necessity. Before they would sell community land that was not in use; now the situation is worse: they sell land already claimed by relatives or neighbours' (Mayangna woman, personal communication, 3 June 2021).

These fragmented and often contradictory ways that Mayangna forest guardians and community members experience struggles over land, nature and governance is the process through which resilience emerges. Forest guardians are beginning to position themselves and Mayangna territorial struggles in ways that shift the practice and scale of authority, and begin to reconfigure the exercise of autonomous rights to sustain human and more-than-human nature. Yet their conditions of possibility are linked to the ways they attend to everyday dynamics of dissent and disagreements, and the emotional and cultural toll of their overall suffering. In the next section, we show how they find anchorage in the past to anticipate future uncertainties.

5.2. Remembering and re-membering: anchors to the past as resilience for anticipating the future

The struggle for the defence of territory and the exercise of autonomy is kept alive by everyday dissent and disagreement among Indigenous community members regarding the meaning and material losses driven by the invasions. These conflicts create spaces to remember the past and strengthen affective relations through a shared sense of loss, and emphasis on recovering historical memory softens intergenerational conflicts.

Forest guardians from two different territorial districts proposed workshops in 2021 where they could share their memories and ways of understanding the losses and injustices they suffer from invasion, deforestation and other related socio environmental changes. Seven workshops took place in total, one with just women. Memories intertwined and refracted through one another as participants began to move from territorial survival in the crucible of conflict to the depth of their emotions. They expressed a deep sense of emotional attachment to the forest and articulated a sustainability sensibility that harkened back to the past.

These memories at first appeared to paralyse participants; 'If I go out, I see fences, I see that rivers are drying, it is not like before and that stresses me … '. Some memories are wistful, recalling times when there were no settlers in the territory and no necessity to defend it; when everyone could use the forests without restriction or fear: 'before the forest guardian job started there was peace' (Workshop with forest guardians, March, 2021), 'Today I went for a walk, and I came back sad, there is a lot of destruction in the hills, stripping of trees, drought' (Workshop with forest guardians, March, 2021) … 'walking upstream my heart was filled with sadness, desolation knowing that people dońt care and not knowing what to do to make a change' (Workshop with forest guardians, March, 2021). While these reflections seem to forget precolonial and colonial domination, they clearly re-member the past to keep alive Indigenous lifeways, pointing to where resilience processes begin.

The reflections among the Mayangna women were particularly insightful. Women discussed with us the suffering they experienced due to conflicts and how it had impacted their everyday lives. They lamented the drying up of the rivers, its contamination, and soil impoverishment due to the creation of pastures. While men and women both expressed that the river is a source of life for them, in practice, women's everyday activities are especially tied to the rivers, where they have bodily connections through washing clothes and dishes, bathing, fishing and canoeing. More importantly, they also function as a key site for social interaction with other women. While these reflections echo prior feminist political ecology claims about agrarian change, social reproduction and nature, we zoom in on what it means in a space/place where territorial rights have formally existed since 1987 and commitment to particular socionatural relations much longer. Maintaining non-capitalist means of production and the autonomy it entails, is not just about food sovereignty or accumulation by dispossession of women's knowledge or labour. When the forest is integral to social reproduction as it is for most Mayangna living in their autonomous territories, hunting, harvesting fruits and collecting medicinal plants and the reciprocal relations they entail is *being* (workshop with forest guardians, June 2021). The process of listening to each other and validating and disputing these sentiments is also integral to navigating the violences of Mayangna territorial conflicts. At the centre of both are maintaining or strengthening relations of co-responsibility between humans and non-humans that are eminently affective and emotional, not transactional (Tynan 2021), affects which from Western eyes appears as an ethics of care (Whyte et al. 2016).

Sometimes, listening and validating occur in private spaces, other times more publicly; but the process solidifies bonds in time and space. As one Mayangna Indigenous youth leader pointed out, re-membering in the sense of actively recovering, reassembling and employing historical memories can rework tensions and frictions between generations. Elders fear that the hard-won struggle for autonomous governance over lands and forests is slipping away.[7] Working with elders, asking, reflecting, and recreating their own history of territorial conflicts, he emphasised, helps them to see themselves not as victims or marginalised; but as people with a long and creative history of preserving their language, cultural practices, cosmovision and territories, against all odds. '[M]y

[7]Also referring to the landmark case of Awas Tingni. The Mayangna Indigenous community of Awas Tingni sued the Nicaraguan state in 1997 for granting a logging concession to private interests in Mayangna traditional territory. The case ended in a landmark Human Rights ruling by the Inter-American Court of Human Rights in 2001 that became a precedent for all Indigenous groups of the world (Gómez Isa 2017) as it established, for the first time in history, Indigenous communities' right to their collective land as a basic Human Right (Inter-American Court on Human Rights 2001).

desire' he said, 'is to bring the same process throughout the nine Mayangna territories' (Mayangna man, personal communication, 30 July 2021). This process of remembering and re-membering (Olsen 2003) suggests the relationship between historical refusals and the possibilities of eroding capitalism: 'the Mayangnas' struggles is not today, not of this decade nor of this millennium, it is a long history that involved our ancestors' (Interview 4, 2 June 2020).

5.3. *Weaving new relations within and beyond capitalism*

Affective relations between people and their land/territory form a foundation for understanding how conflicts reveal the contradictions of Mayangnas as 'agrarian' and 'Indigenous' subjects. Indigenous leaders colluding with agrarian capitalist land use similarly signals a paradoxical collaboration with state territorialisation[8] that ultimately helps retain Indigenous authority and control over land. Some Mayangna men and women have opened themselves to 'meeting *mestizos* halfway' in areas where land encroachment is accentuated and community-level fragmentation expanding, according to the forest guardians. By forging new bonds with settlers, they hope to foster a semblance of peaceful coexistence and lessen the likelihood of violent clashes, yet also maintaining a certain degree of autonomous control over their lifeways.

Settlers form unions with Mayangna women in some communities, purchase goods from community markets, or as is the case of settler youth, attend Mayangna baseball games. Other settlers and Mayangnas have developed trade relationships – such as bartering a mule for a plot of land. One Indigenous leader's conversation with a cattle rancher suggested that he was committed to building a good relationship: 'Look, Madame you are authorities, see we have money and we are cattle ranchers, we want to work and we do not want to be bad, we want to live that is what we want … ' (Mayangna woman leader, personal communication, 2 June 2021). Some Mayangna women saw the creation of new – to some extent caring – relations as necessary, in part due to the fact that their families are now a minority in their own territory, and partially because they have no support from state authorities (Workshop with Mayangna women, 1 June 2021). Crucial to this is an understanding of how these actions are grounded both in economic need and in a deep desire to avoid more conflicts and greater suffering.

In territories that have been dealing with land invasions for a long time, managing conflict has signified accepting settler presence in the territory (due to dynamics of corruption, violence and state irresponsibility), while finding ways to protect territorial autonomy, recover biodiversity, avoid overt violence, and defuse internal dissent (between settlers and Indigenous peoples, and between Indigenous people themselves). The conversations, negotiations, and exchanges entailed, even when they are volatile or painful, reshape social relations and socio-ecological practices.

In one Mayangna territory men and women have agreed to self-regulate land use. Faced with mounting pressure from settlers and the continued lack of state protection on their behalf, they allocated 27% of their territory for lease to settlers but with differentiated rates for land use in order to reverse deforestation and greenhouse gas release,

[8]State territorialisation i.e. the process through which conservation areas consolidate state sovereignty and power spatially through imposed tenure rights and control resource extraction (Sylvander 2021).

while encouraging a shift to agroecological production systems. Leasing for extensive cattle ranching would cost three times more than areas destined for more sustainable production. While regulating land titles (*saneamiento*) can equal state territorialisation (Sylvander 2021), the details about land use conditions and pricing are important in showing how through land markets, Indigenous people contribute to eroding capitalist forces within their territory, or at least reshape them for their own lifeway goals. However, whether or not this manoeuvring contributes to deeper transformation remains uncertain.

Indigenous responses to settler and extractivist practices are thus motivated by a great sense of responsibility between descendants and ancestors, and between the human and non-human worlds. They counter western imaginaries of 'progress' and hierarchies of personhood that drive colonialism and inform alternative ways forward (Thompson and Ban 2021; Whyte 2018). These relations and responsibilities are very different from those that derive from a capitalist logic as they cannot be monetised. They bridge ethnic and gender differences, taking Indigenous lifeways and knowledges seriously to help re-centre the forest within decisions over land exchanges.

5.4. Feeling-knowing-doing resilience

Emotional practices such as keeping silent, suffering and fear as well as solace, are resilience strategies as much as the most visible (legal, patrolling, confrontational) struggles. They do not exist without each other, and one does not prefigure the other; rather, they are the ways through which knowledge of resilience is co-created and contributes to transformation.

Within Indigenous communities themselves, women and men, elders and youth, wrestle with contradictory and ambiguous emotions. Historical suffering and practices that root people in their socionatures inform present-day distrust and dissent. Land invasions, forest destruction, and divergent positions regarding colonisation and extractive practices within and between families shape the silences and solace at the community level. Such collective emotions and relations are not usually perceived as part of resilience practices per se. Yet when examined through their interconnections, they tend to simultaneously move people nearer one another while also creating distance amongst themselves; the experience of being close yet apart (in positions and practice). In this sense, the possibilities within these processes are ambiguous. Here we look at how these play out in specific instances.

The tensions arising when Mayangna communitarians individually enter into agreements with settlers often provoke silence, rather than direct confrontation. When members of an extended family sell or lease community lands, family members often cover it up. If others in the community are aware, they do not challenge it. For them, it is better not to get involved; 'this is not discussed so as not to cause conflicts between families' (Mayangna man, personal communication, 2 June 2021). When asked why elders who have authority cannot stop a nephew, for example, from selling land, the answer is that it 'would deepen family problems'. Instead, the community enters into a pact of silence where tensions may brew or dissipate with time and distance. As another elder commented privately

> when they sell, they do it quietly so that no one notices and when they know that someone finds out, they stay away from the community for seasons to avoid complaints, that is why it is difficult to resolve the situation at the moment. (Mayangna man, personal communication, June 2, 2021)

Collective suffering is expressed differently, but here also divisions within the community shape public responses. Indigenous women and men experience socioenvironmental conflict caused by land encroachment differently. Men can use their own bodies to protect their territory, for instance patrolling long-distances, or they handle violent clashes using rudimentary weapons (machetes, wood-stick, stones); women are considered too vulnerable to do so. Additionally, men are being assassinated while women's bodies can be used to cause triple damage by settlers: physical, psychological, and moral for instance, through sexual violation. In both the women's workshop and in individual interviews, fear of being raped by settlers was constant. Consequently, women (sometimes at the request of their spouse) have relinquished their freedom to walk alone to cultivation areas, rivers for fishing, the forest to collect firewood, seeds or plants used for medicinal purposes, or to bring food home. Many women have moved their cultivation areas into new places in the forest where they feel safer. They go to areas with permaculture crops like fruit trees less frequently, and always in groups of relatives, not alone. In other moments of high risk, women usually move to other communities to protect themselves. During more critical moments of conflict, they rely on their faith; gathering together to pray while the men confront the settlers.

While these stories of suffering are clearly disruptive and traumatising for the community, they also show how collective responses to conflict and violence help maintain community ties and livelihood activities within a context that is rapidly changing due to settlement.

5.5. Re-signifying uncertainty

As the stories of violence and conflict avoidance imply, resilience processes and transformation also unfold under conditions of uncertainty (Mehta, Adam, and Srivastava 2019). This is especially important in authoritarian contexts in which uncertainties are compounded: resilience is about trying to rework relations rather than reversing them (Ojha et al. 2022). The everyday entwinement of dissent and affective relations between Mayangna peoples and *mestizo* peasants illustrate how resilience processes emerge in and through socio-environmental conflict, despite lacking foreseeable outcomes.

For example, some forest guardians called into question the territorial authorities, and in doing so exposed themselves to critique (both from their communities and the authorities), further complicating their job. Some responded to their limited ability to defend their territories by recognising they needed to do something more organised. For them, this meant (i) documenting and publicly revealing to their communities and to regional and national civil society organisations the ways in which the Indigenous territorial authorities under state influence were allowing settler invasions; and (ii) strengthening their organisation. The fact that they gathered to speak outside of the community attracted distrust and criticism: 'they say that when we come to these workshops it is because we are selling land', 'others say that we are getting into politics … but the Mayangnas should not interfere with the parties, we are not politicians' (Forest guardian, personal communication 15 May 2021).

Forest guardians, community members and activists uncover violence (internally and externally), and make connections between different types of invasions through these exposure tactics. Their strategy is simple – to ensure that they are not alone in their struggle; they are not denouncing territorial land grabs and settler colonialism outright, yet neither are they objecting only to the violence. Rather they are attempting to repair the state's relationship with their territory;

> If we denounce what is going on in our territory and what the GTI [Indigenous Territorial Government- *Gobierno Territorial Indígena*] is doing, we can call attention to our situation and get solidarity from other groups to create pressure so that national authorities can take responsibility for this situation. (Forest guardian, personal communication, 23 April 2021)

This process is reshaping subjectivities and relationships of struggle even if the outcomes remain uncertain. First, sharing information with the communities after participation in workshops not only helps to defuse distrust, it also subtly communicates that forest guardians actively communicate about their work, in contrast to the secrecy of the territorial authorities. Moreover, it establishes that their work and actions are necessary for addressing the conflict. Second, they are increasingly responsive to the different needs of community members as opposed to feeling accountable to the territorial authorities. They have already committed to producing printed materials in their native language as well as Spanish. Whether or not these efforts to uncover violence will mobilise dissent against territorial authorities and shift support to the forest guardians most active in denouncing the territorial authorities remains to be seen.

Uncertainty also stems from the fact that collaboration born out of conflict on one level does not imply the absence of tension and internal struggle in other spaces. At the territorial level, the process of collaboration to protect the reserve and exercise autonomy can widen fissures and provoke new tensions. Collaboration practices in the face of conflict often open spaces for navigating other expressions of dissent and exclusion, with the possibility of strengthening resilience processes, questioning capitalist relations and even challenging authoritarian politics.

But how this happens and what it means is disputed. In this sense, while sharing the position to halt invasions, these forest guardians question extractive capitalism as a means of legitimising and reinforcing unequal power relations between community members. Collaborative processes take shape *through* knowledge disputes, agreements and governance practices over how to confront aggressive land use change or deal with other conflicts produced by invasions. Concurrently, as their position solidifies to oppose extractive capitalist invasions, other fissures of dissent appear and/or widen, especially internal dissent along gendered, racialised and generational lines. In this painful and uncertain process lies the seeds of transformation that can erode the entwinement between capitalism, colonialism and climate change.

6. Rethinking resilience: conclusions and beyond

Rethinking resilience through the lens of conflict shows the severe limitations of climate change policies intent on promoting stability and conflict resolution. We need to look outside of dominant visions of climate change in order to be able to put forward valid anti-capitalist approaches to climate change (Paprocki 2022). In Nicaragua, like in many parts of the world, current Indigenous territorial conflicts are rooted in histories of colonialism, relations with the state and capital, and practices with more-than-humans. These roots, however, are uneven across intersectional social relations (gender, age, ethnicity) and emerge in sometimes surprising and uncertain ways as Indigenous people seek to protect their lifeways in the face of agrarian change, climate change interventions and biodiversity protection initiatives. In BOSAWAS, we show how resilience emerges out of the actions of Indigenous forest guardians who challenge not only agrarian and resource frontier settlers, but also Indigenous authorities and local people critical of their efforts to bring into public view violations of both biosphere reserve and local land use regulations. Yet these resistance efforts co-exist with other resilience strategies of silence and acceptance, as when hard pressed young men sell land in order to preserve livelihoods, or people accept *mestizo* neighbours as trading partners. In these practices, we find strategies to reduce violence and conflict and exert a level of control over a monetised land market largely dictated by outsiders, all of which are ultimately aimed at maintaining community cohesion and Indigenous lifeways. While in some sense these practices bring Indigenous people closer to the dynamics of agrarian change, a narrow class reading of them misses efforts at decolonising biosphere reserve governance regimes, intersectionality, and the affective relations with humans and more-than-humans through which such responses unfold.

When Indigenous authorities illegally sign land titles and concessions in collusion with powerful business and government elites, it shakes community foundations. And yet, resilience emerges from the internal struggles that result and commitments to maintaining the core values of the group. It is unlikely that Indigenous men would have gathered together to form a structured response to land encroachment as they have with the forest guardians if their own leaders had not been co-opted, or if violent clashes in the forest did not have embodied affects for women. Similarly, faced with declining control over land, some Indigenous communities have imposed their own conditions on land access, demanding that *mestizo* settlers engage in land use consistent with Indigenous sustainability practices and working at a social level to better integrate them into community social life. Each struggle foregrounds a particular focus such as: biodiversity protection, gender and youth justice, or territorial autonomy, but in practice they evidence intertwined concerns. These multiple interconnections, as well as Indigenous peoples' long history of navigating between isolation and deal-making (Casolo 2011) reflect the *longue durée* of practices of resilience.

Such tactics lay a foundation for 'doing capitalism' differently within Mayangna Indigenous territories. According to Wright (2019), dismantling, escaping, taming and resisting capitalism, when combined strategically can contribute to eroding capitalism from the inside-out. We do not dispute this claim in general, but we recognise that none of these logics sufficiently explain the plurality of ways that Mayangna Indigenous communities wrestle with the class and ethnicity differentiated settlers that lay claim to their territory, the disparate state responses or lack of response, including those labelled as biodiversity conservation or climate change mitigation initiatives, and growing divisions amongst themselves. These are not simple acts of resisting conservation, development, capital accumulation or dispossession. Rather, they are efforts by intersectional and uneven actors to take control and remake the terms under which territorial claims occur. They bring *mestizo* settlers into different socioenvironmental relations, ones that more closely mirror the historical ontologies and practices of Indigenous Mayangna. They help erode the hegemony of outsiders in (illegal) land sales and offer back a modicum of control to Indigenous peoples. These efforts at asserting territorial autonomy become the best way to challenge destructive designs on nature and local lives where state sanctioned capital accumulation goes hand in hand with biodiversity conservation and climate change responses.

Our work shows that authoritarian governance challenges resilience not only in its efforts at top-down control, but also in the micropolitics of co-optation and corruption used as mechanisms of capital accumulation. Mayangna practices of re-membering help ensure these efforts at eroding capitalism are rooted in history and embodied experiences of the more-than-human, but also the transformations of current subjectivities and meanings within agrarian economies. Resilience entails not only identifying and strengthening one's own skills, but also analysing the strategies and endgames of the other. They stimulate processes in which capacity strengthening (learning from elders) and new actions (forest guardians working together) are key. The successes of current Mayangna resilience strategies thus lie not in the actions themselves, but rather in a willingness to understand the subjectivities and positionalities that are emerging, and in relation to what visions, frameworks, and dynamics (past and present). Even if decision-makers and practitioners are conceptually prepared to include complexity in social, cultural and political issues when intervening in development processes (Ensor, Forrester, and Matin 2018), how these issues are incorporated in practice is conditioned by the parameters of authoritarian rule, colonial legacies, and extractive capitalism as well as their reverberations in the territories of agrarian change.

The case here thus sheds light on how the dynamics of conflict and collusion shape abilities to respond to climate change throughout Nicaragua and the region more widely. Ultimately, we have shown how resilience and socioenvironmental conflicts work together dialectically and dialogically. Resilience actions support dissent and the building of affective relations. Conflicts innovate, deepen, reshape, revalue historical and situated resilience practices and/or give birth to new ones. For that reason, even when conflicts and resilience processes are not visible, the silences themselves can speak. Looking for resilience in this manner is not to discount how the dynamics of settlement in the biosphere reserve remain tied to processes of agrarian change, modernisation and the historical legacies of colonial practices. Rather, it is to read the everyday practices of contestation and struggle and the building of affective relations as moments wherein resilience can solidify, relations can be re-written, and community lifeways perpetuated. For us, this is what a process oriented conceptualisation of resilience to climate change means.

Disclosure statement

No potential conflict of interest was reported by the author(s).

Funding

This work was supported by the Swedish Research Council Vetenskapsrådet [Grant Number 2018-05866] and the Swedish Research Council for Sustainable Development Formas [Grant Number 2018-00442]. JC also received support from a Junior Research Grant/The Research Foundation – Flanders (FWO) [Grant Number G0D8620N].

ORCID

Noémi Gonda http://orcid.org/0000-0002-1261-8380

References

100% Noticias. 2020. "Gobierno territorial Mayagna denuncia atroz asesinato de indígenas [Mayagna territorial government denounces atrocious murder of Indigenous people]. YouTube video." https://www.youtube.com/watch?v=HUIYl4elfBw&feature=share&fbclid=IwAR18VQoclhX034GRNBKbCZwa_rHYlmjXqznuYRtjuxjvDaZTQ7l5u-SiYZw.

Abu-Lughod, Lila. 1990. "The Romance of Resistance: Tracing Transformations of Power Through Bedouin Women." *American Ethnologist* 17 (1): 41–55. doi:10.1525/ae.1990.17.1.02a00030

Betteridge, Brittany, and Sophie Webber. 2019. "Everyday Resilience, Reworking, and Resistance in North Jakarta's Kampungs." *Environment and Planning E: Nature and Space* 2 (4): 944–966. doi:10.1177/2514848619853985

Borras, Saturnino M. Jr, Ian Scoones, Amita Baviskar, Marc Edelman, Nancy Lee Peluso, and Wendy Wolford. 2021. "Climate Change and Agrarian Struggles: An Invitation to Contribute to a JPS Forum." *The Journal of Peasant Studies* 49 (1): 1–28. doi:10.1080/03066150.2021.195647.

Boyd, Emily, Henny Osbahr, Polly J. Ericksen, Emma L. Tompkins, Maria Carmen Lemos, and Fiona Miller. 2008. "Resilience and 'Climatizing' Development: Examples and Policy Implications." *Development* 51 (3): 390–396. doi:10.1057/dev.2008.32.

Butler, Judith. 1997. *The Psychic Life of Power: Theories in Subjection*. Stanford, CA, USA: Stanford University Press.

CABEI. 2019. "Bio-CLIMA Nicaragua: Integrated Climate Action for Reduced Deforestation and Strengthened Resilience in the BOSAWÁS and Rio San Juan Biosphere Reserves." Central American Bank for Economic Integration, (Green Climate Fund), March 18. https://www.greenclimate.fund/sites/default/files/document/21710-bio-clima-nicaragua.pdf.

Camargo, Alejandro. 2022. "Imagined Transitions: Agrarian Capitalism and Climate Change Adaptation in Colombia." *The Journal of Peasant Studies* 49 (4): 713–733. doi:10.1080/03066150.2022.2059350.

Campos Cubas, Victor M. 2018. "Cuántos años le quedan a la reserva Indio Maíz? El deterioro de la reserva entre incendios y deforestación." Confidencial, 2018. Accessed 30 May 2022. https://www.confidencial.com.ni/confidencialtv/cuantos-anos-le-quedan-a-indio-maiz-el-deterioro-de-la-reserva-entre-incendios-y-deforestacion/.

Carr, Edward R. 2019. "Properties and Projects: Reconciling Resilience and Transformation for Adaptation and Development." *World Development* 122: 70–84. doi:10.1016/j.worlddev.2019.05.011.

Casolo, Jennifer J. 2011. "Unthinkable Rebellion and the Praxis of the Possible: Ch'orti'Campesin@ Struggles in Guatemala's Eastern Highlands." UC Berkeley.

Casolo, Jennifer J., Selmira Flores Cruz, Noémi Gonda, and Andrea J. Nightingale. 2022. "Choosing to "Stay with the Trouble": A Gesture Towards Decolonial Research Praxis." *Undisciplined Environments*. 8 March 2022. https://undisciplinedenvironments.org/2022/03/08/choosing-to-stay-with-the-trouble-a-gesture-towards-decolonial-research-praxis/.

Correia, Joel E. 2019. "Unsettling Territory: Indigenous Mobilizations, the Territorial Turn, and the Limits of Land Rights in the Paraguay-Brazil Borderlands." *Journal of Latin American Geography* 18 (1): 11–37. doi:10.1353/lag.2019.0001

Cote, Muriel, and Andrea J. Nightingale. 2012. "Resilience Thinking Meets Social Theory: Situating Social Change in Socio-Ecological Systems (SES) Research." *Progress in Human Geography* 36 (4): 475–489. doi:10.1177/0309132511425708

Coulthard, Glen Sean. 2014. *Red Skin, White Masks: Rejecting the Colonial Politics of Recognition. Minneapolis*. Minnesota: University of Minnesota Press.
Cumes, Aura Estela. 2012. "Mujeres indígenas patriarcado y colonialismo: un desafío a la segregación comprensiva de las formas de dominio." *Anuario de Hojas de WARMI* 17: 1–16.
Devine, Jennifer A. 2018. "Community Forest Concessionaires: Resisting Green Grabs and Producing Political Subjects in Guatemala." *The Journal of Peasant Studies* 45 (3): 565–584. doi:10.1080/03066150.2016.1215305
Douwe van der Ploeg, Jan. 2010. "The Peasantries of the Twenty-First Century: The Commoditisation Debate Revisited." *The Journal of Peasant Studies* 37 (1): 1–30. doi:10.1080/03066150903498721.
Ensor, Jonathan, John Forrester, and Nilufar Matin. 2018. "Bringing Rights Into Resilience: Revealing Complexities of Climate Risks and Social Conflict." *Disasters* 42: S287–S305. https://onlinelibrary.wiley.com/doi/abs/10.1111disa.12304. doi:10.1111/disa.12304
Eriksen, Siri H., Andrea J. Nightingale, and Hallie Eakin. 2015. "Reframing Adaptation: The Political Nature of Climate Change Adaptation." *Global Environmental Change* 35: 523–533. doi:10.1016/j.gloenvcha.2015.09.014.
Fash, Benjamin C. 2022. "Redefining Extractivism from Honduras." *Geoforum; Journal of Physical, Human, and Regional Geosciences* 135: 37–48. doi:10.1016/j.geoforum.2022.07.010.
Figueroa Romero, Dolores, and Miguel González Pérez. 2021. "Violencia política en los márgenes del Estado: Autonomía comunitaria afro-indígena en la Costa Atlántica de Nicaragua." *Canadian Journal of Latin American and Caribbean Studies/Revue Canadienne des études Latino-Américaines et Caraïbes* 46 (3): 449–469. doi:10.1080/08263663.2021.1970410
Flores, Selmira, and Sylvia Torres. 2012. "Ganaderas en la producción de leche: una realidad oculta por el imaginario social en dos zonas de Nicaragua." *Encuentro: Revista Académica de la Universidad Centroamericana* 44 (92): 7–28. doi:10.5377/encuentro.v0i92.790.
Folke, Carl, Reinette Biggs, Albert V Norström, Belinda Reyers, and Johan Rockström. 2016. "Social-ecological resilience and biosphere-based sustainability science." *Ecology and Society* 21: 3. doi:10.5751/ES-08748-210341.
Forsyth, Tim. 2018. "Is Resilience to Climate Change Socially Inclusive? Investigating Theories of Change Processes in Myanmar." *World Development* 111: 13–26. doi:10.1016/j.worlddev.2018.06.023.
Gahman, Levi. 2020. "Contra Plantation, Prison, and Capitalist Annihilation: Collective Struggle, Social Reproduction, and the co-Creation of Lifegiving Worlds." *The Journal of Peasant Studies* 47 (3): 503–524. doi:10.1080/03066150.2019.1572606.
Garcia, Alicea, Noémi Gonda, Ed Atkins, Godden Naomi Joy, Paiva Henrique Karen, Meg Parsons, Petra Tschakert, and Gina Ziervogel. 2022. "Power in Resilience and Resilience's Power in Climate Change Scholarship." *Wiley Interdisciplinary Reviews: Climate Change* e762: 1–21. doi:10.1002/wcc.762.
Giraldo, Omar Felipe, and Peter M Rosset. 2018. "Agroecology as a Territory in Dispute: Between Institutionality and Social Movements." *The Journal of Peasant Studies* 45 (3): 545–564. doi:10.1080/03066150.2017.1353496
Gómez Isa, Felipe. 2017. "The Decision by the Inter-American Court of Human Rights on the Awas Tingni vs. Nicaragua Case (2001): The Implementation Gap." *The Age of Human Rights Journal* 8: 67–91. doi:10.17561/tahrj.n8.4.
GoN. 2001. *Ley 407 que Declara y Define la Reserva de BOSAWAS*, edited by Nicaraguan Government. Managua, Nicaragua: La Gaceta.
GoN. 2003. *Ley No. 445. Ley del Regimen de Propiedad Comunal de los Pueblos Indigenas y Comunidades Etnicas de las Regiones Autonomas de la Costa Atlantica de Nicaragua y de los Rios Bocay, Coco, Indio y Maiz*, edited by Nicaraguan Government. Managua, Nicaragua: Gaceta Oficial No.16.
Gonda, Noémi. 2019. "Re-politicizing the Gender and Climate Change Debate: The Potential of Feminist Political Ecology to Engage with Power in Action in Adaptation Policies and Projects in Nicaragua." *Geoforum* 106: 87–96. doi:10.1016/j.geoforum.2019.07.020.
Gonda, Noémi. 2021. "Re-Negotiating Rural Masculinities as Vulnerability: Cattle Ranchers in Climate Change Affected Rural Nicaragua." In *Men, Masculinities, and Earth*, edited by Paul Pulé, and Martin Hultman, 289–307. Cham: Palgrave Macmillan.

Gonda, Noémi, Stephanie Leder, Marien González-Hidalgo, Linley Chiwona-Karltun, Arvid Stiernström, Flora Hajdu, Ildikó Asztalos Morell, Klara Fischer, Alin Kadfak, and Anna Arvidsson. 2021. "Critical Reflexivity in Political Ecology Research: How Can the Coronavirus Pandemic Transform Us Into Better Researchers?" *Frontiers in Human Dynamics* 3: 41. doi:10.3389/fhumd.2021.652968.

Gonda, Noémi, José Pablo Prado Córdova, Frédéric Huybrechs, and Gert T. Van Hecken. 2022. "Exploring the Love Triangle of Authoritarianism, Populism, and COVID-19 Through Political Ecology: Time for a Break-Up?" *Frontiers in Human Dynamics* 4. doi:10.3389/fhumd.2022.653990.

González-Hidalgo, Marien, and Christos Zografos. 2019. "Emotions, Power, and Environmental Conflict: Expanding the 'Emotional Turn' in Political Ecology." *Progress in Human Geography* 44 (2): 235–255. doi:10.1177/0309132518824б..

Gram-Hanssen, Irmelin, Nicole Schafenacker, and Julia Bentz. 2022. "Decolonizing Transformations through 'Right Relations'." *Sustainability Science* 17 (2): 673–685. doi:10.1007/s11625-021-00960-9.

Grove, Kevin, Savannah Cox, and Allain Barnett. 2020. "Racializing Resilience: Assemblage, Critique, and Contested Futures in Greater Miami Resilience Planning." *Annals of the American Association of Geographers* 110 (5): 1613–1630. doi:10.1080/24694452.2020.1715778

Guevara Flores, Ramón. 2017. "Estado de la deforestación y estadísticas socioeconómicas y ambientales recientes en las zonas ocupadas por colonos del el territorio indígena Mayangna Sauni As en BOSAWAS." *Wani* 72: 9–34. doi:10.5377/wani.v0i72.7436.

Haider, L Jamila, Maja Schlüter, Carl Folke, and Belinda Reyers. 2021. "Rethinking Resilience and Development: A Coevolutionary Perspective." *Ambio* 50 (7): 1304–1312. doi:10.1007/s13280-020-01485-8

Harris, Leila M., Eric K. Chu, and Gina Ziervogel. 2018. "Negotiated Resilience." *Resilience* 6 (3): 196–214. doi:10.1080/21693293.2017.1353196.

Inter-American Court on Human Rights. 2001. *Case of the Mayagna (Sumo) Awas tingni community vs. Nicaragua. Judgment of August 31, 2001*, edited by Inter-American Court on Human Rights.

IPCC. 2007. *Fourth Assessment Report of the Intergovernmental Panel on Climate Change*. Cambridge, UK; New York, USA: Cambridge University Press.

IWGIA. 2020. *Attack on Alal Goes Unpunished*. Accessed 22 October 2022. https://iwgia.org/en/nicaragua/4237-iw-2021-nicaragua.html#_ftn26.

Kaika, Maria. 2017. "'Don't Call me Resilient Again!': The New Urban Agenda as Immunology … or … What Happens When Communities Refuse to be Vaccinated with 'Smart Cities' and Indicators." *Environment and Urbanization* 29 (1): 89–102. doi:10.1177/0956247816684763

Kaimowitz, David, Angelica Faune, and Rene Mendoza. 2003. "Your Biosphere is my Backyard: The Story of Bosawas in Nicaragua." Vol. 25: Center for International Forestry Research.

Kautsky, Karl. 1988. *The Agrarian Question*. London: Zwan Publications.

Koshy, Susan, Lisa Marie Cacho, Jodi A Byrd, and Brian Jordan Jefferson. 2022. *Colonial Racial Capitalism*. Durham: Duke University Press.

Kröger, Markus, and Anja Nygren. 2020. "Shifting Frontier Dynamics in Latin America." *Journal of Agrarian Change* 20 (3): 364–386. doi:10.1111/joac.12354.

Larson, Anne M. 2010. "The "Demonization" of Rainforest Migrants, or: What Conservation Means to Poor Colonist Farmers." In *Beyond the Biophysical: Knowledge, Culture, and Politics in Agriculture and Natural Resource Management*, edited by Laura A. German, Joshua J. Ramisch, and Ritu Verma, 49–71. Dordrecht: Springer Netherlands.

Li, Tania Murray. 2014. *Land's End*. Durham: Duke University Press.

Maldidier, Christophe. 2004. "Agricultural Pioneer Fronts, the Crest of a Far-reaching Wave. The Social and Spatial Dimension of Lowland Colonization in Nicaragua." In *Beyond Tropical Deforestation. From Tropical Deforestation to Forest Cover Dynamics and Forest Development*, edited by Didier Babin. Montpellier: Unesco-Cirad.

Martí i Puig, Salvador, and Macià Serra. 2020. "Nicaragua: De-Democratization and Regime Crisis." *Latin American Politics and Society* 62 (2): 117–136. doi:10.1017/lap.2019.64

Matamoros-Chávez, Edwin. 2014. "Micropolíticas de campesinos colonos en territorios indígenas de Nicaragua." PhD, Latin American Studies, The University of Texas at Austin. http://hdl.handle.net/2152/28420.

Mehta, Lyla, Hans Nicolai Adam, and Shilpi Srivastava. 2019. "Unpacking Uncertainty and Climate Change from 'Above' and 'Below'." *Regional Environmental Change* 19 (6): 1529–1532. doi:10.1007/s10113-019-01539-y.

Mehta, Lyla, and Wendy Harcourt. 2021. "Beyond Limits and Scarcity: Feminist and Decolonial Contributions to Degrowth." *Political Geography* 89: 102411. doi:10.1016/j.polgeo.2021.102411.

Mills-Novoa, Megan, Rutgerd Boelens, Jaime Hoogesteger, and Jeroen Vos. 2022. "Resisting, Rejecting, and Reworking Climate Change Adaptation Projects in Ecuador." *Journal of Peasant Studies* 1–29. doi:10.1080/03066150.2022.2144252.

Moore, Donald S. 2005. *Suffering for Territory*. Durham: Duke University Press.

Nicaraguan Ministry of Finance and Public Credit. 2019. "Bio-CLIMA Nicaragua: Integrated climate action for reduced deforestation and strengthened resilience in the BOSAWÁS and Rio San Juan Biosphere Reserves." (Managua, Nicaragua). https://www.greenclimate.fund/sites/default/files/document/21710-bio-clima-nicaragua.pdf.

Nightingale, Andrea J. 2011. "Bounding Difference: Intersectionality and the Material Production of Gender, Caste, Class and Environment in Nepal." *Geoforum; Journal of Physical, Human, and Regional Geosciences* 42 (2): 153–162. doi:10.1016/j.geoforum.2010.03.004.

Nightingale, Andrea J. 2017. "Power and Politics in Climate Change Adaptation Efforts: Struggles Over Authority and Recognition in the Context of Political Instability." *Geoforum; Journal of Physical, Human, and Regional Geosciences* 84: 11–20. doi:10.1016/j.geoforum.2017.05.011.

Nightingale, Andrea J. 2018. "The Socioenvironmental State: Political Authority, Subjects, and Transformative Socionatural Change in an Uncertain World." *Environment and Planning E: Nature and Space* 1 (4): 688–711. doi:10.1177/2514848618816467.

Nightingale, Andrea J., Noémi Gonda, and Siri H. Eriksen. 2022. "Affective Adaptation = Effective Transformation? Shifting the Politics of Climate Change Adaptation and Transformation from the Status quo." *WIRES Climate Change* 13 (1): e740. doi:10.1002/wcc.740.

Nuijten, Monique. 2004. "Peasant 'Participation', Rural Property and the State in Western Mexico." *The Journal of Peasant Studies* 31 (2): 181–209. doi:10.1080/0306615042000224276.

Nygren, Anja. 2004. "Contested Lands and Incompatible Images: The Political Ecology of Struggles Over Resources in Nicaragua's Indio-Maíz Reserve." *Society and Natural Resources* 17 (3): 189–205. doi:10.1080/08941920490270221

Oakland Institute. 2020. "Nicaraguás Failed Revolution; The Indigenous Struggle for Saneamiento." Oakland, CA, USA: Oakland Institute. https://www.oaklandinstitute.org/sites/oaklandinstitute.org/files/nicaraguas-failed-revolution.pdf.

Ojha, Hemant, Andrea J. Nightingale, Noémi Gonda, Benard Oula Muok, Siri Eriksen, Dil Khatri, and Dinesh Paudel. 2022. "Transforming Environmental Governance: Critical Action Intellectuals and Their Praxis in the Field." *Sustainability Science* 17: 621–635. doi:10.1007/s11625-022-01108-z.

Olsen, BjØrnar. 2003. "Material Culture After Text: Re-Membering Things." *Norwegian Archaeological Review* 36 (2): 87–104. doi:10.1080/00293650310000650.

Olsson, Per, Folke Carl, and Moore Michele-Lee. 2022. "Capacities for Navigating Large-Scale Sustainability Transformations: Exploring the Revolt and Remembrance Mechanisms for Shaping Collapse and Renewal in Social-Ecological Systems." In *Applied Panarchy: Applications and Diffusion Across Disciplines*, edited by Lance H. Gunderson, Craig R. Allen, and Ahjond Garmestani. Washington, DC: Island Press.

Paprocki, Kasia. 2022. "Anticipatory Ruination." *The Journal of Peasant Studies* 49 (7): 1399–1408. doi:10.1080/03066150.2022.2113068.

Peet, Richard, and Michael Watts. 2004. *Liberation Ecologies: Environment, Development and Social Movements (2nd Edition). Edited by Richard Peet and Michael Watts.* London and New York: Routledge.

Reyers, Belinda, Michele-Lee Moore, L. Jamila Haider, and Maja Schlüter. 2022. "The Contributions of Resilience to Reshaping Sustainable Development." *Nature Sustainability* 5 (2022): 657–664. doi:10.1038/s41893-022-00889-6.

Rivera Cusicanqui, Silvia. 2010. "Ch'ixinakax utxiwa. Una reflexión sobre prácticas y discursos descolonizadores." Tinta limon.

Rockström, Johan, Will Steffen, Kevin Noone, Åsa Persson, F. Stuart Chapin, Eric F. Lambin, Timothy M. Lenton, et al. 2009. "A Safe Operating Space for Humanity." *Nature* 461 (7263): 472–475. doi:10.1038/461472a.

Santiago Vera, Teresita, Peter Michael Rosset, Antonio Saldívar Moreno, and Victor Ernesto Méndez. 2022. "Peasant Resilience: Decolonization and Re-Conceptualization." *Environmental Justice* 15 (3): 179–184. doi:10.1089/env.2020.0070

Sawyer, Suzanna. 2004. *Crude Chronicles: Indigenous Politics, Multinational Oil, and Neoliberalism in Ecuador*. Durham, NC: Duke University Press.

Sekine, Yukari. 2021. "Emerging 'Agrarian Climate Justice' Struggles in Myanmar." *The Journal of Peasant Studies* 48 (3): 517–540. doi:10.1080/03066150.2020.1839054.

Simpson, Audra. 2007. "On Ethnographic Refusal: Indigeneity,'Voice'and Colonial Citizenship." *Junctures: The Journal for Thematic Dialogue*, 9 December 2007. https://junctures.org/junctures/index.php/junctures/article/view/66/60.

Simpson, Leanne Betasamosake. 2017. *As We have always Done: Indigenous Freedom through Radical Resistance*. Minneapolis: University of Minnesota Press.

Stoler, Ann Laura (ed.). 2013. *Imperial Debris: On Ruins and Ruination*. Durham: Duke University Press.

Stone-Jovicich, Samantha, Bruce E. Goldstein, Katrina Brown, Ryan Plummer, and Per Olsson. 2018. "Expanding the Contribution of the Social Sciences to Social-Ecological Resilience Research." *Ecology and Society* 23: 1. doi:10.5751/ES-10008-230141.

Sultana, Farhana. 2022. "The Unbearable Heaviness of Climate Coloniality." *Political Geography* 102638. doi:10.1016/j.polgeo.2022.102638.

Sultana, Parvin, Paul Michael Thompson, Naya Sharma Paudel, Madan Pariyar, and Mujibur Rahman. 2019. "Transforming Local Natural Resource Conflicts to Cooperation in a Changing Climate: Bangladesh and Nepal Lessons." *Climate Policy* 19 (sup1): S94–S106. doi:10.1080/14693062.2018.1527678

Sylvander, Nora. 2021. "'Territorial Cleansing'for Whom? Indigenous Rights, Conservation, and State Territorialization in the Bosawas Biosphere Reserve, Nicaragua." *Geoforum; Journal of Physical, Human, and Regional Geosciences* 121: 23–32. doi:10.1016/j.geoforum.2021.02.013

Taylor, Lucy, and Geraldine Lublin. 2021. "Settler Colonial Studies and Latin America." *Settler Colonial Studies* 11 (3): 259–270. doi:10.1080/2201473X.2021.1999155.

Thompson, Kim-Ly, and Natalie C. Ban. 2021. ""Turning to the Territory": A Gitga'at Nation Case Study of Indigenous Climate Imaginaries and Actions." *Geoforum; Journal of Physical, Human, and Regional Geosciences* 137: 230–236. https://doi.org/10.1016/j.geoforum.2021.11.006.

Tuana, Nancy. 2013. "Gendering Climate Knowledge for Justice: Catalyzing a New Research Agenda." In *Research, Action and Policy: Addressing the Gendered Impacts of Climate Change*, edited by Margaret Alston, and Kerri Whittenbury, 17–31. Dordrecht: Springer Netherlands.

Tuhiwai Smith, Linda. 2013. *Decolonizing Methodologies: Research and Indigenous Peoples*. London: Zed Books.

Tynan, Lauren. 2021. "What is Relationality? Indigenous Knowledges, Practices and Responsibilities with kin." *Cultural Geographies* 28 (4): 597–610. doi:10.1177/14744740211029287.

Tzul Tzul, Gladys. 2018. *Gobierno comunal indígena y estado guatemalteco*. Guatemala: Instituto Amaq': edicionesbizarras.

Van Hecken, Gert, Vijay Kolinjivadi, Frédéric Huybrechs, Johan Bastiaensen, and Pierre Merlet. 2021. "Playing Into the Hands of the Powerful: Extracting "Success" by Mining for Evidence in a Payments for Environmental Services Project in Matiguás-Río Blanco, Nicaragua." *Tropical Conservation Science* 14: 19400829211020191. doi:10.1177/19400829211020191.

Wainwright, Joel, and Joe Bryan. 2009. "Cartography, Territory, Property: Postcolonial Reflections on Indigenous Counter-Mapping in Nicaragua and Belize." *Cultural Geographies* 16 (2): 153–178. doi:10.1177/1474474008101515

Watts, Michael. 2003. "Development and Governmentality." *Singapore Journal of Tropical Geography* 24 (1): 6–34. doi:10.1111/1467-9493.00140

Watts, Michael. 2021. "The Agrarian Question." In *Handbook of Critical Agrarian Studies*, edited by A. Haroon Akram-Lohdi, Kristina Dietz, Bettina Engels, and Ben M. McKay, 53–66. Cheltenham: Edward Elgar Publishing.

Whyte, Kyle. 2017. "Indigenous Climate Change Studies: Indigenizing Futures, Decolonizing the Anthropocene." *English Language Notes* 55 (1): 153–162. doi:10.1215/00138282-55.1-2.153

Whyte, Kyle. 2018. "Settler Colonialism, Ecology, and Environmental Injustice." *Environment and Society* 9 (1): 125–144. doi:10.3167/ares.2018.090109

Whyte, Kyle, Chris J Cuomo, S. M. Gardiner, and A. Thompson. 2016. "Ethics of Caring in Environmental Ethics." In *The Oxford Handbook of Environmental Ethics*, edited by Stephen M Gardiner, and Allen Thompson, 234. Oxford: Oxford University Press.

Wright, Erik Olin. 2019. *How to be an Anticapitalist in the Twenty-First Century*. London: Verso Books.

Ybarra, Megan. 2018. *Green Wars: Conservation and Decolonization in the Maya Forest*. Oakland: Univ of California Press.

OPEN ACCESS

Resisting, leveraging, and reworking climate change adaptation projects from below: placing adaptation in Ecuador's agrarian struggle

Megan Mills-Novoa, Rutgerd Boelens, Jaime Hoogesteger and Jeroen Vos

ABSTRACT

As climate change escalates, donors, international organizations, and state actors are implementing adaptation projects. Embedded within these adaptation projects are imaginaries of rural resilience. These imaginaries, however, are contested by individuals and collectives targeted by such initiatives. In this article, we draw on Foucault's notion of counter conducts to understand how beneficiaries in Ecuador resist, leverage, and/or rework adaptation interventions and towards what end. We identified five counter conducts: (1) negotiating for control, (2) setting the terms for participation, (3) opting out, (4) subverting the discursive frame, and (5) leveraging longevity. We argue that these counter conducts are generative, enacting multi-scalar counter-hegemonic politics of agrarian transformation.

1. Introduction

Agrarian landscapes and the people who live there are subject to both the impacts of climate change as well as the proliferation of climate change adaptation initiatives. Since the early 2000s, climate change adaptation has become a major focus of policy and rural development projects throughout the Global South through national and regional planning processes as well as concrete projects (Berrang-Ford et al. 2021). These initiatives are often funded through bilateral or multi-lateral climate funds with donations from historical greenhouse gas emitters such as the United States or the European Union. As of 2020, multi-lateral climate funds had approved US$5.8 billion for adaptation (Watson and Schalatek 2021a). In Latin America, US$480 million has been approved for 71 adaptation projects between 2003 and 2020 (Watson and Schalatek 2021b).

Adaptation projects aim to reform or transform agrarian livelihoods and landscapes in response and/or anticipation of climate changes and their effects (IPCC 2007). But the question of what type of agrarian future will be most resilient[1] under climate change is not agreed on and resonates with broader debates about what desirable agrarian transformation entails (Akram-Lodhi and Kay 2010; Paprocki 2018a; Camargo 2022). Project promoters sidestep these contested topics by justifying adaptation interventions based on the urgent nature of the climate crisis (Paprocki 2018b) while advancing an often technocratic rural imaginary that dictates what forms of agricultural production and agrarian lifeways are 'resilient' and need support (Mills-Novoa et al. 2020). Towards this end, climate change adaptation projects restructure access to and use of resources (Borras, Franco, and Nam 2020; Mikulewicz 2021; Borras et al. 2022).

To date, adaptation projects have predominantly made incremental changes that adjust or reform existing systems (Bassett and Fogelman 2013; Berrang-Ford et al. 2021). Beyond their incremental nature, many adaptation projects have explicitly reinforced capitalist forms of agrarian development through interventions such as promoting agro-export development (Mills-Novoa 2020), increasing the agricultural efficiency, productivity, and market access for 'vulnerable' agrarian communities (Zografos 2017; Funder, Mweemba, and Nyambe 2018; Duarte-Abadía, Boelens, and Buitrago 2021; Eriksen et al. 2021), and/or expropriating land and water as part of powerful regimes of accumulation (Borras and Franco 2018; Paprocki 2018a). Among these more capitalist-oriented adaptation approaches, adaptation relies strongly on the 'adaptive and self-organizing capacities of the market above all else' (Watts 2015, 40).

The logic of adaptation projects and the embedded view of agrarian political economy, however, is not always coherent or overtly capitalist. In many cases, adaptation finance and related projects are positioned as 'conciliatory' actions that meld market power with Indigenous or local knowledges and capacities to help these communities endure turbulence wrought by climate change and related forms of capitalism (Abya Yala 2009, 1; Watts 2015; Khan et al. 2019).

Within the literature on climate change adaptation projects, scholars have primarily focused on the project of rule, missing the political contests, negotiations, collaborations, feigned compliance, and refusals that arise on-the-ground. When the focus remains primarily on the strategies and intentions of project promoters, so-called beneficiaries can easily be understood as communities and individuals to whom adaptation projects happen (Keskitalo, Juhola, and Westerhoff 2012; Beck 2017). This conceptualization of beneficiaries, however, obscures the creative, plural ways in which beneficiaries oppose adaptation projects and how their opposition articulates with and builds on broader

[1]While adaptation projects are often framed in terms of increasing resilience to climate change, resilience and adaptation have important differences. Rooted in socio-ecological systems theory (Holling 1974), resilience has been heavily critiqued for its vagueness and depoliticized focus on 'bouncing back' without acknowledging that the starting state may be highly unequal and that the large-scale political, economic and social forces being preserved are precisely the ones that historically produced rampant inequality (Mikulewicz 2019). Adaptation is a broader concept, which encompasses both radical concepts of transformative adaptation that reconsider the political economy of and power relations that produce climate change, as well as more incremental approaches that reinforce the status quo, and transitional adaptation that lies in between (Pelling 2011; O'Brien 2012; Bassett and Fogelman 2013; Watts 2015).

agrarian struggles for territorial control, food sovereignty, or anti-extractivism (McMichael 2008; Borras and Franco 2018; Sekine 2021; Duarte-Abadía 2022).

This article contributes to the JPS forum on 'climate change and critical agrarian studies' by studying the counter-hegemonic politics of climate change adaptation through an examination of how these projects are negotiated on-the-ground and towards what end (van der Ploeg 2008; Tria Kerkvliet 2009; Boelens 2015). We do so through the Foucauldian notion of counter-conducts, which enables us to focus on an analysis of power from the bottom-up reflecting the 'micropolitics' of adaptation (Foucault 2008; Cadman 2010; Dean 2010). In understanding adaptation initiatives as an emerging frontier for advancing agrarian transformation, we use counter conducts to understand the tendencies and processes through which rural relations of domination and subordination are being transformed through varied forms of opposition (Scott 1985; Byres 1991; Bernstein 1996; Akram-Lodhi and Kay 2010). To better understand the strategic logics underlying the identified counter conducts, we draw on Katz's (2004) typology of resistance to capitalism.

To this end, we look across the seven internationally funded projects that constitute Ecuador's adaptation project portfolio to understand if, how and why the so-called beneficiaries resist, leverage, and/or rework adaptation projects. Ecuador is a timely case to study agrarian climate change adaptation projects. As a country, Ecuador has been highly strategic and proactive in accessing international adaptation finance, having received projects from every fund for which they are eligible. The individuals and communities that participated in these climate change adaptation projects as 'beneficiaries' are heterogenous but were universally deemed vulnerable by project implementors due to their reliance on land-based livelihoods and exposure to climatic hazards.

Central to our analysis is an effort to understand how the counter conducts intertwine and relate to broader anti-capitalist social mobilization and agrarian/Indigenous movements in Ecuador. Thus far, climate change adaptation has remained peripheral to large-scale social mobilizations around agrarian or even climate justice. However, we argue that local negotiations of climate change projects should not be understood in isolation, but rather, as part and parcel of broader agrarian justice movements (Colloredo-Mansfeld 2009; Velásquez 2022).

We argue that the counter conducts employed by people negotiating adaptation projects are generative – producing and enacting a counter-hegemonic micropolitics of agrarian transformation that reflect and intertwine with broader goals of grassroots struggles to carve out their own (agrarian) futures (Sekine 2021). By focusing on the imaginaries furthered through bottom-up resistance to adaptation projects, we can understand agrarian transformations more expansively, not just as a material process borne out of class struggle, but also as the product of the 'agencies, desires, and expectations of those who experience climate disasters and adaptation policies in their everyday life' (Camargo 2022, 714).

As the basis for this article, we paired an institutional ethnography of adaptation project governance with multi-sited, community-based research. We conducted 96 semi-structured interviews with adaptation professionals and community leaders, participant observation in communities that participated in one of Ecuador's seven internationally funded adaptation projects, participatory mapping of project sites, and a systematic review of adaptation project documentation. This data was collected by the first author over 12 months of fieldwork in Ecuador (July–August 2018, October 2018–February 2019, and August–December 2019).

The article is structured as follows. We begin by discussing how counter-conducts, as a theoretical lens, provide insight into how climate change adaptation projects are negotiated on-the-ground. Next, we examine the rise of adaptation projects in Ecuador and their entanglements with historical struggles for agrarian justice. We then present the five counter-conducts identified in our research. For each of these counter conducts, we review the dominant rationalities advanced by project funders and implementors as well as the alternative imaginaries of climate change adaptation being advanced by beneficiaries, while also relating these to broader agrarian movements and demands in Ecuador. We conclude by discussing how these identified counter conducts relate to broader struggles for agrarian justice and the possibilities they create for shaping the future of agrarian Ecuador and climate change adaptation.

2. Theoretical framework

2.1 Climate change adaptation governmentalities

When analyzing climate change adaptation projects through the Foucauldian lens of governmentality, these initiatives employ different strategies that aim to conduct-the-conduct of beneficiaries (Oels 2005; Dowling 2010). This is often advanced by project implementors that assert that the entrenchment of capitalist forms of development in the rural agrarian world is *the* means by which to create climate-resilient subjects in an increasingly globalized economy (Bäckstrand and Lövbrand 2006; Borras et al. 2022). This is done through the creation of new socio-economic, political and cultural relations in and among beneficiaries and their broader socio-environmental networks (Foucault 1991, 1991, 2008; Dean 2010).

As analyzed in Mills-Novoa et al. (2020), climate change adaptation projects need and thus create objects (landscapes) and 'vulnerable' subjects (beneficiaries) to justify and target their interventions. Distinct knowledge practices and government techniques (sovereign, disciplinary and neoliberal governmentalities) are employed by project implementors to create an object for intervention in adaptation projects (Foucault 2008; Fletcher 2017; Hommes et al. 2020). Relatedly, the project requirements, recognition politics and associated participatory mechanisms that are enacted simultaneously produce new 'adaptive' and 'resilient' subjects that fit with the adaptation truth regime. In other words, these new beneficiary subjects are instrumental to and constitutive of project implementors' vision of agrarian development and related climate change adaptation (Mosse 2005; Beck 2017). For this, projects need beneficiaries to frame their needs, worldviews, relationships, and approaches in the context of climatic risk, exposure, resilience, vulnerability and adaptive capacity (Li 2007; Eriksen, Nightingale, and Eakin 2015; Webber 2016; Usón, Henríquez, and Dame 2017).

Though beneficiaries are central in the discourse and justification of climate change adaptation projects, their presence in project documentation is peripheral. While analyzing the documentation of Ecuadorian adaptation projects, one only sees fleeting glances of the beneficiary. In climate change adaptation projects' public transcripts such as evaluations, promotional materials, and reports, beneficiaries are enumerated with evaluation references such as '78 drip irrigation systems' (GEF-PNUD 2015, 53). Occasionally, beneficiaries are described as able bodies for communal workdays in progress reports where implementors collaborated 'with human resources in the construction of adequate infrastructure' (ibid, 12). Beneficiaries are sometimes made strategically visible as glossy photos attached to a glowing testimonial. Despite these myriad appearances, in all these cases beneficiaries tend to be portrayed as 'ideal' and 'unproblematic' subjects that unquestionably fit, benefit from, and therefore are in dear need of the projects. This understanding and portrayal of beneficiaries, however, hides, denies, and ignores the agency of the individuals, families and communities that leverage, and at other times resist or rework, an adaptation project through quotidian struggles and engagements in and with these initiatives.

2.2 Negotiating adaptation projects from below through counter-conducts

When beneficiaries consent to a climate change adaptation project, in practice they negotiate and reshape the project to fit their own needs and imaginaries of climate change adaptation (de Certeau 1984; Long and van der Ploeg 1989; Mosse 2004; Camargo 2022). They are not naïve subjects (as often portrayed in project documentation), but rather, active agents that shape and reshape agrarian futures in- and through climate change adaptation projects (Liverman 2015). They do so through strategies such as noncompliance (e.g. Wyborn et al. 2014), ambivalence (e.g. Camargo and Ojeda 2017), fraught negotiation (e.g. Caretta and Borjeson 2015), overt conflict (e.g. Eriksen and Lind 2017), and political use of climate change policies to strengthen territorial and resource control (Funder, Mweemba, and Nyambe 2018). Within adaptation scholarship, opposition has primarily been examined in the context of contentious resettlements in flood-prone or coastal areas (e.g. Broto 2020; Arnall 2018; Henrique and Tschakert 2019; Paprocki 2019).

Opposition to climate change adaptation projects is often positioned as being a reactive, opposing force *against* the imposed dominant rationalities and practices of project implementors (Death 2010; Lindegaard 2016). This simplistic view of opposition, however, does not capture the negotiation, sometimes subtle and sometimes stark, between governments, donors, NGOs, consultants, local communities, and individual beneficiaries within these multi-faceted, multi-scalar projects (Mosse 2004; Horowitz 2011; Rasch and Köhne 2016). There is a tight interrelationship between the techniques of government and the resistance they elicit (Boelens and Gelles 2005; Pieck 2015; Hoogesteger, Boelens, and Baud 2016). Subaltern groups are subject to cross-cutting fields of force even as they resist. This means that there is no autonomous originary space for opposition, but rather it is shaped by and shapes identities, place, and the very terrain of the struggle (Gramsci 1971; Moore 1998; Fletcher 2001).

So how do we understand these 'revolts of conduct' (Foucault 2009, 196)? These acts of opposition are fundamental responses to the power exercised via governmentalities and raise various fundamental questions, which include 'By whom do we consent to be directed? How do we want to be conducted? Toward what do we want to be led?' (ibid, 197). When discordance exists between adaptation project beneficiaries and implementors around these fundamental questions, 'counter conducts' arise as diverse forms of opposition. In many cases, counter conducts may not register as explicitly rejecting government or even as being political in nature (Scott 1998; Odysseos, Death, and Malmvig 2016). Counter conducts, defined as 'the will not to be governed thusly, like that, by these people, at this price,' do not exist outside governmentality but are 'wholly immanent and necessary to the formation and development of governmentality' (Cadman 2010, 540; Foucault 2008, 75).

Counter conducts, themselves, form particular subjectivities and identities through the act of dissent itself (Death 2010). Thus, practices are fundamental, whether consciously deployed or not, because they generate or dissociate forms of identification and belonging to particular knowledge regimes. Dissenting subjectivities are formed by transforming 'one's relation to oneself and others' and represent interventions by groups or individuals into 'both the ethical and political practices and forces that shape us' (Davidson 2011, 32).

Resistance to adaptation can happen on varied terrains of struggles, towards multiple imaginaries, and with diverse tools. Resistance to adaptation can be found in quotidian negotiations over project control (Nightingale 2017) or transnational movements against the commodification of environmental services (Dupuits et al. 2019). As an example of how counter conducts arise and are advanced, Valladares and Boelens (2019), investigated the mining conflict in Ecuador's Quimsacocha highlands through four governmentalities – truth, sovereign, neoliberal and disciplinary governmentality (Foucault 2008, 313; Fletcher 2017). They found repressive mining policy strategies advanced by the state mingled with subtler, capillary powers that attempted to assign identities to human and non-human subjects and set boundaries on peasant political participation. These governmentalities, however, were met with opposition when inhabitants of Quimsacocha defended their territories against mining by disputing each of the four governmentalities enacted by the state to govern them. Valladares and Boelens highlight that the counter-conducts of people in Quimsacocha, 'occur in formal political and legal settings, but above all in communities' day-to-day practices' (Valladares and Boelens 2019, 77). These everyday practices of opposition were also highlighted by Xu, Boelens, and Veldwisch (2022) in a Chinese case in which the state water governmentality scheme was profoundly and permanently contested not through coordinated bottom-up resistance but through ongoing misalignment of institutional structures and interests and actions both among state, local villagers and among villager groups themselves. Consequently, intervention projects were 'continually negotiated through people's everyday practices, which went far beyond and were far more complex than overt resistance or covert weapons of the weak' (Xu, Boelens, and Veldwisch 2022, 9). As these cases exemplify, counter conducts are as multiple and innovative as the varied practices employed by powerful actors to forward their projects.

Within agrarian climate change adaptation projects in Ecuador or elsewhere, not all counter conducts are explicitly anti-capitalist or oppositional. Camargo (2022) found that peasants in Northern Colombia rework and reimagine climate change adaptation, but toward their own vision of an agrarian future in which they act as rural entrepreneurs who actively shape the development of capitalism. In this case, imaginaries of climate change adaptation are remade but not toward an anti-capitalist alternative. Additionally, some counter conducts may not be explicitly oppositional, meaning that those employing the counter conducts are not consciously disrupting or subverting systems of oppression (Benjamin 1978).

Furthermore, in recognizing the open and sometimes implicit nature of opposition, it is important not to exaggerate or romanticize resistance (Abu-Lughod 1990; Horowitz 2011) particularly when examining subtle forms of everyday resistance (Scott 1985). It is therefore important to provide analytical precision to the 'strategic logics' that underlie resistance to adaptation. Such examination enables insight into critical questions regarding the degree to which contemporary agrarian movements have internalized (or not) 'climate change politics as a key context for and object of political struggle' (Borras et al. 2022, 8).

To understand opposition to climate change adaptation projects, Katz's typology of resistance helps to differentiate the strategic logics underlying the counter conducts arising in response to climate change adaptation projects. Cindi Katz distinguishes between resilience, reworking and resistance. *Resilience* refers to autonomous acts that help sustain individuals or communities as they endure capitalism. In this article, we refer to this as 'leveraging' adaptation projects to avoid confusion with mainstream understandings of climate change resilience. *Reworking* entails acts where individuals or collectives seek to alter oppressive and unequal conditions rather than endure them. They seek pragmatic strategies for recalibrating power relations or material distribution, 'retooling themselves as political subjects and social actors' (Katz 2004, 247). These acts can be successful at reordering or occasionally undermining the structural constraints of everyday life under capitalism. *Resistance* entails people seeking to subvert and disrupt the underlying conditions of exploitation and oppression. Resistance requires an 'oppositional consciousness' wherein individuals recognize social relations of power and have a vision for change and utterly different social relations. The categories of leveraging, reworking, and resisting are not exclusionary but overlapping.

Katz's typology for resistance provides analytical precision to examinations of counter conducts, and responds to some critiques of Foucault's theorizations. Foucault has often been criticized for being 'totalizing' and 'power-deterministic'. Contrary to his critics, however, Foucault's concept of power is not that it is 'all-powerful' but 'infinite' (Gordon 1991, 47). Though not always consistently, he elaborated on subjects' resistance to normalizing powers while maintaining the fact of their embeddedness in governmentalizing webs. Subjects are subjectified: simultaneously becoming active subjects and subjected actors. Thereby, actors are not simply determined by powers that cannot be influenced or understood by those who are subject to those powers. Foucault, non-deterministically, remarked that 'there is no power without potential refusal or revolt' (Foucault 1988, 84): but *not* in independent, autonomous ways. By drawing on counter conducts, we can understand resistance as bound together and mutually constitutive with the project of rule in climate change adaptation projects.

Based on these notions, we draw on Katz's typology to better understand the strategic logics underlying the counter conducts we identified and examine in this article namely: (1) negotiating for control over project governance, (2) setting the terms for participation, (3) opting out, (4) subverting the discursive frame of climate change, and (5) leveraging longevity. As we show in our analysis, oppositional acts can integrate various strategic logics (conscious and unconscious) over time, parts of the struggle and at different scales. Therefore important scalar interrelations are at play between the studied 'local' negotiation of adaptation projects and the broader agrarian/Indigenous movements in Ecuador (see Andolina, Laurie, and Radcliffe 2009; Hoogesteger and Verzijl 2015; Hoogesteger et al. 2017; Colloredo-Mansfeld et al. 2018; Velásquez 2022).

3. Climate change adaptation projects and their agrarian entanglements in Ecuador

Since 2009, Ecuador has received seven large-scale adaptation projects from the large multi-lateral climate funds as well as three smaller Green Climate Fund-funded readiness grants for adaptation planning and capacity building, and some limited bilateral adaptation investment from the German Corporation for International Cooperation (GIZ) and Japan International Cooperation Agency (JICA) (Table 1). For this study, we focus on the seven adaptation projects that constitute Ecuador's portfolio of multilaterally funded adaptation projects because these projects are the largest in scale, ambition and finance in the country, and reflect broader trends in adaptation projects. Importantly, the same multi-lateral climate funds and international organizations active in Ecuador are presently funding and shaping climate change adaptation initiatives and their governance across the Global South (Figure 1).

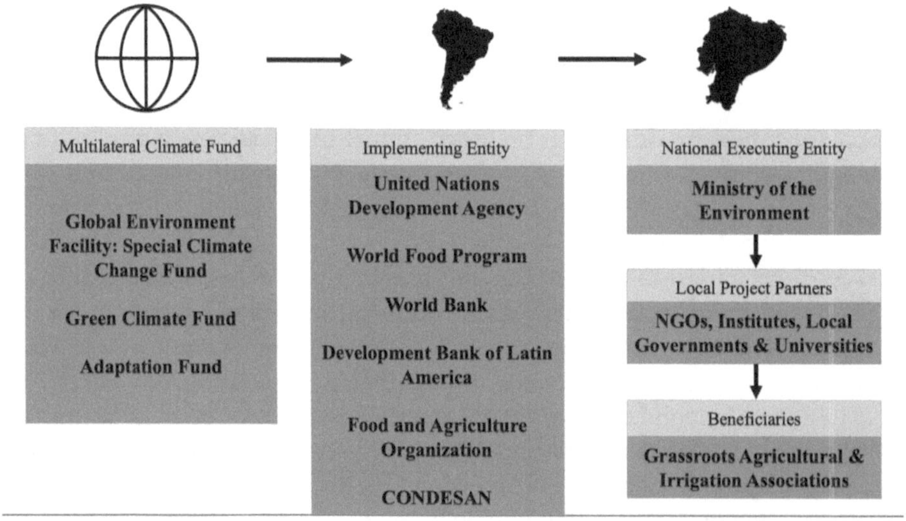

Figure 1. Simplified model of multi-scalar governance of climate change adaptation projects from the global to regional to local scale.

In Ecuador, the bureaucracy governing climate change adaptation finance solidified in direct response to the country's growing portfolio of projects. The rise of climate finance occurred in the context of Rafael Correa's *Revolución Ciudadana* (Citizens' Revolution, 2007–2017), which sought to reassert the role of the state in environmental governance and international development aid. Under Correa, Ecuador built a national policy and organizational architecture to attract international climate dollars. Ecuador's 2008 Constitution, which famously recognized the legal Rights of Nature, explicitly addresses climate change in Article 414 where the Ecuadorian State commits to 'protect at-risk populations' and employ 'adequate, transversal measures' to mitigate climate change (del Ecuador 2008). To bolster the role of the state in climate change governance, then-President Correa issued executive decree 1819 in 2009, creating the Sub-Secretary of Climate

Table 1. Internationally funded climate change adaptation projects in Ecuador.

Project name	Years	Climate fund	Implementing entity	Beneficiaries	Targeted sectors
PACC: Proyecto de Adaptación al Cambio Climático a través de una efectiva gobernabilidad del agua en Ecuador	2008–2015	GEF SCCF	UNDP	4,455 direct beneficiary families and 28,983 indirect individuals	Agro-ecology, sustainable livestock, water saving, efficient irrigation systems, irrigation storage and reforestation
PRAA: Proyecto Regional Andino de Adaptación al Cambio Climático	2008–2014	GEF SCCF	World Bank	Approx. 40,000 direct beneficiaries	Glaciers and conservation of highland wetlands (páramos)
FORECCSA: Proyecto Fortalecimiento de la Resiliencia de las Comunidades ante los efectos adversos del Cambio Climático con énfasis en Seguridad Alimentaria	2011–2018	Adaptation Fund	World Food Programme	19,356 direct beneficiaries and 60,000 indirect beneficiaries	Water, food security, and agriculture
AICAA: Proyecto de Adaptación a los Impactos del Cambio Climático en Recursos Hídricos en los Andes	2018–2021	GEF SCCF	Development Bank of Latin America (CAF)/ CONDESAN	No stated number of expected beneficiaries	Natural resources, hydropower, and conservation
Proyecto Binacional: Construyendo capacidades adaptativas al Cambio Climático a través de la seguridad alimentaria y acciones nutricionales en comunidades afro e indígenas en la zona fronteriza Colombia-Ecuador	2017–2022	Adaptation Fund	World Food Program	Expected 10,144 female and 9,724 male direct beneficiaries	Climate change risk reduction and food security through ancestral knowledge
Proyecto Toachi Pilatón: Aumentar la capacidad de adaptación de las comunidades locales, los ecosistemas y los sistemas hidroeléctricos en la cuenca alta del Río Blanco	Approved in 2018, I 2020–2024	Adaptation Fund	Development Bank of Latin America (CAF)/ FAO	553 direct beneficiary families and 49,367 individual indirect beneficiaries	Water resources and hydropower
Proyecto Regional Chile – Ecuador: Reducir la vulnerabilidad climática y el riesgo de inundación en áreas urbanas y semiurbanas costeras en ciudades	Approved in 2018, 2020–2023	Adaptation Fund	Development Bank of Latin America CAF/ UNDP	700 direct beneficiary families/population of Esmeraldas as indirect beneficiaries	Early warning systems and disaster risk management

Change within the Ministry of the Environment (MAE).[2] This office was created with the explicit goal of capturing resources from the Kyoto Protocol's Clean Development Mechanism and other climate finance sources (Article 2). At the time, Ecuador was already implementing its first set of pilot adaptation projects and Correa's administration recognized the potential of leveraging these dollars for national development priorities. The Sub-Secretary of Climate Change, which was subsequently divided into two offices, one for adaptation and the other for mitigation, is responsible for the coordination of all climate change adaptation and mitigation activities in the country.

MAE is the locus for all adaptation projects in Ecuador. MAE is recognized by multilateral climate funders as the national focal point therefore it represents the national interest in choosing projects, designing initiatives, and implementing funded proposals (Figure 1). The centrality of the state within the administration of climate change adaptation projects means that the projects promoted through these initiatives map closely onto the state's broader politics of agrarian development.

Within Correa's government, there was a highly contested discourse around agrarian futures and food sovereignty. During his rise to power, Correa galvanized a large coalition of agrarian and Indigenous movements with promises of combating hegemonic transnational neoliberalism and reasserting national food sovereignty. Prior to Correa's election in 2006, he signed an agreement with the *Mesa Agraria*, which was a coalition of four prominent peasant/Indigenous organizations. In this agreement he committed to initiating an 'agrarian revolution' based on the tenets of food sovereignty. This revolution would entail the democratization of land and water access and state investment in the revitalization of the 'peasant economy' (Giunta 2014; Henderson 2017; Tilzey 2019). This agrarian revolution, however, never occurred. Correa was unwilling to challenge capitalist social-property relations and never delivered a more radical version of food sovereignty. Instead, he invested in reformist initiatives in small farm productivity and social welfare programs through which he sought to neutralize calls for land and resource redistribution whilst exempting the landed oligarchy from any version of agrarian reform (Herrera Revelo 2017; Tilzey 2019). Importantly, Correa's government funded their social welfare programs and agricultural development initiatives through the extraction of mineral and fossil fuel resources as well as agri-fuels. The contradictions of Correa's post-neoliberal Ecuador – 'earth-extractivism to save Mother Earth' (Valladares and Boelens 2019), wherein explicitly anti-capitalist discourse cloaked social repression and deepening resource exploitation, reflect the evolving capitalisms (and resultant agrarian struggles) visible in South America and beyond (Radcliffe 2012; Silveria et al. 2017; Baud, Boelens, and Damonte 2019).

Adaptation projects in Ecuador have largely focused on agrarian livelihoods and production systems, reconfiguring production through efficient water management, value chain enhancement, and agro-ecology (Table 1). These projects and their constitutive agrarian imaginaries reflect the Ecuadorian state's co-option of food sovereignty, which

[2]When the fieldwork for this paper was undertaken by the first author (2018–2019), the Ministerio de Ambiente (Ministry of the Environment, MAE) was the national executing entity for adaptation projects. MAE, however, has since gone through substantial reorganization. In March 2020, MAE was merged with the Secretariat of Water (SENAGUA) and its name was changed to Ministerio del Ambiente y Agua (Ministry of the Environment and Water, MAEA). In June 2021, MAEA was once again rebranded and became Ministerio del Ambiente, Agua y Transición Ecológica (Ministry of Environment, Water, and Ecological Transition, MAATE). We refer to this Ministry as MAE throughout this article.

has ultimately translated to national food provisioning by productivist means such as expanding agro-exports and value chain enhancement (Tilzey 2019). The dominant imaginaries advanced through adaptation projects in rural Ecuador sideline more radical notions of food sovereignty advanced by peasant and Indigenous movements that seek a post/alternative development model of cooperative social relations founded in the concept of a good life (*buen vivir* or *Sumak kawsay*) (Giunta 2014; Vergara-Camus 2014; Tilzey 2017).

Following Correa's tenure, President Lenin Moreno's government undertook an aggressive austerity campaign to make-up for shortfalls in government revenue due to persistently low oil prices and to satisfy the conditions of a US$4.2 billion loan that Ecuador received from the International Monetary Fund in 2019. These sweeping austerity measures included a 70% cut in MAE's budget between 2015 and 2019. In the face of these budget reductions, the Sub-Secretary of Climate Change has remained relatively insulated because most of its budget comes from international funding. As national environmental and social initiatives have been hollowed out in the name of austerity, internationally funded projects are the sole adaptation initiatives occurring in Ecuador.

In Table 1, we present the seven internationally funded climate change adaptation projects that have been implemented in Ecuador at the time of data collection (see Mills-Novoa et al. (2020) for a detailed examination of two of these projects and their government rationalities). These projects were implemented in varied regions of Ecuador (Figure 2). When speaking with adaptation project implementors and national peasant and Indigenous organizations, we examined dynamics across Ecuador's portfolio of projects. We also analyzed three projects in greater depth: FORECCSA, PACC, and the Binational project. Both FORECCSA and PACC were multi-sited but had the greatest number of beneficiaries in mestizo and Indigenous agrarian communities in the Ecuadorian highlands (Figure 2). During the time of fieldwork, the Binational project in northeastern Ecuador was just beginning and explicitly targets Indigenous and Afro-descended communities on the Colombia/Ecuador border.

In the following sections, we discuss the five counter conducts that we observed across Ecuador's portfolio of climate change adaptation projects.

4. Adaptation counter conducts

4.1 The empowered beneficiary: negotiating for control (resisting adaptation)

Within the dominant imaginary of climate change adaptation projects, the local communities are project partners who participate in projects in prescribed fora, but the state and international agencies retain financial and administrative control over projects and their interventions. Within Ecuador, however, communities are actively negotiating their role in projects to assert greater control in determining the nature of the initiative (Hoogesteger 2013, 2015). Through this counter-conduct, communities seek to transcend the role of beneficiary. In doing so, these communities question the legitimacy of the state as the key interlocutor for adaptation finance and advance their own imaginaries in the process.

Within Ecuador, this negotiation is exemplified by a new multi-lateral adaptation project in which the Grand Family Awá, the Network of Southern Pacific Community councils, and the Afro-Ecuadorian Confederation of Northern Esmeraldas have fought to govern the project by negotiating directly with international agencies and funders. The

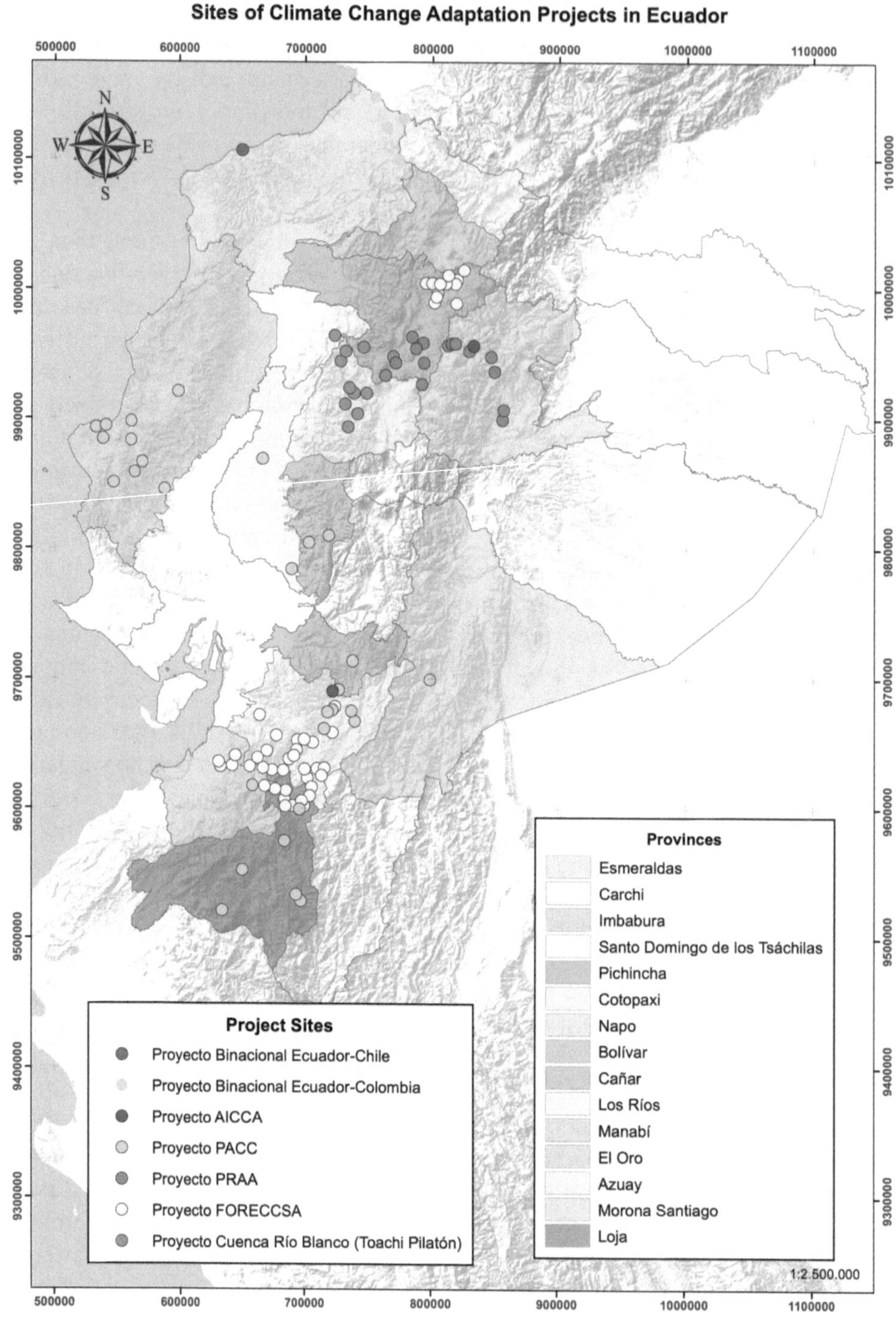

Figure 2. Map of project sites across Ecuador's portfolio of internationally funded climate change adaptation projects.

14-million-dollar project, *Building adaptive capacity to climate change through food security and nutrition actions in vulnerable Afro- and Indigenous communities in the Colombia-Ecuador border areas* which is commonly referred to as the 'Binational' project was developed by the World Food Programme (WFP) and funded by the Adaptation Fund (Table 1).

Unlike previous adaptation projects in the country, project implementors had to get approval from the general assemblies and leaders of these groups to develop and implement the project. Though this project was still in its nascent stages when the first author conducted fieldwork, the Grand Family Awá, which is highly distrustful of the state, had already asserted control in the project by requiring that all significant decisions be presented by WFP to the general assembly for approval (Binational project staff member, Interview, 22 February 2019). Furthermore, the Afro-Ecuadorian and Indigenous organizations have positioned themselves as project executors meaning that they implement the project in their territory and receive resources directly from WFP.

MAE retains a role on the national steering committee, but it has been relegated to the role of advisor. When asked about MAE being sidelined in the Binational project, a former director of MAE's Adaptation Directorate blamed WFP for this, 'We fought until the end and with all the allies of MAE so that this couldn't happen, at the end of the day it is against the sovereignty of the country. We were fighting very strong for the role of the state. The role of the state is above any implementing agency' (Interview, 26 February 2019). WFP's privileging of the sovereignty of the Afro-Ecuadorian and Indigenous project counterparts reflects the growing sensitivity of funders and international agencies to issues of Indigenous sovereignty.

The case of the Binational project reflects how some local actors are countering the state's role in adaptation by asserting a central role in the execution of projects to advance their imaginary of climate change adaptation. In the early stages of the project, the Grand Family Awá and the Network of Southern Pacific Community have reoriented this 'food security and nutrition' themed adaptation project to focus on their imaginary of climate change adaptation including recovering traditional practices and native species with community leaders and members identifying and collecting plants, holding seed fairs, and developing ancestral community-based disaster risk management practices (World Food Programme 2020). These adaptation measures point toward a different vision of adaptation than that being propagated through more top-down adaptation projects in the country where the focus is on improving irrigation efficiency, increasing agricultural productivity, and enhancing value chains. The climate change adaptation imaginary advanced by the Grand Family Awá and the Network of Southern Pacific Communities reflect both an alternative vision of project control and governance as well as material interventions that reflect broader efforts by CONAIE for Indigenous sovereignty, bicultural education, and food sovereignty (Pacari 2020).

While WFP remains a key interlocutor between the Adaptation Fund and these project partners, it is clear the Grand Family Awá and the Network of Southern Pacific Communities councils have employed an effective counter-conduct via direct negotiation to sideline the state and gain greater control over the Binational project. This counter-conduct reflects an overt oppositional consciousnesses where Indigenous and Afro-descended communities actively deny the legitimacy of the state and gradually seek to assume state power through administering their adaptation project thus asserting greater control over their territory, resources, and the nature of adaptation. These political fights over project control are part of CONAIE and its members' long legacy of challenging the state and demanding respect for Indigenous territories and ways of life (Akchurin 2015).

4.2 The conditional beneficiary: setting the terms for being governed (leveraging adaptation)

While the final project evaluations report that FORECCSA benefited 12,585 families and PACC benefited 4,455 households, there exists substantial differences in how those beneficiaries embrace, conditionally accept, and/or abandon adaptation interventions. As summarized by a MAE adaptation official:

> There is 50/50 people that are empowered and have been able to carry forward [agricultural] production for themselves. But there are other people, other beneficiaries that are used to paternalism. When the project ends and there aren't more resources, they leave the initiatives that they participated in on the side. (Interview, 9 November 2019)

This view of beneficiaries is common among project implementors. Namely, that while some beneficiaries are transformed into emerging entrepreneurs, adaptive agriculturalists, and/or resilient irrigators, other beneficiaries will not occupy those subject positions. Or these beneficiaries will only inhabit those roles while the project is ongoing and funds continue to roll into the community. This ambivalence and conditional acceptance of adaptation projects, however, reflects another counter-conduct, where beneficiaries only inhabit the beneficiary subject position if they feel that the project implementors and the state is fulfilling its obligations. Far from becoming an idealized *homo economicus*, fully embodying the rational adaptive, productive agricultural positionality, these beneficiaries make it clear that their support is contingent on ongoing support that helps them endure turbulence in climatic, social, economic, and/or political conditions.

Like the project implementor quoted above, many project implementors attribute the ambivalence of beneficiaries to long-standing paternalism, which project implementors view as something approximating a pathology among beneficiaries and therefore in need of reeducation or sanction. An adaptation expert at MAE describes this view of beneficiaries:

> We have a lot of virtues, but also a lot of flaws and among the defects that we have as Latin Americans is that we are very comfortable. So the Spanish inheritance that the Spaniards brought us was not very good in this sense. I tell you this with shame and sadness, but it is still very ingrained in people the tendency to want things to come easily. And if things come easily for you then why are you going to do difficult things if they can come easily? So, if you got used to the fact that for three or four or five years [the average length of an adaptation project] you have things easy. So are you going to take out the hoe to work for the same thing? You are always waiting for another project to come that will bring you things. Sometimes I think that we have to pass a certain point with the beneficiaries and be more cautious in the selection and identification of beneficiaries, not only for their poverty or lack of services, or certain social or economic indicators for who is the most needy – that is true. But, I also believe we have to work hard to educate the beneficiaries so that they know at the end of the project that they have been beneficiaries. (Interview, 15 January 2019)

This view of the uncooperative beneficiary clearly holds racist and classist undertones, and also reflects the deep frustration that adaptation project implementors feel when beneficiaries don't behave like thankful beneficiaries during or beyond the end of a project.

Expectations of patronage and implicit quid-pro-quo by beneficiaries, wherein adaptation projects are viewed as being in exchange for political support, are highly problematic in the eyes of project implementors because it disrupts the technocratic imaginary of these projects, which they feel should be successful on technical merits alone. This dynamic, however, closely maps onto the state-society relationships reinforced during the technocratic populism of former president Rafael Correa. Under Correa, experts felt they could build a rational and just society without input from social movements or citizens (Torre 2013, 2018). Instead, social welfare payments by the state to Indigenous and other poor Ecuadorians sought to neutralize calls for more radical reforms. These social programs were designed to engender a sense of indebtedness and political allegiance amongst recipients (Torre 2013, 2018). Under Lenin Moreno and deepening economic austerity, these social programs have been incrementally rolled back, but patronage relationships have long been part of the state-society contract in Ecuador and extend to state interventions like adaptation projects. Under the current center-right President Guillermo Lasso, these social programs are likely to be further dismantled.

Adaptation projects, as a form of state intervention, are also being implemented in communities and autonomous, grassroots irrigation or agricultural associations with their own set of social dynamics and governance norms that shape how beneficiaries do or do not relate to an external project. Incursions by project technical staff into the oversight of association members is often met by fierce resistance by local leaders as exemplified by a conflict that arose between a PACC project technician and an autonomous irrigation association. As recounted by a leader of the irrigation association:

> I told him, 'Look Engineer. It is not a problem if the comrades come or do not come. You are directing the project, you are not supervising who comes to work or not. That is what we do. Let's be clear on this, Juaca [an engineer and member of the association] is executing the project and we are the beneficiaries who are lending a hand and you don't get to track who has not been coming. (Interview, 6 December 2019)

This irrigation association leader is strongly asserting that the adaptation project staff member and the state, which that staff member represents, do not have the authority to sanction their members. Within this irrigation association, there is a counter-hegemonic politics that seeks to govern adaptation infrastructure based on autonomous governance structures enacted through grassroots irrigation associations and other local territorial institutions.

The variation in how project beneficiaries relate to project interventions and embody the beneficiary subject position speaks to the messy execution of projects and the varied social dynamics, history, and imaginaries of the individuals and communities that participate in these programs. By viewing beneficiary's willingness to embody that subject position as conditional, we can see how beneficiaries are willing to be governed only under particular terms, and their disengagement is a counter-conduct employed when, from their vantage point, the state-society contract is broken by the state and project implementors. As a counter-conduct, the conditionality of beneficiaries undermines the official narratives of project outcomes as presented in proposals, evaluations, and promotional materials.

4.3 The beneficiary that refuses to be a beneficiary: opting-out (leveraging and resisting adaptation)

Perhaps the most obvious counter-conduct open to potential adaptation project beneficiaries is an out-right refusal to participate in the initiative. This 'right of refusal' to external climate change adaptation projects is a key counter-conduct employed by both Indigenous and mestizo communities in asserting their territorial rights (Simpson 2017). Refusals to climate change adaptation projects are often undocumented and thus hard to find. The communities that do not enter into agreements with MAE for climate change adaptation projects are not documented in formal paperwork and often these refusals happen at a very early stage of project design.

Many project designers avoid these refusals altogether by choosing communities that they know will be amenable to climate change adaptation projects. In our interviews with project designers, they often cited two reasons for why they chose particular sites for intervention. The first was technical with project designers justifying project sites selection based on climate change vulnerability or impact studies. The second site selection criterion was based on the pre-existing relationships, political alliances or the vaguely stated 'willingness to work' of particular communities (Interview, former PACC project staff, 10 September 2019). One former adaptation project designer described their experience of project site selection:

> The majority of the times [project implementors] make a political decision and then figure out how to substantiate it. The majority of time politics function this way, not the reverse. This country is not an open canvas, there is power, there are interests, territories. (Interview, 9 September 2019)

By prioritizing communities with whom state and international actors have pre-existing relationships or political affinities, these refusals are avoided before they can happen.

While it may be hard to document cases of community-wide refusal in climate change adaptation projects, within each community or association that participated in PACC and FORECCSA, there are individuals or families that chose not to participate in a climate change adaptation project. When asked why particular individuals refused, community leaders partially attributed this to both restrictive financial and labor project requirements and the advanced age of beneficiaries. Beyond these constraints, community leaders also cited internal conflicts, mistrust of the state, and unwillingness to provide financial or in-kind contributions as reasons that some individuals chose to opt out.

By opting out, these beneficiaries deny climate change adaptation as a viable field of action and reject its related subjectivities. As a counter-conduct, the opting out of individuals and/or communities are autonomous actions that protect them from enduring climate change adaptation initiatives not by opposing state power directly but sheltering themselves and others from the productivist logics embedded in adaptation initiatives.

The refusals of individuals or communities to participate in climate change projects is also linked to national and international efforts led by the national Confederation of Indigenous Nationalities of Ecuador (CONAIE, *Confederación de Nacionalidades Indígenas del Ecuador*) to challenge the legitimacy of the state in administering climate finance and reject market-based climate change solutions like Reducing Emissions for Deforestation and Forest Degradation (REDD+) and its national manifestation, Socio-Bosque. In 2008,

Humberto Cholango, the then president of CONAIE, wrote directly to Ban Ki Moon and Christiana Figueres, the General Secretary of the UNFCCC, to explicitly condemn REDD+ and denounce the negotiation of the Ecuadorian state and United Nations as against the will of Ecuador's Indigenous peoples. More recently, Jaime Vargas, the president of CONAIE from 2017-2020, denounced Ecuadorian President Lenin Moreno at the 2019 UNFCCC Conference of Parties 25 in Madrid, stating that Moreno's support of REDD+ and payment of ecosystem services was a 'just a little bit of bread to the Indian so they will maintain the forests while we continue industrializing' (Pressenza 2019).

While CONAIE has largely focused their advocacy on contesting market-based mitigation strategies like REDD+, they have openly challenged the legitimacy of the Ecuadorian state in climate change governance. More generally, CONAIE rejects the market-based, capitalist imaginaries of climate change solutions and instead advances an alternative imaginary grounded in food sovereignty, peasant's rights, and Earth-rights.

The trans-environmental Indigenous movement represented by CONAIE and other agrarian movement allies is actively seeking to disrupt the underlying conditions of exploitation, resistance within and through an explicitly anti-capitalist platform. While CONAIE has not made a formal position on climate change adaptation, they have articulated a clear call for 'structural changes of coexistence with Nature through Buen Vivir/Vivir Bien' and a rejection of the 'conciliatory "adaptation" that is the ambition of multinationals [corporations]' (Abya Yala 2009, 1).

4.4 The subversive beneficiary: undermining the discursive frame of climate change adaptation (reworking adaptation)

Adaptation project designers work hard to define and bound how a project responds specifically to climate change impacts. The climate rationality and impact studies that justify adaptation interventions and produce a regime of truth are foundational to how funders select projects and how beneficiaries are schooled to understand projects (Li 2007). For an adaptation project to be labeled, 'just another development project' is a pejorative for many adaptation project implementors (Implementing entity project director, Interview, 12 September 2019). The ability of adaptation projects to respond directly to climate change impacts, cementing their additionality above and beyond baseline development, both justifies projects to funders but also creates a powerful justification for territorial intervention. Considering this careful framing, subverting the discursive frame of climate change adaptation projects represents a key counter-conduct as exemplified by the community of Santa Rosa.[3]

Santa Rosa, like many communities in the southern highlands of Ecuador, has silver, lead and zinc deposits within the parish. To exploit these resources, an underground mine was developed in the 1970s. For twenty years, a Filipino company owned the mine and employed a mixed foreign and local workforce. In the early 1990s, however, the mine was purchased by the Ecuadorian subsidiary of a Canadian mining conglomerate and then closed soon after. This closure was driven by compounding factors including low silver prices, extensive water contamination leading to community resistance, and the efforts of the local labor force to unionize for better wages (Terán 1994). Despite the

[3]Fictionalized name.

closure of the mine twenty years ago, it remains an active specter of the threats of extractivism for the residents of Santa Rosa and neighboring communities.

In 2012, there were reports that staff from the mining company were conducting inspections and visiting families in the vicinity of the mine. This activity stoked fears that the silver mine would reopen prompting protests by communities across the canton[4] and local politicians to denounce this threat to vital water systems . A former employee of the mine and current leader of the local irrigation association remembered this struggle:

> What we didn't want again was those mines because those goddamn mines took our water. [...] We made a war against [the mining company] that wanted to return to mining here. We didn't accept it and we raised up as four parishes, we gathered between 1500 and 2000 people in defense of the water. (Interview, 11 October 2019)

It was in this context that Santa Rosa began preparing their proposal to MAE for a sub-grant from a new adaptation pilot project.

The *Adaptation to Climate Change through Effective Water Governance* (or PACC for its acronym in Spanish) was funded by the Global Environment Facility's Special Climate Change Fund and implemented by the United Nations Development Program between 2009 and 2014 (Table 1). This 3-million-dollar project's objective was to 'reduce vulnerability via the efficient management of water resources' (UNDP 2008, 2). Within the project, MAE officials advanced an imaginary of productivist agricultural reform where campesino farmers increased their agricultural production and water use efficiency through irrigation system improvements, capacity building, and some highland reforestation for improved water availability.

When developing their PACC sub-grant proposal, parish leaders of Santa Rosa created an ambitious project where they would construct seven reservoirs and improve irrigation infrastructure in collaboration with multiple irrigation associations across the parish. While this project was designed to help local farmers respond to changing water availability, it also served an important political purpose in stymying the reactivation of the mine. The parish government decided to site their largest planned reservoir in the tailings of the old mine. Nestled between the abandoned mill house and former company office, community leaders hoped that by constructing the reservoir there they could gain state-recognized rights over related land and water, thus strengthening their claim over their territory and providing them a legal basis by which to oppose the re-opening of the mine. When asked about how his administration chose the site for the largest PACC-funded reservoir, the former president of the parish responded, 'There are many places you can put a reservoir, but only one place we could also stop the mine from re-opening' (Interview, 8 October 2019). Santa Rosa's proposed PACC project thus served a dual purpose of increasing water storage to buffer increasingly erratic seasonal rains and asserting control over land and water rights.

MAE realized that this reservoir site presented political and environmental concerns. Officials even suggested that the reservoir be relocated because of water contamination and instability in the tailings. But in 2014, the Santa Rosa parish government defiantly

---

[4]Within Ecuador, there are three key scales of sub-national state governance – provincial, canton, and parish – with parish being the most local form of government.

informed MAE that it would not be relocating its reservoir. In the official memo, the Santa Rosa parish council credited their decision to the lack of other sites with sufficient untitled water. What was not written in these official documents was the political work that that reservoir was performing in the tailings. The placement of the reservoir represented an overt act of resistance to the neo-extractivist stance of the Ecuadorian state and an explicit statement about a future agrarian landscape free from mining.

The case of Santa Rosa represents a key counter-conduct. Here the local leaders mimic the discourse of climate change adaptation, but toward very different ends. By subverting the discursive frame of the climate change adaptation project, the people of Santa Rosa were able to assert control over their territory and build political subjectivities that are tied to the national movement against extractivism, which faced substantial persecution under former president Correa (Machado Aráoz 2015). The elected leaders of Santa Rosa's parish council and campesino irrigation associations leveraged adaptation projects to reinforce an agrarian future rooted in agricultural lifeways, anti-extractivism, and defense of water.

Additionally, the case of Santa Rosa reflects the skill of local leaders in reworking the adaptation project to support their alternative territorial imaginaries. It is also important to highlight that the reservoir that was constructed in the tailings does still provide important water storage to downstream farmers (though with elevated contaminant levels). Therefore, the subversion of the climate change adaptation rationality is not complete, rather it challenges purist fantasies of project implementors who often wish to separate the risk of climate change from the milieu of long-standing territorial struggles. The strategic articulation of struggles for agrarian climate justice and long-standing anti-extractivism mobilization, highlights the growing convergence between local and national struggles (Borras and Franco 2018; Sekine 2021).

By deploying this counter-conduct, the beneficiaries of Santa Rosa reworked the capitalist logic, seeking to alter the unequal conditions that they have endured under neo-extractivism. Their counter conduct reflects a pragmatic strategy for recalibrating power relations and safeguarding agrarian livelihoods in their territory.

4.5 The long-standing beneficiary: leveraging longevity (reworking adaptation)

Climate change adaptation projects are ephemeral. The implementation of these projects generally lasts between four and five years and then project implementors expect for local governments, individuals, and/or grassroots organizations to sustain these projects into the foreseeable future. The longevity of beneficiaries in their community as compared to the often-episodic presence of international or state actors provides beneficiaries a key counter conduct. By leveraging their longevity, beneficiaries can contest the imposition of dominant territorial projects and reinforce their own imaginaries of climate change adaptation.

The infrastructure installed during adaptation projects in Ecuador such as reservoirs or irrigation systems are left to agricultural or irrigation associations or individuals to maintain. In the hands of grassroots organizations, adaptation projects are subject to customary water management norms, but there remains a fraught give and take with the state over the governance of adaptation interventions and the organizations themselves. On one hand, adaptation projects forge relationships and dependencies between grassroot organizations and the state by requiring water concessions. In both the case of FORECCSA and PACC, water concessions for newly installed reservoirs had to be approved by the secretariat of water (SENAGUA). As one irrigation association board member that participated in PACC explains:

> We have to stay legal with SENAGUA because the water belongs to the state. The concessions do not consider us as owners, but only authorizes the use of the water as an authorization. So that is why we are dependent on SENAGUA. (Lama 2020, 44)

As this board member clearly articulates, PACC fostered ties between the state and autonomous irrigation systems through water concessions, but project implementors stopped short of requiring irrigation associations be legally formalized.

While associations may agree to attain and maintain water concessions for additional security over water access and use, irrigation associations have been reluctant to formally legalize as organizations with the state. The legal formalization of grassroots irrigation or agricultural associations in Ecuador entails registering with SENAGUA, agreeing to ratify and abide by approved water governance statutes, and attaining a tax ID. One parish president explained why he did not require irrigation associations to legalize before participating in PACC:

> It does not matter. It was not necessary to be legally constituted to benefit. We never demanded it from them because we saw that it was a burden. [...] No, we discarded the idea and started to work with organizations that we saw wanted to work ... that had the drive to work. So we said let's work with them. We did not make them legalize. Then came SENAGUA saying they needed to legalize. I don't know how many are (legalized) at the moment, but we did not demand that they be legally constituted. (Interview, 8 October 2019)

Since PACC, very few irrigation associations have completed their legalization process, instead opting to govern their irrigation infrastructure according to customary water governance norms. When asked why so few have legalized, the president of the National Federation of Irrigation Association replied:

> It is about sovereignty [...] There are many communities that aren't legalized and I personally tell them that they don't need to legalize. The truth is that you have to contract a lawyer, which costs a lot, and pay for all the expenses. The other [barrier] is after you are legalized you have to register with the Internal Revenue Service. For that you need an account, a professional to sign for you, another expense. (Interview, 7 September 2019)

While there are clearly monetary and bureaucratic barriers to legalization, these alone are not the only reason. When there have been periods in the past when procedural and financial barriers were simplified or removed by SENAGUA, many associations still chose not to legalize because some irrigation association leaders felt that it was not necessary since they had long been managing water communally and resolving conflicts. The unwillingness to conform with state rationalities about water management following a adaptation project's closure represents a counter conduct through which beneficiaries oppose adaptation rationalities wherein water governance under a changing climate reflects technocratic norms of 'rational' and 'efficient' water use. By not undertaking the legalization process, grassroots irrigation associations defy the desires of the state and support an alternative imaginary wherein newly implemented adaptation projects are manged according to customary norms and collective social relations following the closure of the project.

This counter-conduct by former project beneficiaries, however, does not go unnoticed. When reservoirs or other project interventions need major repairs or other large-scale investments, local leaders often petition the provincial government, which has the constitutional responsibility for rural development and irrigation. As an example, San Bartolomé Central, an irrigation association that participated in PACC, needs to replace the ruptured membrane lining of their PACC-funded reservoir at a cost of US$30,000. As one irrigation association member explains, this assistance is desperately needed, 'When the reservoir [lining] breaks, we need to ask for the support of the regional government otherwise we cannot irrigate anymore' (Lama 2020, 51). The provincial government of Azuay, which encompasses San Bartolomé, however, has withheld support. The irrigation association president recounts the conditions for provincial support, 'We have visited the prefecture asking for help to repair the reservoir, but they won't help us until we are legalized as an irrigation association.' (Interview, 6 October 2019). The ongoing debate over the legalization of irrigation associations following the closure of adaptation projects reflects the governmentalities and counter conducts that arise as the beneficiaries and state continue to negotiate the nature of climate change adaptation and agrarian change into the uncertain future.

The interplay between state actors and local irrigation associations around legalization highlights that the governance of climate change adaptation and its subjects does not terminate with the closing of the project. By leveraging longevity, beneficiaries subvert and seek to turn the formal adaptation project rules and rationalities into myriad projects of their own making: redirecting agrarian livelihoods and asserting control over their own futures.

5. Conclusion

Looking across Ecuador's portfolio of climate change adaptation projects and the opposition that these initiatives have generated, we do not find a burgeoning and coherent trans-environmental, trans-regional, anti-capitalist peasant social movement that places climate justice, much less climate change adaptation, at the core of its political struggle. CONAIE, as one of the strongest and longest-running social movements in Latin America and the hub of a broad agrarian coalition across Ecuador, has played a central role in elevating the rights of Mother Earth (Akchurin 2015) and food sovereignty (Peña 2016). Despite this leadership, CONAIE has remained largely silent on the topic of climate change adaptation beyond dismissing it as a 'conciliatory' discourse driven by the interests of transnational corporations (Abya Yala 2009).

The most recent mass social mobilization in June and July 2022 exemplifies the peripheral nature of climate change to CONAIE's central platform. During the mobilization, CONAIE's leaders alongside their broad coalition of campesino rights and labor-based organizations issued a list of ten demands to President Guillermo Lasso as a response to the 'crisis caused by neoliberalism' (CONAIE 2022). The ten demands included core issues relating to agrarian justice including debt cancelation, fair agricultural prices, protection of national markets, improved labor protection, halting the extractive frontier, bicultural education and Indigenous sovereignty, stopping the privatization of strategic sectors, and price controls on items of basic necessity (ibid). Topping this list, however, was the demand for the increased subsidy of diesel and gasoline to cushion rising prices. This demand highlights how climate justice remains a far second to meeting the immediate needs of Indigenous and poor Ecuadorians in the context of growing domestic austerity and international market turbulence.

Contestation over climate change adaptation is not yet happening on the national stage, but it is happening in agrarian communities across Ecuador. An examination of the counter conducts, as 'revolts of conduct', enable us to examine the micropolitics through which project beneficiaries challenge and negotiate adaptation initiatives on-the-ground and towards what ends (Foucault 2009, 196). Fundamentally, struggles over climate change adaptation are struggles over the future of agrarian people, livelihoods, and landscapes. Therefore these counter conducts should not be seen as separate, dislocated or 'merely local' struggles. Across these varied counter conducts and their underlying strategic logics, communities are also forming new subjectivities, collectivities, and imaginaries wherein vernacular notions and practices of climate change adaptation become incorporated into broader struggles for agrarian justice. We argue that local contestation of agrarian adaptation projects should be seen and theorized as a part of the broader agrarian and Indigenous movement that at different scales pragmatically and eclectically struggles to create an alternative agrarian future for rural communities.

From the Binational project on the northern border to Santa Rosa in the Southern highlands, we see agrarian communities advancing long-standing demands for agrarian justice through their contestation of adaptation projects but with varied logics underlying their counter conducts (Katz 2004). In the Binational Project, Indigenous and Afro-descended communities are actively seeking to disrupt the underlying conditions of oppression by fiercely asserting their sovereignty and demanding control over adaptation finance. As a result of their *resistance*, they are advancing an alternative imaginary of climate change adaptation that resonates with broader struggles to build greater sovereignty, strengthen local territorial control, and foster agrarian production that reflects anti-capitalist notions that forefront food sovereignty, cooperative social relations, and the co-existence and entanglement of people and nature (Abya Yala 2009).

Other individuals are *reworking* the unequal conditions and structures of oppression through their negotiation of climate change adaptation projects. In the case of Santa Rosa, local leaders used their adaptation project to fortify their territorial claims, block the reopening of a silver mine, and promote an imaginary of anagrarian future that is based in smallholder agriculture over extractive industry. As an another example of reworking adaptation, grassroot irrigation associations across the Ecuadorian highlands are outlasting project implementors to bring these projects under local, autonomous governance structures.

Other counter conducts that we identified enable project beneficiaries (or would be beneficiaries) to *leverage* adaptation projects to help them endure turbulence in markets, politics, and even the climate. As an example, some individuals or communities chose to not participate in projects altogether to shelter themselves from state intervention and productivist logics advanced by these initiatives. Others chose to participate in adaptation projects but only under particular conditions or when receiving material benefits.

What do these counter conducts mean for the future of climate change adaptation in agrarian Ecuador? Over the nearly 15 years of adaptation project implementation in Ecuador, beneficiaries have begun to take a more oppositional and proactive role in dictating the terms of climate change adaptation. It is an open question if Indigenous and agrarian movement actors will begin to explicitly incorporate climate change adaptation into their demands, and importantly, if these local counter conducts will enable agrarian communities to position themselves translocally or transnationally in the changing landscape of adaptation finance and the struggle over the nature of climate-resilient agrarian development. It is commonly said by adaptation scholars and practitioners that climate change adaptation is inherently local, yet as we have shown, the negotiation of these local projects is deeply entwined with and rooted in the broader multi-scalar struggles of Indigenous and agrarian movements that aim to carve out space, autonomy, and political leverage to develop alternative agrarian futures.

Acknowledgements

The authors would like to acknowledge Carol Tapia who served as research assistant on this project. The authors would also like to acknowledge Diana Liverman, Andrea Gerlak, Elizabeth Oglesby, and Gökçe Günel for their helpful comments on earlier drafts of this article as well as participants in the 2022 *Climate Change and Agrarian Justice Conference, 2020 EASST+4S Conference*, and 2021 *2nd World Forum on Climate Justice* who provided feedback on earlier versions of this paper.

Disclosure statement

No potential conflict of interest was reported by the authors.

Funding details

This work was supported by University of Arizona's College of Social and Behavioral Sciences Summer Dissertation Writing Fellowship, National Science Foundation Doctoral Dissertation Research Improvement Grant (Grant Number: 2002829), Conference of Latin American Geography PhD Field Study Award, National Science Foundation/Dutch Research Council Graduate Research Opportunities Worldwide Grant, American Association of Geographers Dissertation Fieldwork Grant, American Geographical Society Council Fellowship, University of Arizona - School of Behavioral and Social Sciences Research Institute Dissertation Grant, Ida Grace Guy Endowed PEO Scholarship, American Association of Geographers-Cultural Anthropology & Political Ecology (CAPE) Specialty Group Dissertation Fieldwork Award, American Association of Geographers-Latin American Specialty Group Dissertation Fieldwork Award, and Tinker Foundation Research Travel Award.

ORCID

Megan Mills-Novoa http://orcid.org/0000-0002-1780-9266
Jaime Hoogesteger http://orcid.org/0000-0002-6784-0552

References

Abu-Lughod, L. 1990. "The Romance of Resistance: Tracing Transformations of Power Through Bedouin Women." *American Ethnologist* 17 (1): 41–55. doi:10.1525/ae.1990.17.1.02a00030.

Abya Yala. 2009. *Diálogo de Alternativas y Alianzas de los Movimientos Indígenas, Campesinos y Sociales del Abya Yala*. Quito: *Enlace Indígena*.

Akchurin, M. 2015. "Constructing the Rights of Nature: Constitutional Reform, Mobilization, and Environmental Protection in Ecuador." *Law & Social Inquiry* 40 (4): 937–68. doi:10.1111/lsi.12141.

Akram-Lodhi, A. H., and C. Kay. 2010. "Surveying the Agrarian Question (Part 2): Current Debates and Beyond." *The Journal of Peasant Studies* 37 (2): 255–84. doi:10.1080/03066151003594906.

Andolina, R., N. Laurie, and S. A. Radcliffe. 2009. *Indigenous Development in the Andes*. Durham: Duke University Press.

Arnall, A. 2018. "Resettlement as Climate Change Adaptation: What can be Learned from State-led Relocation in Rural Africa and Asia? *Climate and Development* 11 (3): 253–63.

Bäckstrand, K., and E. Lövbrand. 2006. "Planting Trees to Mitigate Climate Change: Contested Discourses of Ecological Modernization, Green Governmentality and Civic Environmentalism." *Global Environmental Politics* 6 (1): 50–75. doi:10.1162/glep.2006.6.1.50.

Bassett, T. J., and C. Fogelman. 2013. "Déjà vu or Something New? The Adaptation Concept in the Climate Change Literature." *Geoforum* 48: 42–53. doi:10.1016/j.geoforum.2013.04.010.

Baud, M., R. Boelens, and G. Damonte. 2019. "Presentación. Nuevos capitalismos y transformaciones territoriales en la región andina." *Estudios Atacameños (En línea)* 63: 195–208.

Beck, E. 2017. *How Development Projects Persist*. Durham: Duke University Press.

Benjamin, W. 1978. "On the Mimetic Faculty." In *Reflections*, translated by E. Jephcott, edited by P. Demetz, 333–6. New York: Harcourt Brace Jovanovich.

Bernstein, H. 1996. "Agrarian Questions Then and Now." *Journal of Peasant Studies* 24 (1–2): 22–59. doi:10.1080/03066159608438630.

Berrang-Ford, L., A. R. Siders, A. Lesnikowski, A. P. Fischer, M. W. Callaghan, N. R. Haddaway, K. J. Mach, et al. 2021. "A Systematic Global Stocktake of Evidence on Human Adaptation to Climate Change." *Nature Climate Change* 11 (11): 989–1000. doi:10.1038/s41558-021-01170-y.

Broto, V. C. 2020. "Climate Change Politics and the Urban Contexts of Messy Governmentalities." *Territory, Politics, Governance* 8 (2): 241–58.

Boelens, R. 2015. *Water, Power and Identity: The Cultural Politics of Water in the Andes*. London and Washington D.C.: Earthscan.

Boelens, R., and P. H. Gelles. 2005. "Cultural Politics, Communal Resistance and Identity in Andean Irrigation Development." *Bulletin of Latin American Research* 24 (3): 311–27

Borras, S. M., and J. C. Franco. 2018. "The Challenge of Locating Land-Based Climate Change Mitigation and Adaptation Politics Within a Social Justice Perspective: Towards an Idea of Agrarian Climate Justice." *Third World Quarterly* 39 (7): 1308–25. doi:10.1080/01436597.2018.1460592.

Borras, S. M., J. C. Franco, and Z. Nam. 2020. "Climate Change and Land: Insights from Myanmar." *World Development* 129: 104864. doi:10.1016/j.worlddev.2019.104864.

Borras, S. M., I. Scoones, A. Baviskar, M. Edelman, N. L. Peluso, and W. Wolford. 2022. "Climate Change and Agrarian Struggles: An Invitation to Contribute to a *JPS* Forum." *The Journal of Peasant Studies* 49 (1): 1–28. doi:10.1080/03066150.2021.1956473.

Byres, T. J. 1991. "The Agrarian Question." In *A Dictionary of Marxist Thought*, edited by T. Bottomore, 9–11. Oxford, UK: Blackwell.

Cadman, L. 2010. "How (Not) to Be Governed: Foucault, Critique, and the Political." *Environment and Planning D: Society and Space* 28 (3): 539–56. doi:10.1068/d4509.

Caretta, M. A., and L. Borjeson. 2015. "Local Gender Contract and Adaptive Capacity in Smallholder Irrigation Farming: A Case Study from the Kenyan Drylands." *Gender, Place & Culture* 5: 644–61.

Camargo, A. 2022. "Imagined Transitions: Agrarian Capitalism and Climate Change Adaptation in Colombia." *The Journal of Peasant Studies* 49 (4): 713–733.

Camargo, A., and D. Ojeda. 2017. "Ambivalent Desires: State Formation and Dispossession in the Face of Climate Crisis." *Political Geography* 60: 57–65.

Colloredo-Mansfeld, R. 2009. *Fighting Like a Community*. Chicago, Illinois: University of Chicago Press.
Colloredo-Mansfeld, R., A. Ordoñez, H. Paltán López, J. Quick, D. Quiroga, and J. Williams. 2018. "Conflicts, Territories, and the Institutionalization of Post-Agrarian Economies on an Expanding Tourist Frontier in Quilotoa, Ecuador." *World Development* 101: 441–52. doi:10.1016/j.worlddev.2016.11.001.
CONAIE. 2022. Frente a las Crisis Causada por el Modelo Neoliberal: Organizarse es Necesario, Resistir Es Justo, Luchar es un Derecho. https://www.pauta.cl/pauta/site/docs/20220627/20220627171939/demandas_movilizacion_nacional___13_de_junio_2022.pdf
Davidson, A. I. 2011. "In Praise of Counter-Conduct." *History of the Human Sciences* 24 (4): 25–41. doi:10.1177/0952695111411625.
Dean, M. 2010. *Governmentality*. 2nd ed. New York: SAGE Publications.
Death, C. 2010. "Counter-Conducts: A Foucauldian Analytics of Protest." *Social Movement Studies* 9 (3): 235–51. doi:10.1080/14742837.2010.493655.
de Certeau, M. 1984. *The Practice of Everyday Life*. Berkeley: University of California Press.
del Ecuador, G. 2008. *Constitución del Ecuador*.
Dowling, R. 2010. "Geographies of Identity: Climate Change, Governmentality and Activism." *Progress in Human Geography* 34 (4): 488–95. doi:10.1177/0309132509348427.
Duarte-Abadía, B. 2022. *Ríos, Utopias y Movimientos Sociales. Reviviendo flujos de vida en Colombia y España*. Quito, Ecuador: Abyayala.
Duarte-Abadía, B., R. Boelens, and E. Buitrago. 2021. "Neoliberal Commensuration and New Enclosures of the Commons: Mining and Market–Environmentalism Governmentalities." *Territory, Politics, Governance* 00 (00): 1–21.
Dupuits, E., M. Baud, R. Boelens, F. de Castro, and B. Hogenboom. 2019. "Scaling Up but Losing Out? Water Commons' Dilemmas Between Transnational Movements and Grassroots Struggles in Latin America." *Ecological Economics* 172: 106625. doi:10.1016/j.ecolecon.2020.106625.
Eriksen, S., and J. Lind. 2017. "Adaptation as a Political Process: Adjusting to Drought and Conflict in Kenya's Drylands." *Environmental Management* 43: 817–35.
Eriksen, S., A. Nightingale, and H. Eakin. 2015. "Reframing Adaptation: The Political Nature of Climate Change Adaptation." *Global Environmental Change* 35: 523–33. doi:10.1016/j.gloenvcha.2015.09.014.
Eriksen, S., E. L. F. Schipper, M. Scoville-Simonds, K. Vincent, H. N. Adam, N. Brooks, B. Harding, et al. 2021. "Adaptation Interventions and Their Effect on Vulnerability in Developing Countries: Help, Hindrance or Irrelevance?" *World Development* 141: 105383. doi:10.1016/j.worlddev.2020.105383.
Fletcher, R. 2001. "What Are We Fighting for? Rethinking Resistance in a Pewenche Community in Chile." *Journal of Peasant Studies* 28 (3): 37–66. doi:10.1080/03066150108438774.
Fletcher, R. 2017. "Environmentality Unbound: Multiple Governmentalities in Environmental Politics." *Geoforum* 85: 311–5. doi:10.1016/j.geoforum.2017.06.009.
Foucault, M. 1991. *The Foucault Effect*. Chicago: University of Chicago Press.
Foucault, M. 2008. *The Birth of Biopolitics*. London: Picador.
Foucault, M. 2009. *Security, Territory, Population*. London: Picador.
Foucault, M. 1988. *Politics, Philosophy, Culture: Interviews and Other Writings*. New York: Routledge.
Funder, M., C. Mweemba, and I. Nyambe. 2018. "The Politics of Climate Change Adaptation in Development: Authority, Resource Control and State Intervention in Rural Zambia." *The Journal of Development Studies* 54 (1): 30–46. doi:10.1080/00220388.2016.1277021.
GEF-PNUD. 2015. Evaluación Final del Proyecto de Adaptación al cambio climático a través de una efectiva gobernabilidad del agua en Ecuador (No3520 de PIMS) GEF – PNUD Mayo de 2015, 1–88. Panama City: GEF-PNUD.
Giunta, I. 2014. "Food Sovereignty in Ecuador: Peasant Struggles and the Challenge of Institutionalization." *The Journal of Peasant Studies* 41 (6): 1201–24. doi:10.1080/03066150.2014.938057.
Gordon, C. 1991. "Government Rationality." In *The Foucault Effect: Studies in Governmentality*, edited by G. Burchell, C. Gordon, and P. Miller, 1–51. Chicago, Illinois: University of California Press.
Gramsci, A. 1971. *Selections from the Prison Notebooks*. New York: International Publishers.

Henderson, T. 2017. "State-Peasant Movement Relations and the Politics of Food Sovereignty in Mexico and Ecuador." *The Journal of Peasant Studies* 44 (1): 33–55. doi:10.1080/03066150.2016.1236024.

Henrique, K. P., and P. Tschakert. 2019. "Contested Grounds: Adaptation to Flooding and the Politics of (in)visibility in São Paulo's Eastern Periphery." *Geoforum* 104: 181–92.

Herrera Revelo, S. 2017. "Lecciones del Contradictorio Progresismo en el Ecuador." *Lalineadefuego.Info* 31: 1–5.

Holling, C. S. 1974. "Resilience and Stability of Ecological Systems." *Ecology, Evolution, and Systematics* 4: 1–23.

Hommes, L., R. Boelens, S. Bleeker, B. Duarte-Abadía, D. Stoltenborg, and J. Vos. 2020. "Water Governmentalities: The Shaping of Hydrosocial Territories, Water Transfers and Rural–Urban Subjects in Latin America." *Environment and Planning E: Nature and Space* 3 (2): 399–422. doi:10.1177/2514848619886255.

Hoogesteger, J. 2013. "Trans-Forming Social Capital Around Water: Water User Organizations, Water Rights, and Nongovernmental Organizations in Cangahua, the Ecuadorian Andes." *Society & Natural Resources* 26 (1): 60–74. doi:10.1080/08941920.2012.689933.

Hoogesteger, J. 2015. "Normative Structures, Collaboration and Conflict in Irrigation; a Case Study of the Pillaro North Canal Irrigation System, Ecuadorian Highlands." *International Journal of the Commons* 9 (1): 398–415. doi:10.18352/ijc.521.

Hoogesteger, J., R. Boelens, and M. Baud. 2016. "Territorial Pluralism: Water Users' Multi-Scalar Struggles Against State Ordering in Ecuador's Highlands." *Water International* 41: 1.

Hoogesteger, J., Y. Tiaguaro-Rea, E. Rap, and J. P. Hidalgo. 2017. "Scalar Politics in Sectoral Reforms: Negotiating the Implementation Of Water Policies in Ecuador (1990–2008)." *World Development* 98: 300–9. doi:10.1016/j.worlddev.2017.04.036.

Hoogesteger, J., and A. Verzijl. 2015. "Grassroots Scalar Politics: Insights from Peasant Water Struggles in the Ecuadorian and Peruvian Andes." *Geoforum* 62: 13–23. doi:10.1016/j.geoforum.2015.03.013.

Horowitz, L. S. 2011. "Interpreting Industry's Impacts: Micropolitical Ecologies of Divergent Community Responses." *Development and Change* 42 (6): 1379–91. doi:10.1111/j.1467-7660.2011.01740.x.

IPCC. 2007. "Climate Change 2007: Impacts, Adaptation and Vulnerability." In *Working Group II Contribution to the Intergovernmental Panel on Climate Change Fourth Assessment Report Summary for Policy Makers*, edited by M. Parry, O. Canziani, J. Palutikof, P. van der Linden, and C. Hanson, 1–976. Geneva, Switzerland: World Meteorological Organization.

Katz, C. 2004. *Growing up Global: Economic Restructuring and Children's Everyday Lives*. Minneapolis: University of Minnesota Press.

Keskitalo, E., S. Juhola, and L. Westerhoff. 2012. "Climate Change as Governmentality: Technologies of Government for Adaptation in Three European Countries." *Journal of Environmental Planning and Management* 55 (4): 435–52. doi:10.1080/09640568.2011.607994.

Khan, M., S. Robinson, R. Weikmans, D. Ciplet, and J. T. Roberts. 2019. "Twenty-five Years of Adaptation Finance Through a Climate Justice Lens." *Climatic Change* 17 (8): 17–9.

Lama, D. 2020. *Transforming Water Rights and Irrigation Systems Through a Climate Change Adaptation Project in the Southern Andes of Ecuador*. Wagenignan, Netherlands: Wageningen University.

Li, T. M. 2007. *The Will to Improve*. Durham: Duke University Press.

Lindegaard, L. B. 2016. *The Discursive Intersection of the Government of Others and the Government of Self in the Face of Climate Change*, 95–118. City: John Benjamins Publishing Company.

Liverman, D. 2015. "Reading Climate Change and Climate Governance as Political Ecologies." In *The Routledge Handbook of Political Ecology*, edited by T. Perrault, G. Bridge, undefined, and J. McCarthy, 301–19. Oxford: Routledge.

Long, N., and J. D. van der Ploeg. 1989. "Demythologizing Planned Intervention: An Actor Perspective." *Sociologia Ruralis* 29 (3–4): 226–49. doi:10.1111/j.1467-9523.1989.tb00368.x.

Machado Aráoz, H. 2015. "Conflictos socioambientales y disputas civilizatorias en América Latina: Entre el desarrollismo extractivista y el Buen Vivir." *Crítica y Resistencias. Revista de conflictos sociales latinoamericanos* 1: 19–42.

McMichael, P. 2008. "Peasants Make Their Own History, But Not Just as They Please ... " *Journal of Agrarian Change* 8 (2–3): 205–28. doi:10.1111/j.1471-0366.2008.00168.x.

Mikulewicz, M. 2019. "Thwarting Adaptation's Potential? A Critique of Resilience and Climate-Resilient Development." *Geoforum* 104: 267–82. doi:10.1016/j.geoforum.2019.05.010.

Mikulewicz, M. 2021. "Disintegrating Labour Relations and Depoliticised Adaptation to Climate Change in Rural São Tomé and Príncipe." *Area* 53 (3): 422–30. doi:10.1111/area.12630.

Mills-Novoa, M. 2020. "Making Agro-Export Entrepreneurs Out of Campesinos: The Role of Water Policy Reform, Agricultural Development Initiatives, and the Specter of Climate Change in Reshaping Agricultural Systems in Piura, Peru." *Agriculture and Human Values* 37: 667–82. doi:10.1007/s10460-019-10008-5.

Mills-Novoa, M., R. Boelens, J. Hoogesteger, and J. Vos. 2020. "Governmentalities, Hydrosocial Territories & Recognition Politics: The Making of Objects and Subjects for Climate Change Adaptation in Ecuador." *Geoforum* 115: 90–101. doi:10.1016/j.geoforum.2020.06.024.

Moore, D. S. 1998. "Subaltern Struggles and the Politics of Place: Remapping Resistance in Zimbabwe's Eastern Highlands." *Cultural Anthropology* 13 (3): 344–81. doi:10.1525/can.1998.13.3.344.

Mosse, D. 2004. "Is Good Policy Unimplementable? Reflections on the Ethnography of Aid Policy and Practice." *Development and Change* 35 (4): 639–71. doi:10.1111/j.0012-155X.2004.00374.x.

Mosse, D. 2005. *Cultivating Development*. London: Pluto.

Nightingale, A. J. 2017. "Power and Politics in Climate Change Adaptation Efforts: Struggles Over Authority and Recognition in the Context of Political Instability." *Geoforum* 84: 11–20. doi:10.1016/j.geoforum.2017.05.011.

O'Brien, K. L. 2012. "Global Environmental Change II." *Progress in Human Geography* 36 (5): 667–76. doi:10.1177/0309132511425767.

Odysseos, L., C. Death, and H. Malmvig. 2016. "Interrogating Michel Foucault's Counter-Conduct: Theorising the Subjects and Practices of Resistance in Global Politics." *Global Society* 30 (2): 151–6. doi:10.1080/13600826.2016.1144568.

Oels, A. 2005. "Rendering Climate Change Governable: From Biopower to Advanced Liberal Government?" *Journal of Environmental Policy & Planning* 7 (3): 185–207. doi:10.1080/15239080500339661.

Pacari, N. 2020. "Reflexiones sobre el proyecto político de la CONAIE: logros y vigencia." In *¡Así encendimos la mecha! Trienta años del levantamiento indígena de Ecuador: una historia permanente*, edited by M. Martínez Abarca, 13–24. Quito, Ecuador: Ediciones Abya-Yala.

Paprocki, K. 2019. "All That Is Solid Melts into the Bay: Anticipatory Ruination and Climate Change Adaptation." *Antipode* 51 (1): 295–315.

Paprocki, K. 2018a. "Threatening Dystopias: Development and Adaptation Regimes in Bangladesh." *Annals of the Association of American Geographers* 108 (4): 955–973.

Paprocki, K. 2018b. "All That Is Solid Melts Into the Bay: Anticipatory Ruination and Climate Change Adaptation." *Antipode* 51 (1): 295–315. doi:10.1111/anti.12421.

Pressenza. 2019. COP25. Jaime Vargas, CONAIE: "¿Quién está soñando cambiar el mundo? ¿Quién?" Pressenza, 10 Dec.

Pelling, M. 2011. *Adaptation to Climate Change: From Resilience to Transformation*. London and New York, NY: Routledge.

Peña, K. 2016. "Social Movements, the State, and the Making of Food Sovereignty in Ecuador." *Latin American Perspectives* 43 (1): 221–37. doi:10.1177/0094582X15571278.

Pieck, S. K. 2015. "'To Be Led Differently': Neoliberalism, Road Construction, and NGO Counter-Conducts in Peru." *Geoforum* 64 (C): 304–13.

Radcliffe, S. A. 2012. "Development for a Postneoliberal Era? Sumak Kawsay, Living Well and the Limits to Decolonisation in Ecuador." *Geoforum* 43 (2): 240–9. doi:10.1016/j.geoforum.2011.09.003.

Rasch, E. D., and M. Köhne. 2016. "Micropolitics in Resistance: The Micropolitics of Large-Scale Natural Resource Extraction in South East Asia." *Society & Natural Resources* 29 (4): 479–92. doi:10.1080/08941920.2015.1086458.

Scott, J. C. 1985. *Weapons of the Weak: Everyday Forms of Resistance*. New Haven and London: Yale University Press.

Scott, J. C. 1998. *Seeing Like a State*. New Haven and London: Yale University Press.

Sekine, Y. 2021. "Emerging 'Agrarian Climate Justice' Struggles in Myanmar." *The Journal of Peasant Studies* 48 (3): 517–40. doi:10.1080/03066150.2020.1839054.

Silveria, M., M. Moreano, N. Romero, D. Murillo, G. Ruales, and N. Torres. 2017. "Geografías de Sacrificio y Geografías de Esperanza: Tensiones Territoriales en el Ecuador Plurinacional." *Journal of Latin American Geography* 16 (1): 69–92. doi:10.1353/lag.2017.0016.

Simpson, A. 2017. "The Ruse of Consent and the Anatomy of 'Refusal': Cases from Indigenous North America and Australia." *Postcolonial Studies* 20 (1): 18–33. doi:10.1080/13688790.2017.1334283.

Terán, F. A. 1994. *Desarrollo, Explotación Subterránea y Operación de Planta en la Mina San Bartolomé*. Cuenca, Ecuador: Universidad del Azuay.

Torre de la, C. 2013. "Technocratic Populism in Ecuador." *Journal of Democracy* 24 (3): 33–46.

Torre de la, C. 2018. "Ecuador After Correa." *Journal of Democracy* 29 (4): 77–88.

Tilzey, M. 2017. "Reintegrating Economy, Society, and Environment for Cooperative Futures: Polanyi, Marx, and Food Sovereignty." *Journal of Rural Studies* 53: 317–34. doi:10.1016/j.jrurstud.2016.12.004.

Tilzey, M. 2019. "Authoritarian Populism and Neo-Extractivism in Bolivia and Ecuador: The Unresolved Agrarian Question and the Prospects for Food Sovereignty as Counterhegemony." *The Journal of Peasant Studies* 46 (3): 626–52. doi:10.1080/03066150.2019.1584191.

Tria Kerkvliet, B. J. 2009. "Everyday Politics in Peasant Societies (and Ours)." *The Journal of Peasant Studies* 36 (1): 227–43. doi:10.1080/03066150902820487.

UNDP. 2008. *PIMS 3520 – Adaptation to Climate Change Through Effective Water Governance in Ecuador*. Quito, Ecuador: United Nations Development Program.

Usón, T. J., C. Henríquez, and J. Dame. 2017. "Disputed Water: Competing Knowledge and Power Asymmetries in the Yali Alto Basin, Chile." *Geoforum* 85: 247–58. doi:10.1016/j.geoforum.2017.07.029.

Valladares, C., and R. Boelens. 2019. "Mining for Mother Earth. Governmentalities, Sacred Waters and Nature's Rights in Ecuador." *Geoforum* 100: 68–79. doi:10.1016/j.geoforum.2019.02.009.

van der Ploeg, J. D. 2008. *The new Peasantries: Struggles for Autonomy and Sustainability in an era of Empire and Globalization*. London: Routledge.

Velásquez, T. A. 2022. *Pachamama Politics: Campesino Water Defenders and the Anti-Mining Movement in Andean Ecuador*. Tucson, Arizona: University of Arizona Press.

Vergara-Camus, L. 2014. *Land and Freedom: The MST, the Zapatistas, and Peasant Alternatives to Neoliberalism*. London: Zed Books.

Watson, C., and L. Schalatek. 2021a. *Climate Finance Thematic Briefing: Adaptation Finance*. Washington D.C.: Heinrich Boll Stiftung, Climate Finance Fundamentals No. 3.

Watson, C., and L. Schalatek. 2021b. *Climate Finance Regional Briefing: Latin America*. Washington D.C.: Heinrich Boll Stiftung, Climate Finance Fundamentals No. 6.

Watts, M. J. 2015. "Now and Then: The Origins of Political Ecology and the Rebirth of Adaptation as a Form of Thought." In *The Routledge Handbook of Political Ecology*, edited by J. McCarthy and T. Perreault, 19–50. London: Routledge.

Webber, S. 2016. "Climate Change Adaptation as a Growing Development Priority: Towards Critical Adaptation Scholarship." *Geography Compass* 10 (10): 401–13. doi:10.1111/gec3.12278

World Food Programme. 2020. *Project Performance Report*. Washington D.C.: World Food Programme.

Wyborn, C., L. Yung, D. Murphy, and D. R. Williams. 2014. "Situating Adaptation: How Governance Challenges and Perceptions of Uncertainty Influence Adaptation in the Rocky Mountains." *Regional Environmental Change* 15: 669–82.

Xu, Q., R. Boelens, and G. J. Veldwisch. 2022. "Rural Drinking Water Governance Politics in China: Governmentality Schemes and Negotiations from Below." *Political Geography* 97: 102703. doi:10.1016/j.polgeo.2022.102703.

Zografos, C. 2017. "Flows of Sediment, Flows of Insecurity: Climate Change Adaptation and the Social Contract in the Ebro Delta, Catalonia." *Geoforum* 80: 49–60. doi:10.1016/j.geoforum.2017.01.004.

Linking climate-smart agriculture to farming as a service: mapping an emergent paradigm of datafied dispossession in India

S. Ali Malik

Introduction

It is difficult to imagine anything more urgent for critical agrarian scholars and movements than the questions Borras et al. (2022) raise about how to develop an emancipatory mode of confronting climate change in agrarian and rural settings. Unfortunately, we find in India that institutionalized approaches to such challenges in the form of Climate-Smart Agriculture (CSA)/Climate-Smart Villages (CSV) point in the opposite direction and which, as such, are obstacles to the emancipatory approach Borras et al. call on us to imagine. This essay subjects such obstacles to critical scrutiny, arguing that CSA/CSV programs take the form of neoliberal technologies of government which intensify the disenfranchisement and dispossession of Indian farmers. By doing so, these technologies of government reduce farmers' ability and, perhaps, willingness towards the development of 'a sufficiently anti-capitalist, trans-environmental and agrarian approach to confront climate change' (17–18). I suggest that any serious effort to construct the emancipatory project Borras et al. call for will be forced to confront how governmental technologies serve to produce specific forms of agrarian subjectivity antithetical to emancipatory change.

Bridging the agrarian question and the climate question necessitates a contextualization of technologies and politics under information capitalism in India, and increasingly in the Global South at large. This article provides a nascent attempt at such bridging work by pursuing an initial critical mapping of CSA activities in India, while establishing CSA's linkages with an emergent, data-driven service-based farming model. It is based on an analyses of various primary documents: India's *National Action Plan on Climate Change* (2008), NGO Alternative Futures report 'Training of Trainers': Summary Manual on Gender, Climate Change, Agriculture and Food Security', program reports and publications from The Consultative Group on International Agricultural Research (CGIAR) and its associated Research Program on Climate Change, Agriculture, and Food Security (CCAFS), Memorandums of Understanding between the Government of India and various agri-tech firms, and the Indian Institute of Management and a leading American consulting firm Bain & Company's co-authored report *Indian Farming's Next Big Moment: Farming as a Service*. As such, the claims advanced in this article call for further empirical exploration.

CSA must not be dismissed as simply a trojan horse for corporate agro-industrial interests. Rather, CSA and its associated programs should be investigated as mechanisms that produce and accumulate farmer/agricultural/ecological information while reconfiguring agrarian subjectivities towards digital, service-based agriculture in the form of Farming as a Service (FaaS). In doing so, I aim to contribute to the interfacing of Critical Agrarian Studies and Critical Data Studies.

The first section shows how data enacts a unique form of accumulation by dispossession as part of the broader political economy of CSA. The next section begins by showing how India's *National Mission on Sustainable Agriculture* established the policy rationale for the domestic implementation of CSA/CSV programs in India, followed by an exploration of CGIAR's early CSV projects in Haryana. The CSV program, conceptualized as a technology of responsibilization, significantly initiated the inculcation of a 'climate-smart' mentality within participant subjects as a precursor to the broader datafication of Indian agriculture. I demonstrate how CSVs interpellated and prepared Indian farmers to utilize data-driven precision agriculture technologies long before the introduction of big data in Indian agriculture made such technologies more widely incumbent upon farmers in particular regions. The introduction of precision agricultural technologies and training farmers in their use in tandem with the construction of 'climate-smart' agrarian subjectivities established the foundations for the 'Farming as a Service' (FaaS) model introduced in 2017.

Section 3 interrogates FaaS as a neoliberal model that combines data-driven agriculture and financialization in a clearly defined Indian agricultural paradigm that holds great potential for further dispossessing and deskilling Indian smallholders. The Indian state in conjunction with an array of domestic and multinational corporate actors have actively cultivated the infrastructural and financial conditions needed for the actualization of FaaS. I briefly explore its discursive elements in conjunction with its modes of subjectification as exemplified through 'nudge' techniques intended to construct an emergent algorithmic agricultural subject.

The final substantive section further examines two key disciplinary elements of the service-based farming model: contract farming and agricultural insurance. While both have long existed in India, I attempt to show how datafication reconstitutes existing fields of contract and regulation in India through its capacity to expand and intensify the scope and speed of information accumulation. The essay concludes with brief reflections on the future of agrarian struggles in India within the context of climate change and big data technologies.

Data in accumulation by dispossession and the political economy of CSA

Critical data scholars argue that data is never neutral, raw, or objective but provides relational, contingent, and contextual ways of seeing and interpreting the world (boyd and Crawford 2012; Dalton and Thatcher 2014; Kitchin 2013). Data does not pre-exist its generation and it is not fixed. Big data is situated within a power/knowledge matrix that must be politicized through an analysis of the actors and structures that enable its creation, collection, and analysis (Iliadis and Russo 2016). Carolan (2022) accordingly suggests that in addition to understanding what big data *is*, scholars must investigate what data-assemblages *do* in relation to politics and world-making.

In the processes of accumulation by dispossession (Harvey 2004), data provides a 'fix' to overaccumulation without spatial expansion (Thatcher, O'Sullivan, and Mahmoudi

2016). Datafied accumulation by dispossession in agriculture entails a recursive process in which 'digital agriculture firms build their profits on appropriating data produced by (and ostensibly for) farmers' (Duncan et al. 2022, 12; Fraser 2019). This recursive process not only entails the transfer of data from generator-farmers to digital agricultural firms but also creates new objects of property and value, merging 'commodification and theft into one moment' (Nichols 2020, 8). Datafied recursive dispossession facilitates data rentiership (Birch, Chiappetta, and Artyushina 2020), whereby agribusiness and agri-tech firms transform agricultural data into a private asset capable of constructing and extracting value in service-based business models such as Farming as a Service. The neoliberal state accelerates this process by constructing both the requisite digital infrastructure and regulatory conditions for agribusiness and agritech's datafied capital accumulation by dispossession (Stock and Gardezi 2021). Miles (2019) demonstrates how the epistemological relations between agricultural industrialization and information enacted in precision agriculture technologies produce an algorithmic rationality 'that clothes a number of positioned values in the guise of simple, quantifiable, technical facts' (3). Algorithmic rationality circulates in India through both state and corporate discourses that position climate-smart and precision technologies as technical, capital-intensive solutions to both climate and food insecurity (Stock and Gardezi 2021). CSA in India, as I will attempt to show, is a significant conduit to instigate the embedding of this algorithmic rationality and construction of climate-smart subjectivities amenable to the Farming as a Service model.

CSA emerged globally out of contemporary global crises in accumulation in 2007–9 that saw volatility in global food commodity prices in conjunction with changes in international climate policy that favored financialized responses to the climate crisis. The Intergovernmental Panel on Climate Change simultaneously recognized linkages between global agricultural production and climate change, acknowledging the potential for reducing greenhouse gas emissions by linking climate change policies to sustainable development initiatives in agriculture (Smith et al. 2007). As in the financialization of any feature of life, rendering food and agriculture into sites for financial investment and accumulation necessitated the abstraction of its commodifiable subject matter from larger, constitutive material and cultural contexts. This was exemplified in the transformation of agricultural commodities into complex financial instruments such as agricultural commodity derivatives, which have been largely deregulated since the 1990s (Hartmann, Mwaka, and Dannenberg 2021). Several underlying factors constitutive of industrial agricultural practices deepened the incursion of finance into agriculture: the decreasing availability of arable farmland, the conversion of farmlands into biofuel production, and the rise of speculative activity in agricultural commodities, particularly flex crops (Borras et al. 2016; Clapp and Isaakson 2018). The opportunity to financialize both agriculture and the climate crisis by creating new market instruments was reflected in shifts in international climate policy, which resulted in the formation of CSA as a means to increase yields and farmers' incomes while reducing greenhouse gas emissions and incorporate a form of financialization tied to development.

Contemporary studies in agrarian political economy are often bound with political ecology to show how relations between changing forms of global capitalist production, labor, and land/environmental/natural resources produce distinct spatialized forms of agrarian dispossession, reconfigured state-society-environment relations, and resistance

(Baviskar 1995; Escobar 1995; Guha 1989; Hecht and Cockburn 2011; Peluso 2017; Vandergeest and Schoenberger 2019; Watts 1983, 2003). Borras and Franco (2018) demonstrate how CSA is a conduit for land grabs in the name of climate conservation and agricultural efficiency, which also spurs financialized responses to climate mitigation/adaptation such as carbon trading and the commodification of agricultural practices characteristic of agrarian moral economies. Newell and Taylor (2018, 2020) further how CSA's institutional roots in UN agencies such as the World Bank, FAO, and CGIAR have enabled corporate agribusiness to dominate its governance agenda, while positioning financialization of the global food system as a response to climate change through new agricultural technologies and mitigation practices. These features of CSA's political economy, as I attempt to show in the Indian context, are intensifying through the introduction of big data technologies intended to 'translate contentious politics into manageable technocratic responses' (Newell and Taylor 2018, 16). CSA's role in accumulating agricultural information and data is a key aspect in this process. The political economy of CSA under the conditions of information capitalism must contend with the ways in which CSA works as a mechanism for information and data accumulation.

CSA involves a strategy to increase food production under conditions of climate change with three interrelated objectives which its advocates suggest constitute a 'triple-win': (1) increasing agricultural productivity to ensure food security, (2) reducing greenhouse gas emissions while enhancing farmer resilience to climate change, and (3) raising farmers' incomes to accomplish the goals of sustainable rural development. Conceptually, CSA spans production systems, specific agricultural practices and technologies, national policies and institutional arrangements, and financing mechanisms which conjoin agricultural and climate change objectives. No particular practice or technology is absolutely or explicitly designated 'climate-smart'. Instead, agricultural practices and technologies are recognized as climate-smart by fitting within the rubric of the 'triple win' goals. This criterion represents a sliding scale; an irrigation technique acceptable under the CSA branding might be very productive in increasing yields and farmer resilience to climate change but only scantly reduce emissions, leading some to argue that virtually any existing practice might qualify as climate-smart (Neufeldt et al. 2013).

Agricultural activists are particularly concerned about the deployment of biotechnologies as new forms of accumulation by dispossession which will further smallholder dependency upon corporate agriculture. The official CSA paradigm neither mandates nor dismisses the use of chemical inputs or GM-seeds (Lipper and Zilberman 2018, 27), but the FAO's 2016 *State of Food and Agriculture* report advocates for the use of biotechnology to reduce emissions. Activists have frequently highlighted the use of biotechnologies in developing GM-seeds for CSA in the Water Efficient Maize for Africa project, a ten-year project in which new drought-tolerant maize varieties were developed under the banner of 'climate-smart crops', funded by the Gates Foundation (TWN 2017). They argue that the creation and dispersal of these varieties are intended to create smallholder dependency on largely unaffordable patented seed, as alarmingly suggested by an estimated 1,500 patent applications for climate-resilient seeds in 2015 alone (Shiva 2015).

The creation of engineered climate-smart seeds is accelerating under data-driven plant breeding. Plant genomics and phenomics carry copious amounts of data which are complemented by sources containing smaller amounts of data pertaining to spatial and temporal-specific climate conditions, land and soil, irrigation, greenhouse gas emissions, socio-economic conditions, and agricultural markets (Rao 2018). The collection of plant genomic and climatic data provides the 'raw' inputs for data-driven plant breeding to create evermore patent-protected climate-smart seeds.

Making climate-smart villages and climate-smart mentalities in India

In 2008, India released its *National Action Plan on Climate Change* composed of eight categories incorporating various sub-missions broadly linked to its sustainable development objectives. These sub-missions included enacting equitable water distribution and national forestry efforts to increase India's green cover to enhance carbon sequestration. The Plan also included a National Mission on Sustainable Agriculture (NMSA), designed to encourage the development of climate-resilient crops, the creation of weather insurance mechanisms, and to promote changes in agricultural practices (Pandve 2009). The NMSA shared CSA objectives of increasing both agricultural productivity and resilience to climate change.

Much of the NMSA policy focused on achieving greater efficiency in chemical input management, investments in digital infrastructure to improve knowledge delivery to farmers, and the use of biotechnologies to breed high-yielding and climate-resilient crop varieties. The policy framed investment in agro-forestry as a potential spur to increase carbon sequestration (Dept. of Agriculture and Cooperation 2010, 44). In clear anticipation of the CSA model, the NMSA called for:

> … continuous efforts in improving genetic traits to sustain productivity, both in the short and long term, with focus on research and development of resilient genotypes. Simultaneously, reforms in policies, regulatory regimes and conformance to standards [to] promote large scale research and wider adoption of genetically improved varieties by both producers and consumers. International collaborations with Consultative Group on International Agricultural Research (CGIAR) institutions would be critical for fostering these activities. (21–22)

Beginning in 2012, this consortium of actors targeted Haryana as a key site for the expansion of CSA development initiatives. CGIAR created the Research Program on Climate Change, Agriculture and Food Security (CCAFS) to advance research and implementation with the International Maize and Wheat Improvement Centre (CIMMYT) while collaborating with the Haryana government and the Indian Council of Agricultural Research. Haryana's leadership in agricultural production builds upon its history as a primary beneficiary of Green Revolution technologies and agricultural research. Unsurprisingly, this 'success' has also made it a major contributor of carbon emissions and particularly susceptible to climate change (Ahmad et al. 2019, 89) which position it as an ideal site for CSA.

CCAFS initiated Climate-Smart Villages, which are:

> sites where researchers from national and international organizations, farmers' cooperatives, local government leaders, private sector organizations and key policy planners come together to identify which climate-smart agriculture interventions are most appropriate to tackle the climate and agriculture challenges in the village. (CGIAR-CCAFS-CIMMYT 2014, 3)

CGIAR partners with the International Center for Tropical Agriculture (founded by the Rockefeller Foundation), illustrating the ongoing influence of Green Revolution actors in contemporary rural development initiatives. This is a key partnership that positions CGIAR and CCAFS as the lead actor implementing CSA in India. This web of public-private partnerships provides a glimpse of the multi-scalar and multi-national governance structures that subject bodies such as CCAFS to various forms of discipline.

CSVs are local test sites for assessing which CSA technologies and practices most effectively accomplish all three CSA objectives with the eventual intention of scaling-up successful actions in national policies and programs (Vincent et al. 2011). They further operate as sites of governmentality, where Indian farmers were first prepared as subjects to adopt the tools and behaviors needed for data-driven agriculture. The more that they interact with CSV agricultural technologies (instruments, seeds, digital platforms) and become trained in a particular CSV vocabulary, the more 'farmers begin acting like algorithms' (Gardezi and Stock 2021, 2) for use in the FaaS model.

In the earliest Haryana CSV initiatives commencing in 2012, the Karnal district was an attractive site for pilot projects in part due to its proximity to several national agricultural research bodies. These research bodies were Green Revolution-based sites of top-down expert knowledge production, surveillance and agricultural information collection through their control over state-held seedbanks and extension services, as well as laboratories for the production of high-yielding seeds. Karnal district is home to the Central Soil Salinity Research Institute, the National Dairy Research Institute, the Directorate of Wheat Research, and Haryana Agricultural University, all of which were partners with CCAFS and CIMMYT (CCAFS-CIMMYT 2014, 4–5). These actors assisted in creating and dispersing the CCAFS Baseline Household Level Survey in eight randomly selected villages in Karnal district to assess their access to resources, livelihood activities, and cropping practices.

Portfolio creation was a participatory process that involved researchers, farmers, state officials, and private stakeholders in selecting the most appropriate CSA technologies and practices building upon the baseline assessments (Aggarwal et al. 2018). In the pilot CSVs in Haryana and Bihar, consultations with local communities led to the creation of village committees made up of farmers, researchers, and local planners to implement the village portfolio, which contained specific practices, crops, and technologies.

CSVs established farmer learning networks targeted to engage rural women consistent with globally mandated gender inclusive participatory development objectives. CCAFS worked with an Indian development NGO Alternative Futures, to produce a toolkit titled 'Training of Trainers': Summary Manual on Gender, Climate Change, Agriculture and Food Security'. This toolkit established capacity-building workshops to train female leaders to educate other women in CSV villages about the links between climate change, agriculture, and food security relevant to their livelihoods using examples that would resonate with them (Vincent et al. 2011). The objective was 'to empower women by showing them that their own experiences validate the scientific', asking them to 'consider any changes they have observed in climate over their lifetime and their mother's lifetime' with trainers explaining 'how their experiences match the projection of scientists' (56–57). Women were also made aware of government loan and credit programs to reinforce to them their 'need to connect more with government agencies' (Vincent et al. 2011). Feminist theorists have shown how similar programs in various contexts have operated as conduits for gendered accumulation by dispossession (Keating, Rasmussen, and Rishi 2010).

This project represented an attempt to legitimate the CSV as an official state project and to reshape women's subjectivities by teaching them to articulate their experiences in relation to climate change in scientific terms, using the vocabularies of greenhouse gases, climate change indicators, and water cycles and to adapt their understandings and experiences in ways conducive to the adoption of the agricultural technologies introduced by the CSV program. The capacity-building workshops sought to empower female participants to adapt successfully to projected changes by responding to 'available government plans and programs and field-based adaptation models by non-government actors' (Vincent et al. 2011, 6). Similar to participatory rural appraisal projects that operate as technologies of responsibilization with the effect of rendering rural communities technical (Li 2011), CSV capacity-building workshops nurtured 'new ways of thinking and acting' (101). If participant farmers knew how to identify and articulate the effects of climate change on local agriculture, they could then be responsibilized to adapt to such changes more scientifically and thus more effectively. They possessed the potential to be 'empowered' to make rational calculations, which in this case, involved the adoption of precision agricultural technologies.

As Rose and Miller (1992) suggest, 'making people write things down and count them … is itself a kind of government of them, an incitement to individuals to construe their lives according to such norms' (187). CSV participants were first trained to articulate their encounters with the effects of climate change in a modern, technoscientific lexicon. It is through mundane calculative technologies such as these exercises that the subjects of CSVs were constructed, and the rationales of CSA translated into their everyday practices. In learning how to express their experiences in a particular scientific vocabulary, these agrarians were actively engaged in transforming their lived experiences into a legible domain of governmental intervention through the introduction of precision agricultural technologies.

Rather than being condescended to as bearing outmoded and ineffectual traditional knowledges and practices, as they might have been only a generation earlier, women were put into a dialogue of sorts with practitioners of modern science. CSV neoliberal participatory technologies thereby illustrate a significant divergence from the governmental technologies of the Green Revolution, in which peasants were constructed as parochial, traditional, and in need of foreign technological expertise. The neoliberal agricultural subject is, instead, made to feel like a responsible stakeholder in development projects.

This particular dimension of neoliberal development has the capacity to imbue agricultural subjects with new capacities to recast and reshape projects that engage them in forms of neoliberal government. Whereas twentieth century agricultural development projects disregarded, if not ridiculed, the values and understandings of agrarian subjects, neoliberal governmentality recognized, validated, and incorporated these in the way it shapes its subjects and orients them to its projects of government. Algorithmic government extends this project of incorporation through 'nudging' techniques, which will be addressed in my discussion of FaaS.

Particular agricultural technologies and practices became popular in efforts to meet CSA objectives. These became indicia of success and integral to reconfiguring the agrarian subject. Laser Land-Levelers (LLL), a farmer-operated, tractor-towed, laser-controlled device that flattens and evens the soil surface, proved to reduce greenhouse gas emissions by reducing fertilizer usage while increasing crop yields in rice and wheat (CCAFS-CIMMYT 2014). Smallholders rented LLLs at a rate of 6–700 rupees per hour, instead of purchasing them due to their prohibitive cost, which illustrated the growing 'Uberization' of farm machinery in India. While this led to greater accessibility for smallholder farmers, it also largely benefitted private agribusinesses such as Mahindra & Mahindra (India's leading tractor manufacturer and an agri-chemical company), and the state. When such technologies were rented from state-run, custom-hiring call centers, the Ministry of Agriculture was able to deploy an agricultural app described as 'Uber for tractors' to track machinery usage and prices, providing "an invaluable database for policy-makers … to track the usage of new technology that the government wants to promote"' (Jebaraj 2019). This model, as I will show, fits squarely within the imperatives of FaaS.

Levelers were also prerequisite tools for the expanded use of precision-farming technologies, such as the GreenSeeker handheld crop-sensor device. When held above a crop canopy, the GreenSeeker collected agronomic data used to indicate crop health as a prediction of future yields as well as nutrient and nitrogen requirements for a particular plot. The precise readings of these hand-held devices 'allowed farmers to reduce over-application, reducing fertilizer costs and avoiding air and water pollution' (Lapidus et al. 2017, 2). The sensor collected information by sampling the scanned area; when the farmer released the device's trigger, it displayed the measured value of plant health and fertilizer needs, which the farmer recorded and input into an algorithm (Abit and Arnall 2016). This algorithm, significantly, was an essential component of the Nutrient Expert decision-support software, a computer-based tool jointly developed and owned by CIMMYT and the International Plant Nutrition Institute (CIMMYT 2013). This proprietary software analyzed the crop data collected by GreenSeeker and provided tailored information on growing conditions and natural nutrients in the soil. It also sent recommendations for improvements in local practices directly to farmer's phones (Sapkota et al. 2021). Hence, beginning in 2012 and in practices that have been accelerating since 2014, Indian farmers have freely gathered and contributed farm data and information that continues to create ongoing value for actors who hold intellectual property rights in data-based proprietary algorithms and agricultural platforms, while also enriching the state in garnering massive investments from foreign and domestic agribusiness.

These technologies become extensions of a particular algorithmic agricultural subject through capacity-building exercises which support new forms of financialization in which farmers themselves become vectors for information's commodification as data in supply chains. These early CSV enterprises prepared farmers themselves to serve as technological inputs into a flow of data from which others will profit and use as the basis for further financial calculation.

As of 2019, CCAFS began explicitly incorporating big data analytics into CSVs. The 'Big Data Analytics to Identify and Overcome Scaling Limitations to Climate-Smart Agricultural Practices in South Asia' project is a follow up to the pilot Haryana CSV project. CCAFS is working with India's National Research and Extension Systems (NARES) and the domestic private sector:

> The Big Data analytics for climate-smart agriculture in South Asia (Big Data 2 CSA) project responds to the limitations of plot-based agronomy by developing digital data collection systems to source, data-mine and interpret a wide variety of primary agronomic management and socioeconomic data from tens of thousands of smallholder rice and wheat farmers in India … This is carried out in partnership with national research systems and international partners in digital agriculture across the region. (CCAFS 2019)

The project intends to deliver customized CSA recommendations to at least 500,000 farmers. The first completed activity involved capacity building activities to enhance data collection. Ongoing activities, such as the picture-based insurance scheme, range from developing crop advisories transmissible through digital formats, converging existing data from remote sensing and spatial collection into a single dataset and transferring it to a customized cloud platform for processing, and utilizing previously collected data in machine learning to build a predictive capability on crop performance pertaining to CSAs triple goals. It further led to the creation of the Global Agricultural Research Data Innovation & Acceleration Network (GARDIAN), a digital platform that allows to users to search for over 200,000 + reports, 26,000 + datasets, and 12 + terabytes of localized soil and climate data, cultivation practices, and crop and land profiles: 'The ultimate goal is to seamlessly integrate real-world data from farms around the world into algorithms that generate critical insights that can then be shared back with farmers' (Brennan 2018, n.p.) The Indian Council for Agricultural Research and the Open Government Data Platform India are two of the project's key data providers, in addition to the US, UK, and World Bank, while Amazon, Google, IBM, and management firm McKinsey & Co. are among its external partners. GARDIAN emphasizes fair and open access with little concern for actors possessing the power to appropriate and capitalize such data, namely large and powerful corporations. GARDIAN thus represents an agricultural data-as-commons strategy of accumulation (Nye 2019) which naturalizes Indian agricultural data as a natural resource that is available for appropriation and use by corporate actors.

'Farming as a Service': A Model for Datafication & Dispossession

In 2017, the Indian Institute of Management and a leading American consulting firm, Bain & Company, co-authored a report titled *Indian Farming's Next Big Moment: Farming as a Service* (hereinafter the Report). The Report is the first publication to articulate FaaS, an emergent data-driven agricultural paradigm in which various components of farming are transformed into services, available on a subscription or pay-per-use basis. FaaS applies to three categories of farming activity: farm management solutions, production assistance, and access to markets. Farm management solutions rely on data-collecting farm machinery and tools, such as the GreenSeeker, to gather farm data pertaining to soil and seed quality, weather conditions, input use, and monitoring crop output. This data is processed and disseminated to farmers though mobile alerts, digital dashboards, and application platforms. Whereas farmers typically initiate extension services when they determine that they need outside advice, real-time farm advisory under FaaS would allow extension agents to monitor and contact farmers to suggest farming practices. Further, the processed farm data would be shared with financial institutions to facilitate loans, assess risk management, and process crop insurance claims (Kedia 2018). Access to markets involves using platforms to link farmers directly to agribusinesses to purchase seeds and inputs, while eliminating intermediaries (such as commission agents) who are integral to the current Mandi (market) procurement system. As of 2018, FaaS had garnered roughly $115 million USD in private equity investments, much of which is directed to supporting domestic agri-tech startups.

Implementing FaaS is contingent upon an extensive digital infrastructure capable of delivering internet connectivity to farmers. An early project involved ITC, a leading Indian food conglomerate, which created 'e-Choupal', an internet kiosk installed in rural villages through which ITC procures agricultural commodities directly from farmers while also providing advisory services and seed and input suggestions. Kumar (2004) shows how these early projects were primarily driven by the potential to create a more extensive digital agriculture infrastructure: '… the eChoupals … generate value only when they enable the creation of an alternative agricultural infrastructure through the exchange of information … ' (48).

The Modi government launched Digital India in 2015 as a campaign to build a national digital infrastructure in order to electronically deliver nearly all government services, with an emphasis on enhancing broadband connectivity in rural areas. Its 2020 budgetary allocation reached over $500 million USD (The Economic Times 2020) in addition to attracting a monetary pledge from Google to invest $10 billion USD over the next 5–7 years. The digital mapping platform, Bharat Maps, is part of Digital India's National GIS Mission, which provides a tool to visually represent geographic information pertaining to natural resources and governmental sites such as post offices, banks, and schools for location specific planning, decision-making, and monitoring.

Digital India further spans *Aadhaar*, the world's largest national biometric identification system with 1.2 billion enrollees (Rao and Nair 2019). *Aadhaar* stores fingerprints, iris scans, and facial recognition photos along with biographical information for each Indian citizen. Access to basic welfare services all run through *Aadhaar* and the 12-digit number assigned to each registered citizen. *Aadhaar* stores personal data within a centralized database comprised of 7,000 servers located in Haryana and Karnataka, providing a digital foundation for similar initiatives across various sectors. In 2020, the federal government announced plans to create the first *Aadhaar* authenticated database for farmers' data, which will include a broad range of agricultural information from individual farmer socio-economic data to farmland-specific data (Haq 2019). Indian sub-states are beginning the process of verifying landholdings to link these with the farmer beneficiaries whose information is already stored in the central *Aadhaar* database. The collected and processed data will reside in *Agristack*, the first national database that will store data about farmers' and farming practices.

The *Agristack* database is one of several public-private projects intended to build an infrastructure for data-driven digital agriculture. Of India's approximately 630,000 rural villages, it is estimated that less than 1% are connected to the internet (Bellampalli 2018, 33). The *Kisaan Call Center* functions as an information gateway in enabling farmers (who are given mobile phones) to call agricultural scientists for advice on using chemical inputs and hybrid high-yielding seeds. Farmer's inquiries, whether orally communicated over the phone or through SMS, are recorded by Call Center operators. After several years of operation, the Call Center has collected a large amount of data which is now in the process of conversion into a structured dataset amenable to big data analytics (Agarwal and Tripathi 2017). Mobile apps such as *Pusa Krishi* inform farmers about newly available crop varieties, tech-based agricultural tools, and insurance apps that can generate insurance premiums and loan amounts for crops based on farm conditions. Geographic Information Systems are used to test for soil quality, weather forecasting, and crop yield predictability (Bellampalli 2018, 32–35).

In response, Farmer Producer Organizations (FPOs), and state-organized farmer/producer collectives, now express concerns about the growing connections between the state and private actors:

> *Agristack* could strengthen the asymmetry in information flow by providing all information about farmers and their farming easily to corporations who looked at farmers as a consumer base, be it agri-inputs – seed, chemical fertilizer and pesticides, machinery companies or fin-tech companies and to those for whom farmers were suppliers like the food industry. (Rajesh Krishnan, quoted in Kapil 2021)

Without a national data privacy law yet in place, Microsoft could sell this accumulated personal, land, and agricultural data to agribusiness firms, as well as insurance companies and financial firms. The commodification of farmer data represents a new revenue stream within FaaS, while also illustrating how such informational commodification is embedded within the CSV program as well as CSA at large. The state-led creation of national digital infrastructure is crucial in connecting CSA to FaaS. Yet, this must be contextualized alongside the Indian state's engagement with local and multinational corporate actors of various sizes.

India is currently home to over 450 agri-tech startups which attracted investments reaching almost $250 million USD in the first half of 2019 alone (NASSCOM 2019). Indian agri-tech start-ups have launched online services, web/mobile applications, and software to collect agricultural data and apply big data analytics, mimicking the features found in Monsanto's farm data projects. In 2015, India established the Agri Udaan program to provide funding and infrastructural support for selected Indian agri-tech start-ups 'to promote innovation and entrepreneurship in agriculture through rigorous mentoring, networking and helping the startups connect with potential investors' under the management of the Indian Council of Agricultural Research. Agri Udaan's areas of activity include sustainable inputs, precision/smart agriculture, agricultural biotechnology, and agricultural financial technologies.

Various state organizations such as the Ministry of Agriculture and Farmers Welfare and the Ministry of Electronics and Information Technology are key actors in this field of neoliberal governmentality, introducing freely accessible information on crop prices and advisories, but the private sector controls and directs the vast majority of current data-driven projects (Kumar and Basu 2022; Shankarnarayan and Ramakrishna 2020). It is crucial to understand, however, that these projects are the result of a deepening relationship between the state and private corporations. For example, one of India's most influential national think tanks, NITI Aayog, entered a partnership with IBM in 2018 to develop predictive crop yield models to enable real-time advisory services to farmers. State governments are entering into public-private partnerships to further galvanize domestic agri-tech start-ups and implementing agri-tech solutions. The government of Uttar Pradesh is working with the Gates Foundation and Tata Trusts to establish the Indian Agritech Incubation Network at the Indian Institute of Technology-Kanpur (Ernst and Young 2019). CropIn, a burgeoning Indian agri-tech firm, collaborated with the Government of Karnataka's Department of Agriculture to utilize its SmartFarm software for digitizing already recorded farm data, providing real time crop monitoring and yield predictions, 'farm to fork' traceability, and precise weather advisories.

The national government similarly engages with multinational corporate actors. In April 2021 the federal Department of Agriculture, Co-operation and Farmers' Welfare signed a Memorandum of Understanding with Microsoft to jointly collect and digitize farmers' personal information along with land records, fertilizer usage, and crop estimation (Saha 2021). Farmers' data that is collected and stored in the state-owned Agristack database will be shared with Microsoft: 'The government, through this Memorandum of Understanding, aims to provide "required data sets" of farmers' personal information to Microsoft to develop a farmer interface for "smart and well-organised agriculture"' (Kapil 2021). India's agri-tech industry is estimated to reach $24 billion USD in revenues by 2025, while currently holding roughly 1% market penetration (Srivastava 2021). Srivastava (2021) further confirms: 'So far, the government has seeded publicly available data for more than 50 million farmers of the 120 million identified land-holding growers' (n.p.). The state continues to cultivate the conditions for nationally-based firms to become major actors in datafying Indian agriculture, through a development rationale in which the state increasingly becomes the 'owner, supervisor, and promoter of capital' (Alami, Dixon, and Mawdsley 2021, 18).

FaaS identifies smallholder farmers as both the reason for FaaS and as its primary beneficiaries. In assuming that small farms cannot produce enough food to feed a growing population, the Report suggests that small-land holdings (less than 50 hectares) and a lack of access to technology and real-time information are the key obstacles to enhanced productivity and efficiency. Smallholders, the Report suggests, have chemically degraded soil due to their own poor farming practices; they often make bad choices in crop selection due to a lack of information about high-yielding seeds and chemical inputs (7), both of which can be resolved using real-time advisory services and off-site expert monitoring. Smallholders are considered 'low skilled' and must be taught how to select and plan better for demand, crop selection and cropping patterns, achievable both through real-time advisory services and greater use of precision tools (28). Such precision tools along with modern farm machinery are, of course, expensive for smallholders, creating new needs for rental services at custom hiring centers, as well as greater access to credit and loans.

In accessing markets, experts believe that smallholders are often taken advantage of by supply chain middlemen who skim their already meager profits. Brooks (2021) shows how these 'street level bureaucrats' often work with smallholders to negotiate fairer terms for loans and financing, and thus present opportunities for smallholders to exercise greater agency when dealing with financial lenders (385). Aga (2018, 2021), however, demonstrates how small and small-medium farmers in Maharashtra have not only become 'the vehicle or the deepening penetration of agribusiness capital' (14) by becoming retail sellers of agri-inputs, but also receive agribusiness capital to extend credit to local smallholders, effectively creating a downward flow of risk that enables agribusiness to mitigate market and climatic risk 'by passing the risk onto the shoulders of retailers' (16).

Under the FaaS model, smallholders will be left to directly negotiate with large retailers and supermarkets through digital platforms. The smallholder's alleged agricultural ineptitude combined with institutional failures to provide him with access to agricultural technologies and institutional inefficiencies in supply chain mismanagement constitute problems resolvable through FaaS.

In FaaS, acts of 'nudging' guide smallholders towards making correct, rational market-oriented choices (Brooks 2021). Nudging entails making micro-changes in a subject's environment so that they will adjust their behavior without coercion, or even without knowing that they are being induced into altering their behavior. The subject is made to feel that they are making their choices freely, if not intuitively, and for their own well-being. For example, adding calorie counts to food menus is a nudge that steers consumer behavior towards making healthy food choices, rather than banning or taxing the sale of unhealthy foods. As Thaler and Sunstein (2008) comment 'to count as a mere nudge, the intervention must be easy and cheap to avoid. Nudges are not mandates. Putting the fruit at eye level counts as a nudge. Banning junk food does not' (6). Nudging, then, offers a corrective to neoclassical conceptions of the subject as an already formed rational actor by suggesting that the subject often requires soft guidance in making rational choices. In Indian FaaS, nudging unfolds through smallholder use of mobile phones and digital platforms, where service providers and agri-tech firms constantly monitor farmer behavior. Prior to the introduction of big data in agriculture, participant farmers received text messages informing them about local pest outbreaks or expected weather disruptions:

> … texts, voice messages, and specialized call-in radio shows let them [farmers] know about upcoming severe weather, or what to do to prevent pest infestation. Others link them to digital farm development loans tailored to their growing and lean seasons to make repayment easier. Many are simple tips to improve their soil and crop yield, or reminders to help them stay on track. (Grameen Foundation n.d., n.p)

The language of 'tips' and 'reminders' exemplifies the neoliberal discourse of nudging, or as the Grameen Foundation suggests: 'Make self-help a way of life. One text at a time' (n.d.). 'Tips' and 'reminders' present themselves as part of a menu of choices that the farmer, it is implied, can select or ignore. This makes the decision-making process appear to be a matter of the farmer exercising their agency, but, as Brooks notes, nudging actually aims to 'steer farmers towards choices made for them, rather than mobilize their innovative capacity and agency' (2021, 384). By adding big data and predictive technologies into the fold, farmers receive digital recommendations customized to match their preferred language/dialect and literacy levels for chemical inputs suited for their specific soil conditions, long term weather forecasting, and market prices (Fabregas, Kremer, and Schilbach 2019). They will not be penalized if they ignore the recommendations. Yet, they enter a scenario where their failure to adhere to these prescriptions renders them particularly susceptible to individualized blame and could subject them to further interventions that diminish their autonomy and decision-making.

Datafied contract farming and agricultural insurance in faaS

As FaaS utilizes such insidious 'nudging' techniques, it simultaneously deploys disciplinary technologies to control its subjects more explicitly. Contract farming is a neoliberal legal technology that is designed to further manipulate and erode farmer self-sufficiency. Under contract farming, buyers procure agricultural commodities directly from farmers. PepsiCo, for example, has engaged in contract farming in India to produce potatoes for Lay's chips since the early 1990s (Sinha et al. 2021). Buyers and farmers establish the contractual terms of production: a precise quantity of the desired crop, quality standards, time to harvest, and a specific remunerative price for the farmer, with technologies, agricultural inputs, and extension services typically provided to the farmer (Singh 2002).

From the perspective of international financial institutions and institutional economics, contract farming represents an ideal pathway to modernizing agriculture while raising farmers' incomes (World Bank 2007). It provides certainty and predictability for buyers in setting the terms of production, while smallholders can access corporate (patented) seeds and inputs and receive a fixed price for their harvests (Eaton and Shepherd 2011). For critics, however, contract farming is a 'flexible strategy of accumulation' for agribusiness in the transition of smallholder farming towards commercial farming without burdening the buyers with the need to invest in land (Little and Watts 1994). In a recent study in Malawi, Adams et al. (2019) argue that through contract farming, 'corporations can secure access to land, [and] exert control over local communities', while at the same time the practice 'changes rural agrarian relations, transforms local family institutions by carefully selecting a few household members with influence into the scheme and selectively dispossess[es] the poor community members' (1435). Guthman (2017) further suggests that contract farming is characteristic of rent-seeking behavior; the buyer can incorporate nearly all of their products (from inputs to packaging and marketing) into the arrangement or charge fees for the inclusion of such goods 'that build in above-normal returns to the buyer – economic rents' (103).

Recent studies of critical agrarian political economy in India illustrate how contract farming fundamentally represents the corporatization of Indian agriculture (Cohen, Vicol, and Pol 2022; Shrimali 2021; Vicol 2019). Using a case study in Punjab, Shrimali (2021) argues that contract farming is a 'state-mediated capital accumulation strategy that brings together various fractions of capital' (14) which renders farmers as consistent consumers of agri-inputs, facilitates corporate access to state-based financial resources, and depends on the Indian state to establish the regulatory conditions for corporate interventions in domestic agriculture. However, agricultural institutional economists and critical agrarian political economists have yet to consider the impact of big data on contract farming. As Ravis and Notkin (2020) suggest, datafied agriculture is rapidly becoming a key tool for a highly controlled and centralized contract farming. Agricultural platforms enable direct communication between firms and farmers, while the accrued and processed farm data provides proprietary, up-to-date granular information that can be sold to insurance firms, ag-input providers, and food retailers. Microsoft, for example, has been developing digital advisory and predictive farming applications in partnership with the International Crops Research Institute for the Semi-Arid Tropics (ICRISAT), a CGIAR-affiliated organization based in Southern India. With the intensifying datafication of Indian agriculture, the social relations constitutive of the 'moral economy' of contract farming (Narayanan 2012) appear primed for erosion.

Contract farming further spawns depeasantization, whereby farmers directly contribute to agribusiness capital accumulation by working their land for fixed wages and, in essence, alienate themselves from smallholder farming practices (Araghi 1995; McMichael 2013; Shrimali 2021). Relatedly, depeasantization entails agricultural deskilling, in which farming not only becomes automated but the ability of farmers to innovate, experiment, and ultimately make decisions about their crops in interaction with local ecologies is diminished (Stone 2007). Under datafied contract farming, smallholders will receive algorithmically-informed recommendations discursively framed superior, data-based information (Gardezi and Stock 2021). As more smallholders are incorporated into contract farming tied to digital agriculture, the potential for deskilling and ultimately depeasantization is likely to proceed more rapidly than ever before imagined.

Contract farming represents a modern juridical mechanism to standardize, marketize, and discipline what Vicol (2019) designates the 'biological rhythm' (139), or the ecological uncertainties and risks, of smallholder agriculture. A contractual agreement, of course, cannot tame the ecological uncertainty inherent to smallholder agriculture. Buyers often extend loans to farmers to purchase the chemical inputs that the buyer makes available. It is an ideal scenario for buyers because they control the terms of production while the farmer incurs the risk in terms of crop failure or weather damage. These dimensions of contract farming along with the rise of flex crops have spurred the mainstreaming of data-driven agricultural insurance, which have been constitutive of strategies of financial inclusion in international development since 2009 (Soederberg 2013). International donors and financial institutions utilize digital platforms and mobile phones to extend loans and credit to smallholders previously illegible to them, creating new markets for investment of financial capital (Gabor and Brooks 2017). This financial turn attempts to cultivate entrepreneurial behavior in smallholders which institutions such as the World Bank and corporate agribusiness players deem essential to incorporating smallholders into global value chains (Brooks 2021), coinciding with the rise of flex crops highly amenable to financial speculation (Borras et al. 2016). Isakson (2015) argues that in India, 'insurance brokers and other actors who stand to benefit from the marketization of risk management have engaged in far reaching discursive and pedagogical interventions aimed at teaching farmers the "rationality" of insurance and "structurally adjusting culture," with the aim of "creating effective demand"' (269), while Da Costa (2013) details the various campaigns in which the Indian state, development NGOs, and philanthropic organizations have engaged in educating smallholders in financial literacy since 2005, with the aim of producing a demand for weather index-based crop insurance schemes (856).

Datafied contract farming coupled with agricultural insurance exemplifies financialized digital agriculture. Contracts are formulated through records of production coupled with agricultural insurance programs, which are then incorporated into the range of services embedded within the contract (Roberts 2005). Precise and granular agricultural data, then, is a crucial input in contract farming and in the financialization of agriculture more broadly. The amassing and processing of big data in agriculture presents an opportunity to gather and utilize enormous amounts of information in a quick, economical manner. Combined with farm management platforms, this enables direct firm-to-farm communication and oversight while also augmenting buyers' ability to predict and coordinate crop prices with production and demand (Ravis and Notkin 2020). A recent report authored by GRAIN (2021) further comments on this aggregation of agricultural data and platforms, contract farming, and agricultural insurance:

> … they [farmers] must buy the inputs that are promoted and sold on credit (at high interest rates), follow the 'advice' of the chatbot to qualify for crop insurance (which they must pay for), sell their crops to the company (at a nonnegotiable price), and receive payments on a digital money app (for which there is a fee). Any missteps can affect a farmer's credit worthiness and access to finance and markets. It's contract farming on a mass scale. (7)

CCAFS, for example, has engaged in 'picture-based insurance' in Haryana, Odisha, and Tamil Nadu under the banner of 'seeing is believing' since 2018, aiming 'to provide smallholder farmers with low cost, timely and accurate crop insurance by using the novelty of smartphone and remote sensing technologies, sometimes bundled with agricultural credit and advisories' (Rajeesh 2021, 9). Farmers take photos of their crops and download a free mobile application. They then upload these initial photos to the application, beginning at sowing and continuing through harvest. Local agronomists then analyze the photos to verify insurance claims and proceed with payouts for successful claims. Additionally, these photos will be aggregated into a publicly accessible image database so that they may be utilized as inputs for machine learning and artificial intelligence algorithms which generate the e-advisory/extension services prevalent under digital agriculture. As one crop advisory project partner states:

> After a few years, we will have a large database of crops and diseases to build an artificial intelligence system. This will empower plant experts and farmers to detect and match the disease on their crop with the database at our end, and with a click of a button receive precise instant advice. (Quoted in Rajeesh 2021, 21)

The concluded initial phase of the project showed that advisory experts, or 'plant doctors' in this project, can craft highly personalized messages targeted at a farmer's individual situation with the effects of increased adoption of the insurance product while insurers' increased access to monitoring data allowed them to provide recommendations that minimized risk and payouts:

> Results from initial trials in India found that the advisory service became more effective when bundled with insurance services – farmers' engagement increased when the advisory was associated with their insurance policy, and farmers' willingness to pay for the bundled advisory-insurance package was higher than when the two services were offered separately. These synergies between insurance and advisory services could encourage farmers to adopt more climate-smart resilience technologies. (Kramer quoted in McDade 2019, n.p.)

The project's next phase will emphasize the adoption of commercial crops and linking digital crop registries to insurance providers to assess credit worthiness and other financial resources. Such a scenario represents a 'data grab' (Fraser 2019) whereby Indian smallholders actively participate in activities that will contribute to their dispossession.

FaaS, therefore, relies on complementary neoliberal technologies to both shape and regulate the conduct of Indian smallholders. While 'nudging' techniques unfold quietly, neoliberal legal technologies of contract farming and datafied agricultural insurance are explicit attempts to surveil farmers and correct what might be considered 'deviant' farming behavior under FaaS. This arrangement of technologies simultaneously renders farmers more productive in capital accumulation while concurrently enhancing the ability of corporate and state actors to exercise disciplinary power over them. Datafied contract farming enacted in conjunction with datafied agricultural insurance carries immense potential to funnel already economically vulnerable smallholders into debt relations characterized by dependence and dispossession (McMichael 2013, 671). These new modes of subjectification and discipline provide a glimpse into an imminent, bleak era of agricultural dispossession.

Conclusion

This essay has attempted to show how the past and future CSA/CSV projects align with the neoliberal FaaS model, as well their overlap with the imperatives of state-supported corporatized and financialized agriculture at large. These projects, paradigm, and the Indian state-corporate nexus fundamentally revolve around information accumulation as inputs for big data analytics and agricultural algorithms. The datafication of Indian agriculture constitutes the most recent historical iteration of accumulation by dispossession, characterized and reinforced specifically by new modalities of algorithmic subjectification that are well suited to the financialization of agriculture and ultimately the dispossession of Indian smallholders. As Sagari Ramdas (2022) shows, datafied dispossession of indigenous Adivasi agrarians is well underway in Andhra Pradesh. These developments stand as powerful obstacles – though perhaps with some luck and confrontation, also a potential seedbed – for the development an emancipatory agrarian response to climate change.

The recently concluded farmers' protests against the now withdrawn 2020 Farm Bills provided some indication that Indian farmers are increasingly cognizant against the deepening relations between state-domestic corporate agribusiness. Food scholar Raj Patel (2021) commented with respect to the farmers' protests that 'in the camps outside Delhi, you'll hear "Sarkar ki Majboori – Adani, Ambani, Jamakhori," or 'The government is beholden to the hoarders, Adani and Ambani', referencing Indian billionaires Mukesh Ambani and Gautam Adani. These two men, known to have close relationships with Prime Minister Modi, own Reliance Industries Ltd. and Adani Enterprises Ltd., two of India's biggest agribusiness conglomerates. The protests indicated, among other things, an awareness among farmers that the Indian state is proceeding in alliance with both domestic and multinational agribusiness capital.

I echo Fraser's (2020) call for greater convergence between food sovereignty movements and those of data sovereignty. Farmers' movements for food sovereignty and agroecology must reckon with the power of big data technologies, their corporate owners, and enabling state forces. Data sovereignty advocates, particularly those who are the engineers, programmers, and scientists involved in big data analytics and agricultural algorithms and platforms, must be educated in the ethical and political stakes of their work and put into dialogue with food sovereignty movements. While this may seem implausible, we have seen the merging of various distinct but overlapping social movements in the anti-globalization protests of the 1990s (Seoane and Taddei 2002). Movements and scholars critical of the Intellectual Property regime in agricultural/environmental regulation and privacy law scholars must also be included, particularly in light of the datafication of plant and genetic resources that are increasingly amenable to digital biopiracy. India's Traditional Knowledge Digital Library may provide a useful-yet-imperfect starting point in thinking about domestic data governance in the hands of local agrarian communities.

Food sovereignty movements might also find technical support and solidarity from the Global Indigenous Data Alliance and the Indigenous Data Sovereignty Group, both of which are working at the international level towards Indigenous data sovereignty through the UN Declaration on the Rights of Indigenous Peoples (Oguamanam 2020). The UN Declaration on the Rights of Peasants may also open potential avenues for food-data sovereignty engagements. The technologies themselves may also be repurposed as seen in Brazil, where rural communities are utilizing digital agricultural technologies to substantiate land claims, as well as to surveil agribusiness activities (Seufert, Mendonca, and Pitta 2018).

Future battles in Indian agriculture in the context of climate change, and arguably agriculture at-large in the Global South, will be waged through phones, applications and platforms, sensors, and drones, among other information technologies. We may not be able to eat big data (Fraser 2022). However, without a greater sense of urgency and action upon the various issues discussed in this essay, big data may eat the most vulnerable among Indian farmers.

Acknowledgements

Special thanks to Rosemary Coombe and Nate Holdren for their substantive feedback and to Hayley Caldwell for editorial assistance. Thank you to the participants in the Technology in Asia Workshop at Columbia University for commenting on an excerpted version of this essay, as well as my Drake LPS colleagues for doing the same. I also thank the anonymous reviewers for their constructive feedback and the editors who brought together the Forum on Climate Change and Critical Agrarian Studies. I am of course responsible for any errors.

Disclosure statement

No potential conflict of interest was reported by the author(s).

References

Abit, J., and D. Arnall. 2016. "Using the GreenSeeker Handheld Sensor and Sensor-Based Nitrogen Rate Calculator." Oklahoma Cooperative Extension Service, Accessed August 2, 2021. https://shareok.org/bitstream/handle/11244/317978/oksa_pss_2278_2016-01.pdf?sequence=1.

Adams, Timothy, Jean-David Gerber, Michèle Amacker, and Tobias Haller. 2019. "Who Gains from Contract Farming? Dependencies, Power Relations, and Institutional Change." *The Journal of Peasant Studies* 46 (7): 1435–1457. doi:10.1080/03066150.2018.1534100.

Aga, A. 2018. "Merchants of Knowledge: Petty Retail and Differentiation Without Consolidation among Farmers in Maharashtra, India." *Journal of Agrarian Change* 18 (3): 658–676. doi:10.1111/joac.12249.

Aga, A. 2021. *Genetically Modified Democracy: Transgenic Crops in Contemporary India*. New Haven: Yale University Press.

Agarwal, N., and A. Tripathi. 2017. "Big Data Security and Privacy Issues: A Review." *International Journal of Innovative Computer Science & Engineering* 2: 12–15.

Aggarwal, Pramod, K. Andy Jarvis, Bruce M. Campbell, Robert B. Zougmoré, Arun Khatri-Chhetri, Sonja J. Vermeulen, Ana Maria Loboguerrero, et al. 2018. "The Climate-Smart Village Approach: Framework of an Integrative Strategy for Scaling up Adaptation Options in Agriculture." *Ecology and Society* 23 (1): 14. doi:10.5751/ES-09844-230114.

Ahmad, Firoz, Nazimur Rahman Talukdar, Meraj Uddin, and Laxmi Goparaju. 2019. "Climate Smart Agriculture, Need for 21st Century to Achieve Socioeconomic and Climate Resilience Agriculture in India: A Geospatial Perspective." *Ecological Questions* 31: 1–100. doi:10.12775/EQ.2020.008.

Alami, I., A. D. Dixon, and E. Mawdsley. 2021. "State Capitalism and the new Global D/Development Regime." *Antipode* 53 (5): 1294–1318. doi:10.1111/anti.12725.

Araghi, F. A. 1995. "Global Depeasantization, 1945–1990." *The Sociological Quarterly* 36 (2): 337–368. doi:10.1111/j.1533-8525.1995.tb00443.x.

Baviskar, A. 1995. *In the Belly of the River: Tribal Conflicts Over Development in the Narmada Valley*. Oxford: Oxford University Press.

Bellampalli, P. K. 2018. "Role of ICT and Rural Development: A Globalization Prospect." *Wide Spectrum Research Journal* 6 (11): 31–42.

Birch, K., M. Chiappetta, and A. Artyushina. 2020. "The Problem Innovation in Technoscientific Capitalism: Data Rentiership and the Policy Implications of Turning Personal Digital Data into a Private Asset." *Policy Studies* 41 (5): 468–487.

Borras, S. M. Jr., and J. C. Franco. 2018. "The Challenge of Locating Land-Based Climate Change Mitigation and Adaptation Politics Within a Social Justice Perspective: Towards an Idea of Agrarian Climate Justice." *Third World Quarterly* 39 (7): 1308–1325. doi:10.1080/01436597.2018.1460592.

Borras, Saturnino M. Jr., Jennifer C. Franco, S. Ryan Isakson, Les Levidow, and Pietje Vervest. 2016. "The Rise of Flex Crops and Commodities: Implications for Research." *The Journal of Peasant Studies* 43 (1): 93–115. doi:10.1080/03066150.2015.1036417.

Borras, Saturnino M. Jr., Ian Scoones, Amita Baviskar, Marc Edelman, Nancy Lee Peluso, and Wendy Wolford. 2022. "Climate Change and Agrarian Struggles: An Invitation to Contribute to a JPS Forum." *The Journal of Peasant Studies* 49 (1): 1–28. doi:10.1080/03066150.2021.1956473.

boyd, d., and Crawford, K. 2012. Critical questions for big data. *Information, Communication & Society* 15 (5): 662–679. doi:10.1080/1369118X.2012.678878.

Brennan, E. 2018. "Agricultural Intelligence: What AI Can Do for Smallholder Farmers." *Foodtank*. Accessed September 14, 2022. https://foodtank.com/news/2018/10/agricultural-intelligence-what-ai-can-do-for-smallholder-farmers/.

Brooks, S. 2021. "Configuring the Digital Farmer: A Nudge World in the Making?" *Economy and Society* 50 (3): 374–396. doi:10.1080/03085147.2021.1876984.

Carbonell, I. 2016. "The Ethics of big Data in big Agriculture." *Internet Policy Review* 5 (1): 1–13. doi:10.14763/2016.1.405.

Carolan, M. 2022. "Digitization as Politics: Smart Farming Through the Lens of Weak and Strong Data." *Journal of Rural Studies* 91: 208–216. doi:10.1016/j.jrurstud.2020.10.040.

CCAFS. 2019. "Big Data Analytics to Identify and Overcome Scaling Limitations to Climate-Smart Agricultural Practices in South Asia (BigData2CSA)." Accessed January 30, 2021. https://ccafs.cgiar.org/research/projects/big-data-analytics-identify-and-overcome-scaling-limitations-climate-smart.

CGIAR Research Program on Climate Change, Agriculture and Food Security (CCAFS) & International Maize and Wheat Improvement Center (CIMMYT). 2014. "Climate-Smart Villages in Haryana, India." Accessed January 30, 2021. https://cgspace.cgiar.org/handle/10568/42346.

CIMMYT. 2013. "Nutrient Expert™ Decision Support Tools for Maize and Wheat Launched." CIMMYT, 8 July 2013. https://www.cimmyt.org/news/nutrient-expert-decision-support-tools-for-maize-and-wheat-launched/.

Clapp, J., and S. R. Isaakson. 2018. *Speculative Harvests: Financialization, Food, and Agriculture*. New York: Columbia University Press.

Clapp, J., P. Newell, and Z. Brent. 2018. "The Global Political Economy of Climate Change, Agriculture and Food Systems." *The Journal of Peasant Studies* 45 (4): 80–88. doi:10.1080/03066150.2017.1381602.

Cohen, A., M. Vicol, and G. Pol. 2022. "Living Under Value Chains: The new Distributive Contract and Arguments About Unequal Bargaining Power." *Journal of Agrarian Change* 22 (1): 179–196. https://onlinelibrary.wiley.com/doi/full/10.1111joac.12466.

Da Costa, D. 2013. "The 'Rule of Experts' in Making a Dynamic Micro-Insurance Industry in India." *Journal of Peasant Studies* 40 (5): 845–865. doi:10.1080/03066150.2013.857659.

Dalton, J., and J. Thatcher. 2014. "What Does a Critical Data Studies Look Like, and Why Do We Care? Seven Points for a Critical Approach to 'Big Data'." *Society & Space*. https://www.societyandspace.org/articles/what-does-a-critical-data-studies-look-like-and-why-do-we-care.

Dean, M. 2010. *Governmentality: Power and Rule in Modern Society*. London: Sage.

Department of Agriculture and Cooperation. 2010. "National Mission for Sustainable Agriculture: Strategies for Meeting the Challenges of Climate Change." Ministry of agriculture. New Delhi. https://agricoop.nic.in/sites/default/files/National%20Mission%20For%20Sustainable%20Agriculture-DRAFT-Sept-2010.pdf.

"Digital Agriculture." n.d. The Grameen Foundation. https://grameenfoundation.org/solving-poverty/digital-farming. https://grameenfoundation.org/solving-poverty/digital-farming.

Dinesh, D. 2015. "Scaling Up Climate-Smart Agriculture Through Private Sector Engagement." 3 December 2015. https://ccafs.cgiar.org/news/scaling-climate-smart-agriculture-through-private-sector-engagement

Duncan, Emily, Sarah Rotz, André Magnan, and Kelly Bronson. 2022. "Disciplining Land Through Data: The Role of Agricultural Technologies in Farmland Assetisation." *Sociologia Ruralis* 62 (2): 231–249. doi:10.1111/soru.12369.

Eaton, C., and A. Shepherd. 2011. *Contract Farming: Partnerships for Growth*. UN Food & Agriculture Organization, 145.

The Economic Times. 2020. "Govt increases Digital India Program fund by 23% to Rs 3,958 crore." 2020. *The Economic Times*, February 20. https://economictimes.indiatimes.com/news/economy/policy/govt-increases-digital-india-program-fund-by-23-to-rs-3958-crore/articleshow/73872120.cms?from=mdr.

Ernst and Young. 2019. "Agritech – Towards Transforming Indian Agriculture." Accessed February 13, 2022. https://assets.ey.com/content/dam/ey-sites/ey-com/en_in/topics/start-ups/2020/09/ey-agritech-towards-transforming-indian-agriculture.pdf.

Escobar, A. 1995. *Encountering Development: The Making and Unmaking of the Third World*. Princeton: Princeton University Press.

Fabregas, R., M. Kremer, and F. Schilbach. 2019. "Realizing the Potential of Digital Development: The Case of Agricultural Advice." *Science* 366: ay3038. doi:10.1126/science.aay3038

Fraser, A. 2019. "Land Grab/Data Grab: Precision Agriculture and its new Horizons." *The Journal of Peasant Studies* 46 (5): 893–912. doi:10.1080/03066150.2017.1415887.

Fraser, A. 2020. "The Digital Revolution, Data Curation, and the New Dynamics of Food Sovereignty Construction." *The Journal of Peasant Studies* 47 (1): 208–226. doi:10.1080/03066150.2019.1602522.

Fraser, A. 2022. "'You Can't eat Data'? Moving Beyond the Misconfigured Innovations of Smart Farming." *Journal of Rural Studies* 91: 200–207. doi:10.1016/j.jrurstud.2021.06.010.

Gabor, D., and S. Brooks. 2017. "The Digital Revolution in Financial Inclusion: International Development in the Fintech Era." *New Political Economy* 22 (4): 423–436. doi:10.1080/13563467.2017.1259298.

Gardezi, M. and R. Stock. 2021. "Growing Algorithmic Governmentality: Interrogating the Social Construction of Trust in Precision Agriculture." *Journal of Rural Studies* 84: 1–11. doi:10.1016/j.jrurstud.2021.03.004.

Government of India. 2008. "National Action Plan on Climate Change." Accessed January 15, 2022. http://pmindia.nic.in/Pg01-52.pdf.

GRAIN. 2021. *Digital Control: How Big Tech Moves into Food and Farming (And What It Means)*. Barcelona. https://grain.org/system/articles/pdfs/000/006/595/original/Digital20control20EN%2005.pdf?1611163279.

Guha, R. 1989. *The Unquiet Woods: Ecological Change and Peasant Resistance in the Himalaya*. New Delhi: Oxford University Press.

Guthman, J. 2017. "Life Itself Under Contract: Rent-Seeking and Biopolitical Devolution Through Partnerships in California's Strawberry Industry." *The Journal of Peasant Studies* 44 (1): 100–117. doi:10.1080/03066150.2016.1217843.

Haq, Zia. 2019"Govt to take Aadhaar help for new farmers' database." *The Hindustan Times*, October 27. https://www.hindustantimes.com/india-news/govt-to-take-aadhaar-help-for-new-farmers-database/story-2lJTsq202pRdIzU1yGHKjJ.html.

Hartmann, G., I. Mwaka, and P. Dannenberg. 2021. "Large Investments, Small Farmers: A Financialisation Perspective on Value Chains in a Development Corridor." *Development Southern Africa* 38 (1): 122–138. doi:10.1080/0376835X.2020.1799758.

Harvey, D. 2004. "The new Imperialism: Accumulation by Dispossession." *Socialist Register* 40: 62–87. https://socialistregister.com/index.php/srv/article/view/5811.

Hecht, S. B., and A. Cockburn. 2011. *The Fate of the Forest: Developers, Destroyers, and Defenders of the Amazon*. Chicago: University of Chicago Press.

Higgins, W., and K. T. Hallström. 2007. "Standardization, Globalization and Rationalities of Government." *Organization* 14 (5): 685–704. doi:10.1177/1350508407080309.

Iliadis, A., and F. Russo. 2016. "Critical Data Studies: An Introduction." *Big Data & Society* 3 (2): 205395171667423–7. doi:10.1177/2053951716674238.

Isakson, S.R. 2015. "FFS – Small Farmer Vulnerability and Climate Risk: Index Insurance as a Financial Fix." *Canadian Food Studies/La Revue canadienne des études sur l'alimentation* 2 (2): 267–277. doi:10.15353/cfs-rcea.v2i2.109.

Jebaraj, P. 2019. "'Uber for Tractors': Government to Launch App to Aid Farmers." *The Hindu*, August 12. https://www.thehindu.com/business/agri-business/uber-for-tractors-government-to-launch-app-to-aid-farmers/article28985433.ece.

Kapil, S. 2021. "Agristack: The New Digital Push in Agriculture Raises Serious Concerns." *DownToEarth*, June 23. https://www.downtoearth.org.in/news/agriculture/agristack-the-new-digital-push-in-agriculture-raises-serious-concerns-77613.

Keating, C., C. Rasmussen, and P. Rishi. 2010. "The Rationality of Empowerment: Microcredit, Accumulation by Dispossession, and the Gendered Economy." *Signs: Journal of Women in Culture and Society* 36 (1): 153–176. doi:10.1086/652911.

Kedia, S. 2018. "Agri Startups are Bringing in IoT, Big Data and Equipment to Farmer, and Revolutionising the Sector." *Yourstory*, February 26. https://yourstory.com/2018/02/iot-big-data-equipment-farmers-agri-startups/amp.

Kitchin, R. 2013. "Big Data and Human Geography." *Dialogues in Human Geography* 3 (3): 262–267. doi:10.1177/2043820613513388.

Kumar, R. 2004. "eChoupals: A Study on the Financial Sustainability of Village Internet Centers in Rural Madhya Pradesh." *Information Technologies and International Development* 2 (1): 45–73. doi:10.1162/1544752043971161.

Kumar, A., and S. Basu. 2022. "Can end-User Feedback Inform 'Responsibilisation' of India's Policy Landscape for Agri-Digital Transition?" *Sociologia Ruralis* 62 (2): 305–334. doi:10.1111/soru.12374.

Lapidus, Daniel, Latane, Annah, Ortiz-Monasterio, Ivan, Beach, Robert, and María Elena Cárdenas Castañeda. 2017. *The GreenSeeker handheld: A research brief on farmer technology adoption and disadoption*. RTI Press. doi:10.3768/rtipress.2017.rb.0014.1705.

Li, T M. 2011. "Rendering Society Technical: Government Through Community and the Ethnographic Turn at the World Bank in Indonesia." In *Adventures in Aidland: The Anthropology of Professionals in International Development*, edited by D. Mosse, 57–80. Oxford: Berghahn.

Lipper, L. and D. Zilberman. 2018. "A Short History of the Evolution of the Climate Smart Agriculture Approach and its Links to Climate Change and Sustainable Agriculture Debates." In *Climate Smart Agriculture: Building Resilience to Climate Change*, edited by L. Lipper, D. Zilberman., N. McCarthy, S. Asfaw, & G. Branca, Vol. 52, pp. 13–30. Springer International Publishing.

Little, P. D., and M. Watts. 1994. *Living Under Contract: Contract Farming and Agrarian Transformation in sub-Saharan Africa*. Madison: University of Wisconsin Press.

McDade, M. 2019, January 28. *Seeing really is believing: Farmers' photos revolutionize insurance and advisory services*. CGIAR Platform for Big Data in Agriculture. Accepted July 27, 2022. https://bigdata.cgiar.org/blog-post/seeing-really-is-believing-when-farmers-photos-revolutionize-insurance-and-advisory-services/.

McMichael, P. 2013. "Value-chain Agriculture and Debt Relations: Contradictory Outcomes." *Third World Quarterly* 34 (4): 671–690. doi:10.1080/01436597.2013.786290.

Miles, C. 2019. "The Combine Will Tell the Truth: On Precision Agriculture and Algorithmic Rationality." *Big Data & Society*. doi:10.1177/2053951719849444.

Mitchell, Christopher, Shivani Seghal, Priyanka Chopra, and Hemendra Mathur. 2018. *Indian Farming's Next Big Moment: Farming as a Service*. Bain & Company. https://www.bain.com/link/b1b3591fe01b4bf4aecb605db838ffda.aspx.

Narayanan, S. 2012. *Notional Contracts: The Moral Economy of Contract Farming Arrangements in India*. Mumbai: Indira Gandhi Institute of Development Research.

The National Association of Software and Service Companies (NASSCOM). 2019. "Annual Report." https://nasscom.in/sites/default/files/Annual_Report_2019.pdf.

Neufeldt, Henry, Molly Jahn, Bruce M. Campbell, John R. Beddington, Fabrice DeClerck, Alessandro De Pinto, Jay Gulledge, et al. 2013. "Beyond Climate-Smart Agriculture: Toward Safe Operating Spaces for Global Food Systems." *Agriculture & Food Security* 2 (2013): 1–6. doi:10.1186/2048-7010-2-12.

Newell, P., and O. Taylor. 2018. "Contested Landscapes: The Global Political Economy of Climate-Smart Agriculture." *The Journal of Peasant Studies* 45 (1): 108–129. doi:10.1080/03066150.2017.1324426.

Newell, P., and O. Taylor. 2020. "Fiddling While the Planet Burns? COP25 in Perspective." *Globalizations* 17 (4): 580–592. doi:10.1080/14747731.2020.1726127.

Nichols, R. 2020. *Theft is Property! Dispossession and Critical Theory*. Durham: Duke University Press.

Nye, C. 2019. "The Commons as Accumulation Strategy." *Social Text* 37 (2): 1–28. doi:10.1215/01642472-7370967.

Oguamanam, C. 2020. "Indigenous Peoples, Data Sovereignty, and Self-Determination: Current Realities and Imperatives." *The African Journal of Information and Communication* 26 (2): 1–20. doi:10.23962/10539/30360.

Pandve, H. T. 2009. "India's National Action Plan on Climate Change." *Indian Journal of Occupational and Environmental Medicine* 13: 17–19. doi:10.4103/0019-5278.50718.

Patel, R. 2021. "Why Aren't we Talking About Farmers in India?" *Boston Review*, 18 May 2021. https://bostonreview.net/articles/why-arent-we-talking-about-farmers-in-india/.

Peluso, N. L. 2017. "Plantations and Mines: Resource Frontiers and the Politics of the Smallholder Slot." *The Journal of Peasant Studies* 44 (4): 834–869. doi:10.1080/03066150.2017.1339692.

Rajeesh S. 2021. "Picture-based Insurance Project: Stakeholder Perspectives from India." CCAFS Working Paper no. 405. Wageningen, the Netherlands: CGIAR Research Program on Climate Change, Agriculture and Food Security (CCAFS). https://hdl.handle.net/10568/119240.

Ramdas, S. 2022. "Disruptive Technologies: The Case of Indigenous Territories of Andhra Pradesh, India." *Heinrich Boll Stiftung*, 28 March 2022. https://hk.boell.org/en/2022/03/25/disruptive-technologies-case-indigenous-territories-andhra-pradesh-india.

Rao, N. H. 2018. "Big Data and Climate Smart Agriculture-Status and Implications for Agricultural Research and Innovation in India." *Proceedings of the Indian National Science Academy* 96: 625–640. doi:10.16943/ptinsa/2018/49342.

Rao, U., and V. Nair. 2019. "Aadhaar: Governing with Biometrics." *South Asia: Journal of South Asian Studies* 42: 469–481. doi:10.1080/00856401.2019.1595343.

Ravis, T., and B. Notkin. 2020. "Urban Bites and Agrarian Bytes: Digital Agriculture and Extended Urbanization." *Berkeley Planning Journal* 31: 100–122. doi:10.5070/BP331044067.

Roberts, R. A. 2005. *Insurance of Crops in Developing Countries*. Vol. 159. UN Food & Agriculture Organization.

Rose, N., and P. Miller. 1992. "Political Power Beyond the State: Problematics of Government." *The British Journal of Sociology* 43 (2): 173–205. doi:10.2307/591464.

Saha, S. 2021. "Why Microsoft's Role in Digitising India's Land Records is Worrying Farmer Groups." Newslaundry, 2021, June 2. https://www.newslaundry.com/2021/06/02/why-microsofts-role-in-digitising-indias-land-records-is-worrying-farmer-groups.

Sapkota, Tek B., Mangi L. Jat, Dharamvir S. Rana, Arun Khatri-Chhetri, Hanuman S. Jat, Deepak Bijarniya, and Jhabar M. Sutaliya. 2021. "Crop Nutrient Management Using Nutrient Expert Improves Yield, Increases Farmers' Income and Reduces Greenhouse gas Emissions." *Scientific Reports* 11: 1564–1575. doi:10.1038/s41598-020-79883-x.

Seoane, José, and Emilio Taddei. 2002. "From Seattle to Porto Alegre: The Anti-Neoliberal Globalization Movement." *Current Sociology* 50 (1): 99–122. doi:10.1177/0011392102050001008.

Seufert, P., M.L. Mendonca, and F. Pitta. 2018. *When Land Becomes a Global Financial Asset: The MATOPIBA Case in Brazil*. FIAN. https://www.righttofoodandnutrition.org/files/3._eng_when_land_becomes_a_global_financial_asset.pdf.

Shankarnarayan, V. K., and H. Ramakrishna. 2020. "Paradigm Change in Indian Agricultural Practices Using Big Data: Challenges and Opportunities from Field to Plate." *Information Processing in Agriculture* 7 (3): 355–368. doi:10.1016/j.inpa.2020.01.001.

Shiva, V. 2015. *The Violence of the Green Revolution: Third World Agriculture, Ecology, and Politics*. Lexington: University of Kentucky Press.

Shrimali, R. 2021. *Contract Farming, Capital and State*. Singapore: Springer.

Singh, S. 2002. "Contracting out Solutions: Political Economy of Contract Farming in the Indian Punjab." *World Development* 30 (9): 1621–1638. doi:10.1016/S0305-750X(02)00059-1.

Sinha, Ashish Aman, Hari Charan Behera, Ajit Kumar Behura, and Braja Bandhu Swain. 2021. "Land Allocation Choice in Both Contract and Non-Contract Farming: A Study of Potato Growers in West Bengal, India." *Sage Open* 11: 215824402110475. doi:10.1177/21582440211047593.

Smith, Pete, Daniel Martino, Zucong Cai, Daniel Gwary, Henry Janzen, Pushpam Kumar, Bruce Mccarl, et al. 2007. "Policy and Technological Constraints to Implementation of Greenhouse Gas Mitigation Options in Agriculture." *Agriculture, Ecosystems and Environment* 118 (1-4): 6–28. doi:10.1016/j.agee.2006.06.006.

Soederberg, S. 2013. "Universalising Financial Inclusion and the Securitisation of Development." *Third World Quarterly* 34 (4): 593–612. doi:10.1080/01436597.2013.786285.

Srivastava, S. K. 2021. "Agricultural Automation: A Strategic Dilemma of a Rich Indian Farmer." *Journal of Information Technology Teaching Cases* 11 (2): 117–121. doi:10.1177/2043886920951200.

Stock, R., and M. Gardezi. 2021. "Make Bloom and let Wither: Biopolitics of Precision Agriculture at the Dawn of Surveillance Capitalism." *Geoforum; Journal of Physical, Human, and Regional Geosciences* 122: 193–203. doi:10.1016/j.geoforum.2021.04.014.

Stock, R., and M. Gardezi. 2022. "Arrays and Algorithms: Emerging Regimes of Dispossession at the Frontiers of Agrarian Technological Governance." *Earth System Governance* 12: 100137. doi:10.1016/j.esg.2022.100137.

Stone, G.D. 2007. "Agricultural Deskilling and the Spread of Genetically Modified Cotton in Warangal." *Current Anthropology* 48 (1): 67–103. https://www.journals.uchicago.edu/doi/full/10.1086508689.

Thaler, R. H., and C. R. Sunstein. 2008. *Nudge: Improving Decisions About Health, Wealth, and Happiness*. New Haven: Yale University Press.

Thatcher, J., D. O'Sullivan, and D. Mahmoudi. 2016. "Data Colonialism Through Accumulation by Dispossession: New Metaphors for Daily Data." *Environment and Planning D: Society and Space* 34 (6): 990–1006. doi:10.1177/0263775816633195.

Third World Network. 2017. "Philanthrocapitalism: The Gates Foundation's African Programmes are Not Charity." *Third World Resurgence* 328: 2–5. https://www.twn.my/title2/resurgence/2017/328/eco1.htm.

Vandergeest, P., and L. Schoenberger. 2019. *De-centring Land Grabbing: Southeast Asia Perspectives on Agrarian-Environmental Transformations*. London: Routledge.

Vicol, M. 2019. "Potatoes, Petty Commodity Producers and Livelihoods: Contract Farming and Agrarian Change in Maharashtra, India." *Journal of Agrarian Change* 19 (1): 135–161. doi:10.1111/joac.12273.

Vincent, K., T. Cull, A. Kapoor, P.K. Aggarwal, G.D. Bhatta, C. Lau, and S. Nilormee. 2011. *Gender, climate change, agriculture, and food security: A CCAFS training-of-trainers (TOT) manual to prepare South Asian rural women to adapt to climate change*. Cali, Colombia: CGIAR Research Program on Climate Change, Agriculture and Food Security (CCAFS). https://hdl.handle.net/10568/33440.

Watts, M. 1983. *Silent Violence: Food, Famine, and Peasantry in Northern Nigeria*. Athens: University of Georgia Press.

Watts, M. 2003. "Political Ecology." In *A Companion to Economic Geography*, edited by Eric Sheppard and Trevor J. Barnes, 257–274. London: Wiley. doi:10.1002/9780470693445.ch16.

World Bank. 2007. *World Development Report 2008 : Agriculture for Development*. Washington, DC: World Bank. https://openknowledge.worldbank.org/handle/10986/5990 License: CC BY 3.0 IGO.

Prefiguring *buen sobrevivir*: Lenca women's (e)utopianism amid climate change

Benjamin C. Fash, Betty del Carmen Vásquez Rivera and María Sojob

ABSTRACT

Given its utopian orientation, the anti-capitalist, decolonial notion of *buen vivir* understates how precarious life is for those who experiment with practicing its principles. Furthermore, climate change and climate mitigation will exacerbate precarity for the foreseeable future. In response, we develop a conceptual argument that integrates concepts from post-extractivism, climate adaptation, communitarian feminism, and prefigurative politics for a more accurate and more actionable, eutopian concept we term *buen sobrevivir*, or surviving well. We then illustrate the concept through Lenca women's struggles to make alternatives possible in the face of violent extractivism, patriarchy, and climate change impacts in post-coup Honduras.

Introduction

Buen vivir is a utopian, anti-capitalist, decolonial way of life or paradigm situated as an alternative to development (Farah and Vasapollo 2011; Gudynas and Acosta 2011). Inspired by indigenous knowledges, *buen vivir* has translated – with many other names – across cultures and mostly agrarian contexts throughout Latin America. To practice *buen vivir* has been equated to 'live in solidarity, in equality, in harmony, in complementarity, in reciprocity' (Morales 2011), and to pursue 'the land without evil' (Gudynas and Acosta 2011). To advance *buen vivir*, the concept of post-extractivism has been proposed as a paradigm in which resource extraction is drastically reduced, to the point of achieving 'zero poverty' and 'zero extinctions' (Gudynas 2013). Honty and Gudynas (2014, 30) proclaim that implementing alternatives oriented toward *buen vivir* permits humans to 'revert problems, annul the factors that cause them, and the cultural and political conditions that make them acceptable'. This, they argue, can provide change radical enough to achieve a 'secure climate'.

Such proposals have led critics to argue that *buen vivir* literature's 'quasi mystical' utopian orientation has significant dissonance with the realities of many indigenous people who seek it (Stefanoni 2012), and that governments implementing it 'pragmatically' continue to apply developmentalist agendas (Caria and Domínguez 2016). *Buen vivir's* early proponents, meanwhile, have expressed disillusionment with State-level co-optation of the concept, but have highlighted the possibilities of implementing *buen vivir* principles, including 'radical food sovereignty' (Tilzey 2019) at subnational levels, with arguments that it may be best to use a different name for such efforts (Mora 2020). In constructive critiques, decolonial and communitarian feminists have also pressed that beyond painting a future without capitalist/extractivist development, *buen vivir* proposals must set their focus on the already present struggles against oppression, including feminist struggles against patriarchy (Varea and Zaragocin 2017a).

In this paper, responding to both proposals and critiques, we recognize the ways in which *buen vivir* discourse has inspired territorial movements but argue for a reframing based on the ways in which it understates how precarious life is and will continue to be for those who experiment with practicing its principles. We present the struggles of the Lenca women-led *Movimiento Ambientalista Santabarbarense* (MAS, or Santa Barbara Environmentalist Movement) in Honduras, where precarity stems from political economy, patriarchy, climate change, and climate change mitigation. MAS interpret and implement *buen vivir* through such activities as municipal bans of extractivism, collective fish farms, collective water management, artisan training, bartering, native seed exchanges, women's political and economic collectives, medicinal plant knowledge sharing, and healing ceremonies. Though they seek to approach what Lenca leader Berta Cáceres (2015) called 'societies that are just, dignified, and for life', none of them would suggest they live anywhere close to utopia. They're in Honduras: a country with among the highest rates of violence associated with extractivism and patriarchy in the world (Luciano et al. 2019; Global Witness 2020), and significant vulnerability to climate change (Eckstein, Hutfils, and Winges 2018).

We argue that by proposing a paradigm somehow outside of precarious circumstances, the *buen vivir* literature does not do justice to how people live, adapt, and innovate. Therefore, as our analysis of MAS suggests, a more accurate concept to describe their liberatory struggles and aspirations would be *buen sobrevivir*, or surviving well. Firmly rooted in the vision of making other worlds possible, *buen sobrevivir* aspires to *eutopia* – a possible or already existing good place – rather than utopia, which etymologically describes a place that does not and *cannot* exist (Sargent 1994). We define *buen sobrevivir* as follows:

> *Buen sobrevivir* is a eutopian, prefigurative politics that harmonizes more-than-human relations through collective action to create and reproduce societies based on justice, dignity, and peace for all beings, despite precarious circumstances wrought by multiple, intersecting structures of oppression and climate change. Embracing decolonial, feminist, and post-capitalist visions, *buen sobrevivir* rejects modernist notions of development and progress, and promotes the unlearning and deconstruction of all forms of colonization and domination, including dualisms of culture/nature, man/woman, civilized/primitive, and self/other. Just as it promotes finding happiness in simplicity, complementarity, and conviviality in more-than-human communities, *buen sobrevivir* emphasizes the healing and defense of bodies and lands, understood as interdependent territories in a vast web of life under constant threat.

Grown out of Lenca women's struggles to survive well in opposition to patriarchy, capitalism, extractivism, and unjust climate mitigation, *buen sobrevivir* is a politics of hope and possibility for ways of being otherwise to be reinterpreted and applied to contexts around the world in the confrontation with both historic structures of oppression and new challenges from climate change. *Buen sobrevivir* differs from *buen vivir* fundamentally because it reflects present-day political action rather than an imagined future utopia. If *buen vivir* is a 'horizon-concept' (Svampa 2019, 20) to inspire action, *buen sobrevivir* is a pragmatist experiment to make the most livable world possible for ourselves and future generations from where we stand today. To illustrate this, we can relate experiences where a group of mixed gender and ethnic identities marches to live ancestral music while eating locally sourced, traditional, GMO-free foods in a protest against an extractivist project. As these people collectively put their body-territories on the line to defend their land-territory, one might say they are practicing *buen sobrevivir*, but none would say they are in a *buen vivir* utopian context.

This paper will first develop this conceptual argument through a critical review of literatures on *buen vivir*, post-extractivism, climate change, communitarian feminism, and prefigurative politics. We then illustrate our theoretical points by providing an analysis of the Honduran context of life-threatening developmentalism, extractivism, patriarchy, and climate change, in which we situate MAS's diverse efforts to practice communitarian feminism and *buen vivir* principles, which we here reframe as *buen sobrevivir*. We finish with a local case that encapsulates the practice of *buen sobrevivir* in the face of extreme weather events and megadams approved ostensibly as a climate mitigation measure to control flooding.

Our proposals for *buen sobrevivir* offer insights and provocations to debates and gaps in *buen vivir* literature regarding its strategy and implementation, as well as its relationships to climate change and patriarchy. In so doing, we speak to the pressing questions posed for the *Journal of Peasant Studies*' forum on climate change and critical agrarian struggles (Borras et al. 2021). This contributes a grounded response to the question: 'What are the implications of climate change (and climate change mitigation efforts) for rural areas around the world, and how have rural people responded politically to these challenges' (Borras et al. 2021). We also argue that *buen sobrevivir* makes significant contributions to answer Fraser's (2021) call for an anti-capitalist, trans-environmentalism.[1] Thus, we respond to critiques that *buen vivir* and/or post-extractivism are more abstract than operational (Bebbington 2015), incoherent and contradictory (Caria and Domínguez 2016), if not simply romantic and unattainable (Stefanoni 2012). Our critiques of *buen vivir* and post-extractivism are part of an earnest effort to strengthen the utility of these concepts as alternatives to colonialist and capitalist development as well as patriarchy, both in the academy and on the ground.

This critique and reframing of *buen vivir* is, like all arguments and knowledge, situated (Haraway 1988). Our insights are rooted in three complementary trajectories. For the first author, the article is situated within a five-year (2016–2021) activist research agenda exploring the material and discursive forms of post-extractivist agenda in Honduras. He is a founding member of the Copán Environmentalist Coalition, an anti/post-extractivist group in his hometown. Through this work, he met and collaborated with both co-authors in workshops, conducting interviews, and co-directing and co-producing a documentary film with the third author about the women leaders within MAS. The second author is a Lenca woman, a founding member and leader of MAS. Her contributions to

[1]For Fraser (2021, 126), 'trans-environmentalism' consists of an eco-politics that incorporates 'labour rights, feminism, anti-racism, anti-imperialism, class consciousness, pro-democracy, anti-consumerism, anti-extractivism'.

this research are based on her position as an intellectual and strategic leader within the organization, as well as other feminist, environmentalist, and indigenous networks. The third author is a Tsotsil filmmaker whose insights emerge from participation in workshops and other collective learning experiences centered on audiovisual production and politics with indigenous women in Central America, as well as collaborative research and filmmaking with the first author, as mentioned above. This paper primarily draws from participant observation in MAS's activities, 42 informal, 12 semi-structured, and eight in-depth interviews with members of the organization, and archival research undertaken between 2019 and 2022.

To be clear, the term *buen sobrevivir* emerged out of conversations among us, the authors of this manuscript, on how to better describe what we have observed. Prior to this paper, MAS was using the term *buen vivir*, but the dissonance between the literature on *buen vivir* and what is being practiced is significant enough to merit this reframing. We anticipate that focusing on *surviving well* instead of on *living well* may sound like a compromise. Our point is not to curb aspirations, but rather to recognize that *buen vivir's* harmonious, reciprocal, and complementary more-than-human relations do not exist in isolation. Chaos, violence, and instability are realities that extractivist development and patriarchy have wrought on the more-than human lives, particularly in Latin America, where *buen vivir* emerged. With inevitable impacts from climate change set to stay even in best-case scenarios, centering survival better names the state of generalized vulnerability. The question of how 'climate change differ[s] from past exclusions or threats' (Borras et al. 2021, 17) is embedded in our argument that a politics of survival has staying power: while local actions can serve to mitigate and adapt to climate change, none can stop it in the way that extractivism or patriarchy might be stopped. Previously, one could think of *surviving well* in turbulent times as a pathway toward utopian futures, such as *buen vivir*. Now, turbulence related to climate change is here to stay.

Framing *buen sobrevivir*: theorizing five themes

In this section, we frame the concept of *buen sobrevivir* from a critical review of five key themes. We present *buen vivir* as the core concept, which runs transversally through the rest. Then, we provide a constructive critique of post-extractivism, a proposed paradigm in which resource extraction is limited to that which is indispensable for *buen vivir* (Gudynas 2013). We argue that like *buen vivir*, post-extractivism contains both inspirational and unfeasible elements. Third, we discuss how the ongoing, inevitable, and potential impacts of climate change and climate mitigation are inadequately addressed in *buen vivir* and post-extractivism literature as we make the case for a politics centered on survival in the context of intensified precarity. Fourth, we present critical input regarding patriarchy, colonialism, and communalism from communitarian feminism to debates on *buen vivir* in order to frame *buen sobrevivir* as a more just, inclusive, and effective politics. Finally, we discuss the significance of prefigurative politics, which seek to embody, 'within the ongoing political practice of a movement, those forms of social relations, decision-making, culture, and human experience that are the ultimate goal' (Boggs 1977, 100), as a way to unify vision and action in the climate era.

Buen vivir's utopian politics

The genealogy of *buen vivir* is not entirely clear: though commonly traced to a simplistic translation of indigenous conceptions of living in states of harmony, bliss, and balance through practices of reciprocity and complementarity (Macas 2011), it shares many features with Western utopian thinking (Giraldo 2014). Escobar (2018) argues it emerged from the articulation of movements embracing both indigenous and western thinking, which positioned *Sumak Kawsay* and *Suma Qamaña*, Quechua and Aymara terms, as decolonial alternatives to development and modernity, and in particular the ideas of progress, growth, and instrumental rationality. In 2008 and 2009, when ostensibly post-neoliberal regimes in Ecuador and Bolivia enshrined these concepts in new constitutions, many activists celebrated the dawn of a new era. Since then, *buen vivir* has been taken up across Latin America and elsewhere.

While much literature has focused on developing viable proposals and constructive critiques to realize *buen vivir* through national politics (Walsh 2010; Gudynas and Acosta 2011), there has been considerable focus within indigenous intellectual circles on reflecting and systematizing their own concepts of living well for political action at the scale of their territories. Kaqchikel writers in Guatemala, for instance, have analyzed and promoted *Utz'ilaj Kaslemal*, 'a system in which everyone had participation in community organizations in accordance with their abilities, which facilitated time for training, observation, agricultural production, astronomy, mathematics, the exchange of collective harvests, and collective gatherings to create, perform and enjoy art, music, and dance' (Méndez Martínez 2012, 21). Our proposal for *buen sobrevivir* finds resonance with such territorial projects, which neither discount nor rely on national politics, and emphasize praxis over utopianism.

Since utopia etymologically refers to no-place, or a place that does not exist (Sargent 1994), many speak of it as a 'horizon', including Fernando Aguirre, who noted:

> Utopia is on the horizon. I know very well that I will never reach it. If I walk ten steps, she will move ten steps further away. The more I look for her, the less I will find her, because she moves away as I get closer. So, what is utopia for? Utopia is for that - for walking. (Decrecimiento 2012)

Some take issue with the horizon metaphor, arguing that utopian projects have indeed been implemented (Giraldo 2014; Rodríguez Salazar 2016). However, though utopian projects work *toward* utopias, they do not achieve utopias. We position *buen sobrevivir* as *eutopian* (referring to a 'good place' that can or already does exist) rather than utopian, for the sake of drawing a clear line between what MAS leaders describe as 'dreaming while asleep' (utopianism) vs. 'dreaming while awake' (eutopianism). We concur with the Red de Feminismos Decoloniales (Translated by the authors from 2014, 320), who more elaborately state:

> We affirm eutopia in the sense of prefigurations or apparitions that occur in the present time, which allow us to see the emergency of the "alter" sensibility, that is, of rupture with capitalist modernity, enunciating and practicing a radical commitment with life, allowing people to see through diverse small-scale practices, in the realm of production-in-reciprocity, of consensus-based decision-making, of the impartation of justice, of the order of gender, of the inclusion of differences, derivatives of a world that is already possible, recovering from the expropriation of its of communitarian sovereignty.

The common depictions of *buen vivir* in edenic, liberatory terms, seemingly free from disruption, are problematic given that the 'ghosts' of capitalism will haunt even the most non-capitalist future societies, wreaking chaos and havoc through climate change (Gan 2017).

The early hopes that many activists and intellectuals invested in the Ecuadorian and Bolivian constitutional incorporation of *buen vivir* were quickly dashed by the contradictions in government implementation, particularly regarding resource extraction and infrastructure (Bebbington 2009; Gudynas 2009; Acosta 2013), prompting some to drop the term (Mora 2020). In our case, proposing *buen sobrevivir* holds onto many of the same principles, but sheds light on the violent, insecure contexts in which those are prefigured, whether they be from climate change, patriarchy, or resource extraction.

Post-extractivism: limiting resource extraction and promoting food sovereignty

Following observations that human rights violations and environmental impacts of resource extraction continued under Latin American pink tide regimes – including those that co-opted *buen vivir* discourse – many social movement groups and scholars consolidated what they perceived as conceptions of unjust, excessive, and harmful logics and practices of resource extraction in the term *extractivism*. Though popular definitions focus on raw materials extracted for export (Gudynas 2015), especially from the 'Global South' (Acosta 2013; Svampa 2015), here, we follow Fash's (2022) definition:

> Extractivism is a logic and practice of colonialist domination that involves the extraction of natural resource rents without restoration or care of the territories to which the resources belonged nor of the socio-natural relations that they previously sustained. Contemporary extractivism can manifest as development strategies based on synergies between mineral and hydrocarbon extraction, infrastructure (especially energy generation), industrial agriculture, aquaculture and forestry, extensive ranching, even tourism, as well as in the form of smaller scale activities operating with colonialist logics of rent capture without restoration or care.

Analyses of extractivism have thus laid bare the reality that socialism can easily adopt logics and practices that perpetuate the same devastating impacts on the planet, even if the proceeds of extraction are better distributed among humans. But while capitalism structurally *requires* extractivism to operate, socialism does not (Fraser 2021). Thus, just as Fraser argues that eco-politics must be anti-capitalist to be effective, voices from Latin America have made clear that anti-capitalism (and, therefore *buen vivir*) must include alternatives to extractivism if we are to survive (Gudynas 2013; Svampa 2013).

Thus, we embrace and adjust the proposals for post-extractivism (Gudynas 2013), here defined as an alternative paradigm in which extraction is limited to that which is indispensable for *buen sobrevivir*. Building from Fash's (2022) definition of extractivism, post-extractivism should not simply be an alternative regional paradigm to aspire to, but a politics practiced in the midst of extractivist agendas, staging struggles over local policy, questioning energy generation, and accounting for climate change. In this vein, in addition to limiting certain forms of extraction, post-extractivism should focus on unlearning modernist ideals of consumption by decolonizing the horizon of desire, 'a metaphor for what is collectively desirable and feasible' (Gutiérrez Aguilar 2014). For instance, in design and construction, a sector responsible for between

23–40% of greenhouse gas emissions (Olukoya Obafemi and Kurt 2016), unlearning modernist ideals related to concrete and steel makes way for processes to recover adobe construction traditions, which have nearly zero carbon footprint (Olukoya Obafemi and Kurt 2016). Similarly, an integral approach to agroecology and territorial defense involves recovering traditional foods and unlearning habits of consuming sugar and oil palm.

More broadly, on the grounds that post-extractivism entails a reduction in extractivism-derived rents and a move away from agrarian extractivism (McKay, Alonso-Fradejas, and Ezquerro-Cañete 2021), *buen sobrevivir* requires an agricultural shift away from input-intensive corporate monocultures, toward agroecological production (Gudynas 2013) and food sovereignty (Edelman et al. 2014; Martínez-Torres and Rosset 2014). Proponents of agroecology argue that diversified production provides greater economic security against crop losses and market variation, multiple harvesting periods to spread income throughout the year, and higher productivity. By definition, agroecology also supports more wildlife, resists pests – limiting or eliminating contamination from inputs – and reduces soil erosion (Altieri et al. 2015). Finally, agroecology and food sovereignty have been shown to reduce emissions, recycle resources, improve water retention, and prioritize local exchange, making them effective climate mitigation and adaptation strategies (Altieri et al. 2015; Dobler-Morales et al. 2021).

Numerous scholars have contributed valuable critiques of food sovereignty on both theoretical and empirical grounds, bearing the need to position *buen sobrevivir* in such debates. In keeping with *buen sobrevivir*'s response to precarity, following Agarwal (2014), our vision for food sovereignty is not as a singular, romantic one of total national or local self-sufficiency with consensus and collectively owned land, but rather a complex negotiation of the broad principles of the food sovereignty movement with on-the-ground constraints that farmers face, including asymmetrical power in gender relations. We recognize that essentializations of the 'peasant', just like essentializations of the 'Lenca' or 'indigenous', obscure goals and constraints of many farmers employing chemical-intensive practices, monocropping, and/or obtaining income from export (Bernstein 2016; Soper 2020). It is also clear that some of those deploying food sovereignty discourse, like many using *buen vivir* discourse, do so to negotiate or obtain resources in the short term rather than achieve radical change in the long term (Henderson 2017). Nonetheless, these valuable observations do not make food sovereignty and agroecology as currently defined any less possible, relevant, or important to ensuring more-than-human survival (McMichael 2015). Like Tilzey's demonstration that the Bolivian MST practices radical food sovereignty and agroecology amidst national (neo-) extractivist policies (Tilzey 2019), our paper seeks to respond to Bernstein's (2014, 27) call for a 'concrete analysis of a concrete situation'.

Like some food sovereignty and broader *buen vivir* proposals, post-extractivism literature has certain overreaching utopian tendencies, particularly considering climate change. Gudynas (2013, 173) suggests that a post-extractivist paradigm has 'zero poverty, zero extinctions', an example of elements of *buen vivir* and post-extractivism that have 'doubtful coherence' (Caria and Domínguez 2016, 27), given how climate change and other factors will continue to cause the ongoing, major biodiversity extinction event (Novacek and Cleland 2001; IPBES 2019; Román-Palacios and Wiens 2020).

However, revising the goal to be about *reducing* extinctions is no less radical, simply more pragmatic. In general, as we will now argue, trends in climate change give these proposals both grounding and urgency, supporting the case for a eutopian focus on survival.

Climate change: centering survival

Buen vivir literature tends to present its utopian politics as a solution that can resolve climate change, rather than as a strategy to adapt to it. Honty and Gudynas (2014) propose commonplace policy instruments – including decarbonization, energy transitions, organic agriculture, and forest conservation, arguing that these can guarantee a 'secure climate'. Giraldo (2014) suggests climate change's threats to human survival are a cause for the emergence of *buen vivir*, yet he and others present *buen vivir* as a paradigm seemingly unaffected by climate change. While literature on *buen vivir* provides an 'expansive, historically informed analysis' of a wide 'set of environmental struggles in agrarian settings' (Borras et al. 2021, 5), it generally fails to adequately incorporate climate change.

Three postulates regarding climate change motivate our framing of *buen sobrevivir*: First, even in best-case scenarios, climate change's impacts will continue to put survival on the line globally. Second, survival strategies in the near term prefigure societal changes that determine if and for whom there will be conditions for survival. And third, discursively centering the causes and risks of climate change as well as the need to *survive with it* in politics is vital to persuading others to act.

Changing weather patterns that already threaten more-than-human survival call for novel shifts in territorialized life projects. The intensity and time frame of future threats depend largely on the near-term actions of human societies. However, even with the most drastic mitigation efforts, the earth will 'likely be warmer than any other time over the last 800,000 years' (Steffen et al. 2018, 8257), making 'an increasing threat to human life from excess heat now … almost inevitable' (Mora et al. 2017, 508). Mitigation strategies, often equally threatening, are already being used to dispossess climate-vulnerable people of their resources in the present (Paprocki 2018). This urges a politics of *buen sobrevivir* to complement *buen vivir*'s indigenous, place-based traditional concepts with values of experimentation and improvisation. Following communitarian feminists, we argue in favor of performing alternative futures collectively (Paredes 2010), in contrast to top-down technologically-centered interventions or so-called 'survivalism' and 'prepping', which tend toward highly individualistic and reactionary proposals for how to deal with emergencies (Peterson 1984).

In agriculture, traditional practices of what to plant, where, and when are brought into question by the inevitability of rising temperatures, shifting precipitation patterns, increased chances of floods and droughts, and for some, saltwater intrusion. The harmony and reciprocity between humans and nature – as *buen vivir* would have it – face ruptures that beg questions of climate adaptation, which, as Nightingale (2017, 12) notes, needs to be refocused 'to capture the intertwined biophysical and political processes that together shape adaptation needs'. To counteract maladaptation practices (see Dobler-Morales et al. 2021), such as increased uses of external inputs, abandonment of terraces, increased GMO usage, and reduced crop diversity, *buen sobrevivir* includes *campesinx a campesinx* (farmer to farmer) knowledge exchanges at a local level to accompany national and international organizing for food sovereignty (Holt-Giménez 2006; Altieri et al. 2015). Such exchanges are very much aligned with *buen vivir* principles, but bear emphasis considering climate adaptation needs.

Importantly, while accounting for the ongoing and inevitable future impacts of climate change scales back utopian possibilities, it does not, and should not, give into apocalyptic narratives. Our visions of life should be based on the planet that we have, as it is and can be, not as it was. As we move away from trends in *buen vivir* literature and assume the inevitability of climate change, Haraway's (2016, 1) reflections seem especially poignant:

> In urgent times, many of us are tempted to address trouble in terms of making an imagined future safe .... Staying with the trouble requires learning to be truly present, not as a vanishing pivot between awful or edenic pasts and apocalyptic or salvific futures, but as mortal critters entwined in myriad unfinished configurations of places, times, matters, meanings.

This call to navigate precarity finds many complements within communitarian feminism, a grounded movement led by indigenous women fighting to survive well.

Communitarian feminism: defending and healing territories

Communitarian feminism emerged in the 1990s in Bolivia among indigenous women and has found resonance elsewhere in Latin America. Like decolonial feminism, it revindicates indigenous women's embodied knowledges and goals to dismantle colonialism and patriarchy (Millán 2014). Why center patriarchy in a radical politics to confront climate change? As Paredes (2010, 39) puts it, 'there is no revolution as long as women's bodies continue to be a colony for men, governments, and States'. The central goal of communitarian feminism, she says, is to replace patriarchy with non-capitalist, genderless communities focused on collective wellbeing over individual freedom or success. As such, it represents a proposal from below that predates and shares a great deal with Fraser's (2021) proposal for anti-capitalist, trans-environmental politics.

Meanwhile, *buen vivir* is more commonly recognized as alternative to development and extractivism and less as an alternative to patriarchy. Leading thinkers on *buen vivir* as an alternative to development make only superficial mention of feminism as a reaction to 'the patriarchal base', and overlook Latin American feminists' thinking and political agendas when they cite feminism as an example of a western tradition contributing to *buen vivir* (Gudynas and Acosta 2011). As Varea and Zaragocin (2017b) point out, while critiques of how Bolivia and Ecuador continued with a business-as-usual governance of extractive industries echoed far and wide, the feminist critiques that of failures to 'depatriarcalize' have stayed mostly within conversations and small academic circles, some of which don't consider violating women's rights a contradiction with *buen vivir*. Nonetheless, we see *buen vivir* as complementary to decolonial and communitarian feminisms, and look to integrate these frameworks for *buen sobrevivir* (see also Varea and Zaragocin 2017a).

Communitarian feminism provides a framework for extending the defense of territory against extractivist development to a 'second battlefront': ensuring the wellbeing and integrity of women's bodies, through defense of what Cabnal (2019) calls the 'body-territory'. The body-territory faces multiple threats, including stalking, sexual violence, and 'territorial femicide'. These are intensified in humanitarian emergencies (Freedman 2016; Hossain and McAlpine 2017), which climate change will continue to provoke. Women also face mental and spiritual exhaustion from multiple aggressions attributed to patriarchy, racism, colonialism, heteronormativity, and neoliberal capitalism. Thus, liberating the body-territory goes beyond 'defense' and into the realm of collective, reciprocal healing (Cabnal 2017). Notions of reciprocity and collective healing permeate care for the body-territory as well as the 'land-territory' into the broader 'web of life', conceived as the interdependent cosmic network of all living and non-living beings. This is vital for the *buen sobrevivir* framework, because, rather than ideas of an edenic alternative, the defense and healing of body- and land-territories are presented as actions in co-existence with systemic oppression and aggression.

Communitarian feminists embrace indigenous ways and thought in various principles and practices, but do so critically and selectively. While some decolonial feminists suggest the concept of gender arrived with colonialism (e.g. Lugones 2010), Paredes (2010) and Cabnal (2017) state that while ideas fundamental to *buen vivir* like complementarity, reciprocity, and harmony *are* engrained in indigenous cosmovisions, they tend to obscure the ancestral patriarchy that accompanied pre-colonial communities and survives today in tandem with colonial patriarchy.[2] Based on this acknowledgement, a *buen sobrevivir* steeped in communitarian feminism borrows from ancestral ways and thought but does not romanticize them or see them as fixed ideals. Instead, it focuses on that which is applicable to the defense of the body-territory and land-territory, including new ways of thinking, doing, and being. To this effect, communitarian feminism highlights the importance of working *with* rather than against men, arguing that for communities to be well, men and relationships of women with men must be well (Paredes 2010). This expands early feminist contributions to agrarian studies focused on the household-scale, which note that 'peasant households can disintegrate or differentiate themselves out of existence not only because of class relations, but also as a result of tensions in household relations' (Deere 1995, 65), and complements Razavi's (2009) argument that it is vital to understand and intervene in the interconnections between domestic structures and broader economic and political processes.

Prefigurative politics: experimenting with possible futures

The focus of communitarian feminists on everyday actions to defend and heal territory as well as experiment with new practices, is a form of prefigurative politics fundamental for *buen sobrevivir*. Though not explicitly referenced in communitarian feminist circles, prefigurative politics has been an important component of feminisms since the 1960s, precisely because it inspires practices to actualize the vision of a movement in the present moment (Boggs 1977). Prefigurative politics offer an approach open to diverse projects across the ideological spectrum, though the term has mostly referred to leftist anarchist movements (Yates 2015). The counter-futures of pre-figurative

[2]Focusing on both Europe and India, Agarwal provides a similar intervention in arguing that ecofeminist characterizations of the origins of patriarchy problematically paint a picture of a pre-colonial past free of patriarchy when the evidence suggests 'a hierarchical construction of male-female relationships' (Agarwal 1998, 67).

politics are presented in contrast to the pre-emptive futures of 'anticipatory politics', a prescient concept to analyze unjust state-level responses to the climate emergency. As Anderson (2010, 782) puts it, anticipatory politics proceeds 'by (re) making life tensed on the verge of catastrophe in ways that protect, save, and care for certain valued lives, and damage, destroy, and abandon other lives'. The Honduran dams to mitigate flooding but displace vast more-than-human communities offer a prime example of 'anticipatory politics'.

Prefigurative politics are indispensable to *buen sobrevivir*, given the already precarious conditions wrought by extractivism and patriarchy, and the urgency and uncertainty of climate change. Sharing much with Emel's (1991) 'provocative pragmatism', prefigurative politics urge participants to 'start where we are' (Gibson-Graham 2006) with an experimental praxis. Gibson-Graham (2006) highlights the ways in which feminists provide excellent examples of how social transformation can always be under reconstruction and taking place at both interpersonal and societal levels, particularly in cases (like climate change) when we are seeking 'to articulate not only what is, but "what has never been"' (Emel 1991, 387). To this effect, pre-figurative politics typically proceeds through an intensive commitment to improvising with available ideas, materials, spaces, and bodies, and affective states (Jeffrey and Dyson 2021). Prefigurative politics have been central to a variety of movements, from community economies (Gibson-Graham 2006) to Zapatismo to Occupy (Smucker 2014), the reproductive justice organizing project We are BRAVE (Lin et al. 2016), and many more (Cornish et al. 2016).

By transforming a politics of 'nowhere' to a eutopian politics of 'now here' (Gibson-Graham 2006, xxi), the prefigurative elements of *buen sobrevivir* serve as resources for social movements that might despair regarding structures of oppression they face, and analytical resources to answer critics who talk about *buen vivir* or post-extractivism as unobtainable abstractions. As Jeffrey and Dyson (2021, 642) state, 'analysis of oppositional prefigurative politics answers a call among geographers for grounded work on how counter-futures are constructed, discussed, imagined, lived, and defended' (see also Anderson 2010). In this vein, Hondurans Berta Cáceres and Miriam Miranda (2021) argue for the importance of the rights to both dream and experiment, making mistakes and learning from them autonomously.

Surviving well: Lenca women's struggles in Honduras

Honduras is frequently described in terms of extreme violence, corruption, and poverty (Ulku et al. 2021). Here, we situate these very real conditions within a broader interpretation of Honduras as a space in dispute, where activists have been struggling for justice amid brutal repression that constantly poses risks to their survival. Individuals and communities have sought solutions to survive well despite the significant obstacles. This provides the empirical evidence for *buen sobrevivir* as a prefigurative political project inspired by *buen vivir,* post-extractivism, and communitarian feminism in the face of intersecting impacts from climate change, neoliberal extractivism, and patriarchy in Lenca territory. We begin by analyzing ways in which the political economic elite – compounding impacts from climate change – have made the country increasingly uninhabitable through policies and practices of development, extractivism, climate and energy politics, and patriarchy. We then describe MAS's counterhegemonic goals, practices, and achievements, mapping them onto the theoretical framework for *buen sobrevivir* laid out in the previous section. Finally, we present a more detailed examination of Chinda, Santa Barbara, where the struggle to survive well is particularly acute and urgent, given impacts of intensified extreme weather events and extractivist climate mitigation policy.

The brutal implementation of the post-coup agenda

Since the US-backed political-military overthrow of President Mel Zelaya in 2009, activists commonly described Honduras as lacking in governability and the rule of law. Strong evidence for this has hit the global stage as the former authoritarian president Juan Orlando Hernández faces drug trafficking charges in the US (U.S. Department of Justice 2022), after his brother, a former congressman was convicted (Anderson 2021). Meanwhile, social movement organizations, networked with actors from local governments, NGOs, international aid, and multilateral organizations struggle to ameliorate conditions for survival, maintain hope for justice, and celebrate life as an act of resistance. Remarkably, left-leaning populist Xiomara Castro became the first woman elected president in the nation's history in November 2021, sparking hope for decreased risks to women and environmental defenders. Still, the fact remains that climate change and its mitigation efforts will continue to put survival on the line; Article 14 of recent reforms to declare energy a human right explicitly state that hydroelectric dams will be promoted to mitigate flooding and increase energy generation (Poder Legislativo 2022), a retrenchment of extractivist discourse from her predecessor and an alarming reflection of influence from former Ecuadorian President Rafael Correa, now an economic advisor to Xiomara (Torres 2022).

Honduran macroeconomic development has relied on enclaves and those who migrate. These enclaves include *maquilas* (sweat shops), drug trafficking corridors (McSweeney et al. 2018), and *extractivist* projects, including industrial agriculture, forestry, mining, industrial fisheries, megatourism projects, and electricity generation (Fash et al. 2021). 15% of Hondurans live on less than US $1.90/day and income inequality is among the highest in the world (Ulku et al. 2021). Poverty, corruption, and violence have driven hundreds of thousands of Hondurans to risk their lives in migration. Through remittances, migrants provide a 23.5% share of Honduran GDP (World Bank 2021).

The national agenda has never featured *buen vivir*. Instead, a brutal effort to advance extractivism has figured prominently in Honduran development strategy. The post-coup government's plan proclaimed the Honduras's leadership 'in terms of sustainable use of natural resources, generating energy, food, minerals and derivatives of the forest sector, like no other country in the region' (Translated from Congreso Nacional de Honduras 2010, 99). Many organizations have succeeded in stopping local extraction projects – particularly mines and dams – challenging the plan's fulfillment (Fash et al. 2021). From 2009-2021, across the board, extraction expanded (Fash et al. 2021).

In electricity generation – driven by extraction destined primarily for national consumption (Fash et al. 2021) – the state has framed the privatization and expansion of renewables as a solution to economic and climate crises. Beginning in the 1980s, the state rid the national energy company of its most competent staff and resources to create an economic and energy crisis and make the case that the State lacked the funding and competence to generate the energy necessary for national development (Girón Castillo 2007). Since then, the state has approved hundreds of contracts with private energy generation companies at elevated prices. Since 2007, the country has seen an unjust transition to renewables, framed in terms of climate mitigation. To reach its goal of more than 60% renewable generation in 2020, the government attracted investment through generous tax exonerations, high prices, and a legal framework permitting expedited approvals, secrecy, and human rights violations (Fash et al. 2019).

Increasingly entangled with climate politics, dams are the most significant and conflictive renewable energy source (Global Witness 2020; Fash et al. 2019). Based on damage from extreme weather events, Honduras has topped the Germanwatch Climate Change Vulnerability Index (Eckstein, Hutfils, and Winges 2018), and faces precarious conditions from water contamination and scarcity (Trócaire 2015). As a Honduran dam operator explained, droughts already challenge hydropower's viability since insufficient water flow causes costly dam malfunctions and failure to generate at full capacity. Despite the questionable climate credentials for hydropower (Fearnside and Pueyo 2012), and projections that transmission infrastructure maintenance would suffice to meet national demand until 2030, the government has pushed new projects as drivers of development and climate change mitigation. In fierce struggles over dams, former military members, private business, and other actors have been involved in killing numerous activists, including Berta Cáceres (Fash et al. 2021). Berta's murder in 2016 has driven greater attention to the ways in which both environmentalists and women are under attack. Extractivism-related violence has occurred in defense of the interests of the private sector, which has increased its hold over extraction, with considerable US military aid, since the 1980s (Barahona 2005, 2018). Globally, Honduras also has one of the highest rates of violent deaths of women, registering 5,348 murders of women from 2005 to 2017, more than half of them femicides (Luciano et al. 2019).

In 2020, these trends reached a critical point following Hurricanes Eta and Iota. After significant losses of lives, homes, infrastructure, and crops, in the already precarious context of the COVID-19 pandemic, the government's response focused on floods in Honduras's industrial hub, the Sula valley. Claiming an urgency to prevent future floods, and to 'solve the energy crisis', the Honduran president issued executive decree PCM-138-2020, fast-tracking 14 hydroelectric dams as 'megaprojects of national interest' (Poder Ejecutivo 2020). Three of these, as well as others not within the decree, are in Lenca territory in Santa Barbara and have been paused largely due to local resistance. Citing such resistance, one reporter went so far as to blame the flooding from hurricanes in 2020 on Betty Vásquez, MAS's coordinator, for leading groups to stop several major dams over the last decade (IM Defensoras 2021). Against all odds, Betty won a court case against him for defamation, the fruit of a campaign from Betty and MAS's deeply rooted network of relations with feminist, environmentalist, and human rights activists and organizations.

MAS's journey to post-extractivism and communitarian feminism

MAS's struggles against dams form part of a broader agenda to resist extractivism and patriarchy through confrontational politics and the practice of alternatives. Based in

Lenca territory[3] in the department of Santa Bárbara, MAS leaders describe the organization as an umbrella of local individuals, organizations and sectors that come together voluntarily to promote a territorial, environmental, and human rights defense agenda inspired by communitarian feminism. In response to the aggressive national push to expand extractivism after the 2009 coup (Bebbington, Fash, and Rogan 2018; Fash 2022) several anti-mining organizers in the area joined forces to found MAS on December 7, 2011. Since then, MAS has expanded into 18 of the 28 municipalities in Santa Bárbara, primarily in Lenca communities in the central and southern parts of the department. It operates under the modality of a Coordinating Committee made up of 9 directors at the departmental level, while at the community and municipal level it operates through environmental committees.

MAS's membership is open to include urban dwellers, men, and mestizxs, but most of its members are campesinx, and/or women, and/or Lenca. The organization estimates that 80% of its members are *campesinxs* in some regard, while the rest are individuals and organizations from diverse sectors – including from the urban center of Santa Barbara – with environmentalist and/or feminist values. The organization did not begin with a feminist agenda; this emerged through debates as the organization evolved and expanded to be woman-led and majority women. In villages with environmental conflicts, Lenca women were and are often the ones taking lead roles, putting their bodies on the line in defense of territory. While globally, women in environmental organizations have struggled to place their gender interests in the organizations' priorities (Agarwal 1998; Razavi 2009), MAS integrates feminism and environmentalism in the communitarian feminist discourse of 'defense of territory' including the 'land-territory' and 'body-territory' (Cabnal 2019). Beyond Santa Barbara, MAS's leadership has been invited to participate in numerous national and international indigenous, environmental, human rights, and feminist networks. Through these, MAS leaders have adopted concepts such as extractivism, *buen vivir*, communitarian feminism, and climate justice, and applied them to rural Honduran contexts. Here, through a summary of their activities, an example of *buen sobrevivir* emerges.

By way of context: MAS's agenda of resistance and alternatives draws from and responds to the 500-year legacy of colonialist violence and resistance in Lenca territory. From the time of Spanish arrival to today, the Lenca have the largest indigenous group in terms of population and territorial expanse (Chapman 1992; Fash et al. 2021). While the Lenca language(s) is/are no longer spoken fluently, and Christian leaders continue to impose their beliefs, many spiritual, cultural, social, and agricultural pre-colonial traditions survive (Chapman 1992; Herranz 1996; Cáceres and Miranda 2021). MAS practices the preservation and recovery of Lenca traditions and ancestral knowledges through processes of dialogue, archival research, and experimentation. Present day Lenca territory is predominantly rural, with most people practicing some combination of *milpa* farming, coffee farming, small-scale dairy farming, artisanship (making sombreros, shoes, hammocks, piñatas, among other items), though no precise surveys were found to provide a systematic classification of the peasantry in this area. A small number of people work in tourism (around Lake Yojoa and on the newly marketed *ruta Lenca*), construction, or

[3]Lenca territory has been documented as stretching over parts of at least nine of Honduras's 18 departments (UNAH et al. 2021).

mining, given the largest and longest operating mine (El Mochito) in Honduras is in southeast Santa Barbara. More broadly, Santa Barbara has been one of the most sought-after parts of the country for mining companies, and Lenca territory in general the most sought-after region for hydropower investment (Fash et al. 2021).

Anti-extractivism is a prominent expression of MAS's anti-capitalist praxis. When Congress passed a new mining law in 2013, lifting a hard-fought moratorium on mining (Bebbington, Fash, and Rogan 2018), MAS joined in a lawsuit filed by the Coalición Nacional de Redes Ambientales (CONROA, or National Coalition of Environmental Organizations and Networks), which led to the 2017 Supreme Court decision that six articles of the mining law were unconstitutional. This put a pause on concession approvals until late 2019, when Congress passed reforms that more grossly violate the constitution (Fash et al. 2021). Locally, exercising powers granted in the Municipalities Law, MAS – like many other Honduran organizations – has advocated for municipal bans on mining and other forms of extraction. MAS also pioneered municipal bans of extractivism writ large (Fash 2022), and joined a national movement to establish municipal bans on Special Economic Development Zones, or 'ZEDEs', which 'grant [private enterprises] sweeping legal and regulatory autonomy' over Honduran territory (Mason, Peterson, and Cano 2021, 107). As of January 2022, MAS has established bans of mining, dams, ZEDEs and/or extractivism in 14 of the department's 28 municipal districts.

MAS complements its anti-extractivist activities with post-extractivist production practices. In Arenales, Gualala, MAS supported community members to stop the implementation of a mining project on three different occasions. Residents then took to establish a cooperative tilapia farm using abandoned fishponds of the National Agrarian Institute. Timoteo, a local leader, reports that 'after knowing nothing' about fisheries, the 25 associates maintain regular sales to support their families. In 2020, they had 40,000 tilapia and a contract with local schools to provide 120,000 fish for the year. Given their reliance on large quantities of clean water, they actively participate in watershed conservation.

MAS regularly holds meetings at the Arenales collective fishery to establish and strengthen bonds, share stories, and craft resistance strategies. Among these meetings are their seed exchanges, designed to maintain ancestral knowledge of plants, promote agroecological practices, and resist industrial agriculture and GMO's. The organization's leaders invite members to share seeds from their fields and gardens along with knowledges of how and when to plant, make organic fertilizers and pesticides, as well as prepare food, drinks, or medicine. Fittingly, MAS titles these gatherings *intercambios de semillas, saberes y sabores* ('exchanges of seeds, knowledges, and flavors'), forming part of their programs for *buen comer* and *buen beber* (eating well and drinking well) under the umbrella of *buen vivir*. Beyond goals to inform, these gatherings incorporate self-reflection and storytelling with jokes and word play, key pieces in MAS's praxis of weaving senses of trust, joy, friendship, and inclusion into the social fabric of the organization.

At one exchange, Timoteo shared the history of resistance to mining, while Carmen shared how her local women's organization collectively produces and sells a variety of goods, including bread, tamales, and embroidery in Nuevo Celilac. Embroiled in a legal battle over territory, they run an organic vegetable and herb garden. Appreciating the peace in her cool, adobe house with water catchment, some might suggest Carmen's life exemplifies *buen vivir*. But Carmen's day to day is more about survival than utopia.

Though Carmen doesn't eat animal foods, she keeps chickens and ducks to sell or barter out of financial necessity. Her organic gardening started as a response to the expense of chemical inputs and her perception that chemicals are bad for health. Seeking an income, her daughter migrated to the city to work in a *maquila* but was killed in a car accident.

Carmen's town is known for its traditional *caramba* music, artisans, and recreation. The value of collective recreation became an organizing flashpoint when a rancher fenced in the local watering hole, claiming it as an extension of his property. Local women leaders sought and obtained guidance from MAS leaders through meetings, capacity-building processes, and ongoing *acompañamiento* (accompaniment) regarding issues of territory and the commons, after which they filed complaints with their local town hall and police force, ultimately achieving the removal of the fence and the restoration of public access to the public space. Satisfaction and inspiration from collaboration on their local activism led to the locals' greater interest and participation in MAS's broader activities, a common trend in the ongoing pattern of growth within the organization. MAS now holds events there. After performing one day at the pond, *caramba* musicians shared how their music brings communities together to reach accords in the Lenca Guancasco tradition (see also Chapman 1992). Betty Vásquez then juxtaposed the maintenance of Lenca traditions to popular music:

> They are stealing our joy. They are even taking away our hopes??… this system promotes music that summons us to violence, which is calling for inequality, and in some cases music that promotes racism towards indigenous peoples. It is a colonial music that also has to do with seeing women as objects, not as political subjects, actors full of knowledge.

These examples begin to illustrate the spectrum of MAS's *buen sobrevivir* activities. Elsewhere, in La Mica, Ilama, where MAS also organized protests to expel a mining concession, women gather monthly to discuss community issues, administer a credit union, and promote women's participation in local politics. However, in 2022, Dunia, the women's group's convener, migrated to Spain to work, pay off debts and improve living conditions for her family. She is one of several leaders in MAS to migrate either to a Honduran city or internationally, major losses for the organization that light on the question of whether 'the upsurge in rural–urban and international migration undermine place-based struggles' (Borras et al. 2021, 17). Up the hills from La Mica, women from MAS's early anti-mining organizing in Santa Bárbara Mountain National Park now lead a coffee cooperative, credit union, and zero waste initiatives. Across Santa Barbara, such women's groups trace their politics to MAS workshops on issues ranging from climate change to extractivism and women's rights.

Though by no means a feminist utopia, MAS is a space for collective learning and action, in which men and women are prefiguring the vision of more harmonious relations. MAS demonstrates that prefiguring eutopia is much about practicing generosity and making affective ties: in addition to sharing knowledges, skills, and gifts, women and men support each other emotionally, intellectually, politically, and logistically. Though women are the majority in the organization, men also have leadership roles in the organization's environmental activism. At MAS gatherings, members are encouraged to point out patriarchal actions and ideas coming from both men and women in a way that is about collective awareness, rather than disciplining. As a leader in MAS expressed,

> Patriarchy is deeply embedded in our daily lives. All these spaces are patriarchal: family, school, church, political parties, and community organizations. We take on the creation of relationships based on respect as a political project because patriarchy kills just like extractivism kills. So [patriarchy] is something men and women are unlearning together in MAS … we use inclusive language[4] not to change grammar, but to change people. We're clear it's not about a power struggle or a war of the sexes like some say, but rather about weaving convivial relationships without canned prescriptions but rather as processes we construct based on our reality.

Various women members have emphasized the importance of the collective trust within the organization: in a context where the State has treated rampant femicides with apathy and impunity, women are at least a bit safer when traveling with men. Thus, when possible, men in the organizations accompany women to events or provide transportation, even when they have no other role to play at the event. Also, in a context where many women face the risk of domestic violence if they do not ask permission to leave the house, it is increasingly common to see women within MAS managing family finances, making decisions ('especially about our body-territory', one says), taking leadership roles in community organizations, and pursuing opportunities in higher education. In short, MAS represents moves toward 'egalitarian farming systems' (Deere 1995, 56).

Within women-centered activities, men tend to sit and listen quietly, and provide logistical support. This relatively passive stance is achieved by various means, as one woman leader explained:

> Some men respect our space and just want to listen to see what we're up to, and then they talk about things with their partners afterward. Sometimes men start out suspicious or worried about what we're up to, and want to participate a little too much. But we can't let them impose patriarchy on us. So, we let them know this is our space, but welcome them to listen. We can see changes, but it's slow, especially in the older men.

Indeed, the youth groups of the organization are quicker to take interest in 'new masculinities' (Messnerschmidt and Messer 2018) and seek workshops on the topic. On the other hand, church organizations provide resistance, as women leaders from different communities shared that their churches questioned women leaders in MAS and discouraged their feminist agendas.

MAS members carry communitarian feminist values and practices into a variety of other collaborations and political organizations. Supporting the Copán Environmentalist Coalition, Betty Vásquez spoke at a rally and supported them to elevating their petition for a ban on mining to a ban on extractivism in Copán Ruinas (Fash 2022). Betty and others have also represented MAS in the Coalition Against Impunity, the National Network of Human Rights Defenders, Women's Rights Center, and the Foro de Mujeres por la Vida, where women from different contexts, territories, and struggles share experiences, knowledges, and strategies to defend women's rights in north/western Honduras. Regionally, MAS has participated in *Mesoamericanas en resistencia por una vida digna* (Mesoamerican Women in Resistance for a Dignified Life), a movement that has contributed insights and inspiration for MAS's interpretation of communitarian feminism, and, by extension, our theorization of *buen sobrevivir*. Finally, MAS partners with NGOs and international aid organizations to fund many of its activities and expand its reach, part of a growing

[4]In the Spanish language, all nouns are gendered.

trend of NGO partnerships in agrarian struggles (Edelman and Borras 2016). Following Hurricanes Eta and Iota, the organization's leaders solicited and distributed humanitarian aid, boating across the Ulúa river to provide basic supplies to communities isolated from bridge collapses.

Throughout their struggles, MAS leaders interweave cosmopolitical ceremonies (De La Cadena 2015), combining Lenca tradition and political discourse to heal the body-territory and harmonize relations with the land-territory and the web of life (Cabnal 2019). Their regular meetings generally feature an *altar de la vida* (altar of life) including maize, beans, incense, and flowers, around which members gather and share experiences, feelings, and proposals. After the hurricanes, MAS members gathered to perform ceremonies to 'reconcile [themselves] with water'. Members shared the traumas of losing both individually and collectively grown crops that were for survival, for bartering, for sale, or for medicinal use, partly or entirely losing homes they had constructed, and losing contact with loved ones due to infrastructure and communications failures. As a result, numerous members reported feeling 'blocked', with no desire to go near the Ulúa River, or even their local water sources. The ceremonies, then, allowed members to share their experiences, collectively reconnect with the life-giving force of the rivers and streams, and reinvigorate their struggles against extractivism, particularly megadam construction.

Chinda: struggles to survive well in the eye of extractivist climate mitigation

Chinda is a Lenca municipal district of just over 5,000 residents considered 100% rural by the census (Secretaría de Gobernación, Justicia y Descentralización 2022), which lists the primary economic activities as the cultivation of *milpa,* coffee, sugar cane, fruits, and vegetables. According to project files obtained by MAS, more than half of the district could be flooded if the El Tornillito dam project moves forward. With a 160–200 MW generation capacity, the dam would be the third largest in the country. Here, we present the dam project as it relates to Chinda residents' struggles toward food sovereignty, agroecology, gender justice, and post-extractivism. This case elucidates how the eutopian, radical politics of *buen sobrevivir* are practiced through complementary acts to confront oppression and perform alternatives.

The tradition surrounding sardines in Chinda is an ancestral practice. Each November since time immemorial, locals have visited the banks of the Ulúa to fish sardines with the shared understanding that the sardines can be consumed, gifted, and even bartered, but never sold. They say if someone sells sardines or there is conflict within the community, the sardines do not return the following year. When there are sardines, they are prepared in a variety of local dishes, including *pupusas* and eggs.

The role of sardines in the aftermath of Hurricanes Eta and Iota demonstrates how locally controlled, culturally rooted, and diversified food sources (pillars of food sovereignty) are vital for survival. Bridge collapses isolated much of Chinda, and the scarcity of humanitarian aid threatened lives. Waldina, a local MAS leader[5], reports that many lost their crops from intense floods, including MAS members' collective *milpa*. However, that very month, the

[5]Given MAS's model as an umbrella of community-based organizations, the organization operates through various groups in Chinda, including the women's network, water councils, and an environmental organization called *Comunidades Unidas*.

sardines leapt in remarkable numbers, and were an important food source to survive. While some locals expressed anger with the river for destroying homes and taking crops, MAS members articulated counter-hegemonic narratives featuring a political ecological analysis of environmental degradation and vulnerability focused on deforestation, industrial agriculture, and mangrove destruction, all of which increase vulnerability to floods. On the contrary, they suggested, the river blessed people with sardines for survival.

The sardines are a small symbol of the greater stakes in the fierce struggle against the El Tornillito dam. They migrate to, and leap at the very site of the dam project. Since dams disrupt fish migration (Moran et al. 2018) and survival through physical obstruction as well as changes in water temperature, methane, and oxygen content (Galy-lacaux et al. 1997; Wang et al. 2019), the sardines' future is uncertain at best. Meanwhile, the dam company Hidrovolcán has marketed the flooding as a tourism opportunity, saying Chinda will become the 'Venice of Honduras', but, in a town with critical levels of poverty (Secretaría de Gobernación, Justicia y Descentralización 2022), as Waldina put it, 'who is going to own the hotels?' Hidrovolcán is headed by a Honduran energy magnate whose portfolio includes extensive solar farms and one of the largest thermoelectric plants in the country (Radio Progreso 2021). According to a source in the energy sector, the Honduran government has provided him with valuable studies and approved his projects without due bidding.

With support from MAS leaders from outside of Chinda, Waldina has led the effort to stop the dam. In 2017, on a day's notice of a town hall meeting regarding the project, Waldina and her collaborators rallied hundreds of people and managed to establish a ban of hydroelectric dams in Chinda. But Hidrovolcán has continued to operate, dedicating part of its $600M budget to conflict-laden land purchases in the flood zone. Juan Carlos Cerros, a close collaborator of Waldina's and President of the MAS-affiliated group *Comunidades Unidas*, refused to sell his land to the company. Despite being one of 16 members of MAS with government-sponsored 'protective measures', he was killed in March 2021. With his death unsolved, Waldina, echoing some of our personal experiences, reports she's been told she too will 'end up like Berta [Cáceres].'

Nonetheless, Waldina and her collaborators around the district joyfully persevere in practicing *buen sobrevivir*. Sponsored by the Christian Organization for the Integral Development of Honduras (OCDIH)'s 'Women's networks' programs, women collectively learn and practice just and dignified forms to survive well. Their priorities include how to transform or leave toxic relationships, rights to participate in politics, and agroecological production of food and medicine. Waldina and her community garden collaborators demonstrated great pride in developing their own pesticides with chili pepper sprays and growing the 'most delicious cucumbers around'. Unfortunately, they noted, buyers in local markets have grown to value size and shape over taste and production inputs, one more issue in the struggles to decolonize desire toward *buen sobrevivir*.

Despite the municipal ban on dams in Chinda, at the time of writing, the El Tornillito project remains in contention, along with several other major dams in Santa Barbara. Taking advantage of greater opportunities for dialogue with Xiomara Castro's administration, MAS leaders met with the Secretary of Natural Resources in May 2022 to present their deep concerns regarding the projects and request an audience with Xiomara and her cabinet to discuss a path forward. While the leftist agenda has made important strides toward a more just energy system by initiating renegotiations of electricity generation contracts with private generators and providing free electricity to roughly a million families, Xiomara and her Libre party-controlled Congress's decisions regarding new energy generation projects, especially hydroelectric dams, will be a defining feature of their approach to extractivism. Likewise, it is among the highest concerns for MAS.

Throughout this case study, we have illuminated ways in which MAS deploys post-extractivist, communitarian feminist logics and strategies in opposition to authoritarian populism, patriarchy, and extractivist climate change mitigation. In the face of such precarious circumstances, practicing something akin to what elsewhere has been deemed *buen vivir* is, and will continue to be, a means of survival.

Conclusions

The MAS case for *buen sobrevivir* responds to the urgency to understand ways to curb the scale, intensity, and violence of resource extraction in Latin America, and to make other worlds possible. Our theoretical framework and case study also speak to the pressing questions of this *JPS* forum regarding climate change's unique qualities and differentiated impacts, as well as climate narratives and strategies to simultaneously mitigate climate change and erode capitalism (Borras et al. 2021). Here, we summarize our contributions to debates on *buen vivir*, post-extractivism, and the forum.

This paper answers calls to ground *buen vivir* and post-extractivism proposals by focusing on networked experiments at subnational levels, which reveal the ways in which eutopian prefigurative politics to survive well co-exist with precarity wrought by structural violence and injustice related to extractivism, patriarchy, and climate change. We argue that *buen vivir* literature's understatement of precarity justifies its conceptual reframing: given that climate change and climate mitigation efforts will continue to destabilize even the most harmonious nature/society assemblages, we proposed to generalize the survival strategies already in place in spaces such as Lenca territory as *buen sobrevivir*, a politics of 'making multispecies livability possible' (Gan 2017, 5). Considering this precarity, *buen sobrevivir* seeks to do justice to the realities of both ongoing resistance to oppressive structures and adaptation to climate change, which means defending or recovering some indigenous traditions while shedding others and embracing experimentation with new ideas – an approach central to both prefigurative politics and communitarian feminism. This, we argue, responds to critiques that *buen vivir* proposals romanticize the past.

Climate change also begs consideration in reframing the concept of extractivism to better address issues of extraction for national consumption. Extraction that contributes to localized vulnerability (by decreased water quantity/quality or soil water retention) or that which contributes greenhouse gases (e.g. methane from dam reservoirs and biomass energy) is problematic whether it is for export or not. Given the onset of climate change's impacts and a broader diffusion of its future risks, climate change has increasingly entered national politics and social movement agendas. Thus, our case highlights resistance to energy generation, which we include in our definition of extractivism.

Our case illuminates how the state and private sector deploy developmentalist narratives (prioritizing urban areas, electrical energy, and national security) to justify an anticipatory politics of extractivist climate mitigation by way of hydroelectric dams. Rural areas were rendered more vulnerable because hurricane relief efforts were concentrated on urban areas, and the dams proposed ostensibly mitigate flooding in urban areas while deliberately flooding agrarian settings and displacing entire communities.

The prefigurative politics of *buen sobrevivir* operate deliberately 'in the way of development', confronting the exacerbation of social differentiation and trajectories of accumulation (Blaser, Feit, and McRae 2004). Such obstruction has induced defamation and violence in a context where political, economic, and military elite have maintained a constant trajectory of accumulation since the 1980s, through transitions toward privatization and renewables in the Honduran energy sector, manifesting after the hurricanes in the executive decree to fast-track 14 hydroelectric dams. For agrarian politics, the dire conditions of inequality, violence, and uncertainty spawned survival efforts that spanned from anti-extractivist municipal declarations, autonomist politics to create community economies, and electoral politics to erode capitalism.

We can also note that climate change adaptation and resilience programs, though often technocratic or neoliberal, have brought significant funding that social movements channel toward more radical efforts. In meetings such as the seed exchanges, Lenca women recover ancestral knowledge practices linked to the erosion of capitalism and, by extension, the causes of climate change. At the same time, these gatherings provide spaces in which leaders discursively link climate change and extractivism, a vital step in expanding their movement to ban extractivism locally and build support for national policy changes.

Based on Lenca women's experiences, *buen sobrevivir* is a sufficiently radical politics to confront climate change in rural settings that is anti-capitalist, environmental, agrarian, decolonial, and feminist. In the spirit of challenging all forms of domination of the body-territory and land-territory, MAS's experiments involve the eutopian implementation of their vision to survive well locally, while working toward structural change. Rather than the top-down, statist utopian politics that have failed in other contexts, these politics involve the messy groundwork of persuading others to join in processes of learning and unlearning to establish post-patriarchal, post-extractivist, and post-capitalist territories.

This framing of *buen sobrevivir* acknowledges that the time for 'sufficient' action to 'stop' climate change has passed and recognizes that urgent efforts to ensure future survival depend on: obstructing processes that will increase emissions and/or vulnerability; strengthening communities to ensure collective support; and knowledge production to improvise adaptative responses to climate change impacts that are just and equitable. Placing a more-than-human 'community' at the center of a eutopian politics of surviving well, what we make of the future is not a choice between utopia and Armageddon (Carrington 2022), but rather an ongoing struggle to replace structures and practices of oppression with systems and processes to perform societies based on justice, dignity, and interdependence.

Acknowledgements

We would like to thank Tony Bebbington and Jody Emel as well as four anonymous reviewers for their comments, Denise Humphreys Bebbington for inviting us to present the paper at Extractives@Clark, and fellow members of the Movimiento Ambientalista SantaBarbarense and Coalición Ambientalista de Copán for their dedication, camaraderie, and trust.

Disclosure statement

No potential conflict of interest was reported by the author(s).

Funding

Fash's work was supported by the National Science Foundation Graduate Research Fellowship Program [grant number 1945996], the Fulbright Commission, the Institute for Human Geography, Clark University, and the Conference of Latin American Geography. Fash and Sojob's collaborative work was supported by the Instituto Mexicano de Cinematografía.

ORCID

Benjamin C. Fash http://orcid.org/0000-0001-8746-2328

References

Acosta, A. 2013. "Extractivism and NeoExtractivism: Two Sides of the Same Curse." In *Beyond Development: Alternative Visions from Latin America*, edited by M. Lang, and D. Mokrani, 61–86. Amsterdam: Transnational Institute.

Agarwal, B. 1998. "Environmental Management, Equity and Ecofeminism: Debating India's Experience." *Journal of Peasant Studies* 25 (4): 55–95. doi:10.1080/03066159808438684.

Agarwal, B. 2014. "Food Sovereignty, Food Security and Democratic Choice: Critical Contradictions, Difficult Conciliations." *Journal of Peasant Studies* 41 (6): 1247–1268. doi:10.1080/03066150.2013.876996.

Altieri, M. A., C. I. Nicholls, A. Henao, and M. A. Lana. 2015. "Agroecology and the Design of Climate Change-Resilient Farming Systems." *Agronomy for Sustainable Development* 35 (3): 869–890. doi:10.1007/s13593-015-0285-2.

Anderson, B. 2010. "Preemption, Precaution, Preparedness: Anticipatory Action and Future Geographies." *Progress in Human Geography* 34 (6): 777–798. doi:10.1177/0309132510362600.

Anderson, J. L. 2021. "False Friends: How the U.S. Looked Away from Corruption It Helped Create." *The New Yorker*, 58–69.

Barahona, M. 2005. *Honduras en el siglo XX: una síntesis histórica*. Tegucigalpa: Editorial Guaymuras.

Barahona, M. 2018. *Elites, redes de poder y régimen político en Honduras*. Tegucigalpa: Editorial Guaymuras, ERIC-SJ.

Bebbington, A. 2009. "The new Extraction: Rewriting the Political Ecology of the Andes?" *NACLA Report on the Americas* 42 (5): 12–20. doi:10.1080/10714839.2009.11722221.

Bebbington, A. 2015. "Political Ecologies of Resource Extraction: Agendas Pendientes." *European Review of Latin American and Caribbean Studies* 100: 85–98. doi:10.18352/erlacs.10121.

Bebbington, A., B. Fash, and J. Rogan. 2018. "Socio-environmental Conflict, Political Settlements, and Mining Governance: A Cross-Border Comparison, El Salvador and Honduras." *Latin American Perspectives* 46 (2): 84–106. doi:10.1177/0094582X18813567.

Bernstein, H. 2014. "Food Sovereignty via the 'Peasant Way': A Sceptical View." *Journal of Peasant Studies* 41 (6): 1031–1063. doi:10.1080/03066150.2013.852082.

Bernstein, H. 2016. "Agrarian Political Economy and Modern World Capitalism: The Contributions of Food Regime Analysis." *Journal of Peasant Studies* 43 (3): 611–647. doi:10.1080/03066150.2015.1101456.

Blaser, M., H. Feit, and G. McRae, eds. 2004. *In the Way of Development: Indigenous Peoples, Life Projects and Globalization*. London: Zed Books.

Boggs, C. 1977. "Marxism, Prefigurative Communism, and the Problem of Worker's Control." *Radical America* 11 (6): 99–122.

Borras, S. M., I. Scoones, A. Baviskar, M. Edelman, N. L. Peluso, and W. Wolford. 2021. "Climate Change and Agrarian Struggles: An Invitation to Contribute to a JPS Forum." *Journal of Peasant Studies* 49 (1): 1–28. doi:10.1080/03066150.2021.1956473.

Cabnal, L. 2017. "Tzk'at, Red de Sanadoras Ancestrales del Feminismo Comunitario desde Iximulew-Guatemala." *Ecología Política* 54: 100–104.

Cabnal, L. 2019. "El relato de las violencias desde mi territorio cuerpo-tierra." In *En tiempos de muerte: cuerpos, rebeldías, resistencias, Tomo IV*, edited by X. Leyva Solano, and R. Icaza, 113–126. Buenos Aires: CLACSO, RETOS, International Institute of Social Sciences.

Cáceres, B.. 2015. Berta Caceres acceptance speech, 2015 Goldman Prize ceremony. Goldman Environmental Prize. San Francisco, CA. Available at: https://youtu.be/AR1kwx8b0ms

Cáceres, B., and M. Miranda. 2021. "Las luchas de mujeres y pueblos Garífunas y Lencas en Honduras." In *De despojos y luchas por la vida*, edited by X. Leyva Solano, P. Viera Bravo, J. M. Trigueiro de Lima, and A. C. Velázquez Solís, 21–42. Guadalajara: CLACSO, CIESAS, Retos.

Caria, S., and R. Domínguez. 2016. "Ecuador's buen vivir." *Latin American Perspectives* 43 (1): 18–33. doi:10.1177/0094582X15611126.

Carrington, D. 2022. "Hope Amid Climate Chaos: 'We are in a Race between Armageddon and Awesome'." *The Guardian*, 4 October.

Chapman, A. 1992. *Los Hijos del Copal y la Candela: Ritos agrarios de la tradición lenca de Honduras*. México: Editorial UNAM.

Congreso Nacional de Honduras. 2010. *Visíon de País 2010-2038 y Plan de Nación 2010-2022*. Tegucigalpa: Congreso Nacional.

Cornish, F., J. Haaken, L. Moskovitz, and S. Jackson. 2016. "Rethinking Prefigurative Politics: Introduction to the Special Thematic Section." *Journal of Social and Political Psychology* 4 (1): 114–127. doi:10.5964/jspp.v4i1.640.

Decrecimiento. 2012. "Eduardo Galeano: ¿Para qué sirve la Utopía?" Accessed 25 Dec 2021. https://www.youtube.com/watch?v = GaRplBj5xho.

Deere, C. D. 1995. "What Difference Does Gender Make? Rethinking Peasant Studies." *Feminist Economics* 1 (1): 53–72.

De La Cadena, M. 2015. *Earth Beings: Ecologies of Practice Across Andean Worlds*. London: Duke University Press.

Dobler-Morales, C., A. Álvarez Larrain, Q. Orozco-Ramírez, and G. Bocco. 2021. "Grounding Maladaptation: Agricultural Change as a Source of Climatic Risks in Small Farms of the Mixteca Alta, Mexico." *Geoforum* 127: 234–245. doi:10.1016/j.geoforum.2021.11.001.

Eckstein, D., M.-L. Hutfils, and M. Winges. 2018. *Global Climate Risk Index 2019: Who Suffers Most from Extreme Weather Events? Weather-Related Loss Events in 2017 and 1998 to 2017*. Berlin: Germanwatch.

Edelman, M., and S. M. Borras. 2016. *Political Dynamics of Transnational Agrarian Movements*. Canada: Fernwood.

Edelman, M., T. Weis, A. Baviskar, S. M. Borras, E. Holt-Giménez, D. Kandiyoti, and W. Wolford. 2014. "Introduction: Critical Perspectives on Food Sovereignty." *Journal of Peasant Studies* 41 (6): 911–931. doi:10.1080/03066150.2014.963568.

Emel, J. 1991. "Ecological Crisis and Provocative Pragmatism." *Environment & Planning D: Society & Space* 9: 384–390. doi:10.1068/d090384.

Escobar, A. 2018. *Designs for the Pluriverse: Radical Interdependence, Autonomy, and the Making of Worlds*. London: Duke University Press.

Farah, H. I., and L. Vasapollo, eds. 2011. *Vivir bien: ¿Paradigma no capitalista?* La Paz: CIDES-UMSA.

Fash, B.C.. 2022. "Redefining Extractivism from Honduras." *Geoforum* 135 (2022): 37–48. doi:10.1016/j.geoforum.2022.07.010.

Fash, B., E. Velásquez, A. Kaffati, J. Meza, C.N. Mondragón, J.D. Cáceres, R.E. Corrales, A. Carias, C. Molina, and L. Corea. 2019. "Territorios en Riesgo II: Minería, Hidrocarburos y Generación de Energía Eléctrica En Honduras."Tegucigalpa: UNAH, FOSDEH, Oxfam.

Fash, B., E. Velásquez, J. Rivera, J. Cáceres, L. Corea, N. Cuba, S. Sellwood, G. Redman, and M. Burdett. 2021. "Territorios en Riesgo III: Minería, Hidrocarburos y Generación de Energía Eléctrica En Honduras." Tegucigalpa: FOSDEH, UNAH, Oxfam.

Fearnside, P. M., and S. Pueyo. 2012. "Greenhouse-gas Emissions from Tropical Dams." *Nature Climate Change* 2 (6): 382–384. doi:10.1038/nclimate1540.
Fraser, N. 2021. "Climates of Capital: For a Trans-Environmental Eco-Socialism." *New Left Review* 127: 94–127.
Freedman, J. 2016. "Sexual and Gender-Based Violence Against Refugee Women: A Hidden Aspect of the Refugee 'Crisis'." *Reproductive Health Matters* 24 (47): 18–26. doi:10.1016/j.rhm.2016.05.003.
Galy-lacaux, C., R. Delmas, L. Labroue, and P. Gosse. 1997. "Gaseous Emissions and Oxygen Consumption in Hydroelectric Dams: A Case Study in French Guyana." *Global Biogeochemical Cycles* 11 (4): 471–483. doi:10.1029/97GB01625.
Gan, E. 2017. "Haunted Landscapes of the Anthropocene." In *Arts of Living on Damaged Planet*, edited by A. Tsing, H. Swanson, E. Gan, and N. Bubandt, 1–14. Minneapolis: University of Minnesota Press.
Gibson-Graham, J. K. 2006. *A Postcapitalist Politics*. Minneapolis: University of Minnesota Press.
Giraldo, O. F. 2014. *Utopías en la era de la supervivencia: Una interpretación del Buen Vivir*. Chapingo: Editorial Itaca.
Girón Castillo, W. 2007. *Energía eléctrica: Desarrollo-subdesarrollo*. Tegucigalpa: Universidad Nacional Autónoma de Honduras.
Global Witness. 2020. *Defending Tomorrow: The Climate Crisis and Threats Against Land and Environmental Defenders*. London.
Gudynas, E. 2009. "Diez tesis urgentes sobre el nuevo extractivismo. Contextos y demandas bajo el progresismo sudamericano actual." In *Extractivismo, Política y sociedad*, edited by Centro Latinoamericano de Ecología Social, 187–225. Quito: CAAP, CLAES.
Gudynas, E. 2013. "Transitions to Post-Extractivism: Directions, Options, Areas of Action." In *Beyond Development: Alternative visions from Latin America*, edited by M. Lang, and D. Mokrani, 165–194. Amsterdam: Transnational Institute.
Gudynas, E. 2015. *Extractivismos: Ecología, economía y política de un modo de entender el desarrollo y la Naturaleza*. Cochabamba: Centro de Documentación e Información Bolivia.
Gudynas, E., and A. Acosta. 2011. "La renovación de la crítica al desarrollo y el buen vivir como alternativa." *Utopía y Praxis Latinoamericana* 16 (53): 71–83.
Gutiérrez Aguilar, R. 2014. *Rhythms of the Pachakuti: Indigenous Uprising and State Power in Bolivia*. London: Duke University Press.
Haraway, D. 1988. "Situated Knowledges: The Science Question in Feminism and the Privilege of Partial Perspective." *Feminist Studies* 14 (3): 575–599. doi:10.2307/3178066.
Haraway, D. J. 2016. *Staying with the Trouble: Making Kin in the Chthulucene*. London: Duke University Press.
Henderson, T. P. 2017. "State–Peasant Movement Relations and the Politics of Food Sovereignty in Mexico and Ecuador." *Journal of Peasant Studies* 44 (1): 33–55. doi:10.1080/03066150.2016.1236024.
Herranz, A. 1996. *Estado, sociedad y lenguaje*. Tegucigalpa: Editorial Guaymuras.
Holt-Giménez, E. 2006. *Campesino a campesino: Voices from Latin America's Farmer to Farmer Movement for Sustainable Agriculture*. Oakland: Food First Books.
Honty, G., and E. Gudynas. 2014. *Cambio climático y transiciones al Buen Vivir: alternativas al desarrollo para un clima seguro*. Lima: CLAES.
Hossain, M., and A. McAlpine. 2017. *Gender Based Violence Research Methodologies in Humanitarian Settings*. Cardiff: Elhra.
IM Defensoras (Iniciativa Mesoamericana de Mujeres Defensoras de Derechos Humanos). 2021. "HONDURAS / Recrudece la campaña de desprestigio y difamación en contra de la Defensora Betty Vásquez." Accessed 31 March 2022. http://im-defensoras.org/2020/12/alerta-defensoras-honduras-recrudece-la-campana-de-desprestigio-y-difamacion-en-contra-de-la-defensora-betty-vasquez/.
IPBES. 2019. *Global Assessment Report of the Intergovernmental Science-Policy Platform on Biodiversity and Ecosystem Services. Debating Nature's Value*. Bonn: IPBES secretariat.
Jeffrey, C., and J. Dyson. 2021. "Geographies of the Future: Prefigurative Politics." *Progress in Human Geography* 45 (4): 641–658. doi:10.1177/0309132520926569.

Lin, C. S., A. A. Pykett, C. Flanagan, and K. R. Chávez. 2016. "Engendering the Prefigurative: Feminist Praxes That Bridge a Politics of Prefigurement and Survival." *Journal of Social and Political Psychology* 4 (1): 302–317. doi:10.5964/jspp.v4i1.537.

Luciano, D., N. Hidalgo, N. Acuña, and A.-M. Urban. 2019. *Femicidio en Honduras*. Washington, D.C.: BID: División de Género y Diversidad Sector Social.

Lugones, M.. 2010. "Toward a Decolonial Feminism." *Hypatia* 25 (4): 742–759. doi:10.1111/j.1527-2001.2010.01137.x.

Macas, L. 2011. "El Sumak Kawsay." In *Debates sobre cooperación y modelos de desarrollo*, edited by G. Weber, 47–60. Quito: Centro de Investigaciones CIUDAD.

Martínez-Torres, M. E., and P. M. Rosset. 2014. "Diálogo de saberes in La Vía Campesina: Food Sovereignty and Agroecology." *Journal of Peasant Studies* 41 (6): 979–997. doi:10.1080/03066150.2013.872632.

Mason, J., C. Peterson, and D. I. Cano. 2021. "The Honduran ZEDE Law, from Ideation to Action." *Journal of Special Jurisdictions* 1 (2): 107–149.

McKay, B. M., A. Alonso-Fradejas, and A. Ezquerro-Cañete, eds. 2021. *Agrarian Extractivism in Latin America. Agrarian Extractivism in Latin America*. London: Routledge.

McMichael, P. 2015. "A Comment on Henry Bernstein's way with Peasants, and Food Sovereignty." *Journal of Peasant Studies* 42 (1): 193–204. doi:10.1080/03066150.2014.936853.

McSweeney, K., D. J. Wrathall, E. A. Nielsen, and Z. Pearson. 2018. "Grounding Traffic: The Cocaine Commodity Chain and Land Grabbing in Eastern Honduras." *Geoforum* 95 (February): 122–132. doi:10.1016/j.geoforum.2018.07.008.

Méndez Martínez, L. 2012. "Elementos de la constitución antigua del pueblo Maya." *Cambios* 7: 19–23.

Messnerschmidt, J. W., and M. A. Messer. 2018. "Hegemonic, Nonhegemonic, and "New" Masculinities." In *Gender Reckonings: New Social Theory and Research*, edited by J. W. Messerschmidt, P. Y. Martin, M. A. Messner, and R. Connell, 37–72. New York: New York University Press.

Millán, M., ed. 2014. *Más allá del feminismo: caminos para andar*. México: Red de Feminismos Descoloniales.

Mora, A. I. 2020. "Buen vivir / vivir bien: entre crisis política y polifonía de voces subalternas y de movimientos sociales." In *Buenos vivires y transiciones: la vida dulce, la vida bella, la vida querida, la vida sabrosa, la vida buena, la vida plenitud: convivir en armonía*, edited by A. I. Mora, 13–24. Bogotá: Corporación Universitaria Minuto de Dios – UNIMINUTO.

Mora, C., B. Dousset, I. R. Caldwell, F. E. Powell, R. C. Geronimo, C. R. Bielecki, C. W. W. Counsell, et al. 2017. "Global Risk of Deadly Heat." *Nature Climate Change* 7 (7): 501–506. doi:10.1038/nclimate3322.

Morales, E. 2011. "Prólogo." In *Vivir bien: ¿paradigma no Capitalista?*, edited by I. Farah H, and L. Vasapollo, 9–10. La Paz: CIDES-UMSA.

Moran, E. F., M. C. Lopez, N. Moore, N. Müller, and D. W. Hyndman. 2018. "Sustainable Hydropower in the 21st Century." *Proceedings of the National Academy of Sciences* 115 (47): 1–8. doi:10.1073/pnas.1809426115.

Nightingale, A. J. 2017. "Power and Politics in Climate Change Adaptation Efforts: Struggles Over Authority and Recognition in the Context of Political Instability." *Geoforum* 84: 11–20. doi:10.1016/j.geoforum.2017.05.011.

Novacek, M. J., and E. E. Cleland. 2001. "The Current Biodiversity Extinction Event: Scenarios for Mitigation and Recovery." *Proceedings of the National Academy of Sciences* 98 (10): 5466–5470. doi:10.1073/pnas.091093698.

Olukoya Obafemi, A. P., and S. Kurt. 2016. "Environmental Impacts of Adobe as a Building Material: The North Cyprus Traditional Building Case." *Case Studies in Construction Materials* 4: 32–41. doi:10.1016/j.cscm.2015.12.001.

Paprocki, K. 2018. "Threatening Dystopias: Development and Adaptation Regimes in Bangladesh." *Annals of the American Association of Geographers* 108 (4): 955–973. doi:10.1080/24694452.2017.1406330.

Paredes, J. 2010. *Hilando fino desde el feminismo comunitario*. México: Cooperativa El Robozo.

Peterson, R. G. 1984. "Preparing for Apocalypse: Survivalist Strategies." *Free Inquiry in Creative Sociology* 12 (1): 44–46.

Poder Ejecutivo. 2020. "Decreto Ejecutivo Número PCM-138-2020." *La Gaceta*, 6 January, 1–5.

Poder Legislativo. 2022. "Decreto No. 46-2022." *La Gaceta*, 35924, 12–32.

Radio Progreso. 2021. "Energía Sobrevalorada: El consorcio detrás de la crisis de la ENEE." El Progreso, Honduras.

Razavi, S. 2009. "Engendering the Political Economy of Agrarian Change." *Journal of Peasant Studies* 36 (1): 197–226. doi:10.1080/03066150902820412.

Red de feminismos descoloniales. 2014. "Descolonizando nuestros feminismos, abriendo la mirada." In *Más allá del feminismo: caminos para andar*, edited by M. Millán, 319–327. México: Red de Feminismos Descoloniales.

Rodríguez Salazar, A. 2016. "Teoría y práctica del Buen Vivir: Orígenes, debates conceptuales y conflictos sociales." El caso de Ecuador. Universidad del País Vasco.

Román-Palacios, C., and J. J. Wiens. 2020. "Recent Responses to Climate Change Reveal the Drivers of Species Extinction and Survival." *Proceedings of the National Academy of Sciences of the United States of America* 117 (8): 4211–4217. doi:10.1073/pnas.1913007117.

Sargent, L. T. 1994. "The Three Faces of Utopianism Revealed." *Utopian Studies* 5 (1): 1–37.

Secretaría de Gobernación, Justicia y Descentralización. 2022. "Perfil municipal: Índice de Desarrollo Municipal, Chinda, Santa Bárbara." Tegucigalpa.

Smucker, J. M. 2014. "Can Prefigurative Politics Replace Political Strategy?" *Berkeley Journal of Sociology* 58: 74–82.

Soper, R. 2020. "From Protecting Peasant Livelihoods to Essentializing Peasant Agriculture: Problematic Trends in Food Sovereignty Discourse." *Journal of Peasant Studies* 47 (2): 265–285. doi:10.1080/03066150.2018.1543274.

Stefanoni, P. 2012. "¿Y quién no querría 'vivir bien'? Encrucijadas del proceso de cambio boliviano." *Rebelion* 53: 1–6.

Steffen, W., J. Rockström, K. Richardson, T. M. Lenton, C. Folke, D. Liverman, C. P. Summerhayes, et al. 2018. "Trajectories of the Earth System in the Anthropocene." *Proceedings of the National Academy of Sciences of the United States of America* 115 (33): 8252–8259. doi:10.1073/pnas.1810141115.

Svampa, M. 2013. "Resource Extractivism and Alternatives: Latin American Perspectives on Development." In *Beyond Development: Alternative Visions from Latin America*, edited by M. Lang, and D. Mokrani, 117–144. Amsterdam: Transnational Institute.

Svampa, M. 2015. "Commodities Consensus: Neoextractivism and Enclosure of the Commons in Latin America." *South Atlantic Quarterly* 114 (1): 65–82. doi:10.1215/00382876-2831290.

Svampa, M. 2019. "The Latin American Critique of Development." In *Pluriverse: A Post-Development Dictionary*, edited by A. Kothari, A. Salleh, A. Escobar, F. Demaria, and A. Acosta, 18–24. New Delhi: Tulika Books.

Tilzey, M. 2019. "Authoritarian Populism and Neo-Extractivism in Bolivia and Ecuador: The Unresolved Agrarian Question and the Prospects for Food Sovereignty as Counter-Hegemony." *Journal of Peasant Studies* 46 (3): 626–652. doi:10.1080/03066150.2019.1584191.

Torres, M. 2022. ""Honduras debe invertir en hidroeléctricas y renegociar fideicomisos": Parte del diagnóstico de expresidente ecuatoriano Rafael Correa." *HCH*, 30 March.

Trócaire. 2015. "Feeling the Heat: How Climate Change is Driving Extreme Weather in the Developing World." Maynooth, Ireland.

Ulku, H., G. R. Zaourak, G. C. Arenas, N. L. Campora, S. Kasyanenko, C. Lakatos, A. Mulabdic, and S. Yu. 2021. *Unleashing Central America's Growth Potential: Honduras*. Washington, DC: The World Bank.

UNAH (Universidad Nacional Autónoma de Honduras), FOSDEH (Foro Social para la Deuda Externa de Honduras), and O. International. . 2021. "Geoportal Territorios en Riesgo." Accessed 1 August 2022. https://territoriosenriesgo.unah.edu.hn/.

U.S. Department of Justice. 2022. "Juan Orlando Hernandez, Former President of Honduras, Extradited to the United States on Drug-Trafficking and Firearms Charges." Accessed 2 Aug 2022. https://www.justice.gov/usao-sdny/pr/juan-orlando-hernandez-former-president-honduras-extradited-united-states-drug.

Varea, S., and S. Zaragocin. 2017a. *Feminismo y Buen Vivir: Utopias Decoloniales*. Cuenca: PYDLOS Ediciones.

Varea, S., and S. Zaragocin. 2017b. "Introducción." In *Feminismo y buen vivir: utopías decoloniales*, edited by S. Varea, and S. Zaragocin, 5–16. Cuenca: PYDLOS Ediciones.

Walsh, C. 2010. "Development as Buen Vivir: Institutional Arrangements and (de)Colonial Entanglements." *Development* 53 (1): 15–21. doi:10.1057/dev.2009.93.

Wang, B., H. Zhang, X. Liang, X. Li, and F. Wang. 2019. "Cumulative Effects of Cascade Dams on River Water Cycle: Evidence from Hydrogen and Oxygen Isotopes." *Journal of Hydrology* 568: 604–610. doi:10.1016/j.jhydrol.2018.11.016.

World Bank. 2021. "Annual Remittances Data (updated as of May 2021)." Accessed 5 January 2022. https://www.worldbank.org/en/topic/migrationremittancesdiasporaissues/brief/migration-remittances-data.

Yates, L. 2015. "Rethinking Prefiguration: Alternatives, Micropolitics and Goals in Social Movements." *Social Movement Studies* 14 (1): 1–21. doi:10.1080/14742837.2013.870883.

Forest as 'nature' or forest as territory? Knowledge, power, and climate change conservation in the Peruvian Amazon

Maritza Paredes and Anke Kaulard

ABSTRACT

This paper examines the implications of climate crisis governance for rural communities in the Amazon of Peru. It draws the attention to the shared political economy behind the resistance of diverse rural populations particularly, Indigenous and Colono communities. Based on an analysis of two local conservation interventions in the region of San Martín—one involving indigenous communities and the other peasant settlers—this study argues that narrow authorized knowledge obscures the wider historical and agrarian macro context of uneven institutional and ecological arrangements that lead to the reproduction of injustices related to the land, the underlying causes of deforestation, and the authoritarian relationships of these local communities with the state.

1. Introduction

The effects of climate change are catastrophic and are rapidly evolving, putting many populations in imminent danger. However, the effects of greenhouse gas emissions and the policies that try to mitigate them have uncertain and unequal effects among different rural populations. This paper examines the profound implications of climate crisis governance for diverse rural communities in the Amazon of Peru.

Forests have become crucial in the fight against climate change since their conservation saves valuable carbon stocks and reduces greenhouse gas emissions from deforestation (Oldekop et al. 2020). Peru has the second largest share of the Amazonian forests, only after Brazil (NICFI n.d.), but the country also has one of the highest deforestation rates (Ritchie and Roser 2021). Therefore, conservation policies have taken on a new urgency in twenty-first century Peru as part of a new global structure of climate governance (Scoones, Leach, and Newell 2015).

The urgent need to respond to increasing deforestation using market solutions and territorial interventions by states has led to a rapid expansion of national protected areas (Frank, Hironaka, and Schofer 2000; Holmes and Cavanagh 2016), the rising power of renewed conservation regimes (Peluso 2011; Brockington 2002; Larson and Soto 2008), and greater pressure for the adoption of positive incentive programs to reduce emissions from deforestation, such as REDD+ (Mbatu 2016). At the same time, these processes of revitalized conservation are producing resistance from local populations (Merino and Gustafsson 2021; Paredes and Kaulard 2020; Krause 2020; Bocarejo and Ojeda 2016; Rocheleau 2015; Leach and Scoones 2015). In general, conservation programs do not always result in conflicts with local populations and when there are conflicts, they do not necessarily have the same intensity, demands, or forms of struggle. However, information from various grassroots organizations and a growing body of research suggests that there is increasing local resistance to conservation, which expresses the complexity and intertwining of ecological, political, and social struggles related to capitalism in a logic of scales (Fraser 2021).

This paper examines conservation resistance in the Amazon of Peru which is related to growing international support for the Peruvian state to fight deforestation. Peru forms part of multilateral climate change initiatives and is a signatory to several international and bilateral agreements to reduce deforestation and greenhouse gas emissions: it has been involved in Reducing emissions from deforestation and forest degradation Program, REDD+, since 2008; in 2014, it signed a landmark agreement with Norway and Germany to support deforestation efforts in the Amazon; and in 2018, it issued its first framework law on climate change, making Peru's Paris Agreement commitment legally binding. Today, conservation efforts provide new resources for implementing state-led conservation in the Peruvian Amazon. However, at the local level, these efforts may clash with Indigenous and 'Colono' communities sharing a history of the land and agrarian dispossession. Originally, the Peruvian state used the term Colono for Andean peasant settlers who migrated into the Amazon as part of its colonization programs. Over the years, peasant settlers and their indigenous peers appropriated the term to self-name or to differentiate their communities, respectively.

This paper builds on Polanyi (1944) to draw attention to the relationship between capitalism and nature, the complex problems behind resistance to conservation, and how the forest is disembedded from local societies to serve climate change mitigation. The ontological dimension underlying many conservation conflicts has been amply discussed in the context of Indigenous communities (De la Cadena 2015; Descola 1996; Viveiros de Castro 1999). However, this article aims to draw attention to the shared political economy behind the resistance of diverse rural communities employing different 'strategic logics' (Wright 2019) and in particular, Indigenous and Colono communities in the Amazon. Both communities articulate challenges to fundamental capitalist structures, including the role of the state, and these need to be discussed critically in a broad societal context (Fraser 2021). We engage with the discussion of Borras et al. (2022) by studying conservation in the context of historical and multi-scale climate justice and agrarian struggles. Employing the lens of political economy, we seek to contribute to the debate of 'who are the potential social forces behind such political struggles' (Borras et al. 2022, 16) and how cross-identity coalitions can arise.

There have been warnings in the conservation literature about multiple problems related to land dispossession. Those affected are mainly Indigenous communities who claim ancestral territorial rights (West and Brockington 2006), but also local non-Indigenous communities, who obtain their sustenance from the habitat (Colono peasants, for example) and are directly affected by the impact on their ecosystems (Guha and Martínez-Alier 1997). This is because most local rural Indigenous and non-Indigenous communities' social reproduction is closely interrelated with the reproduction of their ecosystems in hybrid multispecies assemblages, frictions, and connections (Tsing 2015). Thus, many of these communities' environmental conflicts are also struggles for their livelihoods (Fraser 2021). Unfortunately, current conservation interventions and programs— which are now embedded in the narrative of fighting climate change with a 'sense of urgency' that pervades current debates about sustainability and justice—reveal the same persistent problems of overlooking the history of land and agrarian dispossession and the creation of ecological inequalities (Scoones, Leach, and Newell 2015, 2).

This paper examines why local implementation of this new wave of conservation policies in the Amazon of Peru as well as the knowledge it is based on, is facing increasing local resistance in rural and agrarian settings. To explain conservation countermovements in the era of climate governance, we focus on how certain conservation 'authorized knowledge' (Li 2007) narrows the scope of understanding when it represents forest as 'nature,' while communities claim the forest as 'territory.' Territory embodies the forest with its inhabitants in a common history, culture, and a system of power relations that regulates life inside and outside the territory. In contrast, the narrow authorized knowledge ('nature') focuses on 'technical fixes' (Nightingale et al. 2020) and obscures the wider historical context of uneven institutional arrangements in the territory claimed by local communities.

We argue that this narrow 'authorized knowledge' overlooks the interconnections between the past of land, agrarian dispossession, and the present with its sense of urgency regarding climate change action. Also, it misses the connections between the macro characterized by state neglect of local agriculture and the micro-actions of sustainable management of the forests. This narrow approach to conservation reproduces inequalities related to land distribution and use, agrarian opportunities, and the state's authoritarian relationship with these rural and local communities in a context of extractive capitalist intensification in rural areas.

Methodologically, the paper focuses on the region of San Martín, where significant efforts have been made to advance forest conservation and implement sustainable management in the last two decades (Gustafsson and Scurrah 2019). The research examines two case studies of conservation where conflicts with local rural communities exist: the Área de Conservación Regional Cordillera Escalera (Cordillera Escalera Regional Conservation Area, CE-RCA), where the conflict involves mostly Indigenous communities, and the National Protection Area, Bosque de Protección Alto Mayo (Protection Forest Alto Mayo, PFAM), where conflicts developed with Colono communities. Information was collected through an extensive review of conservation project documents to construct a timeline and understand the complex history of the cases and the different approaches to conservation. In addition, we conducted in-depth interviews with state environmental authorities at the subnational and national levels, the organizations implementing conservation agreements (NGOs), and the local communities living in the protected areas or buffer zones. Both cases draw on 20 interviews conducted between July and September 2019 and later, in August 2021 and March 2022 in the context of COVID-19. We openly name the cases and the institutions working in both areas, but all information provided by specific individuals has been kept anonymous for ethical reasons.

The paper is organized into five sections. Section two explains the importance of authorized knowledge in narrowing the representations of the forest and its dynamics. Section three introduces the national and subnational context of our conservation cases. Sections four and five examine each of the cases. Finally, section six concludes with remarks on the characteristics of these conservation struggles and the challenges that their similarities pose for climate governance and justice.

2. Forest as 'nature' or territory?

In the twenty-first century, the acceleration of climate change and the repositioning of forests as global value reserves due to their contribution to carbon sequestration and climate mitigation (Katila et al. 2019) is transforming forest landscapes through various forms of 'enclosures' (Borras et al. 2022). Along with this conservation policy, positive incentives to reduce emissions from deforestation (REDD+) in developing countries are being implemented, despite the strong debate they have provoked (Schroeder and McDermott 2014). Countries like Peru are subject to greater international pressure and are receiving significant international funding to permit governments to assume more forest protection commitments and implement sustainable forest management programs.

The current conservation paradigm promotes the global market value of the 'standing' forest to reduce deforestation through larger areas of conservation rather than through the traditional use of the forest by its inhabitants. In this paper, we build on Polanyi's idea of how society embeds economy (1944) to explain the resistance of different rural and agrarian communities in the Amazon to forest commodification trends in conservation policy and the focus on eco-services as tradable goods on the global market (Leach and Scoones 2015; Igoe and Brockington 2007).

Polanyi argues against the narrow knowledge that characterizes the market as separated, autoregulated, and disembedded from society. In the *Great Transformation* Polanyi reveals how the expansion of capitalist markets operated: it expanded not only through the growth of capital but also through the fictitious transformation of labor, land, and money into commodities by social and political forces (none of these goods are produced to be sold in the market). For him, the alienation of these resources from society, particularly labor and land/nature, constitutes a serious threat to society. The fact that they cannot continue to meet their social functions and are reduced to mere articles of exchange or production causes irreparable damage to families, neighborhoods, and public and private life (Polanyi 1944). Fortunately, while social and political forces were able to create these fictitious commodities, these forces' countermovements also led to the creation of the institutions that regulate markets in the global North in the past.

Polanyi enables us to better understand local conservation countermovements in the Peruvian Amazon arising in response to the politics of climate change conservation that disembeds the forests from their local societies. Forests, land, and nature are not produced to be sold on the market. Instead, the forest provides local peoples with eco-services as essential forms of livelihoods that can be related to the markets or not, and in ways that have been historically produced, creating a set of diverse and profound interdependencies between forest ecosystems and their human inhabitants. While livelihoods are capabilities, assets, and activities required for people to earn secure means of living (IFRC 2022), when considering livelihood claims one must recognize cross-scale dynamics and the central place for considerations of knowledge, power, values, and political change (Scoones 2009, 191).

Resistance in the Amazon comes primarily from two sources. First, there are Indigenous communities whose organizations have fought for decades, both nationally and internationally, to protect the forests they inhabit since their cultures and forms of livelihood continue to depend on these ecosystems (De la Cadena 2015; Rasmussen 2018). However, their livelihood also includes small market-oriented agricultural activities which lead them to adapt their uses of the forestland. In addition, Colono communities initially migrated to the Amazon forest following the state's promise of free agricultural lands (and some migration still continues) (Barclay 1991). These local settlers do not have the same ontological relationship with the forest as do Indigenous communities,[1] nor do they have recognized international rights. However, they also have strong historical livelihood interdependencies with the forest and land management. Moreover, these interdependencies were produced by both market and state-led political and rural transformation projects in the past.

In the 60s, state developmentalist strategies considered the Amazon as a region to be tamed by agrarian development and colonization (Santos-Granero and Barclay 2000). Until the massive land reform of 1969, the state attempted to stop conflicts over land in the Andes without modifying the ownership of large land holdings (Brass 1989). Land conflict generated in Peru a significant level of peasant articulation and mobilization in the Andes (Blanco Galdos 1972). However, in the 1980s, both Indigenous and Colono communities were affected by the withdrawal of the state and the abandonment of local agriculture in a context of neoliberal globalization and intensified competitive pressures on Latin American agriculture (Kay 2008). In particular, the growing importance of mining in Peru led to further increases in food imports, which had a detrimental impact on peasant agriculture and food sovereignty (Crabtree 2002). The winner was agrobusiness. Meanwhile, small peasant farmers, in both Indigenous and Colono communities, who mainly participate in local markets, ended up facing unfair competition from subsidized food exports from rich countries (Kay 2008). Moreover, since the state lost interest in local agrarian problems in the 1980s and attempted, unsuccessfully, to connect local production with international agricultural markets, illicit crops, such as coca, became part of the livelihoods of Colonos in the Amazon (Grillo et al. 2021), and gradually also of Amazonian Indigenous communities. This puts at risk their non-commodified collective logic by the intrusion a logic of private property oriented toward participation in these economies (Fraser et al. 2018).

This paper shows how the implementations of conservation programs in these complex rural contexts are still guided by knowledge representations that do not take into account that all nature is socially, culturally and historically constructed (Soper 1995; Smith 1990). The result is insistence on 'technical fixes' that do not work (Myers et al. 2018; Nightingale et al. 2020). The compelling body of scholarship that warns about the social problems of disembedding forests from local society is most often not taken into consideration by actors from state agencies, non-governmental organizations (NGOs), and development projects. Moreover, these actors count on 'epistemic goods' (Montana 2021) to promote and impose practices that originate from other contexts or

[1]Nor do they share the same value equivalence system of nature that extractivist markets try to impose on their ecosystems (Li 2015).

accepted knowledge but often ignore local community ideas, knowledge, and perspectives.

As noted above, the representation of the forest as 'nature' in authorized knowledge contrasts with local peoples' claims of the forest as territory, which embodies both the ecosystem and its inhabitants in a common history of struggle for land and agrarian forms of livelihood, and power relations that regulate rural life within the territory. The claim of the forest as territory not only involves the 'tangible' dimensions of nature and territory for communities. It also includes the 'intangible' elements condensed in the territorial struggle.[2] Donna Haraway (2016, 12, following Strathern 1990) adroitly notes that what ideas we use to think about other ideas, what knowledge knows knowledge, and what stories tell stories matter for rearticulating communities in the context of the climate crisis. At the same time, it is not an easy task to understand and articulate both the tangible and intangible dimensions of territory and struggle together when discussing conservation policies. However, communities are opening new doors for this understanding by reframing forests through the rich and complex meaning of 'territory'. According to Elden, territory is not only a mere administrative demarcation imposed by political authorities or a container of natural resources that can be sold on the global market. Instead, territory is a historically and geographically specific form of organization and thought (Elden 2010, 757). Therefore, by recognizing the forest not as an isolated biological ecosystem but as territory, communities claim a view of the forest that recognizes the historical, cultural, tangible, and intangible interconnections that inhabit and transit the space.

In this context, this paper argues that narrow 'authorized knowledge' obscures both institutional arrangements that connect the present to the past and the macro and the micro and consequently reproduces inequalities and spurs territorial countermovements against conservationist climate policies, even among communities with diverse 'strategic logics' (Wright 2019). This is so because first, a narrow representation of forests as 'nature' ignores the historical alienation of local communities from access to land and sustainable livelihoods and reproduces this alienation. Scholars report that conservation administrators often generate 'green' discourses (West, Igoe, and Brockington 2006) to justify land dispossessions in the name of preservation (Fairhead, Leach, and Scoones 2012; Franco and Borras 2019) and to guide access, use, and management of forest resources.

Second, authorized knowledge overlooks the connections between the macro and the micro.[3] Often, it enables assumptions such as that local and poor people, including Indigenous communities, are the primary motors of deforestation. Such conclusions are reached without inquiring into the macro and historical causes of deforestation and why poor people might engage in such activities (Franco and Borras 2019). Thus, the production of narratives about the misuse of natural resources legitimizes local peoples' alienation from forest governance (Montefrio and Dressler 2016; Rocheleau 2015). In extreme cases, this can lead to increasing militarization of the forest and to local communities being stigmatized as invaders and/or criminals and even to violent evictions (Rasmussen 2018; Schmidt-Soltau and Brockington 2007; Ojeda 2012). Currently, the

[2]We thank our anonymous reviewer for reminding us of the necessity of reflecting on intangible dimensions and complexity of addressing these in processes of policy making.

[3]Chomba et al. (2015), Holmes and Cavanagh (2016), and Schmidt-Soltau and Brockington (2007) are some studies describing these same problems in other contexts.

consequences can be even more serious for non-Indigenous Peoples since they lack internationally recognized rights (Igoe and Brockington 2007).

In sum, authorized knowledge integrated into the global governance of climate change reflects the old dependency problem of Latin America (Ojeda 2014). The representation of forests as 'nature' and their integration into mitigation policies such as REDD+ occurs through multiscale, global, and hierarchical networks (Rocheleau and Roth 2007; Rocheleau 2011) that have the power to promote and impose particular practices. Moreover, it revives the old problem of the developmental and authoritarian state worldwide (Scott 2008). In the Peruvian Amazon authoritarian (Santos-Granero and Barclay 2000) and populist (Crabtree 1997; Coomes 1996) public programs of territorial transformation, land regimes, and ethnic classifications produced the socio-ecological conditions that made many areas more vulnerable to illegal economies (Paredes and Manrique 2021).

3. Mitigation climate policies in the Peruvian Amazon

Peru and other countries in the Amazon basin are increasingly receiving greater international backing to reduce emissions related to land-use change and rising deforestation. For example, during the 14th Conference of the Parties (COP 14) in Poznan, the newly created Ministry of the Environment committed to a very ambitious target: preventing the deforestation of 54 million hectares by 2021, an area representing around 80% of the total forest area (Egerlid 2015). Following this commitment, the country is receiving growing international funding (NICFI n.d; Andes Amazon Fund 2021).

International resources have served to increase the size of the forest sector. In 2010, Peru created the National Forest Conservation Program for Climate Change Mitigation (PNCBMCC) to design and implement forest policies to fulfill the country's commitments. This includes the development of REDD+, which Peru has been involved in since 2008 as a pilot country (Che Piu and García 2011). This backing has also served to revive conservation policy. The number of hectares under some protection (Protected Natural Areas [Áreas Naturales Protegidas, ANP]) has tripled since the 2000s. A now fortified National Service of Natural Protected Areas (SERNANP) manages these areas. In addition, the presence of forest rangers has more than tripled between 2011 and 2019, according to SERNANP.

In 2002, Peru embarked on a process of decentralization. Although this process has been criticized for being incomplete and insufficient, regional governments were given the authority to fulfill various functions in the environmental sector, ranging from preparing environmental plans to granting forest concessions or promoting land use planning. Among the most relevant decentralized functions was the power to grant property titles and rights of use, which was transferred by the Ministry of Agriculture (MINAGRI) to regional governments in 2008 (Art. 51, Organic Law of Regional Governments No. 27867). Additionally, the Ministry of the Environment (MINAM) transferred land-use planning powers to the regional governments, including developing and approving the Ecological and Economic Zoning (ZEE) process.

The territorial policy of the regional government of the San Martín was favorable to forest conservation objectives. Between 2007 and 2014, regional governor César Villanueva promoted a sustainable development agenda, pioneering institutional strengthening, including the creation of a Regional Environmental Authority (ARA). At the same time, the newly created Ministry of the Environment saw this subnational space as something that it could use to achieve its objectives (Augusto Melendez 2018). As a result, San Martín became the first region to receive forestry and environmental competencies to combat deforestation. In particular, its government received the power to authorize rights over forestland and different types of concessions (forestry, conservation, etc.); authorize changes in land use; authorize and approve forest management plans; and promote and control compliance with the national forestry policy. As a result of this transfer of powers, land use planning began through the ZEE process, which reported that 64.42% of the region's territory corresponded to ecological protection and conservation zones and highlighted the potential of San Martín as a provider of ecosystem services. However, it should be noted that these MINAM conservation policies could be at odds with MIDAGRI's competencies of promoting agricultural competitiveness by titling rural individual territories.[4]

4. Indigenous communities' resistance in the Cordillera escalera regional conservation area (CE-RCA)

The Cordillera Escalera Regional Conservation Area (CE-RCA) was created in 2005 and is administered by the regional government of San Martín. It consists of 149,870 ha in the provinces of Lamas and San Martín. It overlaps with the ancestral lands of the Shawi and Kichwa Indigenous Peoples, something that was neglected for years (Paredes and Kaulard 2020). The main complaint of the Kichwa and Shawi is that their claims regarding the recognition of their ancestral lands have been ignored. In addition, there was no prior consultation before creating the CE-RCA and during the elaboration of the area's master management plan (Valderrama 2017).

Amazonian Indigenous organizations, which emerged during the 1960s-70s, first sought to protect their territories from the advance of new extractive activities and the arrival of new settlers by way of state colonization programs (Paredes and Manrique 2021). Finally, in 1974, through Law 20653, the state recognized the collective legal existence of Indigenous Amazonian peoples (Greene 2009); however, it only recognized individual communities rather than Indigenous' Peoples' collective rights. The law used an Andean peasant template and failed to recognize Indigenous territories based on land use patterns, resource needs (obtained through hunting, fishing, and slash-and-burn horticulture), and linguistic and ancestral territory sharing. In addition, the law became more restrictive in 1978, restricting community rights over the forest (only to usufruct) and maintaining only property rights over agricultural land. No land titles were granted in areas suitable for forestry conservation (Forestry Law 1975, DL 21147, and Law 29763).

In more recent years, Indigenous communities are contesting these laws and claiming collective tenure over the forest declared under conservation and demanding the fulfillment of the Prior Consultation Law (PCL) before areas of protection are created. The PCL 29785 was passed in 2011 in compliance with ILO Convention 169, which was signed by Peru in 1994. However, the way this law has been regulated and implemented has been highly criticized for watering down the original aim of protecting the rights of Indigenous Peoples (Schilling-Vacaflor and Flemmer 2015).

[4]See Regional Directorial Resolution N° 212-2017-DRASAM-GRSM.

4.1. Kichwa communities' claims

Communities in the CE-RCA argue that the conservation area's specific objectives refer only to the provision of water for surrounding localities—regarding water as an ecosystem service (see Supreme Decree of creation: DS 045-2005-AG, cited in Valderrama [2017]). Importantly, this decree makes no mention of the Indigenous communities in the area and their use of the forest for their livelihoods: including small-scale agriculture with crops both for self-consumption and the market (cocoa, coffee, and banana), and traditional activities such as fishing, hunting, and gathering.

Furthermore, the communities that are members of CEPKA (Ethnic Council of the Kichwas of the Amazon) accuse the regional government of ignoring the fact that they were in the process of demanding legal land titles when creating the CE-RCA. Later, they were informed by authorities that they could no longer obtain collective property titles to their ancestral territory since titling is not allowed within protected areas in Peru in order to avoid overlapping rights (Interview 30.06.2019). The truth is that Indigenous communities' lands in the Peruvian Amazon region are poorly titled, with almost 50% of them without land titling (Ministerio de Cultura 2019). San Martín is the worst case in the country. While its regional government has developed the most advanced environmental and climate institutions (Augusto Melendez 2018), it presents the lowest rate of land titling of Indigenous communities: only 19% (Ministerio de Cultura 2019). CEPKA's legal demand against the Supreme Decree that created the CE-RCA (Supreme Decree No 45-2005-AG) was rejected in 2019 by the central government (RM N°443-2019-MINAGRI), and they cannot own their ancestral lands due to the titling prohibition in protected areas.

4.2 Authorized knowledge in the CE-RCA

In implementing the CE-RCA, the wider historical context and the uneven arrangements for transformation and distribution of land use for Indigenous communities were overlooked or justified by legal and technical knowledge. In the 1960s, through colonization programs and the expansion of modern agriculture and road infrastructure, the state sought to integrate the Amazon, and especially San Martín, into the national space (Barclay 1991) and to expand the capitalist agricultural frontier in the Amazon (Santos-Granero 1991). In this context, San Martín's Indigenous communities faced (and still face) discrimination from authorities who consider their traditional ways of subsistence and their ancestral cultures as primitive and believe they were wasting the Amazon's resources. Due to this generalized discrimination, many Kichwa communities only recently began to defend their rights and lands and to create an autonomous Kichwa federation (Biffi and Chaparro 2021; Paredes and Kaulard 2020). For example, CEPKA was founded in 2002 and represents the largest number of communities affected by the CE-RCA.

One of the specialists in charge of the design of the CE-RCA pointed out in an interview that only those communities with legal titles had to be taken into account (interview 23.9.2019). Therefore, the CE-RCA perimeter was designed *ad hoc*, with the titled communities remaining outside the protected area. The communities without titles were not considered 'a problem' for its creation (interview 20.9.2019). According to the authorities, these communities could still 'use' the forest and only had to follow the rules established by the park management committee. In contrast, community leaders claim dispossession of their territories. One of our interviewees said that: 'They (state officials) talked about regulations, decrees, many things. And then we looked at each other and never said anything because we did not understand; if we did not do that activity (small agriculture in the CE-RCA), what would we live on? The peasant man, the agricultural man, the native man lives from hunting, fishing, and agriculture' (interview 31.7.2019).

From the technical-legal knowledge approach that has characterized regional bureaucrats' way of dealing with the territorial rights of Indigenous Peoples, the right of 'use' of the forest solves livelihood and food security issues for Indigenous communities. Regional government experts do not understand or do not acknowledge the reasoning of 'recognition' and 'autonomy' (instead of 'use') that lies behind the claims of Indigenous organizations which insist on the titling of their ancestral territory and, as a result, they do not understand why just the right of 'using' the forest is not enough. Some high officials from the environmental branch of the regional government recognize that Indigenous communities had been historically dispossessed of their territories, and there was 'a historical debt to them' (interview 20.0.2021). However, they also maintain that the titling of Indigenous lands is a very complex technical matter, which does not correspond to the environmental but rather to the agricultural sector, 'and that it will be a very complicated process' (interview 20.0.2021). This interview reveals that authorized knowledge is also part of a politics of knowledge, where those in power examine certain things but not others, often based on their interests or goals.

In addition, the authorized knowledge that prevailed in the creation of the CE-RCA did not include an interest in encompassing the protection of the forest to produce global ecosystem services and providing recognition, justice, and securing livelihoods to Indigenous communities in the area. The CE-RCA was justified on the basis of its natural value and the water services it could provide for urban areas. In the justification study for the creation of the protected area (CTAR. No 141-2002-CTAR-SM-PE 2002), environmental strategists do refer to the existence of Indigenous practices in the forest. However, not enough attention is paid to this issue, nor is there concern for finding more information on this subject, to the detriment of the Indigenous communities in the area (Valderrama 2017). For example, the report recognizes that the Kichwa population makes long hunting trips into the protected area, far from the villages where their houses are located. However, when discussing the economic activities of the Kichwa population generally, there is no reference to exactly where they obtain bush meat and medicinal plants 'for subsistence' (Valderrama 2017). Moreover, public and NGO officials have a persistent view that conservation and food security (or Indigenous livelihoods) are incompatible (interview 14.1.2013). This prevents their addressing both issues simultaneously with an approach that accounts for complexity.

Through the creation of the conservation areas, Indigenous practices in the forest have been criminalized, and several Kichwa community members have been accused of entering this park without permission and opening pathways and agricultural fields (interview 20.7.2019). According to the CE-RCA regulations, Indigenous communities must ask for permission to use some limited forest areas for hunting, fishing, or gathering other non-timber resources. They also have to respect agreed-upon rules concerning cultivating temporal crops or opening trails and footpaths (Medina Revilla 2015).

Members of the Alto Pucalpillo community accused park rangers of preventing them from fishing and of burning their homes to evict them. Community members were in turn sued and had to face a two-year trial, which, though they were ultimately found not guilty, was an effective coercion mechanism that forced them to leave their homes (Chaparro 2018). In another case, the Mishkiyakullu community accused the rangers of preventing them from hunting since community members are required to renew entry permits to CE-RCA every three months (Bravo Mego 2019), which implies a bureaucratic procedure (Paredes and Kaulard 2020). Beyond these attacks, the communities deeply resent the recurrent and daily contempt expressed by state authorities. One of the interviewed leaders describes the contempt with which the officials treated them: 'They said: "this is an area belonging to the state, and I can to do what I want there. I am going to enter because you are nobody"; that is how they treated us, "you are nobody"' (interview 31.7.2019).

Ultimately, the regional government used a techno-legalistic tactic to bypass Indigenous rights to Free Prior Informed Consultation. According to CE-RCA specialists, there was no obligation to carry out the consultation stipulated by ILO Convention 169 and ratified by the Peruvian state in 1994 to create the regional protected area. According to them, in 2005, when the CE-RCA was created, a prior consultation law had yet to be approved in Peru (such a law was finally approved in 2011—Law No. 29785). However, CEPKA denounced the government's lack of will: neither the first CE-RCA Master Plan nor the following ones were consulted with communities, despite their repeated requests for consultation even after the 2011 law was approved. Instead, the regional government organized 'participative' workshops for the 'users' of the area to substitute for the consultation stipulated by ILO 169. These were largely informative and sensitization workshops (interview 23.9.2019). Other specialists said that the main aim of these workshops was to ensure that the number of people living inside the protected area was kept to a minimum (interview 23.9.2019). In this case, participation was structured as a procedural step whose purpose was to legitimize authorized knowledge without granting due recognition to the demands of Indigenous Peoples thus leading to a complex escalation in their struggles. While on the international level, it is the initial recognition and the creation of standards of rights, such as prior consultation, which are at stake on the local level, the object of struggle is the implementation and true exercise of these rights.

5. Colono resistance in the protected forest of Alto Mayo

The Alto Mayo Protected Forest (PFAM; acronym in English) is located in the north of San Martín and was created in 1987, long before the Peruvian state agreed to global deforestation commitments to contribute to climate change mitigation. The PFAM pertained to the Ministry of Agriculture at the time, and there was not much conservation enforcement for many years, with Colono communities migrating into the area. With the creation of the Ministry of the Environment in 2008 and the new commitments by the Peruvian state to global conservation, enforcement in the PFAM was revitalized. The national protected areas authority (SERNANP) became responsible for the PFAM and selected the area for the REDD+ pilot project in 2008. However, it was only in 2011 that the state began an intervention in the areas with the implementation of its alliance with the NGO International Conservation to manage the PFAM through the Alto Mayo Conservation Initiative

(ICAM). The ICAM project aims to implement a conservation intervention in the PFAM and is to last until 2028. The project includes funding for the management of the PFAM and preparation of the area for the REDD+ policy. The main intervention instrument is the signature of 'Conservation Agreements'[5] with families inside PFAM and in the buffer area; ICAM also contributes to the enforcement of conservation in the protected area. For example, in 2008, there were only ten park rangers but by 2016, the number had increased to 103, 90% of them paid by the ICAM project (Conservation International Peru 2016).

When these agreements were implemented in 2011, conflicts and resistance developed in the PFAM about who has a right to be in the area, who does not, and since when. According to the authorities, Colonos who entered the area after the early 2000s, when the PFAM changed its legal status,[6] have no right to stay unless they sign conservation agreements with SERNANP (interview 06.07.2022). In this paper, we focus on the complicated relationship between the administration of the area and the Colono communities of *Aguas Verdes*, in the north of the park. These Colonos claim that they inhabited these territories for a long time and that former Ministry of Agriculture officials granted them possession of these lands (interview 9.3.2022).

5.1. Colonos resistance in the PFAM

Starting in 2011, Colonos from the area of Aguas Verdes have been contesting what they see as increasingly strict conservation measures in the PFAM. In 2011, the Peruvian Ombudsman reported on the conflict between the Colonos and SERNANP, particularly the Chief Officer of the Alto Mayo Protected Forest at the time, concerning their settlement and activities permitted within the park (Defensoría del Pueblo 2011). Organized in *rondas*,[7] Colonos complained they were prevented from entering the area and assaulted personnel from the Natural Protected Area. The government responded by increasing the area's militarization, arguing that 'land speculation persists, and traffickers intimidate the local population' (Conservation International Peru 2016, 30). In 2019, the conflict escalated, and the *rondas* asked that the police contingent in the area be removed. More recently, in 2022, the Colonos questioned the approval and updating of the PFAM's New Master Plan (2022), insisting on their right to possession and the implementation of an accurate territorial delimitation.

In response to these claims, the police general who was stationed in the area responded publicly that 'an operation plan will be designed to remove all the people who invaded the park at some point and continue to advance' (ELM 2022). However, he added that before they entered the area by force, a strategic communication process would be carried out to make the population aware of the various penalties for violating the park's laws. Finally, he ended by saying, 'the operation will be carried out peacefully, respecting the rights of everyone.' It is evident that in this case the

[5]Its signatories commit not to expand the agricultural frontier to avoid deforestation and to adopt sustainable practices. In addition, they receive technical assistance, supplies, and work materials from the managing NGO.

[6]SUNARP, Property Registry File No. 6838- R, Volume 40, Folio 1407; Supreme Decree No. 038-2001–AG regulates the Law of Protected Natural Areas (Law No. 26834).

[7]*Rondas* (community patrols) are a form of rural autonomous policing and community justice organization that were initially created in the northern Andes and then spread to other areas in the Andes during the period of political violence (1980–1992), but also to the colonized areas in the Amazon (Thorp and Paredes 2010)

conflict focuses on the legitimacy of Colonos remaining in the PFAM, without conditionalities.

5.2. Authorized knowledge in the PFAM

Narrow authorized knowledge in the PFAM obscured the history of land distribution promoted by the state in San Martín and the context of the agrarian problem, which is crucial to understanding local resistance. Our interviews with PFAM officials gave the impression that the NGO's conservation project was initiated in a forest with vast biodiversity but without a history. According to the PFAM ICAM management office, SERNANP previously thought that Colonos could be expelled, even using military force (interview 18.03.2022), but gradually realized that this could result in a 'massacre.' As result, the park management now allows Colonos to stay if they sign conservation agreements (interview 6.7.2022). However, Colonos claim that from the creation of PFAM in 1987–2011, many settlers moved into the PFAM over the years and did so with the collaboration of local officials from the Ministry of Agriculture until PFAM's borders were finally enforced (interview 9.3.2022). They do not accept conditionalities.

The extensive diagnostic research that served as an input for the Master Plan[8] (INRENA 2008) which was carried out by a broad range of professionals and through many participatory workshops, did not analyze the historical problems related to migration despite the identification of the relationship of migratory flows to certain crop booms: for example, the cultivation of corn in the 70s and 80s, the coca boom in the 80s and 90s, and lucrative coffee cultivation in the 90s and 2000s. Historically, San Martin has been built upon the state-supported promise of a 'land without men for men without land' (Espinosa 2009). The forest was presented to Colonos as a space to be tamed (Santos-Granero and Barclay 2000). The government brought Andean settlers from the highlands to San Martin intentionally and used ambitious colonization plans, credits, and technical and credit programs to expand commercial agriculture (first rice and corn and later livestock) and the construction of considerable infrastructure (Paredes and Manrique 2021; Barclay 1991).

However, instead of organizing and controlling settlement, the state unintentionally encouraged a culture of uncontrolled colonization (Heckmann 2006). The arrival of thousands of people to San Martin, including the Alto Mayo area, was not accompanied by territorial organization and land (use) planning by the state, which would have improved livelihoods for Colonos. In addition, state programs to modernize monoculture agriculture in San Martin failed. Later, erratic agrarian policy and the abandonment of the agrarian sector that commenced at the end of the 1970s created even more vulnerabilities among the Colonos. According to Gootenberg (2018), the failure of colonization programs and global drivers led to coca's rise in the mid-1970s. It turned out that coca production was much more lucrative than most of the low-profit products that had been introduced by the state. San Martín was home to the first large-scale coca hub in Peru —in the Alto Huallaga—but the whole region was militarized in the 1980s due to the conflict that emerged between the state and subversive groups such as the Shining

[8]The Master Plan is the highest level-planning tool for the Alto Mayo Protected Forest. It establishes the general strategies and objectives for the management of the area.

Path (SL) and the Revolutionary Movement of Tupac Amaru (MRTA), which also looked to profit from the growing illegal economy. At the end of the 1990s and the beginning of the 2000s, the region became well known because of the so-called 'San Martín Miracle,' i.e. for overcoming violence and illegal crops with the strong backing of US international cooperation and the promotion of alternative export crops such as coffee and cacao (Manrique López 2016; Kaulard 2021).

In 2008, the study for the Master Plan noted that increasing land trafficking was linked to the boom of these new export crops (INRENA 2008, 54), but it does not analyze this problem and its potential threat to the conservation program. Traffickers sell land in the Amazon to Andean peasants who obtain capital from selling their lands in the highlands to mining or other extractive companies. When they arrive in the Amazon expecting to be able to cultivate crops such as coffee and cocoa, or to raise livestock, they find that these land transactions have no legal basis or that the land is located in protected areas (interview 6.7.2022). Moreover, the study did not identify as a problem the persistent promotion of these crops by other state agencies, including alternative development programs targeting coca and other illegal crops. Finally, the study does not connect this land trafficking to the weak and messy land tenure policy in this region and the country as a whole. Not only has titling been extremely slow (although individual titling is faster than indigenous collective titling), but the state also did not enforce conservation for many years, and the PFAM was a protected area only on paper. Moreover, Colonos claim that the district municipality allowed them to create *caseríos* (hamlets) in the protected forest to respond to education and health demands and to win votes, particularly in local elections.

Adequate knowledge of conservation requires the acknowledgment of this history of land promises, migration, incentives for deforestation, and the growth of illegal crops that became part of the survival patterns of these Colono communities. Representing the forest as pure nature gives the false impression that conservation can be done locally in isolation from the wider and historical agrarian problems facing the region and the country. To the authorized knowledge of the PFAM administrator (Conservation International 2022), Colonos looking for new lands to cultivate coffee, cacao, or coca are the main threat to the PFAM. However, without acknowledging that until the state prioritizes solve family and regional agriculture, conservation will always be under threat, not by the Colonos *per se* but by the lack of rural employment, poverty, and food scarcity.

None of the studies produced for the Master Plan in 2008 nor the Ecological Economic Zoning (EEZ) in 2009 considered the possibility that in a context of agrarian crisis, people would return to the cultivation of coca and other illegal crops. Moreover, recent reports by some NGOs about opium and coca cultivation in the PFAM are minimized by governmental agencies. They argue that most families who live in the PFAM have conservation agreements with the implementing NGO (interview with the Chief of PFAM). Nevertheless, conservation vs. growing illegal crops to earn a living has brought back a trend of militarization of the area (ELM 2022).

This conflict has increased Colonos distrust of the state and of the new 'green' park rangers who have recently become more visible. One of the Colonos we interviewed told us that people increasingly see the park rangers as representing the police and as the 'bad guys,' the 'muscle' of the central state whose purpose is to evict Colonos (interview 9.3.2022). This resident said that the authorities who prepared the Master Plan and oversee its implementation declared the upper zone of the PFAM a red zone without knowing the area. 'The police entered? together with the SERNANP; they created disturbances, knocking down houses [in the hamlets in the PFAM] of poor, humble people.'

While there is a solid international framework for Indigenous Peoples and safeguards regarding the mechanisms for consultation and participation, these laws do not protect other rural communities who live in the forest. There are no Indigenous Peoples inside the FPAM, so there is no legal right to prior consultation. In this complex context, multi-stakeholder tables have been proposed to replace prior consultations. The PFAM Master Plan was developed over several years with the help of international cooperation (INRENA 2008, 20) and was ready in 2008. Fifteen participatory workshops were held with local people (INRENA 2008, 21). However, we wonder how if this could have worked well in a context where local social organization is weak.

On the one hand, many people prefer not to attend participatory meetings and workshops because they feel that these are not representative or imply considerable additional work, according to one representative of the local environmental committee – CAR (interview 14.7.2019). On the other hand, the *rondas* are not seen by conservation authorities as significant interlocutors as they are not 'conservation implementers.' Instead, *ronda* leaders are easily stigmatized in a context of growing militarization and are seen by conservation authorities as a bad influence on their members. One official interviewed said, 'they generate conflicts, ask money from their members to go to Lima, … so that they will remove them [the park rangers] from the area and it will no longer be a protected area' (interview 10.2.2022). In these contexts, how can controversies be solved without imposition and conflict?

6. Discussion and conclusions

Through this paper, we want to emphasize the possibility of cross-ethnic convergence of climate justice and agrarian struggles in the Peruvian Amazon and answer a pivotal question raised in Borras et al. (2022, 17) regarding possible responses. Combining narratives and strategies could contribute to the empowerment of the local countermovements described here, which currently are rather fragmented, and thereby help them to better position themselves in international and local debates and forums.

We demonstrated that a certain type of narrow authorized knowledge approach to climate change conservation faces growing resistance in the Amazon because it reproduces injustices. We looked at two cases, one involving Indigenous communities and the other involving Colonos in the region of San Martín in Peru. Despite clear and undeniable differences between them, both cases can be considered Polanyian countermovements. Although Indigenous communities have different epistemic relationships with the forest than Colonos, when we employ a political economy perspective we also find common claims behind these struggles, particularly their common history of land dispossession and agrarian abandonment. Yet, in both cases, conservation programs focus on 'technical fixes' and overlook redistribution issues and problems related to the recognition of local inhabitants' land and agrarian struggles. We also learned that rather than a universal or abstract conception of nature, these communities experience the forest as a territory, a historically and geographically specific form of organization, culture, and thought. Within the territories, the forest ecosystem provides local peoples

(Indigenous and non-Indigenous) with eco-services as essential livelihood forms and aspirations. These livelihoods may be related to markets of different sizes, to some type of agricultural or farming activity or to another form of economy in ways that have been historically produced or modified. These have sometimes involved much suffering, environmental change and damage, and have created a set of diverse and profound interdependencies between the economy, politics, life, and ecosystems.

In addition, we show how the state reproduces its old strategies. Today the statés 'authorized knowledge' ignores previous institutional arrangements that have disempowered and continue disempowering people, preventing new and fair conservation engagements. For instance, first the state ignored the lack of land tenure or weak and messy land tenure when it created natural parks and REDD+ programs. Second, the neoliberal policies of the 90s led to the abandonment of macro agrarian policy for local farmers which increased the probability of illegal economies in the Andean Amazon, such as coca crops and illegal logging. Though these illicit economies lead to deforestation, a solution within a climate justice framework cannot criminalize communities and avoid addressing the foundational causes of the local agrarian problem. As Fraser argues, ignoring the underlying structural dynamics can lead to a general crisis that affects the local social system and 'threatens the well-being of all, not to mention the planet' (Fraser 2021, 215). The conclusion is that conservation programs do not operate in a vacuum, and meeting deforestation goals is almost impossible without engaging in political debates on the national level to address macro and structural problems driving deforestation crises. Third, the recent creation of processes labeled as inclusive and participatory, such as prior consultations following the ILO 169 Convention, while useful as goals to strive for, when applied on the ground have served to exclude communities from real decision-making and politics in general. Participation must be meaningful and not an administrative step. It must be participation that is active, intercultural, and autonomous, empowering and recognizing socio-political organizations in order to address conflict, tradeoffs, and controversies. But this type of participation demands a change in state logic, participants must go from being seen as program beneficiaries to being seen as citizen and criteria must change from purely technical to informed political.

Finally, when we examine the political economy of these resistance and conservation conflicts, we also learn about the possibilities of their converging on a common transformative agenda. Both cases display their own `strategic logics' (Wright 2019) in fighting against the disembedding of forests from their local societies, and they have different forms of relating to the forest. However, framing forests as territory is key to the potential convergence of these two types of communities as well as many others (Svampa 2019). For both, the forest is embedded in a territory for which they are willing to struggle and defend their livelihoods and identities against the dispossession driven by the state and extractive or green capitalism. Still, it is important to consider the differences of scale. Indigenous organizations who have much more experience in international negotiations, look to 'tame' capitalist and climate change global policies that affect them, asking for more institutional participation and building collective environmental and justice narratives. However, at the national and local levels, international policies continue to be overwhelmed by the power dynamics of the state and capital interests. In this context, Colonos as well as a growing number of Indigenous families (even going against their official organizations' narratives) express a more ambivalent position as they increasingly face a 'complex tradeoff' in the context of the expanding illegal capitalist economies in Peru and Latin America (Gutiérrez-Sanín 2021). Therefore, Wright's (2019) logics of taming, resisting, and embracing a 'forgiven capitalism' can converge only with great difficulty. However, building a common narrative is not impossible, according to our account. The two cases presented here illustrate the 'trans-environmentalist' challenge proposed by Fraser (2021), that is, overcoming the division imposed by the dependent situation of the Global South, including class, racial, and ethnic subordination.

Acknowledgements

We would like to thank Danitza Gil and Nicole Enrico for their excellent research assistance, Miguel Valderrama for the great support during the fieldwork, and Maria-Therese Gustafsson for her generous comments on an early version of the paper. This article is the results of the FONDECYT/PROCIENCIA Project 2020-PI0446 of CONCYTEC/Peru.

Disclosure statement

No potential conflict of interest was reported by the author(s).

Funding

This work was supported by Fondo Nacional de Desarrollo Científico y Tecnológico/PROCIENCIA: [Grant Number PERU- PI0446].

ORCID

Maritza Paredes http://orcid.org/0000-0002-1032-1356

References

Andes Amazon Fund. 2021. Andes Amazon Fund: Safeguarding the Most Diverse Landscapes on Earth." Impact Report, December 2021.

Augusto Melendez, Maria Claudia. 2018. "Cuando el Estado y la sociedad confluyen: la construcción de la política territorial en San Martín (2002–2018)." *Debates en Sociología* 47: 9–39. doi:10.18800/debatesensociologia.201802.001.

Barclay, Frederica. 1991. "Protagonismo del Estado en el proceso de incorporación de la selva." In *Amazonía, 1940–1990: el extravío de una ilusión*, edited by Frederica Barclay, Martha Rodríguez, Fernando Santos-Granero, and Marcel Valcárcel, 43–102. Lima: Terra Nuova / CISEPA PUCP.

Biffi, Valeria, and Anahí Chaparro. 2021. "Introducción. Visiones De San Martín Y Territorios Vividos En La Amazonía Andina." *Anthropologica* 39 (46): 5–9. doi:10.18800/anthropologica.202101.000.

Blanco Galdos, Hugo. 1972. *Tierra o muerte. Las luchas campesinas en Perú*. México: Siglo XXI.

Bocarejo, Diana, and Diana Ojeda. 2016. "Violence and Conservation: Beyond Unintended Consequences and Unfortunate Coincidences." *Geoforum; Journal of Physical, Human, and Regional Geosciences* 69: 176–183. doi:10.1016/j.geoforum.2015.11.001.

Borras, Saturnino M. Jr., Ian Scoones, Amita Baviskar, Marc Edelman, Nancy Lee Peluso, and Wendy Wolford. 2022. "Climate Change and Agrarian Struggles: An Invitation to Contribute to a JPS Forum." *The Journal of Peasant Studies* 49 (1): 1–28. doi:10.1080/03066150.2021.1956473.

Brass, Tom. 1989. "Trotskyism, Hugo Blanco and the Ideology of a Peruvian Peasant Movement." *The Journal of Peasant Studies* 16 (2): 173–197. doi:10.1080/03066158908438389.

Bravo Mego, Sandra Lizeth. 2019. "La afectación directa positiva de la creación de Cordillera Escalera vs el derecho de consulta previa un conflicto sin resolver." Bachelor 's thesis, Pontificia Universidad Católica del Perú. http://hdl.handle.net/20.500.12404/13657.

Brockington, Dan. 2002. *Fortress Conservation: The Preservation of the Mkomazi Game Reserve*. Bloomington: Indiana University Press.

Chaparro, Anahí. 2018. Caminos de crianza Kichwa Lamista: relaciones entre personas y lugares de Alto Pucalpillo." Master diss., Universidad Federal Fluminense.

Che Piu, Hugo, and Tania García. 2011. *Estudio REDD Perú: La situación REDD en el Perú*. Lima: Derecho Ambiente y Recursos Naturales (DAR).

Chomba, Susan, Juliet Kariuki, Jens Friis Lund, and Fergus Sinclair. 2015. "Roots of Inequity: How the Implementation of REDD+ Reinforces Past Injustices." *Land Use Policy* 50: 202–213. doi:10.1016/j.landusepol.2015.09.021.

Consejo Transitorio de Administración Regional (CTAR). 2002. Estudio Justificatorio para el establecimiento del Área de Conservación Regional Cordillera Escalera (ACR – CE) N° 141-2002-CTAR-SM/PE presented to INRENA.

Conservation International Peru. 2016. *Alto Mayo Conservation Initiative Monitoring & Implementation N°3 (2014–2016)*. Monitoring and implementation Report, VCS Version 3, CBB Standards Second Edition. Lima, Peru.

Conservation International Peru. 2022. "Innovando la gestión del Bosque de Protección Alto Mayo." https://www.conservation.org/peru/iniciativas-actuales/innovando-la-gestion-del-bosque-de-proteccion-alto-mayo.

Coomes, Oliver T. 1996. "State Credit Programs and the Peasantry Under Populist Regimes: Lessons from the APRA Experience in the Peruvian Amazon." *World Development* 24 (8): 1333–1346. doi:10.1016/0305-750X(96)00039-3.

Crabtree, John. 1997. "Populismo y neopopulismo: la experiencia peruana." *Apuntes: Revista de Ciencias Sociales* 40: 97–109. doi:10.21678/apuntes.40.457.

Crabtree, John. 2002. "The Impact of Neo-Liberal Economics on Peruvian Peasant Agriculture in the 1990s." *Journal of Peasant Studies* 29 (3–4): 131–161. doi:10.1080/03066150412331311049.

Defensoría del Pueblo. 2011. *Reporte de Conflictos Sociales No. 90*. August 2011. Lima.

De la Cadena, M. 2015. *Earth Beings: Ecologies of Practice Across Andean Worlds*. Durham/London: Duke University Press.

Descola, Philippe. 1996. *In the Society of Nature: A Native Ecology in Amazonia*. Cambridge: Cambridge University Press.

Egerlid, Josefin. 2015. "Governing Indigenous Territories in the Peruvian Amazon: Placing People or Forest First?" Swedish University of Agricultural Sciences. http://urn.kb.se/resolve?urn=urn:nbn:se:slu:epsilon-s-4975.

Elden, Stuart. 2010. "Thinking Territory Historically." *Geopolitics* 15 (4): 757–761. doi:10.1080/14650041003717517.

ELM (En La Mira). 2022. "SERNANP y Policía Nacional Coordinan Plan de Operaciones para la Lucha Frontal contra los Deforestadores del BPAM" (Press Conference Video). Tarapoto, April 8, 2022. https://m.facebook.com/story.php?story_fbid=2063326573827897&id=100004517228627&sfnsn=wa.

Espinosa, Oscar. 2009. "¿Salvajes Opuestos al Progreso?: Aproximaciones Históricas y Antropológicas a las Movilizaciones Indígenas en la Amazonía Peruana." *Anthropologica* 27 (27): 123–168. https://revistas.pucp.edu.pe/index.php/anthropologica/article/view/1602.

Fairhead, James, Melissa Leach, and Ian Scoones. 2012. "Green Grabbing: A new Appropriation of Nature?" *Journal of Peasant Studies* 39 (2): 237–261. doi:10.1080/03066150.2012.671770.

Franco, Jennifer, and Saturnino M. Borras Jr. 2019. "Grey Areas in Green Grabbing: Subtle and Indirect Interconnections Between Climate Change Politics and Land Grabs and Their Implications for Research." *Land Use Policy* 84: 192–199. doi:10.1016/j.landusepol.2019.03.013.

Frank, David Jhon, Ann Hironaka, and Evan Schofer. 2000. "The Nation-State and the Natural Environment Over the Twentieth Century." *American Sociological Review* 65: 96–116. http://www.jstor.org/stable/2657291.

Fraser, Nancy. 2021. "Climates of Capital: For a Trans-Environmental eco-Socialism." *New Left Review* 127: 94–127.

Fraser, Nancy, James Angus, Thiago Cardoso, Angela Steward, and Luke Parry. 2018. "Amazonian Peasant Livelihood Differentiation as Mutuality-Market Dialectics." *The Journal of Peasant Studies* 45 (7): 1382–1409. doi:10.1080/03066150.2017.1296833.

Gootenberg, Paul. 2018. "Introduction: Orphans of Development: The Unanticipated Rise of Illicit Coca in the Amazon Andes, 1950–1990." In *The Origins of Cocaine: Colonization and Failed Development in the Amazon Andes*, edited by Paul Gootenberg, and Liliana Dávalos, 1–18. London: Routledge.

Greene, Shane. 2009. *Customizing Indigeneity*. Redwood City: Stanford University Press. doi:10.1515/9780804771283.

Grillo, Luciana, Allison Kendra, Álvaro Pastor, and Hernán Manrique López. 2021. "Addressing Socio-Environmental Challenges and Unintended Consequences of Peruvian Drug Policy: An Analysis in Two Former Cocalero Valleys." *Journal of Illicit Economies and Development* 3 (1): 97–117.

Guha, R., and J. Martínez-Alier. 1997. *Varieties of Environmentalism: Essays North and South*. London, New York: Earthscan.

Gustafsson, Maria-Therese, and Martin Scurrah. 2019. "Strengthening Subnational Institutions for Sustainable Development in Resource-Rich States: Decentralized Land-use Planning in Peru." *World Development* 119: 133–144. doi:10.1016/j.worlddev.2019.03.002.

Gutiérrez-Sanín, Francisco. 2021. "Tough Tradeoffs: Coca Crops and Agrarian Alternatives in Colombia." *International Journal of Drug Policy* 89, 103156. doi:10.1016/j.drugpo.2021.103156.

Haraway, Donna J. 2016. *Staying with the Trouble: Making Kin in the Chthulucene*. Durham: Duke University Press.

Heckmann, Matthias. 2006. The upper montane forests of the Yuracyacu: Soils and vegetation (Bosque de Protección Alto Mayo, Northern Perú)." Diploma diss., University de Bayreuth.

Holmes, George, and Connor J. Cavanagh. 2016. "A Review of the Social Impacts of Neoliberal Conservation: Formations, Inequalities, Contestations." *Geoforum; Journal of Physical, Human, and Regional Geosciences* 75: 199–209. doi:10.1016/j.geoforum.2016.07.014.

IFRC (International Federation of Red Cross and Red Crescent Societies). 2022. Food security and livelihoods. https://www.ifrc.org/food-security-and-livelihoods.

Igoe, Jim, and Dan Brockington. 2007. "Neoliberal Conservation: A Brief Introduction." *Conservation & Society* 5 (4): 432–449. https://www.jstor.org/stable/26392898.

INRENA (Instituto de Recursos Naturales). 2008. Plan Maestro del Bosque de Protección Alto Mayo 2008–2013. Lima.

Katila, Pia, Carol J. Pierce Colfer, Wil de Jong, Glenn Galloway, Pablo Pacheco, and Georg Winkel, eds. 2019. *Sustainable Development Goals: Their Impacts on Forests and People*. Cambridge University Press. doi:10.1017/9781108765015.

Kaulard, Anke. 2021. "La Construcción de Redes e Infraestructuras Sinérgicas del Cacao en San Martín (2007 - 2018)." Phd diss., Pontificia Universidad Católica del Perú. http://hdl.handle.net/20.500.12404/21232.

Kay, Cristobal. 2008. "Reflections on Latin American Rural Studies in the Neoliberal Globalization Period: A new Rurality?" *Development and Change* 39 (6): 915–943. doi:10.1111/j.1467-7660.2008.00518.x.

Krause, Torsten. 2020. "Reducing Deforestation in Colombia While Building Peace and Pursuing Business as Usual Extractivism?" *Journal of Political Ecology* 27 (1): 401–418. doi:10.2458/v27i1.23186.

Larson, Anne M., and Fernanda Soto. 2008. "Decentralization of Natural Resource Governance Regimes." *Annual Review of Environment and Resources* 33: 213–239. doi:10.1146/annurev.environ.33.020607.095522.

Leach, Melissa, and Ian Scoones, eds. 2015. *Pathways to Sustainability. Carbon Conflicts and Forest Landscapes in Africa*. London: Routledge.

Li, Tania Murray. 2007. "Practices of Assemblage and Community Forest Management." *Economy and Society* 36 (2): 263–293. doi:10.1080/03085140701254308.

Li, Fabiana. 2015. *Unearthing Conflict: Corporate Mining, Activism, and Expertise in Peru*. Durham and London: Duke University Press.

Manrique López, Hernán. 2016. El fin de la guerra de la cocaína: construcción del Estado y desarrollo alternativo en la región San Martín (1978–2015)." Bachelor's thesis, Pontificia Universidad Católica del Perú.

Mbatu, Richard S. 2016. "REDD + Research: Reviewing the Literature, Limitations and Ways Forward." *Forest Policy and Economics* 73: 140–152. doi:10.1016/j.forpol.2016.09.010.

Medina Revilla, Alicia. 2015. "La comunicación para el desarrollo y la gestión participativa-intercultural de áreas naturales protegidas: estudio del caso de Cordillera Escalera en San Martín." Bachelor's thesis, Pontificia Universidad Católica del Perú. http://hdl.handle.net/20.500.12404/6597.

Merino, Roger, and Maria-Therese Gustafsson. 2021. "Localizing the Indigenous Environmental Steward Norm: The Making of Conservation and Territorial Rights in Peru." *Environmental Science & Policy* 124: 627–634. doi:10.1016/j.envsci.2021.07.005.

Ministerio de Cultura del Perú. 2019. "Base de Datos de Pueblos Indígenas u Originarios." Accessed 29 May 2022. http://bdpi.cultura.gob.pe/pueblos-indigenas.

Montana, Jasper. 2021. "From Inclusion to Epistemic Belonging in International Environmental Expertise: Learning from the Institutionalisation of Scenarios and Models in IPBES." *Environmental Sociology* 7 (4): 305–315. doi:10.1080/23251042.2021.1958532.

Montefrio, Marvin Joseph, and Wolfram H. Dressler. 2016. "The Green Economy and Constructions of the "Idle" and "Unproductive" Uplands in the Philippines." *World Development* 79: 114–126. doi:10.1016/j.worlddev.2015.11.009.

Myers, Rodd, Anne M. Larson, Ashwin Ravikumar, Laura F. Kowler, Anastasia Yang, and Tim Trench. 2018. "Messiness of Forest Governance: How Technical Approaches Suppress Politics in REDD+ and Conservation Projects." *Global Environmental Change* 50: 314–324. doi:10.1016/j.gloenvcha.2018.02.015.

NICFI (Norway's International Climate and Forest Initiative). n.d. "Partner Countries: Peru." *Norwegian Ministry of Climate and Environment*. Accessed October 5, 2022. https://www.nicfi.no/partner-countries/peru/.

Nightingale, Andrea Joslyn, Siri Eriksen, Marcus Taylor, Timothy Forsyth, Mark Pelling, Andrew Newsham, Emily Boyd, et al. 2020. "Beyond Technical Fixes: Climate Solutions and the Great Derangement." *Climate and Development* 12 (4): 343–352. doi:10.1080/17565529.2019.1624495.

Ojeda, Diana. 2012. "Green Pretexts: Ecotourism, Neoliberal Conservation, and Land Grabbing in Tayrona National Natural Park, Colombia." *Journal of Peasant Studies* 39 (2): 357–375. doi:10.1080/03066150.2012.658777.

Ojeda, Diana. 2014. "Descarbonización y despojo: desigualdades socioambientales y las geografías del cambio climático." In *Desigualdades socioambientales en América Latina*, edited by Barbara Göbel, Manuel Góngora-Mera, and Astrid Ulloa, 255–290. Bogota: Universidad Nacional de Colombia.

Oldekop, Johan A., Laura Vang Rasmussen, Arun Agrawal, Anthony J. Bebbington, Patrick Meyfroidt, David N. Bengston, Allen Blackman, et al. 2020. "Forest-linked Livelihoods in a Globalized World." *Nature Plants* 6: 1400–1407. doi:10.1038/s41477-020-00814-9.

Paredes, Maritza, and Anke Kaulard. 2020. "Fighting the Climate Crisis in Persistently Unequal Land Regimes: Natural Protected Areas in the Peruvian Amazon." *Journal of Cleaner Production* 265: 121605. doi:10.1016/j.jclepro.2020.121605.

Paredes, Maritza, and Hernán Manrique. 2021. "The State's Developmentalist Illusion and the Origins of Illegal Coca Cultivation in Peru's Alto Huallaga Valley (1960–80)." *Journal of Latin American Studies* 53 (2): 245–267. doi:10.1017/S0022216X21000225.

Peluso, Nancy L. 2011. "Emergent Forest and Private Land Regimes in Java." *Journal of Peasant Studies* 38 (4): 811–836.doi:10.1080/03066150.2011.608285.

Polanyi, Karl. 1944. *The Great Transformation*. New York: Farrar & Rinehart.

Rasmussen, Mattias Borg. 2018. "Paper Works: Contested Resource Histories in Peru's Huascarán National Park." *World Development* 101: 429–440. doi:10.1016/j.worlddev.2017.05.005.

Ritchie, Hannah, and Max Roser. 2021. "Forests and Deforestation." *Our World in Data*. Accessed April 5, 2022. https://ourworldindata.org/deforestation.

Rocheleau, Dianne. 2011. Networks, Territories and Distributed Power in Land Grabs in Latin America." Paper presented at the International Conference on Global Land Grabbing, Sussex, April, 6–8.

Rocheleau, Dianne. 2015. "Networked, Rooted and Territorial: Green Grabbing and Resistance in Chiapas." *The Journal of Peasant Studies* 42 (3e4): 695–723. doi:10.1080/03066150.2014.993622.

Rocheleau, Dianne, and Robin Roth. 2007. "Rooted Networks, Relational Webs and Powers of Connection: Rethinking Human and Political Ecologies." *Geoforum; Journal of Physical, Human, and Regional Geosciences* 38 (38): 433–437. doi:10.1016/j.geoforum.2006.10.003.

Santos-Granero, Fernando. 1991. "Frentes económicos, espacios regionales, y fronteras capitalistas en la amazonía." In *Amazonía, 1940–1990: el extravío de una ilusión*, edited by Frederica Barclay, Martha Rodríguez, Fernando Santos-Granero, and Marcel Valcárcel, 227–287. Lima: Terra Nuova / CISEPA PUCP.

Santos-Granero, Fernando, and Frederica Barclay. 2000. *Tamed Frontiers: Economy, Society, and Civil Rights in Upper Amazonia*. Boulder: Westview Press.

Schilling-Vacaflor, Almut, and Ricarda Flemmer. 2015. "Conflict Transformation Through Prior Consultation? Lessons from Peru." *Journal of Latin American Studies* 47 (4): 811–839. doi:10.1017/S0022216X15000826.

Schmidt-Soltau, Kai, and Dan Brockington. 2007. "Protected Areas and Resettlement: What Scope for Voluntary Relocation?" *World Development* 35 (12): 2182–2202. doi:10.1016/j.worlddev.2007.02.008.

Schroeder, Heike, and Constance McDermott. 2014. "Beyond Carbon: Enabling Justice and Equity in REDD+ Across Levels of Governance." *Ecology and Society* 19 (1): 31. doi:10.5751/ES-06537-190131.

Scoones, Ian. 2009. "Livelihoods Perspectives and Rural Development." *The Journal of Peasant Studies* 36 (1): 171–196. doi:10.1080/03066150902820503.

Scoones, Ian, Melissa Leach, and Peter Newell, eds. 2015. *The Politics of Green Transformations*. London: Taylor & Francis.

Scott, James C. 2008. *Seeing Like a State*. New Heaven: Yale University Press.

Smith, Neil. 1990. *Uneven Development: Nature, Capital, and the Production of Space*. Oxford: Basil Blackwell.

Soper, Kate. 1995. *What Is Nature? Culture, Politics, and the Non-Human*. Oxford: Blackwell.

Strathern, Marilyn. 1990. *The Gender of the Gift: Problems with Women and Problems with Society in Melanesia*. Berkeley: University of California Press.

Svampa, M. 2019. *Neo-extractivism in Latin America: Socio-Environmental Conflicts, the Territorial Turn, and new Political Narratives*. London: Cambridge University Press.

Thorp, Rosemary, and Maritza Paredes. 2010. *Ethnicity and the Persistence of Inequality: The Case of Peru*. Basingstoke: Palgrave Macmillan.

Tsing, Anna Lowenhaupt. 2015. *The Mushroom at the End of the World*. Princeton: Princeton University Press.

Valderrama, Miguel. 2017. "ACR Cordillera Escalera: miopías en los preliminares de una gestión cuestionada." *SERVINDI*, December 14, 2017. https://www.servindi.org/actualidad-noticias/14/12/2017/acr-cordillera-escalera-miopias-en-los-preliminares-de-una-gestion.

Viveiros de Castro, Eduardo. 1999. The transformation of objects into subjects in Amerindian ontologies." In the symposium entitled "Re-animating Religion: A Debate on the New Animism." Paper presented at the 98th Annual Meeting of the American Anthropological Association, Chicago, November.

West, Paige, and Dan Brockington. 2006. "An Anthropological Perspective on Some Unexpected Consequences of Protected Areas." *Conservation Biology* 20 (3): 609–616. https://www.jstor.org/stable/3879216.

West, Paige, Jim Igoe, and Dan Brockington. 2006. "Parks and Peoples: The Social Impact of Protected Areas." *Annual Review of Anthropology* 35: 251–277. doi:10.1146/annurev.anthro.35.081705.123308.

Wright, Erik Olin. 2019. *How to be an Anti-Capitalist in the Twenty-First Century*. New York: Verso Books.

OPEN ACCESS

Whose security? Politics, risks and alternatives for climate security practices in agrarian-environmental perspectives

Corinne Lamain

ABSTRACT

Climate security, albeit highly contested, is moving beyond the discursive realm into policies and practices that implicate the control of land, water and forests. Through a systematic literature review this paper offers a typology of climate security practices. It observes a shift towards human security framing, offering potential for agrarian struggles. However, risks remain: the depoliticisation of scarcity, control-seeking over natural resources, a push for neoliberal approaches, a dominant focus on violent conflict, and knowledge politics. Alternative approaches are suggested, foregrounding place-specific alliances that address the politics of conflict and embrace plurality of knowledges, contributing to (agrarian) climate justice.

Introduction

Debates around climate change being a threat to security, peace and stability are increasingly common in academic, policy and activist circles. These debates, whose topic is referred to as 'climate security', are exemplary of climate politics and marked by stark disagreements about relations between climate change, conflict and security. The perspective on the sense and meaning of this narrative is highly divergent across professional and governance sectors, academic disciplines and geographic regions. Meanwhile, a wide array of initiatives is emerging in which the discourse has been translated into action, across a range of organisations. For example, the United Nations (UN) has seen considerable action; it established its Climate Security Mechanism in 2018 – a cross-UN-departmental unit 'to strengthen the capacity of the UN system to analyze and address the adverse impacts of climate change on peace and security', including a community of practice, training and toolbox (Albrecht 2021). This intensified effort by the UN does not stand alone; rather, it mirrors an overall increase of action around climate security. A broad spectrum of sectors and interests is represented in this agenda: the military, security agencies, governmental bodies, academia, research and think tanks, (international) non-governmental organisations (NGOs), and activist groups. In fact, the rise of action is such that the Planetary Security Initiative[1] released a publication listing all 'climate security practices' in 2021. This presents a wide range of practices along a spectrum of climate-focused interventions on the one end and more '(hard) security' interventions on the other, which in the majority of cases intersect with agrarian struggles. Cases included range from 'Greening the Desert', a mitigation effort that should have 'stability' as a co-benefit, to the energy transition of the French Defence Forces that seeks to limit resource extraction during deployment, and the Juba Peace Agreement of Sudan, in which natural resource management provisions are included (Von Lossow et al. 2021).

The broadness of scope of the practices is indicative of the shift in climate security discourse from a more limited military focus on the effect of climate conflict on state security towards wider concerns around how climate change will affect human and ecological well-being, and thus human security (Adger et al. 2014). This in turn reflects the contestation around the existence of climate-induced conflicts and the political security agendas tied to climate security discourses (Buxton and Hayes 2015; Gemenne et al. 2014). Such agendas have long been accused of being defined by neoliberal tendencies to control the environment, to increase the reach of the security sector, and by determinist and capitalist approaches to scarcities (Buxton 2021; Dalby 2019; Kallis 2019; Scoones et al. 2019). Climate security did not emerge anew; rather, it is rooted in earlier debates around environmental security, that centre on links between environmental degradation and (violent) conflicts. This debate is characterised by a myriad of divergent views and approaches (Floyd and Matthew 2013). It preceded the climate security discourse in pointing towards the complexity of identifying causal links, leading authors to warn that in making that link particular political security discourses may be at play (Dalby 2019; Floyd and Matthew 2013; Hartmann 2010; Peluso and Watts 2001).

Without a doubt, whilst the climate security debate continues to become more heated, the discourses have landed in action. And considering that climate security is conceptualised around a presumed scarcity of natural resources,[2] its actions affect people that are living off land, water and forests. Those can be seen in the light of potential repressive effects, as was found in work on 'green militarisation' or 'green wars'; the engagement of the military in nature conservation, including anti-poaching measures. This has been shown to lead to a greater willingness to use violence, with socially and environmentally detrimental implications (Büscher and Fletcher 2018; Duffy et al. 2019). This makes dissecting the framing and implementation of climate security practices vital to agrarian debates and begs the question of how and to what extent they are reflected in the rapidly increasing body of academic literature around climate security.

This article aims to offer a contribution through a systematic literature review of *the application of* the climate security discourse into *practices*, to identify the framing of these in order to learn from earlier experiences and offer recommendations for future practices. It responds to the first set of questions laid out for this Forum,[3] by dissecting

[1]The Planetary Security Initiative was founded by the Dutch Ministry of Foreign Affairs in 2015 and is run by a consortium of think tanks and research organisations, facilitating a 'community of practice' around climate security.

[2]Scarcity may also be conceptualised in relation to the limited ability of the atmosphere to absorb ever-increasing emissions of greenhouse gases. This fascinating approach is, however, beyond the scope of this paper.

[3]This set of questions is: 'How and in what specific, local and global ways, does climate change differ from past environmental exclusions or threats? What combinations of narratives and strategies frame climate change and the institutionalised responses to it in agrarian settings? What exclusions and inclusions result from this?' (Borras et al. 2021, 17).

how climate security as a 'new' narrative and strategy is framing instutionalised responses that affect agrarian settings. Climate security, as the results presented below suggest, could easily provide another 'corporate-driven, technological narrative' that pushes the assumption that capitalism, although having shaped the current climate crisis, could also offer the way out through innovation and economic growth that serve adaptation (Borras et al. 2021, 9). It matters greatly in what ways climate security is understood, for how its action is defined, and thus who stands to benefit and who does not.

The paper first sketches the climate security debate and its frames. Secondly, it motivates the method and elaborates on how 'climate security action' is understood. This is followed by the results of the review, offering a typology of practices along climate security frames. Six risks in the framing of climate security practices are presented: (1) environmental scarcities, (2) political strategies, (3) control-seeking through top-down governance approaches, (4) a push for neoliberal approaches to economic growth, (5) a dominant focus on violent conflict vis-à-vis conflicts as a clash of interests and (6) knowledge politics. The paper concludes with suggestions for alternative approaches that foreground affected communities through place-specific alliances that address the politics of conflict and embrace plurality of knowledges and degrowth. It therewith aims to contribute to debates on (agrarian) climate justice, as set out by Newell (2022) and Borras and Franco (2018), focusing particularly on the role of militarism and its risks to achieving climate justice.

The rise of the climate security debate

Climate security has solidly landed in academic and policy debates, albeit from rather divergent entry points. Its starting premise is that climate change will have negative effects on peace and stability, through an increase of extreme weather events that will spur scarcity of natural resources, leading to increased conflict, migration, and perhaps even wars. This reasoning strongly resonates with earlier literature on the link between conflict and the environment, that situates 'scarcity' as an instigator of conflict and war (see e.g. Homer-Dixon 1994). This causal chain has been, and is to this day, fiercely critiqued, such as by political ecology authors, who understand these dynamics as more complex, political and deeply embedded in ideology and competing interests around natural resources in which capital dominates (Kallis 2019; Peluso and Watts 2001; Scoones et al. 2019). This paper builds on this work that lays the foundation for understanding 'scarcity politics'. It has been argued that the politics of security and the politics of scarcity have fuelled a stream of interventions around food security, water security, energy security, etc., primarily centred around the interests of the elites. Approaches focused on security are part of a '[military] paradigm that seeks to maintain control rather than address the underlying problems' (Rogers 2009, as cited in Buxton and Hayes 2015, 8). More recently, the academic as well as policy debate around environmental security also shifted in another direction to what came to be known as 'environmental peacebuilding', seeking to shape contestations around natural resources into positive outcomes for collaboration and peace. Examples of these can be found in 'transboundary water management' (Earle et al. 2015; Floyd and Matthew 2013; Krampe 2017). Within climate security debates, the assumed link between climate change and conflict has been called deterministic and Malthusian (Buxton and Hayes 2015; Gemenne et al. 2014; Hartmann 2014) and is refuted in a range of studies that have pointed to the complex of social, political, institutional and ecological factors that, rather than climate change itself, influence whether (violent) conflict erupts or is exacerbated (Adger et al. 2014; Benjaminsen et al. 2012; Bergholt and Lujala 2012; Hartmann 2010). In this, it matters deeply in what manner conflict is defined, also for the way that policy and practice responses are designed. The notion of conflict is really not the same across the domains of peace and conflict studies and political ecology. In the first field, conflict is considered as an event, as an eruption of violence, which may evoke the tendency to suppress it, to avoid violence and disruption. In the political ecology literature, conflicts are instead seen as a process, innate to power structures in society (LeBillon and Duffy 2018). Despite the ongoing debate around the link between conflicts and climate change, the climate security discourse has landed firmly in influential domains, such as the UN Security Council, militaries of the US and other Western nations and major corporations, like Shell (Buxton and Hayes 2015; Buxton 2021). An example: from the Dutch military rose self-proclaimed 'Klimaatgeneraal' (climate general), Tom Middendorp,[4] also the Chair of the International Military Council on Climate and Security, consisting of representatives of the military and security fields. One explanation for this uptake of the concept is the increased urgency that the security element adds to climate change, prioritising it above other agendas, thereby potentially 'opening policy windows' that would otherwise remain shut (Warner and Boas 2017). Framing climate security as a crisis would allow for certain political strategies to be pursued, also paving the way for increased support for technocratic, top-down policies (Boin, 't Hart, and McConnell 2009), fitting in a wider phenomenon of 'disaster capitalism' in which neoliberal policies are pushed in situations of disaster (Klein 2007). It is, as such, an integral part of climate politics: the spheres and frameworks of climate change making, defining and responding, with particular emphasis on relations of power (Franco and Borras 2019). For more elaborations on the climate security debate, see my earlier work (Lamain 2022).

Climate security frames

This paper follows how climate security is framed within the translation from the discursive realm into policies and practices. Climate security has mostly been presented along four frames, as defined by McDonald (2013): state security, international security, human security and ecological security. As a concept, climate security emerged in the mid-2000s from military and security circles, whose primary reference point was the security of the nation state (Lewis and Lenton 2015; Scott 2012). In this state security frame, in which the state is the reference point, conflicts are understood as events – eruptions of violence – that may lead to broader social unrest, perhaps even civil or inter-state wars (Adger 2010). Increasing concerns about the effects of climate change also brought it to the international governance realm as a concern for international security. In this frame,

[4]In 2022, Tom Middendorp published a book, *Climate general: building resilience*, about his experiences in the Dutch so-called 3D (defence, development, diplomacy) missions in Afghanistan, Iraq and other 'fragile states' that taught him how climate change and security are related. He pleads that a steep increase of military effort is needed to address this, requiring reform of the North Atlantic Treaty Organization (NATO) and armies globally.

the international order is to be at risk primarily for conflict, and possibly even war, between countries due to reduced availability of natural resources and heightened migration streams. The UN Security Council, which placed climate change on its agenda following controversy, eventually agreed it to be a 'threat multiplier', a discursive approach that has found much ground in defence and international relations policy circles. It should take into consideration the lack of solid evidence and agreement on direct links between climate and conflict. A much broader understanding of climate security was established through the human security frame with the release of the report of the Intergovernmental Panel on Climate Change (IPCC) in 2014, which included a chapter on human security. 'Human security, in the context of climate change, [is] a condition that exists when the vital core of human lives is protected, and when people have the freedom and capacity to live with dignity' (Adger et al. 2014, 259). This obviously goes beyond military and geopolitical concerns around state building; rather, it intends to bring climate security to the realm of development, democratic institutions and human rights. The fourth frame, ecological security, is a rather transformative proposal that aims to shift the reference point to ecological systems and the position for humans within them. It then moves attention away from the individual and to the system level. This frame is fairly recent and has seen little analysis or translation into practice as yet, aside from its groundwork delivered by McDonald (2021). A sound scientific and empirical base for the assumptions underlying both the state and international security frames is lacking, and both are in fact much critiqued (Gemenne et al. 2014). State security concerns have been argued to hijack the climate change debate, situating it in the realm of the military, rather than as an issue of social and human development (Boas 2015; Hartmann 2010). The impact of climate change is then presented as something that may be controlled by increasing defence as part of security efforts (Selby and Hoffmann 2014). It has been argued that the 'threat multiplier' narrative that was proposed as a response to the contestation continues to limit the understanding of complex dynamics to a linear, deterministic process, with climate change inevitably ending up in conflict. This overlooks other causal chains, such as that the relation may conversely be that conflict leads to increased vulnerability to climate change, that climate change may actually serve to facilitate peaceful arrangements, or that the effects of climate mitigation and adaptation *interventions* may induce or reinforce conflicts (Abrahams and Carr 2017; Dabelko et al. 2013; Hunsberger et al. 2017). Human security is a broad domain with multiple applications: it is actually a field of thought rather than an approach, and knows many interpretations. It is a concept, an objective, an analytical frame, a policy philosophy and a policy planning approach. The policy philosophy distinguishes between 'freedom from want' – e.g. poverty, livelihoods, vulnerability to disaster – and 'freedom from fear' – e.g. crime, violence (Gomez and Gasper 2022). The use of the human security frame has also been critiqued for its tendency to assign risk to populations and individuals and for it serving techno-managerial interventions and promoting depoliticised adaptation (Selby and Hoffmann 2014). It is evident, also in this paper, that human security as framing has gained ground and that climate security efforts are mostly (also) situated within this frame. Due to the limited coverage of ecological security as yet, the first three frames form the framework for the inventory and analysis of practices that this paper presents.

Methodological approach and limitations

Much has been written on the climate security discourse, but given the relatively recent emergence of climate security as a field of action, no systematic reviews are available on its practices. Reviews of policies and governance mechanisms are available, albeit with a scope on specific sectors (e.g. intergovernmental organisations; Dellmuth et al. 2018; Krampe and Mobjörk 2018). This review thus seeks to contribute to the literature by giving an overview of policies and practices that explicitly apply climate security terminology.[5] This calls for a specific and fairly narrow search strategy, as much of the climate security debate, particularly in relation to 'human security', could also be categorised under the much broader frames of environmental peacebuilding (Ide 2020; Krampe 2017), which climate security is considered an element of, and of climate change adaptation. It cuts across debates and practices on environmental cooperation, natural resource management, adaptation, disaster-risk reduction (see e.g. McBean and Ajibade 2009), conservation, development and peacebuilding. Each of these frames encompasses a considerable body of knowledge, which is beyond the scope of this paper. Here I have been particularly interested in how the climate security *discourse* that has risen globally has *permeated arenas of action*. This review therefore follows the simple and straightforward approach of searching for practices and policies that are labelled 'climate security', or are presented in the literature as part of that field. It is therein difficult to identify whether practices are born as a result of the rise of the climate-security narrative and community, or whether they preexisted it and may have integrated concerns on climate-security. It does so by following the methodological approach of the systematic literature review, which is defined by outlining a range of steps that are followed in order for the study to be replicable. This process starts with drawing up a protocol, specifying its objective, the research question, search words, and inclusion and exclusion criteria; the protocol underlying this review is included in the Appendix. The search was conducted in Web of Science, and limited to that database, as the interest was primarily in the emergence of practices in academic debates. This poses a serious limitation, however, considering the rapidly emerging appearance of climate security in realms of action, which has not yet featured in the scientific literature. There is a vast – and quickly growing – body of 'grey literature' available on these practices, which could provide valuable insights in dominant framings. Also, this could shed light on the urgent question of whether and how climate security is shaping climate mitigation and adaptation investments (Dunlap and Fairhead 2014; Lamain 2022). For the sake of manageability, however, this review is concerned only with the representation of climate security action in academic debates. It follows in the example of Oya (2013) in aiming to explore and synthesise a research field that is emerging. The question guiding this review was 'How

[5]In the review, much more diverse definitions were encountered. Some of the those are not linked, it seems, to broader climate security debates. Two notable cases posit the natural environment – respectively marine ecosystems as defence against floods and algae to improve food and fuel security – as vital to protecting against the effects of climate change. Here, climate security refers to the resilience of ecosystems and their contribution to mitigation and adaptation (Meriwether and Forsyth 2018; Walsh et al. 2016). Such readings and related practices are not taken on board, as they stretch beyond the frames that are most prominent (McDonald 2013). They are mentioned here, however, for their demonstrational power of how climate security as a concept has spread widely from its origin.

have climate security discourses shaped policy and practice, and what implications has this translation had for agrarian struggles?'

The search in Web of Science resulted in 31 studies, and after first- (title and abstract) and second-stage (full text review) screening, combined with manual forward search, 16 studies remained (articles and book chapters; see Appendix 1 for a flow chart). Full text review was done using an analytical scheme, leading to organisation in two typologies: the first with regard to the orientation of practice design, the second along McDonald's security frames and their contribution to those. This review is by no means intended to be an exhaustive overview; rather, it is intended to provide insights in the characteristics and articulation of interventions that are explicitly labelled as climate security and why that matters for agrarian struggles.

When applying the frames of McDonald, a strong emphasis on 'human security' becomes apparent; it applies to almost all practices that were encountered[6] (see Table 1). In the majority of cases, human security is interpreted coupled with state security, in which cases the broadness of the effects of climate change, stretching beyond violence and stability, is acknowledged. International security as a frame is less represented, and when it is, primarily in terms of 'regional security', applying to specific geographies. This represents the shift of climate security discourses that moved from being more conflict/defence-oriented towards representing wider concerns around adaptation and resilience, following the lack of empirical soundness with regard to the climate-conflict link. Importantly, the climate security frames are not as separate or mutually exclusive as they may seem, especially in the translation from discourse to policy. As is common in this translation, dominance is exercised by actors in power, so that policies and practice abide the world views of policy actors (Mosse 2005). The self-identified human security practices lean strongly on the 'freedom from need' side of human security, linking them to state security concerns.

One of the papers that is part of this review observes such dynamics in climate security governance, where increased focus on resilience and adaptation amongst practitioners in the UK should help massage the criticised climate-conflict narrative towards greater acceptance with international aid and development communities; 'a developmentalisation of the climate security discourse' (Boas and Rothe 2016, 628). The discursive shift from state security to human security may be understood as well as a broadening move that should redirect attention towards a more social agenda. It is argued that the broadness of the 'human security' discourse is intentional, in order to serve as a boundary concept to cross sectors. This broadness allows for translation according to priorities in specific contexts and as such sees vastly different interpretations (Gasper 2010). Framing of practices in the realm of human security may thus reflect a plurality of meanings, ranging from protection from environmental hazards to risks to livelihoods due to resource conflicts, to interventions to increase productivity of dry land ('freedom from want'), to protection from violence and war ('freedom from need') (see Table 2).

[6]Except for two cases: practices of NATO and the US Department of Defense, aimed at deployment of the military in the context of climate change (see Table 1).

Table 1. Typology of interventions in the review.

Type of intervention	SS	HS	IS	Geographic region	Funded by	Disciplines of authors
Reviews, modelling, frameworks for decision-making						
Review of governance frameworks	■	■	■	Indo-Pacific	Australian Federal Government, Department of Defense	Humanities, arts and social sciences; business, government and law; Flinders government
Modeling exercises to map 'climate vulnerability'	■	■	■	Africa (2012–2014) and Asia (2019)	United States Department of Defense (US DoD)	Public affairs; political science; international security and law; global climate change
Decision-support framework for adopting 'green infrastructure'	■	■		Ontario, Canada	Natural Sciences and Engineering Research Council of Canada	Ministry of the Environment; environmental science and planetary health; physical and environmental sciences
The threat multiplier debate as a tool in policy strategies	■	■	■	US	N/A	Environmental studies; Center for Violence Prevention at USAID
Mathematical assessment of climate security of countries	■	■		183 countries	N/A	Production engineering and management
US Department of Defense Minerva Initiative	■			United States and Africa	US DoD	Conflict and security
Governance mechanisms: policy, statements, units						
China's statements in UNSC, domestic action by Chinese army and General Logistics Department	■	■		China	N/A	International relations
Policy action by four intergovernmental regional organisations: Association of Southeast Asian Nations , Economic Community of West African States , Intergovernmental Authority on Development , SAARC		■		South East Asia, South Asia, Western Africa and Eastern Africa	N/A	International peace research
Climate Resilience Risk Register; Foresight Project on Global Environmental Migration	■	■		Global	United Kingdom Ministry of Defence	?
NATO's Defence and Environment Experts Group	■		■	Global	NATO	Conflict and security
EU's climate diplomacy policy		■		European Union nation states	European Union	Conflict and security
Interventions: missions, projects						
Risk reduction management centres		■		Cuba and five other Caribbean countries	Cuban government	Languages, literature and cultural studies; geography
Humanitarian programmes on climate conflict		■		Karamoja, Uganda	DFID and USAID	Environmental studies, Center for Violence Prevention at USAID
REDD+	■	■		Global South	IPCC, NGOs	Political Science, Sustainability and climate

SS = state security, HS = human security, IS = international security.

Risks in the framing of practices, and alternative approaches

The increase of human security as the dominant frame could be understood as a move towards greater acknowledgement and integration of agrarian struggles in climate security practice, as it centres more around human and ecological well-being. However, across the systematic literature review six risks were identified that persist within this shift – especially in the way in which the frame of human security is understood and applied – that have major implications in agrarian settings. For each of these, reflection on the predominant tendency across the review is provided, as well as alternatives that were encountered or are available elsewhere.

1. *Environmental resources: absolute/relative versus political scarcity*

A first element that cuts across the climate security practices is the premise that climate change will lead to scarcity of resources, and thus to increased (violent) conflicts, including between agrarian groups, such as peasants, herders, and forest dwellers; or to people migrating. This assumption has stood since the Limits to Growth report by the Club of Rome and in academia around that time (se e.g. Homer-Dixon 1994). This kind of scarcity reasoning has been critiqued, and much more nuanced approaches to scarcity are now available. Those move beyond 'absolute' and 'relative' scarcity, which posit resources to be finite, leading to insecurity and disaster, often leading to technological fixes as

Table 2. Intentions of practices per climate security frame.

Frame	Intention of practices
International security	
	International/regional collaboration and diplomacy for reducing pressures on resources, loss of land and lives, migration across borders - Position-taking of states on military security in the United Nations Security Council - Agreements on resource sharing in interregional organisations - Providing overviews of risk to human lives and to military deployment by mapping geographic regions
State security	
	Increasing military security - Securing military bases and installations - Providing support and guidance to the military on its readiness for deployment in the case of climate change events and to address climate change effects - Reducing ecological footprint of army *Increasing political security* - Reducing contestation and conflict due to climate change impacts on natural resources - Reducing forced migration due to climate disasters - Reducing migration due to reduction of livelihood opportunities as a result of reduction of available natural resources - Stimulating collaboration across different segments of society
Human security	
	Freedom from need – violence, crime - Avoiding loss of life due to extreme weather events - Reducing competition for scarce resources among groups and individuals, caused by climate change *Freedom from want – poverty, food security, water security* - Preparing for climate change impacts, such as extreme weather events, that can harm housing, infrastructure, belongings and livelihoods - Improving the capacity to deal with extreme weather events - Contributing to resilience of people and their livelihoods - Evaluating a range of dimensions of climate security to prioritise policy interventions

solutions. Instead, scarcity is increasingly understood as political and ideological and deeply rooted in capitalist dynamics (Kallis 2019; Scoones et al. 2019). 'Absolute' and 'relative' scarcity thinking is evident in the state and international security frames. The response, then, is to suppress unrest with force, as is also observed within 'green militarisation' and 'green wars' (Büscher and Fletcher 2018; Duffy et al. 2019). The human security frame often leans on the 'threat multiplier' narrative, which equally builds on notions of 'absolute' and 'relative' scarcity, in less direct ways. Across the systematic literature review there is generally a recognition of the complexity in causal chains; nonetheless, there is a persistent idea that this linearity will play out, eventually.[7] Prominently lacking are questions around the distributional causes of scarcity, relations with divisions along intersectional lines and the choice-making that is inherent to these. Without posing these questions, any action that is aiming to address scarcity, or its consequences, will be subject to tension and thus face the chance of reinforcing matters of injustice:

> We must always be clear about the political consequences of scarcity-driven interventions on the structural relations within a society, set within the wider political economy that defines these. Scarcity is not universal, given, fixed or determining, but context-specific, socially constructed, politically contested, variable and dynamic; and therefore, always subject to contestation, negotiation and contentious politics, as different groups mobilise around resource issues. (Scoones et al. 2019, 239)

As an example of this, Abrahams (2020) demonstrates the limitations of the 'threat multiplier' discourse that guides interventions of Canada's International Development Research Centre (IDRC) and the UK's Department for International Development (DFID). 'A key challenge for those tasked with addressing this conflict […] is reconciling the simplicity of this discursive framing and the on-the-ground complexities of what is deemed to be resource-related conflict' (Abrahams 2020, 5). After critiques of environmental determinism, environmental peacebuilding was proposed to provide an approach that centres positive peace as an outcome of cooperation, rather than conflict around natural resources. Lessons are available on what Ide (2020) calls 'the dark side' of environmental peacebuilding, which emerges when interventions are marked by depoliticisation and stand to invoke displacement and discrimination. Entirely counterproductive to its objectives, interventions then may in fact exacerbate conflict and damage the environment. Through avoiding the reproduction of these dynamics, environmental peacebuilding offers an alternative to security thinking.

2. *Political strategies: international action versus sovereignty*

The review also highlighted how climate security shapes policy arenas, where the discourse is instutionalised in policy units or entities. The strategic use of climate security narratives by states and state departments demonstrates that they indeed serve geopolitical or domestic political objectives. This draws attention away from the need for socially and ecologically just climate action that would serve marginalised agrarian communities

[7]This is made explicit by Busby et al. (2018, 90): 'Though the effects of climate change and related disasters on conflict are contested […], its accelerative effects on conflict and instability are plausible even if the scope conditions have yet to be fully understood'. In a similar vein, ahead of a review of governance barriers, Tangney et al. (2021, 12) state: '[t]he security implications of these ongoing tensions in the context of climate change remain unclear; however, resource shortages and forced migration as a result of climate disasters are likely to reduce human security in the region and could thereby enhance the possibility of conflict between states with already tense relationships'.

rather than pushing them further into conflict through capitalist undertakings of states and their security conglomerates (Buxton 2021). As such, climate security reinforces neo-colonial relations in international forums and further state enterprises in which the exploitation of agrarian settings is implicated (Borras et al. 2021). Where and how the policy units and entities are positioned is rather diversified across the spectrum of climate security: stretching from more climate change-oriented towards more security-concerned agencies. Climate security policy and practice fulfil an explicit role in political strategising, with divergent objectives, for example to garner support for climate action or to protect security sovereignty. Abrahams (2019) demonstrates the use of the climate security narrative in US policy circles. In the runup to COP21 in Paris, the US government emphasised the concerns of the Pentagon and the US military around climate change being a matter of national security, in order to gain mass for action. Abrahams observes that the climate security community is predominantly populated with military and security actors (more than development actors and especially more than environmental actors), who express that climate security to them is primarily a 'bipartisan issue' and that they do not wish to be too closely associated with the 'left-leaning' communities that environmental organisations relate to (Abrahams 2019). Conversely, climate security narratives are rejected for their interference with sovereignty elsewhere: declaring climate change as a security issue would imply that there is a need for international military action. For example, Bo (2016) analyses how the Chinese government does not recognise climate security at the international (United Nations Security Council (UNSC)) level, to avoid international interference in domestic affairs and to protect its sovereignty on security matters. At the domestic level, however, China does undertake action by launching policy units or committees, albeit to a limited extent. Also, within the South Asian Association for Regional Cooperation (SAARC) , it is observed that the space to manoeuvre towards coordinated climate security action is limited, as streamlining of security efforts may interfere with matters of sovereignty. This 'hampers the capacity for planning', due to tensions between India and Pakistan (Krampe and Mobjörk 2018, 334). Offering alternatives to these geopolitical dynamics is beyond the scope of this paper, but clearly a push for more security efforts stands in the way of 'transform[ing] the relations of production that generate climate change in the first place, through reparation, redistribution and decolonisation' (Borras et al. 2021, 13).

3. *Governance arrangements: top-down coordinated versus bottom-up, place-specific*

Prominent across the literature systematically reviewed here is the tendency to inform or promote coordinated and integrated regional governance arrangements, involving multiple sectors beyond the military and defence (Floyd 2015; Krampe and Mobjork 2018; Tangney et al. 2021). Whilst there is much to say for increased action by states in response to the climate crisis, there may be something to worry about regarding states acting primarily from security interests; this may lead to the military dominating the climate change debate to increase its control (Boas 2015; Hartmann 2010). In their review of barriers in regional governance, for example, Tangney et al. (2021) set out how governance should serve human security; this is nonetheless approached from defence interests.[8]

[8] 'Assuming a whole-of-government approach to effect climate-related policymaking suggests that issues of security governance that may have traditionally been the preserve of sovereign defence forces, borderland security and foreign

This could imply that a slippery governance slope is opened, even when presented under the umbrella of human security. Floyd (2015) analyses how 'institutional and ideational fragmentation' undermine coordinated approaches, resulting from climate security being a debate rather than a concept. Nonetheless, she identifies a range of regional security initiatives[9] intended to address climate security. She also notes how other coordinated governance efforts should serve the 'readiness' of the military: 'Today NATO's primary concern with regard to environmental and climate security is the adverse impact of environmental factors on military readiness and its ability to provide national and international security (Floyd 2015, 130).[10] Security interests then remain focused on the level of control that the military apparatus can maintain, also in a context of climate change. Across the review, however, not all government coordination is observed to lean in on military-centred approaches. Climate security is seen becoming entangled in divergent policy objectives and approaches between sectors (e.g. defence versus environment) and their respective policy communities, its consequent implementation meeting challenges (Abrahams 2019). This could represent the kind of 'inner-class' tension emerging within capitalism's self-reproducing force when meeting environmental limits, as described by Arsel (2023). The state security frame, and related 'threat multiplier' debates, are predominantly observed with Western (intergovernmental) organisations. Their counterparts from South (East) Asia and East and Western Africa as well as UN departments adhere, rather, to the human security framing, particularly to argue for climate change to be treated as a developmental issue, related to peace and stability as well as livelihoods (Dellmuth et al. 2018; Krampe and Mobjörk 2018).[11] An example of such an alternative to top-down approaches was encountered in the review: the Cuban Risk Reduction Management Centres, a successful approach to dealing with hurricanes and other extreme weather events.[12] Its replication in the wider Caribbean region serves as a promising example and powerful counter-narrative. The centres are managed by the 'Cuban Civil Defence'; they build on indigenous environmental and climate knowledge and are place-localised, and each centre involves professionals from different disciplines who manage an information hub at the local level (Jerez Columbié and Morrissey 2020).

4. *Economic interests: capital accumulation versus redistribution*

Climate security debates draw attention away from causes of climate change that urgently need to be addressed, such as the power of corporations and their contribution to carbon emissions, which result from an economic model of perpetual growth, based on fossil fuels and extensive consumption patterns of global elites (Borras et al. 2021; Buxton

affairs overlap substantially with concurrent agendas pursued within domestically focused portfolios such as home affairs, disaster management, environment and energy policy' (Tangney et al. 2021, 3).

[9]Namely, the Organization for Security and Co-operation in Europe or the multiagent Environment and Security Initiative, of which a range of UN departments are partners.

[10]Also, at the national level in China, without much public discourse, a Military Climate Change Expert Committee was installed already in 2008, 'aiming to provide technical support for the military to address climate change and effective guidance for the military's warfare and disaster management training and anti-disaster preparedness' (Bo 2016, 105).

[11]The latter paper was excluded from the systematic literature review selection after a closer read, for not offering practices, but it gives insight relevant to this point.

[12]Between 2005 and 2014, Cuba established 92 Risk Reduction Management Centres across its territory. In the ensuing years, this model was translated and adapted to other contexts in pilot projects: the British Virgin islands, Dominican Republic, Guyana, Jamaica, Trinidad and Tobago (Jerez Columbié and Morrissey 2020).

and Hayes 2015; Dunlap and Fairhead 2014). The evident limitations, or the self-destructive force, of that system has implications in agrarian settings, among which climate change is one major symptom. Buxton (2021) points to how corporate interests are served by security approaches. There is a historical tradition of the military securing trade and investment, which is increasingly anchored in governance institutes across levels. Also, climate change framed as security concern reinforces 'a booming military and security apparatus and industry that has already gained unprecedented wealth and power in the wake of 9/11[13]' (Buxton 2021. 8). This military–industrial complex is integral to neoliberal thinking and stands to benefit from increased relevance, which climate change has been argued to provide it with (Buxton and Hayes 2015). Across the systematic literature review there is a notable presence of departments of defence commissioning practices,[14] which in turn are often directed at continued deployment of the military itself, for example by securing its energy supplies. Moreover, within the review, climate security practices regularly appeared to be grounded in the neoliberal assumption that economic growth will support adaptation. For example, in their 'climate security assessment of countries', Phillis et al. (2018) offer maps in which a level of climate security is measured based on a mathematical model that consists of a set of seven indicators, in which economic resilience is one of seven, and its 'adaptive capacity' indicator heavily leans on gross domestic product per capita. There is, however, also mention within the review of approaches diverging between economic sectors, e.g. defence versus environment, leading to challenges in implementation. Alternatives to this system are explored, for example in degrowth debates, which redeem the current global model of Western-type growth-driven societies and are exploring what is 'the good life', also beyond the material sense (see e.g. Akbulut et al. 2019; Gerber 2020). Degrowth offers a pathway towards climate mitigation by reduced production and consumption patterns, that in turn limits energy needs. Consequently, it also contributes to adaptation by reducing the exploitation of natural resources, which in fact drives most of the socio-ecological conflicts that climate security concerns itself with. More analytical and empirical work on degrowth alternatives to the military would be welcome.

5. *Conflicts – violent events versus structural processes*

Security agendas are discursively driven by the objective to reduce or 'manage' conflicts, which are understood as violent clashes, or even war. Climate security thinking then frames those affected by climate change, especially vulnerable populations also in agrarian settings, into elements of danger or threats that need to be controlled (Buxton 2021). Within the field of political ecology, conflict is understood as a much broader phenomenon and has been defined as 'a clash of interests, values and norms among individuals or groups that leads to antagonism and a struggle for power' (Scheidel et al. 2017, 587). Within this much broader and more nuanced understanding, conflicts are not considered to be 'events' in which violence suddenly erupts, but rather 'processes' with dynamics

[13]This argument was made by Buxton before the Russian invasion of Ukraine in February 2022, which has increased the defence spending of Western countries by staggering amounts. For example, the German government alone increased its defence budget by 100 billion euro (The Guardian 2022).

[14]The US Department of Defense funded the modelling exercises discussed by Busby et al. (2012, 2014, 2018), Busby, Smith, and Krishnan (2014) and Busby et al. (2013), the Australian Department of Defence funded the governance review of Tangney et al. (2021), and NATO's Defence and Environment Experts Group funded that of Floyd (2015).

around structural inequity and injustice (LeBillon and Duffy 2018). It has been argued that narratives around climate conflict strategically ignore this conceptual divergence and that they bypass, or are blind to, the 'everyday structural violence and conflict caused by state and economic forces that are pursuing economic growth' (Dunlap and Fairhead 2014, 947). In the systematic literature review this is encountered likewise, indicating a depoliticisation of conflicts in most of the articles, including in those that are presented as part of a human security framing. Across the review, conflict is hardly defined and covers a broad spectrum: between groups that are rivalling for scarce resources; clashes as a result of migration; intra- and interstate war. Anderson and Gough (2021), for example, apply the metaphor of 'war' as one of the biblical Four Horsemen in reference to the effects of climate change. In most cases, the practice at hand – be it efforts to increase military preparedness, developmental projects aiming to increase resilience or adaptation, or mitigation schemes – is considered a 'solution' to address conflict. How these practices will fall into deeper dynamics of conflict, and specifically how they may *shape and fuel* those dynamics themselves, is not prominently figured. There is, however, ample literature on how (for example) adaptation and mitigation interventions contribute to socio-ecological conflicts around land, water and forests in the Global South (Dabelko et al. 2013; Hunsberger et al. 2017). In fact, an article in the review offers REDD + as an approach for promoting peace and stability, contributing to collaboration, compensation and equity through the payment for ecosystem services element (Tänzler and Ries 2012). A vast body of literature a decade later shows the opposite has been the experience with REDD+: numerous conflicts and injustices have been highlighted (Corbera and Schroeder 2017; Hoang, Satyal, and Corbera 2019; Satyal et al. 2019). Also in relation to involvement of security forces there is the risk that it actually leads to insecurity, as is apparent in contexts with former military invasions such as Afghanistan and Iraq (Buxton 2021). If 'conflict' is not understood in a broader sense, but rather as an 'event' that can be 'controlled' by intervening, possibly even with military force, this is an entirely different starting point for governance than justice and a pursuit of democratic institutions.

6. *Knowledge politics: global, technical, scientific in collaboration with localised, cultured knowledges*

Lastly, a note regarding whose knowledge climate security practices are based on; whose knowledge is considered of value? First, the listing of disciplines of authors (Table 1) sees a dominance of the natural and technical sciences, such as environmental sciences and engineering, and also of international relations and peace and conflict studies. Gemenne et al. (2014) already pointed to the need for more involvement of the social sciences: international relations, economics, political sciences; there seems to be a remaining gap of involvement of demography, development studies, and anthropology. In addition there is a need for philosophy, sociology and implementation sciences, in order to better ground debates and action in ethics and understanding of social change. Preferably, these disciplines do not work separately, but rather collaborate in inter- and transdisciplinary set-ups, as the complexity of questions that are part of climate security debates requires integration of pluriform perspectives (Ledford 2015; Repko and Szostak 2021). The scholar-activism tradition has much to offer in how to engage with social movements and activist groups, that are key in highlighting the priorities for people living in forests, of and with land and water, and in transforming agrarian struggles (Borras 2016). Involving types of knowledge other than academic in the conception as well as the implementation of practices benefits their relevance, for actually being grounded in experiences and realities of those affected by climate change. Interventions could be based on 'the significant environmental knowledge women have built up over generations' (Ide 2020, 10), to increase the chance of achieving their intended goals. Another example is the inclusion of indigenous, localised knowledge and cultures on weather events in the set-up and conduct of the Disaster Risk Reduction Centres in Cuba:

> The success of Cuba's Risk Reduction Management Centres, and their extension across the Caribbean, showcases the import of long-standing, locally-attuned experience in developing resilience to the environmental threats of climate change. The locally-sensitive environmental knowledge of the Global South is not only relevant in developing climate resilience within the framework of South–South cooperation; it is also key to developing integrated, effective responses to climate change in the Global North – via proven strategies of solidarity and cooperation for adaptation. (Jerez Columbié and Morrissey 2020, 28)

This is yet another example showing that where scholars and activists meet, powerful alliances take shape that can engage with spaces in which policies and practices are framed, designed and implemented, so that they contribute to social and climate justice, rather than expand control of the globally dominant economic and military apparatus.

Conclusions and alternatives for moving forward

The troops have taken off, whilst the assumptions underlying the climate security mission have not found common ground. The debate on the climate-conflict link is ongoing, and this becomes evident where climate security practices hit the ground. Climate security is applied for political strategic purposes and its implementation is meeting challenges (Abrahams 2019). There is reason to ask why climate security, even if perceived under the umbrella of human security, would need to be a narrative at all. Security as a frame pushes out alternatives that serve the interests of marginalised populations across the world much better. It tends to lock debates in a 'threat or security' binary, whereas there is a plethora of values that are not attended to in this lock (Buxton 2021). 'Security' as a narrative fits in what Paprocki (2022) has so aptly termed 'anticipatory ruination': the reinforcement of destructive capitalist tendencies by announcing climate disasters, paving the way for forceful responses. These dynamics are thus embedded in larger structures of accumulation by dispossession, affecting natural resources as being expropriated by forces of capital (Harvey 2004).

This paper followed the approach introduced by Bernstein by analysing security narratives along the key questions of critical agrarian studies: 'Who owns what? Who does what? Who gets what? What do they do with it?' (Bernstein 2017, 8). It has looked at frames for security applied in climate security practices, particularly zooming in on who stands to benefit from those practices. It concludes that 'security narratives' subsume alternatives to dealing with climate change such as climate justice and community-based approaches, in which there is acknowledgement and appreciation of pluralism. Through the translation from discourse to practice, power is exercised (Mosse 2005) and the consequences for rural populations are subsumed in intervention logic. Similar observations are found across the review presented in this article; climate security practices are primarily conceptualised around human security, whereas their guiding assumptions remain rooted in the Malthusian arguments on which the climate security narrative was originally founded. Where human security offers opportunity for opening up the scope to more nuanced, place- and conflict-specific thinking and doing (Gasper 2010), the review found that depoliticised, top-down, neoliberal interventions continue to be proposed, now dressed up as 'human security'. In an attempt to address the critiques on the causal chains of climate change and conflict, the 'threat multiplier' framing has been widely adopted, across the UN, defence and development communities. Practitioners tasked with climate security interventions are found to struggle with the continued deterministic, linear thinking that underlies the threat multiplier narrative driving implementation, because the temporal and spatial linkages are unclear and indirect (Abrahams 2020). The discursive shift from climate security as tied to state security towards a wider understanding in terms of human security and related 'resilience' discourses has been found to lead to a greater acceptance of climate security interventions. This has, however, not shifted implementation pathways equally, through the weaving of new narratives into pre-existing state security concerns (Boas and Rothe 2016). This article has outlined six risks that are found in the human security formulation of practices: (1) environmental scarcity thinking, (2) political strategising, (3) control-seeking through top-down governance approaches, (4) a push for neoliberal approaches, (5) a dominant focus on violent conflict and (6) knowledge politics.

In this paper, alternatives are proposed that are not cast from the sky, but rather are promptly available and that will contribute to increased (agrarian) climate justice (Borras and Franco 2018; Newell 2022). The first is environmental justice, which re-politicises socio-environmental relations and conflicts, democratising the use and governance of agrarian spaces and that is conceived and executed in alliances between scholars and activists, integrating disciplinary perspectives (Temper et al. 2018). In fact, agrarian climate justice itself may serve as guiding frame, as also proposed by Calmon, Jacovetti, and Koné (2021). The second alternative proposed is climate action and disaster-risk-reduction efforts, still from a human security lens, but rather aimed at 'freedom from want', people-centred and based on geographic epistemologies. Jerez Columbié and Morrissey (2020) give as an example the Risk Reduction Management Centres in Cuba, which are seated in localised governance structures, involve Indigenous and local knowledges and integrate community members in interdisciplinary groups. It is a bottom-up, locally specific, South–South collaboration-based approach to increasing resilience, in a 'carefully contextualised manner' (ibid. 9). Thirdly, environmental peacebuilding as a strongly related, or even encompassing, field offers an alternative that is locally grounded and oriented towards collaboration and peace rather than suppression. It includes a solid body of knowledge, and lessons on practices are readily available. In addition to its many positive contributions, Ide (2020) lists six types of adverse effects, calling it 'the dark side of environmental peacebuilding': depoliticisation, displacement, discrimination, deterioration into conflict, delegitimization of the state, and degradation of the environment. This offers the chance to improve this field and release the 'security' framing. Lastly, degrowth may serve as a guiding framework: reduction of natural resource extraction through a reorientation of values away from production and consumption would contribute to climate mitigation as well as adaptation (Hickel 2021), and as such address concerns driving the climate security agenda.

Acknowledgements

Gratitude is due to the reviewers that have carefully commented on draft versions, which has improved the quality of the paper significantly. I wish to thank my supervisors Jun Borras, Ashok Swain and Murat Arsel for their patient guidance of my snail-pace progress. Also, the Villagers, in particular Yukari Sekine and Dani Calmon; bless you for giving me the courage to believe I can do this. Dario, rock and anchor, thank you for remaining steady at all times. And lastly, my heart and spirits owe Gabrielle Hubler an amount of gratitude that is beyond words; you continue to be my compass and guide.

Disclosure statement

No potential conflict of interest was reported by the author.

ORCID

Corinne Lamain http://orcid.org/0000-0002-7548-1086

References

Abrahams, D. 2019, July 4. "From Discourse to Policy: US Policy Communities' Perceptions of and Approaches to Climate Change and Security." *Conflict, Security & Development* 19 (4): 323–345. https://doi.org/10.1080/14678802.2019.1637080.

Abrahams, D. 2020. "Conflict in Abundance and Peacebuilding in Scarcity: Challenges and Opportunities in Addressing Climate Change and Conflict." *World Development* 132: 104998. https://doi.org/10.1016/j.worlddev.2020.104998.

Abrahams, D., and E. R. Carr. 2017. "Understanding the Connections Between Climate Change and Conflict: Contributions from Geography and Political Ecology." *Current Climate Change Reports* 3 (4): 233–242. https://doi.org/10.1007/s40641-017-0080-z.

Adger, W. N. 2010. "Climate Change, Human Well-Being and Insecurity." *New Political Economy* 15 (2): 275–292. https://doi.org/10.1080/13563460903290912.

Adger, W. N., J. M. Pulhin, J. Barnett, G. D. Dabelko, U. Oswald Spring, C. H. Vogel, et al. 2014. "Human Security." In *Climate Change 2014: Impacts, Adaptation, and Vulnerability. Part A: Global and Sectoral Aspects. Contribution of Working Group ii to the Fifth Assessment Report of the Intergovernmental Panel on Climate Change*, edited by C. B. Field, V. R. Barros, D. J. Dokken, K. J. Mach, M. D. Mastrandrea, T. E. Bilir, and M. Chatterjee, 755–791. Cambridge, UK and New York, NY, USA: Cambridge University Press.

Akbulut, B., F. Demaria, J.-F. Gerber, and J. Martínez-Alier. 2019. "Who Promotes Sustainability? Five Theses on the Relationships Between the Degrowth and the Environmental Justice Movements." *Ecological Economics* 165: 106418. https://doi.org/10.1016/j.ecolecon.2019.106418.

Albrecht, Patrizia. 2021, August 2023. "Joint Efforts for Sustaining Peace: Meet the UN Climate Security Mechanism." *UN Media Centre / Blog*.https://www.unssc.org/news-and-insights/blog/joint-efforts-sustaining-peace-meet-un-climate-security-mechanism.

Anderson, V., and W. A. Gough. 2021. "Harnessing the Four Horsemen of Climate Change: A Framework for Deep Resilience, Decarbonization, and Planetary Health in Ontario, Canada." *Sustainability* 13 (1): 379. https://doi.org/10.3390/su13010379.

Arsel, M. 2023. "Climate Change and Class Conflict in the Anthropocene: Sink or Swim Together?" *The Journal of Peasant Studies* 50 (1): 67–95. https://doi.org/10.1080/03066150.2022.2113390.

Benjaminsen, T. A., K. Alinon, H. Buhaug, and J. T. Buseth. 2012. "Does Climate Change Drive Land-Use Conflicts in the Sahel?" *Journal of Peace Research* 49 (1): 97–111. https://doi.org/10.1177/0022343311427343.

Bergholt, D., and Päivi Lujala. 2012. "Climate-related Natural Disasters, Economic Growth, and Armed Civil Conflict." *Journal of Peace Research* 49 (1): 147–162. https://doi.org/10.1177/0022343311426167.

Bernstein, H. 2017. "Political Economy of Agrarian Change: Some key Concepts and Questions." *RUDN Journal of Sociology* 17 (1): 7–18. https://doi.org/10.22363/2313-2272-2017-17-1-7-18.

Bo, Y. 2016. "Securitization and Chinese Climate Change Policy." *Chinese Political Science Review* 1 (1): 94–112. https://doi.org/10.1007/s41111-016-0003-5.

Boas, I. 2015. *Climate Migration and Security: Securitisation as a Strategy in Climate Change Politics*. Abingdon, Oxon: Routledge.

Boas, I., and D. Rothe. 2016. "From Conflict to Resilience? Explaining Recent Changes in Climate Security Discourse and Practice." *Environmental Politics* 25 (4): 613–632. https://doi.org/10.1080/09644016.2016.1160479.

Boin, Arjen, Paul 't Hart, and Allan McConnell. 2009. "Crisis Exploitation: Political and Policy Impacts of Framing Contests." *Journal of European Public Policy* 16 (1): 81–106. http://dx.doi.org/10.1080/13501760802453221.

Borras, S. M. 2016. *Land Politics, Agrarian Movements and Scholar-Activism*. The Hague: Inaugural Lecture, ISS.

Borras, S. M., and J. Franco. 2018. "The Challenge of Locating Land-Based Climate Change Mitigation and Adaptation Politics Within a Social Justice Perspective: Towards an Idea of Agrarian Climate Justice." *Third World Quarterly* 39 (7): 1308–1325. https://doi.org/10.1080/01436597.2018.1460592.

Borras Jr, S. M., I. Scoones, A. Baviskar, M. Edelman, N. Lee Peluso, and W. Wolford. 2021. "Climate Change and Agrarian Struggles: An Invitation to Contribute to a JPS Forum." *The Journal of Peasant Studies* 49 (1): 1–28. https://doi.org/10.1080/03066150.2021.1956473.

Busby, J. W., K. H. Cook, E. K. Vizy, T. G. Smith, and M. Bekalo. 2014. "Identifying Hot Spots of Security Vulnerability Associated with Climate Change in Africa." *Climatic Change* 124 (4) (June 2014): 717–731. https://doi.org/10.1007/s10584-014-1142-z.

Busby, J. W., J. Gulledge, T. G. Smith, and K. L. White. 2012. "Of Climate Change and Crystal Balls: The Future Consequences of Climate Change in Africa." *Air & Space Power Journal - Africa and Francophonie* 3 (3): 4–45.

Busby, J. W., T. G. Smith, and N. Krishnan. 2014. "Climate Security Vulnerability in Africa Mapping 3.0."." *Political Geography* 43: 51–67. https://doi.org/10.1016/j.polgeo.2014.10.005.

Busby, J., T. G. Smith, N. Krishnan, C. Wight, and S. Vallejo-Gutierrez. 2018. "In Harm's Way: Climate Security Vulnerability in Asia." *World Development* 112: 88–118. https://doi.org/10.1016/j.worlddev.2018.07.007.

Busby, Joshua W., Todd G. Smith, K. L. White, and S. M. Strange. 2013. "Climate Change and Insecurity: Mapping Vulnerability in Africa." *International Security* 37 (4): 132–172. https://doi.org/10.1162/ISEC_a_00116.

Buscher, Bram, and Robert Fletcher. 2018. "Under Pressure: Conceptualising Political Ecologies of Green Wars." *Conservation and Society* 16 (2): 105. http://dx.doi.org/10.4103/cs.cs_18_1.

Buxton, N. 2021. *Primer on Climate Security: The Dangers of Militarising the Climate Crisis*. Amsterdam: Transnational Institute.

Buxton, N., and B. Hayes. 2015. *The Secure and the Dispossessed: How the Military and Corporations are Shaping a Climate-Changed World*. Transnational Institute.

Calmon, D., C. Jacovetti, and M. Koné. 2021. "Agrarian Climate Justice as a Progressive Alternative to Climate Security: Mali at the Intersection of Natural Resource Conflicts." *Third World Quarterly* 42 (12): 2785–2803. https://doi.org/10.1080/01436597.2021.1965870.

Corbera, E., and H. Schroeder. 2017. "Editorial to the Special Issue "REDD+ Crossroads Post Paris: Politics, Lessons and Interplays"." *Forests* 8 (12): 508. https://doi.org/10.3390/f8120508.

Dabelko, G. D., L. Herzer, N. Schuyler, Meaghan Parker, and R. Sticklor. 2013. *Backdraft: The Conflict Potential of Climate Change*. (Environmental Change & Security Program report vol. 14, issue 2). Washington, DC: Woodrow Wilson international Center for Scholars.

Dalby, S. 2019. "Environmental (In)Security, in International Encyclopedia of Geography: People, the Earth." *Environment and Technology* 2019: 1–11. https://doi.org/10.1002/9781118786352.wbieg0428.pub2.

Dellmuth, L. M., M.-T. Gustafsson, N. Bremberg, and M. Mobjörk. 2018. "Intergovernmental Organizations and Climate Security: Advancing the Research Agenda." *WIRES Climate Change* 9 (1): e496. https://doi.org/10.1002/wcc.496.

Dellmuth, Lisa M, Maria-Therese Gustafsson, Niklas Bremberg, and Malin Mobjörk. 2018. "Intergovernmental Organizations and Climate Security: Advancing the Research Agenda." *WIREs Climate Change* 9 (1). http://dx.doi.org/10.1002/wcc.2018.9.issue-1.

Duffy, Rosaleen, Francis Massé, Emile Smidt, Esther Marijnen, Bram Büscher, Judith Verweijen, Maano Ramutsindela, Trishant Simlai, Laure Joanny, and Elizabeth Lunstrum. 2019. "Why We Must Question the Militarisation of Conservation." *Biological Conservation* 232: 66–73. http://dx.doi.org/10.1016/j.biocon.2019.01.013.

Dunlap, A., and J. Fairhead. 2014. "The Militarisation and Marketisation of Nature: An Alternative Lens to 'Climate-Conflict'." *Geopolitics* 19 (4): 937–961. https://doi.org/10.1080/14650045.2014.964864.

Earle, A., E. Cascao, S. Hansson, A. Jägerskog, A. Swain, and J. Öjendal. 2015. *Transboundary Water Management and the Climate Change Debate*. London and New York: Routledge. ISBN 9780415835152.

Floyd, R. 2015. "Global Climate Security Governance: A Case of Institutional and Ideational Fragmentation." *Conflict, Security & Development* 15 (2): 119–146. https://doi.org/10.1080/14678802.2015.1034452.

Floyd, R., and R. Matthew. 2013. *Environmental Security: Approaches and Issues*. 1st ed.. London and New York: Routledge. ISBN 9780415539005.

Franco, J. C., and S. M. Borras. 2019. "Grey Areas in Green Grabbing: Subtle and Indirect Interconnections Between Climate Change Politics and Land Grabs and Their Implications for Research." *Land Use Policy* 84 (May): 192–199. https://doi.org/10.1016/j.landusepol.2019.03.013.

Gasper, D. 2010. "The Idea of Human Security." *Climate Change, Ethics and Human Security*, 23–46. https://doi.org/10.1017/CBO9780511762475.004.

Gemenne, F., J. Barnett, W. N. Adger, and G. D. Dabelko. 2014. "Climate and Security: Evidence, Emerging Risks, and a new Agenda." *Climatic Change* 123 (1): 1–9. https://doi.org/10.1007/s10584-014-1074-7.

Gerber, J. F. 2020. "Degrowth and Critical Agrarian Studies." *Journal of Peasant Studies* 47 (2): 235–264. https://doi.org/10.1080/03066150.2019.1695601.

Gomez, O. A., and D. Gasper. 2022. *The Position of Crises in Human Development Processes and Thinking - Using the Human Security Perspective in an Era of Transitions*. Background paper for UNDP Special Report on human security. https://hdr.undp.org/content/position-crisis-human-development-processes-and-thinking.

The Guardian. 2022, February 27. Germany to Set Up €100bn Fund to Boost its Military Strength. Accessed August 22, 2022. https://www.theguardian.com/world/2022/feb/27/germany-set-up-fund-boost-military-strength-ukraine-putin.

Hartmann, B. 2010. "Rethinking Climate Refugees and Climate Conflict: Rhetoric, Reality and the Politics of Policy Discourse." *Journal of International Development* 22 (2): 233–246. https://doi.org/10.1002/jid.1676.

Hartmann, B. 2014. "Converging on Disaster: Climate Security and the Malthusian Anticipatory Regime for Africa." *Geopolitics* 19 (4): 757–783. https://doi.org/10.1080/14650045.2013.847433.

Harvey, D. 2004. "The 'New' Imperialism: Accumulation by Dispossession." *Socialist Register* 40: 63–87. https://doi.org/10.1093/oso/9780199264315.003.0007.

Hickel, J. 2021. *Less is More: How Degrowth Will Save the World*. London: Penguin Random House.

Hoang, C., P. Satyal, and E. Corbera. 2019. ""This is my Garden": Justice Claims and Struggles Over Forests in Vietnam's REDD+." *Climate Policy* 19 (sup1): S23–S35. https://doi.org/10.1080/14693062.2018.1527202.

Homer-Dixon, T. 1994. "Environmental Scarcities and Violent Conflict: Evidence from Cases." *International Security* 19 (1): 5–40. https://doi.org/10.2307/2539147.

Hunsberger, C., E. Corbera, S. M. Borras, J. C. Franco, K. Woods, C. Work, R. de la Rosa, et al. 2017. "Climate Change Mitigation, Land Grabbing and Conflict: Towards a Landscape-Based and Collaborative Action Research Agenda." *Canadian Journal of Development Studies / Revue Canadienne D'études du Développement* 38 (3): 305–324. https://doi.org/10.1080/02255189.2016.1250617.

Ide, T. 2020. "The Dark Side of Environmental Peacebuilding." *World Development* 127. Art. 104777. https://doi.org/10.1016/j.worlddev.2019.104777.

Ide, Tobias. 2020. "The Dark Side of Environmental Peacebuilding." *World Development* 127: 104777. http://dx.doi.org/10.1016/j.worlddev.2019.104777.

Jerez Columbié, Y., and J. Morrissey. 2020. "Subaltern Learnings: Climate Resilience and Human Security in the Caribbean." *Territory, Politics, Governance*, 1–20. https://doi.org/10.1080/21622671.2020.1837662.

Jerez Columbié, Yairen, and John Morrissey. 2020. "Subaltern Learnings: Climate Resilience and Human Security in the Caribbean." *Territory, Politics, Governance* 11 (1): 19–38. http://dx.doi.org/10.1080/21622671.2020.1837662.

Kallis, G. 2019. *Limits: Why Malthus Was Wrong and Why Environmentalists Should Care*, 168. Stanford: Stanford University Press. ISBN-10: 1503611558 ISBN-13: 978-1503611559.

Klein, N. 2007. *The Shock Doctrine: The Rise of Disaster Capitalism*. Toronto: Alfred A. Knopf Canada. ISBN-10, 9780141024530.

Krampe, F. 2017. "Towards Sustainable Peace: A new Research Agenda for Post-Conflict Natural Resource Management." *Global Environmental Politics* 17 (4): 1–8. https://doi.org/10.1162/GLEP_a_00431.

Krampe, F., and M. Mobjörk. 2018. "Responding to Climate-Related Security Risks: Reviewing Regional Organizations in Asia and Africa." *Current Climate Change Reports* 4 (4): 330–337. https://doi.org/10.1007/s40641-018-0118-x.

Krampe, Florian, and Malin Mobjörk. 2018. "Responding to Climate-Related Security Risks: Reviewing Regional Organizations in Asia and Africa." *Current Climate Change Reports* 4 (4): 330–337. http://dx.doi.org/10.1007/s40641-018-0118-x.

Lamain, C. 2022. "Conflicting Securities: Contributions to a Critical Research Agenda on Climate Security." *Globalizations*. https://doi.org/10.1080/14747731.2022.2057093.

LeBillon, P., and R. Duffy. 2018. "Conflict Ecologies: Connecting Political Ecology and Peace & Conflict Studies." *Journal of Political Ecology* 25 (1): 239–260. https://doi.org/10.2458.

Ledford, H. 2015. "How to Solve the World's Biggest Problems." *Nature* 525 (7569): 308–311. https://doi.org/10.1038/525308a.

Lewis, K., and T. Lenton. 2015. "Knowledge Problems in Climate Change and Security Research." *Wiley Interdisciplinary Reviews: Climate Change* 6 (4): 383–399. https://doi.org/10.1002/wcc.346.

McBean, G., and I. Ajibade. 2009. "Climate Change, Related Hazards and Human Settlements." *Current Opinion in Environmental Sustainability* 1 (2): 179–186. https://doi.org/10.1016/j.cosust.2009.10.006.

McDonald, Matt. 2013. "Discourses of Climate Security." *Political Geography* 33: 42–51. http://dx.doi.org/10.1016/j.polgeo.2013.01.002.

McDonald, M. 2021. *Ecological Security: Climate Change and the Construction of Security*. Cambridge: Cambridge University Press. ISBN: 9781009024495.

Mosse, D. 2005. *Cultivating Development. An Ethnography of aid Policy and Practice*. London: Pluto Press. https://doi.org/10.2307/j.ctt18fs4st.

Newell, P. 2022. "Climate Justice." *The Journal of Peasant Studies* 49 (5): 915–923. https://doi.org/10.1080/03066150.2022.2080062.

Oya, C. 2013. "The Land Rush and Classic Agrarian Questions of Capital and Labour: A Systematic Scoping Review of the Socioeconomic Impact of Land Grabs in Africa." *Third World Quarterly* 34 (9): 1532–1557. https://doi.org/10.1080/01436597.2013.843855.

Paprocki, Kasia. 2022. "Anticipatory Ruination." *The Journal of Peasant Studies* 49 (7): 1399–1408. http://dx.doi.org/10.1080/03066150.2022.2113068.

Peluso, N. L., and M. Watts. 2001. "Violent Environments." In *Violent Environments*, edited by N. L. Peluso, and M. Watts, 3–38. Ithaca/London: Cornell University Press.

Phillis, Y. A., N. Chairetis, E. Grigoroudis, F. D. Kanellos, and V. S. Kouikoglou. 2018. "Climate Security Assessment of Countries." *Climatic Change* 148 (1-2): 25–43. https://doi.org/10.1007/s10584-018-2196-0.

Repko, A. F., and R. Szostak. 2021. *Interdisciplinary Research: Process and Theory*. 4th ed. Thousand Oaks, VS: Sage Publications. ISBN: 9781544398600.

Rogers, Paul. 2009. "Climate Change and Security." *Clinical Medicine* 9 (2): 108–109. http://dx.doi.org/10.7861/clinmedicine.9-2-108.

Satyal, P., E. Corbera, N. Dawson, H. Dhungana, and M. Maskey. 2019. "Representation and Participation in Formulating Nepal's REDD+ Approach." *Climate Policy* 19 (sup1): S8–S22. https://doi.org/10.1080/14693062.2018.1473752.

Scheidel, A., L. Temper, F. Demaria, and J. Martínez-Alier. 2017. "Ecological Distribution Conflicts as Forces for Sustainability: An Overview and Conceptual Framework." *Sustainability Science* 13 (3): 585–598. https://doi.org/10.1007/s11625-017-0519-0.

Scoones, I., R. Smalley, R. Hall, and D. Tsikata. 2019. "Narratives of Scarcity: Framing the Global Land Rush." *Geoforum; Journal of Physical, Human, and Regional Geosciences* 101: 231–241. https://doi.org/10.1016/j.geoforum.2018.06.006.

Scott, S. V. 2012. "The Securitization of Climate Change in World Politics: How Close Have we Come and Would Full Securitization Enhance the Efficacy of Global Climate Change Policy?" *Review of European Community & International Environmental Law* 21 (3): 220–230. https://doi.org/10.1111/reel.12008.

Selby, J., and C. Hoffmann. 2014. "Rethinking Climate Change, Conflict and Security." *Geopolitics* 19 (4): 747–756. https://doi.org/10.1080/14650045.2014.964866.

Selby, Jan, and Clemens Hoffmann. 2014. "Rethinking Climate Change, Conflict and Security." *Geopolitics* 19 (4): 747–756. http://dx.doi.org/10.1080/14650045.2014.964866.

Tangney, P., C. Nettle, B. Clarke, J. Newman, and C. Star. 2021. "Climate Security in the Indo-Pacific: A Systematic Review of Governance Challenges for Enhancing Regional Climate Resilience." *Climatic Change* 167 (3-4). https://doi.org/10.1007/s10584-021-03197-8.

Tänzler, D., and F. Ries. 2012. "International Climate Change Policies: The Potential Relevance of REDD+ For Peace and Stability." In *Climate Change, Human Security and Violent Conflict: Challenges for Societal Stability*, edited by J. Scheffran, M. Brzoska, G. H. Brauch, P. M. Link, and J. Schilling, 695–705. Hexagon Series on Human and Environmental Security and Peace. Berlin, Heidelberg: Springer. https://doi.org/10.1007/978-3-642-28626-1_35.

Temper, L., F. Demaria, A. Scheidel, D. Del Bene, and J. Martinez-Alier. 2018. "The Global Environmental Justice Atlas (EJAtlas): Ecological Distribution Conflicts as Forces for Sustainability." *Sustainability Science* 13 (3): 573–584. https://doi.org/10.1007/s11625-018-0563-4.

Von Lossow, T., A. Schrijver, M. Van der Kroon, L. van der Schaik, and J. Meester. 2021. Towards a Better Understanding of Climate Security Practices, *Clingendael report*.

Walsh, Michael J., Léda Gerber Van Doren, Deborah L Sills, Ian Archibald, Colin M Beal, Xin Gen Lei, Mark E Huntley, Zackary Johnson, and Charles H Greene. 2016. "Algal Food and Fuel Coproduction Can Mitigate Greenhouse Gas Emissions While Improving Land and Water-Use Efficiency." *Environmental Research Letters* 11 (11): 114006. http://dx.doi.org/10.1088/1748-9326/11/11/114006.

Warner, Jeroen, and Ingrid Boas. 2017. "Securitisation of Climate Change: The Risk of Exaggeration." *Ambiente & Sociedade* 20 (3): 203–224. http://dx.doi.org/10.1590/1809-4422asocex0003v2022017.

Wilson, A. Meriwether W., and Catherine Forsyth. 2018. "Restoring Near-Shore Marine Ecosystems to Enhance Climate Security for Island Ocean States: Aligning International Processes and Local Practices." *Marine Policy* 93: 284–294. http://dx.doi.org/10.1016/j.marpol.2018.01.018.

Appendix 1. Systematic literature review protocol and flow chart

Systematic literature review protocol

10 September 2021

Aim

1. To identify whether and how climate security discourses translated into interventions in policy and practice;
2. To analyse in what manner policy and practice are framed and aim to make a contribution;
3. To identify the implications of the framing.

Research question

"How have climate security discourses shaped policy and practice and what implications has this translation had for agrarian struggles?'

Search keywords

Set	***Category***	***Keyword***
Set 1	*Climate change*	Climate change, global warming
Set 2	*Climate security discourse*	Climate security, environmental peacebuilding
Set 3	*Climate practice*	Climate action, adaptation, mitigation, conservation, climate finance*, energy transition

Terms within each individual set are connected via OR whereas each set is connected with the other sets through AND. All sets search in the title, abstract and keywords of papers. The general search query was:

> ("climate change" OR "global warming") AND ("climate security" OR "environmental peacebuilding") AND ("climate action" OR "adaptation" OR "mitigation" OR "climate finance*" OR "energy transition")

Source databases

- Web of Science Core Collection

Data-inclusion criteria

1) Time frame: from 2006 (when climate security as a term was coined) to the present.
2) Language: English – because the PhD project focuses on climate security discourses in relation to international climate finance.
3) Focus of literature: climate security practice is a key focus area, e.g. it is further developed as a concept or applied as policy/practice.
4) Type of publication: published journal articles.

Exclusion criteria

1) Duplicated items.
2) Items in which climate security is referred to, but not elaborated on as concept or framework, or in which no practices or investments in action are included.

Flow chart of the literature selection process.

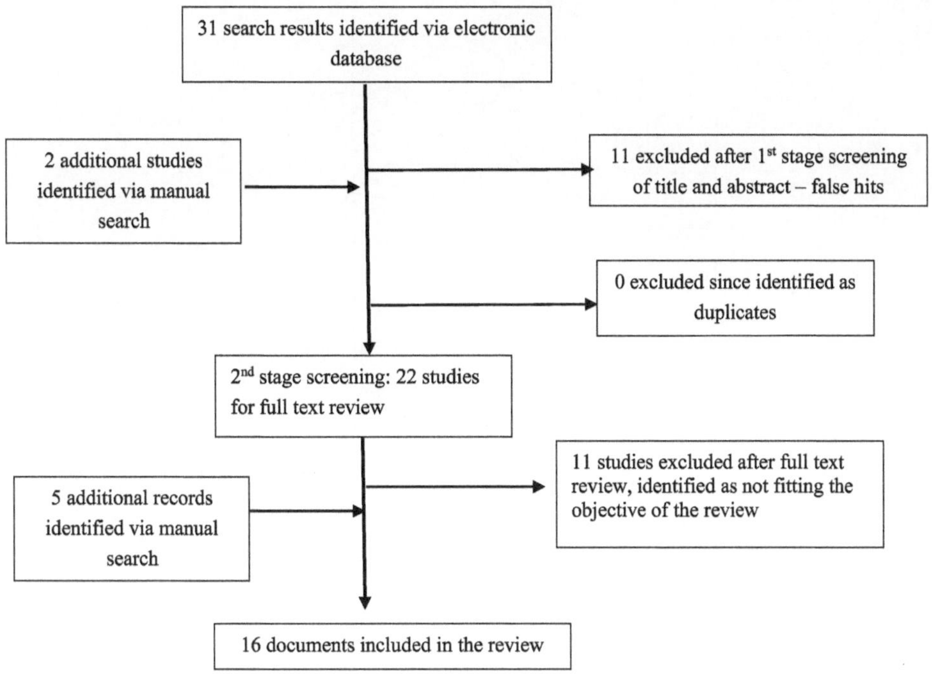
31 search results identified via electronic database
2 additional studies identified via manual search
11 excluded after 1st stage screening of title and abstract – false hits
0 excluded since identified as duplicates
2nd stage screening: 22 studies for full text review
5 additional records identified via manual search
11 studies excluded after full text review, identified as not fitting the objective of the review
16 documents included in the review

Index

Note: **Bold** page numbers refer to tables and *italic* page numbers refer to figures.